Frontispiece

A. *Neohelix albolabris* (Say, 1817), Butler County, PA, diameter 25 mm, this specimen, known as "Hans", has been kept in captivity since 26 April 2002 using the techniques described in Chapter 23, CMNH 71999. B. *Oreohelix idahoensis* (Newcomb, 1866), Salmon River, ID, diameter 13 mm, CMNH 64210. C. *Lampsilis fasciole* Rafinesque, 1820, Little Mahoning Creek, Indiana County, PA, length 77 mm, CFS 71051. D. *Chesapecten jeffersonius* (Say, 1824), Yorktown Formation, Pliocene Epoch, Lee Creek Mine, Aurora, NC, height 73 mm, CFS 71049. E. *Chesapecten septenarius* (Say, 1824), Yorktown Formation, Pliocene Epoch, Lee Creek Mine, Aurora, NC, height 61 mm, CFS 71050. CFS: Sturm Collection, CMNH: Carnegie Museum of Natural History.

Front Cover

Specimens, clockwise from the upper right-hand corner: *Conus marmoreus* Linnaeus, 1758, Indo-Pacific region, CMNH 62.16896. *Liguus fasciatus* (Mueller, 1774), Cape Sable, FL, CMNH 73968. *Liguus virgineus* (Linnaeus, 1767), Dominican Republic, CMNH 73969. *Chiton cumingsii* Frembly, 1827, west coast of South America, CMNH 62.5630. *Spirula spirula* (Linnaeus, 1758), Boynton, FL, CMNH 62.32049. *Elliptio spinosa* (I. Lea, 1836), GA, CMNH 61.1169. *Dentalium grande* Deshayes, 1825, Eocene Epoch, France, CMNH-P 43285. CMNH= Carnegie Museum of Natural History Section of Mollusks, CMNH-P= Carnegie Museum of Natural History Section of Invertebrate Paleontology.

Back Cover

Io fluvialis (Say, 1825), TN, CMNH 62.2167. This freshwater species is the logo for the American Malacological Society. CMNH= Carnegie Museum of Natural History Section of Mollusks.

The Mollusks

A Guide to Their Study, Collection, and Preservation

Edited by
Charles F. Sturm
Timothy A. Pearce
Ángel Valdés

A Publication of the American Malacological Society
Los Angeles and Pittsburgh

Universal Publishers
Boca Raton, Florida
USA • 2006

The Mollusks: A Guide to Their Study, Collection, and Preservation

Copyright 2006   American Malacological Society

All rights reserved. No part of this publication may be reproduced, stored in a retrieval system, or transmitted, in any form or by any means, electronic, mechanical, photocopying, recording or otherwise without the prior permission of the copyright owner.

Acknowledgment of permission to use copyrighted material:
Carl Zeiss, Germany: Figure 7.1
Eastman Kodak Company: Figure 7.4
A. & C. Black, London: Figure 19.2
Daniel Geiger: Figures 20.1 and 20.3
Santa Barbara Museum of Natural History: Figures 27.1-27.9

Acknowledgment of permission to use non-copyrighted material:
David Mulliner: Figure 3.1
Illinois Natural History Survey: Figure 25.1
The Festivus, San Diego Shell Club: Figure 24.4

This volume should be cited as follows:
Sturm, C. F., T. A. Pearce, and A. Valdés. 2006. *The Mollusks: A Guide to Their Study, Collection, and Preservation*. American Malacological Society, Pittsburgh, PA, U.S.A. Pp. xii + 445.

Chapters should be cited as based on the following example:
Sturm, C. F., T. A. Pearce, and A. Valdés. 2006. Chapter 1. The Mollusks: Introductory Comments. *In*: C. F. Sturm, T. A. Pearce, and A. Valdés, eds., *The Mollusks: A Guide to Their Study, Collection, and Preservation*. American Malacological Society, Pittsburgh, PA., U.S.A. Pp. 1-7.

Universal Publishers
Boca Raton, Florida
USA • 2006
ISBN: 1-58112-930-0 (paperback)
ISBN: 1-58112-931-9 (ebook)

Universal-Publishers.com

Library of Congress Cataloging-in-Publication Data

The mollusks : a guide to their study, collection, and preservation / edited by Charles F. Sturm, Timothy A. Pearce, Ángel Valdés.
   p. cm.
 ISBN 1-58112-930-0 (pbk. : alk. paper) -- ISBN 1-58112-931-9 (ebook)
 1. Mollusks--Collection and preservation. 2. Mollusks--Study and teaching. I. Sturm, Charles, F. 1953- II. Pearce, Timothy A., 1954- III. Valdés, Ángel.
 QL406.5.M65 2006
 594.075--dc22
                    2006016238

Preface

This volume promotes the educational mission of the American Malacological Society. The editors and contributors have brought together a broad range of topics within the field of malacology. It is our expectation that these topics will be of interest and use to amateur and professional malacologists.

The chapters in this publication have been peer-reviewed. Each chapter was reviewed by at least one of the editors and by a minimum of two outside reviewers, including at least one amateur and one professional malacologist. Chapters were then revised by authors and again reviewed by the editors, and occasionally, when warranted, by another outside reviewer. Then authors made a second round of changes and editors conducted a final review. While this review process has eliminated some errors and inconsistencies, some may remain.

If anyone does uncover any error or inconsistency in formatting, grammar, or style, the editors would appreciate it being brought to their attention. Constructive suggestions for improving this volume are welcome. Comments can be sent to the editors at <doc.fossil@gmail.com>. Website addresses, cited in this book, were active as of December 2005.

C. F. Sturm
T. A. Pearce
A. Valdés

Table of Contents

Frontispiece ii
Preface vii
Table of Contents ix
1 The Mollusks: Introductory Comments 1
2 Field and Laboratory Methods in Malacology 9
3 Remote Bottom Collecting 33
4 Snorkeling and SCUBA Diving 41
5 Archival and Curatorial Methods 45
6 Digital Imaging: Flatbed Scanners and Digital Cameras 59
7 Applied Film Photography in Systematic Malacology 73
8 Computerizing Shell Collections 101
9 The Molluscan Literature: Geographic and Taxonomic Works 111
10 Taxonomy and Taxonomic Writing: A Primer 147
11 Cladistics and Molecular Techniques: A Primer 161
12 Organizations, Meetings, and Malacology 173
13 Museums and Malacology 181
14 Donating Amateur Collections to Museums 189
15 Fossil Mollusks 197
16 Aplacophora 207
17 Monoplacophora 211
18 Polyplacophora 217
19 Scaphopoda: The Tusk Shells 229
20 Cephalopoda 239
21 Freshwater Gastropoda 251
22 Terrestrial Gastropoda 261
23 Rearing Terrestrial Gastropoda 287
24 Marine Gastropoda 295
25 Unionoida: Freshwater Mussels 313
26 Non-Unionoid Freshwater Bivalvia 327
27 Marine Bivalvia 339
28 The Marine Aquarium: A Research Tool 349
29 An Introduction to Shell-forming Marine Organisms 359
30 Conservation and Extinction of the Freshwater Molluscan Fauna of North America 373
31 Issues in Marine Conservation 385
Appendices 417
Appendix 1: Morphological Features of Gastropod and Bivalve (Pelecypod) Shells 418
Appendix 2: Expanded Table of Contents 427
Glossary 441

Contributors

When two addresses are listed for contributors, the first is their current address while the second is the institution where they were when they submitted the first draft of their chapter.

Frank E. Anderson
Chapter 20
Department of Zoology
Southern Illinois University
Carbondale, IL

Patrick Baker
Chapter 31
Department of Fisheries and Aquatic Sciences
University of Florida
Gainesville, FL

B. R. Bales (1876-1946)
Chapter 2

Arthur E. Bogan
Chapters 25, 30
Curator of Aquatic Invertebrates
North Carolina Museum of Natural Sciences
Raleigh, NC

Thomas A. Burch, MD
Chapter 3
Bremerton, WA

David Campbell
Chapters 11, 15
University of Alabama
Tuscaloosa, AL

Department of Geology
University of North Carolina
Chapel Hill, NC

Eugene V. Coan
Chapter 27
Santa Barbara Museum of Natural History
Santa Barbara, CA

Bobbi Cordy
Chapter 4
Merritt Island, FL

Jim Cordy
Chapter 4
Merritt Island, FL

Clement L. Counts, III
Chapter 17
Department of Natural Sciences
University of Maryland Eastern Shore
Princess Anne, MD

Kevin S. Cummings
Chapters 9, 25
Curator of Mollusks
Illinois Natural History Survey
Center for Biodiversity
Champaign, IL

Robert T. Dillon, Jr.
Chapter 21
Department of Biology
College of Charleston
Charleston, SC

Daniel L. Geiger
Chapters 7, 10, 24
Research Associate
Santa Barbara Museum of Natural History
Santa Barbara, CA

Lucía M. Gutiérrez
Chapter 29
Bióloga Marina y Científica del Medio Ambiente
Guatemala City, Guatemala

Environmental Protection Commission of Hillsborough County
Tampa, FL

Alexei V. Korniushin (1962-2004)
Chapter 26
Institute of Zoology and Zoological Museum
The National Academy of Sciences of Ukraine
Kiev, Ukraine

Ross Mayhew
Chapter 2
Schooner Specimen Shells
Halifax, Nova Scotia, Canada

Fabio Moretzsohn
Chapter 6
Center for Coastal Studies
Texas A&M University - Corpus Christi
Corpus Christi, TX

Department of Zoology
University of Hawaii
Honolulu, HI

Aydin Örstan
Chapters 22, 23
Research Associate - Section of Mollusks
Carnegie Museum of Natural History
Pittsburgh, PA

Timothy A. Pearce
Chapters 1, 9, 14, 22, Glossary
Curator - Section of Mollusks
Carnegie Museum of Natural History
Pittsburgh, PA

Delaware Museum of Natural History
Wilmington, DE

Richard Petit
Chapter 9
North Myrtle Beach, SC

Gary Rosenberg
Chapter 8
Chairman - Department of Malacology
Academy of Natural Sciences
Philadelphia, PA

Patrick D. Reynolds
Chapter 19
Biology Department
Hamilton College
Clinton, NY

Amélie H. Scheltema
Chapter 16
Woods Hole Oceanographic Institution
Woods Hole, MA

Enrico Schwabe
Chapter 9, 18
Department Mollusca
Zoologische Staatssammlung Muenchen
Munich, Germany

Charles F. Sturm
Chapters 1, 2, 3, 5, 9, 12, 13, 28, Glossary
Research Associate - Section of Mollusks
Carnegie Museum of Natural History
Pittsburgh, PA

Ángel Valdés
Chapter 1
Associate Curator of Malacology
Natural History Museum of Los Angeles County
Los Angeles, CA

Paul Valentich-Scott
Chapter 27
Curator of Mollusks
Santa Barbara Museum of Natural History
Santa Barbara, CA

Andreas Wanninger
Chapters 9, 18
Department of Zoomorphology
University of Copenhagen
Copenhagen, Denmark

Beatrice Winner
Chapter 28
North Palm Beach, FL

CHAPTER 1

THE MOLLUSKS: INTRODUCTORY COMMENTS

CHARLES F. STURM
TIMOTHY A. PEARCE
ÁNGEL VALDÉS

1.1 INTRODUCTION

Mollusks have been important to humans since our earliest days. Initially, when humans were primarily interested in what they could eat or use, mollusks were important as food, ornaments, and materials for tools. Over the centuries, as human knowledge branched out and individuals started to study the world around them, mollusks were important subjects for learning how things worked.

Initially, nobility and the wealthy (or scientists with wealthy patrons) carried out such studies on the natural world. Later, in the 19th and 20th Centuries, a professional class of scientists developed. Governments and industry also started supporting scientists. Although in some fields, as professionalism developed, non-professionals took a back seat; in contrast, non-professionals studying mollusks have consistently made important contributions.

Just as bird watchers contribute important observations that allow professionals to study bird migration and changes in populations, so amateurs play important roles to professional malacologists. Amateurs can study the mollusks of a given region over time and see what changes occur in response to interactions with humans and natural forces. The molluscan fauna is inadequately studied in many areas; amateurs can study such areas. You may have the opportunity to describe a new species. While amateurs may not have the resources for undertaking molecular studies, they might collaborate with professionals by providing samples. In addition, many molluscan families and genera have not been subjected to a good revision in many decades. An advanced amateur could review the world literature and summarize it into a well-researched revisionary work. Though this can be a daunting task, it could be the culmination of years of demanding and painstaking work. While such a project may take years or decades, if done correctly, it will be a valuable contribution to practicing malacologists.

This book is intended for three groups of people. The first and foremost is the amateur community. Amateur malacologists are those who study mollusks out of an avocational desire, in contrast to the professional who is employed or was trained as a malacologist or biologist. Some amateurs may be classified as paraprofessionals meaning that their depth of knowledge on a particular subject may be at a professional level, however, they may not have the breadth of malacological knowledge that a professional has. We hope that this book will give amateurs the guidance and skills to deepen their interest in malacology and do so in a professional manner.

Another group of people for whom this book is intended is professional biologists, those individuals whose work, at least in part, relates to mollusks. This group would include, among others, malacologists and ecologists. The techniques outlined in this book will be of use to them in their work. If you are a biologist, and have a need for an up-to-date reference on mollusks, this book is for you. It will be of use to biologists studying both fossil and recent mollusks.

The final group of people is biology students beginning their studies of mollusks. As with the above groups of people, this book will help open up the field of malacology to them. It will provide students with a sourcebook that they can use in their studies.

The terms conchology and malacology need clarification. Conchology has traditionally been thought of as the study of molluscan shells. Malacology is a broader term that includes the study of the animal that made the shell in addition to the molluscan shell. In this book, we use the term malacology to mean the study of mollusks and malacologist to describe anyone interested in mollusks, whether their interest is only the shell or the molluscan fauna in the broadest sense.

What is a mollusk? Among the animals, Mollusca are in a group called Spiralia or Protostomia, which also includes Annelids, Arthropods, and other small phyla. The Spiralia, which make up more than 90% of all living multicellular animals, share several developmental features including spiral cleavage of blastomeres, formation of the mouth from the blastopore, and predictable cell fates such as all mesoderm being formed from the single cell 4d. The circumpharyngeal nerve ring is present in many Spiralia. Within the Spiralia, the group Eutrochozoa includes the Mollusca, Annelida, and other small phyla, which all share a trochophore larva (top shaped larva with an equatorial ciliated band and a dorsal cilia tuft), schizocoely (formation of body cavities from multiple bilaterally paired masses of mesoderm), and paired excretory organs and ducts that open externally. Regarding the Mollusca themselves, despite the wide diversity of body forms, the groups of animals we classify as Mollusca share the following unique features (Eernisse *et al.* 1992, Haszprunar 2000):

- radula (absent in Bivalvia)
- mantle capable of secreting a calcium carbonate-based shell or spicules
- mantle cavity
- ctenidia (specialized gills having countercurrent oxygen exchange)
- osphradia (chemosensory epithelial organs) (absent in Monoplacophora and Scaphopoda)
- pericardium around the heart (not around heart in Scaphopoda)
- mesodermal origin of pericardioducts
- rhogocytes (pore cells) associated with the nephridia (kidneys)
- tetraneury (two pairs of main longitudinal nerve bundles)
- intercrossing dorsoventral muscles
- crystalline style and associated ciliated midgut digestive organs
- esophageal pouches
- broad creeping sole or narrow hydrostatic foot
- large ventral pedal glands that secrete mucus

Mollusks will share some or all of these characteristics. You may be unfamiliar with some of the terms used; many are defined in the glossary found at the end of this book.

Mollusks first appeared at the end of the PreCambrian. Many lineages of mollusks have died out before Recent times. The living mollusks comprise the following classes: Solenogastres and Caudofoveata (together the Aplacophora), Monoplacophora, Polyplacophora, Scaphopoda, Cephalopoda, Gastropoda, and Bivalvia. Mollusks can be found in terrestrial, freshwater, and marine environments.

As indicated in the title, this book focuses on studying, collecting, and preserving mollusks. The study of mollusks can take place in the field, in an aquarium, or in a collection in your home or at a museum. Collection can refer to the collection of field observations or specimens. Preservation can refer to the preservation of specimens in collections, or to conserving living mollusks and their habitats.

It is also important to know what this volume does not cover. It provides only basic information on anatomy. It does not cover the biochemistry, physiology, or genetics of mollusks. Though raising and maintaining mollusks in aquaria or terraria is mentioned in a few chapters, except for Chapter 31, there is very little information on aquaculture of mollusks.

Now we give a brief overview of the remainder of this book. This overview should give you an idea of the information to be found in each chapter and help you use this book effectively.

1.2 CHAPTER REVIEWS

The book is composed of 31 chapters. The first 14 chapters cover basic topics in malacology. The next seventeen chapters cover malacological issues related to specific groups of mollusks.

Chapter 2 presents many techniques for the collection, cleaning, and preservation of mollusks. There are also sections about tagging and narcotizing mollusks. In this chapter, Sturm, Mayhew, and Bales have tried to include not only the latest techniques, but also older and still useful methods. The Walker Dipper, first described in 1904, is illustrated here for the first time.

Chapter 3, primarily written by Burch, is on remote bottom sampling. It updates his paper presented at the annual American Malacological Union meeting in 1941. Burch relates stories about a lifetime of dredging and provides useful insights on this activity. In the second part of this chapter, Sturm discusses several other methods of sampling underwater sediments including grabs, box corers, tangle nets, and the like.

Snorkeling and SCUBA diving are covered by Cordy and Cordy in Chapter 4. This chapter will expose you to new ways of coming face to face with mollusks. While this chapter will not teach you how to dive, the authors have covered the equipment that is needed for snorkeling and SCUBA diving. The inherent risks of these activities are also covered.

Chapter 5 moves us out of the field and into our collections. If you want to learn how to help preserve your collection for the next millennium, this chapter will interest you. This chapter discusses curatorial practices and methods for archival protection of your collection. The chapter starts with a discussion of the risks to stored collections. It continues with a discussion of materials that are used in maintaining collections, stressing archival materials and practices. This chapter concludes with a list of sources for archival materials.

Moretzsohn, in Chapter 6, discusses digital imaging with flatbed scanners and digital cameras. He begins with a basic discussion of the theory behind digital imaging. He goes on to compare digital and film imaging technologies. Digital cameras and flatbed scanners are discussed. Finally, the printing and editing of digital images are reviewed. Moretzsohn also provides an extensive list of Internet sites where one can go for further information.

Geiger completes the topic of imaging with a review of traditional film photography. In Chapter 7, he begins with a discussion of the nature of light and basic photographic theory. He discusses camera equipment and accessories. There is a discussion regarding film. Special topics include infrared and ultraviolet photography, underwater photography, photography of mollusks in aquaria, and storage of images.

In Chapter 8, Rosenberg discusses the use of computer databases in managing collections. In his discussion he includes choosing a software package, the suggested fields to be included, and constructing the database. The discussion is general enough that it will apply to many different operating systems.

Chapter 9 by Sturm, Petit, Pearce, Cummings, Schwabe, and Wanniger provides an introduction to the malacological literature. The first part of the chapter discusses the various types of malacological publications: monographs, iconographies, journals, and separates, just to name a few. The rest of the chapter categorizes over 700 works by biogeographic zones and molluscan families. Thus, if you have an unknown shell from Ecuador, under the Panamic Biogeographic Zone listing, you will find a list of references to help in identifying it. Also listed are books of general interest and others for those interested in taxonomic research.

How to write a taxonomic paper is an art that is handled by Geiger in Chapter 10. Geiger shows the components of scientific papers dealing with both the new description of a genus or species and of a revisionary paper. Geiger also gives a brief overview of the International Code of Zoological

Nomenclature, a set of rules governing the naming of species. Familiarity with the code is essential for anyone describing a new genus or species; for those just interested in cataloging shells, it is useful knowing how names are determined and why a familiar name is sometimes replaced by a less familiar one.

If you are interested in finding out how researchers determine the tree of life, or how closely related different species are, Chapter 11, on cladistic analyses and phylogenetic trees, will interest you. Campbell explains cladistics and then covers some of the major techniques used in a cladistic analysis. He then concludes with a discussion of molecular biology and how it relates to modern phylogenetic analyses.

If you have ever wondered whether or not to join a malacological organization or go to a meeting, Chapter 12 will be of use to you. Here Sturm discusses the basic functions that both amateur and professional malacological associations provide. He describes the functions of such major organizations as the American Malacological Society, Conchologists of America, and others. He discusses the benefits of meetings, both local and national. Six Internet-based discussion groups are listed with directions on how to join them.

Museums are one of the places where we can go to see fossil and Recent mollusks and to learn more about them. In Chapter 13 by Sturm, you will learn about the basic roles of museums in keeping collections and conducting research: they are not just places that exhibit natural history artifacts! In addition, you will take a quick tour of 22 museums in the United States and Canada. These museums all have major malacological collections. So you can learn more about these institutions and what they have to offer, the addresses for their Internet sites are included.

If you ever wondered about museum collections and what museums do with donated specimens, Chapter 14 by Pearce will inform you. Pearce begins by discussing what makes a specimen valuable to a museum. Most specimens will have some value if they have clear and accurate locality data associated with them. Data can be intrinsic (for example color, size, and weight) or extrinsic (for example collected where, when, and by whom). The importance of the extrinsic data is emphasized, as extrinsic data are often what makes specimens of vital importance to a museum. Also covered in this chapter are ways to preserve soft tissue from molluscan specimens and how these may be of importance to a museum. Finally, if you are contemplating donating your collection to a museum, Pearce discusses the way to do this so that the collection will be of greatest use to the museum and other researchers.

With Chapter 15, we will begin exploring different molluscan groups, as well as techniques for rearing mollusks. These chapters are followed by a chapter on the marine biology of non-molluscan organisms.

Chapter 15 by Campbell is a discussion of paleomalacology. Here you will learn about some mollusks that exist only as fossils. Campbell provides a general overview of fossils and ways to collect fossil mollusks.

Chapter 16, by Scheltema, is a review of the Aplacophora (worm mollusks). These organisms comprise a fascinating group of mollusks that, while often under our feet, are rarely collected by professionals or non-professionals. Scheltema discusses the basic biology of these mollusks and their ecologic context. She goes on to discuss ways of collecting aplacophorans and techniques for studying them.

Chapter 17 deals with the smallest class of living mollusks, the Monoplacophora. Counts discusses this group, which was first known from the fossil record; the first living monoplacophoran was identified only in 1957. These mollusks are found in marine environments and at great depths. Counts covers the biology and zoogeography of these organisms. If you are fortunate enough to find a monoplacophoran, directions for their preservation are covered.

Schwabe and Wanniger write about the Polyplacophora (chitons) in Chapter 18. These mollusks have a shell composed of 8 plates and attach tenaciously to the rocks on which they live. Schwabe and Wanniger discuss the external and internal anatomy of these organisms as well as their mode of reproduction and where they live. They include techniques for collecting and preserving the soft tissues and shells of chitons.

The Scaphopoda (tusk shells) are addressed in Chapter 19. Here Reynolds begins with a discussion of the biology of the tusk shells and goes on to review them in an ecological context. He discusses ways to collect, preserve, and maintain collections of tusk shells.

If you ever wanted to know how to preserve a giant squid, this next chapter is for you. In Chapter 20, Anderson discusses the Cephalopoda (i.e., squids and octopuses). He begins with a review of the biology and behavior of these mobile organisms. This review is followed by a taxonomic review of the class. He then goes on to discuss collecting techniques and protocols for preserving cephalopods, including *Architeuthis*, the giant squid. Although you will need hundreds of liters of preservative, an enormous vat, and weeks to months, this chapter will prepare you for the undertaking. Anderson concludes the chapter with information on identifying cephalopods and a section on the difficulties of maintaining cephalopods in an aquarium.

Chapter 21 reviews the freshwater Gastropoda. While some of these gastropods are vectors of disease, others are common aquarium inhabitants. Dillon starts off with a discussion of the biology and ecology of the freshwater gastropods. He also touches on some conservation issues relating to this fauna. He describes techniques for collecting, preparing, and storing freshwater gastropods. Dillon concludes with techniques for maintaining freshwater gastropods in an aquarium.

Pearce and Örstan discuss the terrestrial Gastropoda in Chapter 22. They begin by discussing the biology of this group. There is an extensive discussion of the habitats of the land snails and how to locate them. They continue with a discussion of field methods used in collecting land snails; you will even learn how to collect snails with a leaf blower! They conclude with an exhaustive description of processing and storing land snails as well as a discussion of record keeping.

The discussion of land snails is continued in Chapter 23. Örstan begins with a review of the literature on maintaining land snails in a terrarium. He goes on to discuss short-term maintenance of a terrarium and rearing some North American woodland snails. He concludes with a discussion of factors that can affect the health of land snails. Hans (see the frontispiece) has been raised in captivity for the past 4 years following the advice in this chapter.

The treatment of the Gastropoda concludes in Chapter 24. Here Geiger presents information on the marine Gastropoda. He begins with an extensive discussion of the shell and soft tissue anatomy of the marine gastropods. He continues with a discussion of the major groups of marine gastropods. He then discusses the various habitats where marine mollusks can be found and the ecological aspects of the marine gastropods. He concludes with a discussion of techniques for collecting and preserving marine gastropods, including the preparation of radulae.

The next three chapters cover the Bivalvia. Chapter 25 by Cummings and Bogan reviews the Unionoida or freshwater mussels. The larvae of these mollusks spend part of their lives as parasites of vertebrates. Cummings and Bogan begin with a description of the biology and ecology of the freshwater mussels. This section is followed by a taxonomic treatment of this group. They briefly touch on issues relating to conservation, a topic more fully developed in Chapter 30. They then provide an extensive description of field collecting equipment and techniques, information on identifying freshwater mussels, and methods of curating wet and dry collections of these organisms.

Korniushin, in Chapter 26, provides a similar treatment for the non-unionoid freshwater bivalves. He begins with a treatment of the Sphaeriidae, commonly known as the fingernail, pea, or pill clams because of their small size. This includes the biology and ecology of and methods for collecting and preserving these organisms. He then briefly discusses the Corbiculidae (Asian clams) and Dreissenidae (zebra mussels).

The marine Bivalvia are discussed in Chapter 27. Coan and Scott begin with a discussion of the biology of these organisms. They then give an overview of the five groups (orders) of marine bivalves. They then discuss techniques for the collection, preservation, and study of these animals.

Chapter 28 deals with establishing and maintaining a marine aquarium. Winner begins the chapter with a discussion of the uses of and observations that can be made in an aquarium. The physical components of the aquarium are then discussed: water, plants, rocks, and food. She then describes fourteen marine organisms that she maintained in her aquaria and the observations that she made.

In Chapter 29, Gutierrez gives a brief overview of marine biology. The aim of her chapter is to introduce marine organisms that are sometimes mistaken for mollusks. Some examples are calcareous algae, annelids, brachiopods, and echinoderms.

The last two chapters cover issues of conservation. One covers freshwater mollusks while the other covers marine mollusks.

In Chapter 30, Bogan reviews the sad conservation history of freshwater Gastropoda and Bivalvia in North American streams, rivers, and lakes. He explores the patterns of extinction that have occurred and the reasons behind them. On an upbeat note, he does discuss some reversals in these trends. Bogan includes tables listing the threatened, endangered, and extinct freshwater mollusks of North America.

In the final chapter of this book, Chapter 31, Baker discusses marine molluscan fisheries. He explores the attitudes towards marine mollusks and marine conservation in the 20th and 21st Centuries. He goes on to examine specific examples involving mollusks. He extensively reviews the oyster fisheries on both coasts of the United States. Less extensively, he also discusses the fisheries for abalones, Queen Conchs, scallops, clams, mussels, and cephalopods. He includes tables listing gastropods and bivalves that are covered by state or federal regulations, common edible or commercial mollusks that are not covered by specific regulations, and information for states with marine fishery agencies.

Following Chapter 31 are two appendices and a glossary. Appendix 1 consists of four plates illustrating morphological and anatomic features of gastropods and bivalves. Appendix 2 is an expanded table of contents. It will give you a detailed overview of each chapter and help locating information in the book. In the Glossary, you will find definitions for many of the technical terms used throughout this volume.

This book is meant to be sampled piecemeal and not necessarily read from cover to cover, though one could do this. The book is meant to be sampled in portions relating to one's interests. If you are interested in bivalves, you will concentrate on Chapters 25-27. If you wish to delve into freshwater malacology, Chapters 21, 25, and 26 would be on your reading list. If you are interested in field techniques, read Chapters 2-4, 6-7, and the chapters for the organisms in which you have an interest.

We hope that you will find the information and discussions in this book helpful. As with all books of this type, compromises had to be made. The authors and editors sought to present a balanced handbook of techniques and background information. In preparing each chapter, the authors attempted to be comprehensive in approach while presenting only what they felt would be useful. If there are areas or topics that you feel were slighted or overlooked, let us know.

1.3 ACKNOWLEDGMENTS

The senior editor (CFS) owes a debt of gratitude to many individuals for their involvement with this book. The book is partially an outgrowth of a workshop at the 1999 American Malacological Society (AMS) meeting held in Pittsburgh, Pennsylvania. Opinions as to what topics should be addressed in the workshop were offered by Paul Drez (1947-2004), William Frank, Ross Mayhew, Timothy Pearce, Michael Penziner, Richard Petit, Robert Prezant, Gary Rosenberg, and Evangelos Tzimas. Presenters at that workshop included Kevin Cummings, José Leal, Timothy Pearce, Richard Petit, and Gary Rosenberg. Robert Prezant deserves special thanks, for presenting CFS with the opportunity to organize this workshop.

We would like to extend our thanks to all the contributors to this book: Frank E. Anderson, Patrick Baker, B. R. Bales (1876-1946), Arthur E. Bogan, Thomas A. Burch, David Campbell, Eugene V. Coan, Bobbi Cordy, Jim Cordy, Clement L. Counts, III, Kevin S. Cummings, Robert T. Dillon, Jr., Daniel L. Geiger, Lucia M. Gutierrez, Alexei V. Korniushin (1962-2004), Ross Mayhew, Fabio Moretzsohn, Aydin Örstan, Richard Petit, Patrick D. Reynolds, Gary Rosenberg, Amélie H. Scheltema, Enrico Schwabe, Paul Valentich-Scott, Andreas Wanninger, and Beatrice Winner.

We are thankful for all those individuals who agreed to review submissions. They are: James Albarano, Amir Amiri, Kurt Auffenberg, Glenn Burghardt, Laura Burghardt, Henry Chaney, Eugene Coan, Louise Corpora, Robert Dillon, Douglas Eernisse, James Fetzner, Daniel Geiger, Jose Juves, Albert Koller, Harry Lee, James Lee (1922-2005), Richard Lee, James McLean, Paula Mikkelsen, Paul Monfils, Mohan Paranjpe, Kristin Petersen, Robert Prezant, John Rawlins, Lutfried von Salvini-Plawen, Gerhard Steiner, Patricia Sturm, Michael Vecchione, Janice Voltzow, and Amy Wethington. In addition to individuals already mentioned, we would also like to thank James T. Carlton, Harry A. ten Hove, and Peter J. Wagner for helping to resolve specific questions that arose during the editing process.

The American Malacological Society sponsored this project and provided financial support. Two chairs of the AMS Publications Committee have helped us with this book. The project was started under the guidance of Ronald Toll and came to completion under Janice Voltzow's tenure. Paul Callomon offered valued insight into the intricacies of publishing books.

The quality of this book has been greatly enhanced by the assistance of three librarians who helped track down cited literature and provided help in obtaining interlibrary loans. This work has benefited greatly from the assistance of Bernadette Callery, Sun Xianghu, and Marie Corrado of the Carnegie Museum of Natural History Library.

We would like to thank Emily C. Ullo and Amanda E. Zimmerman for commissioned artwork requested by CFS. Ms. Ullo painted the cover and drew the illustrations in Chapter 21. Ms. Zimmerman drew illustrations in Chapters 2, 3, 9, 11, and 29.

CFS is extremely appreciative of the help given by the two associate editors, Timothy Pearce and Ángel Valdés. They joined the project in an editorial capacity in 2004. Their assistance in copy-editing and layout helped bring this project to its conclusion.

Lastly, CFS thanks his wife Pat for her patience with and encouragement for this project.

1.4 LITERATURE CITED

Eernisse, D. J., J. S. Albert, and F. E. Anderson. 1992. Annelida and Arthropoda are not sister taxa: a phylogenetic analysis of spiralian metazoan morphology. *Systematic Biology* **41**: 305-330.

Haszprunar, G. 2000. Is the Aplacophora monophyletic? A cladistic point of view. *American Malacological Bulletin* **15**: 115-130.

CHAPTER 2

FIELD AND LABORATORY METHODS IN MALACOLOGY

CHARLES F. STURM
ROSS MAYHEW
B. R. BALES (1876-1946)

2.1 INTRODUCTION

Your first exposure to the world of mollusks may have been picking up a seashell on a beach while you were on vacation. It may have been finding a snail or slug in your garden. It may have taken many years before you decided to pursue a more in-depth interest in these organisms. This chapter will introduce you to ways to collect, clean and prepare mollusks.

Part of this chapter was a paper written by B. R. Bales and presented for him by R. Tucker Abbott at the Eleventh Annual Meeting of the American Malacological Union in 1941 (Bales 1942). This paper with minor alterations was reprinted in Abbott *et al.* (1955, 1966) and Jacobson (1974). While much of what Bales had to say is still relevant, some of his suggestions have not stood the test of time and are no longer considered appropriate in the context of current curatorial practices. In this chapter we (CFS and RM) copy and paraphrase from Bales' paper. In addition, we have added information that updates his views and we add information that was not in his original work. Due to our use of Bales' original paper, we have included him as an author; to do otherwise would be tantamount to plagiarism.

In addition to the recommendations in this chapter, you will find additional information in Chapters 3 to 5, and 15 to 27. Bergeron (1971), Lipe and Lipe (1993), and Weil (1998) are other works in which you can find advice and opinions on how to collect, clean, and maintain a molluscan collection.

2.2 COLLECTING BASICS

Other than being given a collection, there are three basic ways to build a collection. You can buy specimens, trade for them, or collect them yourself.

2.2.1 Purchasing. If you chose to purchase shells, you first have to find a dealer. They can be found advertising in publications such as *American Conchologist, La Conchliglia,* and *Of Sea and Shore.* A comprehensive worldwide listing of dealers can be found in *Tom Rice's A Sheller's Directory of Clubs, Books, Periodicals and Dealers* compiled by Tom Rice (2003) and at several good Internet sites such as <www.manandmollusc.net> and <www.conchologistsofamerica.org/home>.

There are several advantages to purchasing shells from a dealer. First, they are already cleaned. Second, they are identified. Most important, you can obtain material from places that you may never be able to collect from yourself, and from fishing boats and other sources inaccessible to the average collector.

The drawbacks to purchasing shells are two-fold. First, having good locality data is usually important, and the locality data of purchased shells may be sparse or suspect. Knowing that you are buying from a reliable dealer helps to minimize this risk. Second, you lose the ability to study the organisms in their natural habitat.

Dealers use a grading system to indicate the quality of their material. The international standard dictates that "Gem" specimens are absolutely perfect to the naked

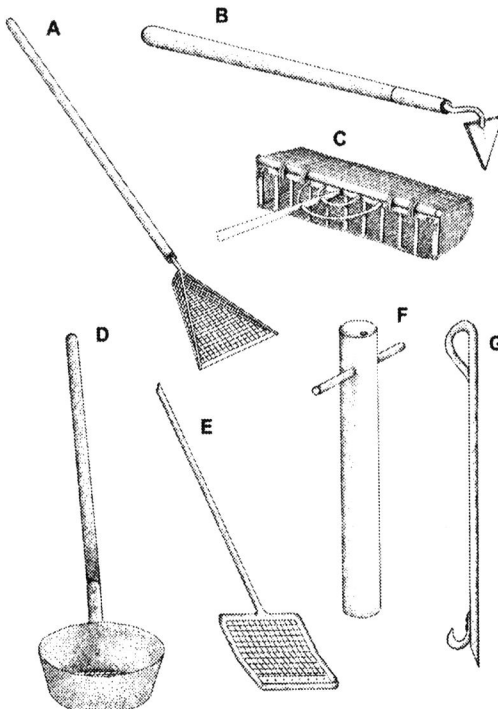

Figure 2.1 Collecting devices and aids.
A. Allison Scoop, B. Ferriss Hoe, C. Davis Rake Drag, D. Walker Dipper, E. van Eeden Scoop, F. Clam tube or clam gun, G. Bales Hook.

eye, "F++ or Gem-", which some dealers call "F+++" specimens, are so close to perfect that any flaws have to be searched for, "F+" indicates a specimen with minor flaws that do not detract much from its aesthetic appeal, "F" (stands for "Fine") specimens have significant flaws, Good specimens are not good at all, and Fair and Poor are recognizable as that species, but just barely. That said, grading is subjective, and there is an understandable temptation to over-grade. If you see a dealer's list with mostly gem designations, beware. *Caveat emptor* is a useful watch-phrase. When placing an order from a dealer you are unfamiliar with, it is best to start small. Make a few modest purchases, and if you agree with the dealer's grading of the shells and the quality of the associated data, you can make larger purchases with confidence. Often dealers will learn your interests and notify you of material in which you might be interested.

2.2.2 Trading. Trading has several rewards. One can trade shells and/or publications for other shells and/or publications. The benefits (and hazards) are the same as with purchasing from dealers. In addition, if you are not trading for a specific shell, you will have the excitement that comes with receiving a shipment of unknown material. As with purchasing from dealers, when you trade with someone, make your first trades modest ones of material that you can afford to lose. If all goes well, mutual trust will develop and larger trades will ensue with confidence. Often, but not always, the material you trade for has been self-collected and therefore the collection data may be more complete and reliable than with purchased material.

2.2.3 Self-collecting. In spite of building a collection by purchasing or trading shells, eventually you may desire to start field collecting yourself. You may decide to collect shells in which the animals have died (dead collecting) or living material (live collecting). Section 2.4 deals with some techniques that you can use to collect mollusks.

The simplest way to self-collect is to look for dead material. You can walk along the beach picking up seashells, an activity known as beachcombing. A particularly good time to try beachcombing is after a storm. Sometimes you will find material from deeper water thrown up on the beach. Also, pelagic material may be blown onto the beach by the storm's winds.

You might walk along the shores of a lake, pond, or river and search for freshwater material. Sometimes you may be lucky and find a shell midden. This is a pile of shells left behind by an animal such as a raccoon or muskrat after it has eaten a meal of mussels. Section 2.4 describes some of the techniques of field collecting.

You might also be interested in collecting terrestrial gastropods. These can be found under rocks or leaf litter. Techniques for collecting these mollusks are covered below and in Chapter 22, Sections 22.4 and 22.6.

2.3 COLLECTING EQUIPMENT

Many different tools or implements may come in handy from time to time, and, while you may achieve a measure of success with very little equipment (such as a spade, a kitchen strainer, some vials and collecting jars, rubber boots, snorkeling equipment, sharp eyes, and plenty of patience), you will eventually find that some specialized tools will be necessary as you progress in the study of malacology (i.e., the study of mollusks as living animals), or conchology (i.e., the study of the calcareous exoskeletons of mollusks, otherwise known as "shells"). What follows now are descriptions of some such tools.

2.3.1 Allison scoop. Allison (1942) described a scoop for collecting *Campeloma* (a freshwater gastropod) from stream bottoms. The scoop was triangular in shape, had a wire basket with a reinforced leading edge and it was attached to a pole (Figure 2.1A). By varying the size of the scoop and the mesh size of the basket, this scoop can be modified to collect a wide variety of mollusks.

2.3.2 Ferriss hoe. Walker (1904) described an implement called the Ferriss hoe. This is a garden hoe with its blade trimmed to 75 mm (3 in) at the top and tapering to a sharp point. The handle is trimmed to the length of a walking cane (Figure 2.1B). This device is a good tool for turning over logs and rocks, breaking up rotting logs, and digging through rotting leaves and around stumps. It is also long enough to pull down tree branches.

2.3.3 Davis rake drag. Davis (1964) described a type of dredge to be used from the shoreline. He took a garden rake and to the crossbar of the rake he attached a 6-8 mm ($1/_4$ in) wire mesh that looped around and was attached to the ends of the tines. He then attached a pipe, filled with sand, to the crossbar to give the device added weight (Figure 2.1C). A rope was attached to the handle of the rake. The device was then thrown out into the water and pulled back so that the wire basket and tines would bite into the substratum. The rope was pulled drawing the device back to shore. The device was then tilted, emptied, and the contents examined for specimens.

2.3.4 Screens, dippers, and nets. To collect the smaller specimens from shallow water, a screen comes in handy. Once you progress beyond kitchen strainers, these may be constructed or purchased in many sizes and forms. The size of the screen and height of the sides depend upon the individual user. It should not be so large that it taxes the strength, for it should be remembered that to be successful in screening, the collector must be persistent and many hours are usually spent in this manner. It is a fascinating form of collecting and the time flies all too soon. Seldom will the collector stop sifting without trying "just one more screenful" of material.

Some collectors prefer several graduated sizes of screen (brass, aluminum, stainless steel, or zinc-coated mesh are best, since they are corrosion-resistant), but most prefer just two; the inside screen to be 6-8 mm mesh ($1/_4$ inch) and the outer one of 3-4 mm mesh ($1/_8$ inch). If you are interested in microshells at all, a third one, of 1 mm or smaller mesh (0.04 in or less) will be necessary. This one will catch most of the smallest shells. The inner screen with the larger mesh should fit snugly into the outer one with the smaller mesh, but not too snugly. Allowances should be made for the natural swelling of the frames, although much swelling will be avoided by painting the frames. Some collectors nail a small cleat to the ends of the outer screen to obtain a firmer grasp. The same results may be obtained by sawing a narrow horizontal slit in the end of the frame. The size of frame and mesh being of individual preference, it is sometimes advisable to try out several before the ideal one is found, and even then, many collectors change from time to time as the occasion demands. You might also find soil sieves useful. These are nested sieves about 20 cm (8 inches) in diameter. They are generally made of brass. They can be purchased from general scientific and forestry supply companies (see the appendix in Chapter 5 for a list of such companies).

One form of screen that is sometimes used has no upright frame on one side and is held in place on the bottom of a body of water by the collector's foot. The sand, mud, marl, or other material is raked or drawn into the screen by the use of a hoe, rake, or other utensil. This type of screen avoids having to lift the material and to deposit it in the screen for those who like to adhere to a less energetic regimen while collecting.

A handy device often used consists of a small round sieve that has been attached to a long handle. It is easily made from a 12-15 cm (5 or 6 inches) gravy strainer to be had at most hardware or discount stores. They have two bent prongs in front that must be bent backward so as not to interfere with the use of the net. No device equals this when working in waist or chest deep water. You might also want to make and use a Walker Dipper (see Figure 2.1D, Chapter 21, Freshwater Gastropoda, and Walker 1904).

When collecting the specimens on various types of aquatic grasses (eel, turtle, etc.), a net made of mosquito or other fine netting sewn around a butterfly net hoop is extremely useful, especially if you wish to be as ecologically sensitive as possible, leaving the grasses in place instead of collecting them and washing the mollusks into fine (the 24-mesh) screens using fresh water.

2.3.5 van Eeden scoop. Another variation of scoop was described in van Eeden (1960) (Figure 2.1E). The scoop was designed for collecting freshwater gastropods. It is a square frame made from a 5 mm iron rod. The square is 25 cm on a side. The frame is angled upwards about 30 degrees two thirds of the way back from the leading edge. The leading edge is reinforced with iron or tin sheeting. A wire screen of appropriate mesh size is attached to the frame. The frame in turn is attached to a handle of appropriate length (2-3 m).

2.3.6 Shovels. Some collectors use a shovel to dig up clams. Commonly, these shovels have a blade 75-100 mm wide (3-4 inches). The blade is longer than it is wide, and it is slightly curved. These shovels are sometimes called clam guns. Insert the shovel so that it curves away from where you believe the clam to be and start digging. Dig and scoop the sediment away from the clam burrow; keeping the shovel blade parallel to the clam burrow. When you see part of the clam reach down and grab it. It is important not to dig towards the clam; doing so may damage it.

2.3.7 Clam tube. There is another device also called a clam gun or sometimes a clam tube. This is a metal or plastic tube, 75-150 mm in diameter (3-6 inches) and 1 to 1.5 meters long (3-4 feet). It is closed on one end with a cap that has a handle and small vent hole (Figure 2.1F). When you locate a clam burrow, place the tube over the burrow. Rock the tube back and forth and twist it so that it drills into the sediment. When you believe that the shell is within the tube, cover the vent hole. Doing so creates a suction effect when you withdraw the tube from the ground. You will pull up a plug of sediment that should include the clam. Empty the contents onto a screen, wash them, and remove the shell. This device works best in muddy and sandy substrates; it does not work well in bottoms composed of gravel and rock.

2.3.8 Hammer. A geologist's or bricklayer's hammer with a chisel end can be quite useful in several respects. You can scrape through leaf litter, dig into the upper layers of soil, turn over rocks (which of course should always be replaced as they were, before leaving the scene), use it in the pursuit of paleomalacology (fossil collecting), chip away the soft rocks in which burrowing species such as many of the family Pholadidae (e.g., *Barnea* and *Zirfaea* spp.) live, and the chisel end comes in handy when taking apart pieces of sunken wood that may harbor ship worms (Families Teredidae and Xylophagidae). It is useful to paint part of these tools a bright yellow or orange. Otherwise, you will learn the hard way how easily such tools can be overlooked when put down in the field and searched for at a later time.

2.3.9 Bales hook. According to Bales, one of the most important tools is a device made from a 15 mm ($^5/_8$ inch) metal rod. One end is looped to make a handle and the other is formed into a point. About 75 or 100 mm (3 or 4 inches) above the point, a curved hook is welded onto the rod giving a form similar to an elephant hook (Figure 2.1G). With such a device, you have an implement that can be used as a walking stick, to turn over rocks, to pull things closer to you, and to pull down branches when looking for arboreal snails. Estwing makes a similar tool called the Gem Scoop®. Instead of a hook, it has a small basket. Of course, many collectors make their own tools as best as they can, given their budget and circumstances, so the implements described in this chapter can be taken as a starting point - an assemblage of ideas and advice, for you to adopt or adapt as you see fit.

2.3.10 Water pumps. When the ocean floor is of rock that is more or less honeycombed with small potholes, it is surprising what fine specimens of the smaller varieties may be obtained by the use of a common bilge pump, which is standard equipment on small boats. The end of the pump is placed in a sand-filled pocket in the rock. The sand as well as the mollusks that have taken refuge in the hole are pumped with the water onto a screen; the sand and water flow through, leaving the specimens all ready for the collector.

2.3.11 Bags and collecting containers. Very important to the collector's outfit is your collecting bag, and this may consist of almost anything from an old tin can, a burlap or nylon sack, a pocket handkerchief or some such makeshift affair, to a real game bag or collecting bag. They are usually made from lightweight canvas and carried in some cases by a strap over the shoulder. Such bags are useful when extra heavy specimens are anticipated. A bag that may be tied or secured about the waist is much handier and has the advantage of always being in place and does not drop in front of the collector when he or she stoops to secure a specimen. Many collectors favor a bag containing partitions: one compartment for tools, vials and other equipment apart from the shells that are collected, and others for samples of various sorts. You might even find that a carpenter's apron makes a useful collecting bag.

When collecting small mollusks, a good supply of glass or plastic vials of various sizes, will prove most useful. These should be cylindrical and of the screw top sort without a narrow neck. Nylon mesh bags are often useful when collecting larger shells such as bivalves. You should be careful to keep shells from different habitats separate, labeling the vials and containers in the field using pencil and strong paper, and/or a grease pencil applied to the container. Basic information will include date, locality, collector, and as a good and brief a description of the habitat as possible. In the heat of collecting, especially when racing the tide, it is easy to mix things up and difficult to take detailed notes, but care nevertheless should be taken to keep facts and specimens straight. Any details that you note regarding the habitat can be jotted down afterwards while they are fresh in the memory.

Freshly taken specimens should never be placed, even temporarily, in a rusty metal container or in contact with rusty chains or other rusty objects for it is remarkable how soon they will become rust stained. It is almost impossible to remove these stains without damaging the shell in the process. Sturdy plastic buckets serve the same purpose at a very reasonable cost, and are easier on fragile species than metal ones.

2.3.12 Glass-bottomed bucket. A useful adjunct to shallow water collecting is the glass-bottomed bucket (first used by sponge collectors), or an equivalent device - sometimes called a water glass or water bucket. Collectors often use one that is square or oblong. An easy way to construct one is to make the frame or box of not too heavy wood and fasten the glass or plexiglas to the open bottom by means of quarter rounds available at any lumberyard. A bead of silicone caulking is placed around the opening between the wood and plexiglas. This ensures a watertight seal. Visualization will be improved if the inside is painted dull black. You can also take a plastic or metal bucket and cut out the bottom leaving a 10 mm ($^1/_2$ inch) rim. To this rim, a circular piece of Plexiglas can be attached with silicone caulking. When in use, frequent wetting of the inside of the glass makes vision clearer and eliminates fogging.

2.3.13 Lights for night collecting. If you have never collected at night, hunting by the aid of artificial light is a revelation. Many mollusks (such as *Conus*) hide wherever they can in the daytime and mainly venture out to feed at night. This is also true of *Marginella, Cypraea, Hydatina,* and the like. While specimens may be taken by the combined use of water glass (see above) and flashlight, you might also try snorkeling or SCUBA (see below and Chapter 4, Snorkeling and SCUBA Diving). Use a bright waterproof flashlight and make certain that you have spare batteries. Extremely good underwater lights can be purchased for less than

$200, and inexpensive models will cost much less. A headlamp with a halogen or LED bulb will give a good cone of light and leave your hands free for collecting. Surprisingly, these can be purchased for less than $100 at most dive shops. Remember, if snorkeling or diving, have a buddy with you, especially if you are collecting at night.

2.3.14 Forceps/tweezers. While collecting small species, a pair of spring forceps comes in handy for small shells like *Caecum*, *Rissoina*, *Cerithiopsis*, Sphaeriidae, etc. Being small, forceps frequently are lost and prudent collectors (especially if they have lost a pair or two) never fail to use a string, tied to both the instrument, and to a wrist or to clothing. However, be sure to make the string long enough so your reach is not restricted.

2.3.15 Loupes. Another device that may prove useful in the field is a small loupe, sometimes called a magnifier or a magnifying glass. These range in magnifying strength from 2X to 20X. Though you might think higher magnification is better, you may be mistaken. For most fieldwork and laboratory work, 5-10X is perfect. This gives sufficient magnification without distortion at the edges of the field, and enough working distance between the specimen and the loupe to allow for adequate lighting (although some loupes can be obtained with a built-in light). With loupes of 15-20X, you will get higher magnification but a very small field of view. If this degree of magnification is necessary, you should consider using a stereoscopic dissecting microscope.

Loupes can be found at craft and jewelry supply shops, at geologist specialty stores, and sometimes even at hardware stores. Two loupes that seem to be good for malacological work are a 7X Hastings Triplet (Bausch and Lomb) and Master Optician's 5X Magnifier (Edmund Optics, Inc., <www.edmundoptics.com/onlinecatalog/displayproduct.cfm?productID=1789>). The former can be attached to a lanyard and worn around your neck while the latter is useful at a workbench. These particular loupes will cost from $30-50 but will last a lifetime. Quite satisfactory 5-10X instruments can be found in the $15-20 range.

Another interesting device that may interest you is the Emoscop SME. This device has been described as an optical Swiss army device. It comprises 3 sets of lenses and depending on how you put them together, you can make several devices. The lenses can be arranged to make a 3X monocular, a 3X telescopic magnifier, a 5, 10, or 15X loupe, or a 30 and 35X microscope. The optical elements compact to 20 mm by 40 mm, and the lenses and microscope base fit in a carrying case that is 30 mm x 65 mm. The device can be purchased from the manufacturer at <www.emoscop.com> and costs $58.

2.3.16 Thread. One of the most useful (and inexpensive) things to have in the collector's kit is a compact roll of rather loosely spun cotton thread or ribbon for holding bivalves together, for tying small chitons to drying boards and other uses that may pop up in the field. Use white thread or ribbon so that you do not have to worry about dyes leaching and staining your specimens.

2.3.17 Tide tables. To attempt intertidal or subtidal (shallow water) collecting without first consulting the tide table would be the height of inefficiency, for all collecting of this type is dependent upon tide conditions. Where there is excellent collecting at a given place at low tide, it would be simply out of the question to do any worthwhile collecting at high tide. Local or regional tide tables may be obtained from the government in most countries (or a fisherman's supply outlet). One can also obtain tide information from the *Old Farmers Almanac* or on the Internet - just look up "tide tables" in any good search engine such as www.google.com or www.dogpile.com. Using the tables is more essential the further north one goes, since ice scouring and sub-freezing air temperatures restrict the number of species able to live in the intertidal and shallow subtidal zone. In many boreal regions such as Eastern Canada, there is decent collecting only at a few new and full moon periods each year, you miss one, and it is months before the next arrives. Further south, conditions are much more congenial to inter- and sub-tidal collecting.

2.3.18 Miscellaneous items. One should not forget items such as maps, a compass, a global positioning system device (GPS), notebooks, field-guides, clothing appropriate for weather conditions, and of course towels and spare clothing for when you stop collecting and wish to get a bit drier and warmer. Also, consider insect repellent, sunscreen and a first aid kit. Additional equipment, such as syringes and dissecting needles, are often used when cleaning specimens. These items are described in the section on cleaning mollusks (Section 2.7).

2.4 FIELD COLLECTING TECHNIQUES

2.4.1 Land snails. Collecting land snails is a fine art and can be infinitely rewarding. The main problem land snail collectors encounter is that of identification. Since many species are endemic to a particular region or even small localities such as isolated valleys or particular portions of mountain ranges, there are few general identification guides, and obtaining regional or national identification literature can be difficult and time-consuming. Nevertheless, the huge variety of forms, shapes, colors, and sculpture (fine-scale superficial detail) make land snail collecting quite worthwhile. See Chapters 9 and 22 for the titles of books that will help you identify land snails.

The first thing to remember when collecting terrestrial mollusks is that mollusks need moisture. Thus, you must look where and when moisture is to be found. After rain and early in the morning when the dew is still present, are the best times to collect from gardens, grasses, bushes, trees, stone walls, and on limestone outcroppings that provide a rich source of calcium for shell formation. Many species will be found in the leaf litter on the forest floor, in localities where the soil is alkaline enough not to dissolve the shells. You will find an assortment of living and dead material this way - some of the dead being very fresh (which saves you the trouble of cleaning them while still providing decent specimens). Bags of leaf littler from damp localities can be collected in the field and many happy hours can be spent at home sifting through it with screens and careful observation especially with the aid of a microscope. Yet another set of species can be found under rocks and old logs. As with marine collecting, the more habitats you check, the more species you come across.

Many terrestrial gastropods enjoy warm, moist areas. Find a site that you believe will contain a molluscan fauna. Take a cotton or burlap sack. Fold it several times and soak it with water. Place it on the ground. On top of the sack place a pile of stones that will somewhat protect the sack from drying out. Check under the sack after several days or a week. You may find a number of slugs and/or snails. Be observant, there will likely be other organisms such as insects, snakes, lizards, and amphibians. In place of the sack, you can also use several layers of corrugated cardboard.

Cleaning and preserving land snails is more difficult than for marine species, since the periostracum, the layer of material that protects the shell from erosion and acidic conditions, must not be removed: it contains the colors and other external characteristics that are vital for identification and appreciation. Chlorine bleach cannot be used as it will remove or damage the periostracum. The best way to go is boiling and picking out the animal with small hooks or pins, followed by vigorous shaking, but leaving them spread out where ants, flies, and other insects can get at them is a very effective method when possible. Small species can be preserved with 80% ethanol.

See Section 2.6 for further cleaning tips for terrestrial mollusks and Chapters 22 and 23 for additional information on land snails.

2.4.2 Freshwater mollusks. Freshwater mollusks include both gastropods and bivalves. May of them can be found by walking along the shallow edges of ponds and lakes, or in shallow streams and rivers. In slightly deeper water one can use an Allison scoop or a van Eeden scoop (see Section 2.3 above) or a Walker dipper (see Chapter 21.4). In yet deeper water, you will need to consider using devices such as dredges, grabs, or bails. More on specific aspects of collecting freshwater mollusks will be found in Chapters 21, 25, and 26.

2.4.3 Marine mollusks. Live material can be collected as well as dead, although most collectors prefer the live material. Live material tends to be preferred as the colors in the shells are more intense (the shells have not been bleached by exposure to the sun) and live shells tend to have less physical damage than ones that are dead and exposed to the action of the surf. Others prefer collecting dead shells since no animal has to be killed and the preparation of the shell requires less work; that is, they do not need to have the dead animal cleaned out of them. Some species can most commonly be found dead - such as deeper water and pelagic species found on the shore after storms or in dredge spoil. Some taxa require special methods of collecting, and these will be mentioned later. When collecting in water, you will be limited by how far you can free dive or reach with your arm or a device like a Walker Dipper (see Chapter 21.4). As a result, many collectors progress to SCUBA diving or dredging to obtain material.

2.4.4 SCUBA diving. The advent of SCUBA diving has opened up a completely new world for collectors, as divers with moderate experience can reach depths up to 30-35 m (around 100 feet) for short periods without special gas mixtures. Beginners would be ill advised to venture beyond 20 m (60-65 feet) until they are able to handle emergencies with fluency and calmness. With this type of collecting, safety is of paramount importance: we do not recommend diving alone or in places where currents or turbulence pose significant hazards. At depths below 15 m (45-50 feet), things can go wrong very quickly, and the results can be dangerous in the extreme. Good, well-maintained equipment and the training to use it properly and with confidence are necessary. Wise divers never push their luck by going deeper than they trained for, or habitually staying down to the last minute of air. Always give yourself a good safety margin: not to do so may cost you your health or even your life!

Much could be written about SCUBA collecting, but the experience of the authors and the scope of this chapter allow only the presentation of a few general principles. In addition to the above cautions, wearing a wet-suit is usually a good idea, especially in cooler waters (hypothermia is insidious and can sneak up on one) or around corals, which can often sting exposed flesh with only a light touch. Sharks are much more benign than generally thought, but they and other carnivorous wildlife should be treated with due caution and respect. You should never tease or relate to them with casual familiarity or carelessness. Jellyfish can be particularly nasty, as can eels and other inhabitants of holes and crevices in rocks or reefs. Use the utmost caution in tropical and sub-tropical waters when trying to find out what is in places one cannot directly see. A special caution is in order for cone snail collectors. Many species, not just the very venomous ones (*Conus geographicus* Linnaeus, 1758, *C. striatus* Linnaeus, 1758, *C. textile* Linnaeus, 1758) can sting you, sometimes with painful results. Living cone snails should always be handled with care.

With shore-based collecting, the more habitats investigated, the more species one is likely to find. Unless you are looking for a particular species or group that you know to be found only in specific habitats, the more inquisitive you are, the more you will find. Sandy bottoms are generally of limited interest. When investigating a reef, it is wise to explore associated rubble fields and dead as well as living coral heads. In places where sponges are abundant, interesting species (triphorids, cerithopsids, etc.) can often be found via the judicious harvest of a few sponges for through dissection on land. For a more extensive treatment of SCUBA diving and snorkeling, see Chapter 4

2.4.5 *Ex pisce* collecting. Sometimes mollusks can be found in the digestive tracts (stomach and intestines) of bottom dwelling fish. The first challenge in *ex pisce* (Latin: from fish) collecting is to obtain the fish or the digestive tracts. The easiest method is to obtain them from the crew of a commercial fishing boat. You can either accompany the crew and get the digestive tracts of the bottom feeding fish they catch or you can supply them with covered buckets in which they can save the digestive tracts for you. They might be more willing to save material for you if you offer to pay for the captain and crew's assistance. When they give you the entrails ask for locality data such as longitude and latitude of the catch, bottom depth, and bottom type if known. Especially provident in the Atlantic Ocean are toadfish and batfish - the latter more often taken by trawling than by angling (H. Lee, pers. comm.). The dover sole can be used in the Pacific Ocean.

Start from the stomach and work your way down the length of the intestine. When you encounter a solid feeling object cut open the organ where you feel it. Wash the object and see if it is a shell. Record where it was found. Shells in the upper portion of the digestive tract are likely to be from the locality where the fish was caught while shells further down may be from a site some distance from where the fish was collected (Clapp 1912).

Take the semi-solid intestinal contents and place them in a strainer or sieve with a fine mesh. Wash the material until the water that drains from the mesh is clear. Allow this material to dry (it will be fairly odorless if washed sufficiently), and then examine it with a loupe or microscope. You may find many micro-shells this way.

While large shells are not found this way, this method has advantages. It can be used for depths where you would not be able to collect by snorkeling or SCUBA diving. Also, it may be productive in areas with rocky bottoms where a dredge would not work. Lastly, the cost of equipment is negligible; definitely less than the cost of dredging equipment and a boat.

2.4.6 Sea stars (starfish). Sea star (formerly known as starfish) stomachs may also be a source for some mollusks. Sea stars can often be obtained as a by-product of commercial fishing. Sea stars have two modes of feeding. Those with long, thin arms [e.g. *Asterias forbesii* (Desor, 1851)] bring the mollusks to their mouth; they then evert their stomachs and eat and digest the mollusks. These types of sea stars do not contain mollusks. Sea stars with short or stiff arms [e.g. *Astropecten articulatus* (Say, 1842) and *Luidia clathrata* (Say, 1825)] swallow mollusks whole and their stomachs may contain shells.

To extract the mollusks, one needs to dissect the digestive tract out of the sea star. The plates from the ventral surface of the sea star are removed from the central disc to a point approximately one third of the way out the arms. The digestive tract is removed and placed in full strength household bleach (5% sodium hypochlorite) for several minutes. When the tissue starts breaking up, it is washed with water in a funnel lined with coarse filter paper. The resultant residue is allowed to dry and then examined with a loupe or dissecting microscope. The shells are picked out with fine forceps.

While this may seem to be a labor-intensive technique, for the collector interested in micro-shells it may be quite productive. Porter (1972, 1974 and references therein) should be consulted for further details regarding this technique.

2.4.7 Tidal pools. You can search tide pools for mollusks. These are generally found in rocky areas. Be forewarned, in many areas tide pool collecting is regulated or illegal. Know what is allowed in your area before you proceed.

2.4.8 Traps. In recent years, mollusk traps have been used to take many of the carnivorous mollusks and shells of non-carnivorous mollusks inhabited by hermit crabs, and reports from those who have used traps have been very encouraging. Some collectors reported success collecting these types of mollusks by simply placing a piece of meat or a dead fish between two sheets of wire mesh and weighting it down with stones overnight. Others have weighed down canvas packets of dried animal dung. For those with limited resources, this remains a very practical collecting method. For more on this technique see Allison (1942). If you can obtain the cooperation of lobster or other trap-based fishermen, all the better! You can even develop your own specialized traps, if the fishermen will agree periodically to collect and re-distribute them for you.

2.4.9 Navigational buoys. Navigational buoys can be searched for evidence of sessile mollusks. If one can gain access to buoys when they are brought in for cleaning, many attached specimens may be recovered. You may be able to obtain the records that indicate when the buoy was placed and recovered. This will allow you to study the colonization of the buoy over a defined time period.

Other avenues of research that occur when studying the fauna attached to buoys include variation in species populations over a geographical range (if multiple buoys are studied) and variations with depth if the anchor chain is also studied. For more on collecting from buoys see Merrill (1974).

2.4.10 Marine grasses and algae. You can also collect mollusks from seaweed and sea grass. Often mollusks will be attached to these plants; however, they may be quite small. Place the seaweed or sea grass in a bucket containing freshwater (not salt water). The freshwater will cause many of the mollusks to loosen their hold and drop to the bottom of the bucket. You can then search through the sediment for these mollusks. An alternative is to decant the freshwater and then replace it with saltwater. As the mollusks recover, they will start moving around. You can then pick them out with your forceps. Also, check the seaweed or sea grass for mollusks that may be attached by a byssus (e.g., *Mytilus*).

Different seaweeds and sea grasses may harbor different mollusks. Check different marine grasses and algae to maximize your yield. Remember to preserve a sample of the plant or algae. You can identify them later and add this habitat data to the specimen label for the appropriate mollusks.

2.4.11 Commercial fishing boats. One of the very best ways to collect deeper water species, especially for collectors without a boat of their own, is to go out on fishing boats of the sort that scrape the bottom seeking scallops, shrimp, groundfish (where you can get the gut contents also), and crabs. This is not for the faint of heart, it must be said, since the smells, often coarse language, accommodations (usually quite spartan), schedule (often 24 hours a day, or at the very least, from dawn 'til dusk), and the constant exposure to massive wastage of marine life by way of by-catch are not the sort of thing easily sustained by everyone. One of us (RM) has found that if you do some work around the ship as you are able, the fishermen are usually extremely friendly and occasionally surprisingly helpful. Safety is a primary consideration. Avoid all moving equipment; get out of the way whenever necessary. Wear a life preserver as a matter of course, and wear footgear that is not only waterproof but also grips the deck securely. Be sure to bring sunscreen, dress for the weather (which can be quite changeable, and considerably cooler than on the shore), and be especially careful in rough weather. It is one thing to fall overboard when it is calm and you can be quickly and easily rescued, but to go over in high waves or even a heavy swell, is quite another matter.

There is certain equipment that should be brought on a fishing boat. Screens for sifting fine or mixed sediments are usually the most productive both qualitatively and quantitatively. You can often make use of the ship's onboard pumps for sifting sediments. Containers of various sizes should be brought; some filled with alcohol (90% or higher concentration of ethanol or isopropyl is best, but be careful with the latter since it is poisonous and you do not wish to contaminate the catch) for preserving the smallest specimens. Have a pair of waterproof gloves (an absolute must, although to pick up the smaller mollusks you usually have to take them off), two changes of clothing, good rain gear, rubber boots, and such materials as you will need to bring home the larger specimens. Most often, the fishermen are glad to share their food, especially if you chip in for your share.

It is important to note that you should check every trawl, and if the operation is 24 hours a day, try to train the crew to spot the best tows for you, so they can awaken you if an especially promising one comes along. As with many things in life, the dictum "You snooze you lose" is especially true here, since, as a general rule, the great majority of your catch, especially of the smaller species, is often found in a surprisingly small number of tows. You may see nothing for hours or even days, then one or two incredible trawls will come up that will make the entire trip worthwhile.

Seldom will you be able to clean your catch onboard. On day trips, this is not a problem, but if you are going offshore for a week or two at a time, you will probably be able to avail yourself of the ice that the crew uses to preserve your catch for prolonged periods. Some specimens can be kept alive in buckets of seawater or on-board tanks, but be sure to refresh the water often to ensure an adequate oxygen supply.

2.4.12 Specific methods for select groups. For each group or type of mollusk, varying search and collection techniques will produce the best results. Techniques must be learned either from an experienced collector or by trial and error, and each collector will develop his or her own favorite techniques. For example, micro-shells can be collected, often in surprisingly shallow water, by lifting medium-size stones and other objects, and gently brushing off the adhering matter into a collecting-bag. Scooping up some of the shell-rich material underneath and at the sides of rocks is certain to yield rich rewards. Terebras are often found in sand, and their trails can be tracked by those who learn what to look for. Cones and many other groups are best collected at night, using a headlamp, perhaps with a red filter to avoid scaring the animals off. Silt and bottoms with mixed clast sizes (i.e., poorly-sorted sediments) are particularly rich in diversity, and taking sediment samples in the most promising spots (a matter of experience and developing an "eye" for such things) will often produce pleasant surprises. For more information on snorkeling and SCUBA, see Chapter 4.

If you do not know how to dive or if you need to collect from depths that are greater than your dive ability, you may have to resort to other techniques. One can dredge for shells, use a bail hook, or use a grab. These techniques are addressed in Chapter 3.

What follows now will be some techniques unique to several special groups. While this is not a comprehensive list, it will give you a good idea of some of the many aspects of collecting. Further information can also be found in Chapters 16-27.

Cyphoma: A useful device for collecting *Cyphoma* consists of a child's toy rake, to be obtained at any hardware or ten-cent store. The teeth of the rake are bent inward on a curve and a small piece of screen wire attached at each side and at the back, thus forming a sort of basket with projecting teeth. By bringing the hook under a strand of *Gorgonia* on which a *Cyphoma* is resting, it may be hooked or scraped off and falls into the basket. By attaching a long handle to this device, it may be used from a boat in deeper water.

Rock borers: If rock borers such as *Petricola* are to be collected, a strong hammer and two or three chisels are needed. One that is 20-25 cm (8-10 inches) in length with a cutting edge of 20-25 mm ($^3/_4$-1 inch) is useful when work must be done in shallow water. Usually most of this type of collecting is done at low tide when the rocks are bare, and a chisel 12-15 cm (5-6 inches) long with a cutting edge of 15-25 mm ($^1/_2$-1 inch) is one that will be most frequently used. When really delicate work is required, a narrower cutting edge is desirable. For more on this group see Chapter 27, Marine Bivalves.

Shipworms: The Teredinidae (shipworms) are wood boring bivalves. To quote Ruth Turner "Most of the characters which differentiate the genera and species in this family are invested in the pallets and largely in the periostracum which covers the calcareous portions of the pallets. It is essential, therefore, that only fresh specimens preserved in glycerin alcohol be used for study as the periostracum sloughs off when the pallets become dry. Distributional records should be taken only from collecting boards or permanent structures as these borers are easily transported from one locality to another by driftwood.

"The best specimens are obtained if the animal is dissected out as soon as the board is removed from the water, or, if this is impossible, the board should be submerged in 70 percent alcohol for a week and then shipped to the laboratory wrapped in a cloth saturated with alcohol. Once the specimens have been extracted from the wood they should be preserved in a mixture of 4 parts alcohol (70%) and one part glycerine. This keeps the periostracal margin of the pallets soft and pliable and, should the alcohol evaporate, the glycerine will keep the pallets moist for some time. The shells and pallets of each specimen should be kept together in a vial and all the specimens from one board should be given the same number" (Turner 1974).

Burrowing clams: When collecting *Cyrtopleura costata* (Linnaeus, 1758), *Ensis directus* Conrad, 1843 and other similar bivalves, it is well to have a bucket of sea water at hand in which to place the specimens as soon as secured, for it is not uncommon for these species to contract their muscles with such force as to fracture the shell. Placing the specimens in water immediately seems to overcome some of this breakage, but even when all precautions are taken, some are broken in this manner.

Cyrtopleura costata and other burrowing mollusks are usually found in colonies, very frequently where there is a mud bottom. Digging out the first specimen or two may cause the water to become so muddy that further collecting is out of the question as the burrows cannot be located. In such cases, a good plan is to secure a supply of small switches from some nearby trees and place a switch in each burrow as it is located and not to start digging until all desired burrows are marked. You may proceed from one switch to another and be able to locate as many specimens as you intended to collect.

Chitons: Smaller chitons may be removed from the rocks with the blade of a small penknife but with the larger ones, a heavier knife is necessary. Care should be taken to not injure the specimens. Injured chitons inmediately roll up and are difficult to flatten subsequently for further preservation. If you are observant, you may determine which end of the chiton is the head, and apply the knife at the other end; they are much easier to remove from the rock in this manner. Chitons are unusual mollusks, and certain procedures that might achieve results with other mollusks do not seem to apply to them. For instance, should the collector desire to study the living animal in action, it is advisable and necessary that the vessel in which it is confined be kept undisturbed. Sometimes if you gently rock the vessel containing the specimens, in a surprisingly short time, even uninjured chitons may unroll and attach themselves to the bottom of the vessel and they can be easily removed and placed upon the stretcher.

For smaller chitons, nothing better can be found on which to stretch them while in the process of preservation than the common wooden tongue depressor or popsicle stick. Some collectors favor the use of narrow strips of glass (such as microscope slides) or plexiglas. The chiton slips into the desired position more easily by reason of the smooth surface. Nonetheless, wood still comes in useful for many species. The chitons are bound to the wood or glass with narrow strips of cloth or ribbon. In cases where collecting time is limited, the living chitons can be placed in a vessel of seawater and set aside. Eventually most specimens will uncurl and they may be placed upon the stretcher and tied firmly with cloth strips. The smaller chitons may be brought back in vials of seawater and mounted on the stretcher at the collector's convenience.

In some instances, the bodies of the larger chitons can be removed from the plates as soon as the specimens are taken and the shells are placed at once upon the stretching board. However, even the largest of specimens may be nicely preserved in a flexible form via the following method (Hanselman 1970).

Chitons, that are tied to a tying board or stretcher, are narcotized by placing them in a solution of 50% isopropyl alcohol and 50% water for 24 hours. Once dead, they are transferred to "chiton goop." This is a mixture of one part glycerin to one part 60% isopropyl alcohol. The chitons are left in this solution for varying lengths of time: 10 days for specimens under 15 mm, 15 days for specimens from 15-35 mm, 20 days for specimens 35-60 mm, 25 days for specimens 60-75 mm, and a proportionally longer time for specimens over 75 mm. While soaking in the chiton goop, they can be removed from the tying board so as to allow the solution to penetrate the specimens adequately.

After the appropriate amount of time soaking, the specimens are removed and allowed to drain on paper towels. They are then reattached to tying boards and allowed to air dry slowly. The time to air dry will roughly be equivalent to the time they were in the chiton goop. You should avoid forced drying in an oven or direct sunlight as the results are not as good as slow air drying. This drying will result in specimens that retain a degree of flexibility.

On the other hand, one can air dry specimens after narcotization. The time to a finished specimen will be shorter but flexibility will be sacrificed. One should remove all organs, leaving only the shell plates and mantle. The soft parts can then be stored in ethanol for future study. Alternately, after narcotization, the whole chiton can be stored in ethanol. These methods do not contaminate the specimen and instead leave the soft parts in a usable condition for researchers (E. Schwabe, pers. comm.). Burghardt and Burghardt (pers. comm.) tie chitons to microscope slides with clear nylon thread. They then place them in vials of ethanol and seal them. The chitons appear to be suspended in the fluid. Additional information on collecting and preparing chitons can be found in Chapter 18, Burghardt and Burghardt (1969: 5-6), and Scheidt (1982).

Pelagic mollusks: One group of mollusks not commonly collected is the pelagic mollusks. These can be collected by using a plankton net. This is a metal hoop to which a net is attached. At the tip of the net is a jar or container. The net is towed through the water and suspended items are collected by the net and concentrated in the container. One then sorts through the contents of the container searching for mollusks. Pelagic organisms will rise and sink in the water column at different times of day. Thus, it is advisable to collect at different times of the day and night to maximize your yield.

2.4.13 Ecological consideratios. Before leaving the subject of collecting, it is well to emphasize that you should leave the ocean floor or wherever you have collected as nearly as possible in the same condition as when found. In the marine or freshwater environment, return overturned rocks to their original orientation. Not doing so is an indication of a thoughtless collector. In days or weeks, the rocks will become bleached and devoid of the organisms that were living on it. It will take many more months for them to be covered by the natural growth necessary to the maintenance of molluscan and other marine life. The undersides of many rocks support what can be considered miniature communities, which are quickly destroyed when the rocks are left overturned.

It is understood that one should also avoid overcollecting. While there are many other far more important causes of population depletion for most mollusk species, it is still common sense to collect only what you need for your own purposes and, where there are enough specimens available, for trading as well.

2.5 TECHNIQUES FOR NARCOTIZING MOLLUSKS

Before beginning the discussion of cleaning techniques, we would like to discuss the topic of narcotization. Sometimes when mollusks are collected, it is important to preserve the animal (soft parts) as well as the shell. The process of narcotizing or anesthetizing will render the animal senseless and relaxed. If the animal is left in the narcotizing agent for a prolonged period of time the animal will die. If anatomical studies are anticipated or if you want to minimize damaging the shell when removing the soft parts, you will want to relax the animal by using a narcotizing agent. While there is no single ideal narcotizing agent, there are some recommendations that can be offered.

For terrestrial gastropods, the simplest way to relax the animal is by placing it in a jar of water sealed so that there are no air bubbles in the container. The animal will die in a relaxed position outside of the shell (if one is present) within 24-48 hours (Pilsbry and Vanatta 1898, Gregg 1944). As soon as the animal is dead, it should be transferred to the appropriate preservative. A good, general-purpose one is 80% ethanol. Some researchers use a mixture of 80% ethanol, 15% water, and 5% glycerin. If this mixture is used, do not use jars with metal lids to store specimens, as glycerin will accelerate the deterioration of the metal lid (see Chapter 5). You can also use formaldehyde but then the tissues cannot be used at a later time for DNA studies (formaldehyde cross-links proteins making the DNA difficult to extract). When DNA studies are anticipated, a 95% or higher concentration of ethanol should be used as the preservative, and the animals should be preserved immediately without drowning (Schander and Hagnell 2003).

Another agent that can be used for terrestrial gastropods is chlorethanone (chlorobutanol) (Hubricht 1951, Clement and Cather 1957). One makes a saturated aqueous solution of chlorethanone and then dilutes one part of this solution with 10-20 parts of water. The snails are put in this solution. Depending on the size of the specimen, it can take from 12-48 hours for complete narcotization. The specimens should then be transferred to the preservative of choice. A. G. Smith (1962) found that chlorethanone also worked well for some marine gastropods.

Nembutal and pentobarbital have been found to be good general narcotizing agents for a variety of freshwater mollusks (van der Schalie 1953, Heard 1965, Runham et al. 1965, Meier-Brooks 1976, Coney 1993, Araujo et al. 1995). These chemicals are considered controlled substances (substances of abuse potential) by the United States Government. You may have difficulty obtaining them unless you have a federal license to possess them or are working with someone who has such a license. Barker (1981) found nembutal a good narcotizing agent for pulmonates and Aquilina and Roberts (2000) found it good for *Haliotis*.

Craze and Barr (2002) used electrical-component freezing spray to kill snails. This material is packaged as an aerosol spray and will cool to -50°C. They found this material quickly froze snails up to approximately 30 mm. The upper limit was mainly dictated by the increasing amount of spray needed to accomplish the task. They found that very small snails might be blown away by the spray. They solved this problem by placing the snails on a sheet of aluminum foil and spraying the under-surface of the foil. Occasionally, they noted a vacuum phenomenon. When they attempted to remove the thawed snail's body from the shell a piece would break off in the apical whorls due to a vacuum created there. They solved this problem by drilling a small hole in the upper whorls to equilibrate the pressure. Electrical-component freezing sprays come in two forms: chlorofluorocarbon (CFC) based and CFC-free. For the protection of the environment, we recommend that the CFC-free forms be used.

Other narcotizing methods utilizing propylene phenoxetal (Owen 1955, Owen and Steedman 1958, Turner 1960, Rosewater 1963, 1965, Mills et al. 1997), methanol (Smith 1996), ethanol (De Winter 1985) Sevin[R] (Carriker and Blake 1959),

magnesium chloride (Smith 1961), amyl chlorohydrin (Smith 1961), and freezing (Carriker and Blake 1959, Bowler et al. 1996) have also been described. Van Eeden (1958) found that a combination of menthol and chloral hydrate (a controlled substance) worked well on freshwater gastropods such as *Physa* and *Bulinus*. Mueller (1972) called this combination Gray's Mixture, and found it useful for relaxing many classes of marine mollusks. Papers by Runham et al. (1965), Crowell (1973), Mueller (1972), Coney (1993), Araujo et al. (1995), Norton et al. (1996), and Aquilina and Roberts (2000) compare multiple agents.

If you must narcotize mollusks, the best way to learn how to do this is by apprenticing yourself to someone with experience. If this is not possible, you should obtain the literature mentioned above and study the intricacies of narcotization. Keep careful records and consider publishing your results and experiences. This is a topic where more in-depth knowledge is needed.

2.6 TECHNIQUES FOR MARKING AND TAGGING MOLLUSKS

Occasionally you will want to study the movements of mollusks. You might want to study how dead shells are moved by the tides. On the other hand, you might need to study the movements of a group of live mollusks. What follows is a list of techniques for marking shelled mollusks. While there are techniques for tracking mollusks without shells (Anderson 1973, Richter 1976, Grimm 1996, Coyer et al. 1999), further discussion of these techniques is beyond the scope of this chapter. Coyer et al. (1999) reviewed methods for tagging various forms of marine life.

One simple way of marking a shell is to write a number directly on it. This can be done with an indelible fine-tipped marking pen or with a fine-tipped drafting pen such as a Rapidograph[R] pen (Koh-I-Noor, Corp., available at art stores and many office supply stores). If the shell is dark, you may have to put on a patch of a quick drying white enamel-like substance such as is found in Liquid Label[R] (Light Impressions, Brea, California). You can also try quick drying enamel paint. It may be necessary to file the periostracum or shell to make it smooth enough to write on. When the shell is numbered, cover the number with a cyanoacrylate adhesive (Superglue, Krazy Glue®) or acrylic polymer (found in Liquid Label®) or dental acrylic. This will protect the number from abrasion (Lonhart 1999 and Lemarié et al. 2000). Some researchers have recommended clear epoxies, however, Lemarié et al. (2000) recommended against using epoxies. They found epoxies to be inferior to cyanoacrylate adhesives.

Another technique is to use pre-printed numbers. These numbered discs or labels are generally 3-8 mm in size. They are attached to the shell using a cyanoacrylate adhesive or acrylic resin. As with writing the number directly on the shells, you may have to file the shell to allow the tag or disc to make better contact. Pre-made tags that have been used for marking include Shellfish Tags (Hallprint Pty., Ltd., Holden Hill, South Australia, Australia), Fingerling Tags (Floy Tag and Manufacturing, Inc., Seattle, WA) and the Queen Marking Kit (bee tags) (E. H. Thorne, Ltd., Wragby, England).

You can also print numbers using a small font, such as a 6 or 8-point font, on waterproof paper with an indelible ink or print them with a drafting pen. Then cut them out or use a hole punch and then attach them to the shell as described above. Young and Williams (1983) used a variant of this technique. They printed numbers on Dymo® tape (a plastic tape). They attached the tags to mussels using a cyanoacrylate adhesive. In studies lasting almost three years, they found that 95% or more of their tags were still attached to the marked mussels.

You can physically alter the shell as a way of marking it. Using a diamond-tipped scribe (available at most hardware stores) or a file, you can scratch a number or code on the shell. If you use a code, remember to record the coding instructions so that others can decode the markings at a later time. Some researchers have used drills to encode shells (Thoma et al. 1959, Wolda 1963, Kleewein 1999). Ropes and Merrill (1970) used either a file or a Dremel Moto-Tool® for notching surf clams.

A method of marking bivalves, not requiring disturbing the specimens to read the mark, has been described by Englund and Heino (1994). This method used a small float attached to a small disc with a cord. The disc was glued to the shell. The middle of the cord was tied to a tag carrying the number or coded message.

Lastly, you can code shells by using colored enamel paints. For instance, a linear arrangement of four dots using five colors will allow you to mark 625 specimens differently. After the paint has dried, cover it with a cyanoacrylate adhesive or acrylic polymer for added protection and durability.

2.7 TECHNIQUES FOR CLEANING AND PRESERVING MOLLUSKS

Methods for cleaning and preserving mollusks are many, varied, and controversial. There are several major methods of cleaning. They include using alcohol, refrigeration or freezing, boiling, cleaning by insects (ants, flies, mealy-worms), microwaving, and ultrasonic cleaning. A technique was even described that uses sea anemones (Fox 1935). Which technique you use will often depend on the type and size of specimen that you are cleaning.

2.7.1 Boiling. A great number of gastropods are cleaned by simply boiling them and removing the animal. Care should be exercised in bringing the animal out with a circular or corkscrew motion. In other words, it should be twisted out and with larger specimens such as *Fasciolaria*, *Busycon*, and the like, an ice pick driven into the body will give a firm hold where it is most needed.

Never boil too great a number of shells at one time, other than bivalves. The bodies of gastropods are much more easily removed while hot, and shells that have been boiled and set aside for a time and have become cold are hard to clean. At a temperature around 65°C (149°F), the columellar muscle relaxes and the body of the animal can be twisted out. If the specimen cools down, the muscle can reharden, making removal of the animal difficult. Also, unless great care and diligence is taken, the liver and other soft parts are often left behind to make themselves very evident by odor at a later time. Very few *Murex* and *Vasum* are perfectly cleaned, as they seem to have a weak connection between the muscular part of the body and the viscera; usually there is a break at this point when drawing out the bodies, leaving portions that are next to impossible to remove.

When boiling, measure the largest shell in the pot and allow approximately one minute of boiling time per 2.5 cm (1 inch) of shell length. This seems to be an adequate amount of time for most gastropods. If bits of tissue remain behind, you can try to flush them out with water, alcohol, or pine oil or place the shells by an anthill (but avoid placing shells in sunlight as they will tend to fade).

2.7.2 Hooks and pins. Hooks useful in drawing the bodies from small gastropod shells may be made by whittling small cylindrical handles from some soft wood, making them 75-100 mm (3-4 inches) long and the diameter of a lead pencil. Into this handle push a needle (with the eye downward into the wood), leaving the point out. This point may be bent to any degree of curvature by heating over a Bunsen burner and bending while white-hot (exercise care when doing this). Some collectors achieve excellent results while using a piece of fairly stiff springy wire such as used as a leader on fishing lines. There are many beautifully prepared specimens for which the only tool used was a safety pin with the point bent into a hook. You may also use a dissecting needle and sometimes a dissecting probe or crochet needle will work.

2.7.3 Flushing with water. While your hook may be a useful implement, you will often leave some tissue behind. At these times, you may resort to water to flush out the remaining tissue. One implement to flush out tissue from gastropod shells is a syringe. The best syringe to use is the one used by dentists and physicians and is of the piston type with a luer-lock to hold on the needle. The common rubber ear syringe may be used. This does not give nearly the amount of force, but is much less expensive and easier to obtain. These work well for small to medium sized shells. You might also want to try an oral irrigating device such as a Water Pik® or Hydro-Pik®. Lastly, for larger gastropods, a reducing nozzle on a garden hose may be appropriate.

To use the alcohol (ethanol) or pine oil treatment, place the shells in a box of sand aperture oriented upward. Fill the shells with the alcohol or a few drops of pine oil. Let them stand for 12 to 24 hours and then rinse them out with water. This method will frequently loosen up retained bits of tissue and allow them to be flushed out. Sometimes a second treatment will be needed. In our experience, most curators in museums do nothing to remove the final traces of tissue. As tissue dries out over time, the smell disappears. Thus, no foreign substances are introduced into the shell. Also, if the humidity in which the collection is stored is kept within appropriate levels, the residual tissue may desiccate.

When at all possible, work underwater. This will prevent the shell from shooting across the room or backyard. It will also help you to avoid losing opercula if they are present in the shells that you are working on. Lastly, you will lessen the risk of splattering yourself with bits of molluscan tissue.

2.7.4 Preservation of tissue. The soft tissues of mollusks are best preserved in a fluid medium. However, if the tissue dries out, it may not be useless. José Leal (pers. comm.), mentioned that he has written more than a couple of papers based on anatomical data drawn from dried soft parts found inside old shells from museum collections, some of them more than 40 years old. He strongly recommends that, if it becomes important that the material should be preserved dry, then odor should not be the emphasis. Instead the preservation of as "artifact-free" a sample as possible should be the aim. Actually, the smell may be minimal if the tissue remains dry; this requires that you control the humidity where the collection is stored (see Chapter 5.2). For methods to reconstitute dried tissue, see Chapter 5.9.

In case it is desired to preserve the specimens entire, ethanol (80-95%) has a decided advantage over most preservatives, especially when the animal is to be used for future dissection, study of molluscan anatomy, and DNA studies. It is very easy to lose the identity of soft parts while preparing shells, so it is well to label each specimen at once. Ordinary bond paper on which the name is written with lead pencil is very satisfactory and may be affixed to the specimen with a short length of thread or string. Although most collectors throw away the soft parts that were extracted from their shells, many museums will be grateful for such material sent to them. In this manner, the amateur collector may, in a small way, partially repay the large amount of free, unselfish, and efficient help given by these institutions (for more on this, see Chapter 14).

Alcohol is often a first step in preserving the dry soft tissue of the smaller species and many are placed in this preservative immediately. Usually a concentration of 80% or higher of ethanol or isopropyl alcohol is used. Isopropyl alcohol is more poisonous but easier to obtain. The smaller operculate shells should be allowed to remain out of water until death occurs and then placed in alcohol before decomposition sets in. In so doing, the opercula are preserved in plain sight. Should these be placed in the preservative before death occurs, the animals retract within the shell, and the opercula may not be seen in the prepared specimens. The specimens should remain in alcohol for at least 24 hours if small and longer if large. After removal, the fluid should be drained from them and they should be allowed to dry for at least a week before packing them away.

2.7.5 Formaldehyde. Some people use formaldehyde for preserving mollusks. This is an acidic substance and in general, if you need to use a preservative, we recommend alcohol. A dilute, well-buffered solution of formalin (a 37-40% aqueous solution of formaldehyde) can be used as a last resort, but be certain to counter its acidity with a neutralizing agent such as limestone chips or borax (sodium borate). These should put in the day before so they have a chance to buffer the solution properly. If you do decide to use formaldehyde do so with caution and in a well-ventilated room. This substance can cause respiratory irritation, and it is believed to be a cancer-causing substance (carcinogen).

2.7.6 Cleaning shells with insects. A method of cleaning long in use by some is the use of blowfly larvae or maggots. This method, while not for the squeamish, is very efficient, and, after a period of time, vigorous rinsing of the shell is all that is necessary. Some other collectors will place shells by anthills and allow the ants to clean out the shells. If you undertake these methods, make sure that animals such as raccoons do not carry off your specimens. Place the shells under a wire box with a weight on top of it. The raccoons will not be able to get to the shells while the insects will.

2.7.7 Vacuum pumps. A novel but quite effective method of cleaning out the last bits of the tissue from old, smelly shells or narrow ones where a bit of the body often remains in the spire (*Terebra* are a major offender here) is the use of a vacuum pump (P. Monfils, pers. comm.). Place the shell, aperture upward, in a bowl of chlorine bleach or hydrogen peroxide. The bottom of the bowl should have a layer of sand or glass beads to support the shell. Place the bowl in a bell jar or vacuum dissector. Using a vacuum pump, decrease pressure slowly so as to allow trapped air bubbles to escape. The bleach or peroxide will now flow into the deepest recesses of the spire and dissolve the remaining fragments of tissue. Use this method only with glassware designed for use with vacuum pumps; to do otherwise risks an implosion. It is also wise to use such an apparatus behind a shatterproof shield in case of an implosion. Use this method only with shells where destruction of the periostracum is of no concern, because the bleach or peroxide will damage the periostracum (see above - Chapter 2.4 Field Collection Techniques: Land Snails, and below - Cleaning the External Shell Surface).

2.7.8 Microwave ovens. Some collectors are now using microwave ovens to clean gastropods. This seems to work best for medium sized shells (2-15 cm). Wrap each shell in a paper towel, place in a covered dish and microwave for one to two minutes. Remove the shells and twist the body out with a corkscrew motion. As microwaves differ in their wattage, you might have to experiment to see what the optimal time will be for a given sized shell in your microwave. As with boiling, if the shells cool down, removal of the body may become difficult. Do not microwave too many shells at a time.

This method can be a messy process, and wrapping the shells in a paper towel serves two purposes. First, it will keep an operculum (if present) associated with its shell. Second, it will keep tissue explosions from soiling the microwave oven. This will cut down on the bad odor left behind and make cleaning the microwave oven easier. As with other methods of cleaning shells, you may also need to flush the shell out with water or use the alcohol or pine oil treatment after microwaving.

2.7.9 Cleaning the exterior shell surface. Some collectors prefer specimens just as taken from the water, but many are more fastidious and try to enhance their beauty by cleaning off all extraneous matter. The periostracum can be removed by a stiff brush. It is much simpler to dissolve it with full-strength household chlorine bleach (5% sodium hypochlorite). This soaking is sometimes followed by brushing to remove what the bleach has not removed.

Where there is a quantity of coralline or calcareous growth, it is necessary to remove this bit by bit and very carefully so as not to injure the shell. Soaking in chlorine bleach will prove extremely useful here as it helps separate the concretions from the shell. A shoemaker's awl is a very efficient tool, as are the very sharp, disposable knives found in hardware stores, however, any sharp pointed instrument will do, including dentists' tools.

Shells that are covered by a mass of vegetation or spongy growth and other encrustations may be readily cleaned by immersion in any of the chlorine bleach solutions found at most grocery stores (5% sodium hypochlorite solution). Do not fear for the integrity of the shell; the chlorine will not attack the lime of the shell, however, the specimen should not be allowed to remain in the solution for too long a time as some loss of color through bleaching may occur. In general, full strength household bleach may be used, but some prefer to use half-strength, especially for more delicate shells. For rugged but encrusted shells, full strength solutions are always best.

When using chlorine bleach, try not to get it on your hands as it can irritate skin (wearing rubber gloves is recommended, but be mindful of latex allergies) or clothing as it tends to eat plastic-based materials such as polyester and acrylic, and can fade the pigments of your jeans or other clothing instantly. Also, remember to rinse the shells well after cleaning to remove all traces of the bleach.

When considering whether to clean the exterior of shells, you must weigh several other factors. In some shells, the color and markings (rays, flammules, etc.) are contained within the periostracum and not the shell itself. Removing the periostracum will leave you with a bleached, white shell. Freshwater bivalves and gastropods and terrestrial gastropods fall into this category. Also in this category are some marine shells such as *Perna viridis* (Linnaeus, 1758) and some brackish water shells. A. G. Smith (1962) found that recently killed freshwater mollusks could be treated with chlorine bleach for one or two minutes. This would remove encrustations and adhered algae without destroying the periostracum. The techniques did not work with old, long-dried out specimens.

Some other shells will show a change in color when treated with chlorine bleach (examples include *Conus ebraeus* Linnaeus, 1758, and *Conus dorreensis* Péron, 1807). When in doubt, treat only one or two shells of a given type or check with someone who has experience with the group of shells in question. Some people also like to leave the encrustations on the shell to show what the shell is associated with in its natural habitat. Other people like to remove all the encrustations and have a clean shell. You can compromise and clean several shells leaving the rest in their natural condition.

It is especially hard to clean shells more or less disfigured by an unsightly mass of barnacles. Often, the mass may be detached in its entirety by applying pressure at just the right point with the cleaning tool, but all too frequently this is not the case and an unsightly white blotch is the result. This may usually be removed by carefully scratching the remains of the barnacle with the tip of a sharp pointed knife and reducing the remains of the barnacle to a powder, care being taken not to injure the shell.

2.7.10 Ultrasonic cleaner. Another method to clean off encrustations is an ultrasonic cleaner. Placing the shell in the tank and turning it on frequently removes the encrustatation without damaging the shell. Try this method on less desirable specimens before applying it to rare or one-of-a-kind shells. The main limit to this technique is the size of the ultrasonic cleaner. The larger the tank (bath) the greater the cost will be. An ultrasound tank with a 10 x 15 cm (4 x 6 inches) tank will be large enough to clean the majority of shells (A. G. Smith 1962, Pojeta and Balanc 1989).

2.7.11 Walnut shell blasting. Recently, a number of reasonably priced machines, operated by compressed air, have become available. They use finely ground walnut shells as a gentle abrasive for cleaning delicate objects. If using this method to clean mollusks, it is best to test the suitability of the method and the appropriate grit size, by first using it on the least valuable specimen. Otherwise, tragedy could easily result.

2.7.12 Techniques for specific groups. What follows are some specific cleaning suggestions for certain groups of marine mollusks.

Janthina: When a crop of *Janthina* is thrown up on the beach, and the bodies are still in the shells, it is well to place the specimens in fresh water overnight, and the bodies may be flushed out the following morning. *Janthina* should not remain in fresh water for more than 12 hours, as longer immersion will soften the periostracum in an irregular manner and the dried shells will present a blotchy appearance.

Xenophora: Removal of the body from *Xenophora* is almost impossible and boiling does not solve the problem. Frequently if the shell is placed with the aperture upward, the living animal will thrust a greater part of its body out and it may be quickly removed. In small specimens of this interesting genus, a needle or pin can be thrust through the body just behind the operculum and allowed to lie crosswise of the aperture. The animal will quickly die and can be removed.

Conidae: The larger *Conus* may be boiled and the animal removed using a straight wire or, as previously mentioned, a crochet hook that has been slightly bent. The wire should be introduced parallel to the long axis of the shell. A firm twist will usually start the body rolling out. Be sure to preserve the operculum of every specimen and see to it that each shell has its own operculum. Place the operculum within the shell and close the aperture with a small wad of cotton or crumpled paper. Later, when preparing the shells for the cabinet, a tightly fitting plug of cotton wedged into the aperture will receive a drop of adhesive on which the operculum is placed and pushed into position in a lifelike manner.

Cypraea: For years, it has been thought that no *Cypraea* should be boiled. It was thought that heating the shell would impair the gloss, and the accepted procedure was to allow the animal to die and decay. The odor would be removed with many rinsings and the shell could eventually appear in respectable company. It has been demonstrated that the idea was fallacious, and now most collectors do not hesitate to boil a *Cypraea*, but it has been recommended that the specimens be placed in tepid water and then brought to a boil, thus avoiding any checking of the shell due to sudden change of temperature. It is important after boiling the larger cowries that the shells are shaken vigorously to loosen the body of the mollusk. It should be shaken until the loosened body may be heard swishing about inside the shell. All moisture should be removed from the shell as any remaining moisture will cause a bluish discoloration in the darker colored species, and this discoloration is often permanent. Flushing with water (Section 2.7.3) is also a good way to clean out *Cypraea* shells.

Large gastropods: With the larger shells such as *Cassis*, *Busycon*, *Pleuroploca*, etc., another technique has proven useful. The shell containing the living mollusk is placed aperture side up for 24 hours or longer. The animal has usually become weakened by this time and a greater portion of it lies out of the shell. A stout cord or wire is tightly wound about the body, just behind the operculum, and then is tied to some convenient object above so that the shell is suspended with all the weight pulling downward. Gradually the body is pulled from the shell by the shell's weight. A bed of soft material should be placed beneath the shell to prevent breakage. Frequently, the cord must be shortened from time to time as the body becomes more and more elongated from the constant traction. Alternately, you can support the shell aperture facing down. As the organism dies, the weight of the tissue mass will cause it slowly to drop out of the shell. You may have to tug at the tissue mass slightly.

Large bivalves: Bivalve shells should be cleaned soon after being collected to avoid discoloration of the interior. A knife is all that is usually needed, however, be careful to avoid scratching the interior when removing the adductor muscles. It is very easy to chip the edge of a bivalve when attempting to open it. The adductor muscles can usually be cut using a razor (for small bivalves) or a very sharp, thin knife (fillet knife or a Finnish puukko) for larger bivalves.

Boiling is not normally necessary, but can be used if desired. Immersion in room temperature water, with gradual heating, will usually cause tightly closed, live bivalves to relax. This causes them to gape and thus be easier to open.

There is a diversity of opinion regarding the position in which bivalves should be dried. Many favor closing the valves in a natural position and holding them in place by a few strands of thread wound about them until they are dry. Others prefer to dry the specimens, wide open, "butterfly fashion" maintaining that the beauty of the shell is enhanced and the inside structure more easily seen and studied. Individual preference should be your guide.

If there is one procedure thoroughly despised by most collectors, it is the practice of holding the valves in position by means of narrow strips of surgical adhesive tape. The adhesive coating of the tape eventually separates from the fabric and adheres to the shell. The residual adhesive is very difficult to remove. Rubber bands should also be avoided. Over time, they will degrade and some leave behind a difficult to remove residue. The same can be said for scotch tape.

Another procedure that is frowned upon by many is to glue the valves of bivalves together. Such adhesives are often hard to remove and fragile specimens may be damaged in the attempt to open them. Tying the valves together with cotton string is the preferred method if keeping them together is desired.

If it is desired that specimens of *Pteria, Anomia*, and the like be preserved on the *Gorgonia* or other object on which they have been found, they should be placed in alcohol for a day or two and then dried *in situ*. *Pododesmus* should be left as found, as attempts to remove the animal portions often prove disastrous. The soft parts are so small that they will dry up and not make themselves unpleasantly evident.

Terrestrial slugs: Terrestrial slugs should be narcotized (see above) and then placed in at least 80% ethanol. One can also freeze-dry slugs but this involves sophisticated techniques and equipment. This technique is described in Crowell (1973).

Miscellaneous taxa: *Dentalium, Rimula*, and the like are left overnight in fresh water and the animal easily removed the following morning. With *Cyphoma, Marginella, Oliva, Olivella*, and *Trivia*, it is necessary to kill the animal with fresh water. They should remain in fresh water for at least 48 hours, with two or more changes of water before the body is sufficiently softened so that it may be removed with a syringe. It might be mentioned that if the shell to be cleaned is held under water during the operation, the collector will obviate all danger of spraying him or herself as well as adjacent territory with none-too-sweet-smelling water.

2.7.13 Coating shells with preservatives. Shells may be found washed up on beaches and they may appear dull and faded. The color may be greatly intensified by a mild application of some greasy substance such as mineral oil, paraffin, silicon oil, or Teflon®. Application of these substances will often brighten up a dull and faded specimen. To preserve the periostracum of such shells as *Sinum*, *Hydatina*, and unionoids, which have a tendency to peel when very dry, an occasional application of one of these substances will prove sufficient. It is very important to remember that the collector is not creating a shell but is simply, in a manner "lifting its face."

Animal and vegetable-based oils will become rancid with time and should be avoided. Silicon oil is a relatively inert substance. Its stability, lack of odor, and safety profile have made it a substance that people are comfortable using. The silicon oil is cut with naphtha or mineral spirits 1 part to 2-4 parts. Shells are dipped into this mixture and then placed on a flat surface. The solvent evaporates leaving behind a thin coating of silicon. If diluted appropriately the amount of silicon oil remaining should not leave the shell feeling greasy or tacky. Too much oil makes the shell slippery and a "dust magnet". Mineral oils can be used in a similar fashion.

A number of collectors have used a solvent called WD-40® to brighten the surface of a shell or to preserve the periostracum. WD-40 is a petroleum distillate. Some collectors dilute it 1 part to 3-4 parts naphtha or mineral spirits and then dip their shells into this solution. Others use the WD-40 full strength. We do not recommend WD-40 for one preservational aspect - it is a proprietary and secret formula. Such mixtures are of unknown composition and can be changed at any time. Thus, you are never certain what you are putting on your shells.

If you are going to apply a coating to your shells, you do not want to apply it full strength. This full strength coating will often make the shell tacky and slippery. Many collectors dilute the oily substance with naphtha or mineral spirits. Diluting the oil one part to three or four parts of solvent is often sufficient. When the solvent evaporates, it leaves behind a thin coating of the oil. Also, the thinned oil is able to penetrate into the shell deeper than the full strength oil would. When using solvent, make sure there is adequate ventilation and no sources of open flames. While naphtha and mineral spirits are relatively low toxicity, they can act as respiratory irritants. They are also flammable, thus no sources of flames or sparking should be present.

Clench (1931) described a process of mixing paraffin with xylene. Place 120 g (4 ounces) of paraffin in 300 ml (10 ounces) of xylene. Warm this in a water bath to dissolve the paraffin in the xylene. Shells are dipped into the solution and then placed on a flat surface to dry. The xylene will evaporate leaving behind a thin coating of paraffin. Remember that xylene is flammable so avoid open flames. Xylene is also a respiratory irritant, so this method should only be performed in a fume hood or a well-ventilated area such as outdoors.

You should also keep in mind that many collectors and most museum workers apply no surface coating to shells. Generally, if you keep the temperature and humidity level of your storage area within a range of 16-21°C (60-70°F) and 50-55% relative humidity, there should be no concern with the periostracum peeling (See Chapter 5, Archival and Curatorial Methods). Also, there is nothing wrong with a slightly faded, pristine shell as opposed to one that is shiny but chemically adulterated.

2.7.14 Acid Treatment. The method of dipping in hot acid followed by a plunge in ice water so as to create a false luster is to be avoided scrupulously. The same may be said of the use of varnish or shellac, or any form of buffing. When you take the necessary time and effort to prepare your specimens properly, you can truthfully say that your collection is "a thing of beauty and a joy forever" (Bales 1942).

2.8 LITERATURE CITED

Abbott, R. T., G. M. Moore, J. S. Schwengel, and M. C. Teskey. 1955. *How to Collect Shells*, 2nd Ed. American Malacological Union, Marinette, Wisconsin. 75 pp.

Abbott, R. T., M. K. Jacobson, and M. C. Teskey (eds.) 1966. *How to Collect Shells*, 3rd Ed. American Malacological Union, Marinette, Wisconsin. 101 pp.

Allison, L. N. 1942. Trapping snails of the genus *Campeloma*. *Science* **95**: 131-132.

Anderson, E. 1973. A method for marking nudibranchs. *Veliger* **16**: 121.

Aquilina, B. and R. Roberts. 2000. A method for inducing muscle relaxation in the abalone, *Haliotis iris*. *Aquaculture* **190**: 403-408.

Araujo, R., J. M. Remón, D. Moreno, and M. A. Ramos. 1995. Relaxing techniques for freshwater molluscs: trials for evaluation of different methods. *Malacologia* **36**: 29-41.

Bales, B. R. 1942. Shore and shallow water collecting. *In: The Eleventh Annual Meeting of the American Malacological Union, held at Rockland and Thomaston, Maine, August 26-29, 1941.* American Malacological Union. Pp. 16-24.

Barker, G. M. 1981. Nembutal for narcotisation of mollusks. *Veliger* **24**: 76.

Bergeron, E. 1971. *How to Clean Seashells*. Great Outdoors Publishing Company, St. Petersburg, Florida. 32 pp.

Bowler, P. A., T. P. Johnson, and W. J. Mautz. 1996. Flash-freezing using cold fluid: Field and laboratory methods for preventing retraction of snails during fixation. *Journal of Molluscan Studies* **62**: 124-126.

Burghartd, G. and L. Burghardt. 1969. *A Collector's Guide to West Coast Chitons*. San Francisco Aquarium Society, Inc., San Francisco, California. 45 pp.

Carriker, M. R. and J. W. Blake. 1959. A method for full relaxation of muricids. *Nautilus* **73**: 16-21.

Clapp, W. F. 1912. Collecting from haddock on George's Bank. *Nautilus* **25**: 104.

Clement, A. C. and J. N. Cather. 1957. A technic for preparing whole mounts of veliger larvae. *Biological Bulletin* **113**: 340.

Clench, W. J. 1931. A preventive for the scaling of the periostracum. *Nautilus* **45**: 30-31.

Coney, C. C. 1993. An empirical evaluation of various techniques for anesthetization and tissue fixation of freshwater Unionoida (Mollusca: Bivalvia), with a brief history of experimentation in molluscan anesthetization. *Veliger* **36**: 413-424.

Coyer, J., D. Steller, and J. Witman. 1999. *The Underwater Catalog: A Guide to Methods in Underwater Research*. Shoals Marine Laboratory, Ithaca, New York. 151 pp.

Craze, P. G. and A. G. Barr. 2002. The use of electrical-component freezing spray as a method of killing and preparing snails. *Journal of Molluscan Studies* **68**: 191-193.

Crowell, H. H. 1973. Preserving terrestrial slugs by freeze-drying. *Veliger* **15**: 254-256.

Davis, J. D. 1964. A rake drag for intertidal or shallow-water dredging. *Turtox News* **42**: 94.

De Winter, A. J. 1985. A new rapid method for the relaxation and killing of slugs. *Basteria* **49**: 71-72.

Englund, V. P. M. and M. P. Heino. 1994. A new method for the identification marking of Bivalvia in still water. *Malacological Review* **27**: 111-112.

Fox, D. L. 1935. A biochemical method for internally cleaning small molluscan shells. *Nautilus* **48**: 99-100.

Gregg, W. O. 1944. Collecting and preserving land slugs. *Bulletin of the Southern California Academy of Sciences* **43**: 41-43.

Grimm, B. 1996. A new method for individually marking slugs (*Arion lusitanicus* (Mabille)) by magnetic transponders. *Journal of Molluscan Research* **62**: 477-482.

Hanselman, G. A. 1970. Preparation of chitons for the collector's cabinet. *Of Sea and Shore* **1**: 17-22.

Heard, W. H. 1965. Comparative life histories of North American pill clams (Sphaeriidae: *Pisidium*). *Malacologia* **2**: 381-411.

Hubricht, L. 1951. The preservation of slugs. *Nautilus* **64**: 90-91.

Jacobson, M. K. (ed.) 1974. *How to Study and Collect Shells*, 4th Ed. American Malacological Union. Wrightsville Beach, North Carolina. 107 pp.

Kleewein, D. 1999. Population size, density, spatial distribution and dispersal in an Austrian population of the land snail *Arianta arbustorum styriaca* (Gastropoda: Helicidae). *Journal of Molluscan Studies* **65**: 303-315.

Lemarié, D. P., D. R. Smith, R. F. Villela, and D. A. Weller. 2000. Evaluation of tag types and adhesives for marking freshwater mussels (Mollusca: Unionidae). *Journal of Shellfish Research* **19**: 247-250.

Lipe, B. and R. Lipe. 1993. *Clean your Shells and other Sea Life*. The Shell Store, St. Petersburg Beach, Florida. 36 pp.

Lonhart, S. I. 1999. Multiple techniques for marking subtidal marine mollusks. *In:* J. H. Heine, D. Caneatro, and G. Wuttken, eds., *Proceeding of the American Academy of Underwater Science 19th Annual Scientific Diving Symposium*. American Academy of Underwater Sciences, Nahant, Massachusetts. Pp. 67-70.

Meier-Brook, C. 1976. An improved relaxing technique for mollusks using pentobarbital. *Malacological Review* **9**: 115-117.

Merrill, A. S. 1974. Collecting from navigational buoys. *In:* M. K. Jacobson ed., *How to Study and Collect Shells*, 4th Ed. American Malacological Union, Wrightsville Beach, North Carolina. Pp. 54-55.

Mills, D., A. Tlili, and J. Norton. 1997. Large-scale anesthesia of the Silver-Lip Pearl Oyster, *Pinctada maxima* Jameson. *Journal of Shellfish Research* **16**: 573-574.

Mueller, G. J. 1972. *Field Preparation of Marine Specimens*. University of Alaska Museum, Fairbanks, Alaska. 44 pp.

Norton, J. H., M. Dashorst, T. M. Lansky, and R. J. Mayer. 1996. An evaluation of some relaxants for use with pearl oysters. *Aquaculture* **144**: 39-52.

Owen, G. 1955. Use of propylene phenoxetal as a relaxing agent. *Nature* **175**: 434.

Owen, G. and H. F. Steedman. 1958. Preservation of molluscs. *Proceedings of the Malacological Society of London* **33**: 101-103.

Pilsbry, H. A. and E. G. Vanatta. 1898. Revision of the North American slugs: *Binneya, Hemphillia, Hesperarion, Prophysaon,* and *Anadenulus. Proceedings of the Academy of Natural Sciences of Philadelphia* **50**: 219-261.

Pojeta, Jr., J. and M. Balanc. 1989. Uses of ultrasonic cleaners in paleontological laboratories. *In:* R. M. Feldmann, R. E. Chapmann, and J. T. Hannibal, eds., *Paleotechniques, The Paleontological Society Special Publication* No. 4. The Paleontological Society, Lawrence, Kansas. Pp. 213-217.

Porter, H. J. 1972. Shell collecting from stomachs of the sea-star genus *Astropecten. New York Shell Club Notes* **180**: 2-4.

Porter, H. J. 1974. Shell collecting from the stomachs of sea-stars. *In:* M. K. Jacobson ed., *How to Study and Collect Shells*, 4th Ed. American Malacological Union, Wrightsville Beach, North Carolina. Pp. 104-105.

Rice, T. 2003. *Tom Rice's A Shellers Directory of Clubs, Books, Periodicals and Dealers*, 26rd Ed. Of Sea and Shore Publications, Port Gamble, Washington. 114 pp.

Richter, K. O. 1976. A method for individually marking slugs. *Journal of Molluscan Studies* **42**: 146-151.

Ropes, J. W. and A. S. Merrill. 1970. Marking surf clams. *Proceedings of the National Shellfisheries Association* **60**: 99-106.

Rosewater, J. 1963. An effective anesthetic for giant clams and other mollusks. *Turtox News* **41**: 300-302.

Rosewater, J. 1965. An effective anesthetic for giant clams. *Indo-Pacific Mollusca* **1**: 394.

Runham, N. W., K. Isarankura, and B. J. Smith. 1965. Methods for narcotizing and anaesthetizing gastropods. *Malacologia* **2**: 231-238.

Schander, C. and J. Hagnell. 2003. Death by drowning degrades DNA. *Journal of Molluscan Studies* **69**: 387-388.

Scheidt, H. E. 1982. Procedures for chiton preparation. *Of Sea and Shore* 12: 141-145.

Smith, A. G. 1962. Notes on cleaning mollusks. *Veliger* **4**: 216.

Smith, D. G. 1996. A method for preparing freshwater mussels (Mollusca: Unionoida) for anatomical study. *American Malacological Bulletin* **13**: 125-128.

Smith, E. H. 1961. Narcotizing and fixing opisthobranchs. *Veliger* **4**: 76.

Thoma, B., G. Swanson, and V. E. Dowell. 1959. A new method of marking fresh-water mussels for field study. *Proceedings of the Iowa Academy of Science* **66**: 455-457.

Turner, R. D. 1960. Some techniques for anatomical work. *The American Malacological Union Annual Report* (for 1959): 6-8.

Turner, R. D. 1974. Collecting shipworms. *In:* M. K. Jacobson ed., *How to Study and Collect Shells*, 4th Ed. American Malacological Union, Wrightsville Beach, North Carolina. Pp. 50-53.

van der Schalie, H. 1953. Nembutal as a relaxing agent for mollusks. *American Midland Naturalist* **50**: 511-512.

van Eeden, J. A. 1958. Two useful techniques in fresh water malacology. *Proceedings of the Malacological Society of London* **33**: 64-66.

van Eeden, J. A. 1960. Key to the genera of South African freshwater and estuarine gastropods (Mollusca). *Annals of the Transvaal Museum* **24**: 1-17.

Walker, B. 1904. Hints on collecting land and fresh-water Mollusca. *Journal of Applied Microscopy and Laboratory Methods* **6**: 2365-2368.

Weil, Art. 1998. *Shell Shocked! A Guide to Sane Shell Collecting*. Evolver srl, Rome. 98 pp.

Wolda, H. 1963. Natural populations of the polymorphic landsnail *Cepaea nemoralis* (L.). Factors affecting their size and their genetic constitution. *Archives Néerlandaises de Zoologie* **15**: 381-471.

Young, M. R. and J. C. Williams. 1983. A quick secure way of marking freshwater pearl mussels. *Journal of Conchology* **31**: 190.

CHAPTER 3

REMOTE BOTTOM COLLECTING

THOMAS A. BURCH, MD
CHARLES F. STURM

3.1 INTRODUCTION

Most people begin their shell collections by walking along the shore and picking up the shells they come across. At some point in time, many of us want to go beyond this technique of collecting, and obtain materials from places we cannot normally reach. This chapter deals with some of the ways to extend our reach. These methods may be as simple as using a dip net to as complicated as using a dredge or box corer. Section 3.2 will discuss dredging from the perspective of Tom Burch. This is followed by Section 3.3, a short discussion of several other methods of remote bottom sampling written by Sturm. For additional treatments of these topics, see Ockelmann (1964), Holme and McIntyre (1971), J. Q. Burch (1974), Skoglund (1990), and Monfils (1999).

3.2 DREDGING

Since I (TAB) wrote one of the first articles in this series in 1941 (Burch 1941), I was very pleased when Dr. Charles Sturm asked me to write again. Actually, he asked because of an article my wife and I wrote (Burch and Burch 2000). It probably did not occur to him that anyone who wrote about dredging in 1941 would still be active sixty years later. The quoted passages in this section are from Burch (1941).

The 1941 article is still pertinent. About the only things that really need revision are costs due to inflation, some technological advances, and fifty years of additional experience with using a dredge.

The opening of the 1941 article was: "I suspect that almost every shell collector who has gotten up at an unearthly hour in the morning to collect at a minus tide has gazed out beyond the narrow bend of shore left uncovered by the retreating waters and wondered what rare treasures he could find, if the sea would but drop a hundred feet or so for awhile. Or what shell collector strolling along the beach after a storm hasn't wished that he could go out beneath the waves and collect live, perfect specimens of some of the shells that lie broken and worn at his feet. The vast majority, however, just sigh and decide that the 'deep stuff' can only be collected with complicated and expensive equipment and is only for institutions and individuals with plenty of finances. A few consider it further and decide that while it would be much easier and more pleasant to dredge from a two-hundred foot cruiser with power winches and a crew to do all the work, if one has a strong back and a few dollars, he too can get some of this same material."

That passage obviously was written before SCUBA became prevalent (see Chapter 4). My first dive was an adventure in a surplus Navy hard hat diving suit that convinced me dredging was the way to go. Twenty years later scuba was popular and I became a certified scuba diver but I still preferred dredging. About the only use I made of scuba was to rescue stuck anchors, retrieve something dropped overboard at anchor or at the dock, or work on the bottom of the boat.

Please consider that the prices shown in the original article, which continues, are those of the Great Depression, which did not really end until we got into World War II. The prices cited need to be multiplied by about 20 to get what they would cost in 2005.

"Whereupon the ambitious collector makes a triangular or rectangular iron frame with a row of holes along one side, or gets a blacksmith to make one for about fifty cents, and gives a friendly fisherman fifty cents for some old fine-meshed fish net. On the way home to sew the net onto the frame he stops at a hardware store and buys about 200 feet of 3/8 inch manila rope for two or three dollars. Then, taking this simple equipment and some containers for the catch, he goes to the beach, rents a skiff for a few hours, rows merrily away from piers and boats and begins to dredge. A few hours later, depending on the physical condition of the dredger, the ambitious collector returns wearily to shore, a tired but happy person, in a couple of hours, having collected more different kinds of shells that are new to his collection than he had dared to hope. I must hasten to say that it does not always turn out as nicely as I have perhaps led my readers to believe. [My first dredge station has the ignominious note, 'Fishing Trip, Dredge did not work.'] The spot chosen in which to dredge makes a great deal of difference. Some who would have become ardent dredging fans have given it up as a bad job after they tried to dredge in a shallow mud slough and got nothing but barren gooey mud, or after they tried to dredge among large rocks and got nothing and perhaps lost their gear. If the embryonic dredger is persistent, however, and not easily discouraged, he will continue, profit from his mistakes, and in time become quite adept at this mode of collecting. Incidentally, if he does continue to dredge, he will soon have many species of shells that less fortunate collectors can never get except by trading, buying, or going dredging themselves. This includes scuba divers even if they wash or brush the shells on algae, stones, etc. into a container or bring up a bag or two of sand from the bottom – unless they get the material from a reef where it is impossible to dredge."

3.2.1 Boats. "While it is sometimes possible to get some very good material by just throwing a dredge off from the end of a pier and dragging it in, a boat of some sort is really essential. I have already suggested that a person with a strong back can dredge a lot of fine material with a skiff. My readers will probably not believe it until they have the experience, but a man can drag just as large a dredge and fill it just as full by rowing as one can with an ordinary small motor boat. Needless to say, the rower will not be able to make as many hauls as the motor boat operator."

"In choosing a dredging boat the most important item to consider, in my estimation, is seaworthiness. If one is going to dredge on the open sea with a small boat, he must always remember that the wind and the sea are treacherous, and if his boat swamps he is in a mighty precarious position. Incidentally in such a case the safest thing to do is to stay with the boat (unless it sinks). If you are going to dredge with a skiff or flat bottomed rowboat, be sure to get one that was made for the ocean and not for some quiet lake or lagoon. It should have fairly wide beam so that you can stand up, if needs be, without tipping it over, the sides and stern should flare a little and there should be some shear. If you have never rowed a boat learn how to, first. Have some old 'salt' show you or read how to in a book. Also, do not try to dredge everything the first day, leave some for other times or your hands will look worse than hamburger, even if you wear gloves – which good oarsmen never do."

I will omit the next few paragraphs in my 1941 account because they were on problems with the pre World War II outboard motors that were nowhere near as reliable as those manufactured after the war. Even then, however, I had a word of caution regarding using a large outboard motor on a small boat. You can get to where you want to go much quicker but if you get stuck while dredging and speed up the engine to break loose, and if you don't break loose, the stern of your boat may be pulled under the water.

I then described our first boat, the JANTHINA. It was an 18 foot (6 m) New England dory with a well for an outboard motor and a shive or pulley in the stern for the dredge line to pass over. The line was wound on a hand windlass forward of the motor operator's seat. We found this a very successful boat and dredged down to 900 feet (300 m). It was very sea worthy, could be launched and landed through the surf so we were not dependent upon the location of boat ramps. Despite my laudatory description I never saw or heard of anyone else using such a boat for dredging. We sold the boat to a fisherman after the War started since the military wouldn't permit us out of the harbor. They considered our dredging activities as suspicious.

After the War and after my various assignments in Central America and Africa with the U.S. Public Health Service, I purchased a Boston Whaler, the JANTHINA V and installed a military surplus winch and powered it with a gasoline engine using a clutch from a Toro ride-around lawn mower. This is a very sea-worthy boat and I used it very successfully off Lewes, Delaware, the northern Gulf of California, and Hawaii. It seemed, however, that whenever I could go dredging in Hawaii there was so much wind that it was unpleasant unless I stayed near shore or in bays, so I purchased a 32-foot (10 m) Salmon Trawler, the JANTHINA VII, that had sailed (or rather chugged) to Hawaii from the mainland. I'm not advocating these particular boats. What I am advocating is that the boat should be sea-worthy and, if large enough, have a derrick or an A-frame to facilitate getting the dredge aboard.

3.2.2 Dredge line. "The simplest type of dredge line is rope. Some prefer to use 1/2 inch or 5/8 inch [12-16 mm] diameter manila rope as it is large and easy on the hands. [Nowadays you would probably get polypropylene or nylon rope.] This is very good, if one is dredging on a shallow, rocky, or shale bottom and does not get it permanently fastened to the bottom. This heavy rope, however, is too bulky for anything but very shallow dredging (5-20 fathoms) and also is much more expensive than the smaller rope which is just as good and sufficiently strong. My father and I have dredged as deep as 50 fathoms with 1/4 inch rope by fastening small weights to the rope at several places. These weights are necessary in deep dredging with rope as it has a tendency to float and pull the dredge off the bottom."

"If one tries to dredge in more than twenty fathoms with rope it is necessary to devise some sort of windlass on which to wind the rope. If you do not use a windlass, it always seemed to me that the rope becomes more tangled when an attempt is made to coil it when dragging in the dredge than if simply allowed to fall naturally ... If you are going to dredge very deep (50 fathoms or more), it is much more convenient to use cable, as rope has a tendency to float and pull the front end of the dredge up. The chief objection to cable is the cost ... Needless to say, a windlass or winch is necessary, if cable is used. If your boat is large enough, connect the winch to the motor, if not crank it by hand."

The cable that I used before the War was twisted wire clothesline that cost about $10.00 for a thousand feet (300 m). I have not seen it since the War. The cable that I used on the Boston Whaler was 3/8 inch (10 mm) stainless steel military surplus cable but the winch I had would only hold enough for very shallow dredging. That was ok off Lewes, Delaware and in the northern Gulf of California but not in Hawaii where the water may be over a hundred fathoms (600 feet, 200 m) a half mile (800 m) from shore. I first got smaller cable and a bigger winch that was powered by a gasoline engine. When I bought the JANTHINA VII, I enlarged the drum capacity so that it would hold over 3000 feet of 3/16 inch (5 mm) steel cable and powered it with an hydraulic motor run from a power takeoff on the boat's diesel engine. The cable was really too small and I discovered that the best time to replace the cable is before departing on a two week dredging trip rather than after returning.

Another thing that I learned the hard way is that one should have a small auxiliary engine to power the hydraulic pump and motor rather than using a power take off from the main engine. The problem is that it is necessary to keep moving forward while pulling up the dredge. If the boat backs up there is slack in the dredge line and the cable will tangle and have to be cut and spliced to get it on board. After discovering this I always tried to dredge with the current rather than against it so the current would keep the cable tight.

3.2.3 Types of dredges. "Dredges can be either very elaborate or very simple. It is possible to make a dredge that will work from a five-gallon oilcan or a

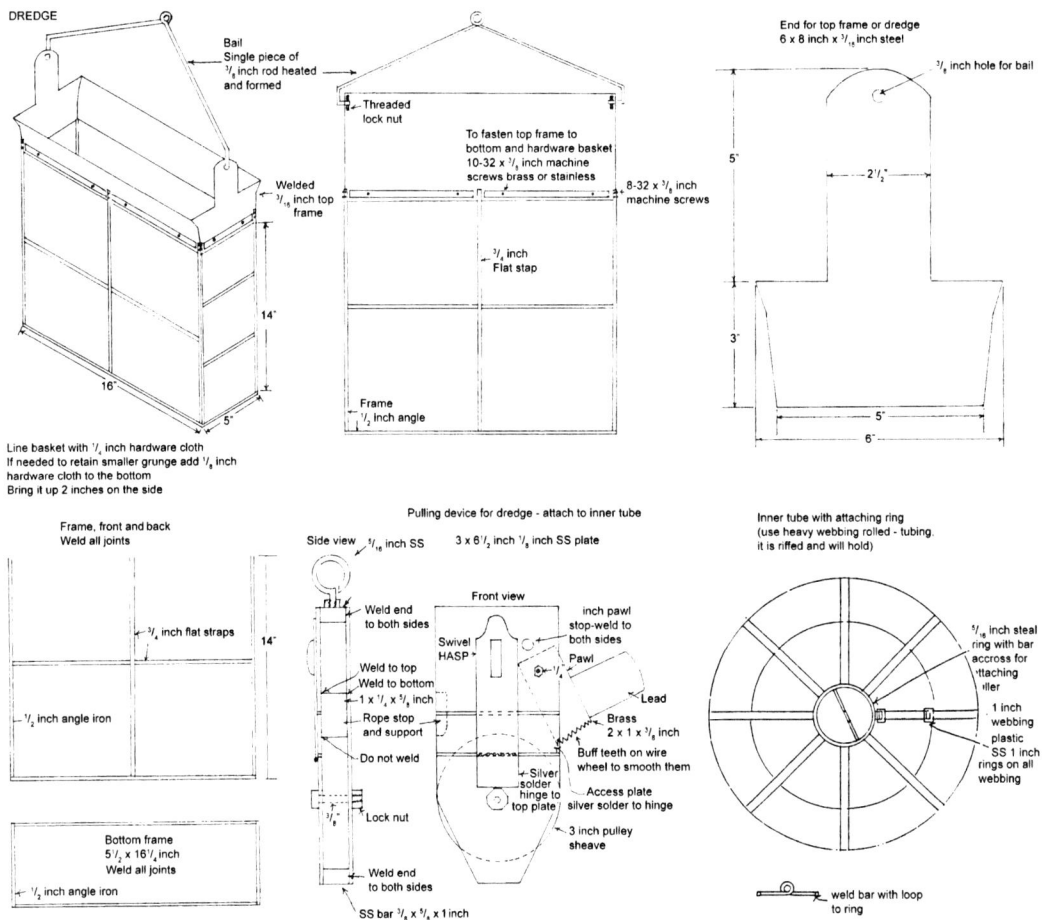

Figure 3.1 Instructions for making a dredge. Plan courtesy of David Mulliner.

piece of iron sewer pipe. These are not as practical as others but can be used in a pinch ... Probably the most efficient cheap dredge consists of a triangular, circular, or rectangular frame with fish net or hardware cloth for a bag ... While the dredge can be considered the standard implement for obtaining marine life from the bottom of the ocean, various other gear, such as trawls, tangles, etc., can be used with success. While there are many different types of trawls, I will only describe a simple beam type that I have used with good results. It consists, briefly, of two iron runners connected to a wooden beam. The size depends on the power of the boat that is used to pull it. A long net is fastened to the beam and the rear end of the runners. The amount of sag to the lead line which drags on the bottom is very important. If it is too little, it digs too much, while if it is too much, it rolls over the material. The right amount must be determined with experience. In addition to a few lead sinkers along the lead line a heavier weight at each end of the line on the runners will help the net stay on the bottom. A heavier net should be placed over the finer net for protection and the end tied shut and not sewed to facilitate removal of the haul."

"The trawl is used on a sand or mud bottom and is used to cover large areas. It picks up only the larger shells, etc., and lets all of the little things go through or under. Tangles are especially useful on rocky bottoms where it is impossible to use a dredge or a trawl. They may be used, however, on any type of bottom. Tangles consist of a beam with a few short pieces of chain to which is fastened unraveled rope" (see Section 3.3.3 below)."

I have also used a 10-foot (3 m) Otter Trawl with the 16-foot (5 m) Boston Whaler in the Northern Gulf of California and the 32-foot (10 m) Salmon Trawler off Hawaii. These are the nets that commercial shrimp boats use to test the bottom to find out if any shrimp are there. I got excellent material from both boats with this gear. It is, however, more complicated and difficult to use an Otter trawl than a dredge or beam trawl. Plans for one type of dredge are shown in Figure 3.1.

3.2.4 Dredging operations. "Probably no two persons who have dredged very much agree on just what equipment is best and how it should be used. My father and I have dredged together quite extensively for the past six years and we do not agree on the proper procedures so no one else will probably agree entirely with me or anyone else."

"The principle of dredging is very simple - you merely throw the dredge overboard, let out about three times as much line as depth, drag it until it is full, pull it up, and remove the contents. The only way to learn to dredge is to actually go out and dredge. You will soon be able to tell by the feel of the line if the dredge is on the bottom and digging; in fact you will soon even be able to tell what type of bottom it is by the way it jerks. Mud hauls usually just get heavier and heavier, sand hauls have many little jerks depending on the character of the sand, while gravel hauls are decidedly jerky and very uneven. The deeper the water, however, the more difficult it is to tell anything about what is happening below."

"It is very important when dredging with cable to keep the line taut by moving the boat forward from the moment the dredge is thrown overboard, otherwise the cable will get full of kinks and break easily. This is not so important when using rope, if one is careful to keep the rope clear of the propeller, but it is usually a good idea."

"A weight of some sort should be placed between the line and the dredge to ensure the cutting edge staying on the bottom. A three or four foot [1-1.3 m] piece of heavy log chain is the most convenient kind of weight, but a sash weight, or anything else will do."

"It is impossible to dredge except at very slow speeds as otherwise the dredge is likely to be pulled clear of the bottom. It is much easier to keep the dredge on the bottom when dredging uphill than down, in fact, if the slope is very marked, it is almost impossible to dredge downhill. The more line you have out the easier it is to keep the dredge on the bottom."

"On dredges with iron bridles it is a good idea to make the dredge line fast to one bridle and tie the other to it with a cord. Then, if the dredge gets caught, the cord [hopefully] will break and the dredge will pull free. As an additional safeguard against breaking the dredge line, it is best to have some arrangement so the line can slip out of the boat, if the dredge gets caught. This is more important with a larger boat."

3.2.5 Locating where you are. "I have suggested that it is a good idea to keep track of the good dredge hauls and the poor, so that you can later return and dredge more of the good material. However, finding the same place again is quite a problem. Trying to find a good dredging spot that you hit by accident is like being the blind man in 'blind man's bluff.' When you get a good dredge haul be sure to take all possible bearings on objects on shore ... The next time you are out maybe you can get within a quarter of a mile of the spot you are looking for."

This is the part of dredging that has changed the most since I started dredging in the 1930s. You want to know exactly where you are so you can come back again, if you get a good haul. At first I took compass bearings of two or more objects on shore that were shown on the nautical chart of the area. I also tried to get two objects on the chart in line one behind the other. Next I used a sextant. My first one was a cheap plastic model but after I took navigation, I invested in a professional one. I used the sextant to determine the angles between object A and B and also between

B and C that were shown on a navigation chart of the area and then using a three arm protractor, I positioned this so that each arm passed over one of the objects and when all three were over their respective objects, the hole in the center of the hub from which the three arms rotated, showed the position of the boat. In the 1960s I used a radio-direction finder to locate the compass bearings of two or three radio beacons and plotted those on the chart. When I got to Hawaii where I used a larger boat, I used Loran. Radar was also available but that cost more than I felt justified in spending. If I were dredging now I would use a Global Positioning Device, which would calculate the latitude and longitude using data from orbiting satellites.

It is also important to know how deep the water is where you want to dredge. It is better to find out you cannot reach the bottom before you put the dredge overboard rather than when you finally pull it up. There have also been a lot of improvements in fathometers (depth gauges). Modern ones will even tell you the type of bottom.

3.2.6 Taking care of the hauls. "When a dredge load is brought up, it should be screened. If the dredge is made of screen, this can be done before bringing it aboard. Otherwise the load must be dumped and screened in the boat. If this is not convenient, just empty the dredge into a sack and take it home to screen."

"The bottom off the West Coast is very spotty with small patches of very good bottom surrounded by large areas of poor bottom with relatively few shells in it. [The same is true but to a lesser extent in most other places in which I have dredged.] For this reason it is very important to keep track of the good hauls and the poor hauls. To do this it is probably best to keep each haul separate until it is dried and sorted. If several hauls, however, are from about the same spot and are apparently the same material, there is no use keeping them separate. After the material is dried and sorted it is up to the individual what he does with it."

"We place everything from the same locality, approximate depth, and character of bottom together.

Thus the material that we have dredged off Redondo Beach, California, is divided into 10, 25, 50, 100, 150 fathoms [18, 45, 90, 180, 275 m]; mud, sand, gravel, or rocks depending upon which it is closest to. Each collector will have to make up his mind according to how much time he has for book keeping."

Since I wrote the last paragraph in 1941 I have had more to do with museums and have changed my ideas and methods. I had assigned numbers to each station whether it were a location on shore, a dredge haul, trawl haul, or haul with a plankton net. The station numbers that I now use start with the last two digits of the year and are then numbered consecutively starting with 001. The advantage of such a system is that you can just put the station number with each set of shells instead of writing it all out. The disadvantage is that when you want to know where it came from, you have to look it up.

"Unless you are not interested in the semi-microscopic shells do not throw any of the material away until after it is dried and re-sorted, as it is practically impossible to see [minute shells] while the material is wet. We usually screen the material, sort out what we can see readily, dry it and then re-sort it at our leisure. This latter sorting can even wait until winter, if you are pressed for time."

Actually, I still have material awaiting the second sorting from dredge hauls made twenty years ago. Somehow, now that I'm in my eighties, I just cannot seem to rush these things.

3.3 OTHER METHODS OF REMOTE BOTTOM SAMPLING

While dredges are efficient devices for sampling the benthos, they do have some drawbacks. They can sample only the upper surface of the sediment. If they sample too deeply they become anchors! They also lose all stratigraphic information. Whatever gets picked up becomes mixed up before the dredge is recovered. Several of the devices mentioned below are designed to avoid these drawbacks.

3.3.1 Grabs. These are devices used to sample sediment at the benthic zone. They obtain a discrete

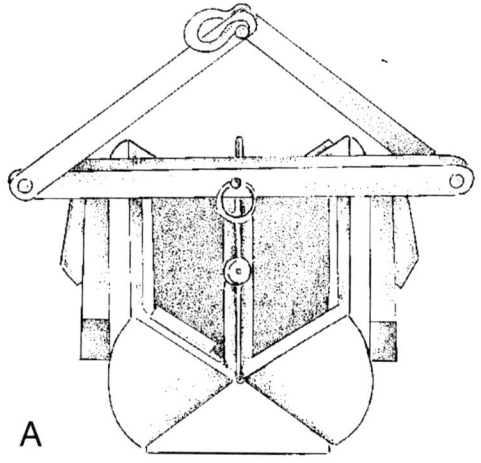

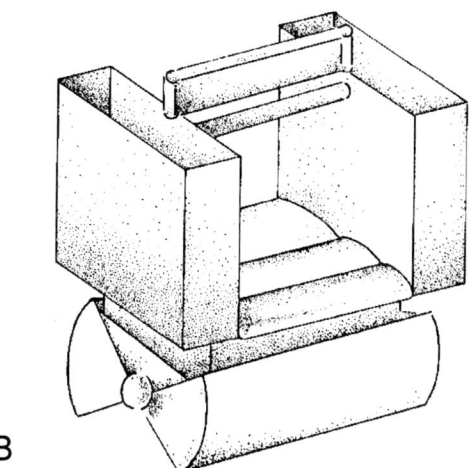

Figure 3.2 Remote bottom sampling devices.
A, Ponar Grab. B, Box Corer.

sample of the bottom that is representative of the general area. This results in a quantitative assessment of the taxa present in the area sampled. You will find that there are several different types of grabs including the Ponar (Figure 3.2A), Ekman, and Peterson Grabs, among others. Grabs vary widely in weight and the type of surface that they are designed to work in. The Ponar Grab can weigh from 6-20 kg (15-45 lbs) and is designed to sample benthic environments that are composed of sand, gravel, or clay. The Ekman Grab is fairly light, weighs in at 4-5 kg (around 10 lbs), and is designed to sample soft sediment that is free of coarse debris and vegetation.

This grab is sometimes attached to the end of a pole as opposed to being lowered on a rope. Lastly, the Peterson Grab is a heavyweight at 40 kg (90 lbs) and is designed for hard bottoms composed of sand, pebbles, clay, and clay compounds.

You can think of a grab as a box that sinks into the bottom substrate of a body of water. As it is pulled up, there are plates or jaws that are released and cover the bottom of the box preventing loss of the sample. The grab is generally heavy enough to sink into the sediment on its own. A messenger is a device that that looks like a donut and slides down the rope or cable attached to the grab. You lower the grab so that it sits on the seafloor, release the messenger and it slides down and triggers the jaws to close. The Ekman Grab is one such grab.

Grabs tend to weigh between 4-100 kg (10-200 lbs). They sample an area from 15 x 15 cm to 30 x 30 cm (6 x 6 inches to 12 x 12 inches) and generally do not penetrate the sediment more than 15 to 30 cm. (6-12 inches). Some grabs are designed so that they can be attached to a pole. Because of the weight of the device, the poles used are not longer than 4-5 meters (12-15 feet). Usually, grabs are deployed by lowering them on a rope by hand or with a winch depending on the depth being sampled and the weight of the grab. If sampling more than several meters deep or with a larger grab, hand deployment becomes problematic. Grabs can be used to sample in depths of several hundred meters (1,000 feet).

There are some drawbacks to using a grab. First, you are not certain how far it penetrated the bottom surface until you bring it up. Second, bottoms covered by large rocks and densely packed clays will defeat a grab. Third, representative sampling will be better for smaller mollusks as opposed to larger ones. Lastly, when the grab is opened, the sample falls out and the sample's integrity is lost. Though you may have a representative sample, you will lose stratigraphic information. The larger, heavier devices called box corers solve some of these drawbacks.

3.3.2 Box cores. Box Corers (Figure 3.2B) are similar to grabs though much larger. The sampling area is from 50 x 50 cm to 100 x 100 cm (20 inches square to 40 inches square) with larger ones being available on special order. A box corer can sample to a depth of up to 100 cm (3 feet). The device is supported in a cage and has weights attached to it. The whole assembly can weigh 750 kg (1,650 lbs) with smaller ones being 200-300 kg (400-700 lbs). These are not devices for the weekend shell collector using a skiff. Also, keep in mind that when retrieving the device there is the weight of the corer and the sample (water soaked sediment) to be dealt with.

You lower a box corer by a winch and when it reaches the bottom, the weight of the device causes it to sink into the sediment. When the winch starts to retrieve it, a set of jaws closes over the opening to the box and prevents the sample from falling out. The corer and the sample are then brought to the deck of the ship and the sample is examined. To use a box corer you need a winch, steel cable, a sizable ship, and appropriate muscle power.

3.3.3 Tangle nets. If you purchase shells, you will occasionally see the phrase "taken in tangle nets." A tangle net is a reasonably fine meshed net made from monofilament nylon. This net is different from the one mentioned by Burch above (see Section 3.2.3). It is lowered to the bottom of a body of water and generally anchored with weights. The net is left in place from a few hours to several days. Bottom dwelling organisms become entangled in the net and are brought to the surface when the net is retrieved. Obviously, organisms that do not crawl about are less likely to be retrieved by a tangle net.

While a tangle net can be quite effective there are some drawbacks. Because they are made from fine monofilament nylon, they can tear easily. They are not selective about what they collect. In addition to some shells, you may also bring up sea stars, fish, and other marine organisms. Lastly, they can be too effective. Some fishermen have stopped using them because of overfishing of mollusks in a given area. For a description of working with a tangle net see Parker (2000).

3.3.4 Bail hooks. Bail hooks are used mainly to collect freshwater mussels. They are described in Chapter 19.6.3, and how to construct them is described in great detail in Nelson (1982).

3.3.5 Dip nets, Walker Dipper, etc. These are devices that you hold in your hand and sweep through vegetation or benthic sediment. A dip net is usually made from a wire hoop to which the net is fastened. The hoop is attached to a stick from 1 to 2 m (3-6 feet) in length. The Walker Dipper is similar, however, the collecting portion is made completely from metal. Further information on this device can be found in Chapter 2.3.4 and 21.4.

3.4 LITERATURE CITED

Burch, J. Q. 1974. Dredging along the California coast. *In:* M. K Jacobson, *How to Study and Collect Shells*, 4th Ed. American Malacological Union. Wrightsville Beach, North Carolina. Pp. 27-30.

Burch, T. 1941. Dredging for everyone. *In: American Malacological Union 11th Annual Meeting.* Pp. 24-29.

Burch, B. L. and T. A. Burch. 2000. *Xenophora* as bioindicators in the Hawaiian Archipelago, a preliminary study. *Western Society of Malacologists (Annual Report for 1999)* **32**: 2-9.

Holme, N. A. and A. D. McIntyre (eds.). 1971. *Methods for the Study of Marine Benthos*, International Biological Programme Handbook No. 16. Blackwell Scientific Publications, Oxford. xii + 334 pp.

Monfils, P. 1999. The old shell game: Extend your reach beyond the beach - dredging for shells. *American Conchologist* **27**: 30-31.

Nelson, D. 1982. Sampling for mussels. *In:* A. C. Miller, compiler, *Report of Freshwater Mollusks Workshop, 19-20 May 1981.* U.S. Army Engineer Waterways Experimental Station, Vicksburg, Mississippi. Pp. 41-60.

Ockelmann, K. W. 1964. An improved detritus-sledge for collecting meiobenthos. *Ophelia* **1**: 217-222.

Parker, R. 2000. Philippine tangle nets. *Of Sea and Shore* **23**: 19-21, 55.

Skoglund, P. 1990. Small boat dredging. *Festivus* **22**: 106-109.

CHAPTER 4

SNORKELING AND SCUBA DIVING

JIM CORDY
BOBBI CORDY

4.1 INTRODUCTION

Immersing ourselves in the environment and habitat of mollusks allows us great advantage over intertidal or beach collecting of mollusks. The advantages of collecting by snorkeling or using SCUBA (Self Contained Underwater Breathing Apparatus) far outweigh the disadvantages. SCUBA and snorkeling give you a chance to see mollusks crawling, swimming, eating, being eaten, mating, laying eggs, and even sitting on their eggs. Reading books that discuss the actual animals that live in the shells will help you greatly in locating them in their habitat. Some mollusks live on grass, sea fans, and algae, in sand, under rocks, on pilings and sunken ships, or in rock crevices. Some mollusks are more active at night than during the daylight hours, so night diving is also in order.

We have learned more by observing mollusks ourselves than we have by reading about them in books. The experience of seeing mollusks in their natural environment and locating their habitat is exciting. It took us many years of collecting to locate the habitat of some mollusks. After 40 years of active collecting, there are some species that still have us puzzled.

Going back to the same areas to collect never gets boring. We find that the areas where we collect frequently change seasonally and with storms. We often find shells in an area only during certain times of the year. There is always something new to be found.

We collected intertidally for several years before we took the plunge of learning to dive. Diving has given us a big advantage in quantity and quality of shells that we collect. We are always looking for the best quality shells and are careful to avoid over collecting. If we are in an area where there are several specimens of one species, we frequently check many of them before we select one to be collected. We also spend a lot of time turning over rocks and dead coral. We carefully turn the rocks back over after examining what is under them. This is appropriate to protect the sites where organisms live. You can actually spend hours in a small locality finding and collecting many different shells.

Underwater photography is another area in which you may develop an interest. Photographing the animals underwater is a great hobby. Studying the animals and their habitat becomes a natural secondary activity as you collect underwater. It is exciting to catch them eating, laying eggs, mating, etc. and capturing it on film (See Chapter 7).

4.2 LEARNING TO SNORKEL

You must be a good swimmer, but not necessarily an expert. To snorkel, you must be confident in water. You can snorkel in many places including water that is not over your head. If you can swim 6-8 laps in a pool, you should be fine. We started our daughter snorkeling when she was 10 years old and she did quite well.

You can learn to snorkel from a friend or shell collector who already knows how to snorkel. Have them take you to a local pool to practice. You can also check out a local dive shop for snorkeling classes. We suggest you borrow or rent equipment before you purchase any. You will see if you are going to enjoy snorkeling and you will have the opportunity to try out different types of gear.

Start in nice shallow, calm water the first time you get in the ocean. Stay with a friend! This increases the safety factor and you can work together to collect.

4.3 EQUIPMENT NEEDED TO SNORKEL

You will need a well-fitting dive mask. Masks come in various sizes and shapes. Everyone's face is shaped differently. Take time to find a mask that fits. There is nothing worse than a leaking mask. You can get a mask with prescription lenses built in or you can wear your mask over your contact lenses (just do not rub your eyes)! One of our last masks was purchased at Wal-Mart and has been one of the best masks we have ever had. You will need a snorkel and a snorkel holder. Some of the new snorkels are very complex and keep water out completely. A pair of dive fins will be next on your list. If you have not used a pair of fins, you will soon rely on these for great strength in swimming, especially against currents. Fins come in various sizes and shapes according to your swimming abilities and the strength in your legs. The larger professional fins are for the more advanced swimmer or those in good physical shape.

You will want boots or socks to wear under the fins. Do not try wearing your fins with bare feet. It will not be long before they will irritate exposed skin and cause a blister. Socks work well if you do not have to come in and walk on a rocky surface. Boots will protect your feet the best. A dive knife or tool is a useful implement. These are not for protection but to dig into holes, fan the sand, or to pop a shell off a rock. We use ours constantly. If they are not made of stainless steel make sure that you dry them after diving. This will prevent them from rusting.

You will want to protect your body from the sun and scratches. You can invest in skins (dive suits) made of Lycra that completely cover your body or go the inexpensive way and dig out a pair of those old polyester or knit pants and a long sleeved T-shirt. Covering all exposed areas of your body with a good waterproof sunscreen is essential. If you plan to snorkel in colder water, a neoprene wet suit is essential to stay warm.

Collecting bags and/or bottles will come in handy when you dive. Always put venomous cone snails in solid containers, not bags, to protect against being stung. Mesh dive bags can be purchased at most dive shops. We prefer a homemade version. We make ours of a lightweight canvas or heavy cotton fabric (about 15-18 cm^2 or 6-7 $inch^2$) with a Velcro closure. We sew a cord on one corner to go around the wrist or hook onto a belt. Tabs sewn on the top to pull the bag open when under water are also a great help. Another great collecting item is a plastic Parmesan cheese bottle. They have a screw-on lid with small holes on one side and a larger hole on the other with a flip top. We put holes in the bottle with an ice pick and feed a cord through it and then tie a good knot in the cord on the inside of the bottle. This can then be attached to a wrist or hung from a belt. Fill the bottle with water before you dive. This prevents it from floating up when you jump into the water. We have also used nylon tummy packs with zippers. Be sure the zipper is nylon and not metal to avoid its rusting.

Gloves will protect your hands from many sources of injury. You can invest in expensive dive gloves at a dive shop or use a pair of canvas garden gloves. Either way, they soon wear out from turning rocks or snagging them on coral.

A weight belt will be needed to help you stay submerged. You will need some help with selecting this item. It is a good idea try one out in a swimming pool to see the amount of extra weight you will need to dive to the bottom. Some people only need 5-6 kg (10-13 pounds) while others may need as much as 10-15 kg (22-33 pounds). As you become more confident in the water and start diving at night, an underwater dive light will be indispensable.

Snorkeling is a wonderful way to view not only the environment where mollusks are found but also the beautiful fish, lobster, crabs, sea stars, and other sea creatures that abound. Still, there are a few precautions to keep in mind. These are covered later in this chapter.

4.4 SCUBA DIVING

SCUBA will enable you to go deeper and stay underwater for long periods but it requires professional training. Locate a local dive shop to sign up for a Basic Dive Course. This course usually involves a swimming test, 20 hours of pool training, 20 hours of classroom work, and two checkout dives in the ocean. Dive shops frequently offer classes on a two for one price deal or as a group. You will need a dive friend for this sport. Minimum age for SCUBA is 15 years old.

SCUBA is more involved as far as equipment and investment than snorkeling. More advanced classes can teach you to use a compass underwater, perform search and rescue activities, or become a Dive Master.

4.5 EQUIPMENT NEEDED FOR SCUBA DIVING

Some of the equipment needed for SCUBA diving is the same as for snorkeling. In addition, some pieces of equipment are unique to SCUBA. You will need a good dive mask, snorkel and snorkel holder, swim fins, dive knife or tool, and dive gloves and boots. You will have all of these if you started out as a snorkeler.

Equipment that you will need to get specifically for SCUBA includes a buoyancy compensator. This piece of equipment is vital to keep you buoyant with your SCUBA tank and weight belt. It also functions as a safety flotation device. You will also require a dive computer. This is used to calculate the depth and duration of each dive. You will need aluminum SCUBA tanks, two are recommended, and these can be rented or purchased. The dive regulator attaches to the tanks and supplies the air in them to your mouth. Lastly, is the sea view gauge. This device displays the air pressure in tank, and gives an indication of how long you can remain under water. If you plan on taking deep dives or diving at night, you will also need an underwater dive light.

4.6 DRAWBACKS TO SCUBA DIVING

There are some drawbacks to SCUBA. First, there is the high cost of equipment. If you belong to a dive club or rent your gear, this may help to keep the costs down. Second, some commercial dive boats do not allow you to take live shells. If you are interested in collecting specimens, as opposed to observational or photographic studies, be sure to ask the boat owner what the policy is about collecting. Also, check with the captain whether collecting is allowed where he or she is taking you. If you are diving in a nature preserve or some other protected area, collecting may be illegal. Two such areas would be the Tortugas in the Gulf of Mexico and around the Galapagos Islands.

You must be in good health. Ask your doctor about your health before you try SCUBA. Some of the medical risks include ear, heart, and lung problems. Diving with a cold, symptomatic allergies, or sinus problems can be risky. Many divers take a decongestant before diving. Decompression sickness (the bends) is usually not a problem as long as you follow the precautions as taught in the required SCUBA course. Proper decompression should be undertaken by anyone who dives deep enough to have to worry about the bends. For more on this topic see Kizer (2001) and Brubakk and Neuman (2002).

4.7 UNDERWATER HAZARDS

Both snorkeling and SCUBA diving can be a wonderful way to add to your collection and your knowledge of shell collecting. It might even add a few lobster or fish to that collecting bag for dinner at night. However, some caution must be exercised. Besides the inherent dangers of the activities of snorkeling and SCUBA, there are certain environmental hazards (Halstead *et al.* 1990).

Fire coral is a white tipped coral that can cause a severe irritation to bare skin. Other corals can be sharp and cut exposed skin. Sea urchins, especially those with the very long spines, can be dangerous if they pierce the skin causing puncture wounds. Moray eels can swim out from under rocks and from inside crevices. They are not poisonous, however, they aggressively protect their territory. Their sharp teeth can inflict a serious bite.

Jellyfish can sting with their long tentacles. The box jellyfish of Australia is among the most venomous of all animals in the world. All jellyfish should be approached with caution. Stonefish are poisonous and are well camouflaged in their environment. In the Indo-Pacific region, you have to watch out for sea snakes, lionfish, and the blue-ringed octopus. Poisonous cone shells, especially in the Indo-Pacific region, are also of concern. Beware of picking these up by hand. Use tongs or very heavy gloves and handle them with caution! There are several types whose stings have been fatal to humans.

We suggest that you study up on local creatures in an area before going there to dive. It is also wise to discuss with the ship's captain or the dive master what dangers are likely to be found in the region where you are diving.

4.8 CONCLUSIONS

Snorkeling and SCUBA diving can be great ways to explore marine and freshwater molluscan habitats. With the appropriate training and prudent caution, it is also a safe activity. Jump in and get your feet wet; you will learn more than you can from handling only dry conchological specimens.

4.9 LITERATURE CITED

Brubakk, A. O. and T. S. Neuman. (eds.) 2002. *Bennett and Elliott's Physiology and Medicine of Diving*, 5th Ed. Saunders, Philadelphia. 779 pp.

Halstead, B. W., P. S. Auerbach, and D. Campbell. 1990. *A Color Atlas of Dangerous Marine Animals*. CRC Press, Inc., Boca Raton, Florida. 192 pp.

Kizer, K. W. 2001. Chapter 57. Diving medicine. *In:* P. Auerbach, ed., *Wilderness Medicine*, 4th Ed. Mosby, Inc., St. Louis, Missouri. Pp. 1366-1401.

CHAPTER 5

ARCHIVAL AND CURATORIAL METHODS

CHARLES F. STURM

5.1 BASIC PRINCIPLES

Collectors often do not give enough thought to the preservation of their shell and fossil collections. A shell is a hard, durable item, or so we think. Unfortunately, that shells are hard durable items is not the case. Many forces are conspiring to destroy our collections. If we want our collections to maintain maximum longevity, we must apply basic principles of good curation (Solem et al. 1981, Rose and de Torres 1992, Rose et al. 1995). Applying these principles will not only help preserve our collections but will also make it easier to incorporate them into a museum's collection. They will also help to maximize the value of a collection if it is to be sold.

The first principle of good curation is to do no harm. Whatever else, do not treat or handle a specimen in a way that makes it worth less after curation than it was worth before. One should strive to apply treatments that can be reversed easily. For example, it is better to use an adhesive that can be easily removed as opposed to one that will defy the use of a hammer and chisel to remove it.

The second principle dictates that techniques should be used that cause as little permanent change to a specimen as possible. It should be assumed that whatever we do to a specimen will cause some change, and sometimes the change may not be completely reversible. Thus, we should minimize what the permanent change to the specimen will be. An example of this is the way some people preserve the periostracum of naiads (Unionidae) with a paraffin/xylene solution. The shell is dipped in the solution and the xylene then evaporates. Paraffin is left behind and this helps to preserve the periostracum from drying out and flaking off. If one wants to remove the paraffin, successively soaking the shell in xylene will leach the paraffin out. However, in addition to the paraffin, other xylene soluble substances will also be removed. This will cause a permanent change to the shell. If morphometric studies of the shell are the objective of a study, then the paraffin/xylene treatment will be of no consequence. However, if we wish to undertake some biochemical studies of the shell, the effects of the treatment will be of importance. One way out of this dilemma would be to treat some but not all of the shells, and record which ones were treated.

The third of the basic principles is to record what techniques you apply to a specimen, and to record them on the specimen's label. In this way, others will be able to tell what was done to a given specimen, whether it can be undone, and what effect it will have on future analyses that might be applied to the specimen.

Techniques that follow these principles will be discussed in this chapter. Sources for many of the products mentioned in the next few pages are listed in the appendix at the end of the chapter.

5.2 DANGERS TO A COLLECTION

I would like to address six basic dangers to collections. They include risks that result from acid exposure, temperature, humidity, light, pests, and shock/abrasion. There are other dangers such as flooding, storms and other natural disasters, fire, theft, and being exposed to armed conflicts, but these catastrophic dangers are beyond the scope of this chapter.

The first aim should be to avoid exposing a collection to these dangers. If avoidance has not been done consistently then you have to fall to a second line of defense, which is to detect incipient problems and block their subsequent damage. Least satisfactory is detecting damage that has already occurred and trying to stop further degradation. Rarely can one reverse damage that has already occurred.

5.2.1 Acid, temperature, and humidity. The first three dangers, acid, temperature, and humidity (often measured as relative humidity, RH) all contribute to a condition called Bynesian Decay (also called Bynes Disease) (Tennent and Baird 1985, Shelton 1996). The calcium carbonate of the shell decomposes in this condition. Bynesian Decay was thought to be due, in part, to a process initiated by bacteria, but is now known to be a chemical process. Bynes originally investigated this process and published papers on it between the years 1899 to 1907. Since it is a chemical process and not a bacterial one, I advocate the term decay as opposed to the older designation as a disease.

This condition can affect collections large and small. Sometimes a few shells are affected, occasionally whole collections. Once the decay has started, the part of a specimen that has been affected cannot be restored. However, further deterioration can be halted.

For Bynesian Decay to occur, acid must be present in the microenvironment of the shell. The acids most commonly involved are formic and acetic acids. These acids are produced by the wood used in cabinets, cardboard trays and boxes, and labels, as well as from the adhesives used in the construction of cabinets and in the repair of specimens. Along with the acidic fumes, temperature is also a concern. The higher the temperature the faster acids can react with the calcium carbonate of the specimens and change it into calcium acetate-formate salts. Thus, at lower temperatures, Bynesian Decay will progress more slowly. Lastly, the humidity must be high enough to provide moisture for the acids to dissolve and precipitate out onto the specimens to cause the decay.

Thus, we see that Bynesian Decay is due to the misfortune of several conditions being present to allow this process to occur. As mentioned above, wood can be a source of the acids. Some woods are more acidic than others and the acidic woods should be avoided when constructing cabinets. The worst offender is oak. Some of the better woods are spruce, mahogany, walnut, birch, basswood, poplar, and balsa. If the woods being used are veneers, then one has to be concerned with the adhesives used in making the veneers and whether they will offgas acidic fumes, as well as the type of wood underneath the veneer.

Plywoods are often constructed using urethane adhesives which can offgas formaldehyde, an acidic compound. If plywood has to be used, one designed for exterior use is preferred. The adhesives used are less harmful than those used in interior plywoods, and the offgasing problem is less intense. Particleboard and pressed wood should be avoided because of the adhesives used in their production (Hatchfield 1995).

The ideal cabinet construction is a metal cabinet that is painted using a powder coat process. Here the metal is painted by a process of electrostatically coating the surface with pulverized polymers (the paint) and then fusing it to the metal with heat. There will be no offgasing of organic solvents or other substances. Two manufacturers of such cabinets are Lane Scientific Equipment Corp. and Steel Fixture Manufacturing Co.

R. Tucker Abbott described how to build a wooden cabinet (Abbott 1954). He recommended a standard size cabinet 40 inch high (101 cm), 22 inch (56 cm) wide, and 32 inch (81 cm) deep. Runners for the drawers should be 30 inch long and set $2\text{-}1/_4$ inch apart. The internal dimensions of the drawers were $20 \times 30 \times 1\text{-}5/_8$ inch. The door should be hinged so that a drawer can be pulled out when the door is open only 90 degrees, and an additional feature is a door that can be completely lifted off its hinge.

If a wooden cabinet is being painted, water based paints and varnishes are preferred. Oil based products can give off formaldehyde and other volatile organic compounds (VOC). These can contribute to Bynesian Decay. Allow four weeks for the cabinet to completely air out and offgas before putting specimens into it. Remember, other sources of acids are the paper you use in the labels and trays, and the inks you use. These will be addressed later.

Temperature is a concern in storing a collection. At higher temperatures, chemical reactions occur faster, thus the decay process will be more likely to occur at a higher than at a lower temperature. In addition, offgasing of acid vapors is faster at a higher temperature. An additional complication is that the items that we are trying to preserve have different ideal temperatures for storage. The ideal temperature for storing photographic images will differ from that for books and papers, which will be different from that for shells. Thus, there is no easy answer to the question, "What is the best temperature to maintain a room that houses a mixed collection?" In general, for shells, a temperature in the range of 16-21°C (60-70°F) appears to be a good compromise between the needs of the collection and a reasonable working environment for people.

Relative humidity is the amount of moisture in the atmosphere at a given temperature relative to how much the air could possibly hold at that temperature. As temperature increases, the atmosphere can hold more water and therefore dissolve more acidic gases. That is the reason that humidity is of such concern in collection management. Fluctuations in humidity can also be problematic. Mineral specimens such as shales, clays, and amber can crack and flake if the humidity gets too low. Some shells, such as the gastropod genus *Paryphanta* and the bivalve genus *Pyganodon* will get too dry at low humidity and crack or shatter. Paper products may become brittle at low humidity. Inks may flake off the page and book covers may warp at high humidity levels.

You might have the occasion to collect fossils that are composed of Pyrite. Pyrite will start degrading at elevated humidity and give off acidic gases. This condition is called Pyrite Disease (or more recently Marcasite Disease), though as in the above discussion on Bynesian Decay, it is not a bacterial disease. I prefer the term Pyrite Decay (Marcasite Disease has not yet gained widespread use). In Pyrite Decay, iron sulfide combines with oxygen and water and forms iron sulfates and sulfuric acid. The result is that the integrity of the specimens is destroyed. If any calcium carbonate shells are present, the sulfuric acid can precipitate Bynesian Decay.

Many techniques have been proposed for preventing Pyrite Decay. Their basis is to prevent the specimen from coming in contact with air either by encasing the specimen with an artificial resin or submerging it in a liquid such as glycerin, paraffin oil, kerosene, or silicone oil. None of these techniques work well (Howie 1992).

The best prevention is to maintain the pyrite specimens in a low relative humidity, around 30% (Howie 1992, Waller 1992). If some degradation has already occurred, neutralization of the degradation products with ethanolamine thioglycollate may help (Cornish and Doyle 1984). Some researchers advocate placing specimens in an oxygen free environment (anoxic environment). This technique is beyond the scope of this chapter but you can read about it in Burke (1996).

There are causes of variation in temperature and RH that we may not even think of. If a cabinet sits in sunlight for part of the day, it may experience fluctuations in temperature and RH on a daily basis. If a cabinet is against a sun-exposed outside wall, it may experience fluctuations that are not seen in other cabinets against inside walls. Basements and lower floors of a building tend to be more humid. Thus, some people recommend that collections are better stored on the upper levels of buildings.

For the above diversity of problems, universal recommendations are difficult to come by. For a collection of a given type of specimens and storage conditions, you may find several different recommendations of temperature and RH. In general, if the relative humidity of a shell collection is maintained within the range of 50-55%, one should not be concerned. This, with a temperature range of 16-21°C (60-70°F) is well within the abilities of home heating/ventilation/air conditioning systems and may need only occasional help from a dehumidifier. The best way to monitor the RH is with a wet-dry bulb thermometer or a hygrometer. I use a mechanical Abbeon Model HTAB 169B hygrometer/thermometer combination while the Section of Invertebrate Zoology at the Carnegie Museum uses a battery operated Oakton Digital Maximum/Minimum Thermohygrometer. I like the fact that I do not have to worry about the battery running out; they like the recording capability of the battery operated device. Hygrometers cost $35-200 depending on the sophistication and accuracy one seeks in the instrument; the Abbeon is approximately $150 while the Oakton is approximately $40.

There are also devices called dataloggers. These are small devices that can cost from $100 to $1000. They are electronic devices that take multiple measurements of temperature and relative humidity; every few minutes, every few hours, or daily. The data are stored on a microchip within the datalogger. You then download the information to a computer and can graph out the results. Some dataloggers contain a readout so that you can see what the current measurements are. Dataloggers may also have alarms to alert you when preset parameters have been exceeded. Most dataloggers will run for at least a year on a set of batteries. Arenstein (2002) provided more information on dataloggers including general datalogger features, comparative data for fourteen models, and addresses for manufactures of these devices. Dataloggers can be purchased from several of the suppliers listed at the end of this chapter.

5.2.2 Light. Light provides energy for chemical reactions to occur. If we limit exposure to light, we can slow down these chemical processes. The type of light is also important. While visible light can cause problems, ultraviolet light (UV) is more problematic. UV light has more energy than visible light and, therefore, it causes more damage than visible light. Fluorescent lights give off more ultraviolet radiation than do incandescent bulbs.

Light causes several problems in natural history collections (Weintraub and Wolf 1995:194). Exposure to visible and ultraviolet light can cause specimens to fade. Non-archival inks will fade on exposure to light. Sunlight and incandescent bulbs may cause an increase in temperature and variations in RH. Short-term exposure should not be problematic. If fluorescent lights are on for a long period of time, UV filter tubes should be placed around the bulbs. If there is a large amount of sunlight in the room, UV filters can be placed on the windows. Neither of these types of products causes noticeable changes in light intensity. They do need to be changed periodically as they lose their filtering ability. Replacement, which is dependent upon the intensity of light exposure, should not be necessary more than every 5-10 years.

5.2.3 Pests. Pests may cause many problems. Silverfish and cockroaches can eat the adhesives that are used in bookbindings. They can also eat the paper used in labels and books. Mice and rodents can also chew up paper materials. Mold will grow on paper and even on some porous specimens making removal difficult or impossible. Prevention is better than treatment. Mold and most insects will be kept under control at the levels of temperature and RH recommended above. If vermin are a problem, mechanical traps are useful. In regards to insecticides, not much is known on how their components will affect a collection. Pyrethrums or permethrin should afford the safest alternative among the insecticides. A powder form is preferred, as it would not contain the solvents and other volatile organic compounds found in an aerosol spray. Consult the paper by Jessup (1995) for more details on pest management.

5.2.4 Shock and abrasion. Shock and abrasion can also wreak havoc on a collection. Specimens can be dropped, crushed, or abraded. They can be crammed together in tight spaces. The headroom above a drawer may be insufficient and the drawer above it may hit the specimen. Specimens can abrade labels and make them difficult to read. There are several ways to decrease damage due to shock and abrasion.

Foam liners in specimen trays can be effective. The foam must be archival and polyethylene foam is commonly used (the archival nature of plastics is discussed below). Avoid polyurethane foam (foam rubber). It is not archival and it will degrade over time giving off acidic vapors (formaldehyde). Labels can be placed in polyethylene tetraphthalate (Mylar®) or polyethylene bags or sleeves. This prevents them from being abraded by the specimen and also decreases contact between specimens and acidic labels. Specimens should not be placed in a tray that is too small. The tray should be big enough to allow space around each of the specimens in a lot. One can use a standup label. This allows all the pertinent information on the label to be read without having to handle the specimen. One can view the specimen and read the label with minimal handling of either. Drawers should slide smoothly to minimize jostling. Lastly, one can fill all empty spaces in a drawer with empty trays. This prevents trays with specimens from sliding around when a drawer is moved.

A collection should be organized in a manner that minimizes the handling of specimens. By arranging a collection systematically according to some system (e.g. Thiele 1929-1935, Vaught 1989, Thiele 1992-1998, Millard 2003), a specimen can be located without shuffling through the whole collection. In arranging a fossil collection, one would store specimens using a combination of stratigraphic and systematic criteria.

If a collection includes any type material, these should be segregated from the research collection. By segregation, I mean that they should not be with the regular collection, but stored in a separate drawer or cabinet. Type specimens are specimens that were used by someone in describing a new genus or species. Type specimens are discussed more fully in Chapter 10.5.2.

In some collections, other classes of specimens will also be segregated from the main reference collection. Non-type material that might be segregated includes rare, threatened, endangered, or recently extinct species. Figured specimens, sometimes known as hypotypes, are specimens used for the purpose of illustration in a scientific publication. These specimens are also segregated in some collections. This allows them an additional degree of protection from excessive handling. Topotypes, material from the same locality as the name bearing types but not used in the original description are sometimes segregated into the type collection.

5.3 PAPER

Paper is used to construct trays and make labels and collection catalogs. The paper used should be archival. There are several points to consider when selecting paper products. The first is that they should be acid free. Acids will cause paper to yellow, become brittle, and decompose. These effects are due to acids breaking down the bonds between cellulose fibers leading to a loss of integrity of the paper. Second, paper should be low in lignin content. Lignin can break down to form acids that can in turn cause the paper to disintegrate. Most papers made from wood pulp are high in lignin while those made from cotton or linen are low in lignin. A third consideration is a paper that is buffered. This means that the manufacturer adds a chemical to the paper that will help to neutralize acids. Calcium carbonate is the most common buffer used. Lastly, the paper should be alkaline sized and not acid-rosin sized.

These points suggest that the optimal paper would be one with a pH of 7-8, a lignin content less than 0.3%, buffered with 2-3% calcium carbonate, and alkaline sized. When buying paper, look for paper for which the manufacturer is willing to supply full specifications. There are also indicator pens made that will determine the pH or lignin values of paper. These work best on white or light colored paper. For example, using a pH indicator pen, you would make a mark on the paper in question and the color that the mark turns indicates the approximate pH of the paper. The Section of Invertebrate Zoology at the Carnegie Museum uses Mohawk Superfine, 65 pound, smooth white paper. The malacology groups at the Natural History Museum of Los Angeles County and the Delaware Museum of Natural History use Perma/Dur paper, which can be obtained from University Products, while the Section of Mollusks at the Carnegie Museum uses Permalife Bond Writing Paper, a 20 pound 25% cotton, acid free, archival paper from Fox River Paper Co. For more information on cellulose based papers see Burgess (1995).

There is also paper made from Tyvek®, a form of polypropylene. This product is virtually indestructible. It does have several drawbacks. Inks tend to bleed on Tyvek® and it does not accept pencil markings as well as cellulose based paper. In addition, it cannot be used with a laser printer, as it will melt. Teslin® is another polymer based paper that has been developed. Like Tyvek®, it is waterproof and quite resistant to wear and tear. Some forms of it seem to work well with ink jet and laser printers. How well it will stand up as an archival product has not yet been determined.

5.4 INKS AND COMPUTER PRINTERS

Most of us give very little thought to the inks that we use. They are, however, complex chemical mixtures and their compositions are often proprietary secrets. Inks are mixtures of pigments, dyes, binders, and vehicles. Dyes stain the paper fibers, while pigments settle out on the fibers. India inks are pigment type inks that use carbon black as the pigment. Several India inks that have been found to be relatively stable and are considered archival are Rotring 17 Black, Hunt Speedball Super Black Ink, Pelikan 17 Black, Higgins T-100, and Pelikan 50 Special Black (Williams and Hawks 1986). In addition, some brands of disposable pens are considered archival. They are convenient to use and come in various point widths. Two brands that fall into this category are Pigma and ZIG.

There are three main types of computer printers: dot matrix, ink jet, and laser. Dot matrix printers are generally considered archival (depending on the ink). The ribbon is coated with a carbon pigment and is applied to the paper by impact. If the pigment is worn away, the impact of the print may still be read on the label by oblique illumination. Ink jet printers should be avoided. The inks used in these printers are rarely archival and often are soluble in water and organic solvents. This solubility prevents them from being used in wet collections. In addition, the inks tend to fade on exposure to light.

Some laser printers may be used for archival purposes. Mainframe laser printers apply ink and then set it with heat and pressure. This process appears to give an archival result. Desktop laser printers do not use the same amount of heat and pressure, and thus the result is not archival. Because of the heat involved, Tyvek® paper should not be used with laser printers.

Another printing method that is gaining popularity in museums is one using thermal transfer printing. Two different methods are used. In direct thermal printing, the printing is burned into the medium, a plastic material such as polyester. The second method, thermal wax transfer, uses heat to melt the ink which is then applied to the medium, paper or plastic. The labels hold up well in ethanol and thus are well suited for wet collections. These printers will print down to a 4 point font size. The limiting factor is cost. These printing systems cost around $1,300. Bentley (2004) discusses these methods in greater detail.

Photostatic or xerographic printing is also a cause for some concern. Some of these inks are not archival, some are not solvent resistant, and flaking of the print can be problematic. Since one frequently does not know the particulars of the ink, paper, and process involved with these methods of printing, I cannot recommend them as being archival. One should not print labels or documents this way unless you can obtain assurance from the manufacturer of your equipment that it uses an archival process.

Recently, CD-ROM's have been used to publish books. With the advent of writable and rewritable CD-ROM's, some collectors have been storing their collection catalogs in this format. Not all CD-ROM's are archival. There have been improvements in long-term stability and some are rated at life spans of 25-100 years if stored at 25°C (77°F). If you plan to use a digital format for long-term storage, make sure you are using an archival product and store it under archival ranges of temperature and RH. It is advisable to have a printed copy for backup (see below).

5.5 VIALS AND JARS

Vials and jars are used for holding small specimens or in storing material in a wet collection. Generally, glass is used because of its stability and clarity. Borosilicate glass, such as Pyrex, Kimax, and Wheaton 800 are brands to look for. Flint glass (soda-lime glass) is a less desirable form of glass, as it is more likely to degrade over time. Using ultraviolet light, one can identify the type of glass. Under ultraviolet light, flint glass fluoresces a bright yellow-green color (Simmons 1995).

5.5.1 Glass decay. I would like to discuss the inherent weaknesses of flint (soda-lime) glass and a condition known as Glass Disease. As with Bynesian Decay, Glass Disease is not a disease but a decay process that affects glass so a more appropriate term might be Glass Decay. Glass tends to be thought of as a relatively stable and inert substance; however, under certain conditions this is not true, especially for soda-lime glass.

Glass Decay is a deterioration of glass that is exposed to water and carbon dioxide. Non-silicate elements of the glass, such as sodium oxide, leach out, react with water vapor, and form alkaline solutions. Over time, the glass weakens and a white powder deposits on the glass.

The chemical reaction appears to be one of sodium oxide in the glass reacting with water in the air and forming sodium hydroxide. The sodium hydroxide reacts with carbon dioxide to form sodium carbonates. This reaction seems to be able to occur at RH's as low as 20%, but more commonly at a RH of 40% or higher.

Note, that this process requires water (moisture) in order to develop. Also recall that chemical reactions occur at faster rates at higher temperatures than at lower temperatures. Therefore, the same controls to prevent Bynesian Decay should also prevent or slow down the development of Glass Decay: keeping a low relative humidity and temperature in the collection storage area.

A similar reaction occurs in potash glass; potassium oxide reacts with water to form potassium hydroxide. This in turn reacts with carbon dioxide and results in the formation of potassium carbonate. The result is that the soda-lime and potash glass crazes, cracks, and develops a frosted appearance over time. The resulting container is weakened.

Since Glass Decay results in basic chemicals being formed (not acidic chemicals), it does not cause the same risks to a collection as seen in Bynesian Decay. Most specimens stored in glass vials are small. If the glass container turns from a clear one to a frosted one, specimens will have to be removed more frequently to be viewed. This causes risks from physical handling to these smaller, more fragile specimens. Also, the salts formed in Glass Decay can precipitate out on the specimens making fine details hard to view.

If the process of decay occurs only on the surface (a process more common with glass manufactured in recent centuries) the process that occurs is called devitrification. This surface process occurs on account of recent soda-lime glass being more stable than those of previous times.

Besides controlling storage conditions, you can also consider using borosilicate glass. Borosilicate glass is preferable to soda glass. While more expensive, it appears to be a more stable form of glass, and is less susceptible to the processes of Glass Decay and devitrification. See Hamilton (1998) for more on issues involving glass and its conservation.

5.5.2 Vials, jars, and closures. When closing vials there are several options. Cork should be avoided as an option; it tends to be acidic and decomposes over time. It also exposes the collection to acidic vapors. A second option is cotton. This tends to be acceptable if a high quality grade of cotton is used. Cheaper grades of cotton may be acidic. Before using cotton, test its pH with a pH-testing pen. The use of polyester fiber or batting is becoming more common. It is cheaper than cotton and just as easy to work with. Lastly, some vials come with polyethylene or polypropylene snap caps. These are archival, but more costly than using polyester fiber.

When it comes to sealing jars, there is a combination of a lid and a liner. There are several different lids and liners available. The first type of lid, and one of the poorer choices, is made from a plastic called Bakelite®. Bakelite is usually a black plastic that is very rigid. It often comes with a cardboard liner that is foil coated. This type of lid is not considered archival on three accounts. First, the lids tend to crack if over tightened. Second, they are susceptible to backing off, a process where a lid loosens due to fluctuations in temperature or exposure to vibration. Lastly, the cardboard liner absorbs fluid that causes it to swell and deteriorate over time.

A second type of lid is a metal lid, with or without a liner. Metal lids tend to rust, and in the presence of glycerin or formaldehyde, they tend to corrode rapidly. These chemicals are frequently found in wet collections. For these reasons, metal lids are generally unacceptable.

Polypropylene lids with polyethylene liners are an excellent combination. The softer polyethylene molds to the irregularities of the jar opening and the polypropylene lid can be tightened without undue fear of cracking. If one is planning to maintain a small wet collection, it may be easier to contact a university or museum with a malacology collection to see if you can obtain a few jars and closures from them.

A final option is a bail top jar or Mason jar. These tend to work well but have some drawbacks. The lids snug down nicely with the wire snap; however, the rubber gaskets tend to deteriorate over time and need to be changed periodically. While some gaskets seem to be more resistant than others, they can be difficult to track down or purchase in small quantities. Thus, the main drawbacks of these jars are their cost, the difficulty in finding them, and the difficulty in finding or replacing the gaskets. If you acquire a few of these jars, a museum or university may have appropriate sized gaskets.

When storing tissue in fluid (wet collection), ethanol is the best all around preservative. A concentration of 80-90% should be used. Specific recommendations for preservation of soft tissues are made in the taxonomic chapters, in Chapter 14, and below in Section 5.9.

Many collectors are using archival plastic boxes and polyethylene ziplock bags for storing dry specimens. These containers have several advantages. They tend to be less expensive than glass vials, have good clarity, and come in a large range of sizes. They also help to control the microenvironment surrounding the specimen. The types of plastics used in their manufacture are considered archival and are discussed in the next section.

5.6 PLASTICS

I have mentioned the use of plastics at several points in this chapter. Some plastics are considered archival while others are not. In general, one wants to use plastics that have no or low levels of plasticizers. The fewer UV inhibitors and dyes that are put into the plastic the better. Lastly, one wants a plastic that is free of surface coatings. With these considerations in mind, there are five plastics that are considered to be archival. They are polyethylene, polypropylene, polycarbonate, polyethylene tetraphthalate, and polytetrafluoroethylene (Teflon®). There is some question of the long-term stability of polystyrene. Some people feel that as long as you avoid exposing it to organic solvents, there should be few problems using polystyrene. Other curators and collection managers suggest avoiding it completely. If you chose to use polystyrene, I recommend that you periodically check the specimens stored in these containers. Polyvinylchloride and polyurethane should be avoided as they can degrade and give off acidic vapors.

Some people like to reuse plastic containers that originally had some other use, for example, margarine tubs or 35mm film containers. The question is how to decide if they are made out of archival plastics. This has become quite easy with the advent of recycling. Plastic products that can be recycled have a code on them. The code is either a triangle with a number in it, an abbreviation of the plastic's name or both. This is generally located on the bottom of the container. Table 5.1 shows the code and which products are archival. Again, watch for possible contamination from colors added to the plastic or the product being coated with other substances. Note that polycarbonate and polytetra-

Table 5.1 Recyclable plastics.

Number	Abbreviation	Plastic	Archival
1	PETE	Polyethylene tetraphthalate Polyester, Dacron®, Mylar®	Yes
2	HDPE	High density polyethylene	Yes
3	PVC	Polyvinyl chloride	No
4	LDPE	Low density polyethylene	Yes
5	PP	Polypropylene	Yes
6	PS	Polystyrene	Uncertain
7	Others	Other plastics	No

fluoroethylene are not recyclable plastics and thus do not appear in the chart.

Williams *et al.* (1998) discussed a sequence of tests that are useful in distinguishing twelve clear plastics. Their algorithm includes polyethylene, polypropylene, polystyrene, polycarbonate, and polyethylene tetraphthalate.

5.7 CONSOLIDANTS AND ADHESIVES

I will begin with some definitions. An adhesive is a substance used to bind items together. Glues are adhesives that are derived from animal origins; for example, rabbit hide glue. A consolidant is a substance used to permeate and strengthen a specimen. Just because a glue or consolidant was described as being useful in the past literature, do not assume that it is still an appropriate substance to use. Shellac was once widely used, but it is no longer considered an acceptable consolidant. With adhesives and consolidants, reversibility is an important consideration. We should be able to undo what we did, with minimal effect on the specimen. Along with reversibility, a non-acidic nature and dimensional stability are also important and desired attributes.

Among the adhesives and consolidants that are considered to be safe and archival are polyvinyl butyral (Butvar 76 and Butvar 98), polyvinyl acetate (Vinac), and acrylic copolymer (Lucite, Acryloid B72, and Paraloid). My favorite is Butvar 76. It is soluble in acetone and I make a thick mixture by dissolving 2.2 kg (one pound) of Butvar 76 in 4 L (one gallon) of acetone. This mixture is used as an adhesive. I dilute this slurry 1:1 with acetone and use this solution as a consolidant. It has the properties of fast penetration, a non-adhesive finish (non-tacky when dry), a brief drying time, and low toxicity.

If Butvar is applied to a specimen that contains moisture, a white haze will result on the surface of the specimen that was consolidated. Therefore, be sure specimens are dry when using Butvar. The main precaution is to use it with adequate ventilation. The solvent, acetone, can cause respiratory depression. Good ventilation prevents this. Acetone is also flammable and should not be used around an open flame.

Some adhesives and consolidants are no longer considered archival. These should be avoided. This group includes polyvinyl alcohol (shrinkage with age), cellulose nitrate (glyptal), and commercial mixtures. The problem with commercial mixtures (at least those that do not disclose the ingredients) is that one does not know what components are in the mixture. Also, without any warning, the manufacturer can reformulate the mixture.

You should also realize that some people avoid adhesives completely. They do not glue pieces of a specimen back together unless there is a specific reason to do so. They just put the pieces in a polyethylene bag and store them that way. This avoids any possible contamination to the specimen. If a specimen needs to be repaired in the future, a decision can be made then as to the best method to use. For further information on adhesives and consolidants see Elder *et al.* (1997).

5.8 RECORDS

A complementary activity to using archival methods is maintaining adequate records for the collection. There are several aspects of record keeping of which the collector should be made aware. The first point concerns the data that accompanies the specimen. All data should be connected with a specimen in an unambiguous manner (see Chapter 14). The second point is that a catalog should be kept. A hardbound, lined journal is a good choice. These can be purchased at most office supply stores. The data for each lot of shells is recorded on one line. In addition, this information should be on a label that is stored with the specimen(s). Many collectors are maintaining their catalogs digitally. If this is done, frequent backups should be undertaken, and the software that allows the catalog database to be read should also be backed up. This prevents data from being lost if the computer crashes, or a floppy disc or CD-ROM becomes corrupted and unreadable (it happens!).

Computer databases are excellent for searching the records of a collection. However, one must make sure that the data can be accessed now and in the future. Whether a database that is stored and not used for ten or twenty years will remain accessible is uncertain because of rapid technological changes. How many collectors still have a computer capable of reading a $5\text{-}1/_4$ inch floppy disc? On account of these uncertainties, some collectors are continuing to maintain a paper-based catalog along with their digital one.

Such a paper-based catalog can be a journal as mentioned above or a printout of the digital catalog. If one chooses the printout version, it should be printed on acid free paper and eventually bound. This will protect the data and prevent pages from becoming lost. This also allows the collection to remain useful if the digital form becomes corrupted or unreadable. For more information on databases, see Chapter 8.

How to label or identify a shell is a matter of personal preference. Most museums place the catalog number directly on the specimen with an archival India ink. If the shell is porous or friable, the number can be written on a small strip of paper and glued to the shell. Another alternative is to treat the surface of the shell with a consolidant, painting over this with an opaque paint, and write the number on this surface. One such product, called Liquid Label™, is available from Light Impressions. You paint on the opaque layer and let it dry. You then write your number on it and paint over it with the acrylic copolymer. If you wish to remove the number, just wash off the Liquid Label™ with acetone. If the shells are small, they may be put in a vial or plastic bag along with a slip of paper with the catalog number recorded on it. This can also be done with larger shells if one does not wish to write numbers on them.

Some collectors will measure a shell to the nearest millimeter or tenth of a millimeter, recording the size and the data in a ledger. The size can be used as the means of identifying the specimen. This works well if there are only one or a few specimens of a given taxon in a lot or collection. This method becomes impractical if there are tens or hundreds of specimens of a given taxon.

Do not use codes for recording localities and other data. Even if you have a key to decipher the code, it may get lost. Write the information out in full. I have worked with a collection donated to the Carnegie Museum where much of the locality data was recorded in code. For many of the specimens, the code could not be deciphered. When a key was finally found, a number of the codes were not on it. Consequently, these specimens have been stripped of their scientific value.

5.9 WET COLLECTIONS

Sometimes you may find it necessary to preserve mollusks that do not have a shell. You may find a need to preserve the soft tissues of a shelled mollusk. For this, ethanol is the best all around preservative. For general purposes an 80% concentration should be used. If the tissue is being saved for DNA studies, then a 95-100% concentration of ethanol should be used. Sometimes an 80% ethanol, 15% water, 5% glycerin solution is used. This allows the tissues to retain some flexibility if the preservative evaporates. If this solution is used, do not use metal lids. As mentioned above, glycerin will accelerate the deterioration of the metal lid.

If the preservative has evaporated and left behind desiccated tissue, you do have some options to reconstitute the tissue. Also, when a bivalve shell dries, the ligament and resilifer become hard and inflexible. These occasionally need to be softened so that the valves can be repositioned. You can accomplish this by one of several methods.

The point must be stressed, unless you need to rehydrate a specimen, do not do so. Rehydrating solutions all have an effect on a specimen and the effect is not always benign (Beccaloni 2001, Simmons 2002: 102-103). If a specimen that was stored in alcohol has dried out, leave it dry. A better method of rehydration may come along in the future. However, if you need a soft specimen again try the following techniques.

You can try soaking the specimen in a 0.5% aqueous solution of trisodium phosphate (Van Cleave and Ross 1947). Thompson *et al.* (1966) used a method that involved soaking the dried specimen in a mixture of 50% ethylene glycol and 50% water. Lastly, you can try an aqueous detergent solution. Presnell and Schreibman (1997: 484) described a method using a 1% aqueous solution of Trend® while Taylor (2003: 8) used a 5% solution of Decon 90®. If the tissue to be re-hydrated is small, you may have to soak it only overnight; larger specimens may require several days.

5.10 CONCLUSIONS

Following the above principles will cost more than following non-archival practices. However, I feel that the benefits derived from archival practices far outweigh their additional costs. The basic principles are doing no harm, using archival substances as often as possible, and recording, on its label, what you do to a specimen. Controlling aspects of the collection environment, temperature, humidity, volatile acids, light exposure, pests, and shock/abrasion, will help to preserve a collection for posterity. Following good curatorial and record keeping practices will assure that the scientific integrity of a collection will be preserved. If you develop an interest in archival practices, I strongly recommend that you join the Society for the Preservation of Natural History Collections. This group was created to investigate archival matters and publishes a newsletter, journal, and several books on archival and curatorial practices. For more information go to their web site at <www.spnhc.org>.

5.11 LITERATURE CITED

Abbott, R. T. 1954. Collecting American seashells. *In: American Seashells*. Van Nostrand Reinhold, New York. Pp. 56-69.

Arenstein, R. P. 2002. Comparison of temperature and relative humidity dataloggers for museum monitoring. *SPNHC Leaflets* **4**: 1-4 (Spring 2002). [Published by the Society for the Preservation of Natural History Collections].

Beccaloni, J. 2001. A comparison of trisodium phosphate and Decon 90 as rehydrating agents for Arachnida and Myriapoda dry specimens. *Biology Curator* **22**: 15-23.

Bentley, A. C. 2004. Thermal transfer printers - Applications in wet collections. *SPNHC Newsletter* **18**: 1-2, 17-18.

Burgess, H. D. 1995. Other cellulose materials. *In:* C. L. Rose, C. A. Hawks, and H. H. Genoway, eds., *Storage of Natural History Collections*, Vol. I: A Prevention Conservation Approach. Society for the Preservation of Natural History Collections, Iowa City, Iowa. Pp. 291-303.

Burke, J. 1996. Anoxic microenvironments: A simple guide. *SPNHC Leaflets* **1**(1): 1-4 (Spring, 1996). [Published by the Society for the Preservation of Natural History Collections].

Cornish, L and A. Doyle. 1984. Use of ethanolamine thioglycollate in the conservation of pyritzed fossils. *Paleontology* **27**: 421-424.

Elder, A., S. Madsen, G. Brown, C. Herbel, C. Collins, S. Whelan, C. Wenz, S. Alderson, and L. Kronthal. 1997. Adhesives and consolidants in geological and paleontological conservation: a wall chart. *SPNHC Leaflets* **1**(2): 1-4 (Spring, 1997). [Published by the Society for the Preservation of Natural History Collections].

Hamilton, D. L. 1998. File 5: Glass conservation. In: *Methods of Conserving Underwater Archaeological Material Culture*. Conservation Files: ANTH 605, Conservation of Cultural Resources I. Nautical Archaeology Program, Texas A&M University. Available at: <nautarch.tamu.edu/class/ANTH605/File0.htm>.

Hatchfield, P. 1995. Wood and wood products. *In:* C. L. Rose, C. A. Hawks, and H. H. Genoway, eds., *Storage of Natural History Collections*, Vol. I: A Prevention Conservation Approach. Society for the Preservation of Natural History Collections, Iowa City, Iowa. Pp. 283-290.

Howie, F. M. 1992. Pyrite and marcasite. *In:* F. M. Howie, ed., *The Care and Conservation of Geological Materials: Minerals, Rocks, Meteorites and Lunar Finds.* Butterworth-Heinemann, LTD., Oxford. Pp. 70-84.

Jessup, W. C. 1995. Pest management. *In:* C. L. Rose, C. A. Hawks, and H. H. Genoway, eds., *Storage of Natural History Collections*, Vol. I: A Prevention Conservation Approach. Society for the Preservation of Natural History Collections, Iowa City, Iowa. Pp. 211-220.

Millard, V. 2003. *Classification of Mollusca. A Classification of World Wide Mollusca*, 3rd Ed. 3 Volumes. Privately published, South Africa. 1992 pp.

Presnell, J. K. and M. P. Schreibman. 1997. *Humason's Animal Tissue Technique*, 5th Ed. The John Hopkins University Press, Baltimore, Maryland. 572 pp.

Rose, C. L., C. A. Hawks, and H. H. Genoway. (eds.) 1995. *Storage of Natural History Collections*, Vol. I: A Prevention Conservation Approach. Society for the Preservation of Natural History Collections, Iowa City, Iowa. 448 pp.

Rose, C. L. and A. R. de Torres. (eds.) 1992. *Storage of Natural History Collections*, Vol. II: Ideas and Practical Solutions. Society for the Preservation of Natural History Collections, Iowa City, Iowa. 346 pp.

Shelton, S. Y. 1996. The shell game: Mollusks shell deterioration in collections and its prevention. *The Festivus* **28**: 74-80.

Simmons, J. E. 1995. Storage in fluid preservatives. *In:* C. L. Rose, C. A. Hawks, and H. H. Genoway, eds., *Storage of Natural History Collections*, Vol. I: A Prevention Conservation Approach. Society for the Preservation of Natural History Collections, Iowa City, Iowa. Pp. 161-186

Simmons, J. E. 2002. *Herpetological Collecting and Collections Management*, Revised Ed. Society for the Study of Amphibians and Reptiles Herpetological Circular Number 31. Society for the Study of Amphibians and Reptiles, Salt Lake City, Utah. 153 pp.

Solem, A., W. K. Emerson, B. Roth, and F. G. Thompson. 1981. Standards for malacological collections. *Curator* **24**: 19-28.

Taylor, D. W. 2003. Introduction to Physidae (Gastropoda: Hygrophila). Biogeography, classification, morphology. *Revista de Biología Tropical* **51** (Supplement 1): 1-287.

Tennent, N. H. and T. Baird. 1985. The deterioration of Mollusca collections: Identifcation of shell efflorescence. *Studies in Conservation* **30**: 73-85.

Thiele, J. 1929-1935. *Handbuch der Systematischen Weichtierkund.* Fischer, Jena. 1154 pp.

Thiele, J. 1992-1998. *Handbook of Systematic Malacology.* Smithsonian Institution Libraries, Washington, D.C. 1690 pp. [Scientific editors of translation R. Bieler and P. Mikkelsen, translator J. S. Bhatti].

Thompson, R. J., M. H. Thompson, and S. Drummond. 1966. A method for restoring dried crustacean specimens to taxonomically usable condition. *Crustaceana* **10**: 109.

Van Cleave, H. J. and J. A. Ross. 1947. A method of reclaiming dried zoological specimens. *Science* **105**: 318.

Vaught, K. C. 1989. *A Classification of the Living Mollusca.* American Malacologists, Inc., Melbourne, Florida. 189 pp.

Walker, R. 1992. Temperature- and humidity- sensative mineralogical and petrological specimens. *In:* F. M. Howie, ed., *The Care and Conservation of Geological Materials: Minerals, Rocks, Meteorites and Lunar Finds.* Butterworth-Heinemann, LTD., Oxford. Pp. 25-50.

Weintraub, S. and S. J. Wolf. 1995. Environmental monitoring. *In:* C. L. Rose, C. A. Hawks, and H. H. Genoway, eds., *Storage of Natural History Collections*, Vol. I: A Prevention Conservation Approach. Society for the Preservation of Natural History Collections, Iowa City, Iowa. Pp 187-196

Williams, R. S., A. T. Brooks, S. L. Williams, and R. L. Hinrichs. 1998. Guide to the identification of common clear plastic films. *SPNCH Leaflets* **3**: 1-4 (Fall, 1998). [Published by the Society for the Preservation of Natural History Collections]

Williams, S. L. and C. A. Hawks. 1986. Inks for documentation in vertebrate research collections. *Curator* **29**: 93-108.

APPENDIX 5.1

General and archival laboratory suppliers:

Acme Vial and Glass
1601 Commerce Way, Paso Robles, CA 93446
<www.acmevial.com>
glass containers

Althor Products
PO Box 640, Bethel, CT 06801
<www.thomasregister.com/olc/althor/>
plastic boxes

Bioquip Products
17803 LaSalle Ave., Gardena, CA 90248-3602
<www.bioquip.com>
collecting equipment

Carolina Biological Supply Company
2700 York Place, Burlington, NC 27215
<www.carolina.com>
general field and laboratory equipment

Conservation Resources International, Inc.
8000-H Forbes Place, Springfield, VA 22151
<www.conservationresources.com>
archival supplies

Fisher Scientific
585 Alpha Drive, Pittsburgh, PA 15238
<www.fishersci.com>
general laboratory equipment

Forestry Suppliers
PO Box 8397, Jackson, MS 39284-8397
<www.forestry-suppliers.com>
field equipment

Gaylord Bros.
PO Box 4901, Syracuse, NY 13221-4901
<www.gaylord.com>
archival supplies

Lane Scientific Equipment Corp.
225 West 34th Street, New York, NY 10122-1496
<www.lanescience.com>
metal cabinets

Light Impressions
PO Box 940, Rochester, NY 14603-0940
<www.lightimpressionsdirect.com>
archival supplies

Sigma (Chemicals)
PO Box 14508, St. Louis, MO 63178
<www.sigma-aldrich.com>
chemicals

Steel Fixture Manufacturing Co.
PO Box 917, Topeka, KS 66601-0917
<www.steelfixture.com>
steel cabinets

Thomas Scientific
PO Box 99, Swedesboro, NJ 08085-0099
<www.thomassci.com>
general laboratory supplies

University Products, Inc.
517 Main Street, PO Box 101, Holyoke, MA 01041-0101
<www.universityproducts.com>
archival supplies

Wheaton Scientific Products
1501 N. 10th Street, Millville, NJ 08332-2093
<www.wheatonsci.com>
glass containers

CHAPTER 6

DIGITAL IMAGING: FLATBED SCANNERS AND DIGITAL CAMERAS

FABIO MORETZSOHN

6.1 INTRODUCTION

There have been earlier books that discuss techniques in malacology, however, this is one of the first in which the use of computers is mentioned. When *How to Study and Collect Shells* (Jacobson 1974) was published, computers were very large and used primarily by institutions and businesses. Since then the size of computers has been greatly reduced while the processing speed and capacity have increased dramatically leading to the use of computers in homes, schools and workstations, and more recently, portable and wireless systems. Due to the general use of computers, digital imaging has become popular for personal and business applications, and perhaps has even contributed to the expansion of the Internet.

One of the main advantages of digital imaging is the instant review of the results, unlike film photography, which requires the purchase of film and the cost of processing for even unwanted photographs. Digital photography permits selective printing and the sharing of images by way of e-mail.

6.2 PRIMER ON DIGITAL IMAGING

A digital image is an image that has been digitized to be used in a computer. There are basically two ways to represent an image in a computer: vector or raster (also known as bitmap). Lines of a single color represent vector images; they can be straight, curved, or irregular lines. Formulas describe the lines and shapes; colors and textures can be assigned to fill in the spaces. This is the method usually employed in computer-generated images, text, and 3D-models, such as those seen in video games. Two advantages of vector graphics are: (1) the theoretically infinite resolution and (2) small file size. Although vector images can be transformed into bitmap, transforming bitmap images into vector graphics can be very complex and computer-intensive (Kay and Levine 1995).

Bitmap graphics are most widely used to represent images from the real world, such as photographs and scanned images. The entire image is divided into a uniform grid of spots called *pixels* (from picture element), and each pixel represents a single, solid color. The image file has a table with the location of each pixel in the image and its light value (i.e., lightness and color). Bitmap graphics usually result in large image size, but because they can represent almost any image, program developers use them widely. Bitmap graphics are more easily edited than vector images (Kay and Levine 1995).

6.2.1 Pixels, resolution, and image size. Bitmap images contain grids of pixels; the more pixels the sharper the image and the higher the resolution. The resolution for input devices is expressed in how many pixels are captured. Recently, camera makers started to use the word *megapixel* (millions of pixels) to describe cameras; e.g., a camera that captures images in 1,600 x 1,200 pixels is rated as 2 megapixels (MP). Monitors, on the other hand, are rated on the screen size (in diagonal, but actual size is smaller than advertised) and resolution that it can display, e.g., a 17 inch (43 cm) monitor capable of displaying 1,600 x 1,200 pixels. However, screen

Table 6.1 File size at different resolutions and color-depths in a 4 x 6 inch (10 x 15 cm) image.

Resolution (dpi)	Pixel count	File size (Kbytes) at color-depth			
		1-bit (black and white)	8-bit (256 shades grayscale)	24-bit (16.7 million colors)	42-bit (4 trillion colors)
72	124,416	15	124	373	653
100	240,000	30	240	720	1,260
200	960,000	120	960	2,880	5,040
300	2,160,000	270	2,160	6,480	11,340
400	3,840,000	480	3,840	11,520	20,160
500	6,000,000	750	6,000	18,000	31,500

resolution should be expressed in *ppi* (pixels per inch). The screen resolution is surprisingly low; typically 72 or 96 ppi, yet the images displayed can be very realistic.

Output resolution (e.g., in printers) is commonly expressed in *dpi* (dots per inch), but this can be misleading. Color inkjet printers rated as 4,800 dpi resolution may actually produce 4,800 droplets of ink per inch of paper. However, to represent one color pixel using a combination of four or six colors (the usual number of inks in most inkjet printers), many droplets are necessary, and the actual resolution is less. On the other hand, black pixels in inkjet and laser printers can print at high resolution. The same is true for pixels in the same color as the inks used (usually cyan, magenta, and yellow), because they do not need a combination of different droplets to produce that particular color.

The resolution needed for a picture will depend on the final use of the image. If you want to display the image in your computer or post it on the Internet, then a low resolution is all that is needed: 640 x 480 pixels is enough to fill about half or more of most computer screens used today, and the file size is small enough to be easily emailed. Remember that not everyone uses a 19-inch (48 cm) monitor; if you post large images on the Internet, some viewers will have to scroll to see the entire image. Generally, output on paper at a resolution around 300 dpi is good enough to simulate continuous tones. A 2 MP-level camera can produce near photo-quality 4 x 6 inches (10 x 15 cm) prints. More pixels are needed for photo-realistic images printed at larger sizes. A 35mm color Ektachrome slide is equivalent to approximately 4,500 x 3,000 pixels at 32-bit color depth (Kilbourne 1991). This resolution is currently matched only by digital backs on high-end [medium format, 4 x 5 inches (10 x 12.7 cm)] professional cameras (e.g. Ihrig and Ihrig 1995), but soon digital SLR will match and surpass this resolution.

You can always take photographs at high resolution and then edit them to reduce the final resolution to that which fits your needs, e.g. to send a photograph as an email attachment. To photograph a protoconch or a detail on the animal, use the macro function and the highest resolution possible; then crop the image to focus attention on the detail (W. Thorsson, pers. comm.). The number of pixels and color depth (see below) directly affect the size of the image file. File size increases exponentially with resolution and linearly with color depth (Table 6.1).

6.2.2 Color depth and graphic types. The color depth expresses the maximum number of colors that can be represented. The number of colors equals 2^n, where n is the color depth in bits. Therefore, a 1-bit image can have two colors (usually black and white), and is used for line drawings; an 8-bit image has 256 possible colors (or tones of gray in grayscale images); while a 24-bit image has up to 16.7 million colors (used for photographic images). The human eye can discriminate color on the order of 24-bit color depth (which is why 24-bit color is also known as *truecolor*) (Murray and van Ryper 1996). Higher color depths are found in some scanners and other input devices. However, because current output devices cannot produce that many colors, high color depths such as 42-bit colors (theoretically more than four trillion colors), are impressive, but meaningless for most users.

Graphics types: There are many different algorithms that encode graphics data, each one with its own set of characteristics and advantages. Some considerations should be taken into account before choosing which graphics format is best for you: programs that support the desired format, compression scheme, flexibility, cross-platform (i.e., can be used both in PCs and Macs), and so forth. The most widely used cross-platform formats for continuous-tone images (e.g., photographs) are currently TIFF (or TIF) and JPEG (or JPG). TIFF is a *lossless* format, meaning that its compression does not cause any loss of information, but the resulting image size is large. Most authors recommend saving an original image as TIFF before editing it.

JPEG was designed to offer great compression (smaller file size), but at the expense of some loss of redundant data at high compressions (referred to as *lossy*). The tradeoff is acceptable in most cases, with no perceptible loss in quality, if a moderate compression is used. For this reason, JPEG is the standard for photographic images on the Internet, and most digital cameras only use the JPEG format. See Kay and Levine (1995), and Murray and van Ryper (1996) for an in-depth discussion on graphics formats. Some other useful formats exist. Adobe Photoshop's PSD format allows images with several layers (like superimposed slides) that can be edited separately, but file sizes are greater. Kodak's FlashPix format offers multiple resolutions, but not many programs and cameras support it. Portable Network Graphics (PNG, pronounced "ping") is a new lossless graphics format that supports a greater range of color depths, text storage, and other improvements over TIFF and JPEG, but currently few programs support it (Niederst 1999). Other graphics formats are under development particularly for Internet use.

6.3 DIGITAL VS. FILM PHOTOGRAPHY

Analog (film) photography and digital imaging are basically similar, but there are some subtle differences and advantages to each medium (Table 6.2). Early digital imaging output used to have a pixelated (blocky) appearance, with hard edges and unnatural color transitions. However, recent developments in equipment and supplies (inks, paper) now may render near photographic quality results. It will not be long until consumer-level photographic film image quality is matched and surpassed by digital photography. However, because of higher prices for digital equipment and the extensive existing network of film photofinisher labs and makers, film photography will continue to be the choice for many consumers.

6.4 USES OF DIGITAL PHOTOGRAPHY IN MALACOLOGY

- Using images in reports and manuscripts; preparing posters for exhibitions and shell shows;
- Incorporating images into computer programs to prepare slide or computer multimedia presentations;
- Archiving images in a collection database, where images can be linked to each shell lot (see Chapter 8). Databases can be saved on a disc (CD-ROM) or posted world-wide on the Internet;
- Send images via email, even from remote locations, to experts for help with identification, or to friends as electronic postcards;
- Enhancing images for study (e.g., by removing a distracting background, making composite plates, etc.);
- Using the image to perform accurate measurements for morphometric study (e.g., aperture/shell length ratios, measuring angles, etc.);
- Printing shell images on a label to help match individual specimens with their labels (e.g., for a special specimen such as a holotype);
- Micromollusk voucher slides, glass slide mounts, and scanning electron microscopy (SEM) stubs can be photographed to help as a map would when studying specimens under a microscope;
- Other uses of digital images include printing your own postcards with your favorite shells, making calendars, and other personalized gift ideas available through many photofinishers (e.g., T-shirts, mouse pads, mugs, etc.).

Table 6.2 Comparative advantages of film-based vs. digital imaging.

Film Photographic	Digital
Pros: • Highest image quality (to date) • Widely available, great range of options (e.g. lenses) and equipment • Lower equipment cost (usually) • Does not require a computer Cons: • Costly and time-consuming processing • High film and processing costs • Relatively short life-span of unexposed films, difficult archiving of negatives/slides • Need to finish the film to change film type/roll (most 35mm cameras but not with APS format) • *Lossy* image copy/reproduction • Time-consuming trials with no previews • More physical space needed for storage of films, prints and negatives than for digital images • Scanner required for electronic distribution/editing	Pros: • Instant results, increased productivity • Easy editing, color and contrast correction • No running cost associated with films and film processing • Ready for Internet posting • Easy and cheap trials: view/delete pictures instantly • *Lossless* duplication/reproduction • Long-term storage stability (e.g. CD-ROM) • Unlimited storage capacity • Image editing can be done with great accuracy • Environment-friendly Cons: • Higher initial equipment costs • Image quality (resolution) not yet as good as film (but gap is closing fast) • Image output not yet as sharp and color-correct as from film (but approaching same quality as film) • Steep learning curve to become fluent with software and technology

6.5 INPUT DEVICES

There are basically two ways to obtain a digital image: make your own or use a pre-existing image. To produce your own digital images, the most common devices are the following:

6.5.1 Scanners. There are several types of scanners, which range in quality from poor to professional. Scanners are usually used to digitize flat images such as photographs or printed media, but they can also be used to capture images from 3D objects such as shells, as we will see below. Some of the main types of scanners currently used are listed below.

Handheld scanners: This type of device used to be the cheapest digitizing device, and was popular in the late 1980s and early 1990s. New versions, pen-like scanners for text, might be making a comeback. Handheld scanners have a single array of CCD (coupled-charged device) that captures the image as the user glides the scanner over an image. Problems associated with handhelds include low resolution and the requirement that the user maintain a steady speed and straight path or else the image becomes distorted.

Flatbed scanners: These are currently the most widely used, popularized by dropping prices and increasing quality. An array of CCDs is arranged under a glass platen over which the document or image is placed. The equipment works very much like a xerographic copying machine: the CCD array moves across the document while a powerful light illuminates the image and the CCD captures each bit of image, one small step at a time. High-end models can also scan translucent media, such as slides and offer high optical resolution. Flatbed scanners can be the cheapest and most versatile type of digitizer for amateurs.

Slide scanners: These are dedicated scanners that digitize slides or negative film at high optical resolution but small scan sizes. The slide scanner is a good solution for users with many slides to scan. Translucent, thin-mounted specimens can also be scanned.

Drum scanners: Professional machines used to scan photographs and artwork for professional printing. These scanners used to cost up to $200,000 a few years ago (Ihrig and Ihrig 1995); now flatbed scanners almost rival drum scanners in quality, at a fraction of the cost; however, drum scanners still offer the highest resolution (up to more than true

8,000 dpi) and high dynamic range (up to 4.0, compared to the average 3.0 on midrange level flatbed scanners) required for color separation in the publishing business and fine arts. The dynamic range measures the ability to capture gradations from the lightest to the darkest parts of an image.

3D-scanners: Currently under development, industrial scanners that use laser beams to scan objects for 3D model reconstruction. Flatbed scanners can simulate 3D effects (Schubert 2000). (To complement these scanners, an emerging technology uses a transformed ink-jet type printer to print layers of a plastic material, thus making it possible to print in 3D. Soon, you will be able to print your own 3D shell models at home).

6.5.2 Digital cameras. Currently, the fastest growing segment of the imaging market is digital photography. Quality and prices range from entry-level to professional, with some now rivaling the quality of film cameras. Digital video cameras can also shoot still photos, but usually offer lower resolution than digital still cameras.

Consumer market: At the low end are cameras at the 640 x 480 pixels level, which is good enough for the Internet. The high-end consumer models are referred to as *Prosumer* models (*Pro*fessional/ Con*sumer*) and now offer resolution of several megapixels. Resolution is increasing fast and soon will equal and exceed that of consumer film photography. At present, there is no need in the consumer market for a camera with more than 6 MP, but the industry will likely create the demand for it, in the same way the computer market is industry-driven. The list of features available in consumer digital cameras is constantly increasing. Digital cameras are the easiest way to capture photos, and arguably the most fun digitizer to use.

Professional market: At the high end of professional digital photography are digital camera backs that are also known as scanning cameras. As of February 2005, the highest resolution was around 140 MP, for a *Super 10K-HS* digital scanning back mounted on a 4"x 5" (10 cm x 12.7 cm) camera <www.betterlight.com>.

6.5.3 Other sources.
Video capture: Computer cards or plug-in attachments allow turning any video source such as video cameras, VCR, television, cable TV, etc. into a still or video image in the computer. These cards capture an analog image and transform it into a digital one, usually increasing the resolution by interpolation, rendering a good quality image.

Service bureaus: Many photo labs now offer services to develop photographic film, scan it, and have the images stored in a CD-ROM or posted on a website for downloading. You can also get your old photographs scanned there if you do not have access to a scanner.

Internet: The Internet is a good place to look for digital images on any topic. Google <www.google.com> has an excellent image search engine (searching 330 million images as of February 2002; quadrupled to nearly 1.2 billion images in Feb. 2005). As of November 2004 there were an estimated eight billion Internet sites worldwide and millions more added everyday (Zetter and McCracken 2000). Many sites on mollusks and shell collecting contain photos (see some examples under *Resources*), but images posted on the Internet are usually copyrighted and cannot be used without prior arrangement with the copyright owner. Other alternatives are to use CD-ROM image banks or other commercially available products.

6.6 THE FLATBED SCANNER: AN AFFORDABLE SOLUTION TO SHELL PHOTOGRAPHY

As the name implies, a flatbed scanner has a flat bed or glass platen suitable to scan flat (two dimension or 2D) media such as photographs, printed images, and text. The latter is very useful for OCR (optical character recognition) to recognize text from a scanned image, transforming it into editable text. Scanned images are useful for a range of uses discussed above.

Within certain size and depth limits, three-dimensional objects can also be scanned with surprisingly good quality by a flatbed scanner. Indeed, the resulting image quality can be better than from most entry- to medium- level consumer digital cameras. There are some limitations and disadvantages over digital cameras (Table 6.3). Note that not all scanners are suitable for 3D-scanning, because of the small depth of field in some models (models with LED as light sources usually are not suitable for 3D scanning). Try to test a particular model before you buy it. In general, the flatter the shell (or object), the better the scanning. Shells that are small (or too big to fit on the scanner platen) are not suitable for scanning. Since most shells that collectors are likely to have fall between the extremes, the flatbed scanner may be an affordable and a useful tool for the collector and researcher.

Some current flatbed scanner models have true optical resolution of 4,800 dpi; this means more than 23 MP/in^2 (3.6 MP/cm^2). A scanner with a letter-sized platen at this resolution has more than 2,154 megapixels, which is much higher than that of any professional digital camera available today. The highest current resolution is around 140 MP for a digital back on medium format cameras.

6.6.1 How to scan shells with a flatbed scanner.

Before you start scanning shells on your scanner, remember to be careful with the glass platen; any scratch on the glass may appear in all subsequent images. Also, do not let shells or other objects drop on the glass platen, and avoid putting too much weight on the scanner (go easy with that *Strombus goliath*!). The scanner captures the image or view of the shell facing the glass platen. Although there may be a number of different ways to scan shells, I present here a simple technique that I developed over the years through trial and error (described in Moretzsohn 1998). The most difficult part of scanning shells is holding the specimen in position. In my experience, using *silly putty* (or modeling clay) is the easiest technique for most shells. I study cowries, the shell of which is basically a flattened ellipsoid, and it is easy to secure the shell in any position. If you want to scan muricids (or other irregular shells) you may have to improvise alternative means of keeping the shell in place long enough for the scanner to complete the scan (depending on the resolution, it can take several minutes).

I have designed two specimen holders that allow changes in the height depending on the shell used.

Table 6.3 Comparative advantages of flatbed scanners and consumer digital cameras.

Flatbed scanners	Digital cameras
Pros: • Large capacity (limited only by the computer's memory) • Images available for editing/printing much faster than from camera • Better brightness/contrast/color correction controls • Larger preview image than cameras • Much higher total resolution • Cheaper than most cameras • Higher color depth and wider dynamic optic range than cameras • More graphics file format options and flexibility • Scan text for OCR (some high-end cameras can also do it) Cons: • Not portable, computer-dependent • Limited depth of field • Limited size range (especially for 3D objects) • More difficult to position specimens for scanning than photographing	Pros: • Portable; convenient • Unlimited subject size range • Larger depth of field; focus can be very accurate • Independent of computer to acquire image, preview, display on TV (and more recently, to print) • Some models can also save a short sound tag with each image, or video Cons: • Higher price than most scanners • Removable storage media still expensive and limited in size • Batteries do not last long, especially when using the LCD preview screen • Uploading images via USB and FireWire connections may take a long time for large capacity cards

The first one is a simple device that can be made from cardboard or plates of acrylic or glass glued together, and staggered in a zigzag fashion. Make two of these columns as high as you need to, and then get a clear glass slide or piece of acrylic that is the same thickness as the material used for the zigzag column. Use a small lump of *silly putty* to attach the shell onto the glass slide so that the shell will hang above the scanner glass. Adjust the height sliding the glass slide into the appropriate slots in the zigzag columns (Figure 6.1). In my experience, it is better for the shell not to touch the glass platen, but to be as close as possible to the glass (a few millimeters is usually good - experiment and discover what works best with your scanner). This avoids Newton rings (concentric annular artifacts formed when polished surfaces are placed close together) as well as too much light on the lowest part of the shell.

The second specimen holder is a little more difficult to produce, but be creative and improvise your own. I used the column of an old dissecting microscope. I removed the scope head, and in its place I secured an acrylic plate to which the specimen is attached with *silly putty* (Figure 6.2). This contraption works in the same way as the zigzag columns described above, but it has the advantage of smooth and easy changes in the height of the acrylic plate.

Once you have figured out how to secure your specimen above the scanner, either remove the scanner lid (document cover) or prop it open (be careful not to slam it closed accidentally, as this will smash your shell and damage the scanner). Place your specimen such that it hovers above the glass platen, and adjust its position for scanning. Remember that the side of the shell facing the scanner is the one that will be scanned. You can scan different views of the shell and make a composite image. You can experiment with different backgrounds; however, if the background is far enough (i.e., a few centimeters away from the glass platen), it will become dark. It is important

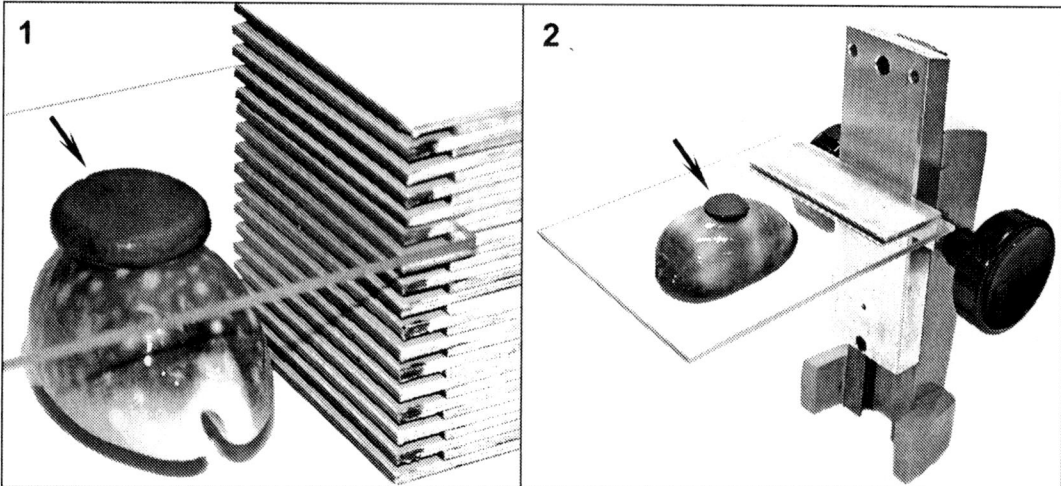

Figure 6.1 Specimen holder version one.
Specimen holder using a zigzag column made from cardboard. A glass (or acrylic) slide fits into the slots in the column, allowing the height of the glass to be changed as needed. A small lump of *silly putty* (arrow) secures the shell in place. Both the zigzag column and the shell are placed on the scanner platen for scanning. In this figure, the aperture of the shell is being scanned.

Figure 6.2 Specimen holder version two.
Specimen holder made using the column of an old dissecting microscope, fitted with an attachment to hold an acrylic plate to which the shell is secured with *silly putty* (arrow). The specimen holder sits outside of the scanner platen (not shown), and the height of the acrylic plate is adjusted so that the shell secured underneath the plate is close to the scanner platen. The main advantage of this specimen holder over that illustrated on Fig. 6.1 is the easier and continuous height adjustment.

that the background is parallel to the glass platen to be evenly illuminated by the scanner. In my experience, a pure black background is the easiest to produce and to edit, but it will depend on the contrast with the shell colors. I use a cardboard box cut open on one side, and with the interior lined with matte black paper. Then I put this box over the specimen being scanned, isolating it from the ambient light. The result is a deep black background, and a shell that may be nicely or poorly illuminated by the scanner. Experiment with your scanner settings, such as brightness and contrast to find out what works best for that particular shell. Shells that are too high may show too dark in the scans, and the far edges will be distorted and out of focus. You can later correct some of the color/contrast using editing software, but it is better to get well balanced scans to start with.

Some scanners have only automated settings. If available, use a manual setting, and adjust it to your needs. A good resolution to start your experiments is around 200-300 dpi. If your scanner optical resolution is higher, you may want to go with higher resolutions in order to get better details on the shell. Interpolation does not usually help much with resolving details, and too high a resolution wastes disk space and time. Save the scanned image in TIFF format. After you edit the image, you can save it as a graphics format such as JPEG that offers better compression than TIFF (see *Editing* below). Schubert (2000) describes another alternative to hold small 3D objects: turn the scanner upside-down, resting it on books, and place the objects in a tray under the scanner.

6.7 DIGITAL CAMERAS

Electronics makers (e.g., Sony, Epson) produce good cameras, loaded with features. However, traditional camera makers (e.g., Nikon, Canon) produce cameras with better lenses, which can take higher quality photos. Before making a decision on which model best suits your needs and budget, I recommend that you take a look at consumer reviews and photo magazines for the latest models and Internet sites dedicated to digital photography (see *Resources*).

6.7.1 Desirable functions on a digital camera (or buying tips).

Resolution: First of all, get the highest resolution you can afford, but do not sacrifice lens quality for higher resolution. Second, beware of interpolation - many makers advertise high-interpolated resolution, but what matters is true optical resolution.

Storage: New options and higher capacity removable storage are available nearly every month. An ideal storage option would be one that offers storage of many photographs, yet it is widely used and not expensive. Currently CompactFlash (CF) cards are cheaper and have higher capacity than SmartMedia cards, and more camera models are compatible with CF cards than SmartMedia. The recent CF-II-compatible microdrives promise a good capacity/price solution (currently up to 8 Gb). Cameras usually are sold with bare minimum storage capacity so you will have to buy some extra storage if you want to take more than 6-10 photos at a time. Professional digital camera backs that are used in studios have internal hard drives to store photos, or can be tethered to a computer, thus eliminating the limiting size of removable media.

Lenses: High quality, glass lenses produce the best quality images. Professional models have interchangeable lenses compatible with film cameras. Consumer cameras currently do not offer interchangeable lenses, but add-ons can be used with some models.

Zoom: These are very useful for general situations, as well as when photographing specimens of different sizes. A good macro capability is also useful to photograph details or small shells.

LCD screen: Indispensable feature if the camera does not have a good viewfinder (some prosumer models do not have an LCD preview screen, but work as single lens reflex (SLR) cameras, in which the image you see in the viewfinder is what you get on film/pixels). More desirable is an LCD screen that is large, bright (especially outdoors - a rarity), and that can be turned off when the composition can be done with the optical finder alone. The ability to turn off the screen is particularly useful to save batteries.

Controls: It is better to have manual as well as automatic controls of functions such as focus, exposure, shutter speed, film sensitivity, etc. Consumer models offer automated programs ("point-and-shoot"); prosumer models also offer manual and customizable programs.

Connection: Cameras can download photos to a computer using different types of connections. Older models use a serial port connection, which is very slow. In this case, a card reader is strongly recommended. Wireless infrared connection is also slow. Newer models use USB or FireWire connections, which are fast.

Batteries: Rechargeable, non-proprietary batteries are recommended. Buy an extra set of batteries if working away from a studio or home. An AC adapter makes possible long shootings, wherever it is possible to use one. Some vendors sell AC adapters separately.

File formats: The standard today is to have photos saved in JPEG format, which is a compressed, lossy graphics format. High-end cameras offer uncompressed TIFF or RAW formats, but file sizes are much larger. It is useful to have different options for compression/image quality/resolution to suit each use of the camera.

Audio tag and video: Recent models offer these options, which can be useful, e.g., to save time when taking notes on each photo, but are not absolutely necessary to photograph shells.

Other features to consider: Software included with camera, shutter speed, light requirements, color depth, compatibility (both PC and Mac formats), price, etc.

6.7.2 Recommended accessories.
- USP 2.0 card-reader, especially if your camera does not have a USB or FireWire port connection;
- A good color "photo" inkjet or dye sublimation printer. Other types of printers include color laser (expensive), solid ink, and thermal wax printers (require special media and ribbons). The most widely used (and least expensive) ones are inkjet printers (but ink can be expensive). You should also consider buying coated or glossy paper for optimal results. You can skip the printer if you prefer to get your prints done through a commercial digital photofinisher (retail or on the Internet, such as Ofoto.com, Shutterfly.com, etc.);
- Because digital imaging produces large files, it is also a good idea to buy some type of removable storage media, such as CD-R (650 - 700 Mb), DVD-R (4.7 Gb to more) and tape drives (many sizes available) (Sawalich 2000). Currently CD-Rs offer long-term storage at the cheapest cost per megabyte and can be more widely used than are other options;
- Image editing software (most cameras and scanners come bundled with entry-level image editors) to edit your photos, as well as image cataloging software (Figure 6.3) (e.g. CompuPic, Picasa 2, etc.) to help you find the right image in a haystack of files you will soon have in your computer;
- Extra memory (RAM) for your computer. Image editing is computer-intensive and extra memory may improve performance considerably. More RAM is useful. The same is also true for computer speed (CPU clock speed).

6.7.3 Tips on photographing shells.
General tips for film photography also apply to the digital medium. See Chapter 7 on photography for more on this topic.

In the digital darkroom (your computer), you can edit pretty much any image, but bear in mind that it is better to go through a little effort to take a good photo than to spend hours editing a photo on a computer. An image with a good contrast between the shell and the background makes selection (masking) and editing easier than an image with an uneven background that blends in with the shell.

Also, when printing, a black (or other color) background will use a lot of ink, while a white background saves the most ink. I personally prefer black backgrounds, because they usually produce a good contrast with the shell, and also look good on the computer monitor as well as on paper. However, Blaker (1989) recommends a white background in most cases, because black backgrounds occasionally do not reproduce well in publications. Gray should be avoided because it tends to appear muddy in print and blends with the object image.

6.8 OUTPUT DEVICES

The most common output devices for digital imagery are printers and film recorders. The latter are still expensive equipment that exposes traditional film, usually slides, to images or text to create material for a slide presentation. In recent years, there has been an explosion in the market for digital projectors, allowing you to make an audiovisual presentation directly from your computer or removable storage (Gordon 2000).

The most commonly used output device, however, is the printer. Printers come in a variety of sizes, resolutions, and prices. Color inkjet printers are very popular in the consumer market because of the low equipment prices, high quality output, and flexibility with the type of media handled. Quality varies from cheap, entry level, basic prints to excellent, professional photographs. Epson has a line of archival inks with an expected print life of more than 200 years (Wedding 2000). Other more specialized printers such as continuous tone thermal wax printers are more expensive than consumer inkjets and, therefore, have a smaller market share, but produce photographic quality prints, albeit with narrow options of paper sizes. Color laser printers are becoming more accessible but quality is usually inferior to good photo ink-jet printers.

6.9 BASICS OF DIGITAL IMAGE EDITING

A computer with the proper imaging software is more powerful than a traditional darkroom, allowing you to perform tasks that are lengthy or even impossible in a traditional setting. These include resizing the image, correcting colors, modifying or distorting parts (or the whole) of the image, applying special filters, editing single pixels, etc. In summary, you can do magic with images, and basically you can edit almost any image. However, poor originals are not worth editing, because of the lengthy editing required. It is better to start with a good image that needs little editing. There

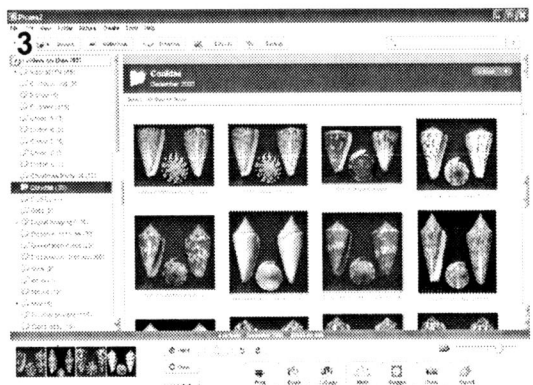

Figure 6.3 Screen shot of Google's Picasa 2.
Screen shot of Google's Picasa 2, photo organizer software. This free program scans your computer for photos (and movies), and produces thumbnails of your images. You can write captions and keywords to help search the right image. It also allows for simple image editing, sending images via email, producing simple webpages, etc.

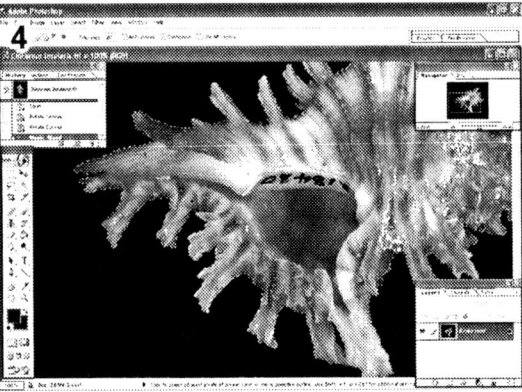

Figure 6.4 Screen shot of Adobe Photoshop.
Adobe Photoshop is the industry's standard for digital image editing. It is powerful, with many resources such as a navigator tool (shown upper right), layers and channels (shown lower right), history (upper left), and plug-in filters made by third-party companies. Most of the editing tools are intuitive, with an icon on the tool bar (lower left). The muricid photo shown in the main window was zoomed in for accurate masking selection.

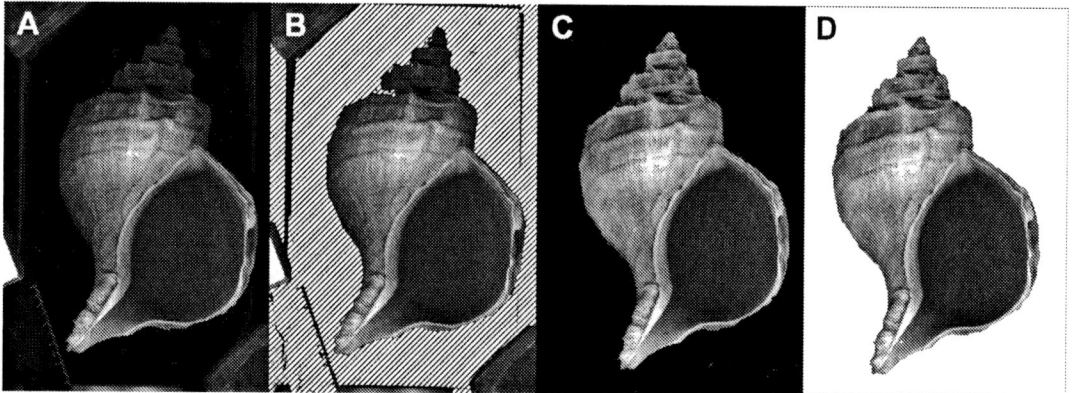

Figure 6.5 Basic steps of image editing.
Basic steps of image editing: flatbed scanning of a large shell (over 15 cm in length) of *Neptunea* sp.
A. A raw scanned image shows an irregular background, with the support for the black box showing in the photo (corners). B. An automatic selection of the background was made on the image on Fig. 6.5A. Note that because of the unevenness of the background, the apex and parts of the shell were also selected, and had to be deselected manually, or the editing applied to the background would also apply to the shell. C. Background was made even, and shell color corrected and rotated. D. Background changed to white for increased contrast.

are good references and sites on how to edit images (see *Resources* below).

For the sake of brevity, I will illustrate a simple project to make a composite image. First obtain the image (e.g., by scanning a shell), and open it in an editing program (Figure 6.4) such as Adobe Photoshop™ (the industry's standard; countless other programs perform similarly, sometimes at a fraction of the price). Use the magic wand or similar selection tool to select the background to isolate the shell. If the background is irregular or uneven (Figure 6.5A), it may be easier to rescan it or you will have to select at least parts of it manually (Figure 6.5B). Using the lasso tool (or similar), carefully drag the cursor around the edges of the shell. Making a careful selection around an irregular object can be the most time-demanding part of editing. To make sure that the edge is accurately selected, it is often recommended that you zoom in to 100% or even closer, and inspect the whole contour of the shell for possible dark areas that the automatic selection missed.

After selecting the whole background, use the bucket tool to paint the background in the same color to make it homogeneous (Figure 6.5C). Then use a filter to soften your selection edges (e.g., feather set to 3 to 5 pixels). Check the foreground and background colors in the color picker, and select white as background, and then press the delete key (Figure 6.5D) and you should have a nice white background. If you make a mistake you can undo this action. Some programs offer unlimited *undo*'s ("actions" function in Photoshop 7.0 and later); others are limited to the last action, so decide if you want to undo your last action before doing the next one. When you close the image, the history list is reset. It is a good idea to save the image often, especially after completing a difficult task.

The edges of the shell should look natural. If they are hard, try to use a wider pixel range for the feather filtering selection. Then select the background, invert selection to select only the shell to correct the colors, brightness, and contrast, thus adjusting the shell image to look realistic. Note, however, that what you see on your monitor is not necessarily what you will get from your printer (you should do a monitor and printer calibration). Also, PC monitors are darker than Macs, and individual monitor screens vary in gamma (brightness).

When you use the paint tool, play close attention to the edges, because subtle pixel changes may "eat the shell away" if you are not careful. If you want to add another image to make a composite, you can enlarge the size of the image (canvas in Photoshop), and move the edited shell to one side. Then open the second image and repeat the steps above. When you are done editing it, select the second shell, copy it, and paste it into the first image, now with some space open on the right side. When you paste an image into another, Photoshop creates a new layer, which can be edited independently from other layers. You can move the pasted image by dragging it around, until the composition is good. You can resize or rotate the image to match other images of the same shell (e.g. Figure 6.6). If your program supports layers, you can save the image with layers, or flatten them to save the image as TIFF and also JPEG (if no further editing).

6.10 CONCLUSIONS

Digital imaging is a powerful tool for both amateurs and professionals, and it is becoming increasingly popular. The quality of digital images already nearly matches that of film photography, and the future of photography is likely to be digital. The current explosion in digital cameras will likely establish them as the main input device into computers for images. Scanners offer an affordable alternative for capturing images of shells. However, if price were not an issue, I would prefer a good quality camera because of its portability and flexibility.

6.11 RESOURCES ON DIGITAL IMAGING AND MALACOLOGY

Popular photography magazines (such as *Popular Photography*) and computer magazines (e.g., *PC Magazine*) often have columns in which digital photography is discussed. More recently, there are magazines now dedicated to the topic, such as *PC Photo*.

The Internet has many sites on digital photography. Below are just a few sites to get you started. You can use search engines to find updated information on a topic in which you have a specific interest. Digital imaging tutorials, product reviews, and resources:
- Photoshop Tutorials at Adobe: <www.adobe.com/products/tips/photoshop.html>
- Kodak's Digital Learning Center: <kodak.com/US/en/digital/dlc/>
- Imaging Resource: <www.imaging-resource.com/>
- Wayne Fulton's "A Few Scanning Tips:" <www.scantips.com/>
- Steve's DigiCams - excellent equipment reviews: <www.steves-digicams.com/ >

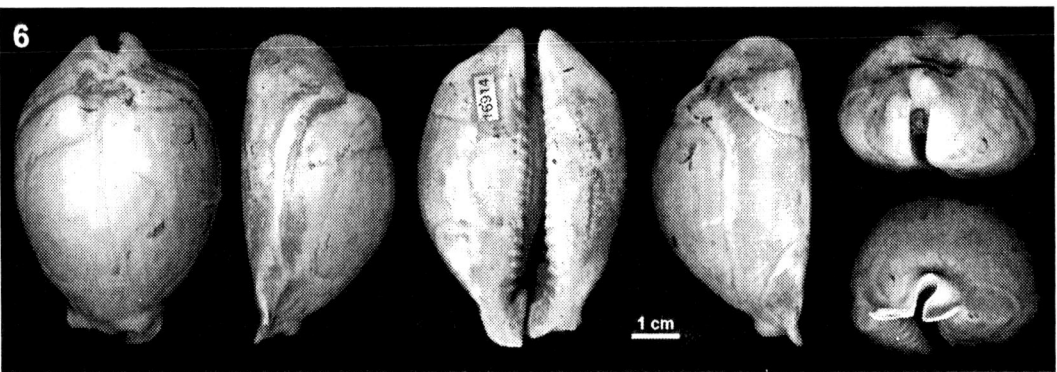

Figure 6.6 Composite image.
Composite image showing six views of a fossil cowry, *Muracypraea* sp., represented to scale. A plate like this one enables the reader to see different parts of the shell at once. These photos were made with a flatbed scanner (an old HP ScanJet IIcx) and edited in Photoshop to produce a seamless plate. Images can be resized, rotated, and edited individually in layers, which can be flattened to be saved as TIF (or JPG), or saved in Photoshop's PSD format (with layers).

- D. Geiger's Zootaxa Digital Imaging Guide: <www.mapress.com/zootaxa/imaging/index.html>

Mollusks and shell clubs - both the American Malacological Society (AMS) and Conchologists of America (COA) maintain good websites with links to many molluscan resources:

- AMS: <www.malacological.org>
- COA's ConchNet: <www.conchologistsofamerica.org/home/>

Shell photo galleries - just a couple of examples of sites with excellent photos:

- Femorale website (shell dealer from Brazil): <www.femorale.com.br/>
- Richard Goldberg's Worldwide Conchology: <www.worldwideconchology.com/>

Photo organizer software - there are several such software, but recently Google made available a very useful yet free software, *Picasa 2*: <www.google.com/options/index.html>

6.12 ACKNOWLEDGMENTS

I am indebted to two remarkable women, my advisor Dr. E. Alison Kay and my wife, Heather Stanton Moretzsohn, for their constant support and encouragement, as well as suggestions on the manuscript. I also thank Gary Rodwell for his computer advice, Wes Thorsson for frequent discussions on photography, Dr. Daniel Geiger for providing updated information on the latest digital camera models, and Dr. Charles Sturm for inviting me to contribute to this project. I thank the Department of Zoology, University of Hawaii for access to computers, scanners, and the digital camera used in the development of this manuscript.

6.13 LITERATURE CITED

Blaker, A. A. 1989. *Handbook for Scientific Photography*, 2nd Ed. Focal Press, Boston. 287 pp.

Jacobson, M. K. ed. 1974. *How to Study and Collect Shells*, 4th Ed. American Malacological Union. 107 pp.

Gordon, A. 2000. Five ways to get prints from your digicam. *The Photo Times* [Online newsletter], <photohighway.com/pt/article.asp?article=40726>. [October 2000].

Ihrig, S. and E. Ihrig. 1995. *Scanning The Professional Way*. Osborne McGraw-Hill, Berkeley, California. 148 pp.

Kay, D. C. and J. R. Levine. 1995. *Graphics File Formats*, 2nd Ed., Windcrest/McGraw Hill, Blue Ridge Summit, Pennsylvania. 476 pp.

Kilbourne, S. 1991. A primer on digital imaging - post production for still photography: Part I. *Journal of Biological Photography* **59**: 43-48.

Moretzsohn, F. 1998. Digital photography in malacology: The flatbed scanner as a great research tool. *In:* R. Bieler and P. Mikkelsen, eds., *Abstracts of the World Congress of Malacology Washington D.C., 25-30 July 1998*. Unitas Malacologica, Washington D.C. p. 232

Murray, D. and W. van Ryper. 1996. *Encyclopedia of Graphics File Formats*. 2nd Ed. O'Reilly and Associates, Inc., Sebastopol, California. 1116 pp.

Niederst, J. 1999. *Web Design in a Nutshell*. O'Reilly, Sebastopol, California. 560pp.

Sawalich, W. 2000. Desperately seeking storage. *PC Photo* for September 2000: 76-80.

Schubert, R. 2000. Using a flatbed scanner as a stereoscopic near-field camera. *IEEE Computer Graphics and Applications* for March/April 2000: 38-45.

Wedding, G. 2000. Epson Stylus Photo 2000P. Epson's new archival inks and papers make their debut in this stylish, extremely capable photo printer. *DigitalFoto* for September 2000: 76-77.

Zetter, K. and H. McCracken. 2000. How to stop searching and start finding. *PC World* for September 2000: 129-143.

CHAPTER 7

APPLIELD FILM PHOTOGRAPHY IN SYSTEMATIC MALACOLOGY

DANIEL L. GEIGER

7.1 INTRODUCTION

Photography has two applications in systematic malacology. A photograph can provide a record of a specimen without collecting the animal itself. It can be a proof of the organism's existence, and may further contain information on habitat, lifestyle, and abundance. In that case, the photographic record takes on the function of a specimen, and all the considerations regarding record keeping and loans apply to it as well.

On the other hand, imaging an existing specimen from a collection is a branch of reproduction photography in which the appearance of the specimen is to be recorded as faithfully as possible. The data of the photograph are the same as those on the specimen record, which may be supplemented with technical photographic details. This chapter covers both aspects of photography as they apply to systematic malacology.

From a technical perspective the chapter is limited to silver halide, or film photography; some remarks on the possible application or limits with respect to digital imaging are given where appropriate (see Chapter 6 for more on digital imaging). The chapter concentrates on 35 mm single lens reflex (SLR) photography, which offers the optimum balance between performance and user friendliness. Photography is as much a science as an art form. I treat it here as a branch of the natural sciences, but fully recognize that striking photographs can be produced by violating every one of the recommendations I will give in this chapter. I will emphasize understanding of optical, chemical, and physical principles that can be applied to any photographic situation. I will introduce some concepts with which most working malacologists are unfamiliar, but the deeper understanding will lead to superior results in the end.

I will first cover the nature of light in terms of intensity and color balance. I proceed to give a quick overview on equipment useful for scientific photography. A review on film types precedes the practical aspects, namely reproduction photography including ultraviolet and infrared applications, photography through water and glass, and underwater photography. Finally, archival concerns will be addressed. As with any technical field, photography is evolving at a fast pace. In order to make this chapter useful beyond the shelf life of any particular product mentioned, I will introduce the reader to some standard metrics used to evaluate the performance of competing products.

I assume you will have a basic familiarity with SLR photography. Terms such as f-stop, focal length, shutter speed, ISO/ASA, and depth of field should at least sound familiar. Otherwise I recommend an introductory text to general photography such as the technical treatment of Sturge (1976), the comprehensive volumes of Jacobson *et al.* (1988) and Stroeble *et al.* (1986, 2000), and the very readable and richly illustrated work of London and Upton (1998). Selected terms can be looked up in Stroeble and Zakia (1993).

7.2 LIGHT

7.2.1 Light intensity. The amount of light is one of the most important factors in photography and determines many camera related settings. Light intensity is measured as photons per area and time. In photography, changes in light intensity are measured on the f-stop scale. One increment (increase or decrease) in this scale denotes a change in light intensity by a factor of two; therefore, the f-stop scale is exponential to the base 2. The term f-stop is taken from the diaphragm settings on a lens, i.e., the numbers like 2, 2.8, 4, 5.6, 8 etc. The difference between two stops is a factor of the $\sqrt{2}$, because of the relation between the *linear diameter* of the diaphragm to the *area* it circumscribes.

Generally, photographic products are made to an accuracy of ±10%: absolute accuracy with the settings is not necessary. The smallest change of intensity you may start to take into account is 1/3 of an f-stop (=1 DIN: see below). A distinction must be made between two kinds of light intensities: incident light and reflected light.

All built-in camera light meters measure light intensity through-the-lens (TTL) and are adjusted to measure reflected light levels. They assume that the object is neutral gray, which by definition reflects 18% of the incident light (see Ray 1999, for details). Some hand held light meters can be used to measure incident or reflected light. Camera light meters can operate in three fundamental ways: integral, spot, and matrix mode.

In integral mode, each point of the image is taken as equivalent and the average of the light intensity of the scene is measured; often the center brightness of the image is given somewhat more weight than that of the corners (center weighted integral). In spot metering, a small circle in the center of the image is measured, whereas the remainder of the image is not considered at all. In matrix metering, the image is subdivided into several areas (5-30 depending on the particular camera), each area is measured independently, the contrast of the scene is evaluated, outliers are identified, and the 'ideal' exposure is determined by some camera algorithm. Matrix metering enhances the success rate of snapshots, but is entirely unpredictable and is of no utility for scientific applications.

Kodak (Anonymous 1987) produces the Neutral Gray Cardboard (NGC), which is very helpful in order to determine accurate exposure values independent of the reflective properties of the object or scene. The reflective properties of either black or white differ from neutral gray by approximately 2 to 2.5 f-stops (for half tone-films: see below). An alternative approach to quantify shades of gray is implemented with the Zone System (e.g., Johnson 1999). This system is geared more towards black and white photography and takes into consideration some of the particularities of black and white developers (Anchell and Troop 1998), hence, is more artistic than scientific.

In order to obtain a reliable reading of the light meter, use the NGC in the place of the object to be photographed. The simplest procedure is to adjust the exposure of the NGC in manual mode and to keep these settings when photographing the actual object. Alternatively (particularly with TTL flashes), note the meter reading for the NGC (say, 1/8 s @ f/16, ±0) and adjust the exposure of the object (say, 1/4 s @ f/16, ±0) with the exposure correction so that it is identical to that of the NGC (1/8 s, @ f/16, -1 exposure correction). It is important that the NGC is metered in spot or integral mode, and that the object is metered in integral mode. If you cannot place the neutral gray card in place of the object, or if the object is not in the same focal plane as the majority of the picture (for instance in extreme close-up photography), take an incident light reading with an appropriate meter, or estimate neutral gray correction.

7.2.2 Estimating Exposure Correction. With some practice, the deviation from neutral gray can be reliably estimated. For the first few rolls, keep very precise records of how you adjusted the exposure and critically examine your slides for detail in the highlight and shadow areas. Initially you may want to bracket widely by ±2/3 (or ±0.5) and ±4/3 (or ±1). Eventually you will develop a feel for exposure correction and the narrower bracketing will be sufficient. In order to estimate nominal exposure correction (±0 for bracketing) follow the steps below. The general procedure with dark objects is to force the camera to underexpose (minus correction), with bright objects to overexpose (plus correction). Remember, the camera *assumes* neutral gray; therefore, it has to be tricked to properly expose very dark or very bright objects:

1. Is the overall image more or less neutral gray? Check the tonality against NGC. If not continue to 2.
2A. Are there pure black elements (minus correction)? If not, continue to 2B or 3A. Estimate how much area is black. Multiply it by -2 f-stops: 50% of the image is black, 50% x -2 f-stops = -1 f-stop. 1/3 of the image is black, 1/3 x -2 f-stops = -2/3 f-stops.
2B. Are there pure white elements (plus correction)? If not, continue to 3A. Estimate how much area is white. Multiply it by 2 f-stops: 50% of the image is white, 50% x 2 f-stops = +1 f-stop. 1/3 of the image is white, 1/3 x 2 f-stops = +2/3 f-stops.
3A. Are there dark, non-black elements? If not, continue to 3B. Estimate how much area is dark. Estimate how far off neutral gray it is. Multiply both factors by -2 f-stops: 50% of the image is dark; dark is in between neutral gray and black: 50% darker, 50% x 50% x -2 f-stops = -0.5 f-stops.
3B. Are there bright, non-white elements? Estimate how much area is bright. Estimate how far off neutral gray it is. Multiply both factors by 2 f-stops: 50% of the image is bright; bright is in between neutral gray and white: 50% brighter, 50% x 50% x 2 f-stops = +0.5 f-stops.
4. Bracket. For 2B, the nominal exposure correction is +2/3. Therefore, for the +2/3 bracket set the dial to +4/3; for the -2/3 bracket set the dial to zero.

7.2.3 The color of light. In color photography, the color balance is of major importance. Both the properties of the film and the light source have to be taken into account. The color of the light source is described by its temperature, measured in Kelvin. An alternative metric is the microreciprocal-degree (mired), defined as 10^6 divided by the color temperature in Kelvin. The color of a light source (i.e., a particular mixture of light of various wavelengths), has to be distinguished from the (spectral) color of light with a particular wavelength (λ) measured in nm.

The former refers to the actual temperature of a tungsten wire when emitting light of a certain color balance. Because human vision readily adapts to various color temperatures, a lighting setup that looks good to the human eye may reproduce an object with unacceptable colors. Therefore, careful attention should be given that the light source and the film material are compatible within very narrow tolerances. For further information on measuring and reproducing color see Hunt and Gainer (1987) and Hunt (1998).

The color temperature of different light sources vary tremendously from reddish (warm) to bluish (cold): indoor incandescent light (2,400-2,800 K), regular photofloods and fiber optic lights (3,200 K), special photoflood lamps (3,400 K), daylight and electronic flashes (5,000-6,000 K), skylight (>10,000 K). Although it seems counterintuitive, a low color temperature (2,000 K) is perceived as a warm light source (reddish), whereas a high color temperature (10,000 K) is perceived as a cold light source (bluish).

Fiber optics and microscope illuminators produce light of their nominal temperature (3,200 K) only if the lamp is given the nominal voltage for that particular bulb. Because light intensity is adjusted by varying the voltage, the proper color temperature is only achieved at one setting. Some fiber optic and microscope light sources have a photo setting; others indicate the nominal voltage on the dial, which is equivalent to a photo setting. Fluorescent tubes should be avoided as light source for color photography, because their emission spectra contain square peaks, which interact unpredictably with the absorption curves of every film emulsion. This lighting results in aberrant coloration, usually a bile green cast (see Chapter 5 in Stroeble *et al.* 1986).

Mixed lighting situations are commonplace, as when using photofloods (3,400 K) in a room with fluorescent tubes (square profile), or using a fiber optics light source (3,200 K) as a focusing light in macro flash photography (5,500 K). Because a film is calibrated for a specific color temperature, multiple light sources of different color temperatures may present a problem. As a rule, you do not need to worry about the unbalanced light source, if it alone would produce an exposure at least four to five f-stops longer than with the balanced light source. The unbalanced light source will only contribute $1/2^4$ to $1/2^5 = 1/16$-$1/32 \approx 5\%$ of the total exposure, which is negligible.

In the fiber optics/flash example, the precise parameters depend on the flash synchronization time of the camera, i.e., the shortest exposure time at which a flash can be used. For example, if the fiber optics light would result in an exposure of 1/2 s with a flash synchronization time of 1/60 s, then the two exposure times are five f-stops apart. With a flash synchronization time of 1/250 s, a 1/8 s exposure of the fiber optics light is admissible, affording a brighter image when focusing. This is one of the practical advantages of a short flash synchronization time. For color critical applications either turn the focusing light off during exposure, or use color conversion filters for the fiber optics light source.

Films are adjusted to one particular color temperature, in the vast majority of cases either 3,200 K (tungsten) or 5,500 K (daylight). The color of the light source can be adjusted with color conversion filters. For instance, the 80A filter converts light from a 3,200 K light source for use with a daylight film (5,500 K). An 80B filter converts light from a 3,400 K light source for use with a daylight film. The 85B filter converts light from a 5,500 K light source to be used with a 3,200 K balanced film. The filter can be placed either over the light source or over the camera lens. For non-standard color temperature, adjustments can be calculated using mired shift values (Stroeble *et al.* 1984, Anonymous 1990, Anonymous 1998, and Anonymous 1999a).

7.3 EQUIPMENT

7.3.1 Body. New features are introduced with every camera and lens model. Many of these features have little value for documentary photography including flash synchronization on first or second curtain, shutter speeds shorter than 1/1000s, predictive auto focus, and ultrasonic motors. However, other features deserve careful attention for reasons given in parentheses; further engineering details are given in Goldberg (1992):

- TTL flash metering (proper exposure for bellows photography).
- TTL flash cable socket (connection of flash or flashes off camera body: macro, through glass/water).
- Flash synchronization speed (fill flash with telephoto lens, focusing light for flashed macro).
- Interchangeable viewfinders: Canon F1, Pentax LX, Nikon F3, F4, F5 (sport/high eye point viewfinder for underwater photography).
- User interchangeable viewfinder screens (bright/clear screens for extreme macro).

Figure 7.1 Comparison in performance of three lenses.
Columns 1-3: Zeiss MacroPlanar 100 mm f/2.8. Chromatic and aberration correction for close focus (1:10 optimum) is provided by floating element. Column 1: at infinity focus. Column 2: at 1:10 magnification: optimum correction of lens. Column 3: at 1:1 magnification (= life size), focus at 45 cm. Column 4: Zeiss Planar 100 mm f/2 at infinity focus. Aberration corrections are optimized at infinity focus. Minimum focus is 1 m (~1:8 magnification). Column 5: Zeiss VarioSonnar 35-135 mm f/3.3-4.5 at 135 mm and infinity focus. Aberration corrections are optimized at infinity focus and f = 70 mm. Macrosetting for 1:4 magnification. Row 1: longitudinal section through lens. Note the more balanced design of the macro lens as compared to the two non-macro lenses. Row 2: Modulation transfer at 10 lines/mm (top), 20 lines/mm (middle), and 40 lines/mm (bottom), measured as a function of distance from center of image. The chart is given for the optimum f-stop for the respective lens, both in sagittal (solid lines) and tangential (stippled lines) target orientation. The more even center to corner (0-24 mm) performance of the macro lens is evident as compared to the zoom lens with macrofocusing capability. Row 3: The relative brightness (= illuminance) from center to corner of the image is given at full aperture (solid line) and at optimal f-stop (stippled line). The performance of the fix-focal length lenses is superior to the zoom lens. Row 4: Distortion (positive values = pincushion; negative values = barrel) is graphed from center to corner. The best performance has the non-macro lens at infinity, whereas the zoom lens is worst. Compiled from material available on the Zeiss web site<www.zeiss.de> and reproduced with kind permission of Carl Zeiss. i.s. = image scale. k = set f-stop.

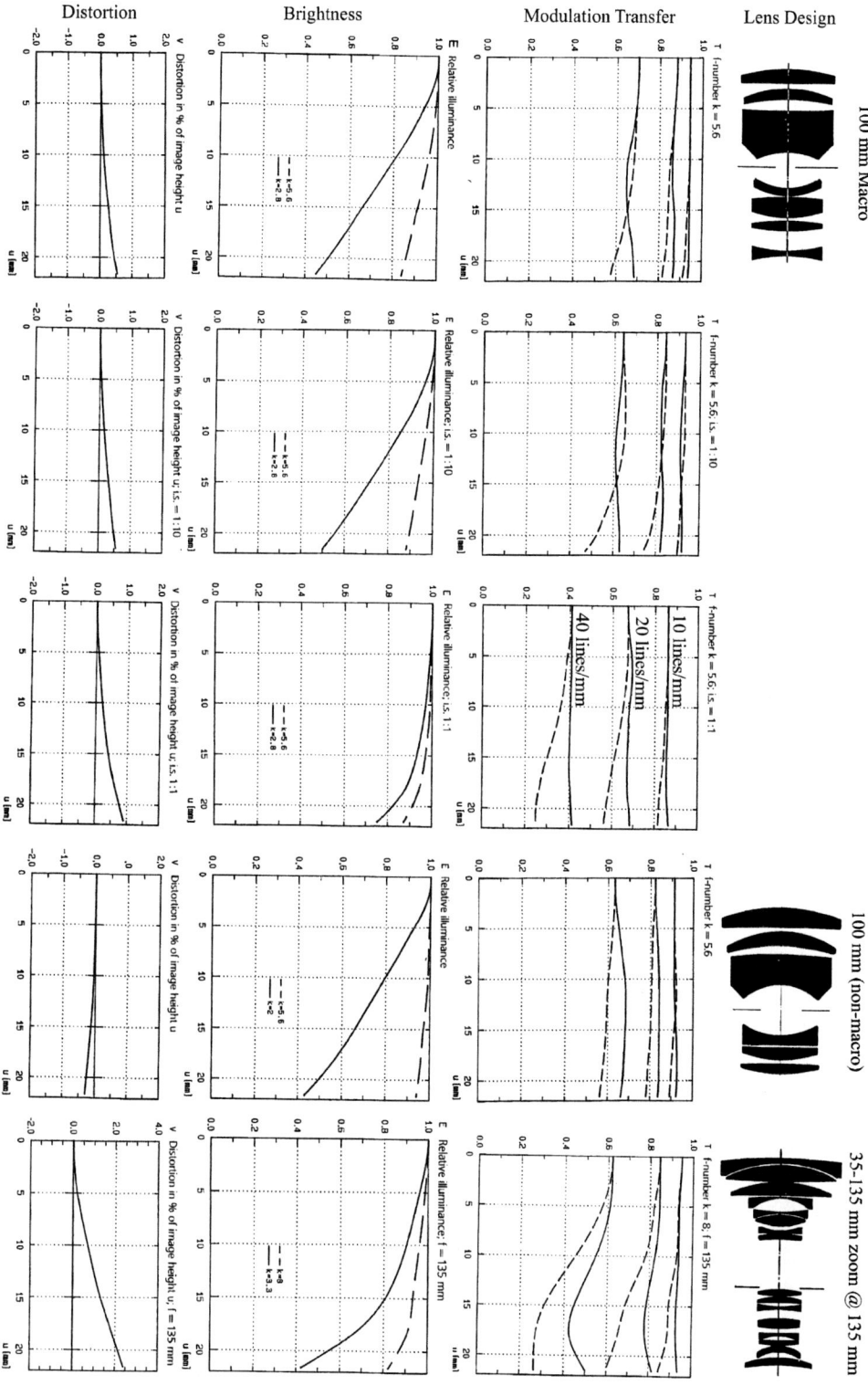

- Depth-of-field preview by body or lens (judging depth of field).
- (Pseudo-) mirror lock-up (reduces vibrations in long time exposures).
- Easy exposure correction/± correction (adjust exposure for non-neutral gray objects).
- Easy turn off auto focus (macro).
- Vacuum film flattener: Contax RTS III (overall image quality: sharpness, chromatic aberration).

Most SLRs can produce very acceptable results. Yet I recommend you chose a major camera manufacturer with a full line of macro and flash components. If underwater photography with a housing is considered, Nikon, Canon, and Pentax are strong contenders. For superior optics Contax (Zeiss lenses), Leica, Nikon, and Canon are potential choices. Although it may sound paradoxical, pay more attention to the available lenses, than to that of the features of the bodies (exception: interchangeable viewfinders if desirable). The line of camera bodies is evolving much faster than that of the lenses.

7.3.2 Motor drive. Many modern cameras have a built-in motor drive. For fieldwork, the possibility of battery failure must be considered. Either carry sufficient spare batteries with you or make sure that your camera allows manual advance and manual rewind of the film. Alternatively, separate motor drives or winders can be attached to most system bodies. Often a grip handle is part of the motor drive and offers greater stability for handheld photography. For copy stand work, motorized film transport will not impose the torque produced by manual film advance and will keep the camera position more stable. On the other hand, when mounting the camera plus a separate motor drive on a tripod or a copy stand, one more mechanical connection with inherent play is found in the set-up, and may amplify vibrations from the mirror swing or the shutter (Keppler 1999). For photographic purists, film flatness is decreased with motorized film advance. The Contax RTSIII is the only SLR with a vacuum film flattener system (Goldberg 1992).

7.3.3 Lenses. A readable overview of SLR lenses has been provided by Landt (1993). Full details including mathematical derivation can be found in Ray (1994). The performance of lenses can be characterized by a number of metrics and attributes. Traditional metrics include focal length and f-stop range.

Certain optical errors can be corrected with special lens elements. The surface of a sphere segment does not focus all rays in one plane. This lens error is called spherical aberration. Aspherical elements, which have surfaces that are not a segment of a sphere, can correct for spherical aberration and field flatness particularly with wide-angle lenses. In uncorrected lenses, the rays of various wavelengths do not focus at the same distance along the lens axis; this error is called chromatic aberration. Uncorrected (chromatic) lenses can only properly focus light of one wavelength.

Most photographic lenses are corrected for two wavelengths (achromatic), whereas the remainder of the spectrum focuses slightly in front or behind the film plane. In telephoto lenses, chromatic aberration is most pronounced. Apochromatic telephoto and zoom lenses contain special glasses, which correct chromatic aberration for three wavelengths and ensure minimal deviation over the entire spectrum. The glasses are variously called low dispersion, extra low dispersion (ED), or super low dispersion (SD) elements. Both of these special lens elements are found only in lenses that are more expensive.

A lesser known metric is the 'Modulation Transfer Function' (MTF). It is obtained from an experimental procedure in which an image composed of black and white parallel bars of variable width (the target) is imaged through a lens and the normalized brightness ratio of the black and white bars is graphed (Figure 7.1). The resulting graph indicates the possible fine detail resolution of a particular lens. If the resolution limit measured in lines/mm is quoted in a chart, usually the value at 50% brightness on the MTF is given.

Much of reproduction photography is in the close-up ($0.1 < m < 0.5$) to macro range ($m > 0.5$). Although many modern lenses (particularly zooms) have what is called a macro-setting, these produce

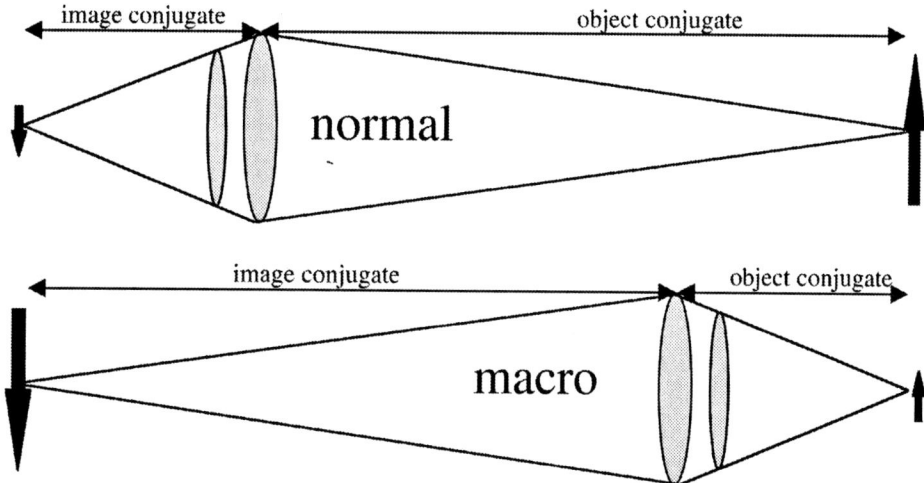

Figure 7.2 Normal and macro photography with respect to the geometry of image and object conjugates. In normal photography, a large object is shown reduced in size, whereas in macro a small object is magnified. Accordingly, the image conjugate is longer in macro photography, hence, a normal, non-macro lens should be reversed (retro position) for macro applications.

significantly lower quality images than dedicated, fix-focal length macro lenses (Figure 7.1). Compare the graphs of the macro lens at infinity to the zoom lens. Macro as well as regular macro-range lenses focus from infinity to the close-up range, however, the optics of regular lenses are corrected for infinity, whereas chromatic and spherical aberration are corrected for the close-up range in true macro lenses. Therefore, true macro lenses will provide a superior image in the close-up range. Macro lenses are manufactured in two focal length ranges: 50-60 mm, and 90-105 mm. I recommend opting for the 90-105 mm range because it allows for greater freedom in positioning the camera, and has a narrower collecting angle (see Section 7.5 Reproduction photography and 7.8 Through glass and water photography).

The lenses for regular photography are designed for the object conjugate - the distance from the lens to the object - to be much greater than the image conjugate - the distance from the lens to the film. This geometry is reversed in macrophotography with much greater than life-size magnification (1:1, m = 1) when using bellows. Therefore, a normal lens should be reversed (= retro position) for photography with m > 1 (Figure 7.2). It can be achieved by reversing the front standard of the bellows or with a reversal ring. It leaves the rear lens unprotected and usually the auto diaphragm function is lost; either the lens has to be stopped down manually before the film is exposed (working aperture) or a double cable release can be introduced depending on the specific system.

Alternatively, macro head lenses with microscope thread mount instead of the usual bayonet, can be mounted on a lens board. These lenses are specifically designed for macrophotography on bellows, should not be reversed, and cannot be used for regular photography. They are usually not listed in the normal catalog of system cameras, but can be obtained for most major camera brands (e.g., OM 20 mm, 38 mm, 80 mm; Leitz Photar series; Zeiss Luminar and Microtar series).

7.3.4 Focus and depth of field. The focal plane is always a plane, and cannot be made deeper by any setting of the diaphragm. Depth of focus is the equivalent of depth of field (DoF) at the film plane. What is changed with the f-stop setting is the depth of field. DoF is a result of the reduction of the blur circle of an out of focus object due to a reduced diameter of the diaphragm. The out of focus object remains out of focus, but it appears sharp. The gain in DoF is counterbalanced by blurring introduced through diffraction of the light at the blades of the diaphragm. Diffraction will increase the diameter of the blur circle (Airy disk) of all image points, including those of an object in the focal plane. Hence, given a sufficiently small aperture, the entire image will appear blurred. Two factors influence the blurring: the grain of the film and the enlargement of the picture.

The more important of the two factors is the enlargement. For 35 mm film, a standard guideline is for an 20 x 25 cm (8 x 10 inch) enlargement held at arms length, which corresponds approximately to viewing conditions of projection slides. Under these conditions, an image appears blurred to the average person if two dots, which are separated by approximately 1/30 mm, touch each other. This happens at an effective f-stop f/32 focused at infinity. If the information gained by increased DoF much exceeds the information loss due to image blur, you may exceed that rule by up to two f-stops with acceptable results; beyond this point, the blurring effect will be excessive. Because the sharpness depends on the magnification, contact prints are sharp even at extremely small f-stops (f/128).

Macrophotography needs further considerations, because the light is spread as it passes through the extension tubes or bellows: the effective f-stop at the film plane is greater than that indicated by the diaphragm setting. As a rule, the maximum advantageous f-stop (f_{max}) set at the diaphragm is a function of the chosen maximum f-stop for infinity (f_∞, usually 32) and the magnification (m): $f_{max} = f_\infty / (m + 1)$. At 1:2, f_{max} = f/22, and at life size (1:1), f_{max} = f/16. Still, as you set f/4 at the lens for a photograph with m = 7, the effective f-stop at the film plane is f/32. Some macro lenses (Nikon Mircronikkor series) correct for this effect in the range over which the lens itself focuses. However, if you use such a lens in combination with an extension tube or bellows, you must take into consideration the magnification *and* the focus setting of the lens, when you determine the proper f-stop setting on the lens.

As DoF is extremely limited in macro photography, it is important that the focal plane is placed as desired. Two factors need to be considered: distance and spatial orientation of the plane. Focal distance can be adjusted with the focus setting of the lens, but note that the magnification of the image is also affected when changing focus of the lens. Alternatively, the entire camera set-up can be moved relative to the specimen. Several options are available. The column of the repro stand or the tripod (particularly with geared columns) can be moved. With bellows, a focusing rail is often integrated in the design. Adorama, Kirk, and ReallyRightStuff produce macro sliders that offer geared movement in two axes. For most SLR applications, the focal plane is parallel to the film plane. Some camera manufacturers offer swing/tilt and/or raise/fall bellows that allow you to place the focal plane at an angle to the film plane (Contax, old Minolta manual focus bellows, discontinued Nikon PB-4). The position of the focal plane is then governed by the "Scheimpflug rule;" for details see Ray (1994) or any book on medium or large format photography (e.g., Merklinter 1993).

The depth of field is distributed asymmetrically to either side of the focal plane in non-macro shots: 1/3 to the front, and 2/3 to the back. In the macro range (>1:1) the distribution of the DoF becomes more or less equal (1/2 to the front and back: see Ray 1994). Many lenses indicate the DoF for certain f-stops on the lens, though many autofocus lenses lack this useful feature. With some cameras, you can close the diaphragm before you take the picture and you can check the DoF with a matt focusing screen. Clear screens allow precise focusing at low light levels in macrophotography, but DoF preview is virtually impossible. When in doubt, focus to the closest part of the object; for animal photographs, focus on the eyes.

7.3.5 Filters. The summary performance of an optical system is always less than that of its best member, but can be further degraded with the addition of a single inferior element. Filters used on lenses are weak points in the optical pathways. I strongly recommend purchasing quality multicoated filters (Kodak Wratten, B&W, Schott, Heliopan, and Lee) and to resist adding a cheap version of any filter.

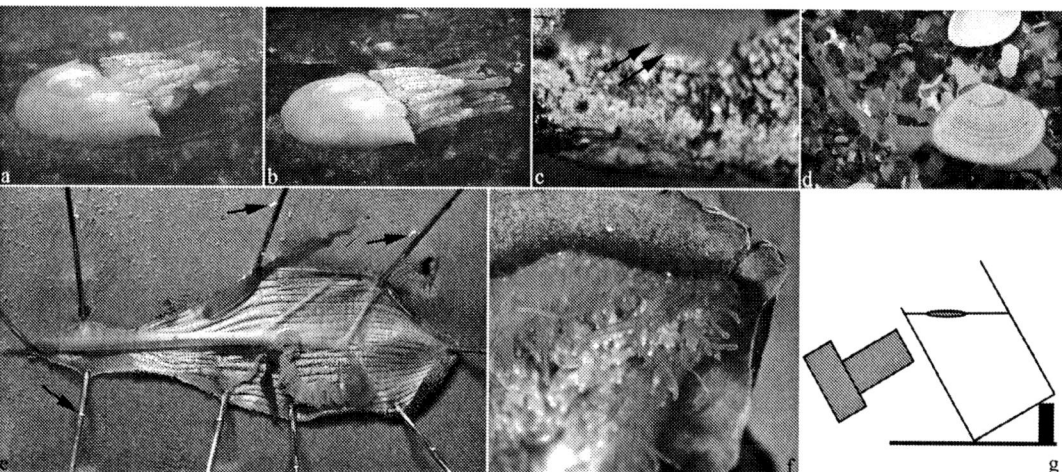

Figure 7.3.
Figures 7.3a and 7.3b: Reduction of diffuse reflection on water with a polarization filter. Note the difference in contrast and the overall gray veil in Figure 7.3 a and b. *Rhizostoma pulmo = octopus* (Macri, 1778). OM4Ti, 65-200 @ 200 mm, circular polarization filter, available light, Kodachrome 64. Worms Head, Rhossili, Gower, Wales, 2/viii 1992. DLG#2583. Figure 7.3a-Without a polarization filter. Figure 7.3b-With a polarization filter. Figure 7.3c-Blow-up showing twig enlarged. Camera shake in mixed light situation with flash. Note the two outlines (ghosting) of twig (arrows). OM4Ti (flash synchronization 1/60 s), 90 mm macro, T-32, Fuji Velvia. Joshua Tree National Park, California, ii 1995. DLG#3766. Figure 7.3d-Nested petri dish technique. *Tellina tenuis* da Costa, 1778, in petri dish nested in second petri dish with substrate. OM4Ti, 90 mm macro, 2x T-32, Kodachrome 64. Millport, Isle of Cumbrae, Scotland, 12-22/v 1992. DLG#2524. Figure 7.3e-Underwater dissection of sipunculid [*Phascolosoma granulatum* (Sato, 1930)]. Note needles at angle and small specular highlight at menisci surrounding needles (arrows). OM4Ti, 80 mm macro, extension tube 65-113 mm, T-32, Kodachrome 64. DLG#2222. Figures 7.3f and 7.3g: Photography in aquarium at angle. Figure 7.3f-Underside of *Velella velella* Linnaeus, 1758. Note shallow depth of field. OM4Ti, 90 mm macro, T-32. Mumbles, Swansea, Wales, 17/ix 1992. DLG#2623. Figure 7.3g- Schematic drawing of set-up in Figure 7.7. Note that film plane is parallel to glass of aquarium.

There are numerous filters used in photography. Usually they are referred to by the Kodak Wratten number. A few general ones are discussed here. Others will be mentioned in the section of their particular application. For technical information on filters (optical density, transmission spectrum, stability) consult the *Kodak Photographic Filters Handbook* (Anonymous 1990).

The most commonly used one is the skylight filter (1A or 1B). It protects the front lens from dirt and scratches and cuts out the near UV, which otherwise can produce a blue tint in the photographic image.

Polarization filters remove glare and reflections from some shiny objects (Figures 7.3a and 7.3b). They can also intensify color saturation in hazy outdoor scenes. Two kinds of polarization filters need to be distinguished: linear and circular. A linear one allows only the portion of the light ray with an e-vector (direction of amplitude) in a particular plane to pass the filter. With a beam of randomly oriented rays, theoretically half the light energy will be absorbed by the filter. However, due to the iodine in the filter material necessitating further absorptive color correction, the actual filter factor of a polarization filter is up to two f-stops.

A circular polarization filter is a linear filter with an additional birefringent $\lambda/4$ filter between the linear polarizer and the camera lens. The $\lambda/4$ plate is a beam splitter that rotates one of the resultant beams by 90° and retards it by $\pi/2 = \lambda/4$. Interference of the two rays makes the e-vector of the emergent ray rotate 360° along its axis over one full wavelength ($2\pi = \lambda$) resulting in a spiraling motion of the light ray. Newer auto focus cameras and those with more intricate, built-in light meters have half-reflecting mirrors and beam splitters integrated in the measuring mechanisms. These optical elements are themselves linear polarizers, hence, will give erroneous readings if presented with a linearly polarized beam from the lens polarizing filter (Goldberg 1992). For these cameras, it is imperative to use the somewhat more expensive circular polarization filters. If in doubt, use a circular polarization filter, because a camera that can accept a linear version will work fine with a circular one, but not vice versa.

Close-up filters are sold as cheap alternatives to the more expensive macro lenses. The strength of these lenses is measured in positive diopters (+1 to +3D). Unless specifically designed for a lens (e.g., Olympus f = 170 mm for Zuiko 80 mm macro, Leica Elpro 1:2-1:1 for 100 mm Apo-Macro-Elmarit-R) these close-up lenses seriously degrade the image. Noticeable is blurring through spherical aberration and loss of field flatness in the corners of the image. Particularly damaging is the combination of a zoom lens with a diopter close-up filter.

7.3.6 Flashes. Flashes are useful as a color accurate light source, particularly in macro and reproduction photography. The strength of a flash is measured by the guide number (GN). It allows calculation of the f-stop for an object at a given distance and a given film speed (f-stop @ ISO 100 = GN/distance). In my experience, metric GN 32 is sufficient for most practical applications. If some tele-photography is envisioned, you may consider a GN 45 - 60 strobe, or make sure that a tele-extender is available to fit your unit. With some experimentation, a Fresnel lens on a rail can be made into a flash focusing unit. Flashes built into the camera body have very limited utility for snapshot photography, are disastrous for through water and glass photography, and cannot substitute for a stand alone unit.

Some flashes are constructed in a modular fashion and allow the use of various heads, e.g., an infrared head (Sunpak 622 Super Pro). When purchasing a flash, make sure it is TTL compatible with your camera and purchase a TTL synchronization cable with it. The most useful ones are the approximately 30-60 cm spiral cable and the 1.5-2 m straight cable.

If you have more than one flash unit, they should be usable in concert. To accomplish this, either one unit has more than one socket for cables, or a multiconnector can be used.

Flashes can also be fired remotely with IR or radio slave sensors. Only a few manufacturers offer remote TTL flash control (Canon EX series); most slaved flashes will be fired in manual mode. Some flashes have built-in continuous focusing lights (e.g., Olympus T-28), which are very helpful in situations with low ambient light, and if reflections may have to be evaluated. Alternatively, a flashlight can be taped to the unit.

When considering flash photography, take the flash synchronization time of the camera body into account. The flash synchronization time is that shutter speed at which the entire film surface is exposed to light at one time; at shorter shutter speeds a slit of variable width moves over the film surface in order to restrict exposure to light. Modern cameras have a flash synchronization time of 1/125 s or 1/250 s, whereas older cameras have a slow 1/60 s; check your camera's manual for specifications.

A regular flash emits a short burst of light between 1/1,000 s and 1/10,000 s duration. Olympus (F-280) and Canon (EX series) produce linear flashes, which have a continuous output of light for the duration of the flash synchronization time, or for the time it takes the slit between the two curtains of the shutter to travel over the entire surface of the film (1/60 s to 1/250 s). These flash units can be used with some of the later models (OM-4T, OM-3T) and allow any flash synchronization up to 1/4,000 s. Although technically interesting, the nominal power of the flash - say GN 28 - is reduced proportionally in f-stop values for exposure times shorter than the nominal flash synchronization time, say 1/60s. The GN at 1/1000s then is 2, which severely limits the utility of these units.

7.3.7 Monopods, tripods, copy stands. An underrated category of equipment is the camera support. Hand held photography is limited to the slowest shutter speed before image blur due to camera movement becomes apparent. In a number of photographic books and articles a fixed time, usually 1/30 s, is indicated. This is complete, utter nonsense! The critical parameter is the angular deviation due to your movement during exposure, which is still within the blur circle of no more then 1/30 mm in the final picture as viewed at standard distance. It is apparent that the value depends upon the collecting angle of the lens.

With a wide-angle lens, you can tilt the film plane much more until the same angular deviation is reached on the film plane than with a telephoto lens. A 1° tilt in the film plane will result for a 180° format filling fisheye lens in a movement on the film plane of a 1/180 times the diagonal of the film format (43 mm): 0.24 mm. A 1° tilt with a 600 mm telephoto lens (4°) will move a point 1/4 of the film diagonal: 10.75 mm. As a general rule, the longest shutter speed is 1/focal length of the lens: for a 16 mm lens 1/15 s, for a 200 mm lens 1/250 s. This rule is only applicable under ideal circumstances; when taking pictures from a moving object (car, boat, continental plate) the longest shutter speed is considerably shorter. Some newer lenses have vibration reduction technology, which allows them to extend the longest time by up to three f-stops.

Take your body posture into consideration. The right hand is on the shutter, and the right arm is held against the chest. The palm of the left hand forms a platform for camera and lens, adjusts focus and diaphragm, while the left arm rests against the chest. Press the shutter very lightly; otherwise, the impact of your finger can introduce significant camera movement.

Surprisingly, even the use of a flash does not guarantee absence of motion blur also called ghosting (Figure 7.3c). The duration of the exposure is determined by the duration of the flash only if the contribution of the available light is minimal (<5%). The duration of a flash depends somewhat on the proportion of the capacitor being emptied and ranges from 1/1,000 s for a full discharge to 1/10,000 s for a fractional discharge. Accordingly, the limits for hand-held flash photography are at f >200 mm. However, for mixed light situations with fill flash, the duration of the flash synchronization time (see above, Section 7.3.6) between 1/60 s and 1/250 s limits the focal length of the lens. The choice of lens for hand-held photography in mixed light situations is limited to focal lengths less than 60-250 mm, respectively. Only with linear flashes (see section *Flashes* above) shutter speeds of up to 1/4,000 s can be used.

For fieldwork, a monopod can allow you to use a shutter speed one f-stop longer than would be appropriate for a particular lens. This applies only for situations where the ground is stable; monopods are problematic on rocking boats; on motorboats a monopod can transmit the engine vibrations to the camera system and will significantly degrade the image. A beanbag may provide suitable stabilization. In extreme situations, a gyroscope may be necessary

For longer exposure times and for careful composition of a picture a tripod is most handy. A tripod can also be used as a copy stand when traveling; make sure that the center column can be reversed or that the ball head can be attached on both ends of the column for close-up work. Select a sturdy kind (Gitzo, Bogen) with a ball head (Bogen, Linhof, ArcaSwiss). Carbon fiber tripod legs are now on the market, which are approximately 30% lighter than a comparable aluminum version without compromise in stability. A combination that has been proven in the field is a set of Gitzo Mountaineer carbon fiber legs with a Linhof Profi II ball head and an ArcaSwiss-style quick release plate. Such a combination costs approximately $1,000, but is well worth the investment.

A copy stand can be purchased in the store or can be home made. The most important characteristics are the flat copy surface, the column at an exact right angle to the plane of the copy surface, and a height adjustable camera support, which guarantees that the film plane is parallel to the copy surface. To check for parallel orientation of the film plane to the copy surface, place a piece of graph paper on the copy board and make sure that the pattern is symmetrical; note that the lines will not be fully parallel to the frame due to residual distortion of the lens. Optionally, a lighting system can be mounted to the copy stand. Alternatively, allowing more flexibility, the lighting set-up can be left separate from the copy stand (Young *et al.* 1996).

When the exposure time is 4-10 f-stops longer than allowed for free hand, it is advisable to swing up the camera internal reflex mirror before the shutter opens in order to minimize vibrations and blurring (Keppler 1999). On some cameras (e.g., Nikon F3, F4, F5, Pentax LX, Contax RTS III) a special lever is available, with other cameras, you may use the self-timer for the same effect (pseudo mirror lock-up). For even longer exposures, mirror lock-up is less of a concern, because the proportion of the exposure under the influence of vibration is negligible.

7.4 FILM

7.4.1 Types of film. Films can be broadly categorized into the following groups.

Half-tone vs. line art: Half tone films produce images that resemble a gray scale or color image as accurately as possible (e.g., Kodak T-MAX, Ektachrome), whereas line-art film show all light grays as white, and dark grays as black (Kodalith).

Black and white (B&W) vs. color: B&W films reproduce a color scene as a gray scale or line-art image, whereas color film records also the color values of the original scene as gray scale values in three different emulsion layers. For B&W films, the spectral sensitivity is further specified: unsensitized = only blue sensitive (Eastman fine grain release positive 5032); orthochromatic = blue and green sensitive, no red sensitivity (Agfa Ortho), panchromatic = blue, green, and red sensitive (Kodak T-MAX, Fuji Neopan).

Negative vs. positive (= reversal = slide = transparency): A negative film has its primary application for producing prints as the final output, whereas slide films produce a transparency from which prints can be obtained.

Fast vs. slow: The sensitivity of a film to a given amount of light is indicated by the speed, measured either in DIN or in ISO/ASA. The most frequently encountered reference speed is ISO 100 = 21 DIN (also 21°). With the ISO scale, a two fold change in film speed (= 1 f-stop = two fold change in exposure time) is reflected in a two fold change in the numerical ISO/ASA-value. On the DIN-scale, a change of one f-stop is indicated by adding or subtracting three DIN-units: ISO 400 = 27 DIN; ISO 50 = 18 DIN.

E6 vs. Kodachrome: Transparency films are produced in two basic types. E6 films can be processed rather easily in the kitchen sink and in the field, whereas Kodachrome films can be factory processed only. E6 films once developed are more stable when exposed to light, but are less stable for long-term dark storage as compared to Kodachrome film (Wilhelm 1993). In E6 film, the color particles are added during development, which in the past meant lesser detail resolution and larger grain. Kodachrome film emulsions have the color dye built into the emulsion. Advances in grain shape (T-grain technology) have allowed E6 films to be at least on par with Kodachrome films in terms of detail resolution, granularity, and archival stability. Kodachrome is being phased out.

Daylight vs. tungsten: Daylight film (= Type S) is balanced for a 5,500 K light source, whereas tungsten film (= Type L) is balanced for a 3,200 K light source. Daylight films have a straight reciprocity behavior (see below) at shorter exposure times than tungsten film. For instance the daylight Fuji Velvia (ISO 50) requires no time or color correction from 1/4,000 s to 1 s, and exposures of longer than 64 s are not recommended; the tungsten Fujichrome 64T is not recommended for use with shutter speeds shorter than 1/15 s and requires no time or color balance adjustment between 1/15 s and 64 s.

7.4.2 Specialty films. Films with a high possible range of speeds include the B&W negative Ilford XP2 (ISO 50-1,600), which is processed like a color negative film (C-41 process); it is a monochrome color (or chromogenic) negative film. A high-speed range and best high-speed reversal film is the Fuji MS 100/1,000 (ISO 100-1,000: RMS 15 @ ISO 800). Alternatively, some ISO 200 to ISO 400 films can be push processed by one to two f-stops to obtain ISO 800-1,600. The films have low RMS (Root Mean Square) values for their nominal speed (RMS 11), which will increase an unknown amount when push developed (see Section 7.4.4).

Some color slide films (e.g., Ektachrome E100VS [professional] or EXB [consumer]) and color print films (e.g., Agfa HDC series) have extremely high color saturation. These films are primarily intended for the tourist market, where more vivid colors associate better memories with places visited. Such films have an interesting application in photo microscopy and aerial photography, where they can compensate for the inherently low contrast scenes. For lower than normal contrast films consider slide duplication (e.g., Kodak 5071, 7121, Fuji CDU II) and internegative films (e.g., Kodak Vericolor III ID, Fuji IT-N). Pre-exposing film with uniform exposure can also lower contrast (Young et al. 1996).

There is only a single B&W reversal film, the Agfa Scala II, which can be factory processed only. For infrared (IR) applications B&W films with extended sensitivity (Kodak High Speed Infrared, Konica Infrared 750) and false color material is available (Kodak EIR).

7.4.3 Color print or slide film? Print films are the consumer choice, because the main viewing form is a large print for which no specialized viewing equipment is necessary. Although print films are cheaper, processing and printing make them more expensive in the end. Print films have greater exposure latitude; a ± 1 f-stop deviation from the proper exposure can still lead to a quite acceptable print. Why then do professionals mostly use transparencies? The dynamic range of the film measure in optical densities (OD) is greater ($\Delta OD_{print} \approx 2.3$, $\Delta OD_{slide} \approx 3.3$), meaning that more shadow and more highlight detail can be recorded, despite the fact that the exposure latitude is only approximately 1/3 f-stop. Higher color fidelity and higher color saturation are further plus points for slide film. Last but not least, the best and most archival print paper (Ilfochrome: expensive) is produced for use with slides and not with negatives. Even the use of an internegative from a slide to print on regular color print paper yields superior results (Delly 1988: 103). For purists, the influence of the color print lab on color balance, exposure, and cropping are further concerns. For publication purposes, slides can be scanned in and transformed into B&W files. In some instances, the color information in the alpha channels can be further exploited even for a final B&W image. Hence, I strongly recommend the use of color slide film, unless special circumstances warrant something else.

7.4.4 Comparative metrics. The performance of a film can be measured by several criteria. The metrics mentioned in this section can be obtained only from the technical data sheets available for most films. Most film manufacturers have material published on the World Wide Web. Kodak and Fuji produce abbreviated versions as small booklets (Anonymous 1999a, 1999b). The grain size of film is measured with the Root Mean Square (RMS) granularity. The lower the RMS number the finer the grain. For color film the RMS granularity is a film specific metric because of the standardized processing. For B&W, however, the developer, temperature and time of development, and nominal sensitivity can strongly influence the granularity (and contrast value "gamma") of the final negative (Anchell and Troop 1998). Note that push development will increase the grain of the film (Fujichrome MS 100/1000: RMS 10 @ ISO 100, RMS 16 @ ISO 1000).

The resolving power of a film is measured with the MTF (see Section 7.3.3; Fig. 7.4 right side), which is obtained by contact printing the target on the film. Due to edge effects during development values of greater than 100% are possible at low spatial frequencies (Figure 7.4 Kodachrome 64). Because the MTF is a normalized metric, its value is independent of the contrast value gamma of the film. Edge effects are due to the over development of the margin with a strong latent image (white bar) with developer from the adjacent area with no latent image (black bar).

Color accuracy can be assessed with the 'characteristic curve' (Figure 7.4 left side). The curve

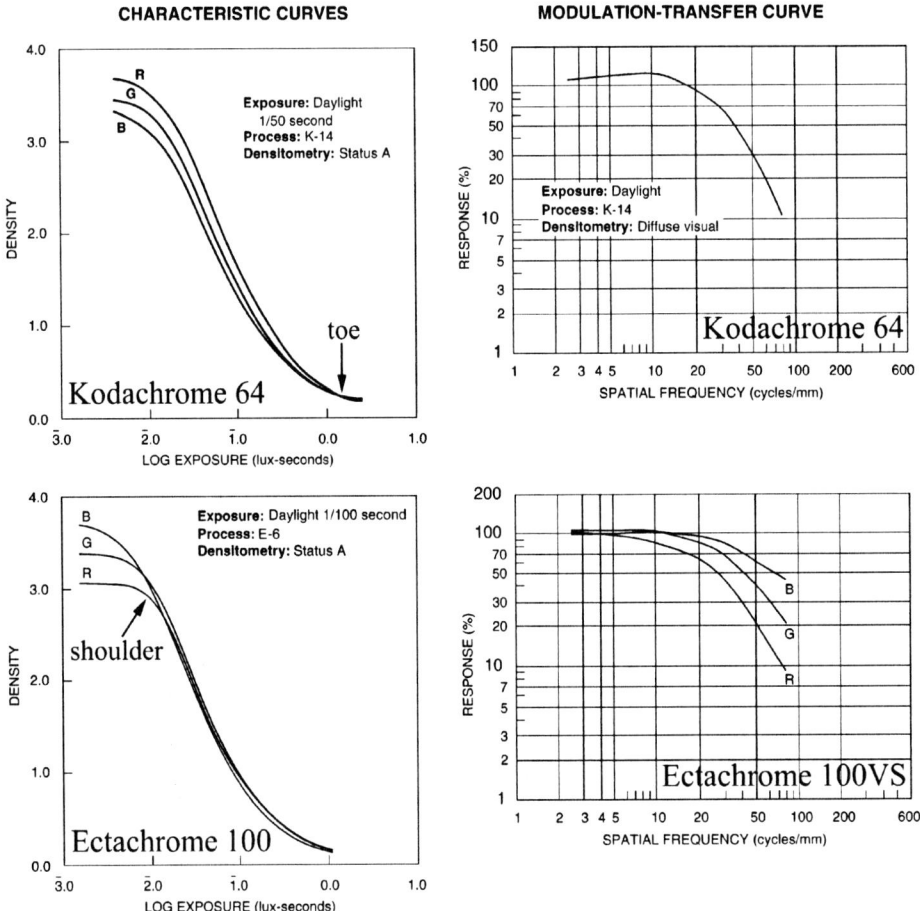

Figure 7.4 Characteristic curves (left) and modulation transfer functions (MTF: right) of some films.
The characteristic curve shows both a shoulder and a toe. In Kodachrome 64, the red sensitive layer is consistently more sensitive than the green and the blue sensitive layers, which accounts for the Kodachrome typical red cast. In the Ektachrome 100, all three curves are more or less superimposed over the entire linear part between the shoulder and the toe, resulting in a more neutral image. In the MTFs, Kodachrome 64 shows greater than 100% response at low spatial frequencies, which is caused by edge effects during development. For the Ektachrome 100VS, note that the MTFs depend on the wavelength of the light: shorter wavelengths afford higher resolution. Sources of illustrations: Kodak Publications E-27 (Ektachrome 100), E-55 (Kodachrome 64), and E-163 (Ektachrome 100VS) available on the web (www.kodak.com). Figures reproduced courtesy of Eastman Kodak Company. KODAK is a trademark.

describes the response of the film measured in terms of its optical density (OD) to varying light intensities. Each film has a toe and a shoulder in which the tonality of the shadow and highlight areas will no longer be accurately reproduced. In the middle, the linear part of the curve is found. The placement of the shoulder (the position of the toe is standardized), and the linearity of the linear part can be compared (Figure 7.4 Kodachrome vs. Ektachrome 100). For color film, one can additionally see how parallel the lines of the three-emulsion layer for the three primary colors are. If they are not parallel, then a color shift due to different exposure will be found (Figure 7.4 Kodachrome 64). For slide film, if the lines are parallel, but not on top of one another, then a constant color tint is to be expected. In color negative material the lines are always separated, which compensates in combination with the orange filter in the base of the film for filtration needs in order to make prints.

Despite all the technical information on a film, the exact performance is difficult to predict. In order to perform a standardized test, either photograph the very same subject with different films; particularly telling differences between films are shown with hues of yellow. Alternatively, photograph a Macbeth ColorChecker (GretagMacbeth, New Windsor, New York), a card with 24 color squares representing the most common colors, and compare the photographs to the actual card. The Macbeth card can also be used to calibrate digital input devices. The hue of a color is also influenced by the exposure, where an underexposure results in denser and darker color, and an overexposure in a washed-out color.

7.4.5 Professional and consumer film.
For many films, two grades are available: the consumer and the professional version. The best quality is found in the professional films and the effective film speed will be determined for every batch of film emulsion; deviation will be noted by 1/3 f-stop increments on the leaflet or on the box of the film. Professional films should be stored below 13°C (see below). For almost all common applications in science, the consumer version is adequate, but for certain films no consumer version is produced (Fuji Provia, Kodak Vericolor).

7.4.6 Storage of films.
Films are composed of a mixture of chemicals that mature or ripen. After approximately a year, they have gone bad. The speed with which the films and other photographic chemicals decay depends on the temperature at which the material is stored, where the standard time refers to storage at low room temperature (18-20°C). If the material is tightly sealed, it can be kept in a refrigerator or even in a freezer for almost indefinite times, i.e., several years to a decade. Infrared films are sensitive to gamma radiation, hence will decay even in the freezer. When taking material out of the freezer, let it warm up for at least 1 h for a 35 mm roll, longer for larger quantities or items such as 30 m bulk rolls (5 h), otherwise water will condense on the film.

7.4.7 Error of reciprocity.
There is a general linear relation between light intensity, shutter speed, and diaphragm (f-stop), called reciprocity. If one is increased, the others have to be decreased by the same amount, calculated in f-stops, in order to keep the light level on the film plane constant. However, there are limits to the linearity of this relation, because the latent picture is formed in a three-step process, where the first two photons excite the silver halide molecule and only the last one completes the transformation to metallic silver. If between the three steps a considerable amount of time elapses, the excited molecule may spontaneously fall down to a lower energy level again, and then more than the minimum number of three photons will be needed, requiring a longer exposure. The same effect may also apply to high photon densities, as two photons may hit one molecule at the same time, but only excite it by one energy level. The time range (shutter speed) in which the reciprocity holds, depends on the type of film and will be indicated on the technical data sheets obtainable from the manufacturer. As a general guideline, for daylight films it is between 1/4000 and 1/10 s, for tungsten films it is between 1/100 s and 10 s (see also Section 7.4.2 Specialty films above). For B&W films which are composed of one light sensitive layer, a simple adjustment of the effective exposure time will be sufficient. Color films, however, are composed of three light sensitive layers, each of which reacts slightly differently to changes in light intensities, therefore, changes in the color balance will also occur. These can be equalized with color compensation filters.

7.5 REPRODUCTION PHOTOGRAPHY

Reproduction photography attempts to show a specimen or object in as natural a condition as possible. This type of photography exercises the greatest degree of control over the set-up conditions. One important factor is camera position; tripods and copy stands were discussed in Section 7.3.7 Equipment above. The various approaches to placement and orientation of the focal plane are discussed in Section 7.3.3 Equipment: Lenses above. Much of reproduction photography is in the macro range; general introductions to macrophotography are given by Brück (1984), White (1987), and Bracegirdle (1995). Other factors are discussed below.

7.5.1 Even illumination. For reproduction photography, even illumination of a flat surface is often desirable. The most effective means is the use of two light sources placed at an angle of 45° to the center of the image. The difference in the light intensity across the frame is inversely correlated to the distance of the light source to the center of the image. If the center light intensity is set at 100%, the illumination of the margins of a square surface is 120% towards the light source and 90% on the sides away from the light source, if the light sources are placed at a distance from the center equal to the width of the square. If the light sources are placed at approximately twice the width of the square, then the values are 110% and 96%, respectively. If four light sources are placed in the diagonals of the square at a distance of the width of the square, the corner illumination is 105%. The light fall-off can be matched to the f-stop dependent loss of corner brightness of a lens (Figure 7.1). Photographing a uniformly colored sheet of paper at various f-stops can help to reveal uneven brightness in the lighting plus lens set-up. Alternatively, a hand-held light meter with a diffuser head can provide readings across the board, which should not vary more than 20%. For glossy objects or for document photography under glass, the collecting angle of the lens should be <45° (f >50 mm) because the incident angle of the light source should be 45° for even illumination (Young *et al.* 1996, Ray 1999: chapter 15, and Figure 7.5). The requirements for slide duplication are discussed in Young *et al.* (1996).

7.5.2 Surface texture. Elements of the surface texture of a shell such as ribbing, cancellation,

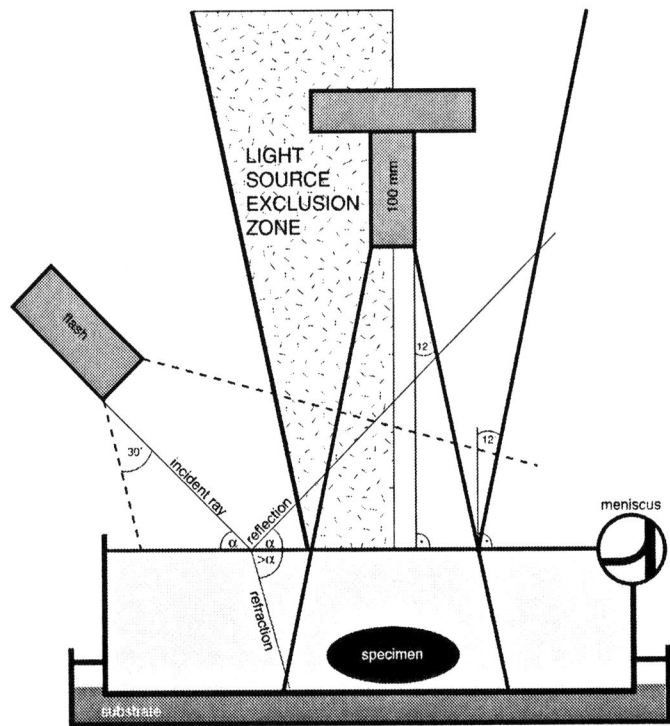

Figure 7.5 Geometry and optics of through liquid photography. An SLR camera with a 100 mm lens is mounted with the film plane parallel to the surface of the liquid. The light source exclusion zone is delimited by a line at half the collecting angle of the lens (12°) reflected at the air-liquid interface at the same angle (12°). The light source (flash) is positioned outside the exclusion zone. Light from the flash is partially reflected and partially transmitted through the water. The transmitted light is additionally refracted at the air-liquid interface. The inner container is over-dimensioned so that the meniscus of the liquid will not reflect any light into the lens. The outer container is filled with water-submerged substrate to produce a suitable background.

and pustules can be enhanced by three different methods. First, the specimen can be coated with black opaque (Sakamoto 1973). Second, the lighting angle can be decreased. The surface texture will then produce more of a shadow, emphasizing the structural elements with highlights. Third, the lighting intensity on each side can be varied. I find between 1/2 and 2/3 of an f-stop difference pleasing. Change either the intensity or the distance of one light source. Because the illumination intensity at the object is measured as an area unit, whereas distance is a linear measurement, the position of the light is changed by a factor of $2^{0.5}$ to $2^{0.66} = 1.41$-1.59. These three methods can be freely combined.

If only a single light source is available such as an electronic flash, use a white or silver reflector (paper, aluminum foil) to brighten the other side of the specimen.

Often selective blocking of light can enhance the aesthetic quality of the image ("subtractive lighting" of R. Meyer, pers. comm.). Anything from white paper, cardboard, wood blocks, aluminum foil, clay, to a finger, can be employed. The only limit is your imagination. Consider the reflective property of the object and its potential influence on the color rendition of the object.

7.5.3 Shadow-free illumination. Ring flashes are often sold for the specific purpose of shadow-free illumination. However, they are usually not strong enough, still produce strong vertical shadows (exception: Olympus indirect ringflash T8), and cannot be used in higher magnification macro shots as they will nicely illuminate the area surrounding the specimen, but not the specimen itself. Due to their limited application, they are rarely used. I prefer the paper cone method over ring flashes, as I can utilize my normal flash or a fiber optics light source for this purpose.

A cone made of white paper may be used in order to minimize reflection on metallic objects. However, 'white' paper is not white and will cause some alteration of the light color, usually towards the blue-magenta. This is due to the UV fluorescent brighteners in paper, hence, causing an increase in the blue area of the spectrum. Because every type of paper is different, no general filtration can be indicated. For critical shots, an unfiltered test photograph is viewed on a 5,500 K calibrated light table/x-ray viewer. The slide is then overlayed with (gelatin) color correction filters until the color balance is right (say 30 cyan). The color correction filtration (30 cyan) must be divided by 1.5 (20 cyan) because the color values (say 0, 1) of the original are not reproduced on the film as 0 and 1, but as 0 and 1.5, i.e., the slope or gamma of the slide films is usually 1.5. The specimen is then re-photographed with the same settings but with the color correction filters (20 cyan) between the lens and the object.

There is another, somewhat more elaborate method. The light cone is replaced by a half dome with an opening for the camera lens. The specimen is placed on a glass plate, under which a suitable background is placed. From the side a light source brings light to a mirror at a 45-degree angle. This mirror reflects the light upwards, which will be reflected on the half dome. The problem with the change in color temperature due to the half dome is as for the transparent cone. However, an even, shadow free illumination can be achieved with a single light source. Note, that at high magnification macro shots (>5:1) a regular flash without paper cone gives surprisingly even illumination, because the flash head is proportionally large compared to the object, hence, acts as a large light source (Hunter and Fuqua 1997).

7.5.4 3D-objects, top view. The nicest images are produced if the background is invisible and just a plain color. There are a few different techniques to be described briefly in the following paragraphs. All of these techniques have one goal in common: the maximum differentiation between object and background (Figure 7.6).

One possibility is to move the background far away from the object, such that the background is completely out of focus. Additionally, for black backgrounds, the background will be underexposed due to the much larger distance between light source and background as compared to light source and object. The background is only a color, free of texture and shadows. The object can be placed on a clean glass plate, which is elevated sufficiently from the background. The glass plate must be extremely clean because any dust particles and water stains will be reproduced. Alternatively, a board with one or multiple holes for pegs can be used. In this case, the object is placed on top of the peg with some modeling clay. As background use finely woven tissue, particularly velvet, or any smooth paper. The easiest background colors are black or white. For any other background color, the illumination is a little more involved, because no shadows should be visible on the background. With black background, the shadows are usually invisible.

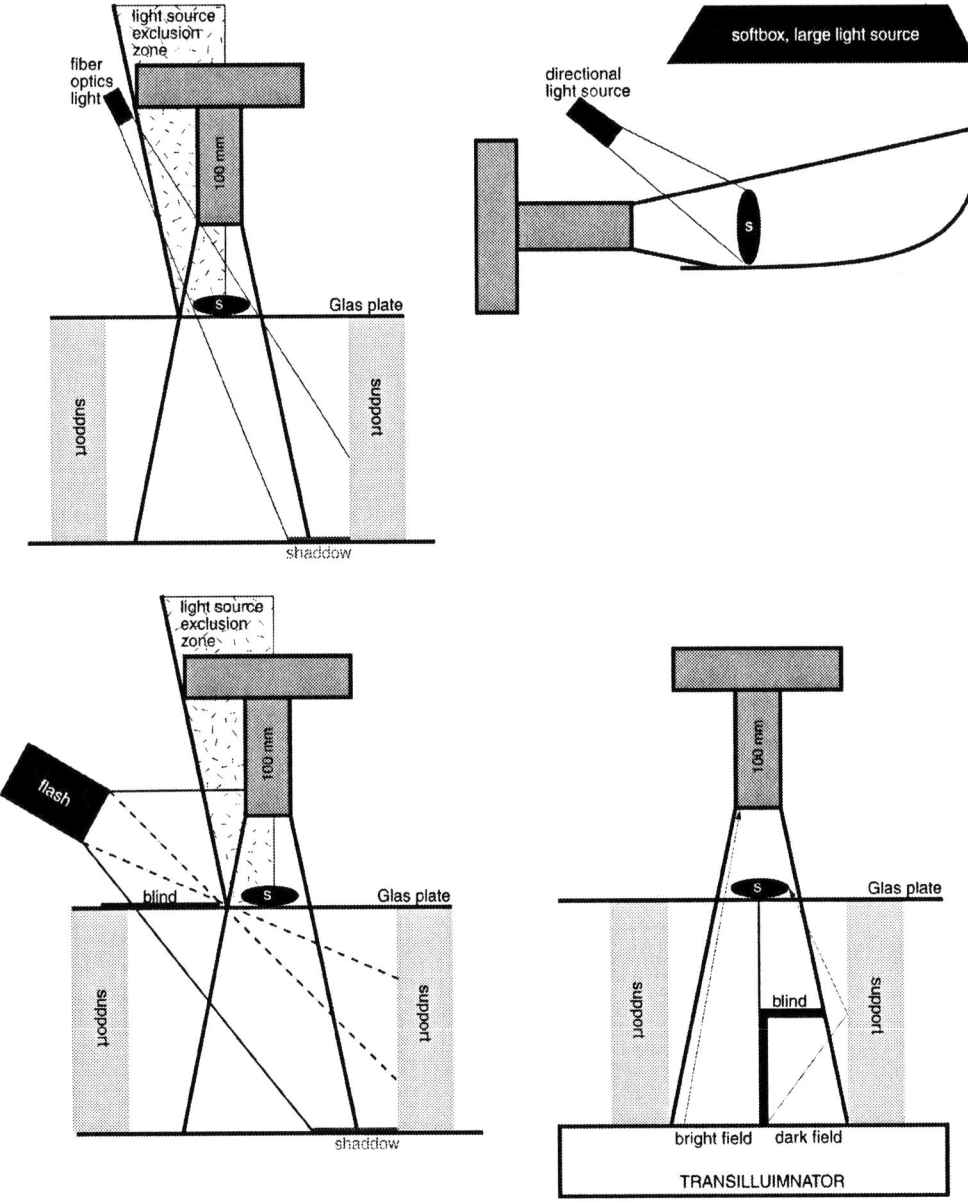

Figure 7.6 Various techniques for reproduction photography.
Top left: top view of specimen on dark background. The glass plates hold the specimen above the background so that the latter is out of focus. The focal length of the lens determines the light source exclusion zone (see also Figure 7.7). The dark shadow produced by the specimen will not show on the black background. Top right: lateral view of a specimen. The specimen is placed on a continuous background pulled up gently in the far end. A large light source provides even illumination over the entire distance. A directional light source can provide a local highlight and emphasize sculptural elements. Bottom left: Top view with background other than black. The blind placed on the glass surface outside the image area will prevent a shadow from being cast on the uniform background. Bottom right: Transillumination set–ups. On the left, a light box provides background illumination producing a dark outline of the specimen on a light background (bright field). Bright field illumination may introduce excessive lens flare, reducing edge definition and overall image contrast. On the right, direct light is blocked with a blind. The margin of the specimen is illuminated with oblique rays. It will result in a bright outline of the otherwise dark specimen on a dark background (dark field).

The position of the light source does not matter for a black background. For white the problems can be minimized by overexposing the background with an additional light source. The background should still not be overexposed by more than 3 f-stops, because otherwise lens flare increases and the edges of the object will appear soft and washed out. For colored backgrounds, place the light source at an angle so that the shadow of the object falls outside the background area to be incorporated in the image. Blinds are often useful to restrict light from one light source from illuminating another area. A frame cut from black paper or cardboard can provide overall restriction of light falling on the background at undesired places. In place of a frame of fixed size, two L-shaped pieces can be superimposed to make an adjustable frame. Three factors afford more liberty in the placement of the light source: the larger the object is shown, the longer the focal length of the lens, and the farther the background from the object.

Alternatively, place a mirror at a 45° angle under the object on a glass plate. The mirror will then be illuminated with a colored light source. Use either an electronic flash with a color filter, a slide projector with a color filter over the lens, or a colored slide in the slide tray. For uniform background, defocus the projected image. Intensity can be controlled with neutral gray filters, or a fine grain B&W negative mounted as a slide, or by moving the projector to or away from the set-up.

7.5.5 3D-objects, lateral view. Although a top view is generally preferred, it might be necessary to take side view pictures of objects. In order to make the background color uniform or softly changing, place the objects on a large area of either paper or fine cloth, which will be pulled up gently in the background (Figure 7.6). Avoid sharp lines on the background. Alternatively, the object can be placed on a transilluminator with a frosted but smooth glass plate.

7.5.6 Glossy objects. Glossy specimens abound in malacology. Photographs will show glare and unappealing highlights. Such highlights are caused by direct reflection of incident light. For shells, usually the reflection is polarized. These reflections can be reduced with dulling spray, ammonium chloride, or black opaque; supposedly, these coatings can be removed (Sakamoto 1973). These techniques will obscure some of the color patterns of the specimen, but may be suitable to enhance morphological details, and so may be applicable to many fossil specimens. If loaned specimens are to be treated, first ask the responsible curator for permission to do so. Alternatively, use a single or cross-polarizing filter (Young et al. 1996), or use a large diffuse light source (see Section 7.5.3).

7.5.7 Microphotography. A microscope or dissecting scope may be fitted with a trinocular head which allows mounting a camera system. Either a dedicated microscope camera without viewfinder or a general purpose SLR camera can be used. Both approaches can produce quality results. Consider the color temperature of the light source and the expected exposure for reciprocity failure. For short exposures consider using a color conversion filter in conjunction with a daylight film; for long exposures rather use a tungsten film. For histology, contrast and color saturation can be increased with a film with high color saturation.

As an alternative, some third party manufacturers (Hama) produce eyepiece photo adapters. These consist of a tube to be slid over the eyepiece and an inner filter ring to be screwed into the filter ring of a 50 mm standard lens. Avoid macro lenses, as the diaphragms tend to cause vignetting. Reasonable photographs can be produced, though a trinocular head is by far the superior solution. More detailed accounts of microphotography are found in White (1987) and Delly (1988).

7.6 ULTRAVIOLET (UV) PHOTOGRAPHY

Ultraviolet photography should be carried out with all necessary safety precautions being taken. When using continuous UV light sources such as a UV illuminator or a black light, all skin and particularly the eyes need to be protected with clothes and an acrylic face shield. Furthermore, UV light forms harmful ozone (O_3) from the oxygen (O_2) in the air. Proper ventilation of the work area is mandatory.

For an overview of techniques see Anonymous (1972a) and Young et al. (1996). Wilson (1975) discussed applications to molluscan specimens.

7.6.1 UV reflectance photography.

Here the actual UV light ray is the image-forming ray. Applications include physical characterizations of materials, pattern detection in non-visible areas of the spectrum, and enhanced fine detail resolution due to the shorter wave length of the UV light (cf. Figure 7.4 MTF of Ektachrome 100 VS). The chief problem is that regular glass and the cement used to combine lenses absorb UV light. Two solutions are available. Mirror lenses without glass (catadioptric lenses) can be used. Alternatively, special silica, fluorite, and quartz refractive (dioptric) lenses are employed. The latter are somewhat more common, but only a limited number (Zeiss 60 mm Planar f/4 UV, Nikon Micro-Nikkor 105 mm f/4.5 UV) are available and they are extremely expensive. The silver halide of the photographic emulsion is inherently UV sensitive so no special film is necessary. However, a pure UV light source is difficult to obtain, hence, the non-UV light from an electronic flash or even a UV illuminator must be filtered out with a filter either over the light source or over the lens (Kodak Wratten 18A, 18B; Schott UG-11). If the UV light source produces any visible light, then a filter is necessary.

7.6.2 UV fluorescence photography.

Many materials fluoresce under UV light. The fluorescence is in the visible spectrum, so no special lenses are needed, and the focus does not need to be adjusted. Only a UV light source, or a flash fitted with a UV filter, are required.

7.7 INFRARED (IR) PHOTOGRAPHY

Infrared photographs taken in the 700-900 nm range can reveal hidden structure in specimens

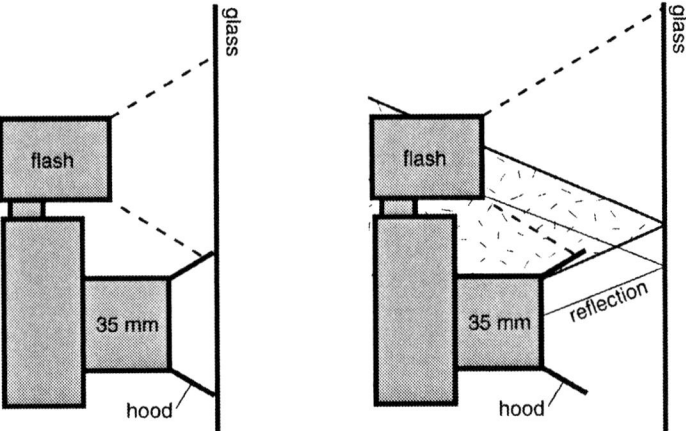

Figure 7.7 No reflection photography in aquaria.
On the left, the hood of a 35 mm lens is held flush against the glass, barring any reflection. There is no light source exclusion zone; hence, the flash can remain camera mounted. On the right, the camera is removed a short distance from the glass. The camera mounted flash is now in the stippled exclusion zone. Strong reflection is apparent in the final image.

and documents (Young et al. 1996). A number of films are available that can be utilized for this purpose (e.g., Kodak EIR: 700-900 nm; Konica Infrared 750: 700-800 nm). Photographic lenses transmit freely in the near IR, but their chromatic correction is carried out for the visible spectrum (400-700 nm). Focus needs to be readjusted, where the visual focal distance is transferred to the IR dot or IR line, usually marked in red with the letter 'R' on the barrel of the lens. A visually opaque IR pass filter (Kodak Wratten 12) is then added before the photograph is taken. Due to the longer wavelength of light being imaged, diffraction effects are enhanced. In most instances, the detection of relatively large-scale patters is sufficient. However, if detail resolution is desired with a practical limit of approximately 10 lines/mm (compared to 40-50 lines/mm in the visible spectrum), the photograph should be taken with open f-stop. Focus bracketing will be advisable. Exposure is a matter of experimentation.

Alternatively, specialized semiconductor elements are produced for the IR spectra up to 20,000 nm, though they are not used in SLR type digital cameras (Ray 1999). The more cost effective approach to IR photography is still with silver halides.

7.8 PHOTOGRAPHY THROUGH GLASS AND WATER

7.8.1 Optics and geometry. The most apparent problem when taking pictures through glass or water is the reflections caused on the surface; a second problem is distortion. Two properties of glass and water have to be distinguished: reflection and refraction. The angle of reflection is equal to the angle of incidence and can vary between >0° and <180°. Reflection increases steadily with decreasing angle of incidence less than 45°. Refraction is the bending of light due to the passage of light through the interface of two media of different densities. As the light source usually is in air and the subject behind glass or under water, the bending of the light is towards the line at a right angel to the surface of the glass/water, i.e., towards the lens axis. Light rays in seawater are at a maximum angle of 48° out of plumb according to Snell's Law (see Figure 7.5, and Sathyendrath and Platt 1990).

Distortion is caused by the unevenness of the surface and in the case of glass, by the inherent variability in the thickness of the glass. Distortion can be minimized by one simple measure: keep the film plane parallel to the surface of glass or water (Figures 7.3g, 7.5, and 7.7). Thus, the unevenness of the surface, and also the thickness of the glass plate are minimized. If the picture appears blurry through the viewfinder of the camera, chances are close to 100% that the picture will be entirely unacceptable; do not hope for mitigating effects. Despite all the considerations given to minimize reflection from glass and water, the surface may be too dirty or uneven to allow successful photography.

In order to eliminate reflection, we can calculate or estimate a suitable position of the light source (e.g., flash), from the following parameters: lens axis at right angle to glass surface (given); collecting angle of lens; scattering angle of light source. The collecting angle of the lens is negatively correlated to its focal length, i.e., a short focal length lens (wide angle lens) has a large collecting angle, whereas a long focal length lens (telephoto lens) has a narrow collecting angle. The scattering angle of most flashes is 60° to illuminate a 35 mm lens.

Zoom heads will vary the scattering angle of the light: 28 mm ~80°, 135 mm ~20°.

As an example, consider a 100 mm macro lens (24°) with a standard flash (60°) as shown in Figure 7.5. Both these angles refer to the total angle, i.e., the off-axis angle is half that value: 12° and 30°. Reflection from the water will be collected by the lens, if the light source is placed in the light source exclusion zone. It is given by half the collecting angle of the lens extended away from the lens until it meets the water surface, where it is reflected at the same angle (12°): incident angle = reflected angle. In Figure 7.5 only the light source exclusion zone on the left half is shown for clarity. The light source can be placed anywhere outside this exclusion zone, but a healthy safety margin is advisable, translating into a low illumination angle. The incident light ray from the light source will be partially reflected and partially refracted at the air-water interface. Because incident angle = reflected angle (α), the reflected light from the light source cannot enter the lens. The transmitted light will also be refracted towards the lens axis; hence, the illumination is not as flat as it would be in air; Snell's Law shows that the illumination must be at 48° or steeper.

The light source exclusion zone is defined by the collecting angle of the lens, which is a function of focal length. Wide-angle lenses (= short focal length) have a larger exclusion zone than telephoto lenses (= long focal length). The latter allow more flexibility in the placement of the light source. Suitable lenses are in the 90-135 mm range. As light sources use either continuous output incandescent lights or TTL flashes. Manual flashes are very difficult to use, because part of the light is reflected on the water surface before it reaches the subject. The precise amount of transmitted light is virtually impossible to determine without taking a test roll, hence, setting the proper f-stop is impossible; wide bracketing will be necessary. Any camera-mounted or built-in flashes need to be turned off.

These principles hold for any photography through reflective material: glass, water, ethanol, formalin. Some typical applications are zoo and aquarium photography through glass windows, specimens mounted in glass jars or embedded in acrylic, tide pool photography, and dissections of animals under water/ethanol (Figure 7.3e). For the last applications, when using needles in wax trays, pin them at a flat angle so that the needle will penetrate the liquid at a distance from the specimen, or use short pins that are entirely submerged. Otherwise, the reflected light from the meniscus formed around the needle will interfere with the specimen (Figure 7.3e).

7.8.2 Aquarium set-up. Aquarium (or other fluid filled container) photographs can be taken either from the top through the air-liquid interface, or from the side or from underneath through the air-glass/acrylic interface. For the former, the specific liquid (fresh water, seawater, alcohol, formalin, and glycerol) necessitates little practical considerations, except for the presence of a meniscus along the wall of the container (Figure 7.5). A container that is oversized for the specimen to be photographed should be used. The frame of the image should be at least 1 cm from the wall of the container, otherwise the light source will be reflected on the curved surface of the meniscus, which acts as a parabolic mirror. For critical applications with water, filter the water through a coffee filter or similar device to remove any suspended particles. In order to remove particles from the surface of the water, use an artist's brush. The aquarium itself should be neutral in color: transparent, white, gray, black. Reflection of balanced light from a colored surface will introduce a color tint.

The internal structure of the aquarium set-up is mostly an artistic consideration. If a white or black tray is selected, be aware that the overall tonality of the image is not neutral gray; make appropriate exposure compensation and bracket if necessary. Some popular set-ups are: pure white tray, pure black tray, container lined with black velvet, glass, or acrylic container on black velvet (consider reflection of light off the bottom of the container). One of my favorite arrangements is the nested container (Figure 7.5), which is particularly suitable for burrowing and sand inhabiting animals (Figure 7.3d). A larger outer container is filled with a suitable substrate (gravel, sand, mud), which is entirely submerged in water. A flat bottom inner container with water only is placed on the substrate. Set it in with the bottom at an angle, so that no air bubbles are trapped underneath the bottom. The animal in the top container is now in clean water on proper substrate and cannot burrow in it. The bottom of the glass or acrylic container is sufficiently thin to allow depth of field to render the background sufficiently sharp.

Lateral or bottom-up aquarium photography allows a creative circumvention of the problem of placing the light source. You can hold the rim of the lens or the lens hood flush against the glass (Figure 7.7), which prevents any light being reflected into the lens. You can use a wide-angle lens, keep the flash mounted on the camera, and still have an image free of any reflections. Additionally, any unevenness and scratches in the glass are as far away from the focal plane as possible.

If an animal needs to be photographed at an oblique angle (bottom view of the neustonic *Janthina*), the film plane must still be parallel to the glass surface. Accordingly, the container should be tilted (Figures 7.3f and 7.3g).

An interesting variation is extreme tele-macro photography in aquaria. Many mollusks are small and prefer to sit in the far corner of the aquarium. I have used a 300 mm f/4.5 with a variable length extension tube (65-113 mm) giving approximately a 1:2 magnification at close focus. Because depth of field needs to be maximized, light loss due to the extension tube is considerable. A camera mounted flash would be very far from the subject, hence, the flash is connected with a TTL synchro cord and held next to the lens flush against the glass of the aquarium. Because exposure is controlled by the duration of the flash (1/1,000 s) hand held photos are possible even with such an unyielding set-up.

7.8.3 Outdoor applications. Photography in tide pools adds further challenges to proper lighting and problems with reflections because the sun or skylight as uncontrolled light sources have to be considered. The film plane has to remain parallel to the surface of the water; hence, the noon sun may be positioned in the light source exclusion zone even when using telephoto lenses. In such situations, it may be necessary to produce a shadow over the area to be photographed and to use an appropriately positioned flash in broad daylight. For macrophotography, the photographer's body is usually sufficient, but an umbrella or a tent may have to be installed for larger areas. With diffuse skylight or light being reflected off the surrounding environment, mirror reflections may be apparent on the surface. These can be attenuated or eliminated with a polarization filter (see Section 7.3.5 for correct type of polarization filter; Figures 7.3a and 7.3b). Any ripples in the water from current or from wind make successful photography through the water impossible. Occasionally minor disturbances can be sufficiently reduced by immersion of a foot, hand, arm, or leg in a strategic position. One may also photograph through a tube or box placed in the water to create an even surface.

7.9 UNDERWATER (UW) PHOTOGRAPHY

I will provide below a very short overview on underwater photography. More detailed information can be found in Edge (1999).

7.9.1 Range of Equipment. General rule: the pictures are as good as the equipment, which is reflected in the price. The simplest version is the Kodak and Fuji instamatic cameras, which consist of a disposable camera with film. They can be used for surface photographs and up to a depth of 3 m in well-lit waters, as no artificial light source (flash) is available for these cameras. The cheap plastic molded lens produces images only suitable for emergency documentation. The cost of a disposable camera is US$10-15.

The Minolta Weathamatic (and a few other similar brands) can be used to a depth of up to 10 m, and has a small built-in flash. As the flash is more or less in the optical axis of the lens, heavy reflection from any suspended particles must be accepted ('snowed' pictures). This camera is only useable in transparent waters of the tropical oceans. It is affordable (US$200-300) and gives reasonable results under optimal conditions.

Ewamarine produces PVC bags for most still cameras and video cameras. Supposedly, they work to a depth of 10 m. However, as the bags are compressible you have the choice of two trade-offs. Either, you put little air in the bag and adjust the camera at the surface as such is impossible at depth, because the bag clings firmly to the camera (special problems with autofocus cameras arise!). Alternatively, the camera bag is filled with air on the surface to make it operable at depth; however, it is then difficult to submerge a strongly positively buoyant bag. In my opinion, these bags are perfectly suitable for surface shots and in bad weather outdoors, but are of no value for true UW photography.

Some 135-format (24 x 36 mm negative) viewfinder cameras are specially designed for UW use, and can withstand a depth of 30-40 m. The best-known brands are Sea & Sea, Motormarine, and the Nikonos family. Interchangeable lenses and various supplementary flashes are available, of which many work in TTL mode (Nikonos V plug). The whole system is handy and with a weight of 1 to 2 kg on land, they are not too heavy. However, as these cameras are rangefinder cameras, the viewfinder image is not the same as the image taken by the camera lens (parallax problem). On certain dive bases, Nikonos cameras can be rented for an exorbitant fee. Church and Church (1986) provided an introductory text to the Nikonos system.

The next possibility is to take an SLR-camera and have a housing built around it. The advantage is that one can frame exactly the picture one wants to take. However, the housings are bulky and therefore impose considerable drag UW. Although UW they are usually -0.2 to -0.3 kg buoyant, on land they weigh 4-10 kg. Housings are available for most cameras and most lenses, lenses being interchangeable only on land. However, cameras with exchangeable viewfinders are preferred (Nikon F3, F4, Canon F1, and Pentax LX), as the distance between the eye and the viewfinder is long due to the facemask and the housing: a sport viewfinder or the Pentax LX 45° viewfinder FB-1 come in handy. One option costing approximately $1,000 is a plexiglas housing from Ikelite. The working depth is limited to 60 m. The better housings are made of aluminum among others by Aquatica, AquaVision, and Swiss Hugyfot with working depth of 70-100 m. Such cameras have become more commonplace and more affordable ($1,300-2,000). Flashes ($400-1,500) are connected by the standard Nikonos V plug. Many TTL-systems are supported by various flashes.

Nikon had introduced a Nikonos-style SLR camera, the Nikonos RS. It claimed to be the ultimate camera for UW use and to supersede housings. However, operation with five to seven millimeter gloves is virtually impossible, it is 1 kg negatively buoyant, just three lenses are available with no new ones being developed, and the price is comparable to a housing system. The only advantage is that the size is comparable to the other Nikonos cameras. Features such as synchronization of the flash on first or second shutter are of no practical use whatsoever. It is no longer produced, but may be found in the second hand market.

Underwater cameras need a lot of care. Many O-rings are found in an UW camera, and particularly in housings. They have to be greased regularly with silicone grease (mineral greases destroy rubber) and have to be kept dust free in order to prevent leakage. Highly viscous silicon grease is preferable, such as those used for high-vacuum systems; environmental grease of SCUBA equipment manufacturers will be washed off immediately. Although the cameras are pressure resistant, they are rather shock sensitive. Flash cables must not be bent, otherwise the camera may be flooded through the cable. Never let seawater dry on an UW camera; if you cannot rinse it immediately in fresh water, keep it moist in a wet towel.

7.9.2 Photography in water. Additional to the differential absorption of colors with depth, water is also approximately 800 times denser than air. In surface photography, the amount of air between the camera and the object is generally neglectable, with underwater photography it plays an important role. In order to frame the same picture at a closer distance, wide-angle lenses are more commonly used. The 35 mm lens is comparable to the 50 mm lens on the surface, the 21 mm is comparable to the 35 or 28 mm lens on the surface. Special rules apply to the housings: The lens is covered with a thick walled glass lens housing called the port. The light beam is broken when it is passing the port and causes a narrowing of the collection angle of the lens: the 50 mm lens becomes an 80 mm lens, the 35 mm is approximately a 50 mm. For wide-angle lenses (f <50 mm), the glass is no longer plane, but bell shaped (dome port); the effect of longer focal lengths only applies to an insignificant amount.

Further considerations with lighting arise. As the light passes the water, the red part of the spectrum is filtered out, rendering the light colder (more blue) and increasing the color temperature. For this reason, most UW flash tubes are of a warm 5,000 K, and with some flashes even the reflector can be exchanged (Hardenberger down to 4,000 K) in order to adjust the color temperature to the working distance. Another problem underwater is the correct f-stop calculation. As the water absorbs much more light than air, the indication of the strength of the flash (guide number) has to be adjusted. Most UW flashes have UW guide numbers of eight to sixteen (the Hardenberger 500 TTL might have 22). Most manufacturers' indications exaggerate the power of their flashes by approximately one f-stop (Frei 1992). When used in air, the UW guide number can be multiplied by a factor of three to four. The light scattering angle of an UW flash is mostly 45-90°, and not in excess of 100° as often listed in manufacturer's documentation. Therefore, the use of two flashes is highly recommended with super-wide angle lenses (f <24 mm).

Underwater the many suspended particles are of great concern to the photographer. Our brain is very well suited to filter out any disturbing particles, but the film is relentless in its ability to record every last one of the suspended particles. In most instances, a flash has to be used for UW photography. It will send light in an axis close to the optical axis of the lens, and, therefore, the lens will collect a lot of the reflected light from suspended particles. For any given object distance (say 2 m) and distance of flash to optical axis (say 0.5 m) the light source is closer to the optical axis with wide angle lenses (16-35 mm) than with a long focal length (100 mm macro); the snowing effect is more pronounced with wide angle than in telephoto lenses. As the particles are mostly in the foreground, they will be overexposed, producing the infamous snow on UW photographs. At a narrow distance to the object the flash can be mounted in such a fashion, that a side or top illumination is realized, either using a long flash arm, or by handholding the flash (point shooting).

The diver him/herself is the single most important contributor to the amount of suspended particles in the water column, because any movement has the tendency to stir up sediment. UW photographers tend to move very slowly, sometimes even appearing unconscious. Much of the movement is done by using the lungs for fine regulation of the buoyancy as opposed to fin work. Dive buddies of UW photographers should be mindful of natural and fin induced currents that can transport suspended particles into the scene. Making all the adjustments, without stirring up any dirt, takes considerable time, typically for a very easy shot 1 minute. For more complicated situations, e.g., macrophotographs on a vertical wall, 3-5 minutes may be needed.

For underwater photography, the same films as for surface photography can be used. Kodak offers Sea processing aimed at snapshot, no-flash UW photography carried out by tourists. It consists of electronically removing the blue cast (reduce blue and green alpha channels, intensify red alpha channel). It is an interesting marketing strategy without merits for scientific applications. If color accuracy of available light shots is important, shoot a Macbeth ColorChecker in situ and devise the accurate filtration in Photoshop.

7.10 STORAGE AND ARCHIVAL CONSIDERATIONS

7.10.1 Processing. Most photographic material can be processed by the user with some technical expertise, with the exception of Kodachrome and Agfa Scala films. For B&W film and paper development, the two most crucial steps for archival considerations are fixation and washing. Underfixation leaves unexposed silver halide in the emulsion which leads to brownish deposits in the emulsion. Over fixation removes some of the desired image. For film, determine the time at which the entire film including the perforated area is transparent, and keep the film for the same time period after the clearing point in the fix bath. Check the film fix bath regularly for silver content with commercially available hypocheck solutions. For paper, follow manufacturer's instructions and check silver content. Fiber papers are considered more archival than the newer resin papers, but fiber papers change size more due to processing. Ilfochrome is the most archival of the color reversal papers. A thorough wash is essential in order to remove any traces of the processing chemicals. Hence, appropriate fixation time, good condition of the fixation bath, and sufficient washing are vital for long-term stability of the image. Selenium toning can convey additional image stability for B&W prints.

Color development for negative (C41), slide film (E6), or reversal paper (R4) is often considered to be tricky for non-specialized facilities. This is positively overstating the sophistication of the procedure. Most likely, the most attractive of the processes, E6, can be carried out in any kitchen sink, using a 10-15 l bucket, a thermometer, and a watch. A rotary processor is not necessary. Most processing kits can be used over a temperature range of 34-42°C with ideal temperature at 39°C. The bucket is filled with 40-42°C water in which the roll of film and the other solutions are pre-warmed for 10 minutes. The water will have cooled by then to approximately 40°C. The subsequent development of approximately 6 minutes is accompanied by a further cooling of 1-2°C. The time in the middle of the development, say 39.5°C after 3 minutes, is taken as the temperature for which the total developing time has to be adjusted according to the included table. For equipment check in the field or on research vessels, a single use kit is suitable (e.g., Kodak Professional single-use chemistry kit: 5 l). For regular, multiroll processing the Beseler CS6 1 liter kit has produced quite satisfactory results. Yet its proclaimed shelf life of up to 4 months is doubtful if not all necessary precautions such as replacing the oxygen containing air with protective gas (N_2), and storing all solutions in amber bottles are taken. Ideally, the entire kit should be used within a couple of weeks. The kit should be used according to the included instruction with one exception. I found that I had to bang the container twice on the counter in order to dissipate any adhering bubbles. This kit, like its competitor's, allows one and two f-stop push and pull processing. The color cast of the image is somewhat affected by the temperature at which development takes place. The possible color casts are much smaller than differences in color temperature of morning and noon sunlight. For color critical application, I recommend processing by a professional lab. Most home processing kits do not include a hardener bath, which makes film more susceptible to scratches. Separate hardeners can be used at the end of the processing sequence.

7.10.2 Mounting. Slide mounts come in a variety of styles. Most commonly, simple plastic slip-in, snap, or contact glued mounts are found. I do not recommend glass mounting, because of Newton rings produced by the slide adhering to the glass, the thickness of the mount using more space, and the greater likelihood of fungus developing between the gelatin emulsion and the glass, particularly in more humid areas. In the US, most Kodachrome slides and older Ektachrome slides are mounted in cardboard. Cardboard as an organic material also harbors the possibility for microbial and fungal activity. Remounting slides is a difficult decision, because of the handling involved. Glass mounted slides are usually easy to disassemble and to remount, but cardboard mounted slides are usually glued together and are often difficult to dissemble. For remounting slides, I recommend snap mounts such as GEPE.

7.10.3 Labeling. Labeling of slides bears the same considerations as labeling specimens. Self-adhesive labels can become dislodged or can lose some of their adhesive property, potentially smudging the photographic image. This worst-case scenario is particularly worrisome with any negative/slide material, because the original image can become damaged, as opposed to a print for which the negative/slide may still be available. I favor a handwritten serial number that references the image to supplementary data in a collection database. For plastic mounts a sharpie or overhead transparency marker are suitable, for cardboard mounts pencil works better, because the ink from the felt tip markers is absorbed by the cellulose fabric. Photographic prints can be marked on the reverse in pencil, pen, or with a self-adhesive label.

7.10.4 Storage. Storage of unexposed silver halide products is detailed under section 7.4.6 Films. Exposed but unprocessed material is most sensitive to heat and high moisture. Whenever possible keep such material in the refrigerator or freezer, and process the material as soon as possible. The conversion of silver halide in the unexposed film to metallic silver during exposure to light is a three step process requiring at least three photons (see also section 7.4.7 Error of Reciprocity above). The two first steps are reversible and this decay of the latent image is temperature dependent. Loss of clarity of the image as well as color shifts may ensue under adverse storage conditions.

Labeled material may be stored in archival polypropylene or polyethylene plastic sleeves (not polyvinyl chloride = PVC). Boxes and sleeves should be made of acid free cardboard and paper. A variety of sleeve designs accommodate up to twenty 35 mm mounted slides, approximately seven 35 mm negative or slide strips of five images, or one to four prints depending on size. Storage of transparencies in clear plastic sleeves allows easy viewing on a light box with minimal handling of the actual images. Openings are either on top or on the side of the sheet, depending on whether the sheet is stored in portrait format in binders, or in landscape format in hanging registers of filing cabinets. Avoid styles in which the openings of two pockets are facing one another, because the slides tend to fall out of one side; all openings should be in the same direction. Any photographic material should be stored in a dry (RH 30-50%), cool (<18°C), and foremost dark place, because even a developed image will fade due to exposure to light. For particularly valuable images consider duplicating or scanning the image, and sealed frozen storage. For technical details on storage and fading of color images see Keefe and Inch (1990) and Wilhelm (1993); for historical materials and restoration see Anonymous (1972b), Weinstein and Booth (1977), Anonymous (1985), and Hendriks (1991).

7.11 ACKNOWLEDGMENTS

R. Gschwind and R. Heilbronner (Department of Physical Chemistry, University of Basel, Switzerland) introduced me to the theory and practice of scientific photography. Jim McLean (LACM), Christine Thacker (LACM), Russel Zimmer (USC), Fabio Moretzsohn (UH), and Anna Bass (USF) read the manuscript and made valuable comments. Carl Zeiss GmBH and Eastman Kodak Company kindly permitted to use material available on their web sites.

7.12 LITERATURE CITED

Anchell, S. G. and B. Troop. 1998. *The Film Developing Cookbook*. Focal Press, Boston. 163 pp.
Anonymous. 1972a. *Ultraviolet & Fluorescence Photography*. Kodak Publication No. M-27. Eastman Kodak, Rochester. 32 pp.
Anonymous. 1972b. *Caring for Photographs: Display, Storage, Restoration*. Time-Life Books, New York. 192 pp.
Anonymous. 1985. *Conservation of Photographs*. Kodak Publication No. F-40. Eastman Kodak Company, Rochester, New York. 156 pp.
Anonymous. 1987. *Kodak Gray Cards*. Kodak Publication No. R-27. Eastman Kodak, Rochester, New York.
Anonymous. 1990. *Kodak Photographic Filters Handbook*. Eastman Kodak Company, Rochester, New York. 161 pp.
Anonymous. 1998. *Kodak professional photoguide*, 6th Ed. Silver Pixel Press, Rochester, New York. 56 pp.
Anonymous. 1999a. *Fujifilm Professional Data Guide '99*. Fujifilm, Tokyo. 110 pp. [This publication can be obtained without charge from you local photoretailer, or from Fuji: 800-788-3854 x73].
Anonymous. 1999b. *Kodak Professional Reference Dataguide*. Eastman Kodak Company, Rochester, New York. 93 pp. [This publication can be obtained without charge from you local photoretailer, or from Kodak: 800-242-2424]
Bracegirdle, B. 1995. *Scientific Photomacrography*. Bios Scientific Publishers, Oxford. 105 pp.
Brück, A. 1984. *Close-up Photography in Practice*. David and Charles, Newton Abbot. 144 pp.
Church, J. and C. Church. 1986. *The Nikonos Handbook*. Privately published, Gilory. ix, 167 pp.
Delly, J. G. 1988. *Photography Through the Microscope*, 9th Ed. Kodak Publication No. P-2. Eastman Kodak Company, Rochester, New York. 104 pp.
Edge, M. 1999. *The Underwater Photographer*, 2nd Ed. Focal Press, Boston. 252 pp.
Frei, H. 1992. Die Stunde der Wahrheit. *Unterwasserfotographie* **92**: 46-57.
Goldberg, N. 1992. *Camera Technology: The Dark Side of the Lens*. Academic Press, Boston. 309 pp.
Hendriks, K. B. 1991. *Fundamentals of Photograph Conservation: A Study Guide*. Lugus Publications, Toronto. 560 pp.
Hunt, R. W. G. 1998. *Measuring Colour*, 3rd Ed. Fountain Press, Tolworth, U.K. 344 pp.
Hunt, R. W. G. and R. W. Gainer 1987. *The Reproduction of Colour: In Photography, Printing & Television*. Fountain Press, Tolworth, U.K. 640 pp.
Hunter, F. and P. Fuqua. 1997. *Light: Science & Magic*, 2nd Ed. Focal Press, Boston. 344 pp.
Jacobson, R. E., S. F. Ray, and G. G. Attridge. 1988. *The Manual of Photography*, 8th Ed. Focal Press, Boston. 394 pp.
Johnson, C. 1999. *The Practical Zone System*, 3rd Ed. Focal Press, Boston. 192 pp.
Keefe, L. E. and D. Inch. 1990. *The Life of a Photograph: Archival Processing, Matting, Framing, Storage*, 2nd Ed. Focal Press, Boston. 384 pp.
Keppler, H. 1999. For sharpest focus do you really need a mirror lockup? If so why don't all top cameras have it? *Popular Photography* **1999**: 18-22, 24, 64.
Landt, A. 1993. *Lenses for 35 mm Photography*. Silver Pixel Press, Rochester, New York. 112 pp.
London, B. and J. Upton. 1998. *Photography*, 6th Ed. Longman, New York. 399 pp.
Merklinger, H. M. 1993. *Focusing the View Camera*. Privately published, Dartmouth, Canada. 128 pp.
Ray, S. F. 1994. *Applied Photographic Optics*, 2nd Ed. Focal Press, Boston. 586 pp. [3rd Ed. of 2002].
Ray, S. F. 1999. *Scientific Photography and Applied Imaging*. Focal Press, Boston. 559 pp.
Sakamoto, K. 1973. Techniques for photographing modern mollusks. *Veliger* **16**: 140-142, 1 pl.
Sathyendranath, S. and T. Platt. 1990. The light field in the ocean. *In*: P. J. Herring, A. K. Campbell, M. Whitfield, and L. Maddock, eds., *Light and Life in the Sea*. Cambridge University Press, Cambridge. Pp. 3-18.
Stroebel, L., J. Compton, I. Current, and R. Zakia. 1986. *Photographic Materials and Processes*. Focal Press, Boston. 585 pp.
Stroebel, L., J. Compton, I. Current, and R. Zakia. 2000. *Basic Photographic Materials and Processes*, 2nd Ed. Focal Press, Boston. 410 pp.
Stroeble, L. and R. Zakia (eds.), 1993. *The Focal Encyclopedia of Photography*, 3rd Ed. Focal Press, Boston. 914 pp.
Sturge, J. M. 1976. *Neblette's Handbook of Photography and Reprography. Material, Processes and Systems*. 7th Ed. Van Nostrand Reinhold Company, New York. 641 pp.
Weinstein, R. A. and L. Booth. 1977. *Collection, Use, and Care of Historical Photographs*. American

Association for State and Local History, Nashville, Tennessee. 222 pp.

White, W. 1987. *Photomacrography, an Introduction*. Focal Press, Boston. 221 pp.

Wilhelm, H. 1993. *The Permanence and Care of Color Photographs: Traditional and Digital Color Prints, Color Negatives, Slides, and Motion Pictures*. Preservation Publishing Company, Grinnell, Iowa. 744 pp.

Wilson, E. C. 1975. Light show from beyond the grave. *Terra* **13**: 10-13.

Young, W. A., T. A. Benson, G. T. Eaton, and J. Meehan. 1996. *Copying and Duplicating: Photographic and Digital Imaging Techniques*. Silver Pixel Press, Rochester, New York. 143 pp.

C. F. Sturm, T. A. Pearce, and A. Valdés. (Eds.) 2006. The Mollusks: A Guide to Their Study, Collection, and Preservation. American Malacological Society.

CHAPTER 8

COMPUTERIZING SHELL COLLECTIONS

GARY ROSENBERG

8.1 INTRODUCTION

Once a shell collection reaches a certain size it becomes difficult to keep in order without a formal system. Specimens stray from labels and unrecorded information fades from memory. One solution is to create a catalog, numbering the specimens and recording associated information under those numbers in a ledger or index card file. The catalog ensures that the data for each specimen are retrievable. A numerical catalog, however, is not particularly useful except as an archive. Lists sorted systematically, geographically, or alphabetically are more convenient than numerical ones. Therein lies the advantage of an electronic database over the traditional catalog. Information is typed only once, but can then be sorted and manipulated into many products, such as labels, catalogs, have-lists, and trade-lists. The challenge in computerizing a shell collection is devising a system complex enough to generate these products, but simple and efficient enough to encourage routine use.

This article introduces database conventions and describes fields and tables that will allow individuals to design conchological databases appropriate to their needs. Institutional scale collections will have more complex needs, including tracking loans and recording published information about specimens they hold, although the basic principles will be the same.

8.2 CHOOSING A PROGRAM

Three kinds of computer program are capable of creating electronic databases: spreadsheets, such as Microsoft Excel, Corel Quattro Pro, and Lotus 1-2-3, word processors that include rudimentary database features, such as Microsoft Word or Corel WordPerfect, and dedicated "relational" database programs such as FileMaker Pro, Microsoft Access, or Corel Paradox. The dedicated programs are called relational because relationships between tables of data can be defined, which gives more control over the data. Relational databases use space efficiently, prevent duplications, enforce formats if desired, provide for vocabulary control, and have sophisticated querying and reporting functions. Use of word processors or spreadsheets for creating databases for shell collections is not recommended if more than a couple of hundred lots will be involved.

After deciding to computerize a shell collection, you must choose what software to use. The current versions of many off-the-shelf database programs, such as FileMaker, Access, or Paradox, will perform well. More important than which of these to use will be how it is configured, and what resources are available to assist in that task. So, if you like the database program you use at work, or a helpful friend or relative is versed in a particular one, that in itself may be sufficient reason to choose it. If, however, you are considering older or less widely used software, first familiarize yourself with database approaches and then ask a number of questions. Some of these questions are also relevant if you are considering using a word processor or spreadsheet instead of a database program.

- Is the program fully compatible with your computer's operating system?
- Can it import and export standard formats, such as delimited ASCII (American Standard Code for Information Exchange), DBase

Table 8.1 Database table showing a subset of the fields that would appear in a typical catalog book.

Catalog number	Identification	Locality	Date collected	Collector	Quantity
1537	Chicoreus florifer (Reeve, 1846)	North Bimini, Bahamas	12 Dec 1973	W. Loman	3
1538	Cypraea spurca acicularis Gmelin, 1791	Key West, Florida U.S.A.	15 Apr 1999	U. Samuels	12
1539	Zoila friendii friendii (Gray, 1831)	Rottnest Island, Western Australia	30 Nov 1984	W. Gretsky	1

(for databases), or Lotus 1-2-3 (for spreadsheets)?
- What kind of technical support is available from the manufacturer? Preferably, there will be toll-free phone line, email support, and FAQs (frequently asked questions) available via the Internet.
- Does the program accept images, *e.g.*, from a digital camera or flatbed scanner, and can they be easily imported?
- How many records can be handled before performance becomes unacceptably slow, especially if images are involved? This will also depend on the operating speed and memory capacity of the computer.
- Are there restrictive limits on the total number of records, the number of fields or images per record, or the number of characters per field or per record?
- Can fields be added midstream and their properties, such as length or type, be easily changed?
- Can reports be printed using italic and bold typefaces, various font sizes, and proportionally spaced fonts? Can these be previewed on screen?
- Is it easy to insert letters with diacritic marks such as in Linné, Röding, Müller, and Bruguière? What about those in Møller or São Paulo, which are not among the 255 characters of extended ASCII?
- Is the program "Web-ready", that is, does it support HTML (Hyper-Text Markup Language)? This is desirable if data will be made available on a website on the Internet.
- Is the program object-oriented? That is, can objects such as "Genus + Species + Author +,+ Date" be created and used repeatedly, or must they be created anew each time?

- Is the program SQL (Structured Query Language) compliant? This is important only if scripts or macros are used for repetitive tasks; if they are written in SQL, they can be exported to another compliant database if needed.

8.3 DATABASE CONVENTIONS

An electronic database contains tables with columns corresponding to fields and rows to records (Table 8.1). Some database programs treat each table as a separate database, whereas others allow a collection of tables to be named as a single database. In its simplest form, the database for a shell collection could be constructed as a "flat file" to emulate the pages of a catalog (Table 8.1). Each field would correspond to a single column in the catalog and each row would correspond to a single lot (a lot contains all the specimens of a species collected in one place at one time and obtained from a particular source).

Although some people will find such a layout adequate, it is not particularly functional, because it violates a fundamental rule of database design: a field should contain only a single kind of information. In Table 8.1, the Identification and Locality fields contain multiple kinds of information. This prevents precise sorting, restricts options for formatting output, and impedes use of the relational properties of the database. For example, the Locality field cannot be sorted by country or state. If the order of entry were made largest to smallest ("USA, Florida, Cedar Keys" instead of "Cedar Keys, Florida, USA"), it would be possible to sort by country, but still not by state or any other hierarchical level. Locality is therefore better split into a number of fields, as in Table 8.2. One immediate

Table 8.2 Database table showing some of the fields that can be used specifying localities.

Locality ID	Continent	Ocean	Country/Territory	Primary Division	Location
345	Australia	Indian	Australia	Western Australia	Rottnest Island
346		Atlantic	Bahamas		North Bimini
347	North America	Atlantic	United States	Florida	Key West, Florida Keys

Table 8.3 Database table showing atomized fields for handling species names.

Species ID	Generic name	Subgeneric name	Specific name	Subspecific name	Author	Year	Parentheses
152	Chicorus		florifer		Reeve	1846	y
153	Cypraea	Erosaria	spurca	acicularis	Gmelin	1791	n
154	Zoila		friendii	friendii	Gray	1831	y

benefit is that items such as Ocean or Continent that normally would not be included in the Locality field, since they are not usually printed on labels, become available for querying and sorting.

Similarly, splitting the Identification field into a series of narrowly defined fields as in Table 8.3 gives more control over the data. Most database programs will allow formats or ranges of acceptable values to be set for individual fields, for example, "initial letter lower case" for Specific Name, or "greater than 1757" for Year. Misspellings can be prevented by linking the text fields to appropriate master lists, such as of authors or generic names. Such master lists are called "authority files". They can be accumulated by the user as names are encountered, or adopted intact or modified from existing lists. Each authority file is a separate table in the database. Atomized fields also provide more flexibility in reporting. Table 8.1 allows only "*Zoila friendii friendii* Gray, 1831", as output, with the species name lacking italics. Table 8.3 allows "*Zoila friendii friendii* (Gray, 1831)", Zoila friendii (Gray, 1831)", "*Zoila friendii friendii*", "*Zoila friendii*", and "*friendii, Zoila*" depending on what is wanted in a given context.

8.4 SUGGESTED FIELDS

Getting optimal performance from a database requires careful choice and definition of its fields. More than 50 fields that might be used in a conchological database are discussed here. Any given database will use only a subset of these, depending on the emphasis of the collection. Some of the recommended fields may not be of immediate use to the collector, but could enhance the value of the collection were it ever sold, or donated to a museum. The fields also serve as mnemonic devices to ensure that some kinds of information are recorded that might be easily overlooked in a standard catalog format such as in Table 8.1. Essential fields are marked with two asterisks (**), recommended fields with one (*).

For each field or set of fields, I describe the intended use and possible field types. Field types include alphanumeric (A) for normal text; numeric (N) or integer (I) for fields on which calculations might be done; date (D) for exact dates; currency (C) for monetary values; logical (L) for yes/no fields; memo (M) for long blocks of text; and picture (P) for digital images. For alphanumeric fields, I append typical maximum lengths; for example, A22 means an alphanumeric field 22 characters in length. Most current database programs have variable length fields, so the user does not need to specify the maximum length. Date fields can be used only for precise dates; alphanumeric should be used for imprecise items such as "pre 1960", "spring 1981", or "April 27-29 2000". Fields are grouped in related blocks that correspond in large part to those that would appear together in a table. A summary of fields appears in Table 8.4.

Table 8.4 Suggested configuration of fields in tables.
Underlined items are unique identifiers for each record in the table. Items in italics link to the unique identifiers of other tables.

Specimen Table	Identification Table	Classification Table
Catalog Number	Species ID	Child
Species ID	Genus (or higher taxon)	Rank
Qualifier	Subgenus	Parent
Identified by	Species	Systematic number
Date identified	Subspecies	
Number of specimens	Author	or
Locality ID	Year	
Collecting Event ID	Parentheses	Genus (or higher taxon)
Microhabitat		Subfamily
Size	and (if Classification Table is	Family
Condition	not used)	Superfamily
Live/Dead		Order
Preservative/Treatment	Subfamily	Class
Source	Family	Systematic number
Transaction	Superfamily	
Cost/Value	Order	
Trade	Class	
Date obtained	Systematic number	
Date cataloged		
Specimen Comments		
Collecting Event Table	**Locality Table**	**Image Table**
Collecting Event ID	Locality ID	Image ID
Station	Ocean	Image
Habitat (macrohabitat)	Continent	*Catalog Number*
Depth	Region	*Locality ID*
Elevation	Country/Territory	*Collecting Event ID*
Collection Method	Primary Division	*Species ID*
Date collected	Secondary Division	Subject
Time	Location	Photographer
Tide	Drainage	Image Comments
Collector	Coordinates	
Vessel	Source of Coordinates	
Expedition	Locality Comments	
Collecting Comments		

8.4.1 Specimen

** Catalog Number (I or A6): A unique number or code assigned to each lot. This number serves to link various tables of data pertaining to the lot.

**Number of Specimens (I or A15): An integer field can be used if the collection contains only intact specimens, which allows the total number of specimens in a collection to be calculated. An alphanumeric field should be used if integers would be ambiguous, e.g., "6 valves" if valves are not paired, "3 fragments" if the fragments might be from more than one individual, or "100 +", for a number too large to count conveniently. Two fields might also be used, an integer for the count and a text field specifying what was counted, which preserves the ability to total the integer field.

Size (N or A15): May be numeric if a single number (e.g., maximum size) is entered and is recorded in standard units (e.g., mm), or the units are specified in a separate field; must be alphanumeric for a range of measurements, multiple measurements of one specimen, or if units are included. Recording sizes of specimens, preferably to a tenth of a millimeter, is an excellent way to ensure that they can be reassociated with their labels, especially if catalog numbers are not written on them, if, for example, a collection drawer is dropped.

Condition (A30): For comments about condition, e.g., gem, fine, with periostracum, with operculum, with protoconch, sinistral, etc.

*Live/Dead (A4): Used to indicate reliability of depth data, since dead specimens can be transported by currents or hermit crabs. This may be expanded to include fossil and subfossil designations, which might indicate that a species no longer lives at a site. This can also be an indicator of condition.

*Preservative/Treatment (A30): For materials or procedures used to maintain condition of specimen: ethanol, glycerin, silicone, etc.

*Source (A30): Person, business, or institution from which a lot was obtained.

Transaction (A10): Gift, exchange, purchase, etc.

Cost/Value (C or N): Price paid or value assigned.

Trade (L or A1): A yes/no field that allows list of items for trade to be generated.

Date Obtained (D or A11): Useful for tax purposes in the event of a donation, for establishing period of ownership. Also useful for showing a latest date at which a specimen could have been collected if date of collection is not known.

Date Computerized (D or A11): Useful for tracking progress in computerization. Some programs can automatically track the date and time a record was created, so the information doesn't need to be typed in. A separate Date cataloged field can be used if part of a collection had been cataloged before it was computerized.

Images (P): Used for digital images of live animals, specimens, habitats, collecting sites, and collectors. If more than one image will be included in a record, then a separate table is needed for images. Additional fields can be used to describe images, as shown in Table 8.4.

**Comments (A255 or M): For data that do not readily fit in the field structure. If a particular comment is often repeated, it suggests the need for another field. Comments fields may appear in more than one table if desired, specific to the subject of the table.

8.4.2 Identification

*Species ID: Provides a unique identifier for records if a separate table is used for identifications.

**Genus (A22): For the generic part of a species name.

*Subgenus (A22): For the subgeneric part of a species name.

**Species (A22): For the specific part of a species name, i.e., the trivial epithet.

*Subspecies (A22): For a subspecific name.

**Author (A34): For the author of the specific or subspecific name.

**Year (A4 or I): For the year in which the specific or subspecific name was introduced.

*Parentheses (L or A1): Indicates the need for parentheses around author and year if the generic placement has changed since the species was named.

*Qualifier (A4): For expressing doubt about the species identification (?, cf., aff. near, like).

*Identified by (A30): Name of the person who identified the lot, which can be useful for determining the reliability of the identification.

*Date Identified (D or A11): Date on which the lot was identified, which can be useful for assessing

Table 8.5 Dynamic structure for higher classification that accepts an unlimited number of ranks.

Child	Rank (of child)	Parent
Bernayinae	Subfamily	Cypraeidae
Chicoreus	Genus	Muricinae
Cypraea	Genus	Cypraeinae
Cypraeidae	Family	Cypraeoidea
Cypraeinae	Subfamily	Cypraeidae
Cypraeoidea	Superfamily	Neotaenioglossa
Gastropoda	Class	Mollusca
Muricinae	Subfamily	Muricidae
Muricidae	Family	Muricoidea
Muricoidea	Superfamily	Neogastropoda
Neogastropoda	Order	Gastropoda
Neotaenioglossa	Order	Gastropoda
Zoila	Genus	Bernayinae

the likely accuracy of an identification when the concept of a species has changed.

8.4.3 Classification

Child (A22): Used as a general category for names of higher taxa (see Table 8.5). If a structure like that in Table 8.6 is used additional ranks are possible, e.g., Tribe, Infraorder, Suborder, Superorder, Subclass.

Parent (A22): Used as a general category for names of higher taxa (see Table 8.5).

Rank (A11): For ranks of higher taxa (see Table 8.5).

Systematic Number (N or I): A number assigned to each Child to allow it to be sorted into systematic order. Generally, all the taxa in a family or a superfamily would be assigned the same number. The list would thus be sorted systematically at family or superfamily level and above, and alphabetically below.

8.4.4 Locality

*Locality ID: Provides a unique identifier for records if a separate table is used for locality fields.

*Ocean (A16): Used for marine species, and for land and freshwater species from oceanic islands.

Continent (A13): Needed only if database is likely to be queried or sorted by continent.

Region (A20): Used as desired for areas such as Polynesia that contain more than one country.

**Country/Territory (A25): Countries often possess remote territories, but since the country name can be misleading as to geography, the territory name might be preferred. For example, the Falkland Islands are part of Great Britain, but lie off the coast of Argentina. It may be more convenient therefore to enter Falkland Islands in this field than Great Britain. Consistency is also important. If England is entered in some cases and Great Britain or United Kingdom in others, a search will not find all desired items.

*Primary Division (A25): For main subdivisions of countries such as states, provinces, prefectures, and departments.

Secondary Division (A80): For lesser subdivisions of countries, such as counties or parishes in the U.S.A.

**Location (A120): For the remainder of the locality data that does not fit in the above fields, written in the order desired for printing on the label.

Drainage (A20): For freshwater species, the drainage in which the body of water lies.

Table 8.6 Static structure for higher classification, which must be modified to add ranks.

Genus	Subfamily	Family	Superfamily	Order	Class
Cypraea	Cypraeinae	Cypraeidae	Cypraeoidea	Neotaenioglossa	Gastropoda
Chicoreus	Muricinae	Muricidae	Muricoidea	Neogastropoda	Gastropoda
Zoila	Bernayinae	Cypraeidae	Cypraeoidea	Neotaenioglossa	Gastropoda

*Coordinates (A23): For latitude and longitude written as a single string of characters, e.g., 24°59'58"S, 130°05'30"W, or, using decimal minutes, 24°59.97'S, 130°05.50'W. If mapping or calculations are to be done with the coordinates, two numeric fields for decimal degrees are preferable: Latitude: 24.9994 and Longitude: 123.0917. North and East are shown as positive numbers, South and West as negative numbers.

In reporting coordinates, consider the precision to which they are actually known. Atlases and gazetteers typically record the position of the center of a geographic feature, but the collector is generally not at the center, for instance, of a town on the seacoast. Rounding off to the nearest minute is appropriate if coordinates are obtained from a general or secondary source, one minute being about one mile. Particularly with self-collected material, if the collector (or another reliable source) has determined the coordinates via a topographic map, navigational chart, or GPS (Global Positioning System) receiving unit, then reporting seconds is appropriate.

Online sources of coordinates include the Geonet Names Server (GNS) for international locations <gnswww.nga.mil/geonames/GNS/index.jsp>, and the Geographic Names Information System (GNIS) for the U.S.A. and its territories <geonames.usgs.gov>. Google Earth is a free program that zooms into satellite and aerial photographs worldwide <earth.google.com>. It can be used to determine coordinates when you know where you were, but don't know the name of the precise place.

Another system of coordinates that might be useful to those who self-collect is UTM (Universal Transverse Mercator). Modern topographical maps and navigational charts often have a UTM grid at a one-kilometer scale. It is usually easier to determine your position from the UTM grid than the latitude and longitude grid. GPS receivers can be set to read either system and to convert between them.

*Source of Coordinates (A60): Because maps and GPS units use different models (datums) for the shape of the earth, which is not precisely spherical, coordinates are not all on the same grid system. Thus, the name of the atlas, gazetteer, topographic map, or navigational chart used to determine the coordinates should be given. If the coordinates came from a GPS unit, the model should be recorded, along with the datum to which it was set. The most widely used datum currently is WGS-84 (World Geodetic System).

8.4.5 Collecting event

*Collecting Event ID: Provides a unique identifier for records if a separate table is used for collecting event fields. Station numbers do not serve this purpose because not all specimens are collected at formal stations.

*Station (A10): Used mainly for self-collected material. Stations should generally be designated by a code combining letters and numbers and should be unique for each site visited by the collector. Some collectors use their initials followed by consecutive numbers, e.g., RTA151, RTA152, etc. The station code, if unique, makes it easy to generate species lists for each station once the material has been identified and computerized.

**Habitat (A80): Habitat data can be difficult to categorize because of its fractal nature. It can be a geographic feature at a large scale (e.g., a beach), and reasonably entered as part of the locality data, or be the niche of a particular species on a small scale (e.g., seawhips). The field might also be split in two, as Macrohabitat, for description of a collecting site, and Microhabitat, for ecological observations on a particular species. If particular measurements such as salinity or temperature are routinely made, fields can be added for them.

Depth (A12): Because it is rarely used for calculations, the depth field is normally alphanumeric, which allows ranges and original units to be recorded, e.g., "10-30 fms." A comment such as "shallow" or "intertidal" would normally go under Habitat. Converting all depths to standard units such as meters can give a false appearance of precision, for example, 100 fathoms converts to 183 meters, so it is good to preserve a record of the original measurement.

Elevation (A12): The comments about Depth apply here as well. Note that a specimen can have both a depth and an elevation, for example, a freshwater snail at a 5 m depth in a lake 500 m above sea level. Some marine species might also have elevations. For example, it makes more sense to say that a periwinkle living two meters above the high tide line is at "2 m" elevation than at "-2 m" depth, which is ambiguous.

*Collecting Method (A50): For data such as "trawled by shrimp boat", "SCUBA at night", "screened from 1 liter of coral sand", or "tangle nets."

Date Collected (A15): Important for establishing when a species lived in an area, especially if the environment has changed.

Time (A20): Useful for showing how much time was spent at a collecting site, the time of day at which live-collected specimens are most likely to be found, or determining the state of the tide retrospectively.

Tide (A10): Used for recording the state of the tide (spring, neap, low, -0.4 m, etc.).

Collector (A30): Important for establishing reliability of locality data, and sometimes in interpreting ambiguous data. In fields such as collector in which the usual entry is a person's name, you must decide the format for the name. For example, "Pilsbry, H.A." is preferable for sorting, but "H. A. Pilsbry" looks better printed on labels. Separate fields for Last Name and Other Names would allow both formats, but still could not handle multiple names such as "H. A. Pilsbry & A. A. Olsson". A still more complicated structure can solve this, but the gain in functionality is probably worthwhile only for museum-scale collections.

Vessel (A30): Name of the boat, ship, or submersible used for collecting.

Expedition (A30): Provides a name by which all the stations on a collecting trip can be identified.

8.5 DATABASE STRUCTURES

Once the fields for a database have been defined, they must be arrayed into a set of tables. A simple structure would be a single table containing all the main database fields, with appropriate authority files attached to particular text fields as desired for vocabulary control. Such a structure would be functional, but would not take full advantage of the relational properties inherent in databases because it would violate another rule of database design: avoidance of redundancy. Redundancy can lead to inconsistency if, for example, duplicate information is updated in one place but not in another. Eliminating redundancy requires splitting the data into two or more tables with relationships defined between them. So, for example, Table 8.1 could be reformatted to link to Table 8.3 for the identification, via the Species ID field, and to Table 8.2 for the locality data as shown in Table 8.7.

Each species name would be represented only once in Table 8.3 and any changes to it would propagate to all the specimen records in the database linked to it. For example, if *Cypraea acicularis* were treated as a full species instead of as a subspecies of *Cypraea spurca*, only a single species record would

Table 8.7 Species ID and Locality ID are used to link to the Identification and Locality Tables.

Catalog number	Species ID	Locality ID	Date Collected	Collector	Quantity
1537	152	346	12 Dec 1973	W. Loman	3
1538	153	347	15 Apr 1999	U. Samuels	12
1539	154	345	30 Nov 1984	W. Gretsky	1

need to be changed, which is preferable to searching the database to change many records individually. Similarly, each locality would be listed only once in Table 8.2.

Structuring the data to avoid redundancy is called "normalization". As the data become more normalized, consistency and ease of maintenance increase and storage space generally decreases. At the same time, however, tables multiply, which makes entering and manipulating data more complicated. Therefore, a balance must be struck between usefulness and normalization. This can be illustrated with an example from higher classification. The rules of normalization say that the fields for higher classification should not be added into Table 8.3, but rather linked from a separate table (Table 8.6).

If the systematic placement of a genus changes, the correction need be made only in one place. However, if the collector routinely uses family names but rarely uses higher ranks, having the family name included in Table 8.3 would be more convenient for easy sorting and querying. Also, Table 8.6 itself is not normalized, since there is redundancy above the genus level. A child-parent structure (Table 8.5) is more efficient.

This structure has several advantages: the "child" field gives the option for instituting vocabulary control on the "parent" field (since every item in Parent except the listhead Mollusca also appears in Child); no additional fields need be created to add ranks such as suborder; and ranks are optional (some scientists regard ranks as artificial, so an increasing number of taxa are being named without ranks). It is also more space-efficient, which becomes more apparent as the number of taxa listed grows, but this is a minor concern since computer storage space is cheap. This structure is more difficult to query, however, so Table 8.6 might be preferable to many users.

When the process of normalization is carried out to a reasonable extent, the structure shown in Table 8.4 results. Depending on the needs and preferences of the individual, it may be carried out to a greater or lesser extent than in Table 8.4. The tables are linked through the ID fields and other "key" fields such as Catalog Number. In most cases, a key field contains a unique number or code, but in the Classification Table the names in Genus or Child are themselves unique since homonyms are not allowed among names for animals. If the database treated plants and animals then a Taxon ID field would be necessary in the Classification Table to distinguish, for instance, *Ficus* the mollusk from *Ficus* the plant.

8.6 TESTING AND USING THE SYSTEM

Once the table and field structure has been implemented, it must be tested in order to catch problems early. Select lots representing a broad array of taxonomy, geography, sources, condition, and method of preservation and try entering data from them into the system. Fields can be left blank if not applicable or the data are unknown. Sometimes unexpected problems crop up. These are sometimes solved by reconfiguring the tables, and sometimes by clarifying or redefining the fields.

Testing will show, for example, that the structure of Table 8.4 seems to assume that all specimens are identified to the species level, because a Species ID number is needed in the Specimen Table. Items identified only to genus or subgenus level are handled easily by entering "sp." or "sp. A" in Species and leaving Author and Year blank in the Identification table. But what about a specimen identified only to family level or above? Species ID could be left blank in the Specimen Table, but then no identification could print in a report. Better is to allow names of higher taxa in the Genus field in the Identification Table and call it "Genus or higher taxon". It still links as planned to the Child field in the Classification Table, since that field contains names of all ranks. It could also link to the Genus field if the alternate structure (Table 8.6) were used, in which case that field also would be renamed "Genus or higher taxon" as in Table 8.4.

The next stage in testing is to make sure that you can get the output desired from the system. Try querying the database in various ways to make sure that it behaves as expected. Try creating reports with the desired format for labels and lists of various sorts. Proper formatting in reports might require the use of if-then statements to account for varying degrees of completeness of the data. For example printing "Location + ',' + Primary Division + ',' + Country" from Table 8.2 yields "Rottnest Island, Western Australia, Australia" (which is fine), "North Bimini,, Bahamas" (with two commas in a row since Primary Division is blank), and "Key West, Florida Keys, Florida, United States" (with United States perhaps viewed as unnecessary on a label by someone living in the United States). Instead try (depending on the database program): "Location + nothing if Primary Division is blank, else ',' + Primary Division + nothing if Country/Territory equals 'United States', else ',' + Country/Territory". It can take a lot of tinkering to get reports formatted satisfactorily, but once it is done, they run automatically.

After testing has been completed, and the table and field structures of the database are set, screens can be constructed that present the fields more conveniently for data entry than the default horizontal or vertical arrays provided by most programs. Next, add restrictions on the format, range of values, and the vocabulary for various fields. Even at this stage, restructuring the database is still possible, but reports and data-entry screens might also need to be reformatted correspondingly.

During development and after each data-entry session, the database should be backed up on media that can be removed from the computer, such as a Zip or compact disk or a flash drive. Printed catalog pages provide another form of backup, as do original labels, which should be retained so data can be confirmed, even if the new computer-generated labels are more uniform or esthetically pleasing.

Although a substantial effort is required to computerize a collection, the results are worthwhile. Information is easily shared among collectors, especially through electronic media such as email and websites. Summary lists of species can help the collector avoid duplicate purchases, especially in the heat of the bourse. If a collection is appraised, it is much less expensive to have the appraiser work from an existing list than compile one from scratch; also, a lot-by-lot appraisal will realize a higher value than a summary appraisal. An appraisal or computerized inventory, especially if accompanied by photographs or a videotape of the collection, could bolster an insurance claim in the case of a disaster. Copies of the database and images should be stored outside the home to guard against their loss. Ultimately the database will enhance the value of the collection, monetarily should it ever be sold, and scientifically by ensuring the integrity of the data.

8.7 ACKNOWLEDGMENTS

I thank Paul Callomon and Paul J. Morris for valuable comments on drafts of the manuscript, and Charlie Sturm for editorial support and continuing patience despite missed deadlines.

CHAPTER 9

THE MOLLUSCAN LITERATURE: GEOGRAPHIC AND TAXONOMIC WORKS

CHARLES F. STURM
RICHARD PETIT
TIMOTHY A. PEARCE
KEVIN CUMMINGS
ENRICO SCHWABE
ANDREAS WANNIGER

9.1 INTRODUCTION TO THE MALACOLOGICAL LITERATURE

Few non-professional malacologists know the vastness of the malacological literature. This literature includes works in systematics, taxonomy, biogeographic studies, ecology, aquaculture, anatomy, physiology, and paleontology, just to name a few. In this chapter we are focusing on the first three areas; systematics, taxonomy, and biogeographic studies.

When viewing some malacological libraries for the first time, many people ask if one reads all of the books. This is like asking if a mechanic uses all the tools in his or her garage. The answer is, of course, no. However, like the mechanic who needs a 20 mm socket for the first time, one never knows when he or she will need to refer to a work that he or she has never before had reason to consult. Thus, many of us acquire malacological literature that may encompass the broad themes as well as our specific interests. The term "malacological literature" encompasses a variety of types of work. In general terms, some of them are listed below.

9.1.1 Iconographies. The classical iconographies (Kiener, L. and P. Fischer, *Spécies Général et Iconographie des Coquilles Vivantes*: 1834-1879; Sowerby, G. B. *Thesaurus Conchyliorum*: 1842-1887; Reeve, L. and G. B. Sowerby, *Conchologia Iconica*: 1843-1878) are composed of monographs of various families or genera of mollusks and are finely illustrated with hand-colored plates. Now extremely expensive, they are still good investments. Old iconographies are important as they contain many descriptions of species and genera. It is difficult to do any type of work on molluscan taxonomy without referring to one or more of the old works. Some modern monographs are called iconographies but they are simply lavishly illustrated monographs of limited scientific value and are really identification books.

9.1.2 Monographs. There are two types of monographs, one being a study of the entire molluscan fauna of a restricted area and the other a study of all of the species in a given genus or family. In order to be a monograph as used herein, a work must give a full taxonomic treatment of included taxa. Examples of monographs are the serial publications *Johnsonia*, *Monographs of Marine Mollusca*, and the *Indo-Pacific Mollusca*. There are also non-serial publications that publish monographs. One such example is Bieler (1993).

9.1.3 Nomenclators. Nomenclators are lists of taxa enabling workers to find the original sources for taxa (species, genera). This category includes the *Zoological Record* (Anonymous 1865-present), Sherborn (1902, 1922-1932), Neave (1939-1940), and Ruhoff (1980).

9.1.4 Handbooks. This term is used in the absence of a better term for those important works that list molluscan taxa in systematic order. The best known of these are the *Treatise on Invertebrate Paleontology* (Moore and Teichert, 1953-1981), Wenz (1938-1944), Thiele (1929-1935), Zilch (1959-1960), Vaught (1989), and Millard (2003).

9.1.5 Classic books. This category is used for the many pre-1860 shell books that have hand-colored plates or are otherwise rare or exotic. Many books of this category are of great importance in molluscan systematics [Perry, G. (1811), *Conchology*, Martyn, T. (1784-1787), *The Universal Conchologist*, and Sowerby, G. B. (1839), *A Conchological Manual*]. Some books of the mid-19th Century are not particularly expensive but others are very costly. As with many collectibles, the ones that cost the most usually increase in value the fastest.

9.1.6 General books. This category could be broken down into numerous parts. Some books cover the shells of a given faunal area, for examples see the following works in the bibliography: Habe (1964a), Keen (1971), and Abbott (1974). Others cover one genus or family worldwide, for example Walls (1979a) for Conidae and Rombouts (1991) for Pectinidae. Sometimes referred to as identification books, their quality varies from excellent to abysmal.

9.1.7 Journals. Over 150 current journals and newsletters are devoted entirely to mollusks (Bieler and Kabat (1991). It is in these journals that you will find the descriptions of most shells named by professional malacologists. As few people (or institutions) subscribe to all of these, extra copies of papers in journals are printed and often distributed by the authors. These are properly termed "separates" although commonly referred to as reprints.

9.1.8 Separates. As stated above, these are papers that have appeared in various journals. They are very important as many malacological works are printed in journals other than those that are strictly on malacology. Publications such as the *Journal of the Academy of Natural Sciences of Philadelphia* and similar publications of other institutions such as the Alfred Wegener Institute for Polar and Marine Research (Germany) often contain important papers on mollusks. If you do not subscribe to their publications, there are two main ways of obtaining them: through correspondence with authors or the institution or by interlibrary loan. Some malacologists have files containing thousands of separates. In the course of a year, they will probably refer to a few hundred of them at the most, but they never know which ones will be needed! If you see a reference to a paper that you think will be important to you in your study of mollusks, write a letter to the author or the institution, and ask if a separate is available.

9.1.9 Miscellaneous printed works. There are dozens of categories that could be listed, such as theses, government reports, and published environmental surveys, but it is impractical to attempt to divide them up. Do not bother to categorize the literature, just acquire and use it!

9.1.10 Internet resources. The Internet can be a two edged sword for malacologists. There are problems associated with accuracy and permanence of sites.

While there are many sites, not all present accurate information. You may have to check with individuals who know more than you to find out which sites can be trusted. If using a site to identify a shell, try to confirm the identification with another site or a printed source.

Permanency of web sites falls into two different areas. The first is whether the address (URL) changes frequently or disappears. While sites for museums and larger organizations tend to be stable, sites hosted by individuals may change frequently. It may be difficult to track down the new address. Some individuals keep the old site running for a while and direct you to the new site. This is very helpful, especially if you refer to the site often. If a web site is no longer supported by its creator, it will be gone for good. Out of print publications can often be found in libraries or through used book dealers; old websites cannot.

The second permanence problem is with the information presented. In a printed publication, the information is stable until the next edition of the publication. You can also track changes through time by reviewing different editions of a work. The ability to track changes is rarely the case with web sites. When a web page or site is updated, the old page or site is usually replaced and there is no record of the old one. Only the new information is to be found.

These problems should not drive us away from using the Internet. The Internet is an evolving medium and we should take advantage of it despite any inherent weaknesses. For example, two websites maintained by the Academy of Natural Sciences of Philadelphia are very useful in selecting the proper name for a species from among synonyms, and as a reference to the original publication of the name and to other pertinent information. One site is for western Atlantic gastropods <erato.acnatsci.org/wasp/> and the other is for Indo-Pacific gastropods <data.acnatsci.org/obis/>. These sites, especially the western Atlantic site, are kept up-to-date and corrections are made when discovered. Yet a third good site is the Check List of European Marine Mollusca <www.somali.asso.fr/clemam/index.clemam.html>.

When we are asked what molluscan works a person should acquire, our reply is "everything affordable," with preference being given to works that apply directly to your area of interest. There are not many works on mollusks that cannot contribute a little to one's knowledge and you never know in advance when a particular work will be of importance. Many of the books recommended here are available only on the used book market. Sometimes an early edition can be purchased for much less that a later one. One such example is *American Seashells* by Abbott; the first edition was published in 1954, while the second edition was published in 1974. If you cannot locate (or afford) the latest editions, you will find the early editions to still be very useful and certainly better than not having either.

9.2 REGIONAL AND TAXONOMIC GUIDES

The rest of this chapter deals with an introduction to the malacological literature. It is a bibliography of more than 700 works. Only one citation is earlier than 1900. This preference for literature of the last 100 years is not meant to discount the importance of older literature. However, we had to limit what was included, and references to the important works of the 1700's and 1800's will be found in the works cited in this bibliography. While we do not list journals *per se*, many of the listed works are published in various malacological and general biological journals.

The works cited are divided into several categories. The first is "Mollusks - General References". These are works that deal with mollusks in general. These works cover the biology and ecology of mollusks. The next category is "Mollusks - Research Tools". These are works that will be of interest to those undertaking taxonomic and systematic research.

The next several categories deal with various marine and non-marine biogeographic zones. Figure 9.1 shows the marine biogeographic zones of the world, while Figure 9.2 shows the world's non-marine biogeographic zones. These zones are not hard and fast. For example, some would include Australia and the islands east of the Wallace Line in the Australian Province for non-marine mollusks. Others would divide these into two zones, the Australian and the Oceanic Zones. Some divide the Californian Province into more than one zone and others separate the waters of Antarctica from those of southern South America. By checking the maps you will be able to determine how we chose to define the various biogeographic zones.

The final categories deal with important works for certain higher taxonomic categories ranging from families to classes. We selected families that we thought would be of interest to the greatest number of readers. Input was also solicited from members of the on-line discussion group Conch-L.

Finally, not all recent works are included. A book or paper that you might think is very important or useful may not be on this list. With tens of thousands of possible works, we had to be selective. However, we are always willing to consider additional books for future inclusion. Please e-mail or send recommendations to the senior author at csturmjr@pitt.edu.

Other useful bibliographies can be found in:
- Taxonomic groups in Chapters 16 to 28 of this volume
- Worldwide marine - Abbott and Dance (1986: 379-390)
- Worldwide tropical land snails - Parkinson *et al.* (1987: 77-107, 265-268)
- Worldwide marine bivalves - Coan *et al.* (2000: 574-705)
- Worldwide land snails - Abbott (1989: 195-229)

9.2.1 Mollusks - general references and research tools.

General references: Barker 2001, Bauer and Wächtler 2000, Beesley *et al.* 1998, Claassen 1998, Cox 1979, Dance 1966, 1969, 1976, 1986, Dillon 2000, Evans 1972, Fretter and Graham 1994, Harper *et al.* 2000, Holme and McIntyre 1971, Hyman 1967, Moore and Teichert 1953-1981, Morton 1979, Ponder *et al.* 1988, Purchon 1977, Solem 1974, Taylor 1996, Thompson 1976, Vermeij 1993, Wade *et al.* 2001, Wilbur 1983-1988, Wilbur and Yonge 1964-1966, Yonge and Thompson 1976.

Research tools: Anonymous 1865-present, Arnold 1965, Bieler and Kabat 1991, Bieler and Petit 1990, Bouchet and Rocroi 2005, Brown 1956, Burch 1990, 1991, Callomon 1999a, 1999b, Callomon and Petit 2004, Coan *et al.* 2005, Edwards and Hopwood 1966, Edwards and Tobias 1993, Edwards and Vevers 1975, Edwards *et al.* 1996, Goto and Poppe 1996, International Commission on Zoological Nomenclature 1999, Jaeger 1955, Kabat and Boss 1992, Mayr and Ashlock 1991, Millard 2003, Moore and Teichert 1953-1981, Neave 1939-1940, 1950, Petit and Bieler 1996, Radwin and Coan 1976, Rosenberg and Petit 2003, Ruhoff 1980, Schuh 2000, Scudder 1882, Sherborn 1902, 1922-1932, Simpson 1990, Solís 2002, Thiele 1929-1935, Turgeon *et al.* 1998, Vaught 1989, Vokes 1980, 1990, Wenz 1938-1944, Winston 1999, Zilch 1959-1960.

9.2.2 Marine biogeographic zones.

Worldwide: Abbott and Dance 1986, Dance 1974, Eisenberg 1981, Kaicher 1973-1992, Lalli and Gilmer 1989, Rosenberg 1992.

Abyssal/bathyal: Bouchet and Marshall 2001, Bouchet and Warén 1979, 1980, 1985, 1986, 1993, Clarke 1961, 1962, Dijkstra 1995, Knudsen 1970, 1985, Sirenko 1988, Turner 1985, Warén and Bouchet 1993.

Aleutian: Akimushkin 1965, Baxter 1987, Bernard 1967, 1983, Berry 1917, 1919, Bogdanov and Sirenko 1993, Coan and Scott 1997, Coan *et al.* 2000, Griffith 1967, Higo *et al.* 1999, 2001, Keen and Coan 1974, La Rocque 1953, Nakayama 2003, Okutani 2000, Oldroyd 1924, 1924-1927, Quayle 1970, White 1976, Yakovleva 1965, Yoo 1976.

Antarctic: Dell 1960, 1964a, 1990, Hain 1990, Linse 2002, Narchi *et al.* 2002, Nicol 1966, Numanami 1996, Powell 1960, Troncoso *et al.* 2001.

Arctic: Abbott 1991b, Akimushkin 1965, Baxter 1987, Bogdanov and Sirenko 1993, Coan *et al.* 2000, Golikov 1994, Knudsen 1985, La Rocque 1953, Lubinsky 1980, MacGinitie 1955, MacGinitie 1959, MacPherson 1971, Richling 2000, Yakovleva 1965.

Australian: Abbottsmith 1969, Allan 1959, Bail and Limpus 1998, Beesley *et al.* 1998, Cotton 1959, 1961, 1964, Hinton 1977, Hodgkin *et al.* 1966, Iredale and McMichael 1962, Jansen 1995, Lamprell and Healy 1998, Lamprell and Whitehead 1992, Lorenz 2001, MacPherson and Gabriel 1962, May and MacPherson 1958, Norman and Reid 2000, Richmond 1992, 1997, Wells and Bryce 1984, 1986, 1993, Wilson 1993, 1994, Wilson and Gillett 1974.

Australian (New Zealand only): Dell 1952, Penniket and Moon 1970, Powell 1976, 1979.

Boreal: Abbott 1991b, Bouchet and Warén 1979, 1980, 1985, 1986, 1993, Bousfield 1960, Emerson and Jacobson 1976, Johnson 1934, La Rocque 1953, Poppe and Goto 1991-1993, Wagner 1984.

Californian: Bernard 1983, Berry 1917, 1919, Burghardt and Burghardt 1969, Coan and Scott 1997, Coan *et al.* 2000, Fitch 1953, Hedgpeth 1971, Keen and Coan 1974, MacFarland 1966, McDonald and Nybakken 1981, McLean 1978, McLean and

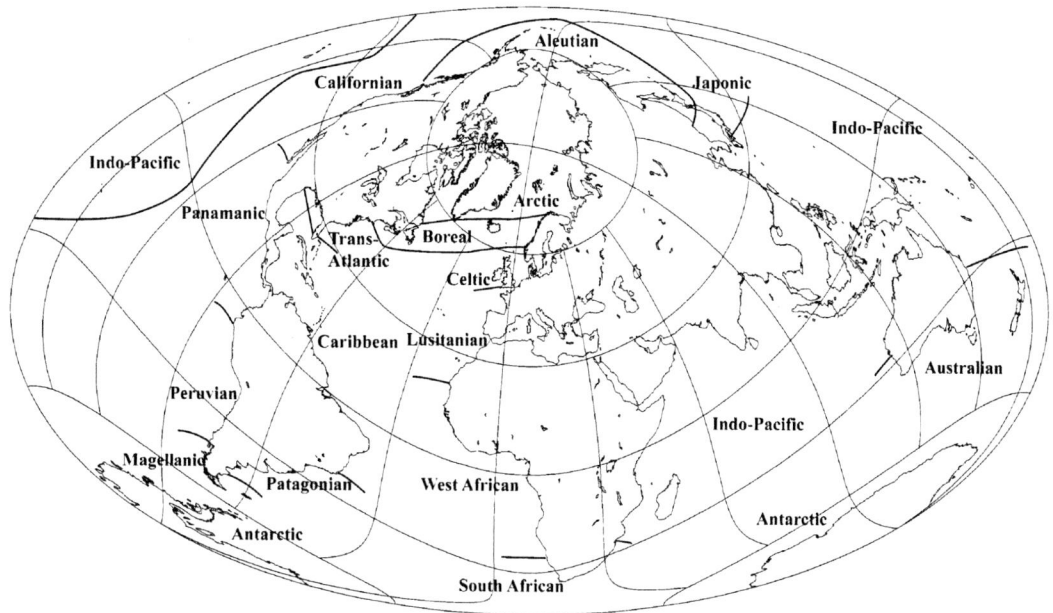

Figure 9.1 Marine biogeographic regions.

Gosliner 1996, Oldroyd 1924, 1924-1927, Packard 1918, Rice 1971, Scott and Blake 1998, Smith, A. G. and Gordon 1948, Young 1972.

Caribbean: Abbott 1958, Bayer and Voss 1971, Coomans 1989, Corsi 1900-1901, De Jong and Coomans 1988, Diaz and Puyana 1994, Emerson and Jacobson 1976, Fernández-Milera 1997, Ferreira 1984, García-Cubas 1981, Humfrey 1975, Johnson 1934, Kaas 1972, Lange de Morretes 1949, 1953, Lipe and Abbott 1991, Macsotay and Campos Villarroel 2001, Mikkelsen *et al.* 1993, Moscatelli 1987, Olsson and McGinty 1958, Redfern 2001, Rios 1994, Rosewater 1975, Vokes and Vokes 1984, Waller 1973, 1993, Warmke and Abbott 1961, Weisbord 1962, 1964, Work 1969.

Celtic: Abbott 1991b, Bouchet and Warén 1979, 1980, 1985, 1986, 1993, Graham 1971, Lellák 1975, McMillan 1975, Picton and Morrow 1994, Poppe and Goto 1991-1993, Tebble 1966, Thompson and Brown 1976, Wagner [1991].

Indo-Pacific: Abbott 1991a, Apte 1998, Bernard *et al.* 1993, Beu 1998, Bieler 1993, Bons 1984, Bosch and Bosch 1982, 1989, Bosch *et al.* 1995, Bouchet and Marshall 2001, Brost and Coale 1971, Cernohorsky 1971, 1972, 1978, Coulombel 1994, Dall *et al.* 1938, Debelius 1998, Dharma 1988, 1992, Dijkstra 1989, 1995, 1998, Drivas and Jay 1988, Garcia 2003, Green and Chouhfeh 1994, Habe 1964b, Habe and Kosuge 1974, Higo *et al.* 1999, 2001, Hinton 1972, [1978], Hylleberg and Kilburn 2003, Jarrett 2000, Kay 1979, Kay and Schoenberg-Dole 1991, Kira 1962, Kirtisinghe 1978, Knop 1996, Kuroda 1941, Lorenz 1998, Maes 1967, Mayissian 1974, Morris and Purchon 1981, Oliver 1992, Pechar *et al.* [1984], Pinn 1990, Purchon and Purchon 1981, Qi 2004, Rehder 1980, Robba *et al.* 2002, 2004, Salvat and Rives 1980, 1984, Salvat *et al.* 1988, Sharabati 1981, 1984, Springsteen and Leobrera 1986, Spry 1968, Subba Rao 2003, Swennen *et al.* 2001, Tan and Chou 2000, von Regteren Altena 1969, 1971, 1975, Voss 1963, Voss and Williamson 1971, Waller 1972, Way and Purchon 1981, Wells *et al.* 1990, Yoo 1976.

Indo-Pacific (Australia only): Hinton 1977, Jansen 1996, 2000, Lamprell and Healy 1998, Lamprell and Whitehead 1992, Lorenz 2001, Rippingale and

McMichael 1961, Short and Potter 1987, Wells and Bryce 1984, 1986, 1993, Wilson 1993, 1994, Wilson and Clarkson 2004, Wilson and Gillett 1974.

Japonic: Azuma 1960, Habe 1961, 1964b, 1977, Habe and Ito 1974, Higo *et al.* 1999, 2001, Hirase and Taki 1951, Kira 1955, 1962, Kuroda and Habe 1952, 1981, Kuroda *et al.* 1971, Kwon 1990, Kwon and Park 1993, Lai [1986], 1987, Nakayama 2003, Okutani 2000.

Lusitanian: Burnay and Monteiro 1977, Dell'Angelo and Smriglio 2001, Demir 2003, Grossu 1955, 1956, [1962], 1981-1987, Lellák 1975, Macedo *et al.* 1999, Nobre 1938-1940, Plá [2000], Poppe and Goto 1991-1993, Sabelli *et al.* 1990-1992, Wagner [1991].

Magellanic: Alamo Vásquez and Valdivieso Milla 1997, Bernard 1983, Carcelles 1953, Carcelles and Williamson 1951, Forcelli 2000, Linse 1997, 2002.

Panamic: Alamo Vásquez and Valdivieso Milla 1997, Bayer and Voss 1971, Bernard 1983, Berry 1917, 1919, Finet 1994a, 1994b, Hickman and Finet 1999, Keen 1971, Keen and Coan 1974, Olsson 1961, 1964, Skoglund 1990, 1991, 1992, Smith, M. 1944.

Patagonian: Barattini and Ureta 1960, Carcelles 1944, 1950, Castellanos 1967, Corsi 1900-1901, Leal 1990.

Pelagic (non-Cephalopoda): Lalli and Gilmer 1989.

Peruvian: Alamo Vásquez and Valdivieso Milla 1997, Basly Santa Maria 1982, Bernard 1983, Dall 1909, Marincovich 1973.

South African: Barnard [1951], 1958, 1959, 1963a, 1963b, 1964, 1969, 1974, Bartsch 1915, Dijkstra and Kilburn 2001, Kennelly 1964, Kensley 1974, Kilburn and Rippey 1982, Liltved 1989, Steyn and Lussi 1998.

Trans-Atlantic: Abbott 1991b, Andrews 1971, 1977, Emerson and Jacobson 1976, Jacobson and Emerson 1971, Johnson 1934, Long Island Shell Club 1988, Maury 1971, Moscatelli 1987, Perry and Schwengel 1955.

West African: Ardovini and Cossignani 2004, Bernard 1984, Burnay and Monteiro 1977, Edmunds 1978, Nicklès 1950, Pasteur-Humbert 1962a, 1962b, Rosewater 1975.

9.2.3 Terrestrial and freshwater biogeographic zones.

Australian - freshwater: Allan 1959, Baker 1924, Cotton and Gabriel 1932, Cowie 1998, Cowie *et al.* 1995, Dell 1953, Haynes 1988, 1993, 2001, Iredale 1934, 1943, Korniushin 2000, Kuiper 1966a, 1983, McMichael and Hiscock 1958, McMichael and Iredale 1959, Pointier and Marquet 1990, Ponder 1997, Powell 1979, Prashad 1928, Smith, B. J. 1992, Smith, B. J. and Kershaw 1979, 1981, Starmühlner 1976, 1993, Taylor 2003, van Benthem Jutting 1963a, Webb 1948, Winterbourn 1973.

Australian - terrestrial: Abbott 1989, Allan 1959, Baker 1924, 1938-1941, Cowie 1998, Cowie *et al.* 1995, Hemmen 2004, Hemmen and Groh 1991, Iredale 1937a, 1937b, 1938, Parkinson *et al.* 1987, Powell 1979, Schileyko 1998-2004, Smith, B. J. 1992, Smith, B. J. and Kershaw 1979, 1981, van Benthem Jutting 1963b, 1964, 1965, Webb 1948.

Ethiopian (Afrotropic) - freshwater: Appleton 1979, 1996, Barnard [1951], Brown 1967, 1994, Brown and Gallagher 1985, Brown and Mandahl-Barth 1987, Brown *et al.* 1992, Connolly 1912, 1925, 1939, Curtis 1991, Fischer-Piette and Vukadinovic 1973, Kuiper 1964, 1966b, Mandahl-Barth 1954, 1988, Pilsbry and Bequaert 1927, Taylor 2003, van Eeden 1960, Verdcourt 1972, Webb 1948, West *et al.* 2003.

Ethiopian (Afrotropic) - terrestrial: Abbott 1989, Barnard [1951], Bequaert 1950, Connolly 1912, 1925, 1939, Fischer-Piette *et al.* 1993, 1994, Gerlach 1987, Hemmen 2004, Hemmen and Groh 1991, Herbert and Kilburn 2004, Mead 1961, Parkinson *et al.* 1987, Pilsbry 1919, Schileyko 1998-2004, van Bruggen 1966, Verdcourt 1972, Warui *et al.* 2001, Webb 1948.

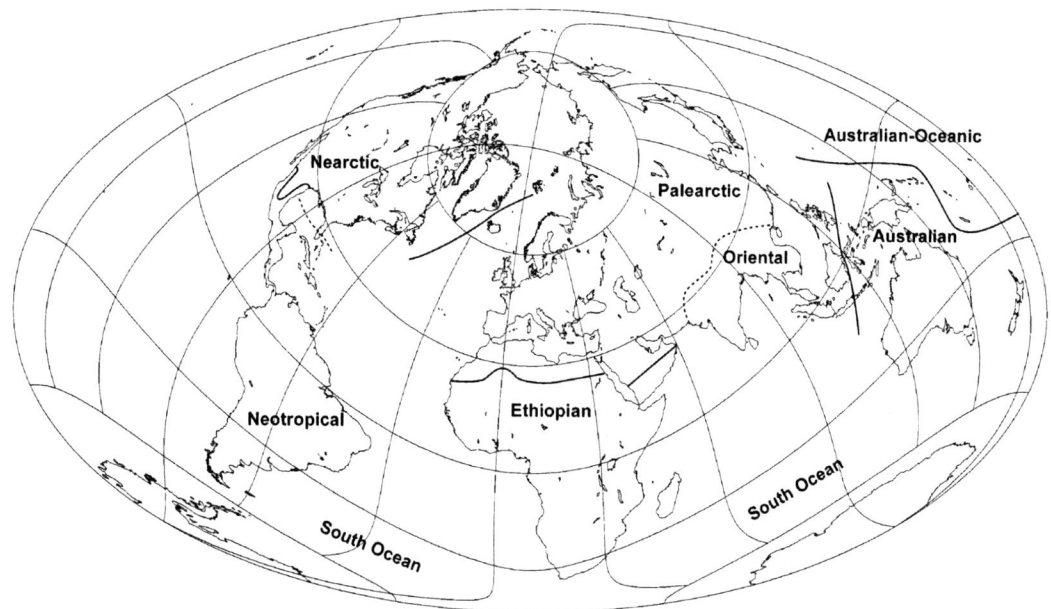

Figure 9.2 Non-marine biogeographic regions.

Nearctic - freshwater: Baker, F. C. 1911, 1928, Baxter 1987, Bleam *et al.* 1999, Bogan 2002, Burch 1975a, 1975b, 1982, 1988, 1989, Burch and Jung 1987, 1992, Burch and Tottenham 1980, Chamberlain and Jones 1929, Clarke 1973, 1981, Couch 1997, Cowie and Thiengo 2003, Cummings and Mayer 1992, Dall 1905, Emerson and Jacobson 1976, Fichtel and Smith 1995, Fuller and Brynildson 1985, Goodrich and van der Schalie 1944, Graf 2001, Hannibal 1912, Henderson 1924, 1929, 1935, Herrington 1962, Howells *et al.* 1996, Johnson 1972, 1999, Jokinen 1983, 1992, Korniushin 2001, La Rocque 1953, Leonard 1959, Martin 1997, 1998, Mathiak 1979, McKillop 1996, Murray and Leonard 1962, Nedeau [2003], Nedeau *et al.* 2000, Oesch 1984, Parmalee and Bogan 1998, Prashad 1928, Robertson and Blakeslee 1948, Strayer and Jirka 1997, Taylor 1970, 2003, Thompson 1984, 1999, Wu and Brandauer 1978, Wu 1989, Wu *et al.* 1997.

Nearctic - terrestrial: Abbott 1989, Baker, F. C. 1939, Baker, H. B. 1928-1930, Baxter 1987, Bequaert and Miller 1973, Burch 1962, Burch and Jung 1988, Burch and Pearce 1990, Chamberlain and Jones 1929, Chichester and Getz 1973, Cocke 1995, Cuezzo 2003, Dall 1905, de la Torre *et al.* 1942, Dirrigl and Bogan 1996, Emberton 1988, 1991, 1995, Emerson and Jacobson 1976, Forsyth 2004, Franzen and Leonard 1947, Goodrich and van der Schalie 1944, Grimm 1971, Hanna 1966, Henderson 1924, 1929, 1935, Hemmen 2004, Hemmen and Groh 1991, Hubricht 1985, La Rocque 1953, MacMillan 1950, Martin 2000, Metcalf and Smartt 1997, Oughton 1948, Pilsbry 1939-1948, Robertson and Blakeslee 1948, Roth and Sadeghian 2003, Schileyko 1998-2004, Smith, A. G. *et al.* 1990, Taft 1961, Walker 1928.

Neotropical - freshwater: Baker H. B. 1922-1923, 1924, Clench 1964, Corsi 1900-1901, Cowie and Thiengo 2003, Emerson and Jacobson 1976, Haas 1949, Harry 1962, Ituarte 1996, Johnson 1981, Lange de Morretes 1949, Mansur 1970, Mansur and Valer 1992, Ortmann 1921, Pain 1960, Parodiz 1968, Parodiz and Bonetto 1963, Pilsbry 1911, Prashad 1928, Quintana 1982, Taylor 2003, van der Schalie 1948.

Neotropical - terrestrial: Abbott 1989, Baker 1922-23, 1924, 1928-1930, Bieler and Slapcinsky 2000, Burch and Pearce 1990, Clench 1946, Close 1995, 2000, Corsi 1900-1901, Emerson and Jacobson 1976, Fernández 1973, Fernández-Milera 1997, 1999, Haas 1949, Hemmen 2004, Hemmen and Groh 1991, Jacobson and Boss 1973, Lange de Morretes 1949, Parkinson *et al.* 1987, Parodiz 1957, Pilsbry 1911, 1912, Quintana 1982, Schileyko 1998-2004, Tillier 1981, van der Schalie 1948, Webb 1948.

Oceanic - freshwater: Clench 1958, Cowie 1998, Cowie *et al.* 1995, Franc 1956, Haynes 1988, 1993, 2001, Kay and Schoenberg-Dole 1991, Pointier and Marquet 1990, Prashad 1928, Solem 1959, 1961, Starmühlner 1976, 1993, Taylor 2003, Webb 1948.

Oceanic - terrestrial: Abbott 1989, Baker 1938-1941, Clench 1958, Cowie 1998, Cowie *et al.* 1995, Franc 1956, Hemmen 2004, Hemmen and Groh 1991, Kay and Schoenberg-Dole 1991, Parkinson *et al.* 1987, Riedel 1980, Solem 1959, 1961, 1976-1982, Webb 1948.

Oriental (Indo-Malaysian) - freshwater: Annandale 1920, Brandt 1974, Burch 1980, 1984, Habe 1964a, Horikawa 1935a, 1935b, Keawjam 1986, Khatoon and Ali 1978, Kuroda 1941, Pace 1973, 1975, Prashad 1928, Subba Rao 1989, Taylor 2003.

Oriental (Indo-Malaysian) - terrestrial: Abbott 1989, Bartsch 1917, Brandt 1974, Dance 1970, Hemmen 2004, Hemmen and Groh 1991, Hemmen and Hemmen 2001, Kuroda 1941, Parkinson *et al.* 1987, Ramakrishna and Mitra 2002, Schileyko 1998-2004, Solem 1966a, Springsteen and Leobrera 1986, Thompson and Dance 1983, Vermeulen and Whitten 1998, Webb 1948.

Palaearctic - freshwater: Adam 1960, Anderson 2005, Backhuys 1975, Bank *et al.* 2001, Beran and Horsák 1998, Burch *et al.* 1987, Dhora and Welter-Schultes 1996, Ellis 1978, Falkner *et al.* 2001, 2002, Graham 1971, Grossu 1955, 1956, [1962], 1981-1987, Haas 1917, Janssen and De Vogel 1965, Kira 1955, Korniushin 1994, 2001, Kruglov and Starobogatov 1993a, 1993b, Kuiper *et al.* 1989, Kwon 1990, Lozek 1964, Mandahl-Barth 1938, Mouthon and Kuiper 1987, Piechocki 1989, Prashad 1928, Prozorova 1995, Stadnichenko 1984, Taylor 2003, Tsaldiikhin 2004, van Damme 1984, van Loen and Van Goethem 1997, Zhadin 1952.

Palaearctic - terrestrial: Abbott 1989, Adam 1960, Anderson 2005, Backhuys 1975, Bank *et al.* 2001, Cameron 2003, Cameron *et al.* 1983, Cameron and Redfern 1976, Cossignani and Cossignani 1995, Dhora and Welter-Schultes 1996, Falkner *et al.* 2001, 2002, Graham 1971, Grossu 1955, 1981-1987, Heller 1993, Hemmen 2004, Hemmen and Groh 1991, Kerney 1999, Kerney and Cameron 1979, Kerney *et al.* 1983, Kira 1955, Kwon 1990, Likharev and Rammel'meier 1962, Lozek 1964, Mandahl-Barth 1938, Schileyko 1998-2004, Schütt 2001, Solem 1979, Webb 1948.

Southern Ocean Islands - freshwater and terrestrial: Dell 1964b, Pugh and Scott 2002.

9.2.4 Marine molluscan groups.

Information is included under specific families and several other taxonomic groups such as Aplacophora, Cephalopoda, Monoplacpophora, Opistobranchia, Ostreoidea, Polyplacophora, Scaphopoda, and Trochoidea.

Angariidae: Poppe and Goto 1993.

Aplacophora: Barnard 1963b, Hyman 1967.

Architectonicidae: Bieler 1993.

Arcidae: Rost 1955.

Buccinidae: Cernohorsky 1981, Egorov and Barsukov 1994.

Bursidae: Beu 1998, Cossignani 1994, Smith, M. 1948.

Calyptraeidae: Hoagland 1977.

Cardiidae: Fischer-Piette 1977, Hylleberg 2004.

Cassidae: Abbott 1968, Kreipl 1997, Smith, M. 1948.

Cephalopoda: Adam and Rees 1966, Akimushkin 1965, Berry 1917, 1919, Dell 1952, Kuroda and Habe 1981, Nesis 1987, Norman 2000, Norman and Reid 2000, Robson 1929-1932, Roper *et al.* 1969, 1984, Sweeney *et al.* 1992, Voss 1963, Voss and Williamson 1971, Voss *et al.* 1998, Young 1972.

Cerithiidae: Houbrick 1978, 1992.

Conidae: Filmer 2001, Kohn 1992, Marsh and Rippingale 1974, Monteiro *et al.* 2004, Röckel *et al.* 1995, Tomlin 1937, Walls 1979a.

Coralliophilidae: Clover 1982, Kosuge and Suzuki 1985.

Costellariidae: Robin and Martin 2004, Turner 2001.

Cypraeidae: Allan 1956, Bradner and Kay 1996, Burgess 1970, 1985, Liltved 1989, Lorenz 1998, 2001, 2002, Lorenz and Hubert 2000, Schilder and Schilder 1938, 1971, Taylor and Walls 1975, Walls 1979b, Wilson and Clarkson 2004, Wilson and McComb 1967.

Epitoniidae: Bouchet and Warén 1986, Garcia 2003, Nakayama 2003, Weil *et al.* 1999.

Eratoidae: Cate 1977, Keen 1977.

Fasciolariidae: Snyder 2003.

Ficidae: Poppe and Verhaeghe 2000.

Haliotidae: Geiger 1998, Geiger and Poppe 2000.

Harpidae: Poppe *et al.* 1999, Rehder 1973, Smith, M. 1948, Walls 1980.

Littorinidae: Reid 1996, Rosewater 1970, 1972.

Marginellidae: Clover 1968, Coovert 1988, Coovert and Coovert 1995, Lipe 1991.

Mitridae: Cernohorsky 1976, 1991, Pechar *et al.* [1984], Robin and Martin 2004.

Monoplacophora: Hyman 1967, Menzies *et al.* 1959, Warén and Gofas 1996.

Muricidae: Egorov 1993, Emerson and Cernohorsky 1973, Fair 1976, Houart 1992, 1994, Ponder and Vokes 1988, Radwin and D'Attilio 1976, Smith, M. 1939, Vokes 1971.

Nassariidae: Cernohorsky 1984.

Olividae: Kilburn 1981, Petuch and Sargent 1986, Streba 2004, Tursch and Greifeneder 2001, Zeigler and Porreca 1969.

Opisthobranchia: Debelius 1998, Farmer 1980, MacFarland 1966, McDonald and Nybakken 1981, Picton and Morrow 1994, Russell 1971, 1986, Skoglund 1991, Thompson 1976, Thompson and Brown 1976, Wells and Bryce 1993.

Ostreoidea: Harry 1985, Moore and Teichert 1953-1981 [*Part N, Vol. 3:*1971], Shirai 1994.

Ovulidae: Cate 1973.

Patellidae: Powell 1973.

Pectinidae: Cox 1957, Dijkstra 1989, 1994, 1995, 1998, Dijkstra and Kilburn 2001, Grau 1959, Jonkers 2003, Rombouts 1991, Wagner [1991], Waller 1972, 1993.

Personidae: Beu 1998, Henning and Hemmen 1993, Piech 1995.

Pinnidae: Rosewater 1961.

Pleurotomariidae: Anseeuw and Goto 1996.

Polyplacophora: Barnard 1963b, Burghardt and Burghardt 1969, Cotton 1964, Dell'Angelo and Smriglio 2001, Ferreira 1984, Hyman 1967, Kaas 1972, Kaas and Van Belle 1985a, 1985b, 1987, 1990a, 1990b, 1998, MacPherson 1971, Sirenko 1988, Slieker 2000, Way and Purchon 1981, Yakovleva 1965.

Ranellidae: Beu 1998, Henning and Hemmen 1993, Piech 1995.

Scaphopoda: Barnard 1963b, Bernard 1967, Habe 1977, Kuroda and Habe 1981, MacPherson 1971, Reynolds 1997, Steiner and Kabat 2001, 2004.

Spondylidae: Lamprell 1986.

Strombidae: Abbott 1960, 1961, Jung and Abbott 1967, Kreipl *et al.* 1999, Moscatelli 1987, Walls 1980.

Terebridae: Aubry 1999, Bratcher and Cernohorsky 1987.

Teredinidae: Turner 1966.

Tridacnidae: Knop 1996, Rosewater 1965.

Triviidae: Cate 1979.

Trochoidea: Hickman and McLean 1990, Robertson 1985.

Turbinellidae: Harasewych 1983.

Turbinidae: Alf *et al.* 2003.

Turridae: Bouchet and Warén 1980, Powell 1964, 1967, 1969, Taylor *et al.* 1993, Tucker 2004.

Typhidae: D'Attilio and Hertz 1988.

Vasidae: Abbott 1959.

Volutidae: Abbottsmith 1969, Bail and Limpus 1998, Bail and Poppe 2001, 2004, Bail *et al.* 2001, Poppe and Goto 1992, Smith, M. 1942, Weaver and DuPont 1970.

Xenoporidae: Kreipl and Alf 1999, Ponder 1983.

9.2.5 Terestrial gastropods.

General terrestrial: Baker, H. B. 1922-1923, 1928-1930, Bank *et al.* 2001, Burch and Pearce 1990, Clench 1958, Dirrigl and Bogan 1996, Falkner *et al.* 2001, Fretter and Graham 1994, Grimm 1971, Hanna 1966, Herbert and Kilburn 2004, Kerney and Cameron 1979, Lozek 1964, Metcalf and Smartt 1997, Oughton 1948, Pilsbry 1939-1948, Schütt 2001, Smith, B. J. and Kershaw 1981, Smith, A. G. *et al.* 1990, Solem and van Bruggen 1984, Tillier 1989, van Bruggen 1966, Warui *et al.* 2001, Zilch 1959-1960.

Acavidae: Emberton 1990, 1999, Groh *et al.* 2002, Hausdorf and Perera 2000.

Achatinidae: Bequaert 1950, Mead 1961, Schileyko 1998-2004 [part 4].

Agriolimacidae: Chichester and Getz 1973, Runham and Hunter 1970, Schileyko 1998-2004 [part 11], Wiktor 2000.

Arionidae: Chichester and Getz 1973, Runham and Hunter 1970.

Bradybaenidae: Richardson 1983, Schileyko 1998-2004 [part 12].

Camaenidae: Bartsch 1917, Cuezzo 2003, Laidlaw and Solem 1961, Richardson 1985, Schileyko 1998-2004 [part 11], Solem 1966b, 1979-1997, van Benthem Jutting 1965, Wurtz 1955.

Cerionidae: Richardson 1992, Schileyko 1998-2004 [part 4].

Charopidae: Pugh and Scott 2002, Schileyko 1998-2004 [part 7].

Clausiliidae: Nordsieck 1998, Schileyko 1998-2004 [part 5], Szekeres 1984, Thompson and Dance 1983, Welter-Schultes 1998.

Cyclophoridae: de la Torre *et al.* 1942.

Endodontidae: Solem 1976-1982.

Gastrocoptidae: Franzen and Leonard 1947, Schileyko 1998-2004 [part 2].

Haplotrematidae: Richardson 1989, Schileyko 1998-2004 [part 6].

Helicidae: Kerney and Cameron 1979, Likharev and Rammel'meier 1962, Richardson 1980, Steinke *et al.* 2004.

Helicinidae: Richling 2004, Thompson and Dance 1983.

Helminthoglyptidae: Richardson 1982, Schileyko 1998-2004 [part 12].

Hygromiidae: Hausdorf 1988.

Limacidae: Chichester and Getz 1973, Runham and Hunter 1970, Schileyko 1998-2004 [part 11].

Milacidae: Chichester and Getz 1973, Runham and Hunter 1970, Schileyko 1998-2004 [part 10].

Oreohelicidae: Richardson 1984.

Orthalicidae: Clench 1946, Close 1995, 2000, Pilsbry 1912, Schileyko 1998-2004 [part 3].

Partulidae: Crampton 1916, 1925, 1932, Richardson 1990, Schileyko 1998-2004 [part 3], Smith, H. H. 1902.

Philomycidae: Chichester and Getz 1973, Fairbanks 1998, Runham and Hunter 1970.

Polygyridae: Emberton 1988, 1991, 1995, Richardson 1986.

Pupillidae: Franzen and Leonard 1947, Schileyko 1998-2004 [part 1], Thompson and Dance 1983.

Rhytididae: Richardson 1989, Schileyko 1998-2004 [part 6].

Sagdidae: Richardson 1986, Schileyko 1998-2004 [part 2].

Spiraxidae: Schileyko 1998-2004 [part 6].

Streptaxidae: Dance 1970, Richardson 1988, Schileyko 1998-2004 [part 6], van Bruggen 1967.

Succineidae: Rundell *et al.* 2004, Patterson 1971, Tillier 1981.

Truncatellinidae: Franzen and Leonard 1947, Schileyko 1998-2004 [part 2], Thompson and Correa S 1991, 1994.

Urocoptidae: Richardson 1991, Schileyko 1998-2004 [part 3], Thompson and Correa S 1991, 1994.

Valloniidae: Gerber 1996, Schileyko 1998-2004 [part 1].

Vertiginidae: Franzen and Leonard 1947, Schileyko 1998-2004 [part 2].

Zonitidae: Baker 1938-1941, Riedel 1980.

9.2.6 Freshwater mollusks.

Ampullariidae: Alderson 1925, Annandale 1920, Burch 1989, Cowie and Thiengo 2003, Keawjam 1986, Pain 1960, 1972, Perera and Walls 1996, Turner and Mikkelsen 2004.

Hydrobiidae: Hershler and Thompson 1992, Kabat and Hershler 1993, McKillop 1996, Perez *et al.* 2005.

Lymnaeidae: Baker 1911, Burch 1989, Hubendick 1951, Kruglov and Starobogatov 1993a, 1993b, McKillop 1996, Wu 1989.

Physidae: Burch 1989, McKillop 1996, Taylor 2003, Wu 1989.

Planorbidae: Burch 1989, Baker 1946, Baker and van Cleave 1945, Harry 1962, McKillop 1996, Wu 1989.

Pleuroceridae: Burch 1989, Graf 2001, Minton and Lydeard 2003.

Sphaeriidae: Bank *et al.* 2001, Burch 1975a, Falkner *et al.* 2001, Herrington 1962, Ituarte 1996, Korniushin 1994, 2000, 2001, Kuiper 1964, 1966a, 1966b, 1983, Kuiper *et al.* 1989, Mandahl-Barth 1988, Mansur and Meier-Brook 2000, Martin 1998, Mouthon and Kuiper 1987, Piechocki 1989, Prozorova 1995, Wu 1989, Wu and Brandauer 1978.

Unionoida: Appleton 1979, Baker 1922-1923, Bank *et al.* 2001, Bauer and Wächtler 2000, Bleam *et al.* 1999, Bogan 2002, Brim Box and Williams 2000, Burch 1975b, Cotton and Gabriel 1932, Couch 1997,

Cummings and Mayer 1992, Dell 1953, Ellis 1978, Falkner *et al.* 2001, Fichtel and Smith 1995, Fuller and Brynildson 1985, Haas 1917, Habe and Ito 1974, Howells *et al.* 1996, Iredale 1934, 1943, Johnson 1972, 1981, 1999, Mandahl-Barth 1988, Mansur 1970, Mathiak 1979, Martin 1997, McMichael and Hiscock 1958, Murray and Leonard 1962, Nedeau [2003], Nedeau *et al.* 2000, Oesch 1984, Ortmann 1921, Parmalee and Bogan 1998, Parodiz 1968, Parodiz and Bonetto 1963, Shirai 1994, Stadnichenko 1984, Strayer and Jirka 1997, van Damme 1984, Wu 1989, Wu and Brandauer 1978.

Viviparidae: Prashad 1928.

9.3 MALACOLOGICAL BIBLIOGRAPHY

Abbott, R. T. 1958. The Marine mollusks of Grand Cayman Island, British West Indies. *Monographs of the Academy of Natural Sciences of Philadelphia* **11**: 1-138, pls. 1-5.

Abbott, R. T. 1959. The family Vasidae in the Indo-Pacific. *Indo-Pacific Mollusca* **1**: 15-32, pls. 1-10.

Abbott, R. T. 1960. The genus *Strombus* in the Indo-Pacific. *Indo-Pacific Mollusca* **1**: 33-146, pls. 11-117.

Abbott, R. T. 1961. The genus *Lambis* in the Indo-Pacific. *Indo-Pacific Mollusca* **1**: 147-174, pls. 118-134.

Abbott, R. T. 1968. The helmet shells of the World (Cassidae). Part 1. *Indo-Pacific Mollusca* **2**: 15-201, pls. 1-187.

Abbott, R. T. 1974. *American Seashells: The Marine Mollusks of the Atlantic and Pacific Coasts of North America*, 2nd Ed. Van Nostrand Reinhold Company, New York. 663 pp., 24 pls.

Abbott, R. T. 1989. *Compendium of Landshells*. American Malacologist, Inc, Melbourne, Florida. vii + 240 pp. [2,000 species of terrestrial shells in color; see Hemmen and Groh 1991 and Hemmen 2004 for corrections].

Abbott, R. T. 1991a. *Seashells of South East Asia*. Tynron Press, Dumfriesshire, Scotland. 145 pp., 52 pls.

Abbott, R. T. 1991b. *Seashells of the Northern Hemisphere*. Gallery Books, W. H. Smith Publishers, Inc., New York. 191 pp., text illustrations.

Abbott, R. T. and S. P. Dance. 1986. *Compendium of Seashells*, 3rd Printing, Revised. American Malacologists, Inc, Melbourne, Florida. ix + 411 pp.

Abbottsmith, F. 1969. *Multiform Australian Volutes*. Privately published, Manly, New South Wales, Australia. 109 pp.

Adam, W. 1960. *Faune de Belgique, Mollusques*, Tome I, Mollusques Terrestres et Dulcicoles. Institute Royal des Sciences Naturelles de Belgique, Bruxelles. 402 pp., 4 pls., 163 text illustrations.

Adam, W. and W. J. Rees. 1966. A Review of the Cephalopod Family Sepiidae. *The John Murray Expedition 1933-34 Scientific Reports* **11**(1): i-iv + 1-165, pls. 1- 46.

Akimushkin, I. I. 1965. *Cephalopods of the Seas of the USSR*. Israel Program for Scientific Translation, Jerusalem. viii + 223 pp., 60 figs. [Originally published in Russian in 1963].

Alamo Vásquez, V. and V. Valdivieso Milla. 1997. *Lista Sistemática de Moluscos Marinos del Perú*, Segunda Edición, Revisada y Actualizada. Instituto del Mar del Perú, Callao, Perú. xv + 183 pp., 2 pls., 406 figs.

Alderson, E. G. 1925. *Studies in* Ampullaria. W. Heffer and Sons, Cambridge, England. xx + 102 pp., 19 pls.

Alf, A., K. Kreipl, and G. T. Poppe. 2003. *A Conchological Iconography: The Family Turbinidae, Subfamily Turbininae, Genus* Turbo. ConchBooks, Hackenheim, Germany. 68 pp., 95 pls.

Allan, J. 1956. *Cowry Shells of World Seas*. Georgian House Pty. Ltd., Middle Park, Victoria, Australia. x + 170 pp., 15 pls.

Allan, J. 1959. *Australian Shells, with Related Animals Living in the Sea, in Freshwater and on the Land*, Revised Ed. Georgian House Pty., Ltd., Melbourne, Australia. 487 pp., 40 pls., 110 figs.

Anderson, R. 2005. An annotated list of the non-marine mollusca of Britain and Ireland. *Journal of Conchology* **38**: 607-638.

Andrews, J. 1971. *Sea Shells of the Texas Coast*. University of Texas Press, Austin, Texas. xvii + 298 pp. [300+ taxa illustrated].

Andrews, J. 1977. *Shells and Shores of Texas*. University of Texas Press, Austin, Texas. xx + 365 pp. [Expanded version of Sea Shells of the Texas Coast, 360+ taxa illustrated].

Annandale N. 1920. The apple-snails of Siam. *Journal of the Natural History Society of Siam* **4**: 1-24, pls. 1-2.

Anonymous. 1865 - present. *Zoological Record*. Biosis, Philadelphia, Pennsylvania. [Issued annually, Section 9 covers mollusks; series previously published by the Zoological Society of London in cooperation with the British Museum (Natural History)].

Anseeuw, P. and Y. Goto. 1996. *The Living Pleurotomariidae. A Synopsis of the Recent Pleurotomariidae Including Colour Plates of All Extant Type Specimens*. Elle Scientific Publications, Osaka, Japan. 202 pp. [All taxa illustrated].

Appleton, C. C. 1979. The Unionacea (Mollusca: Lamellibrachiata) of south-central Africa. *Annals of the South African Museum* **77**: 151-174, 8 figs.

Appleton, C. C. 1996. *Freshwater Molluscs of Southern Africa*. University of Natal Press, Pietermaritzburg, South Africa. 64 pp., 75 figs.

Apte, D. 1998. *The Book of Indian Shells*. Bombay Natural History Society, Oxford University Press, Oxford. ix + 115 pp., 13 pls.

Ardovini, R. and T. Cossignani. 2004. *West African Seashells*. Mostra Mondiale Malacologia, Cupra

Marittima, Italy. 320 pp. [Checklist of 2300 species, 1210 species illustrated].

Arnold, W. H. (compiler). 1965. A glossary of a thousand-and-one terms used in conchology. *Veliger* **7** (Supplement): iii + 50 pp., 155 figs.

Aubry, U. 1999. *Nuove Terebre e Antichi Versi. New Terebras and Ancient Verses*. L'Informatore Piceno Ed. Ancona, Italy. 47 pp.

Azuma, M. 1960. *A Catalogue of the Shell-Bearing Mollusca of Okinoshima, Kashiwajima and the Adjacent Area (Tosa Province) Shikoku, Japan*. Tosa Bay Malacological Literature Publication Committee, Japan. 1, 7, 102, 17 pp., 5 pls.

Backhuys, W. 1975. *Zoogeography and Taxonomy of the Land and Freshwater Molluscs of the Azores*. Backhuys and Meesters, Amsterdam. xii + 350 pp., 16 pls.

Bail, P. and A. Limpus. 1998. *Revision of* Cymbiola *(Cymbiolacca) from East Australian Coast: The "pulchra complex"*. Evolver, Rome. 79 pp., 266 figs.

Bail, P., A. Limpus, and G. T. Poppe. 2001. *A Conchological Iconography: The genus* Amoria. ConchBooks, Hackenheim, Germany. 50 pp., 93 pls.

Bail, P. and G. T. Poppe. 2001. *A Conchological Iconography: A taxonomic introduction to the Recent Volutidae*. ConchBooks, Hackenheim, Germany. 30 pp., 5 pls.

Bail, P. and G. T. Poppe. 2004. The tribe Lyriini. *In:* G. T. Poppe and K. Groh, eds., *A Conchological Iconography*. ConchBooks, Hackenheim, Germany. Pp. 2-93, pls. 1-68.

Baker, F. C. 1911. The Lymnaeidae of North and Middle America; Recent and fossil. *The Chicago Academy of Sciences Special Publication* **3**: i-xvi + 1-539, pls. 1-57.

Baker, F. C. 1928. The Freshwater Mollusca of Wisconsin. *Wisconsin Geological and Natural History Survey Bulletin* **70** [Vol. 1: Gastropoda. xx + 507 pp., 28 pls.; Vol. 2: Pelecypoda. vi + 495 pp., 77 pls.]

Baker, F. C. 1939. *Fieldbook of Illinois Land Snails,* Manual 2. Illinois Natural History Survey, Urbana, Illinois. 166 pp., text figures.

Baker, F. C. and H. J. van Cleave (collation, revisions, and additions). 1945. *The Molluscan Family Planorbidae*. The University of Illinois Press, Urbana, Illinois. xxxvi + 530 pp., 141 pls.

Baker, H. B. 1922-1923. The Mollusca collected by the University of Michigan-Walker Expedition in Southern Vera Cruz, Mexico I-IV. *Occasional Papers of the Museum of Zoology, University of Michigan* **106**: 1-61, 17 pls. [Parts I-III, 1922]; **135**: 1-18, 1 pl. [Part IV, 1923].

Baker, H. B. 1924. Land and freshwater molluscs of the Dutch Leeward Islands. *Occasional Papers of the Museum of Zoology, University of Michigan* **152**: 1-159, pls. 1-21.

Baker, H. B. 1928-1930. Mexican mollusks collected for Dr. Bryant Walker in 1926. I-II. *Occasional Papers Museum of Zoology, University of Michigan* **193**: 1-54, pls. 1-6 [Part I, 1928]; **220**: 1-45, pls. 7-11 [Part II, 1930].

Baker, H. B. 1938-1941. Zonitid snails from Pacific Islands. Parts 1-4. *Bernice P. Bishop Museum Bulletins* **158** (Part 1): 3-102, pls. 1-20; **165** (Part 2): 105-201, pls. 21-42; **166** (Parts 3/4): 205-370, pls. 43-65.

Baker, H. B. 1946. Index to F. C. Baker's "The Molluscan Family Planorbidae." *Nautilus* **59**: 127-141.

Bank, R. A., P. Bouchet, G. Falkner, E. Gittenberger, B. Hausdorf, T. Von Proschwitz, and T. E. J. Ripken. 2001. Supraspecific classification of European non-marine Mollusca (CLECOM Sections I+II). *Heldia* **4**: 77-128.

Barattini, L. P. and E. H. Ureta. 1960. *La Fauna de las Costas Uruguayas del Este (Invertebrados)*. Museo Damaso Antonio Larrañaga, Montevideo, Uruguay. 204 pp. [Mollusks treated in pp. 76-195, pls. 17-52].

Barker, G. M. 2001. *The Biology of Terrestrial Molluscs*. CABI Publishing, New York. xiv + 558 pp.

Barnard, K. H. [1951]. *A Beginner's Guide to South African Shells*. Maskew Miller Limited, Cape Town, South Africa. ii + 215 pp., 32 pls.

Barnard, K. H. 1958. Contributions to the knowledge of South African marine Mollusca. Part I. Gastropoda: Prosobranchiata: Toxoglossa. *Annals of the South African Museum* **44**: 77-163, pls. 1-2, 30 figs.

Barnard, K. H. 1959. Contributions to the knowledge of South African marine Mollusca. Part II. Gastropoda: Prosobranchiata: Rhachiglossa. *Annals of the South African Museum* **45**: 1-237, 52 figs.

Barnard, K. H. 1963a. Contributions to the knowledge of South African marine Mollusca. Part III. Gastropoda: Prosobranchiata: Taenioglossa. *Annals of the South African Museum* **47**: 1-199, 37 figs.

Barnard, K. H. 1963b. Contributions to the knowledge of South African marine Mollusca. Part IV. Gastropoda: Prosobranchiata, Rhipidoglossa, Docoglossa, Tectibranchiata. Polyplacophora. Solenogastres. Scaphopoda. *Annals of the South African Museum* **47**: 201-360, 30 figs.

Barnard, K. H. 1964. Contributions to the knowledge of South African marine Mollusca. Part V. Lamellibranchiata. *Annals of the South African Museum* **47**: 361-593, 39 figs.

Barnard, K. H. 1969. Contributions to the knowledge of South African marine Mollusca. Part VI. Supplement. *Annals of the South African Museum* **47**: 595-661, pls. 1-2, 30 figs.

Barnard, K. H. 1974. Contributions to the knowledge of South African marine Mollusca. Part VII. Revised faunal list. *Annals of the South African Museum* **47**: 663-781.

Bartsch, P. 1915. Report on the Turton collection of South African marine mollusks, with additional notes on other South African shells contained in the United States Museum. *United States National Museum Bulletin* **91**: i-xii + 1-305, pls. 1-54.

Bartsch, P. 1917. The Philippine land shells of the genus *Amphidromus*. *United States National Museum Bulletin* **100**: Vol. 1, Part 1: 1-47, pls. 1-21.

Basly Santa Maria, J. 1982. *Moluscos Marinos del Norte de Chile. Catalogo Ilustrativo*. W. Morales A, Vina del Mar, Chile. 49 pp., 12 pls.

Bauer, G. and K. Wächtler, eds. 2000. Ecology and Evolution of the Freshwater Mussels Unionoida. *Ecological Studies Analysis and Synthesis* **145**: i-xxii + 1-394.

Baxter, R. 1987. *Mollusks of Alaska*. Shells and Sea Life, Bayside, California. 163 pp.

Bayer, F. M. and G. L. Voss. 1971. *Studies in Tropical American Mollusks*. University of Miami Press, Coral Gables, Florida. 236 pp., text figs.

Beesley, P. L., G. J. B. Ross, and A. Wells, eds. 1998. *Mollusca: The Southern Synthesis. Fauna of Australia*, Vol. 5. CSIRO Publishing, Melbourne, Australia. Part A: xvi + 1-563 pp., pls. 1-21; Part B: viii + 565-1234 pp., pls. 22-37.

Bequaert, J. C. 1950. Studies in the Achatininae, a group of African land snails. *Bulletin of the Museum of Comparative Zoology, Harvard University* **105**: 4-216, pls. 1-81.

Bequaert, J. C. and W. B. Miller. 1973. *The Mollusks of the Arid Southwest, with an Arizona Check List*. University of Arizona Press, Tucson, Arizona. 271 pp.

Beran, L. and M. Horsák. 1998. Aquatic molluscs (Gastropoda, Bivalvia) of the Dolnomoravský úval lowland, Czech Republic. *Acta Societatis Zoologicae Bohemicae* **62**: 7-23.

Bernard, F. R. 1967. Prodrome for a distributional check-list and bibliography of the Recent Mollusca of the west coast of Canada. *Fisheries Research Board of Canada Technical Report* **2**: i- xxiv + 1-261. [Checklist].

Bernard, F. R. 1983. Catalogue of the living Bivalvia of the eastern Pacific Ocean: Bering Strait to Cape Horn. *Canadian Special Publication of Fisheries and Aquatic Sciences* **61**: i- vii + 1-102.

Bernard, F. R., Ying-Ya Cai, and B. Morton. 1993. *Catalogue of the Living Marine Bivalve Molluscs of China*. Hong Kong University Press, Hong Kong. 146 pp.

Bernard, P. A. 1984. *Coquillages du Gabon / Shells of Gabon*. Privately published. Libreville, Gabon. 140 pp., 74 pls.

Berry, S. S. 1917. Notes on west American chitons - 1. *Proceedings of the California Academy of Sciences* (Series 4) **7**: 229-248, 4 figs.

Berry, S. S. 1919. Notes on west American chitons - 2. *Proceedings of the California Academy of Sciences* (Series 4) **9**: 1-36, 6 figs.

Beu, A. G. 1998. Indo-West Pacific Ranellidae, Bursidae and Personidae (Mollusca: Gastropoda). A monograph of the New Caledonian fauna and revision of related taxa. Résultats des Campagnes MUSORSTOM, Vol. 19. *Mémoires du Muséum National d'Histoire Naturelle* **178**: 1-255, 70 figs.

Bieler, R. 1993. Architectonicidae of the Indo-Pacific (Mollusca: Gastropoda). *Abhandlungen des Naturwissen schaftlichen Vereins in Hamburg (NF)* **30**: 1-376, 286 illustrations.

Bieler, R. and A. R. Kabat. 1991. Malacological journals and newsletters, 1773-1990. *Nautilus* **105**: 39-61.

Bieler, R. and R. E. Petit. 1990. On the various editions of Tetsuaki Kira's "Coloured Illustrations of the Shells of Japan " and "Shells of the Western Pacific in Colour Vol. I" with an annotated list of new names introduced. *Malacologia* **32**: 131-145.

Bieler, R. and J. Slapcinsky. 2000. A case study for the development of an island fauna: Recent terrestrial mollusks of Bermuda. *Nemouria: Occasional Papers of the Delaware Museum of Natural History* Number **44**: 1-99, 49 figs.

Bleam, D. E., K. J. Couch, and D. A. Distler. 1999. Key to the unionid mussels of Kansas. *Transactions of the Kansas Academy of Science* **102**: 83-91, text figures.

Bogan, A. 2002. *Workbook and Key to the Freshwater Bivalves of North Carolina*. North Carolina Museum of Natural Sciences, Raleigh. 101 pp., 10 pls.

Bogdanov, I. and B. Sirenko. 1993. *Seashells of Russia in Colour*. Editrice La Conchiglia, Roma, Italia. 76 pp., 73 pls.

Bons, J. 1984. *Mollusques Marins de l'Océan Indien: Comores, Mascareignes, Seychelles*. Agence de Coopération culturelle et technique, Paris. 109 pp., 19 pls.

Bosch, D. and E. Bosch (K. Smythe, ed.). 1982. *Seashells of Oman*. Longman Group Ltd., New York. 206 pp., text figures.

Bosch, D. and E. Bosch. 1989. *Seashells of Southern Arabia*. Motivate Publishing, United Arab Emirates. 95, 28 pp. [In English and Arabic].

Bosch, D. T., S. P. Dance, R. G. Moolenbeek, and P. G. Oliver (S. P. Dance, ed.). 1995. *Seashells of Eastern Arabia*. Motivate Publishing, Dubai, UAE. 296 pp. [1000+ taxa illustrated].

Bouchet, P. and B. A. Marshall, eds. 2001. Tropical deep-sea benthos, Vol. 22. *Mémoires du Muséum National d'Histoire Naturelle* **185**: 9-407, pls. 1-4. [Compilation of 12 papers on Indo-Pacific deep water molluscan taxa].

Bouchet, P. and J. P. Rocroi. 2005. Classification and nomenclature of gastropod families. *Malacologia* **47**: 1-397.

Bouchet, P. and A. Warén. 1979. The abyssal molluscan fauna of the Norwegian Sea and its relation to other faunas. *Sarsia* **64**: 211-243, 55 figs.

Bouchet, P. and A. Warén. 1980. Revision of the northeast Atlantic bathyal and abyssal Turridae (Mollusca, Gastropoda). *Journal of Molluscan Studies Supplement* **8**: 1-119, figs. 1-281.

Bouchet, P. and A. Warén. 1985. Revision of the northeast Atlantic bathyal and abyssal Neogastropoda excluding Turridae (Mollusca, Gastropoda). *Bolletino Malacologico* Supplemento **1**: 121-296, figs. 282-723.

Bouchet, P. and A. Warén. 1986. Revision of the northeast Atlantic bathyal and abyssal Aclididae, Eulimidae, Epitoniidae (Mollusca, Gastropoda). *Bolletino Malacologico* Supplemento **2**: 297-576, figs. 724-1267.

Bouchet, P. and A. Warén. 1993. Revision of the northeast Atlantic bathyal and abyssal Mesogastropoda. *Bolletino Malacologico* Supplemento **3**: 579-840, figs. 1268-1953.

Bousfield, E. L. 1960. *Canadian Atlantic Sea Shells*. National Museum of Canada, Ottawa, Ontario, Canada. 72 pp., 13 pls.

Bradner, H. and A. Kay. 1996. An atlas of cowrie radulae (Mollusca: Gastropoda: Cypraeoidea: Cypraeidae). *The Festivus* **28** (Supplement): 1-179, figs. 1-238.

Brandt, R. A. M. 1974. The non-marine aquatic Mollusca of Thailand. *Archiv für Molluskenkunde* **105**: i-iv + 1-423, pls. 1-30.

Bratcher, T. and W. O. Cernohorsky. 1987. *Living Terebras of the World*. American Malacologists, Inc., Melbourne, Florida. 240 pp., 74 pls.

Brim Box, J. and J. D. Williams. 2000. Unionid mollusks of the Apalachicola Basin in Alabama, Georgia and Florida. *Bulletin of the Alabama Museum of Natural History* **21**: i-vii + 1-143, 95 figs.

Brost, F. B. and R. D. Coale. 1971. *A Guide to Shell Collecting in the Kwajalein Atoll*. Charles E. Tuttle Company, Inc., Rutland, Vermont. xii + 157 pp., 33 figs.

Brown, D. S. 1967. A review of the freshwater Mollusca of Natal and their distribution. *Annals of the Natal Museum* **18**: 477-494.

Brown, D. S. 1994. *Freshwater Snails of Africa and Their Medical Importance*, 2nd Revised Ed. Taylor and Francis, London. x + 605 pp., 147 figs.

Brown, D. S. and M. D. Gallagher. 1985. Freshwater snails of Oman, southeastern Arabia. *Hydrobiologia* **127**: 125-149., 18 figs.

Brown, D. S., B. A. Curtis, S. Bethune, and C. C. Appleton. 1992. Freshwater snails of East Caprivi and the lower Okavango River Basin in Namibia and Botswana. *Hydrobiologia* **246**: 9-40, 16 figs.

Brown, D. S. and G. Mandahl-Barth. 1987. Living molluscs of Lake Tanganyika: A revised and annotated list. *Journal of Conchology* **32**: 305-327, pl. 28.

Brown, R. W. 1956. *Composition of Scientific Words: A Manual of Methods and a Lexicon of Materials for the Practice of Logotechnics*. Smithsonian Institution Press, Washington, D.C. 882 pp.

Burch, J. B. 1962. *How to Know the Eastern Land Snails: Pictured-Keys for Determining the Land Snails of the United States Occurring East of the Rocky Mountains*. William C. Brown Company Publishers, Dubuque, Iowa. 214 pp. [Text illustrations].

Burch, J. B. 1975a. *Freshwater Sphaeriacean Clams (Mollusca: Pelecypoda) of North America*, Revised Ed. Malacological Publications, Hamburg, Michigan. xi + 96 pp., 35 figs.

Burch, J. B. 1975b. *Freshwater Unionacean Clams (Mollusca: Pelecypoda) of North America*. Revised Ed. Malacological Publications, Hamburg, Michigan. xviii + 204 pp., 252 figs.

Burch, J. B. 1980. A guide to the freshwater snails of the Philippines. *Malacological Review* **13**:121-143, figs. 1-49.

Burch, J. B. 1982. North American freshwater snails. V-X. V. Keys to the freshwater gastropods of North America, VI. Generic synonymy, VII. Supplemental notes, VIII. Glossary, IX. References, X. Index. *Walkerana* **1**: 217-365.

Burch, J. B. 1984. Freshwater snails of the Philippines. *Walkerana* **2**: 81-112, figs. 1-61.

Burch, J. B. 1988. North American freshwater snails. I-III. I. Introduction, II. Systematics, nomenclature, identification and morphology, III. Habitats and distribution. *Walkerana* **2**: 1-80, figs. 1-20.

Burch, J. B. 1989. *North American Freshwater Snails*. Malacological Publications, Hamburg, Michigan. 365 pp. [this work is a compilation of Burch 1982, 1988, and Burch and Tottenham 1980].

Burch, J. B. 1990. Thomas Say's glossary for conchology. *Walkerana* **4**: 279-306.

Burch, J. B. 1991. Glossary for North American freshwater malacology. I. Gastropoda. *Walkerana* **5**: 263-288.

Burch, J. B., P. R. Chung, and Y. Jung. 1987. A guide to the freshwater snails of Korea. *Walkerana* **2**: 195-232, figs. 1-53.

Burch, J. B. and Y. Jung. 1987. A review of the classification, distribution and habitats of the freshwater gastropods of the North American Great Lakes. *Walkerana* **2**: 233-291, figs. 1-61.

Burch, J. B. and Y. Jung. 1988. Land snails of the University of Michigan Biological Station area. *Walkerana* **3**: 1-177, figs. 1-109.

Burch, J. B. and Y. Jung. 1992. Freshwater snails of the University of Michigan Biological Station area. *Walkerana* **6**: v + 1-218, 113 figs.

Burch, J. B. and T. A. Pearce. 1990. Terrestrial gastropods. *In:* D. L. Dindal, ed., *Soil Biology Guide*. John Wiley and Sons, New York. Pp. 201-309, 215 figs.

Burch, J. B. and J. L. Tottenham. 1980. North American freshwater snails. IV. Species lists, ranges and illustrations. *Walkerana* **1**: 81-215, figs. 21-771.

Burgess, C. M. 1970. *The Living Cowries*. A. S. Barnes and Company, Inc., Cranbury, New Jersey. 389 pp., 44 pls.

Burgess, C. M. 1985. *Cowries of the World*. Seacomber Publications, Cape Town. xiv + 289 pp., 20 pls.

Burghardt, G. E. and L. E. Burghardt. 1969. *A Collector's Guide to West Coast Chitons*. San Francisco Aquarium Society, Inc. and California Academy of Sciences, San Francisco. 45 pp., 4 pls.

Burnay, L. P. and A. A. Monteiro. 1977. *Seashells from Cape Verde Islands*. Privately published, Lisbon, Portugal. 88 pp., 65 figs.

Callomon, P. 1999a. Card Catalog of World-wide Shells by S. D. Kaicher. Pack Contents Index Version 1.0 November 1999. Available at: <www.conchologistsofamerica/org/info/kaicher2.txt>.

Callomon, P. 1999b. Card Catalog of World-wide Shells by S. D. Kaicher. Index of Species Version 1.1 December 1999. Available at: <www.conchologistsofamerica/org/info/kaicher.txt>.

Callomon, P. and R. E. Petit. 2004. Tadashige Habe's 'Coloured Illustrations of the Shells of Japan (II)' and 'Shells of the Western Pacific in Color vol. 2': Comparison of printings and treatment of included taxa. *Venus* Supplement **3**: 1-61.

Cameron, R. 2003. *Keys for the Identification of Land Snails in the British Isles*. FSC Publications, Shropshire, Great Britian. 82 pp., 4 pls., 166 figs.

Cameron, R. A. D., B. Eversham, and N. Jackson. 1983. A field key to the slugs of the British Isles. *Field Studies* **5**: 807-824, 4 pls.

Cameron, R. A. D. and M. Redfern. 1976. British Land Snails (Mollusca: Gastropoda): Keys and Notes for Identification of the Species. *Synopses of the British Fauna (New Series)* **6**: 1-64, 31 figs.

Carcelles, A. R. 1944. Catálogo de los moluscos marinos de Puerto Quequén. *Revista del Museo de la Plata (NS) Sección Zoología* **3**: 233-309, 15 pls.

Carcelles, A. R. 1950. Catálogo de los moluscos marinos de la Patagonia. *Anales del Museo Nahuel Huapi* **2**: 41-100, 6 pls.

Carcelles, A. R. 1953. Catálogo de la malacofauna Antárctica Argentina. *Anales del Museo Nahuel Huapi* **3**: 155-250, 5 pls.

Carcelles, A. R. and S. I. Williamson. 1951. Catalogo de los moluscos marinos de la provincia Magellanica. *Revista del Instituto Nacional de Investigacion de las Ciencias Naturales anexo al Museo Argentino de Ciencias Naturales "Bernardino Rivadavia," Ciencias Zoológicas, Buenos Aires* **2**: 225-383.

Castellanos, Z. J. A. d. 1967. Catalogo de los moluscos marinos Bonaerenses. *Anales de la Comisión de Investigación Científica* **8**: 9-365, 26 pls. [Dated 1967 but published in 1970].

Cate, C. N. 1973. A systematic revision of the Recent cypraeid family Ovulidae (Mollusca: Gastropoda). *Veliger* **15** (Supplement): i-iv + 1-116, 51 pls.

Cate, C. N. 1977. A review of the Eratoidae (Mollusca: Gastropoda). *Veliger* **19**: 341-366, 15 pls. [see Keen 1977 for commentary].

Cate, C. N. 1979. A review of the Triviidae (Mollusca: Gastropoda). *San Diego Society of Natural History Memoir* **10**: 1-16, figs. 1-177.

Cernohorsky, W. O. 1971. *Marine Shells of the Pacific*, Revised Ed. Pacific Publications, Sidney, Australia. 248 pp., 444 figs.

Cernohorsky, W. O. 1972. *Marine Shells of the Pacific*, Vol. II. Pacific Publications, Sydney, Australia. 411 pp., 68 pls.

Cernohorsky, W. O. 1976. The Mitridae of the World. Part 1. The subfamily Mitrinae. *Indo-Pacific Mollusca* **3**: 273-528, pls. 248-466.

Cernohorsky, W. O. 1978. *Tropical Pacific Marine Shells*. Pacific Publications, Sydney, Australia. 352 pp., 68 pls.

Cernohorsky, W. O. 1981. The family Buccinidae Part 1: The genera *Nassaria*, *Trajana*, and *Neoteron*. *Monographs of Marine Mollusca* **2**: 1-52, pls. 1-42.

Cernohorsky, W. O. 1984. Systematics of the family Nassariidae (Mollusca: Gastropoda). *Bulletin of the Auckland Institute and Museum* **14**: 1-356, 51 pls., 173 figs.

Cernohorsky, W. O. 1991. The Mitridae of the World. Part 2: The subfamily Mitrinae concluded and subfamilies Imbricariinae and Cylindromitrinae. *Monographs of Marine Mollusca* **4**: 1-164, pls. 1-155.

Chamberlain, R. V. and D. T. Jones. 1929. Descriptive catalog of the Mollusca of Utah. *Bulletin of the University of Utah* **19**: 1-203, 86 figs.

Chichester, L.F. and L.L. Getz. 1973. The terrestrial slugs of northeastern North America. *Sterkiana* **51**: 11-42, figs. 1-8.

Claassen, C. 1998. *Shells*. Cambridge University Press, New York. xiv + 266 pp. [Part of the Cambridge Manuals in Archaeology series].

Clarke, A. H. 1961. Abyssal mollusks from the south Atlantic Ocean. *Bulletin of the Museum of Comparative Zoology at Harvard College* **125**: 345-387, 4 pls.

Clarke, A. H. 1962. Annotated list and bibliography of the abyssal marine molluscs of the world. *National Museum of Canada Bulletin* **181**: i-vi + 1-114.

Clarke, A. H. 1973. The freshwater molluscs of the Canadian interior basin. *Malacologia* **13**: 1-509, 28 pls.

Clarke, A. H. 1981. *The Freshwater Molluscs of Canada*. National Museum of Natural Sciences, National Museums of Canada. Ottawa, Canada. 446 pp., 176 figs.

Clench, W. J. 1946. A catalogue of the genus *Liguus* with a description of a new subgenus. *Occasional Papers on Mollusks, Museum of Comparative Zoology of Harvard University* **1**: 117-128 [for the supplement see Clench, W. J. 1946. ibid. pp. 442-444].

Clench, W. J. 1958. The land and freshwater Mollusca of Rennell Island, Solomon Islands. *Natural History of Rennell Island, British Solomon Islands, Copenhagen* **2**: 155-202, pls. 16-19.

Clench, W. J. 1964. Land and freshwater Mollusca of the Cayman Islands, West Indies. *Occasional Papers on Mollusks, Museum of Comparative Zoology of Harvard University* **2**: 345-380, pls. 62-63.

Close, H. 1995. *Lure of the* Liguus: *The Florida Tree Snails*. Of Sea and Shore Publications, Port Gamble, Washington. 136 pp., 30 pls.

Close, H. 2000. *The Liguus Tree Snails of South Florida*. University Press of Florida, Gainesville, Florida. xiii + 161 pp., 8 pls.

Clover, P. W. 1968. *A Catalog of Popular* Marginella *Species*. Privately published. 15 pp., 117 figs. [Also

Clover, P. W. [no date]. *Supplement to a Catalog of Popular* Marginella *Species*. 4 pp. Figures 118-137].

Clover, P. W. 1982. Latiaxis *Catalog and Illustrated Check List of Coralliophilidae Family*. Privately published. 38 pp., 18 pls.

Coan, E. V., A. R. Kabat, and R. E. Petit. 2005. 2400 Years of Malacology. Available at: <erato.acnatsci.org/ams/publications/2400_malacology.html>.

Coan, E. V. and P. H. Scott. 1997. Checklist of the marine bivalves of the northeastern Pacific Ocean. *Santa Barbara Museum of Natural History Contributions in Science* Number **1**: 1-28.

Coan, E. V., P. V. Scott, and F. R. Bernard. 2000. *Bivalved Seashells of Western North America*. Santa Barbara Museum of Natural History, Santa Barbara, California. viii + 764 pp., 124 pls.

Cocke, J. 1995. *Common Land Snails of Los Angeles County California*. Privately published. 136 pp. [Text illustrations].

Connolly, M. 1912. A revised reference list of South African non-marine Mollusca; with descriptions of new species in the South African Museum. *Annals of the South African Museum* **11**: 59-306, 1 pl.

Connolly, M. 1925. The non-marine Mollusca of Portuguese East Africa. *Transactions of the Royal Society of South Africa* **12**: 105-220, pls. 4-8, 30 figs.

Connolly, M. 1939. A monographic survey of South African non-marine Mollusca. *Annals of the South African Museum* **33**: i-iii + 1-660, 19 pls., 58 figs.

Coomans, H. E. 1989. *Antillean Seashells, the 19th Century Watercolours of Caribbean Molluscs Painted by Hendrik van Rijgersma*. De Walburg Pers, Zutphen, The Netherlands. 191 pp., 74 pls.

Coovert, G. A. 1988. A bibliography of recent Marginellidae. *Marginella Marginalia* **5**: 1-43.

Coovert, G. A. and H. K. Coovert. 1995. Revision of the supraspecific classification of marginelliform gastropods. *Nautilus* **109**: 43-110, 79 figs.

Corsi, A. F. 1900-1901. Moluscos de la República Oriental del Uruguay. *Anales del Museo Nacional de Montevideo* **2** (XV - 1900): 291-368; **2** (XVI - 1900): 369-443; **2** (XVII - 1901): 449-528.

Cossignani, T. 1994. *Bursidae of the World*. L'Informatore Piceno, Ancona, Italy. 119 pp. [64 taxa illustrated].

Cossignani, T. and V. Cossignani. 1995. *Atlante delle Conchiglie Terrestri e Dulciacquicole Italiane*. L'Informatore Piceno, Ancona, Italy. 208 pp.

Cotton, B. C. 1959. *South Australian Mollusca: Archaeogastropoda*. British Association for the Advancement of Science, Adelaide, Australia. 449 pp., 215 figs.

Cotton, B. C. 1961. *South Australian Mollusca: Pelecypoda*. British Association for the Advancement of Science, Adelaide, Australia. 363 pp., 350 figs.

Cotton, B. C. 1964. *South Australian Mollusca: Chitons*. British Association for the Advancement of Science, Adelaide, Australia. 151 pp., 131 figs.

Cotton, B. C. and C. J. Gabriel. 1932. Australian Unionidae. *Proceedings of the Royal Society of Victoria* **44** (N.S.): 155-161, 1 pl.

Couch, K. J. 1997. *An Illustrated Guide to the Unionid Mussels of Kansas*. Privately published, Olathe, Kansas. ix + 126 pp. [48 taxa illustrated].

Coulombel, A. 1994. *Coquillages de Djibouti*. Edisud, Aix-en-Provence, Paris. 143 pp. [Text figures].

Cowie, R. H. 1998. Catalog of the nonmarine snails and slugs of the Samoan Islands. *Bishop Museum Bulletin in Zoology* **3**: i-viii + 1-122.

Cowie, R. H., N. L. Evenhuis, and C. C. Christensen. 1995. *Catalog of the Native Land and Freshwater Molluscs of the Hawaiian Islands*. Backhuys Publishers, Leiden, The Netherlands. vi + 248 pp.

Cowie, R. H. and S. C. Thiengo. 2003. The Apple Snails of the Americas (Mollusca: Gastropoda: Ampullariidae: *Asolene, Filipponea, Marisa, Pomacea, Pomella*): A nomenclatural and type catalog. *Malacologia* **45**: 41-100.

Cox, I., ed. 1957. *The Scallop. Studies of a Shell and its Influences on Humankind*. The Shell Transport and Trading Company, Ltd., London. 135 pp.

Cox, J. A. 1979. *Shells. Treasures from the Sea*. Larousse and Company, Inc., New York. 254 pp. [Popular text with numerous figures].

Crampton, H. E. 1916. Studies on the variation, distribution, and evolution of the genus *Partula*. The species inhabiting Tahiti. *The Carnegie Institute of Washington Publication* **228**: 1-312. [Illustrated].

Crampton, H. E. 1925. Studies on the variation, distribution, and evolution of the genus *Partula*. The species of the Mariana Islands, Guam and Saipan. *The Carnegie Institute of Washington Publication* **228A**: 1-116. [Illustrated].

Crampton, H. E. 1932. Studies on the variation, distribution, and evolution of the genus *Partula*. The species inhabiting Moorea. *The Carnegie Institute of Washington Publication* **410**: 1-335. [Illustrated].

Cuezzo, M. G. 2003. Phylogenetic analysis of the Camaenidae (Mollusca: Stylommatophora) with special emphasis on the American taxa. *Zoological Journal of the Linnean Society* **138**: 449-476, figs. 1-8.

Cummings, K. S. and C. A. Mayer. 1992. Field guide to freshwater mussels of the Midwest. *Illinois Natural History Survey, Manual* **5**: i-xiii + 1-194, figs. 1-78.

Curtis, B. A. 1991. Freshwater macro-invertebrates of Namibia. *Madoqua* **17**: 163-187. [Checklist of freshwater invertebrates including mollusks].

Dall, W. H. 1905. Land and fresh water mollusks of Alaska and adjoining regions. In: *Alaska*, Vol. 13. Doubleday, Page and Company, New York. Pp. 1-171, 2 pls., figs. 1-118.

Dall, W. H. 1909. Report on a collection of shells from Peru, with a summary of the littoral marine Mollusca of the Peruvian Zoological Province. *Proceedings of the United States National Museum* **37**: 147-294, pls. 20-28.

Dall, W. H., P. Bartsch, and H. A. Rehder. 1938. A manual of the Recent and fossil marine pelecypod mollusks of the Hawaiian Islands. *Bernice P. Bishop Museum Bulletin* **153**: i-iv + 1-233, pls. 1-58.

Dance, S. P. 1966. *Shell Collecting. An Illustrated History*. Faber and Faber, Ltd., London. 344 pp., 35 pls.

Dance, S. P. 1969. *Rare Shells*. Faber and Faber, Ltd., London. 128 pp., 24 pls.

Dance, S. P. 1970. Non-marine molluscs of Borneo, I. Streptaxacea: Streptaxidae. *Journal of Conchology* **27**: 149-162, pl. 6. [For parts 2 and 3, see Thompson and Dance 1983].

Dance, S. P. 1974. *The Collector's Encyclopedia of Shells*. McGraw-Hill Book Company, New York. 288 pp. [1500 taxa illustrated].

Dance, S. P. 1976. *The World's Shells. A Guide for Collectors*. McGraw-Hill Book Company, New York. 192 pp.

Dance, S. P. 1986. *A History of Shell Collecting*. E. J. Brill, Leiden, The Netherlands. 265 pp., 33 pls.

D'Attilio, A. and C. M. Hertz. 1988. An illustrated catalogue of the family Typhidae Cossmann, 1903. *The Festivus* **20** (Supplement): 1-73, 109 figs.

Debelius, H. 1998. *Nudibranchs and Sea Snails. Indo-Pacific Field Guide Second revised edition*. Ikan-Unterwasserarchiv, Frankfurt, Germany. 321 pp. [1000+ figures].

De Jong, K. M. and H. E. Coomans. 1988. *Marine Gastropods from Curaçao, Aruba and Bonaire*. E. J. Brill, Leiden, The Netherlands. v + 261 pp., 47 pls.

de la Torre, C., P. Bartsch, and J. P. E. Morrison. 1942. The cyclophorid operculate land mollusks of America. *United States National Museum Bulletin* **181**: i-iv + 1-306, pls. 1-42.

Dell, R. K. 1952. The Recent Cephalopoda of New Zealand. *Dominion Museum Bulletin* **16**: 1-157, pls. 1-35.

Dell, R. K. 1953. The freshwater Mollusca of New Zealand. Part I. The genus *Hyridella. Transactions of the Royal Society of New Zealand* **81**: 221-237, pls. 17-19.

Dell, R. K. 1960. Antarctic and subantarctic Mollusca. *Records of the Auckland Institute and Museum* **5**: 117-193. [Checklist].

Dell, R. K. 1964a. Antarctic and subantarctic Mollusca: Amphineura, Scaphopoda and Bivalvia. *Discovery Reports* **33**: 99-250, pls. 2-7.

Dell, R. K. 1964b. Land snails from sub-Antarctic islands. *Transactions of the Royal Society of New Zealand (Zoology)* **4**: 167-173.

Dell, R. K. 1990. Antarctic Mollusca: With special reference to the fauna of the Ross Sea. *Royal Society of New Zealand Bulletin* **27**: 1-311, figs. 1-482.

Dell'Angelo, B. and C. Smriglio. 2001. *Living Chitons from the Mediterranean Sea*. Edizioni Evolver S.r.l., Roma. 255 pp., 68 pls., 130 figs.

Demir, M. 2003. Shells of Mollusca collected from the seas of Turkey. *Turkish Journal of Zoology* **27**: 101-140. [Checklist].

Dharma, B. 1988. *Siput dan Kerang Indonesia I: Indonesian Shells I*. P. T. Sarana Graha, Jakarta, Indonesia. xv + 111 pp., 35 pls. [In Indonesian and English].

Dharma, B. 1992. *Siput dan Kerang Indonesia II: Indonesian Shells II*. Verlag Christa Hemmen, Wiesbaden, Germany. 135 pp., 38 pls. [In Indonesian and English].

Dhora, D. and F. W. Welter-Schultes. 1996. List of species and atlas of the non-marine molluscs of Albania. *Schriften zur Malakozoologie* **9**: 90-197, pls. 5-16.

Diaz, J. M. and M. Puyana. 1994. *Moluscos del Caribe Colombiano. Un catálogo ilustrado*. Fundación Natura, Invemar, Colombia. 291 pp., 78 plates.

Dijkstra, H. H. 1989. Pectinidae from French Polynesia (a preliminary report). *Xenophora* **48**: 11-19. [In French and English, illustrated].

Dijkstra, H. H. 1994. Type specimens of Recent species of Pectinidae described by Lamarck (1819), preserved in the Muséum d'Histoire Naturelle of Geneva and the Muséum d'Histoire Naturelle of Paris. *Revue Suisse de Zoologie* **101**: 465-532, 30 pls.

Dijkstra, H. H. 1995. Bathyal Pectinoidea (Bivalvia: Propeamussiidae, Entoliidae, Pectinidae) from New Caledonia and adjacent areas. *Mémoires du Muséum national d'Histoire naturelle* **167**: 9-73, figs 1-154.

Dijkstra, H. H. 1998. Pectinoidea (Mollusca: Bivalvia: Pectinidae: Propeamussiidae) from Hansa Bay, Papua New Guinea. *Molluscan Research* **19**: 11-52, 9 pls.

Dijkstra, H. H. and R. N. Kilburn. 2001. The family Pectinidae in South Africa and Mozambique (Mollusca: Bivalvia: Pectinoidea). *African Invertebrates* **42**: 263-321, 54 figs.

Dillon, Jr., R. T. 2000. *The Ecology of the Freshwater Molluscs*. Cambridge University Press, New York. xii + 509 pp.

Dirrigl, F. J., Jr. and A. E. Bogan. 1996. Revised checklist of the terrestrial gastropods of New Jersey (Mollusca: Gastropoda). *Walkerana* **8**: 127-138.

Drivas, J. and M. Jay. 1988. *Coquillages de La Réunion et de l'Ile Maurice*. Delachaux and Niestlé, Lausanne, Switzerland. 159 pp., 58 pls.

Edmunds, J. 1978. *Sea Shells and other Molluscs Found on West African Shores and Estuaries*. Ghana Universities Press, Accra, Ghana. xii + 146 pp., 27 figs.

Edwards, M. A., P. Manly, and M. A. Tobias. 1996. *Nomenclator Zoologicus*, Vol. 9, 1978-1994. The Zoological Society of London, London. 747 pp.

Edwards, M. A. and A. T. Hopwood. 1966. *Nomenclator Zoologicus*, Vol. 6, 1946-1955. The Zoological Society of London, London. 329 pp.

Edwards, M. A. and M. A. Tobias. 1993. *Nomenclator Zoologicus*, Vol. 8, 1966-1977. The Zoological Society of London, London. 620 pp.

Edwards, M. A. and H. G. Vevers (eds.) 1975. *Nomenclator Zoologicus*, Vol. 7, 1956-1965. The Zoological Society of London, London. 374 pp.

Egorov, R. 1993. Trophoninae (Muricidae) of Russian and adjacent waters. *Ruthenica* Supplement **1**: 1-48, 39 figs.

Egorov, R. and S. Barsukov. 1994. *Recent Ancistrolepidinae (Buccinidae)*. Moscow, Russia. 47 pp., 30 figs.

Eisenberg, J. M. [William E. Old, Jr., consulting editor.] 1981. *A Collector's Guide to Seashells of the World*. McGraw-Hill, New York. 239 pp., 158 pls.

Ellis, A. E. 1978. British freshwater bivalve Mollusca: keys and notes for the identification of the species. *Synopses of the British Fauna (New Series)* **11**: 1-109, pls 1-15.

Emberton, K. C. 1988. The genitalic, allozymic, and conchological evolution of the eastern North American Triodopsinae (Gastropoda: Pulmonata: Polygyridae). *Malacologia* **28**: 159-273, figs. 1-51.

Emberton, K. C. 1990. Acavid land snails of Madagascar: Subgeneric revision based on published data (Gastropoda: Pulmonata: Stylommatophora). *Proceedings of the Academy of Natural Sciences of Philadelphia* **142**: 101-117.

Emberton, K. C. 1991. The genitalic, allozymic and conchological evolution of the Tribe Mesodontini (Pulmonata: Stylommatophora: Polygyridae). *Malacologia* **33**: 71-178, figs. 1-60.

Emberton, K. C. 1995. When shells do not tell: 145 million years of evolution in North American polygyrid land snails, with a revision and conservation priorities. *Malacologia* **37**: 69-110.

Emberton, K. C. 1999. New acavid land snails from Madagascar. *American Malacological Bulletin* **15**: 83-96.

Emerson, W. K. and W. O. Cernohorsky. 1973. The genus *Drupa* in the Indo-Pacific. *Indo-Pacific Mollusca* **3**: 1-40, pls. 1-35.

Emerson, W. K. and M. K. Jacobson. 1976. *The American Museum of Natural History Guide to Shells: Land, Freshwater and Marine, from Nova Scotia to Florida*. Alfred A. Knopf, New York. vii + 482 pp., 47 pls.

Evans, J. G. 1972. *Land Snails in Archeology*. Seminar Press, Inc., New York. xii + 436 pp.

Fair, R. H. 1976. *The Murex Book: An Illustrated Catalogue of Recent Muricidae (Muricinae, Muricopsinae, Ocenebrinae)*. Privately published, Honolulu, Hawaii. 138 pp., 23 pls.

Fairbanks, H. L. 1998. Clarification of the taxonomic status and reproductive anatomy of *Philomycus batchi* Branson, 1968 (Gastropoda: Pulmonata: Philomycidae). *Nautilus* **112**: 1-5.

Falkner, G., R. A. Bank, and T. Von Proschwitz. 2001. Check-list of the non-marine molluscan species-group taxa of the states of Northern, Atlantic and Central Europe (CLECOM I). *Heldia* **4**: 1-76.

Falkner, G., T. E. J. Ripken, and M. Falkner. 2002. Mollusques continentaux de France. Liste de référence annotée et bibliographie. *Patrimoines Naturels* **52**: 1-350.

Farmer, W. M. 1980. *Sea-slug Gastropods*. W. M. Farmer Enterprises, Inc., Tempe, Arizona. 177 pp. [157 taxa illustrated].

Fernández, D. 1973. *Catálogo de la Malacofauna Terrestre Argentina*. Monografía 4. Comisión de Investigaciones Científicas de la Provincia de Buenos Aires, La Plata, Argentina. 197 pp.

Fernández-Milera, J. 1997. *Joyas de Cuba: Moluscos Marinos*. Editorial Oriente, Santiago de Cuba, Cuba. 230 pp. [Gastropods only, 298 illustrations].

Fernández-Milera, J. 1999. Polymita, 2nd Ed. Editorial Científico-Técnica, Ciudad de la Habana, Cuba. 147 pp.

Ferreira, A. J. 1984. Chiton (Mollusca: Polyplacophora) fauna of Barbados, West Indies, with the description of a new species. *Bulletin of Marine Science* **36**: 189-219, 16 figs.

Fichtel, C. and D. G. Smith. 1995. The freshwater mussels of Vermont. *Nongame and Natural Heritage Program, Vermont Fish and Wildlife Department Technical Report* **18**: 1-54. [Illustrated].

Filmer, R. M. 2001. *A Catalogue of Nomenclature and Taxonomy in the Living Conidae 1758-1998*. Backhuys Publishers, Leiden, The Netherlands. 388 pp.

Finet, Y. 1994a. Marine Molluscs of the Galapagos: Gastropods - A monograph and revision of the families Haliotidae, Scissurellidae, Fissurellidae and Lottiidae. *Monograph on Galapagos Mollusca* No. 1. L'Informatore Piceno, Ancona, Italy. 110 pp., 26 pls.

Finet, Y. 1994b. *The Marine Mollusks of the Galapagos Islands: A Documented Faunal List*. Muséum d'Histoire naturelle de Genève, Genève, Switzerland. 180 pp.

Fischer-Piette, E. 1977. *Revision des Cardiidae (Mollusques Lamellibranches)*. Éditions du Muséum, Paris. 212 pp., 12 pls. [Also published as Mémoires du Muséum National d'histoire Naturelle Nouvelle Série, Série A, Zoologie 101].

Fischer-Piette, E., C. P. Blanc, F. Blanc, and F. Salvat. 1993. Gastéropodes terrestres prosobranches. *Muséum national d'Histoire naturelle, Paris, Faune de Madagascar* **80**: 1-281, pls. 1- 16, figs. 1-114.

Fischer-Piette, E., C. P. Blanc, F. Blanc, and F. Salvat. 1994. Gastéropodes terrestres pulmonés (excl. Veronicellidae et g. *Elisolimax*). *Muséum national d'Histoire naturelle, Paris Faune de Madagascar* **83**: 1-551, pls. 1-46, figs. 1-194.

Fischer-Piette, E. and D. Vukadinovic. 1973. Sur les mollusques fluviatiles de Madagascar. *Malacologia* **12**: 339-378, 24 figs.

Fitch, J. E. 1953. Common marine bivalves of California. *Fish Bulletin* **90**: 1-102, figs. 1-63.

Forcelli, D. O. 2000. *Moluscos Magallánicos - Guía de los Moluscos de la Patagonia y del Sur de Chile*. Vázquez Mazzini Editores. Buenos Aires, Argentina. 200 pp. [627 taxa illustrated].

Forsyth, R. G. 2004. *Land Snails of British Columbia*. Royal British Columbia Museum, Victoria, British Columbia, Canada. iv + 188 pp. [92 taxa illustrated].

Franc, A. 1956. Mollusques terrestres et fluviatiles de l'Archipel Néo-Calédonien. *Mémoirs du Muséum National d'Histoire Naturelle, Nouvelle Série, Série A. Zoologie* **3**: 1-200, pls. 1-24.

Franzen, D. S. and A. B. Leonard. 1947. Fossil and living Pupillidae (Gastrocopta: Pulmonata) in Kansas. *University of Kansas Science Bulletin* **31**(2): 311-411, pls. 17-22.

Fretter, V. and A. Graham. 1994. *British Prosobranch Molluscs: Their Functional Anatomy and Ecology.* The Ray Society, London. xix + 820 pp., figs. 1-343.

Fuller, S. and I. Brynildson (revisor). 1985. *Freshwater Mussels of the Upper Mississippi River.* Wisconsin Department of Natural Resources, Madison, Wisconsin. 64 pp. [Text figures].

Garcia, E. F. 2003. New records of Indo-Pacific Epitoniidae (Mollusca: Gastropoda) with the description of nineteen new species. *Novapex* **4**: 1-22, figs. 1-70.

García-Cubas, A. 1981. Moluscos de un sistema Lagunar Tropical en el sur del Golfo de México (Laguna de Términos, Campeche). *Instituto de Ciencias del Mar y Limnología, Universidad Nacional Autónoma México Publicaciones Especiales* **5**: 1-182, figs. 1-176.

Geiger, D. L. 1998. Recent genera and species of the family Haliotidae Rafinesque, 1815 (Gastropoda, Vetigastropoda). *Nautilus* **111**: 85-116, 33 figs.

Geiger, D. L. and G. T. Poppe. 2000. *A Conchological Iconography: The Family Haliotidae.* ConchBooks, Hackenheim, Germany. 135 pp., 83 pls.

Gerber, J. 1996. Revision der Gattung *Vallonia* Risso 1826 (Mollusca: Gastropoda:Valloniidae). *Schriften zur Malakozoologie* **8**: 1-227, figs. 1-85.

Gerlach, J. 1987. *The Land Snails of Seychelles, a Field Guide.* Privately published, Northamptonshire, U.K. 43 pp. [Text illustrations].

Golikov, A. N. 1994. *Shell-Bearing Gastropods of the Arctic.* Moscow, Russia. 108 pp., 139 figs.

Goodrich, C. and H. van der Schalie. 1944. A revision of the Mollusca of Indiana. *American Midland Naturalist* **32**: 257-320.

Goto, Y. and G. T. Poppe. 1996. *A Listing of Living Mollusca.* L'Informatore Piceno, Ancona, Italy. Vol. 1, Parts 1-2, 868 pp, Vol. 2, Parts 1-2, 1031 pp.

Graf, D. L. 2001. The cleansing of the Augean stables, or a lexicon of the nominal species of the Pleuroceridae (Gastropoda: Prosobranchia) of Recent North America, north of Mexico. *Walkerana* **12**: 1-124.

Graham, A. 1971. British prosobranch and other operculate gastropod molluscs: keys and notes for the identification of the species. *Synopses of the British Fauna (New Series)* **2**: 1-112, figs. 1-119.

Grau, G. 1959. Pectinidae of the Eastern Pacific. *Allan Hancock Pacific Expeditions* **23**: i-viii + 1-308, pls. 1-57.

Green, S. and N. Chouhfeh. 1994. *Bahrain Sea Shells.* Arabian Printing and Publishing House, W. L. L., Bahrain. 183 pp. [Illustrated].

Griffith, L. M. 1967. *The Intertidal Univalves of British Columbia*, Handbook 26. British Columbia Provincial Museum, Department of Recreation and Conservation, Victoria, Canada. 101 pp. [Illustrated].

Grimm, F. W. 1971. Annotated checklist of the land snails of Maryland and the District of Columbia. *Sterkiana* **41**: 51-57.

Groh, K., G. T. Poppe, and M. Charles. 2002. *A Conchological Iconography: Family Acavidae (excluding Ampelita).* ConchBooks, Hackenheim, Germany. 66 pp., 44 pls.

Grossu, A. V. 1955. *Fauna Republicii Populare Romîne. Mollusca*, Vol. 3, Fascicula 1 - Gastropoda Pulmonata. Academiei Republicii Populare Romîne, Bucharest, Romania. 518 pp., 282 figs.

Grossu, A. V. 1956. *Fauna Republicii Populare Romîne. Mollusca*, Vol. 3, Fascicula 2 - Gastropoda Prosobranchia si Opisthobranchia. Academiei Republicii Populare Romîne, Bucharest, Romania. 220 pp., 101 figs.

Grossu, A. V. [1962]. *Fauna Republicii Populare Romîne. Mollusca*, Vol. 3, Fascicula 3 - Gastropoda Bivalvia. Academiei Republicii Populare Romîne, Bucharest, Romania. 426 pp., 221 figs.

Grossu, A. V. 1981-1987. *Gastropoda Romaniae.* Editura Litera, Bucharest, Romania. Vol. 1 (1986): 524 pp. - Prosobranchia si Opisthobranchia; Vol. 2 (1987): 443 pp. - Subclass Pulmonata, Ordo Basommatophora si Stylommatophora; Vol. 3 (1981): 269 pp. - Subclass Pulmonata, Ordo Stylommatophora, Vol. 4 (1983): 564 pp. - Subclass Pulmonata, Ordo Stylommatophora. [903 figures].

Haas, F. 1917. Estudio para una monografia de las náyades de la Peninsula Ibérica. *Anuari de la Junta de Ciències Naturales* **2**: 131-190.

Haas, F. 1949. Land and fresh-water mollusks from Peru. *Fieldiana Zoology* **31**: 235-250, figs. 50-59.

Habe, T. 1961. *Coloured Illustrations of the Shells of Japan*, II. Hoikusha Publishing Co., Ltd., Osaka, Japan. ix + 182 pp., 66 pls. [In Japanese; from 1961 to 1994 there have been several editions and printings of this work; there may be minor differences among printings, see: Callomon and Petit (2004) for additional details].

Habe, T. 1964a. Freshwater molluscan fauna of Thailand. *In:* T. Kira and T. Umesao, eds., *Nature and Life in Southeast Asia* **3**: 45-66, 2 pls.

Habe, T. 1964b. *Shells of the Western Pacific in Color*, Vol. 2. Hoikusha Publishing Co., Ltd., Osaka, Japan. 233 pp., 66 pls. [From 1964 to 1975 there have been several editions and printings of this work; there may be minor differences among printings, see: Callomon, P. and R. E. Petit (2004) for additional details. This has been considered an English edition of Habe 1961, however, it contains more extensive treatments of some taxa].

Habe, T. 1977. *Systematics of Mollusca in Japan: Bivalvia and Scaphopoda*. Zukan-no-Hokuryukan Co., Ltd., Tokyo. xiii + 372 pp., 72 pls. [In Japanese].

Habe, T. and K. Ito. 1974. *Shells of the World in colour*. Vol. 1: The Northern Pacific. Hoikusha Publishing Co., Ltd., Osaka, Japan. vii + 176 pp., 56 pls. [In Japanese; between 1965 and 1991 there have been 11 printings of this work; there may be minor differences among printings, see: Petit and Bieler 1996 for additional details].

Habe, T. and S. Kosuge. 1974. *Shells of the World in colour*. Vol. 2: The Tropical Pacific. Hoikusha Publishing Co., Ltd., Osaka, Japan. vi + 194 pp. 68 plates. [In Japanese; between 1966-1991 there have been 10 printings of this work; there may be minor differences among printings, see: Petit and Bieler 1996 for additional details].

Hain, S. 1990. Die beschalten benthischen Mollusken (Gastropoda und Bivalvia) des Weddellmeeres, Antarktis. *Berichte zur Polarforschung* **70**: 1-181, 30 pls.

Hanna, G. D. 1966. Introduced mollusks of western North America. *Occasional Papers of the California Academy of Sciences* **48**: 1-108, pls. 1-4, figs. 1-85.

Hannibal, H. 1912. A synopsis of the Recent and Tertiary freshwater Mollusca of the Californian Province, based upon an ontogenic classification. *Proceedings of the Malacological Society of London* **10** (Part 2 - June): 112-165, **10** (Part 3 - September): 165-211, 8 pls.

Harasewych, M. G. 1983. A review of the Columbariinae (Gastropoda: Turbinellidae) of the western Atlantic with note on the anatomy and systematic relationships of the subfamily. *Nemouria* **27**: 1-42, figs. 1-59.

Harper, E. M., J. D. Taylor, and J. A. Crame. 2000. *Evolutionary Biology of the Bivalvia*. Geological Society Special Publication No. 117. The Geological Society of London, Bath, United Kingdom. vii + 494 pp.

Harry, H. W. 1962. A critical catalogue of the nominal genera and species of neotropical Planorbidae. *Malacologica* **1**: 33-53.

Harry, H. W. 1985. Synopsis of the supraspecific classification of living oysters (Bivalvia: Gryphaeidae and Ostreidae). *Veliger* **28**: 121-158, 30 figs.

Hausdorf, B. 1988. Zur Kenntnis der systematischen Beziehungen einiger Taxa der Helicellinae Ihering, 1909 (Gastropoda: Hygromiidae). *Archiv für Molluskenkunde* **119**: 9-37, figs. 1-19.

Hausdorf, B. and Perera, K. K. 2000. Revision of the genus *Acavus* from Sri Lanka (Gastropoda: Acavidae). *Journal of Molluscan Studies* **66**: 217-231.

Haynes, A. 1988. The gastropods in the streams and rivers of five Fiji islands: Vanua Levu, Ovalau, Gau, Kadavu and Taveuni. *Veliger* **30**: 377-383.

Haynes, A. 1993. The gastropods in the streams and rivers of four islands (Guadalcanal, Makira, Malaita and New Georgia) in the Solomon Islands. *Veliger* **36**: 285-290.

Haynes, A. 2001. *Freshwater Snails of the Tropical Pacific Islands*. Institute of Applied Sciences, Suva, Fiji. 116 pp. [Text illustrations].

Hedgpeth, J. W. 1971. *Introduction to Seashore Life of the San Francisco Bay Region and the Coast of Northern California*. California Natural History Guide No. 9. University of California Press, Berkeley, California. 136 pp., 8 pls., 85 figs.

Heller, J. 1993. *Land Snails of the Land of Israel. Natural History and a Field Guide*. Ministry of Defense, Tel Aviv. 271 pp. [In Hebrew, all taxa illustrated].

Hemmen, J. 2004. Further corrections to R. Tucker Abbott "Compendium of Landshells (1989). *Schriften zur Malakozoologie* **21**: 19-20.

Hemmen, J. and K. Groh. 1991. Preliminary list of corrections and additions for R. Tucker Abbott (1989) "Compendium of Landshells." *Schriften zur Malakozoologie* **4**: 39-54.

Hemmen, J. and C. Hemmen. 2001. Aktualisierte liste der terrestrischen Gastropoden Thailand. *Schriften zur Malakozoologie* **18**: 35-70, 18 figs.

Henderson, J. 1924. Mollusca of Colorado, Utah, Montana, Idaho and Wyoming. *University of Colorado Studies* **13**: 65-223, pls. 1-3, figs. 1-96. [Supplement ibid. Vol. 23: 81-145 (1936)].

Henderson, J. 1929. The non-marine Mollusca of Oregon and Washington. *University of Colorado Studies* **17**: 47-190, 186 figs. [Supplement ibid. Vol. 23: 251-280 (1936)].

Henderson, J. 1935. *Fossil Non-Marine Mollusca of North America*. Geological Society of America Special Paper No. 3. Geological Society of America. vii + 313 pp.

Henning, T. and J. Hemmen. 1993. *Ranellidae and Personidae of the World*. Verlag Christa Hemmen, Wiesbaden, Germany. 263 pp., 30 pls.

Herbert, D. and D. Kilburn. 2004. *Field guide to the Land Snails and Slugs of Eastern South Africa*. Natal Museum, Pietermaritzburg, South Africa. 340 pp. [678 illustrations].

Herrington, H. B. 1962. A revision of the Sphaeriidae of North America (Mollusca: Pelecypoda). *Miscellaneous Publications Museum of Zoology, University of Michigan* **118**: 1-74, 7 pls.

Hershler, R. and F. G. Thompson. 1992. A review of the aquatic gastropod subfamily Cochliopinae (Prosobranchia: Hydrobiidae). *Malacological Review* Supplement **5**: 1-140, figs. 1-74.

Hickman, Jr., C. P. and Y. Finet. 1999. *A Field Guide to Marine Molluscs of Galápagos*. Sugar Spring Press, Lexington, Virginia. ix + 150 pp. [256 taxa illustrated].

Hickman, C. S. and J. H. McLean. 1990. Systematic revision and suprageneric classification of Trochacean gastropods. *Natural History Museum of Los Angeles County Science Series* **35**: i-vi + 1-169, figs. 1-100.

Higo, S., P. Callomon, and Y. Goto. 1999. *Catalogue and Bibliography of the Marine Shell-Bearing Mollusca of Japan. Elle Scientific Publications, Osaka, Japan. 749 pp.

Higo, S., P. Callomon, and Y. Goto. 2001. *Catalogue and Bibliography of the Marine Shell-Bearing Mollusca of Japan: Type Figures*. Elle Scientific Publications, Osaka, Japan. 208 pp. [2653 taxa illustrated].

Hinton, A. G. 1972. *Shells of New Guinea and the Central Indo-Pacific*. The Jacaranda Press, Milton, Australia. xviii + 94 pp., 44 pls. [Published in the United States by Charles E. Tuttle in 1975.]

Hinton, A. G. 1977. *Guide to Australian Shells*. Robert Brown and Associates Pty. Ltd., Port Moresby, Papua New Guinea. 82 pp., 77 pls.

Hinton, A. [n.d., circa 1978]. *Guide to Shells of Papua New Guinea*. Robert Brown and Associates Pty. Ltd., Port Moresby, Papua New Guinea. 74 pp., 68 pls.

Hirase, S. and I. Taki (revised and enlarged by). 1951. *An Handbook of Illustrated Shells, in Natural Colors, from the Japanese Islands and Their Adjacent Territory*. Revised and Enlarged Ed. of "A Collection of Japanese Shells". Maruzen, Tokyo. xxvii + 134, 46 pp., 134 pls.

Hoagland, K. E. 1977. Systematic review of fossil and recent *Crepidula* and discussion of evolution of the Calyptraeidae. *Malacologia* 16: 353-420, 28 figs.

Hodgkin, E. P., G. Kendrick, L. Marsh, and S. Slack-Smith. 1966. *The Shelled Gastropoda of South Western Australia. Handbook* 9. Western Australian Naturalist's Club, Perth, Australia. 60 pp., 21 pls.

Holme, N. A. and A. D. McIntyre, eds. 1971. *Methods for the Study of Marine Benthos*. Blackwell Scientific Publications, Oxford, U.K. xii + 334 pp.

Horikawa, Y. 1935a. A list of fresh water shells of Taiwan. *Venus* 5: 26-33, 2 pls. [In Japanese].

Horikawa, Y. 1935b. Distribution of fresh water shells of Taiwan. *Transactions of the Natural History Society of Taiwan* 25: 226-231. [In Japanese, checklist].

Houart, R. 1992. The genus *Chicoreus* and related genera (Gastropoda: Muricidae) in the Indo-West Pacific. *Mémoires du Muséum National d'Historie Naturelle Série A, Zoologie* 154: 1-188, pls. 1-4, figs. 1-140.

Houart, R. 1994. *Illustrated Catalogue of Recent Species of Muricidae Named Since 1971*. Verlag Christa Hemmen, Wiesbaden, Germany. 179 pp., 28 pls.

Houbrick, R. S. 1978. The Family Cerithiidae in the Indo-Pacific Part 1: The Genera *Rhinoclavis*, *Pseudovertagus* and *Clavocerithium*. *Monographs of Marine Mollusca* 1: 1-130, pls. 1-98.

Houbrick, R. S. 1992. Monograph of the genus *Cerithium* Bruguière in the Indo-Pacific. *Smithsonian Contributions to Zoology* 510: 1-211.

Howells, R. G., R. W. Neck, and H. D. Murray. 1996. *Freshwater Mussels of Texas*. Texas Parks and Wildlife Press, Austin, Texas. iv + 218 pp. [52 taxa illustrated].

Hubendick, B. 1951. Recent Lymnaeidae. Their variation, morphology, taxonomy, nomenclature, and distribution. *Kungligasvenska Vetenskapsakademiens Handlingar Fjärde Serien* 3: 1-225, pls. 1-5, figs 1-369.

Hubricht, L. 1985. The distribution of the native land mollusks of the eastern United States. *Fieldiana: Zoology (New Series)* 24: i-viii + 1-191.

Humfrey, M. 1975. *Sea Shells of the West Indies, A Guide to the Marine Molluscs of the Caribbean*. Taplinger Publishing Company, New York. 351 pp., 32 pls.

Hylleberg, J. 2004. Lexical approach to Cardiacea. *Phuket Marine Biological Center Special Publication* 29: Part 1: 1-352, 149 pls; Part 2: 353-644; Part 3: 645-939, 92 pls.

Hylleberg, J. and R. N. Kilburn. 2003. *Marine Molluscs of Vietnam - Polyplacophora, Gastropoda, Cephalopoda, Bivalvia, Scaphopoda*. ConchBooks, Hackenheim, Germany. 300 pp., 10 pls.

Hyman, L. H. 1967. *The Invertebrates*. Vol. 6: Mollusca I. Aplacophora, Polyplacophora, Monoplacophora, Gastropoda. The coelomate Bilateria. McGraw-Hill, Inc., New York. vii + 792 pp.

International Commission on Zoological Nomenclature. 1999. *International Code of Zoological Nomenclature, Fourth Edition*. The International Trust for Zoological Nomenclature, London. xxix + 306 pp.

Iredale, T. 1934. The freshwater mussels of Australia. *Australian Zoologist* 8: 57-78, pls. 3-6.

Iredale, T. 1937a. A basic list of the land Mollusca of Australia. *Australian Zoologist* 8: 287-333.

Iredale, T. 1937b. A basic list of the land Mollusca of Australia - Part II. *Australian Zoologist* 9: 1-39, pls. 1-3.

Iredale, T. 1938. A basic list of the land Mollusca of Australia - Part III. *Australian Zoologist* 9: 83-124, pls. 12-13.

Iredale, T. 1943. A basic list of the fresh-water Mollusca of Australia. *Australian Zoologist* 10: 188-230.

Iredale, T. and D. F. McMichael. 1962. *A Reference List of the Marine Mollusca of New South Wales*. The Australian Museum, Sydney. 109 pp.

Ituarte, C. F. 1996. Argentine species of *Pisidium* Pfeiffer, 1821, and *Musculium* Link, 1807 (Bivalvia: Sphaeriidae). *Veliger* 39: 189-203.

Jacobson, M. K. and K. J. Boss. 1973. The Jamaican land snails described by C. B. Adams. *Occasional Papers on Mollusks, Museum of Comparative Zoology of Harvard University* 3: 305-519, pls. 54-91.

Jacobson, M. K. and W. K. Emerson. 1971. *Shells from Cape Cod to Cape May with Special Reference to the New York City Area*. Dover Publications, Inc., New York. xviii + 152 pp. [Originally published as *Shells of the New York City Area* (1961); the Dover edition has updated nomenclature and an appendix containing 15 species not included in the 1961 version].

Jaeger, E. C. 1955. *A Source-book of Biological Names and Terms*. 3rd Ed. Charles C. Thomas, Publisher, Springfield, Illinois. xxxv + 323 pp.

Jansen, P. 1995. *Seashells of Central New South Wales*. Privately published, Townsville, Australia. xi + 129 pp, 484 figs.

Jansen, P. 1996. *Common Seashells of Coastal Northern Queensland*. Privately published, Townsville, Australia. 56 pp., 4 pls. [200 taxa illustrated].

Jansen, P. 2000. *Seashells of South-East Australia*. Capricornica Publications, Lindfield, New South Wales, Australia. 118 pp, 414 figs.

Janssen, A. W. and E. F. De Vogel. 1965. *Zoetwatermollusken van Nederland*. Nederlandse Jeugdbond voor Natuurstudie, Amsterdam, The Netherlands. 160 pp. 17 pls.

Jarrett, A. G. 2000. *Marine Shells of the Seychelles*. Carole Green Publishing, Cambridge, U.K. xiv + 149 pp, 649 figs.

Johnson, C. W. 1934. List of marine Mollusca of the Atlantic Coast from Labrador to Texas. *Proceedings of the Boston Society of Natural History* **40**: 1-204.

Johnson, R. I. 1972. The Unionidae (Mollusca: Bivalvia) of peninsular Florida. *Bulletin of the Florida State Museum, Biological Sciences* **16**: 181-249, 12 figs.

Johnson, R. I. 1981. Recent and fossil Unionacea and Mutelacea (Freshwater Bivalves) of the Caribbean Islands. *Occasional Papers on Mollusks, Museum of Comparative Zoology of Harvard University* **4**: 269-288, pls. 38-39.

Johnson, R. I. 1999. Unionidae of the Rio Grande (Rio Bravo del Norte) system of Texas and Mexico. *Occasional Papers on Mollusks, Museum of Comparative Zoology of Harvard University* **6**: 1-65. pls. 1-7.

Jokinen, E. H. 1983. *The Freshwater Snails of Connecticut*. State Geological and Natural History Survey of Connecticut, Hartford, Connecticut. vii + 83 pp., 35 figs.

Jokinen, E. H. 1992. *The Freshwater Snails (Mollusca: Gastropoda) of New York State*. University of the State of New York, State Education Dept., New York State Museum Biological Survey. Albany, New York. vi + 112 pp., 28 figs.

Jonkers, H. A. 2003. Late Cenozoic-Recent Pectinidae (Mollusca: Bivalvia) of the Southern Ocean and neighbouring regions. *Monographs of Marine Mollusca* **5**: i-viii + 1-125, pls. 1-16.

Jung, P. and R. T. Abbott. 1967. The genus *Terebellum* (Gastropoda: Strombidae). *Indo-Pacific Mollusca* **1**: 445-454, pls. 318-327.

Kaas, P. 1972. Polyplacophora of the Caribbean Region. *Studies on the Fauna of Curaçao and other Caribbean Islands* **41**: 1-162, pls. 1-9, figs. 1-247.

Kaas, P. and R. A. Van Belle. 1985a. *Monograph of Living Chitons (Mollusca: Polyplacophora)*, Vol. 1, Order Neoloricata: Lepidopleurina. E. J. Brill/Dr. W. Backhuys, Leiden, The Netherlands. 240 pp., 95 pls.

Kaas, P. and R. A. Van Belle. 1985b. *Monograph of Living Chitons (Mollusca: Polyplacophora)*, Vol. 2, Suborder Ischnochitonina: Ischnochitonidae: Schizoplacinae, Callochitoninae and Lepidochitoninae. E. J. Brill/Dr. W. Backhuys, Leiden, The Netherlands. 198 pp., 76 pls.

Kaas, P. and R. A. Van Belle. 1987. *Monograph of Living Chitons (Mollusca: Polyplacophora)*, Vol. 3, Suborder Ischnochitonina: Ischnochitonidae: Chaetopleurinae, Ischnochitoninae (pars), additions to Vol. 1 and 2. E. J. Brill/Dr. W. Backhuys, Leiden, The Netherlands. 302 pp., 117 pls.

Kaas, P. and R. A. Van Belle. 1990a. *Monograph of Living Chitons (Mollusca: Polyplacophora)*, Vol. 4, Suborder Ischnochitonina: Ischnochitonidae: Ischnochitoninae (continued), additions to Vol. 1, 2 and 3. E. J. Brill/Dr. W. Backhuys, Leiden, The Netherlands. 298 pp., 117 pls., 141 figs.

Kaas, P. and R. A. Van Belle. 1990b. *Monograph of Living Chitons (Mollusca: Polyplacophora)*, Vol. 5, Suborder Ischnochitonina: Ischnochitonidae: Ischnochitoninae (continued), additions to Vols. 1-4. E. J. Brill/Dr. W. Backhuys, Leiden, The Netherlands. 402 pp.

Kaas, P. and R. A. Van Belle. 1998. *Catalogue of Living Chitons (Mollusca: Polyplacophora)*. 2nd Revised Ed. Backhuys Publishers, Leiden, The Netherlands. 208 pp.

Kabat, A. R. and K. J. Boss. 1992. An indexed catalogue of publications on molluscan type specimens. *Occasional Papers on Mollusks, Museum of Comparative Zoology, Harvard University* **5**: 157-336. [Addendum to this work: Kabat, A. R. and K. J. Boss. 1997. *Occasional Papers on Mollusks, Museum of Comparative Zoology, Harvard University* **5**: 337-370].

Kabat, A. R. and R. Hershler. 1993. The prosobranch snail family Hydrobiidae (Gastropoda: Rissooidea): review of classification and supraspecific taxa: *Smithsonian Contributions to Zoology* **547**: 1-94. [Checklist].

Kaicher, S. D. 1973-1992. *Card Catalog of World-wide Shells*. Packs 1-60.

Kay, E. A. 1979. *Hawaiian Marine Shells. Reef and Shore Fauna of Hawaii* Section 4: Mollusca. Bishop Museum Press, Honolulu, Hawaii. xviii + 653 pp., 195 figs.

Kay, E. A. and O. Schoenberg-Dole. 1991. *Shells of Hawai'i*. University of Hawaii Press, Honolulu, Hawaii. 89 pp., 141 figs. [Marine, freshwater and terrestrial].

Keawjam, R. S. 1986. The apple snails of Thailand: Distribution, habitats and shell morphology. *Malacological Review* **19**: 61-81, figs. 1-12.

Keen, A. M. 1971. *Sea Shells of Tropical West America. Marine Mollusks from Baja California to Peru*, 2nd Ed. Stanford University Press, Stanford, California. 1064 pp., 22 pls.

Keen, A. M. 1977. Comment on "A Review of the Eratoidae" by Crawford N. Cate. *Veliger* **19**: 446-448.

Keen, A. M. and E. V. Coan. 1974. *Marine Molluscan Genera of Western North America*, 2nd Ed. Stanford University Press, Stanford, California. 208 pp.

Kennelly, D. H. 1964. *Marine Shells of South Africa*. Thomas Nelson and Sons (Africa) (Pty) Ltd., Johannesberg, South Africa. 92 pp., 32 pls.

Kensley, B. F. 1974. Contributions to the knowledge of South African marine Mollusca. Index: Parts 1-7. *Annals of the South African Museum* **47**: 783-830.

Kerney, M. P. 1999. *Atlas of the Land and Freshwater Molluscs of Britain and Ireland*. Harley Books for the Conchological Society of Great Britain and Ireland, Colchester, U.K.. 264 pp. [208 taxa illustrated].

Kerney, M. P. and R. A. D. Cameron. 1979. *A Field Guide to the Land Snails of Britain and North-West Europe*. William Collins Sons and Company, Ltd., London. 288 pp., 24 pls.

Kerney, M. P., R. A. D. Cameron, and J. H. Jungbluth. 1983. *Die Landschnecken Nord- und Mitteleuropas*. Verlag Paul Parey, Hamburg, Germany. 384 pp., 24 pls., 890 figs.

Khatoon, S. and S. R. Ali. 1978. Freshwater molluscs of Pakistan. *Bulletin of Hydrobiology Research* Series 1 **24/25**: 518-525, 2 pls.

Kilburn, R. N. 1981. Revision of the genus *Ancilla* Lamarck, 1799 (Mollusca: Olividae: Ancillinae). *Annals of the Natal Museum* **24**: 349-463, 262 figs.

Kilburn, R. and E. Rippey. 1982. *Sea Shells of Southern Africa*. Macmillan South Africa, Ltd. xi + 249 pp., 45 pls., 230 figs.

Kira, T. 1955. *Coloured Illustrations of the Shells of Japan*. Hoikusha Publishing Co., Ltd., Osaka, Japan. 224 pp., 67 pls. [In Japanese, from 1954 to 1989 there have been two editions and 38 printings of this work; there may be minor differences among printings, see: Bieler and Petit (1990) for additional details].

Kira, T. 1962. *Shells of the Western Pacific in Color*, 3rd Ed. Hoikusha Publishing Co., Ltd., Osaka, Japan. 224 pp., 72 pls. [From 1962 to 1975 there have been three editions and nine printings of this work; there may be minor differences among printings, see: Bieler and Petit (1990) for additional details. This is an English edition of Kira 1955].

Kirtisinghe, P. 1978. *Sea Shells of Sri Lanka: Including forms scattered throughout the Indian and Pacific Oceans*. Charles E. Tuttle Company, Inc., Rutland, Vermont. 202 pp., 61 pls.

Knop, D. 1996. *Giant Clams*. Dahne Verlag, Ettlingen, Germany. 255 pp. [Illustrated].

Knudsen, J. 1970. The systematics and biology of abyssal and hadal Bivalvia. *Galathea Reports* **11**: 7-236, pls. 1-20, figs. 1-132.

Knudsen, J. 1985. Abyssal Mollusca from the Arctic Ocean. *Journal of Conchology* **32**: 97-107, 2 pls.

Kohn, A. J. 1992. *A Chronological Taxonomy of* Conus, *1758-1840*. Smithsonian Institution Press, Washington, D.C. x + 315 pp., 26 pls.

Korniushin, A. V. 1994. Review of the European species of the genus *Sphaerium* (Mollusca, Bivalvia, Pisidioidea). *Ruthenica* **4**: 43-60, 12 figs.

Korniushin, A. V. 2000. Review of the family Sphaeriidae (Mollusca, Bivalvia) of Australia, with description of four new species. *Records of the Australian Museum* **52**: 41-102, 58 figs.

Korniushin, A. V. 2001. Taxonomic revision of the genus *Sphaerium sensu lato* in the Palaearctic Region, with some notes on the North American species. *Archiv für Molluskenkunde* **129**: 77-122, figs. 1-29.

Kosuge, S. and M. Suzuki. 1985. Illustrated catalogue of *Latiaxis* and its related groups. Family Coralliophilidae. *Institute of Malacology of Tokyo Special Publication* **1**: 1-83, pls. 1-50.

Kreipl, K. 1997. *Recent Cassidae*. Verlag Christa Hemmen, Wiesbaden, Germany. 151 pp., 24 pls.

Kreipl, K. and A. Alf. 1999. *Recent Xenophoridae*. ConchBooks, Hackenheim, Germany. 148 pp., 28 pls.

Kreipl, A., G. T. Poppe, L. M. in't Velt, and K. de Türck. 1999. *A Conchological Iconography: The Family Strombidae*. ConchBooks, Hackenheim, Germany. 59 pp., 130 pls.

Kruglov, N. D. and Y. I. Starobogatov. 1993a. Guide to Recent molluscs of northern Eurasia. 3. Annotated and illustrated catalogue of species of the family Lymnaeidae (Gastropoda Pulmonata Lymnaeiformes) of Palaearctic and adjacent river drainage areas. Part 1. *Ruthenica* **3**: 65-92, 15 figs.

Kruglov, N. D. and Y. I. Starobogatov. 1993b. Guide to Recent molluscs of northern Eurasia. 3. Annotated and illustrated catalogue of species of the family Lymnaeidae (Gastropoda Pulmonata Lymnaeiformes) of Palaearctic and adjacent river drainage areas. Part 2. *Ruthenica* **3**: 161-180, 10 figs.

Kuiper, J. G. J. 1964. Contribution to the knowledge of the South Africa species of the genus *Pisidium*. *Annals of the South African Museum* **48**: 77-95, 32 figs.

Kuiper, J. G. J. 1966a. Critical revision of the New Zealand Sphaeriidae clams in the Dominion Museum, Wellington. *Records of the Dominion Museum* **5**: 147-162, 27 figs.

Kuiper, J. G. J. 1966b. Les espèces africaines du genre *Pisidium*, leur synonymie et leur distribution (Mollusca: Lamellibranchiata: Sphaeriidae). *Annales du Musée Royal de l'Afrique Centrale, Série in 8°, Sciences Zoologiques* **151**: i-viii + 1-78, pls. 1-15.

Kuiper, J. G. J. 1983. The Sphaeriidae of Australia. *Basteria* **47**: 3-52, figs. 1-100.

Kuiper, J. G. J., K. A. Økland, J. Knudsen, L. Koli, T. von Proschwitz, and I. Valovirta. 1989. Geographical distribution of the small mussels (Sphaeriidae) in northern Europe (Denmark, Faroes, Finland, Iceland, Norway and Sweden). *Annales Zoologici Fennici* **26**: 73-101. [Checklist].

Kuroda, T. 1941. A catalogue of molluscan shells from Taiwan (Formosa), with description of new species. *Memoires of the Faculty of Science and Agriculture, Taihoku Imperial University* **22**: 65-216, 7 pls. [Freshwater, land, and marine].

Kuroda, T. and T. Habe. 1952. *Check List and Bibliography of the Recent Marine Mollusca of Japan*. Leo. W. Stach, Publisher, Tokyo. 210 pp.

Kuroda, T. and T. Habe. 1981. A catalogue of molluscs of Wakayama Prefecture, The Province of Kii. I. Bivalvia, Scaphopoda and Cephalopoda. Based on the Kuroda's manuscript and supervised by Tadashige Habe. *Seto Marine Biological Laboratory, Special Publication Series* 7: i- xx + 1-301, pls. 1-13.

Kuroda, T., T. Habe, and K. Oyama. 1971. *Sagami-wan san Kairui (The Sea Shells of Sagami Bay)*. Maruzen, Tokyo. xix + 741(Japanese), 489 (English), 51(index) pp., 121 pls. [In Japanese and English].

Kwon, Oh Kil. 1990. Mollusca (I). *Illustrated Encyclopedia of Fauna & Flora of Korea* 32: 1- 446, pls. 1-23 (bivalves), pls. 1-25 (gastropods). [In Korean].

Kwon, Oh Kil. and Gab-Man Park. 1993. *Coloured Shells of Korea*. Academy Publishing Company, Seoul, Korea. 445 pp., 114 pls. [In Korean].

Lai, Kin-Yang. [1986]. *Marine Gastropods of Taiwan (1)*. Taiwan Museum, Taiwan. 49 pp., 23 pls.

Lai, Kin-Yang 1987. *Marine Gastropods of Taiwan (2)*. Taiwan Museum, Taiwan. 116 pp., 52 pls.

Laidlaw, F. F. and A. Solem. 1961. The land snail genus *Amphidromus* - a synoptic catalogue. *Fieldiana Zoology* 41: 505-677, figs. 15-40.

Lalli, C. M. and R. W. Gilmer. 1989. *Pelagic Snails: The Biology of Holoplanktonic Gastropod Mollusks*. Stanford University Press, Stanford, California. xiv + 259 pp., 91 figs.

Lamprell, K. 1986. *Spondylus: Spiny Oyster Shells of the World*. Robert Brown and Associates Pty. Ltd., Bathurst, Australia. 84 pp., 36 pls.

Lamprell, K. and J. Healy. 1998. *Bivalves of Australia*, Vol. 2. Backhuys Publisher, Leiden, The Netherlands. 288 pp., 112 pls.

Lamprell, K. and T. Whitehead. 1992. *Bivalves of Australia*, Vol. 1. Crawford House Press Pty. Ltd., Bathurst, New South Wales, Australia. xiii + 182 pp., 77 pls.

Lange de Morretes, F. 1949. Ensaio de catálogo dos moluscos do Brasil. *Arquivos do Museu Paranaense* 7: 5-216.

Lange de Morretes, F. 1953. Adende e corrigenda ao ensaio de catálogo do moluscos do Brasil. *Arquivos do Museu Paranaense* 10: 37-76.

La Rocque, A. 1953. *Catalogue of the Recent Mollusca of Canada*. Queen's Printer and Controller of Stationary. Ottawa, Canada. ix + 406 pp.

Leal, J. H. 1990. *Marine Prosobranch Gastropods from Oceanic Islands off Brazil: Species Composition and Biogeography*. University of Miami, Coral Gables, Florida. x + 418 pp., 25 pls.

Lellák, J. 1975. *Shells of Britain and Europe*. Hamlyn Publishing Group, Ltd., New York. 235 pp. [Text illustrations].

Leonard, A. B. 1959. Handbook of gastropods in Kansas. *Kansas Museum of Natural History Miscellaneous Publication* 20: 1-224, pls. 1-11.

Likharev, I. M. [as Likhachev in translation] and E. S. Rammel'meier. 1962. *Terrestrial Mollusks of the Fauna of the U.S.S.R*. Academy of Sciences of the U.S.S.R. Zoological Institute 43: 1-511, figs. 1-420. [Translated from the Russian publication of 1952. Jerusalem: Israel Program for Scientific Translations. Available from National Technical Information Service (NTIS), Springfield, Virginia 22161 Document No. TT 60-21816].

Liltved, W. R. 1989. *Cowries and their Relatives of Southern Africa*. Seacomber Publications. 208 pp., 298 figs.

Linse, K. 1997. Die verbreitung epibenthischer Mollusken im Chilenischen Beagle-Kanal. *Berichte zur Polarforschung* 228: i-vi + 1-131, 2 pls.

Linse, K. 2002. *The shelled Magellanic Mollusca: With special reference to biogeographic relations in the Southern Ocean*. A. R. G. Gantner Verlag KG, Rugell, Liechtenstein. vii + 252 pp., 21 pls.

Lipe, R. 1991. *Marginellas*. The Shell Store, St. Petersburg Beach, Florida. 40 pp. 18 pls.

Lipe, R. E. and R. T. Abbott. 1991. *Living Shells of the Caribbean and Florida Keys*. American Malacologists, Inc., Melbourne, Florida. 80 pp. [Illustrated].

Long Island Shell Club. 1988. *Seashells of Long Island, New York*. The Long Island Shell Club. 209 pp. [Illustrated].

Lorenz, F. 1998. Kauris von Ostafrika. *Schriften zur Malakozoologie* 11: 1-150, pls. 1-26, figs. 1-121.

Lorenz, F. 2001. *Monograph of the Living Zoila: A fascinating group of Australian endemic Cowries (Mollusca: Prosobranchia: Cypraeidae)*. ConchBooks, Hackenheim, Germany. 187 pp., 54 pls.

Lorenz, F. 2002. New worldwide cowries. Description of new taxa and revisions of selected groups of living Cypraeidae (Mollusca: Gastropoda). *Schriften zur Malakozoologie* 20: 1-292, pls. 1-40.

Lorenz, F. and A. Hubert. 2000. *A Guide to Worldwide Cowries, Second, Enlarged and Completely Revised Edition*. ConchBooks, Hackenheim, Germany. 584 pp., 128 pls.

Lozek, V. 1964. *Quartärmollusken der Tschechoslowakei*. Prague: Herausgegeben von der Geologischen Zentralanstalt im Verlag der Tschechoslowakischen Akademie des Wissenschaften. 374 pp., 32 pls.

Lubinsky, I. 1980. Marine bivalve molluscs of the Canadian central and eastern Arctic: Faunal composition and zoogeography. *Canadian Bulletin of Fisheries and Aquatic Sciences* 207: i-vi + 1-111, pls. 1-11.

Macedo, M. C. C., M. I. C. Macedo, and J. P. Borges. 1999. *Conchas Marinhas de Portugal: Seashells of Portugal*. Editorial Verbo, Lisbon, Portugal. 516 pp. [Introduction in English, illustrated].

MacFarland, F. M. 1966. Studies of opisthobranchiate mollusks of the Pacific Coast of North America. *California Academy of Sciences Memoir* 6: 1-546, pls. 1-72.

MacGinitie, G. E. 1955. Distribution and ecology of the marine invertebrates of Point Barrow, Alaska.

Smithsonian Miscellaneous Collections **128**: i-iv + 1-201, pls. 1-8.

MacGinitie, N. 1959. Marine Mollusca of Point Barrow, Alaska. *Proceedings of the United States National Museum* **109**: 59-208, pls. 1-27.

MacMillan, G. K. 1950. The land snails of West Virginia. *Annals of the Carnegie Museum* **31**: 89-238, 15 pls.

MacPherson, E. 1971. The marine molluscs of Arctic Canada. Prosobranch gastropods, chitons and scaphopods. *National Museum of Natural Science Publication in Biological Oceanography* **3**: i-viii + 1-149, pls. 1-7.

MacPherson, J. H. and C. J. Gabriel. 1962. *Marine Molluscs of Victoria.* Melbourne University Press, Melbourne, Australia. xv + 475 pp., 486 figs.

Macsotay, O. and R. Campos Villarroel. 2001. *Moluscos Representativos de la Plataforma de Margarita - Venezuela - Descripción de 24 Especies Nuevas.* Privately published, Valencia, Venezuela. iii + 230 pp., 32 pls.

Maes, V. O. 1967. The littoral marine mollusks of Cocos-Keeling Islands (Indian Ocean). *Proceedings of the Academy of Natural Sciences* **119**: 93-217, 26 pls.

Mandahl-Barth, G. 1938. Land and freshwater Mollusca. *Zoology of Iceland* **4**: 1-31.

Mandahl-Barth, G. 1954. The freshwater mollusks of Uganda and adjacent territories. *Annales du Musée Royal du Congo Belge, Série in 8°, Sciences Zoologiques* **32**: 1-206, figs. 1-96.

Mandahl-Barth, G. 1988. *Studies on African Freshwater Bivalves.* Danish Bilharziasis Laboratory. Charlottenlund, Denmark. 161 pp., 330 figs. [A compendium of three earlier publications from 1982, 1983, and 1985, covering Unionacea and Sphaeriidae].

Mansur, M. C. D. 1970. Lista dos moluscos bivalves das famílias Hyriidae e Mycetopodidae para o estado do Rio Grande do Sul. *Iheringia Série Zoologia* **39**: 33-95.

Mansur, M. C. D. and C. Meier-Brook. 2000. Morphology of *Eupera* Bourguignat 1854, and *Byssanodonta* Orbigny 1846 with contributions to the phylogenetic systematics of Sphaeriidae and Corbiculidae (Bivalvia: Veneroida). *Archiv für Molluskenkunde* **128**: 1-59, figs. 1-151.

Mansur, M. C. D. and R. M. Valer. 1992. Moluscos bivalves do Rio Uraricoera e Rio Branco, Roraima, Brazil. *Amazoniana* **12**: 85-100, 10 figs.

Marincovich, Jr., L. 1973. Intertidal Mollusks of Iquique, Chili. *Natural History Museum of Los Angeles County Science Bulletin* **16**: 1-49, figs. 1-102.

Marsh, J. A. and O. H. Rippingale. 1974. *Cone Shells of the World*, 3rd Ed. Jacaranda Press Pty., Ltd., Melbourne, Australia. 185 pp., 24 pls.

Martin, S. M. 1997. Freshwater mussels (Bivalvia: Unionoida) of Maine. *Northeastern Naturalist* **4**: 1-34, 2 pls.

Martin, S. M. 1998. Freshwater Fingernail and Pea Clams (Bivalvia: Veneroida: Sphaeriidae) of Maine. *Northeastern Naturalist* **5**: 29-60, pl. 1.

Martin, S. M. 2000. Terrestrial snails and slugs (Mollusca: Gastropoda) of Maine. *Northeastern Naturalist* **7**: 33-88.

Mathiak, H. A. 1979. *A River Survey of the Unionid Mussels of Wisconsin 1973-1977.* Sand Shell Press, Horicon, Wisconsin. 75 pp., 11 pls.

Maury, C. J. 1971. *Recent Molluscs of the Gulf of Mexico and the Pleistocene and Pliocene Species from the Gulf States.* Paleontological Research Institution, Ithaca, New York. iv + 282 pp. [Originally published in the *Bulletins of American Paleontology* Volume 8 (34) 1920 and Volume 9 (38) 1922, Checklist].

May, W. L. and J. H. Macpherson (revisor). 1958. *An Illustrated Index of Tasmanian Shells.* L. G. Shea, Government Printer, Tasmania, Australia. 72 pp., 50 pls. [Revised edition of May's 1923 Index, including additional plates and text].

Mayissian, S. 1974. *Coquillages de Nouvelle-Caledonie et de Melanesie.* Privately published. 72 pp., 28 pls.

Mayr, E. and P. D. Ashlock. 1991. *Principles of Systemic Zoology.* McGraw-Hill, Inc., New York. xx + 475 pp.

McDonald, G. R. and J. W. Nybakken. 1981. *Guide to the Nudibranchs of California.* R. Tucker Abbott, ed. American Malacologists, Inc., Melbourne, Florida. 67 pp. [112 taxa illustrated].

McKillop, B. 1996. Geographic and environmental distribution of freshwater gastropods in Manitoba, Canada. *Manitoba Museum of Man and Nature Occasional Series* **1**: 1-34.

McLean, J. H. 1978. Marine shells of southern California. Revised edition. *Natural History Museum of Los Angeles County Science Series 24, Zoology* **11**: 1-104, pls. 1-54.

McLean, J. H. and T. M. Gosliner. 1996. *Taxonomic Atlas of the Benthic Fauna of the Santa Maria Basin and Western Santa Barbara Channel*, Vol. 9. The Mollusca Part 2 - The Gastropoda. Santa Barbara Museum of Natural History, Santa Barbara, California. vii + 228 pp. [155 taxa illustrated].

McMichael, D. F. and I. D. Hiscock. 1958. A monograph of the freshwater mussels (Mollusca: Pelecypoda) of the Australian region. *Australian Journal of Marine and Freshwater Research* **9**: 372-508, 19 pls.

McMichael, D. F. and T. Iredale. 1959. The land and freshwater Mollusca of Australia. *Monographiae Biologicae* **8**: 224-245, 2 pls.

McMillan, N. F. 1975. *British Shells.* Revised Ed. Frederick Warne, London. 196 pp. [355 taxa illustrated].

Mead, A. R. 1961. *The Giant African Snail: A Problem in Economic Malacology.* University of Chicago Press, Chicago, Illinois. xvii + 257 pp., 15 figs.

Menzies, R. J., M. Ewing, J. L. Worzel, and A. H. Clarke, Jr. 1959. Ecology of the Recent Monoplacophora. *Oikos* **10**: 168-182, 10 figs.

Metcalf, A. L. and R. A. Smartt, eds. 1997. Land Snails of New Mexico. *New Mexico Museum of Natural History and Science Bulletin* **10**: 1-145, figs. 1-4.

Mikkelsen, P. M., R. Bieler, and R. E. Petit. 1993. A bibliography of Caribbean malacology 1826-1993. *American Malacological Bulletin* **10**: 267-290.

Millard, V. 2003. *Classification of Mollusca. A Classification of World Wide Mollusca*, 3rd Ed. Privately published, South Africa. 1992 pp. [Issued in 2004, though title page lists 2003].

Minton, R. L. and C. L. Lydeard. 2003. Phylogeny, taxonomy, genetics and global heritage rank of imperiled, freshwater snail genus *Lithasia* (Pleuroceridae). *Molecular Ecology* **12**: 75-87.

Monteiro, A., M. J. Tenorio, and G. T. Poppe. 2004. *A Conchological Iconography: The Family Conidae-The West African and Mediterranian Species of* Conus. ConchBooks, Hackenheim, Germany. 102 pp., 164 pls.

Moore, R. C. and C. Teichert, eds. 1953-1981. *Treatise on Invertebrate Paleontology*. Geological Society of America and the University of Kansas, Boulder, Colorado.
  Part I: Mollusca 1 (Mollusca - General Features - Scaphopoda - Amphineura - Monoplacophora - Gastropoda General Features - Archaeogastropoda, mainly Paleozoic Caenogastropoda and Opisthobranchia). 1960. xxiii + 351 pp., 216 figs.
  Part K: Mollusca 3 (Cephalopoda - General Features - Endoceratoidea - Actinoceratoidea - Nautiloidea - Bactritoidea). 1964. xxviii + 519 pp., 361 figs.
  Part L: Mollusca 4 (Cephalopoda - Ammonoidea). 1957. xxii + 490 pp., 558 figs.
  Part L (Revised): Mollusca 4 (Cretaceous Ammonoidea). 1996. xx + 362 pp., 216 figs.
  Part N, Vol. 1: Mollusca 6 (Bivalvia). 1969. xxxviii + 1-489., 301 figs.
  Part N, Vol. 2: Mollusca 6 (Bivalvia). 1969. 491-952., 311 figs.
  Part N, Vol. 3: Mollusca 6 (Bivalvia). 1971. iv + 953-1224, 153 figs.

Morris, S. and R. D. Purchon. 1981. The marine shelled Mollusca of West Malaysia and Singapore, Part 3: Bivalvia. *Journal of Molluscan Studies* **47**: 322-327.

Morton, J. E. 1979. *Molluscs*, 5th. Ed. Hutchinson & Co. Ltd., London. 264 pp., 60 figs.

Moscatelli, R. 1987. *The Superfamily Strombacea from Western Atlantic*. Privately published. 91 pp., 38 pls.

Mouthon, J. and J. G. J. Kuiper. 1987. Inventaire des Sphaeriidae de France. *Inventaires de Faune et de Flore* **41**: 1-60. [7 taxa illustrated].

Murray, H. D. and A. B. Leonard. 1962. Handbook of unionid mussels in Kansas, *University of Kansas Museum of Natural History, Miscellaneous Publication* **281**: 1-84, pls. 1-45.

Nakayama, T. 2003. A review of northwest Pacific epitoniids (Gastropoda: Epitoniidae). *Monographs of Marine Mollusca* **6**: i-vii + 1-143, pls. 1-20.

Narchi, W., O. Domaneschi, and F. D. Passos. 2002. Bivalves antárticos e subantárticos coletados durante as Expedições Científicas Brasileiras à Antártica I a IX (1982-1991). *Revista Brasileira de Zoologia* **19**: 645-675, 53 figs.

Neave, S. A. (ed.). 1939-1940. *Nomenclator Zoologicus: a List of the Names of Genera and Subgenera in Zoology from the Tenth Edition of Linnaeus, 1758, to the end of 1935*, Vols. 1-4. The Zoological Society of London, London. xiv + 3805 pp.

Neave, S. A. (ed.). 1950. *Nomenclator Zoologicus*, Vol. 5, 1936-1945. The Zoological Society of London, London. 308 pp.

Nedeau, E. J. [2003]. *A Field Guide to the Freshwater Mussels of Connecticut*. Bureau of Natural Resources, Wildlife Division, Connecticut Department of Environmental Protection, Hartford, Connecticut. 32 pp. [Illustrated].

Nedeau, E. J., M. A. McCollough, and B. I. Swartz. 2000. *The Freshwater Mussels of Maine*. Maine Department of Inland Fisheries and Wildlife, Augusta, Maine. 118 pp. [16 taxa illustrated].

Nesis, K. N. 1987. *Cephalopods of the World: Squids, Cuttlefishes, Octopuses, and Allies*. Translated by B. S. Levitov, edited by L. A. Burgess. TFH Publications, Neptune City, New Jersey. 351 pp., 88 pls.

Nicklès, M. 1950. *Mollusques Testacés Marins de la Côte Occidentele d'Afrique. Manuels Ouest-Africains*, Vol. 2. Libraire pour les Sciences Naturelle, Paris. 269 pp., 459 figs.

Nicol, D. 1966. Description, ecology, and geographic distribution of some Antarctic pelecypods. *Bulletins of American Paleontology* **51**: 1-102, 10 pls.

Nobre, A. 1938-1940. *Fauna Malacológica de Portugal: I. Moluscos Marinhos e das Águas Salobras. Annaes de Sciencias Naturaes* i-xiii + 1-806, 87 pls.

Nordsieck, H. 1998. Critical revision of the system of the Japanese Phaedusinae, proposed by Minato (1994) (Gastropoda: Stylommatophora: Clausiliidae). *Archiv für Molluskenkunde* **127**: 21-32.

Norman, M. 2000. *Cephalopods - A World Guide*. ConchBooks, Hackenheim, Germany. 320 pp. [800+ illustrations].

Norman, M. and A. Reid. 2000. *A Guide to Squid, Cuttlefish, and Octopuses of Australasia*. CSIRO Publishing. Victoria, Australia. 96 pp. [Illustrated].

Numanami, H. 1996. Taxonomic study on Antarctic gastropods collected by Japanese Antarctic Research Expedition. *Memoirs of the National Institute of Polar Research, Series E (Biology and Medical Science)* **39**: 1-244.

Oesch, R. D. 1984. *Missouri Naiades: A Guide to the Mussels of Missouri*. Missouri Department of Conservation, Jefferson City, Missouri. 270 pp. [illustrated].

Okutani, T. ed. 2000. *Marine Mollusks in Japan*. Tokai University Press, Tokyo. xlviii + 1173 pp. [In Japanese and English, 5100 + figures].

Oldroyd, I. S. 1924. Marine shells of Puget Sound and vicinity. *Publications of the Puget Sound Biological Station of the University of Washington* **4**: 1-272, pls. 1-49.

Oldroyd, I. S. 1924-1927. *The Marine Shells of the West Coast of North America*. Stanford University Press, Stanford, California. Vol. 1 (1924): 247 pp., 57 pls. Vol. 2 Parts I-III (1927): 941 pp., 108 pls.

Oliver, P. G. 1992. *Bivalved Seashells of the Red Sea*. Verlag Christa Hemmen, Wiesbaden, Germany. 330 pp., 46 pls.

Olsson, A. A. 1961. *Mollusks of the Tropical Eastern Pacific, Particularly from the Southern Half of the Panamic-Pacific Faunal Province (Panama to Peru). Panamic-Pacific Pelecypoda*. Paleontological Research Institution, Ithaca, New York. 574 pp., 86 pls.

Olsson, A. A. 1964. *Neogene Mollusks from Northwestern Ecuador*. Paleontological Research Institution, Ithaca, New York. 256 pp., 38 pls. [37 Recent taxa are discussed along with many fossil taxa].

Olsson, A. A. and T. L. McGinty. 1958. Recent marine mollusks from the Caribbean coast of Panama with the description of some new genera and species. *Bulletins of American Paleontology* **39**: 1-58, pls. 1-5. [A list of 534 species with descriptions of 33 new taxa].

Ortmann, A. E. 1921. South American naiades; a contribution to the knowledge of the fresh-water mussels of South America. *Memoirs of the Carnegie Museum* **8**: 451-670. Plates 34-48.

Oughton, J. 1948. A zoogeographical study of the land snails of Ontario. *University of Toronto Studies Biological Series* **57**: 1-128. [Annotated checklist].

Pace, G. L. 1973. The freshwater snails of Taiwan (Formosa). *Malacological Review* (Supplement 1): 1-118. Plates 1-19, figures 1-17.

Pace, G. L. 1975. The fresh water mussels (Bivalvia: Unionidae) of Taiwan (Formosa). *Bulletin of Malacology, Republic of China* **2**: 47-61. 3 pls.

Packard, E. L. 1918. Molluscan fauna from San Francisco Bay. *University of California Publications in Zoology* **14**: 199-452. 47 pls.

Pain, T. 1960. *Pomacea* (Ampullariidae) of the Amazon River system. *Journal of Conchology* **24**: 421-432. [Annotated checklist].

Pain, T. 1972. The Ampullariidae, an historical survey. *Journal of Conchology* **27**: 453-462.

Parkinson, B., J. Hemmen, and K. Groh. 1987. *Tropical Landshells of the World*. Verlag Christa Hemmen, Wiesbaden, Germany. 279 pp., 77 pls.

Parmalee, P. W. and A. E. Bogan. 1998. *The Freshwater Mussels of Tennessee*. The University of Tennessee Press, Knoxville, Tennessee. xii + 328 pp. 133 pls.

Parodiz, J. J. 1957. Catalog of the land Mollusca of Argentina, Parts 1-3. *Nautilus* **70**: 127-135, **71**: 22-30, 63-66. [Checklist].

Parodiz, J. J. 1968. Annotated catalogue of the genus *Diplodon* (Unionacea - Hyriidae). *Sterkiana* **30**: 1-22.

Parodiz, J. J. and A. A. Bonetto. 1963. Taxonomy and zoogeographic relationships of the South American naiades (Pelecypoda: Unionacea and Mutelacea). *Malacologia* **1**: 179-213, figs. 1-17.

Pasteur-Humbert, C. 1962a. Les mollusques marins testacés du Maroc. Catalogue non critique. I – Les Gastéropodes. *L'Institut Scientifique Chérifien Série Zoologie* **23**: 1-245, pls. 1-42.

Pasteur-Humbert, C. 1962b. Les mollusques marins testacés du Maroc. Catalogue non critique. II – Les Lamellibranches et les Scaphopodes. *L'Institut Scientifique Chérifien Série Zoologie* **28**: 1-184, pls. 1-39.

Patterson, C. M. 1971. Taxonomic studies of the land snail family Succineidae. *Malacological Review* **4**: 131-202.

Pechar, P. C. Prior, and B. Parkinson. [1984]. *Mitre Shells from the Pacific and Indian Oceans*. Robert Brown and Associates, Pty. Ltd. Bathurst, Australia. 56 pls.

Penniket, J. R. and G. J. H. Moon. 1970. *New Zealand Seashells in Colour*. A. H. and W. Reed, Ltd., Wellington, Australia. 112 pp., 51 pls.

Perera, G. and J. Walls. 1996. *Apple Snails in the Aquarium*. T. H. F. Publishing, Inc., Neptune City, New Jersey. 121 pp. [Illustrated].

Perez, K. E., W. F. Ponder, J. Donald, D. J. Colgan, S. A. Clark, and C. Lydeard. 2005. Molecular phylogeny and biogeography of spring-associated hydrobiid snails of the Great Artesian Basin, Australia. *Molecular Phylogenetics and Evolution* **34**: 545-556.

Perry, L. M. and J. S. Schwengel. 1955. *Marine Shells of the Western Coast of Florida*. Paleontological Research Institution, Ithaca, New York. 318 pp., 55 pls.

Petit, R. E. and R. Bieler. 1996. On the new names introduced in the various printings of "Shells of the World in Colour" [Vol. I by Tadashige Habe and Kiyoshi Ito; Vol. II by Tadashige Habe and Sadao Kosuge]. *Malacologia* **38**: 35-46.

Petuch, E. J. and D. M. Sargent. 1986. *Atlas of the Living Olive Shells of the World*. Coastal Education and Research Foundation, Charlottesville, Virginia. xiii +253 pp., 39 pls.

Picton, B. E. and C. C. Morrow. 1994. *A Field Guide to the Nudibranchs of the British Isles*. Immel Publishing, London. 143 pp. [Illustrated].

Piech, B. J. 1995. *Ranellidae and Personidae: A Classification of Recent Species*. Delaware Museum of Natural History, Wilmington, Delaware. 60 pp.

Piechocki, A. 1989. The Sphaeriidae of Poland (Bivalvia, Eulamellibranchia). *Annales Zoologici (Warsaw)* **42**: 249-320, 94 figs.

Pilsbry, H. A. 1911. Non-marine Mollusca of Patagonia. *Reports of the Princeton University Expedition to Patagonia* **3**: 513-633, 15 pls.

Pilsbry, H. A. 1912. A study of the variation and zoogeography of *Liguus* in Florida. *Journal of the Academy of Natural Sciences, Philadelphia* **15**: 427-472, pls. 37-40.

Pilsbry, H. A. 1919. A review of the land mollusks of the Belgian Congo chiefly based on the collection of the American Museum Congo Expedition, 1905-1915. *Bulletin of the American Museum of Natural History* **40**: i-x + 1-370, pls. 1-23, figs. 1-163.

Pilsbry, H. A. 1939-1948. Land Mollusca of North America (north of Mexico). *Academy of Natural Sciences, Philadelphia Monograph* **3**. Vol. 1, Part 1: i-xvii + 1-573 + i-ix, figs. 1-377. Part 2: i-vi + 575-994 + i-ix, figs. 378-580. Vol. 2, Part 1: i-vi + 1-520, figs. 1-281. Part 2: i-xlvii + 521-1113, figs. 282-585.

Pilsbry, H. A. and J. Bequaert. 1927. The aquatic mollusks of the Belgian Congo. With a geographical and ecological account of Congo malacology. *Bulletin of the American Museum of Natural History* **53**: 69-602, pls. 10-77.

Pinn, F. 1990. *Sea Snails of Pondicherry*. Nehru Science Centre, Pondicherry, India. xv + 116 pp., 215 figs.

Plá, E. [2000]. *Moluscos Gasterópodos y Bivalves de la Marina Alta y Baleares*. Institut de Cultura "Juan Gil-Albert", Portugal. 180 pp. [218 taxa illustrated].

Pointier, J-P. and G. Marquet. 1990. Taxonomy and distribution of freshwater mollusks of French Polynesia. *Venus* **49**: 215-231, 4 pls.

Ponder, W. F. 1983. A revision of the recent Xenophoridae of the world and of the Australian fossil species. *The Australian Museum Memoir* **17**: 1-126, pls. 1-42.

Ponder, W. F. 1997. Freshwater molluscs of northeast Tasmania. *Records of the Queen Victoria Museum and Art Gallery* **103**: 185-191.

Ponder, W. F., D. J. Eernisse, and J. W. Waterhouse, eds. 1988. Prosobranch phylogeny: Proceedings of a symposium, 9th International Malacological Congress, Edinburgh, 1986. *Malacological Review* Supplement **4**: 1-346.

Ponder, W. F. and E. H. Vokes. 1988. A revision of the Indo-West Pacific fossil and Recent species of *Murex* s. s. and *Haustellum* (Mollusca: Gastropoda: Muricidae). *Records of the Australian Museum* Supplement **8**: 1-160, figs. 1-89.

Poppe, G. T., S. P. Dance, and T. Brulet. 1999. *A Conchological Iconography: The Family Harpidae*. ConchBooks, Hackenheim, Germany. 17 pp., 51 pls.

Poppe, G. T. and Y. Goto. 1991-1993. *European Seashells*. Verlag Christa Hemmen, Wiesbaden, Germany. Vol. 1 (1991) Polyplacophora, Caudofoveata, Solenogastres, Gastropoda, 352 pp., 40 pls.; Vol. 2 (1993) Scaphopoda, Bivalvia, Cephalopoda, 221 pp., 32 pls.

Poppe, G. T. and Y. Goto. 1992. *Volutes*. [published for] Mostra Mondiale Malacologia by L'Informatore Piceno, Ancona, Italy. 348 pp., 107 pls.

Poppe, G. T. and Y. Goto. 1993. *Recent Angariidae*. [published for] Mostra Mondiale Malacologia by L'Informatore Piceno, Ancona, Italy. 32 pp., 10 pls.

Poppe, G. T. and M. Verhaeghe. 2000. *A Conchological Iconography: The Family Ficidae*. ConchBooks, Hackenheim, Germany. 31 pp., 27 pls.

Powell, A. W. B. 1960. Antarctic and subantarctic Mollusca. *Records of the Aukland Institute and Museum* **5**: 117-193. [Checklist].

Powell, A. W. B. 1964. The family Turridae in the Indo-Pacific. Part 1. The subfamily Turrinae. *Indo-Pacific Mollusca* **1**: 227-345, pls. 172-262.

Powell, A. W. B. 1967. The family Turridae in the Indo-Pacific. Part 1a. The subfamily Turrinae concluded. *Indo-Pacific Mollusca* **1**: 409-443, pls. 298-317.

Powell, A. W. B. 1969. The family Turridae in the Indo-Pacific. Part 2. The subfamily Turriculinae. *Indo-Pacific Mollusca* **2**: 215-415, pls. 188-324.

Powell, A. W. B. 1973. The patellid limpits of the World (Patellidae). *Indo-Pacific Mollusca* **3**: 75-205, pls. 60-182.

Powell, A. W. B. 1976. *Shells of New Zealand: An Illustrated Handbook. 5th revised edition*. Whitcoulls, Christchurch, New Zealand. 1976. 154 pp., 45 pls. [Formerly titled The Shellfish of New Zealand].

Powell, A. W. B. 1979. *New Zealand Mollusca. Marine, Land and Freshwater Shells*. William Collins Publishers, Ltd., Auckland, New Zealand. xiv + 500 pp., 82 pls.

Prashad, B. 1928. Recent and fossil Viviparidae. A study in distribution, evolution and paleogeography. *Memoirs of the Indian Museum* **8**: 153-251, 1 pl.

Prozorova, L. A. 1995. Species composition and classification of the genus *Pisidium* (Bivalvia, Pisidiidae) of the Russian Far East. *Zoologicheskii Zhurnal* **74**: 32-36. [In Russian].

Pugh, P. J. A. and B. Scott. 2002. Biodiversity and biogeography of non-marine Mollusca on the islands of the Southern Ocean. *Journal of Natural History* **36**: 927-952. [Annotated list].

Purchon, R. D. 1977. *The Biology of the Mollusca*. 2nd Ed. Pergamon Press, New York. xxv + 560 pp.

Purchon, R. D. and D. E. A. Purchon. 1981. The marine shelled Mollusca of West Malaysia and Singapore, Part 1: General Introduction and an account of the collecting stations. *Journal of Molluscan Studies* **47**: 290-312.

Qi, Zhongyan, ed. 2004. *Seashells of China*. China Ocean Press, Beijing. viii + 418 pp., 193 pls.

Quayle, D. B. 1970. *The Intertidal Bivalves of British Columbia*. Handbook 17, British Columbia Provincial Museum. British Columbia Provincial Museum, Department of Education, Victoria, British Columbia, Canada. 104 pp. [Text figures].

Quintana, M. G. 1982. Catologo preliminar de la malacofauna del Paraguay. *Revista del Museo Argentino de Ciencias Naturales "Bernardino Rivadavia" et*

Instituto Nacional de Investigacion de las Ciencias Naturales Zoologica **11**: 61-158.

Radwin, G. E. and E. V. Coan. 1976. A catalogue of collations of works of malacological importance. *Western Society of Malacologists Occasional Paper* **2**: 1-34.

Radwin, G. E. and A. D'Attilio. 1976. *Murex Shells of the World: An Illustrated Guide to the Muricidae.* Stanford University Press, Stanford, California. x + 284 pp., 32 pls., 192 figs.

Ramakrishna and S. C. Mitra. 2002. Endemic land molluscs of India. *Records of the Zoological Survey of India Occasional Paper* **196**: 1-65, pls. 1-13.

Redfern, C. 2001. *Bahamian Seashells: A Thousand Species from Abaco, Bahamas.* Bahamianseashell.com, Inc., Boca Raton, Florida. ix +280 pp., 124 pls.

Rehder, H. A. 1973. The family Harpidae of the World. *Indo-Pacific Mollusca* **3**: 207-274, pls. 183-247.

Rehder, H. A. 1980. The marine mollusks of Easter Island (Isla de Pascua) and Sala y Gómez. *Smithsonian Contributions to Zoology* **289**: 1-167, pls. 1-14.

Reid, D. G. 1996. *Systematics and Evolution of Littorina.* Ray Society, London. x + 463 pp., 131 figs.

Reynolds, P. D. 1997. The phylogeny and classification of Scaphopoda (Mollusca): an assessment of current resolution and cladistic reanalysis. *Zoologica Scripta* **26**: 13-21.

Rice, T. 1971. *Marine Shells of the Pacific Northwest.* Ellison Industries, Inc., Edmonds, Washington. 102 pp., 40 pls.

Richardson, L. 1980. Helicidae: Catalog of species. *Tryonia* **3**: Part 1: 1-350; Part 2: 351-697.

Richardson, L. 1982. Helminthoglyptidae: Catalog of species. *Tryonia* **6**: 1-117.

Richardson, L. 1983. Bradybaenidae: Catalog of species. *Tryonia* **9**: 1-253.

Richardson, L. 1984. Oreohelicidae: Catalog of species. *Tryonia* **10**: 1-30.

Richardson, L. 1985. Camaenidae: Catalog of species. *Tryonia* **12**: 5-49.

Richardson, L. 1986. Polygyracea: Catalog of species (Parts 1, Polygyridae; 2, Corillidae; 3, Sagdidae). *Tryonia* **13**: 1-139, 1-40, 1-38.

Richardson, C. L. 1988. Streptaxacea: Catalog of species, Part I. Streptaxidae. *Tryonia* **16**: 1-326.

Richardson, C. L. 1989. Streptaxacea: Catalog of species, Part II. Ammonidellidae, Chlamydephoridae, Haplotrematidae, Rhytididae, Systrophiidae. *Tryonia* **18**: 1-154.

Richardson, C. L. 1990. Partulidae: Catalog of species. *Tryonia* **19**: i + 1–96.

Richardson, C. L. 1991. Urocoptidae: Catalog of species. *Tryonia* **22**: 1-245.

Richardson, C. L. 1992. Cerionidae: Catalog of species. *Tryonia* **25**: 1-121.

Richling, I. 2000. Arktische Bivalvia - eine taxonomische Bearbeitung auf Grundlage des Materials der Expeditionen Transdrift 1 und ARK IX/4 (1993) in das Laptevmeer. *Schriften zur Malakozoologie* **15**: 1-93, figs. 1-84.

Richling, I. 2004. Classification of the Helicinidae: Review of morphological characteristics based on a revision of the Costa Rican species and application to the arrangement of the Central American mainland taxa (Mollusca: Gastropoda: Neritopsina). *Malacologia* **45**: 195-440, figs. 1-340.

Richmond, M. H. 1992. *Tasmanian Sea Shells*, Vol. 2. Richmond Printers, Devonport, Tasmania, Australia. 111 pp. [139 taxa illustrated].

Richmond, M. H. 1997. *Tasmanian Sea Shells Common to other Australian States*, Revised Ed. Richmond Printers, Devonport, Tasmania, Australia. 80 pp. [170 taxa illustrated].

Riedel, A. 1980. *Genera Zonitidarum: Diagnosen supra spezifischer Taxa der Familie Zonitidae (Gastropoda, Stylommatophora).* Dr. W. Backhuys, Publisher, Rotterdam, The Netherlands. 197 pp., 2 pls., 294 figs.

Rios, E. 1994. *Seashells of Brazil*, 2nd Ed. Universidade do Rio Grande, Rio Grande, Brazil. 492 pp., 113 pls.

Rippingale, O. H. and D. F. McMichael. 1961. *Queensland and Great Barrier Reef Shells.* The Jacaranda Press Pty., Ltd., Brisbane, Australia. 210 pp., 29 pls.

Robba, E., I. DiGeronimo, N. Chaimanee, M. P. Negri, and R. Sanfilippo. 2002. Holocene and Recent shallow soft-bottom mollusks from the northern Gulf of Thailand area: Bivalvia. *Bollettino Malacologico* **38**: 49-132, 21 pls.

Robba, E., I. DiGeronimo, N. Chaimanee, M. P. Negri, and R. Sanfilippo. 2004. Holocene and Recent shallow soft-bottom mollusks from the northern Gulf of Thailand area: Scaphopoda, Gastropoda, additions to Bivalvia. *La Conchiglia* **35** (Supplement to No. 309): 5-288, 37 pls. [In Italian and English].

Robertson, I. C. S. and C. L. Blakeslee. 1948. The Mollusca of the Niagara frontier region and adjacent territory. *Bulletin of the Buffalo Society of Natural Sciences* **19**: i-xi + 1-191, pls. 1-14.

Robertson, R. 1985. Archaegastropod biology and the systematics of the genus *Tricolia* (Trochacea: Tricoliidae) in the Indo-West Pacific. *Monographs of Marine Mollusca* **3**: 1-103, pls. 1-96.

Robin, A. and J. C. Martin. 2004. *Mitridae Costellariidae.* ConchBooks, Hackenheim, Germany. 34 pp., 32 pls.

Robson, G. C. 1929-1932. *A Monograph of the Recent Cephalopoda Based on the Collections in the British Museum (Natural History).* British Museum, London. Part I (1929): Octopodinae. xi + 236 pp., 7 pls.; Part II (1932): The Octopoda (excluding the Octopodinae). xi + 359 pp., 6 pls.

Röckel, D., W. Korn, and A. J. Kohn. 1995. *Manual of the Living Conidae.* Vol. 1: Indo-Pacific Region. Verlag Christa Hemmen, Wiesbaden, Germany. 516 pp., 84 pls.

Rombouts, A. 1991. *Guidebook to Pecten Shells: Recent Pectinidae and Propeamussiidae of the World*. Edited and revised by H. E. Coomans, H. H. Dijkstra, R. G. Moolenbeek, and P. L. van Pel. Universal Book Services/Dr. W. Backhuys, Oegstgeest, The Netherlands. xiii + 157 pp., 29 pls.

Roper, C. F. E., M. J. Sweeney, and C. Nauen. 1984. Cephalopods of the world. An annotated and illustrated catalogue of species of interest to fisheries. *FAO Fisheries Synopsis* **125**, Vol. 3: i-viii + 1-277.

Roper, C. F. E., R. E. Young, and G. L. Voss. 1969. An illustrated key to the families of the order Teuthoidea (Cephalopoda). *Smithsonian Contributions to Zoology* **13**: 1-32, pls. 1-16.

Rosenberg, G. 1992. *The Encyclopedia of Seashells*. Robert Hale, Ltd., London. 224 pp. [Text illustrations].

Rosenberg, G. and R. E. Petit. 2003. Kaicher's card catalog of World-wide shells: A collation, with discussion of species names therein. *Nautilus* **117**: 99-120.

Rosewater, J. 1961. The Family Pinnidae in the Indo-Pacific. *Indo-Pacific Mollusca* **1**: 175-226, pls. 135-171.

Rosewater, J. 1965. The family Tridacnidae in the Indo-Pacific. *Indo-Pacific Mollusca* **1**: 347-393, pls. 263-293.

Rosewater, J. 1970. The family Littorinidae in the Indo-Pacific. Part 1. The subfamily Littorininae. *Indo-Pacific Mollusca* **2**: 417-506, pls. 325-387.

Rosewater, J. 1972. The family Littorinidae in the Indo-Pacific. Part 2. The subfamilies Tectariinae and Echininae. *Indo-Pacific Mollusca* **2**: 507-529, pls. 388-408.

Rosewater, J. 1975. An annotated list of the marine mollusks of Ascension Island, South Atlantic Ocean. *Smithsonian Contributions to Zoology* **189**: 1-41, figs. 1-24.

Rost, H. 1955. A report on the family Arcidae (Pelecypoda). *Allan Hancock Pacific Expedition* **20**: 173-249, pls. 11-16.

Roth, B. and P. S. Sadeghian. 2003. Checklist of land snails and slugs of California. *Santa Barbara Museum of Natural History Contributions in Science* **3**: 1- 81, figs. 1-92.

Ruhoff, F. A. 1980. Index to the species of Mollusca introduced from 1850 to 1870. *Smithsonian Contributions to Zoology* **294**: 1-640.

Rundell, R.J., B. S. Holland, and R. H. Cowie. 2004. Molecular phylogeny and biogeography of the endemic Hawaiian Succineidae (Gastropoda: Pulmonata). *Molecular Phylogenetics and Evolution* **31**: 246-255.

Runham, N. W. and P. J. Hunter. 1970. *Terrestrial Slugs*. Hutchinson University Library, London. 184 pp., 57 figs.

Russell, H. D. 1971. *Index Nudibranchia: A Catalog of the Literature 1554-1965*. Delaware Museum of Natural History, Greenville, Delaware. iv + 141 pp.

Russell, H. D. 1986. Index Nudibranchia supplement I, 1966-1975. *Department of Mollusks, Harvard University Special Occasional Publication* **7**: i-ii + 1-100

Sabelli, B., R. Giannuzzi-Savelli, and D. Bedulli. 1990-1992. *Catalogo Annotato dei Molluschi Marini del Mediterraneo/Annotated Check-list of Mediterranean Marine Mollusks*. Edizioni Libreria Naturalistica Bolognese, Bolognese, Italia. Vol. 1 (1990): i-xiv + 1-348. Vol 2 (1992): 349-500. Vol. 3 (1992): 501-781.

Salvat, B. and C. Rives. 1980. *Coquillages de Polynésie*. 2nd Ed. Éditions du Pacifique, Papeete, Tahiti. 391 pp., 446 figs.

Salvat, B. and C. Rives. 1984. *Shells of Tahiti*. Les Editions du Pacifique, Bruat, Papeete, Tahiti. 159 pp., 40 pls.

Salvat, B., C. Rives, and P. Reverce. 1988. *Coquillages de Nouvelle-Caledonie*. Les Editions du Pacifique, Singapore. 142 pp., 28 pls.

Schilder, F. A. and M. Schilder. 1938. Prodrome of a monograph on living Cypraeidae. *Proceedings of the Malacological Society of London* **23**: 119-231.

Schilder, M. and F. A. Schilder. 1971. A catalogue of living and fossil cowries. *Institute Royals des Sciences Naturelles de Belgique Memoires, Deuxième Série* **85**: 1-246.

Schileyko, A. A. 1998-2004. Treatise on Recent terrestrial pulmonate molluscs. Parts 1-12. *Ruthenica* Supplement **2**: 1-1763, figs. 1-2259. [This work, which discusses terrestrial gastropods to the generic level, is not completed, several more parts are planned. The following families are found in the following parts: Achatinidae (part 4), Agriolimacidae (part 11), Bradybaenidae (part 12), Camaenidae (part 11), Cerionidae (part 4), Charopidae (part 7), Clausiliidae (part 5), Gastrocoptidae (part 2), Haplotrematidae (part 6), Helminthoglyptidae (part 12), Limacidae (part 11), Milacidae (part 10), Orthalicidae (part 3), Partulidae (part 3), Pupillidae (part 1), Rhytididae (part 6), Sagdidae (part 2), Spiraxidae (part 6), Streptaxidae (part 6), Truncatellinidae (part 2), Urocoptidae (part 3), Valloniidae (part 1), Vertiginidae (part 2)].

Schuh, R. T. 2000. *Biological Systematics: Principles and Applications*. Comstock Publishing Associates, a division of Cornell University Press, Ithaca, New York. ix + 236 pp.

Schütt, H. 2001. Die türkischen Landschnecken. *Acta Biologica Benrodis* Supplementband **4**: 1-549. [531 taxa illustrated].

Scott, P. V. and J. A. Blake, eds. 1998. *Taxonomic Atlas of the Benthic Fauna of the Santa Maria Basin and the Western Santa Barbara Channel*. Vol. 8. The Mollusca Part 1 - The Aplacophora, Polyplacophora, Scaphopoda, Bivalvia, and Cephalopoda. Santa Barbara Museum of Natural History, Santa Barbara, California. viii + 250 pp. [95 taxa illustrated].

Scudder, S. H. 1882. Nomenclator Zoologicus. I. Supplemental list. II. Universal index. *Bulletin of the United States National Museum* **19**: i-xix + 1-376, 1-340.

Sharabati, D. 1981. *Saudi Arabian Seashells. Selected Red Sea and Arabian Gulf Molluscs*. VNU Books International. 119 pp.

Sharabati, D. 1984. *Red Sea Shells*. KPI Limited, London. 127 pp., 49 pls.

Sherborn, C. D. 1902. *Index Animalium*. Sectio Prima, 1758-1800. British Museum (Natural History), London. lix + 1195 pp.

Sherborn, C. D. 1922-1932. *Index Animalium*. Sectio Secunda, 1801-1850. British Museum (Natural History), London. cxlvii + 7056 pp., 1098 pp. [Published in 33 parts, with corrections and additions].

Shirai, S. 1994. *Pearls and Pearl Oysters of the World*. Marine Planning Co., Ishigaki, Okinawa, Japan. 108 pp. [Text figures].

Short, J. W. and D. G. Potter. 1987. *Shells of Queensland and the Great Barrier Reef. Marine Gastropods*. Robert Brown and Associates, Bathurst, New South Wales, Australia. vi + 135 pp., 60 pls.

Simpson, G. G. 1990. *Principles of Animal Taxonomy*. Columbia University Press, New York. x + 247 pp.

Sirenko, B. I. 1988. A new genus of deep sea chitons *Ferreiraella* gen. n. (Lepidopleurida, Leptochitonidae) with a description of a new ultra-abyssal species. *Zoologichesky Zhurnal* **67**: 1776-1786. [In Russian].

Skoglund, C. 1990. Additions to the Panamic Province bivalve (Mollusca) literature 1971 to 1990. *Festivus* **22** (Supplement 2): i-v + 1-74.

Skoglund, C. 1991. Additions to the Panamic Province Opisthobranchia (Mollusca) literature 1971 to 1990. *Festivus* **22** (Supplement 1): i-iii + 1-27.

Skoglund, C. 1992. Additions to the Panamic Province gastropod (Mollusca) literature 1971 to 1992. *Festivus* **24** (Supplement): i-viii + 1-169.

Slieker, F. J. A. 2000. *Chitons of the World: An Illustrated Synopsis of Recent Polyplacophora*. Mostra Mondiale Malacologia, Cupra Marittima, Italy. vi + 154 pp., 50 pls.

Smith, A. G. and M. Gordon, Jr. 1948. The marine mollusks and brachiopods of Monterey Bay, California, and vicinity. *Proceedings of the California Academy of Sciences (Series 4)* **26**: 147-245, 2 pls.

Smith, A. G., W. B. Miller, C. C. Christensen, and B. Roth. 1990. Land Mollusca of Baja California, Mexico. *Proceedings of the California Academy of Sciences (Series 4)* **47**: 95-158, figs 1-40.

Smith, B. J. 1992. Non-marine mollusca. *Zoological Catalogue of Australia* **8**: i-xii + 1-405.

Smith, B. J. and R. C. Kershaw. 1979. *Field Guide to the Non-Marine Molluscs of South Eastern Australia*. Australian National University Press, Canberra, Australia. 285 pp. [Illustrated].

Smith, B. J. and R. C. Kershaw. 1981. Tasmanian land and freshwater molluscs. *Fauna of Tasmania Handbook Series* **5**: 1-148. [Illustrated].

Smith, H. H. 1902. An annotated catalogue of the genus *Partula* in the Hartman Collection belonging to the Carnegie Museum. *Annals of the Carnegie Museum* **1**: 422-485.

Smith, M. 1939. *An Illustrated Catalog of the Recent Species of the Rock Shells: Muricidae, Thaididae and Coralliophilidae*. Tropical Laboratory, Lantana, Florida. ix + 83 pp., 28 pls.

Smith, M. 1942. *A Review of the Volutidae: Synonymy, Nomenclature, Range and Illustrations*. Published by the authority of the Beal-Maltbie Shell Museum of Rollins College, Winter Park, Florida. 127 pp., 26 pls.

Smith, M. 1944. *Panamic Marine Shells, Synonymy, Nomenclature, Range and Illustrations*. Beal-Maltbie Shell Museum of Rollins College, Winter Park, Florida. xiii + 127 pp., 912 figs.

Smith, M. 1948. *Triton Helmet and Harp Shells. Synonymy, Nomenclature, Range and Illustrations*. Tropical Photographic Laboratory, Winter Park, Florida. v + 57 pp., 16 pls.

Snyder, M. A. 2003. Catalog of the marine gastropod family Fasciolariidae. *Academy of Natural Sciences Special Publication* **21**: i-iv + 1-431, pl. 1.

Solem, A. 1959. Systematics and zoogeography of the land and freshwater Mollusca of the New Hebrides. *Fieldiana Zoology* **43**: 1-359, pls. 1-33.

Solem, A. 1961. New Caledonian land and fresh-water snails: an annotated checklist. *Fieldiana Zoology* **41**: 415-501.

Solem, A. 1966a. Some non-marine mollusks from Thailand, with notes on classification of the Helicarionidae. *Spolia Zoologica Musei Hauniensis* **24**: 7-110, 3 pls.

Solem, A. 1966b. The Neotropical land snail genera *Labyrinthus* and *Isomeria* (Pulmonata: Camaenidae). *Fieldiana Zoology* **50**: 1-226, figs. 1-61.

Solem, A. 1974. *The Shell Makers: Introducing Mollusks*. Wiley-Interscience, New York. xiii + 289 pp., 12 pls.

Solem, A. 1976-1982. *Endodontoid Land Snails from Pacific Islands (Mollusca: Pulmonata, Sigmurethra). Part 1: Family Endodontidae. Part 2: Families Punctidae and Charopidae, Zoogeography*. Field Museum of Natural History, Chicago, Illinois. Part I: xii + 508 pp., 208 figs.; Part II: ix + 336 pp., 143 figs.

Solem, A. 1979. Some mollusks from Afghanistan. *Fieldiana Zoology New Series* **1**: i-vi + 1-89, figs. 1-32.

Solem, A. 1979-1997. Camaenid land snails from Western and central Australia (Mollusca: Pulmonata: Camaenidae), I-VII. *Records of the Western Australian Museum* **1**: suppl. 10: 1-142, figs. 1-35, pls. 1-11 (1979); **2**: suppl. 11: 147-320, figs. 36-73, pls. 12-14 (1981); **3**: suppl. 11: 321-425, figs. 74-110, pls. 15-18 (1981); **4**: suppl. 17: 427-705, figs. 111-180, pls. 19-63 (1984); **5**: suppl. 20: 707-981, figs. 181-256, pls. 64-94 (1985); **6**: suppl. 43: 983-1459, figs. 257-367, pls. 95-176 (1993); **7**: suppl. 50: 1461-1906, figs. 368-453, pls. 177-238.

Solem, A. and A. C. van Bruggen, eds. 1984. *World-Wide Snails: Biogeographical Studies on Non-Marine Mollusca*. E. J. Brill/W. Backhuys, Leiden, Netherlands. 289 pp., 102 figs.

Solís, R. M. 2002. *Diccionario Etimológico de Malacología*. Reseñas Malacológicas **12**: 1-316.

Springsteen, F. J. and F. M. Leobrera. 1986. *Shells of the Philippines*. Carfel Seashell Museum, Manila, Philippines. 377 pp., 100 pls.

Spry, J. F. 1968. *The Sea Shells of Dar es Salaam*. The Tanzania Society, Dar es Salaam, Tanzania. Vol. 1: Gastropods, Revised and Enlarged, 40 pp., 8 pls.; Vol. 2: Pelecypoda. 41 pp., 8 pls. [Reprinted from Tanganyika Notes and Records 56 (March 1961) and 63 (September 1964) respectively].

Stadnichenko, A. P. 1984. Freshwater bivalves (Bivalvia: Unionidae) of the fauna of the Ukrainian-SSR. *Vestnik Zoologii* **1**: 32-38, 2 pls. [In Russian].

Starmühlner, F. 1976. Beiträge zur Kenntnis der Süsswasser-Gastropoden pazifischer Inseln. *Annalen des Naturhistorischen Museum in Wien* **80**: 473-656, 21 pls.

Starmühlner, F. 1993. Beiträge zur Kenntnis der Süss- und Brackwasser-Gastropoden der Tonga- und Samoa- Inseln (S. W. Pazifik). *Annalen des Naturhistorischen Museum in Wien* **94/95**: 217-306, 11 pls.

Steiner, G. and A. R. Kabat. 2001. Catalogue of supraspecific taxa of Scaphopoda (Mollusca). *Zoosystema* **23**: 433-460.

Steiner, G. and A. R. Kabat. 2004. Catalog of species-group names of Recent and fossil Scaphopoda (Mollusca). *Zoosystema* **26**: 549-726.

Steinke, D., C. Albrecht, and M. Pfenninger. 2004. Molecular phylogeny and character evolution in the Western Palaearctic Helicidae *s.l.* (Gastropoda: Stylommatophora). *Molecular Phylogenetics and Evolution* **32**: 724-734.

Steyn, D. G. and M. Lussi. 1998. *Marine Shells of South Africa, An Illustrated Collector's Guide to Beached Shells*. Ekogilde Publishers, Hartebeespoort, South Africa. ii + 264 pp. [1006 taxa illustrated].

Strayer, D. L. and K. J. Jirka. 1997. The pearly mussels of New York State. *New York State Museum Memoir* **26**: i-xiii + 1-113, pls. 1-27.

Streba, G. H. W. 2004. *Olividae: A Collector's Guide*, Revised and Enlarged. ConchBooks, Hackenheim, Germany. 172 pp., 62 pls. [A version in German is also available].

Subba Rao, N. V. 1989. *Handbook Freshwater Molluscs of India*. Zoological Survey of India, Calcutta. xxiii + 289 pp., 103 pls.

Subba Rao, N. V. 2003. *Indian Seashells (Part-1) Polyplacophora and Gastropoda*. Zoological Survey of India, Kolkata, India. x + 416 pp, 96 pls.

Sweeney, M. J., C. F. E. Roper, K. M. Mangold, M. R. Clarke, and S. V. Boletzky. 1992. "Larval" and juvenile Cephalopoda: A manual for their identification. *Smithsonian Contributions to Zoology* **513**: 1-282, figs. 1-277.

Swennen, C., R. G. Moolenbeek, N. Ruttanadakul, H. Hobbelink, H. Dekker, and S. Hajisamae. 2001. *The Molluscs of the Southern Gulf of Thailand*. Biodiversity Research and Training Program, Bangkok, Thailand. ix + 210 pp. [531 taxa illustrated].

Szekeres, M. 1984. Some notes of the distribution of the South American Clausiliidae (Gastropoda, Pulmonata). *In:* Solem, A. and A. C. Van Bruggen, eds., *World-Wide Snails. Biogeographical Studies on Non-Marine Mollusca*. E. J. Brill, Leiden, The Netherlands. Pp. 172-177.

Taft, C. 1961. The shell-bearing land snails of Ohio. *Bulletin of the Ohio Biological Survey, New Series* **1**: i-ix + 1-108. [85 taxa illustrated].

Tan, K. S. and L. M. Chou. 2000. *A Guide to Common Seashells of Singapore*. Singapore Science Center, Singapore. 168 pp. [161 taxa illustrated].

Taylor, D. W. 1970. *West American Freshwater Mollusca, 1: A Bibliography of Pleistocene and Recent Species*. Memoir 4. Society of Natural History, San Diego, California. 73 pp.

Taylor, D. W. 2003. Introduction to Physidae (Gastropoda: Hygrophila). Biogeography, classification, morphology. *Revista de Biologia Tropical* **51**(Supplement 1): 1-287, pls. 1-11, figs. 1-191.

Taylor, J. and J. G. Walls. 1975. *Cowries*. T. H. F. Publications, Inc., Neptune City, New Jersey. 288 pp. [All taxa illustrated].

Taylor, J. D., ed. 1996. *Origin and Evolutionary Radiation of the Mollusca*. Oxford University Press, Oxford, England. xiv + 392 pp.

Taylor, J. D., Y. I. Kantor, and A. V. Sysoev. 1993. Foregut anatomy, feeding mechanisms, relationships and classification of the Conoidea (=Toxoglossa) (Gastropoda). *Bulletin of the Natural History Museum of London (Zoology)* **59**: 125-170.

Tebble, N. 1966. *British Bivalve Seashells: A Handbook for Identification*. 2nd Ed. Royal Scottish Museum, Edinburgh, Scotland. 212 pp., 12 pls. [There was a revised edition in 1976].

Thiele, J. 1929-1935. *Handbuch der systematischen Weichtierkunde*. Gustav Fischer Verlan, Jena. 1154 pp., 895 figs. [Published in 4 parts; A translated version, edited by R. Bieler and P. Mikkelsen, translated by J. S. Bhatti, is available under the title Handbook of Systematic Malacology, Translation Publishing Program, Smithsonian Institution Libraries, Washington, D.C. 1992-1998. 1690 pp. Available from National Technical Information Service (NTIS), Springfield, Virginia 22161 Document No. TT 87-600185].

Thompson, F. G. 1984. *The Freshwater Snails of Florida: A Manual for Identification*. University Presses of Florida, Gainesville, Florida. 94 pp., 193 figs.

Thompson, F. G. 1999. An identification manual for the freshwater snails of Florida. *Walkerana* **10**: i-v + 1-96, figs. 1-209.

Thompson, F. G. and A. Correa S. 1991. Mexican land snails of the genus *Hendersoniella*. *Bulletin of the Florida Museum of Natural History* **36**: 1-23.

Thompson, F. G. and A. Correa S. 1994. Land snails of the genus *Coelocentrum* from northeastern Mexico. *Bulletin of the Florida Museum of Natural History* **36**: 141-173.

Thompson, F. G. and S. P. Dance. 1983. Non-marine mollusks of Borneo. II Pulmonata: Pupillidae, Clausiliidae. III Prosobranchia: Hydrocenidae, Helicinidae. *Bulletin of the Florida State Museum of Natural History, Biological Sciences* **29**: 101-152, figs. 1-75. [For part 1, see Dance 1970].

Thompson, T. E. 1976. *Biology of the Opisthobranch Molluscs*, Vol. 1. Ray Society, London. 207 pp., 20 pls.

Thompson, T. E. and G. H. Brown. 1976. British Opisthobranch Molluscs. Mollusca: Gastropoda. Keys and Notes for the Identification of Species. *Synopsis of the British Fauna* **8** (New Series): 1-203, figs. 1-203.

Tillier, S. 1981. South American and Juan Fernández succineid slugs (Pulmonata). *Journal of Molluscan Studies* **47**: 125-146, figs. 1-26.

Tillier, S. 1989. Comparative morphology, phylogeny and classification of land snails and slugs (Gastropoda: Pulmonata: Stylommatophora). *Malacologia* **30**: 1-303, figs. 1-704.

Tomlin, J. R. l'B. 1937. Catalogue of Recent and fossil cones. *Proceedings of the Malacological Society of London* **22**: 205-330, 333.

Troncoso, N., J. L. van Goethem, and J. S. Troncoso. 2001. Contributions to the marine Mollusca fauna of Kerguelen Islands, South Indian Ocean. *Iberus* **19**: 83-114.

Tsaloikhin, S. J. ed. 2004. *Key to Freshwater Invertebrates of Russia and Adjacent Lands*. Nauka, St. Petersburg. 526 pp., 163 pls. [In Russian, Mollusks pp. 9-491, pls. 1-159].

Tucker, J. K. 2004. Catalog of Recent and fossil turrids (Mollusca: Gastropoda). *Zootaxa* **682**: 1-1295.

Turgeon, D. D., J. F. Quinn, Jr., A. E. Bogan, E. V. Coan, F. G. Hochberg, W. G. Lyons, P. M. Mikkelsen, R. J. Neves, C. F. E. Roper, G. Rosenberg, B. Roth, A. Scheltema, F. G. Thompson, M. Vecchione, and J. D. Williams. 1998. *Common and Scientific Names of Aquatic Invertebrates from the United States and Canada: Mollusks*, 2nd Edition. American Fisheries Society Special Publication 26. American Fisheries Society, Bethesda, Maryland. ix + 526 pp.

Turner, H. 2001. *Katalog der Familie Costellariidae MacDonald 1860 (Gastropoda: Prosobranchia: Muricoidea)*. ConchBooks, Hackenheim, Germany. 100 pp., 22 figs.

Turner, R. D. 1966. *A Survey and Illustrated Catalogue of the Teredinidae (Mollusca: Bivalvia)*. Harvard University, Museum of Comparative Zoology, Cambridge, Massachusetts. vii + 265 pp., 64 pls.

Turner, R. D. 1985. Notes on mollusks of deep-sea vents and reducing environments. *American Malacological Bulletin* Special Edition **1**: 23-34, 8 figs.

Turner, R. L. and P. M. Mikkelsen. 2004. Annotated bibliography of the Florida Applesnail, *Pomacea paludosa* (Say) (Gastropoda: Ampulariidae), from 1824 to 1999. *Nemouria, Occasional Papers of the Delaware Museum of Natural History* **48**: 1-188.

Tursch, B. and D. Greifeneder. 2001. Oliva *Shells. The Genus* Oliva *and the Species Problem*. L'Informatore Piceno, Ancona, Italy. x + 570 pp., 77 pls.

van Benthem Jutting, T. 1963a. Non-marine Mollusca of west New Guinea part 1 - fresh and brackish water. *Nova Guinea, Zoology* **20**: 409-521, pls. 24-25, figs. 1-55.

van Benthem Jutting, T. 1963b. Non-marine Mollusca of West New Guinea part 2 - operculated land snails. *Nova Guinea, Zoology* **23**: 653-726, pls. 27-30.

van Benthem Jutting, T. 1964. Non-marine Mollusca of West New Guinea part 3 - Pulmonata I. *Nova Guinea, Zoology* **26**: 1-74, pls. 1-2, 62 figs.

van Benthem Jutting, T. 1965. Non-marine Mollusca of West New Guinea part 4 - Pulmonata II. *Nova Guinea, Zoology* **32**: 205-304, pls. 7-10.

van Bruggen, A. C. 1966. The terrestrial Mollusca of the Kruger National Park: A contribution to the malacology of the Eastern Transvaal. *Annals of the Natal Museum* **18**: 315-399, figs. 1-70.

van Bruggen, A. C. 1967. An introduction to the pulmonate family Streptaxidae. *Journal of Conchology* **26**: 181-188.

van Damme, D. 1984. The freshwater Mollusca of northern Africa. Distribution, biogeography and palaeoecology. *Developments in Hydrobiology* **25**: i-xi + 1-164, figs. 1-144.

van der Schalie, H. 1948. The land and fresh-water mollusks of Puerto Rico. *University of Michigan Museum of Zoology Miscellaneous Publications* **70**: 1-134, pls. 1-14.

van Eeden, J. A. 1960. Key to the genera of South African freshwater and estuarine gastropods (Mollusca). *Annals of the Transvaal Museum* **24**: 1-17, figs. 1-44.

van Loen, H. and J. L. van Goethem. 1997. Distribution of neritid, viviparid and valvatid freshwater gasteropods in Belgium (Mollusca, Prosobranchia). *Bulletin de l'Institut Royal des Sciences Naturelles de Belgique* **67**: 31-38, 8 figs.

Vaught, K. C. 1989. *A Classification of the Living Mollusca*. Edited by R. T. Abbott and K. J. Boss. American Malacologists, Inc., Melbourne, Florida. xii + 195 pp.

Verdcourt, B. 1972. The zoogeography of the non-marine Mollusca of East Africa. *Journal of Conchology* **27**: 291-348.

Vermeij, G. J. 1993. *A Natural History of Shells*. Princeton University Press, Princeton, New Jersey. viii + 207 pp., 22 pls.

Vermeulen, J. J. and A. J. Whitten. 1998. *Fauna Malesiana, guide to the land snails of Bali*. Backhuys Publishers, Leiden, The Netherlands. ix + 164 pp., 130 figs.

Vokes, E. H. 1971. Catalog of the genus *Murex*, Linné. *Bulletin of American Paleontology* **61**: 5-141.

Vokes, H. E. 1980. *Genera of the Bivalvia: A Systematic and Bibliographic Catalogue.* Revised and Updated. Paleontological Research Institution, Ithaca, New York. xxvii + 307 pp.

Vokes, H. E. 1990. Genera of the Bivalvia: A systematic and bibliographic catalogue - addenda and errata. *Tulane Studies in Geology and Paleontology* **23**: 97-120.

Vokes, H. E. and E. H. Vokes. 1984. Distribution of shallow-water marine Mollusca, Yucatan Peninsula, Mexico. *Mesoamerican Ecology Institute Monograph 1 and Middle American Research Institute Publication* **54**: 1-183, pls. 1-50.

von Regteren Altena, C. O. 1969. The marine Mollusca of Surinam (Dutch Guiana) Holocene and Recent. Part I. General introduction. *Zoologische Verhandelingen* **101**: 1-49, pls. 1-4.

von Regteren Altena, C. O. 1971. The marine Mollusca of Surinam (Dutch Guiana) Holocene and Recent. Part II. Bivalvia and Scaphopoda. *Zoologische Verhandelingen* **119**: 1-100, pls. 1-10.

von Regteren Altena, C. O. 1975. The marine Mollusca of Surinam (Dutch Guiana) Holocene and Recent. Part III. Gastropoda and Cephalopoda. *Zoologische Verhandelingen* **139**: 1-104, pls. 1-11.

Voss, G. L. 1963. Cephalopods of the Philippine Islands. *United States National Museum Bulletin* **234**: 1-180, figs. 1-36.

Voss, G. L. and G. R. Williamson. 1971. *Cephalopods of Hong Kong.* Hong Kong Government Press. 138 pp., 35 pls., 68 text fig.

Voss, N. A., M. Vecchione, R. B. Toll, and M. J. Sweeney, eds. 1998. Systematics and biogeography of cephalopods. *Smithsonian Contributions to Zoology* **586**: Vol. 1: 1-276, Vol. 2: 277-599.

Wade, C. M., B. Clarke, and P. B. Mordan. 2001. A phylogeny of the land snails (Gastropoda: Pulmonata). *Proceedings of the Royal Society, Biological Sciences* **268**: 413-422.

Wagner, F. J. E. 1984. *Illustrated Catalogue of the Mollusca (Gastropoda and Bivalvia) in the Atlantic Geosciences Centre Index Collection.* Geological Survey of Canada, Ottawa, Canada. 76 pp. [Text figures].

Wagner, H. P. [1991]. Review of the European Pectinidae. *Vita Marina* **41**: 3-48. [In Dutch and English, illustrated].

Walker, B. 1928. The terrestrial shell-bearing Mollusca of Alabama. *University of Michigan Museum of Zoology, Miscellaneous Publication* **18**: 1-180, figs. 1-278.

Waller, T. R. 1972. The Pectinidae (Mollusca: Bivalvia) of Eniwetok Atoll, Marshall Islands. *Veliger* **14**: 221-264, pls. 1-8.

Waller, T. R. 1973. The habits and habitats of some Bermudian marine mollusks. *Nautilus* **87**: 31-52, 33 figs.

Waller, T. R. 1993. The evolution of "*Chlamys*" (Mollusca: Bivalvia: Pectinidae) in the tropical western Atlantic and eastern Pacific. *American Malacological Bulletin* **10**: 195-249, pls. 1-14.

Walls, J. 1979a. *Cone Shells. A Synopsis of the Living Conidae.* T. F. H. Publications, Inc., Neptune City, New Jersey. 1011 pp. [All taxa illustrated].

Walls, J. 1979b. *Cowries,* Revised Ed. T. H. F. Publications, Inc., Neptune City, New Jersey. 286 pp. [All taxa illustrated].

Walls, J. 1980. *Conchs, Tibias and Harps.* T. F. H. Publications, Inc., Neptune City, New Jersey. 191 pp. [Text figures].

Warén, A. and P. Bouchet. 1993. New records, species, genera and a new family of gastropods from hydrothermal vents and hydrocarbon seeps. *Zoologica Scripta* **22**: 1-90, figs. 1-59.

Warén, A. and S. Gofas. 1996. A new species of Monoplacophora, redescription of the genera *Veleropilina* and *Rokopella*, and new information on three species of this class. *Zoologica Scripta* **25**: 215-232, 15 figs.

Warmke, G. L. and R. T. Abbott. 1961. *Caribbean Seashells. A Guide to the Marine Mollusks of Puerto Rico and other West Indian Islands, Bermuda and the Lower Florida Keys.* Dover Publications, New York. xx + 348 pp., 44 pls.

Warui, C. M., P. Tattersfield, and M. B. Seddon. 2001. Annotated checklist of terrestrial molluscs of Mount Kenya, Kenya. *Journal of Conchology* **37**: 291-300.

Way, K. and R. D. Purchon. 1981. The marine shelled Mollusca of West Malaysia and Singapore, Part 2: Polyplacophora and Gastropoda. *Journal of Molluscan Studies* **47**: 313-321.

Weaver, C. S. and J. E. duPont. 1970. *The Living Volutes. A Monograph of the Recent Volutidae of the World.* Delaware Museum of Natural History, Greenville, Delaware. xv + 375 pp., 79 pls.

Webb, W. F. 1948. *Foreign Land and Fresh Water Shells from All Parts of the World Except the United States and Canada.* Privately published, St. Petersburg, Florida. 183 pp., 73 pls.

Weil, A., L. Brown, and B. Neville. 1999. *The Wentletrap Book. Guide to the Recent Epitoniidae of the World.* Evolver srl, Rome. 244 pp., 507 figs.

Weisbord, N. E. 1962. Late Cenozoic gastropods from northern Venezuela. *Bulletins of American Paleontology* **42**: 7-672, pls. 1-48. [95 of the 288 gastropods discussed are Recent].

Weisbord, N. E. 1964. Late Cenozoic pelecypods from northern Venezuela. *Bulletins of American Paleontology* **45**: 5-564, pls. 1-59. [51 of the 172 bivalves discussed are Recent].

Wells, F. E. and C. W. Bryce. 1984. *A Guide to the Common Molluscs of South-Western Australian Estuaries.* Western Australian Museum, Perth, Australia. 112 pp. [84 taxa illustrated].

Wells, F. E. and C. W. Bryce. 1986. *Seashells of Western Australia.* Western Australian Museum, Perth, Australia. 207 pp., 74 pls.

Wells, F. E. and C. W. Bryce. 1993. *Sea Slugs of Western Australia*. Western Australian Museum, Perth, Australia. viii + 184 pp., 226 figs.

Wells, F. E., C. W. Bryce, J. E. Clark, and G. M. Hansen. 1990. *Christmas Shells. The Marine Molluscs of Christmas Island (Indian Ocean)*. Christmas Island Natural History Association, Christmas Island. ii + 98 pp. [377 taxa illustrated].

Welter-Schultes, F. W. 1998. Albinaria in central and eastern Crete: distribution map of the species (Pulmonata: Clausiliidae). *Journal of Molluscan Studies* 64: 275-279.

Wenz, W. 1938-1944. Gastropoda. Algemeiner Teil und Prosobranchia. In: O. H. Schindewolf, ed., *Handbuch der Paläozoologie*, Band 6. Gebrüder Borntraeger, Berlin. Pp. 1-1639. [2500+ figures].

West, K., E. Michel, J. Todd, D. Brown, and J. Clabaugh. 2003. *The Gastropods of Lake Tanganyika*. International Association of Theoretical and Applied Limnology at the University of North Carolina, Chapel Hill, North Carolina. 128 pp., 101 fig.

White, J. S. 1976. *Seashells of the Pacific Northwest*. Binford and Mort, Portland, Oregon. vii + 127 pp. [78 taxa illustrated].

Wiktor, A. 2000. Agriolimacidae (Gastropoda: Pulmonata) - a systematic monograph. *Annales Zoologici (Museum and Institute of Zoology, Polish Academy of Sciences)* 49: 347-590, figs. 1-831.

Wilbur, K. M. ed. 1983-1988. *The Mollusca*. Academic Press, Inc., New York. [Vol. 1. (1983). Metabolic Biochemistry and Molecular Biomechanics. xviii + 510 pp.; Vol. 2. (1983). Environmental Biochemistry and Physiology. xviii + 362 pp.; Vol. 3. (1983). Development. xx + 352 pp.; Vol. 4. (1983). Physiology, Part 1. xx + 522 pp.; Vol. 5. (1983). Physiology, Part 2. xx + 500 pp.; Vol. 6. (1983). Ecology. xx + 695 pp.; Vol. 7. (1984). Reproduction. xix + 486 pp.; Vol. 8. (1985). Neurobiology and Behavior, Part 1. xvii + 415 pp.; Vol. 9. (1986). Neurobiology and Behavior, Part 2. x + 499 pp.; Vol. 10. (1985). Evolution. xx + 491 pp.; Vol. 11. (1988). Form and Function. xxviii + 504 pp.; Vol. 12. (1988). Paleontology and Neontology of Cephalopds. xxiv + 335 pp.

Wilbur, K. M. and C. M. Yonge, eds. 1964-1966. Physiology of Mollusca. Academic Press, New York. [Vol. 1: xii + 473 pp.; Vol. 2: xiii + 645 pp.]

Wilson, B. and P. Clarkson. 2004. *Australia's Spectacular Cowries*. Odyssey Publishing, El Cajon, California. ix + 396 pp. [Illustrated].

Wilson, B. R. 1993. *Australian Marine Shells. Prosobranch Gastropods, Part One*. Odyssey Publishing, Kallaroo, W. A., Australia. 408 pp., 44 pls.

Wilson, B. R. 1994. *Australian Marine Shells. Prosobranch Gastropods*, Part Two (Neogastropods). Odyssey Publishing, Kallaroo, Western Australia. 370 pp, 53 pls.

Wilson, B. R. and K. Gillett. 1974. *Australian Shells*. A. H. and A. W. Reed Pty, Ltd., Sydney, Australia. 168 pp, 106 pls. [Also published by Charles E. Tuttle, Rutland, Vermont (1971)].

Wilson, B. R. and J. A. McComb. 1967. The genus *Cypraea* (subgenus *Zoila* Jousseaume). *Indo-Pacific Mollusca* 1: 457-484, pls. 329-343.

Winston, J. 1999. *Describing Species: Practical Taxonomic Procedure for Biologists*. Columbia University Press, New York. xx + 518 pp.

Winterbourn, M. J. 1973. A guide to the freshwater Mollusca of New Zealand. *Tuatara* 20: 141-159, pls. 1-5.

Work, R. C. 1969. Systematics, ecology and distribution of the mollusks of Los Roques, Venezuela. *Bulletin of Marine Sciences* 19: 614-711. [Checklist].

Wu, Shi-Kuei. 1989. Colorado freshwater mollusks. *Natural History Inventory of Colorado* 11: 1-117, figs. 1-126.

Wu, Shi-Kuei and N. Brandauer. 1978. *The Bivalvia of Colorado*. Natural History Inventory of Colorado 2. University of Colorado Museum, Boulder, Colorado. 60 pp., 87 figs.

Wu, Shi-Kuei, R. D. Oesch, and M. E. Gordon. 1997. *Missouri Aquatic Snails*. Natural History Series No. 5. Missouri Department of Conservation, Jefferson City, Missouri. iv + 97 pp., 135 figs.

Wurtz, C. B. 1955. The American Camaenidae (Mollusca: Pulmonata). *Proceedings of the Academy of Natural Science of Philadelphia* 107: 99-143, pls. 1-19.

Yakovleva, A. M. 1965. *Shell-bearing Mollusks (Loricata) of the Seas of the USSR*. Israel Program for Scientific Translation, Jerusalem. viii + 127 pp., 11 pls. [Originaly published in Russian in 1952].

Yonge, C. M. and T. E. Thompson. 1976. *Living Marine Molluscs*. William Collins Sons and Co., Ltd., London. 288 pp., 16 pls.

Yoo, Jong-Saeng. 1976. *Korean Shells in Colour*. Il Ji Sa Publishing Co., Seoul, Korea. 196 pp., 30 pls. [In Korean].

Young, R. E. 1972. The systematics and areal distribution of pelagic cephalopods from the seas off southern California. *Smithsonian Contributions to Zoology*, Number 97: i-iii + 1-159, pls. 1-38.

Zeigler, R. F. and H. C. Porreca. 1969. *Olive Shells of the World*. Privately published, West Henrietta, New York. 96 pp., 13 pls.

Zhadin, V. I. 1952. *Mollyuski Presnykh i Solonovatykh vod SSSR. [Mollusks of Fresh and Brackish Waters of the USSR]*. Zoological Institute of the Academy of Sciences of the U.S.S.R. 368 pp. [Translated by A. Mercado, under the title Mollusks of Fresh and Brackish Waters of the U.S.S.R. Jerusalem: Israel Program for Scientific Translations, 1965. xvi + 368 pp. 339 figures. Available from National Technical Information Service (NTIS), Springfield, Virginia. Document No. TT 65-50019].

Zilch, A. 1959-1960. Euthyneura. In: W. Wenz, ed., *Gastropoda*, Vol. 2. Gebrüder Borntraeger, Berlin. Pp. i-xii + 1-834. [In German, 2515 figures].

CHAPTER 10

TAXONOMY AND TAXONOMIC WRITING: A PRIMER

DANIEL L. GEIGER

10.1 INTRODUCTION

Taxonomic issues arise in many writing situations in malacology. You may be preparing an account of some life-history aspect of a single species, where that species needs to be properly identified. You may write a travel account with a list of all the species that you encountered. Alternatively, you may refer to specimens in a collection. In all these cases, the main taxonomic aspect of the paper is that of reference. The organism(s) should be identified in an unambiguous fashion and should be universally understandable.

A second category of manuscripts are taxonomic accounts, in which the taxonomy is the main objective of the treatment. Two main types should be distinguished: descriptions and revisions. When preparing a description, the specimen at hand is considered to be formally unknown to science. For revisions, an already known taxon is treated in a comprehensive fashion. Existing knowledge is summarized and some original observations may be added. Occasionally, some taxonomic decisions, particularly with respect to type specimens, may be taken.

10.2 REFERENCING A SPECIES

10.2.1 Basics. In order to refer to a species in an unambiguous and universal fashion the following points should be kept in mind. The universal reference system or convention is that of binominal nomenclature (the Linnean system). A species is referred to with two names: the genus name and the species name, also called the species epithet. These two names are always distinguished from the main text in a roman font by either putting them in italics or underlining them. The genus name always starts with a capital letter; the species epithet is spelled in all lower case letters.

In old works some of the species epithets were capitalized, a convention that is no longer practiced today. Furthermore, the full binomen is followed by the author of the species epithet, separated by a comma from the year in which it was described. As an example, consider the common black mussel *Mytilus edulis* Linnaeus, 1758. The authority is usually mentioned only the first time that particular species is used in a manuscript. Taxa of family and higher rank are single words that are spelled with an initial capital letter and are not italicized: Conidae, Heterodonta, Mollusca.

10.2.2 Further considerations. The binomen can be expanded by further elements, namely the subgenus, the subspecies, and in some cases the variety or form. The subgenus is placed after the genus in parentheses and with a capital initial letter; the subspecies is placed after the species name but is not placed in parentheses. The authority of the full name is that of the subspecies. Example: *Strombus (Doxander) vittatus vittatus* Linnaeus, 1758 and *Strombus (Doxander) vittatus japonicus* Reeve, 1851.

The use of names for units below the subspecies, so-called infra-subspecific names (varieties, forms), are somewhat disputed. They are not accepted as taxonomically valid under the *International Code of Zoological Nomenclature* (ICZN), but are sometimes used to further differentiate among morphs (e.g., Reid 1996). The form name should be separated from the binomen with an indication that a form name is used: *Littorina littorea* var. *conica* Hamer, 1920.

The genus name in a binomen is usually abbreviated to a single letter after the first mention of that genus, unless it is the first word of a sentence. For example, in a paper that mentioned *Littorina brevicula*, a subsequent sentence could read, "It was a *L. brevicula*." If your writing mentions genera with the same initial letter, use a sensible abbreviation. For *Cypraea*, *Conus*, *Charonia*, and *Chiton*, use *Cy.*, *Co.*, *Cha.*, and *Chi.*. If only the genus name is used, it is always spelled out: "Members of the genus *Littorina* often occur on rocks."

The placement of the author plus year in parentheses has a very specific meaning, although this convention is often unappreciated and ignored. If the parentheses are not used, the species level name preceding that of the author was originally named in the same genus as written in the present contribution. In the example above, Linnaeus, 1758 described a species *vittatus* in the genus *Strombus*. On the other hand, *brevicula* (Philippi, 1844) was described by Philippi in the genus *Turbo*. Hence, neither always using nor always omitting parentheses for reasons of consistency is appropriate, because the parentheses convey a specific piece of information. As previous authors often ignored the function of parentheses (or parentheses were adulterated by publishers), copying from old publications will often be incorrect. The only way to make sure that the parentheses are set appropriately is to check in the original description.

Occasionally you may want to tell the reader what kind of an organism you are talking about. This can be achieved with a coarse-grained classification of that organism. Such a classification is given in parentheses and each level is separated with either a colon (:), semicolon (;), or hyphen (-) starting with the highest ranked taxon: *Orthotheres haliotidis* Geiger and Martin, 1999 (Decapoda: Pinnotheridae) was found in *Haliotis asinina* Linnaeus, 1758. Most readers of a malacological paper will be unfamiliar with the genus *Orthotheres*. The inclusion of (Decapoda: Pinnotheridae) will tell the reader that it is a higher crustacean (Decapoda) in the pea-crab family (Pinnotheridae). How fine-grained the indications should be is somewhat subjective, where the likely target audience should be kept in mind. It is open to debate whether the above example should read (Arthropoda: Crustacea: Malacostraca: Decapoda: Brachyura: Pinnotheridae) or simply (Decapoda: Pinnotheridae). Try to be economical with words. Most readers will be familiar with the term Decapoda; hence, the three preceding terms are superfluous. However, it may be helpful to indicate that Pinnotheridae is a kind of a crab (Brachyura).

10.2.3 New names. New names should be mentioned only in the context of a proper species description as detailed below. Careless mention of new names in sales catalogs, activity reports, or meeting abstracts should be avoided. It leads to so-called *nomina nuda* (nude names), which will have to be sorted out later at great expense of time and effort. Authors-to-be who want to discuss potential new species with colleagues should not mention the intended new name to the correspondent. Use "sp. A", "the undescribed brown one", or some similar term.

10.2.4 Uncertainty statements. When compiling a faunal list, the identification of some specimens may be tentative. Several levels of uncertainty can be employed starting with the least severe. If you are more or less but not 100% sure about the identity, use the abbreviation cf. for the Latin *confer* = compare: *Littorina* cf. *brevicula*. If you encounter a species that is somewhat similar but certainly not that species, use the abbreviation aff. for Latin *affinis* = affinity: *Littorina* aff. *brevicula*. In order to indicate that the genus is known, but the species could not be determined, use the abbreviation sp. or use species in roman font: *Littorina* sp., *Littorina* species. For general statements about multiple species in a genus use the abbreviation spp: all *Littorina* spp. were found on rocks; all *Haliotis* spp. have a nacreous interior.

10.3 DESCRIPTION OF A NEW TAXON

10.3.1 Quality. The description of new taxa weighs more heavily than any other subject on the literature of organismal biology. The reason is that the description of a new taxon is eternal. Whereas a physiological measurement may be refined or retracted because of some fault in the original experiments, a new name is destined for eternity. In order to revoke a name, great expenses in time and effort have to be made. Hence, greatest care should be afforded to describing new taxa. This caveat applies to all publishing authors, be they amateurs or professionals; there are stellar examples and sinners in both camps. The brief rules outlined below should be considered as only a starting point for original taxonomic endeavors. The novice is strongly encouraged to read more detailed accounts on how to describe taxa (Mayr and Ashlock 1991, Winston 1999), to consult a scientific style manual (e.g., Council of Biology Editors 1994), and to purchase a personal copy of the current (4th) edition of the *International Code of Zoological Nomenclature* (ICZN 1999) to be read cover to cover. Furthermore, examine recently published descriptions by renowned workers, and follow their example. After a first draft has been written, I strongly encourage the writer to circulate the manuscript to other specialists in the field and experienced professionals, particularly in museum environments. Authors do not need to fear that their new species are "stolen" in the review process. For one, most professionals regard the description of new species as a tedious burden. Additionally, the ICZN code of ethics (see section 10.6.14) affords protection.

10.3.2 Publication platform. The conchological community has a diverse array of publications, all of which are produced by dedicated volunteers in countless hours and to good quality standards thanks to recent technological advances in desktop publishing. Many of the contributions, such as travelogues and shell show accounts, do not need to be highly scrutinized for scientific content. Yet, due to the eternal validity of taxonomic descriptions and related decisions, a higher standard applies to contributions describing new taxa. The process of peer review, although far from perfect, will help to improve the quality of the work, and will prevent some mistakes from being published. Hence, in the end, it benefits the author, the periodical, and the community at large to undergo this process. The quality of descriptions as measured by the synonymy ratio of the proposed names has increased steadily, but is not quite perfect yet (Rosenberg 1996). Many molluscan species remain to be discovered and to be described (Bouchet 1997).

Accordingly, I suggest that authors of taxonomic manuscripts take the high road and submit their work to peer-reviewed publications, and that editors of non peer-reviewed journals should not accept any taxonomic manuscripts for publications. The latter condition is already maintained by such respected publications as *The Festivus* and *American Conchologist*, and others are advised to revise their policy one way or the other. A review of your manuscript will point out some inconsistencies, some areas that need to be clarified, and may even provoke some further thought. The arguments are usually presented in a pointed fashion but usually remain courteous even in cases of major disagreement. Any disrespectful language only reflects back on the reviewer hiding behind anonymity. Although a source for annoyance, it need not be addressed because it lacks substance. Also note, the opinion of a reviewer needs not be accepted at face value and can be rebutted when submitting the revised version of the manuscript.

10.3.3 Illustrations. When describing new taxa or reporting unusual specimens, the old saying of "a picture says a thousand words" is still true. With your text you should provide quality illustration(s) that show all the relevant features of the specimen(s) depicted.

Most journals will accept black and white images only, while you will have to pay in order to get color images printed. There are several illustration methods including drawings, film photography (see Chapter 7), and digital photography (see Chapter 6). Use the appropriate technique for the project at hand. The following are some guidelines that have worked well in the past but are by no means the only acceptable fashion of producing quality images.

In 90% of the cases, regular light photography (on film, or digital) will suffice. For regular photography, cut out the shells to remove the background and glue the cutouts onto white paste-up board. In my opinion, the images should be spaced as tightly as possible, because any background space not used for the image is wasted space that could have contained image information. Artistic considerations are of little relevance for scientific publications. Place transfer lettering on the board. For scanning electron microscope images, place the cutouts on a black background and use white transfer letters. For digital imaging, place the cutouts into separate layers over a white or black background layer. For specimens with little color or tonal differentiation, consider a line drawing. Alternatively, you can experiment with lighting techniques for photography, or various digital manipulations in PhotoShop. Consider providing partial enlargements of particular features such as the protoconch of gastropods or the hinge of bivalves.

10.3.4 How to recognize a new species. This is the most difficult task of describing a new species and is fundamental in the practice of taxonomy and nomenclature. You will have to spend considerable time measured in years to familiarize yourself with the group. You must be able to recognize each species and be sure that you know all described species in that group. Merely knowing that you cannot identify the species is insufficient reason to call it new. Often very few specimens of a potentially new species are initially available. Sometimes, the specimen at hand is not a prime exemplar, either.

With a few specimens at hand, the question arises whether it is a mere form of a known species, a teratological form, a hybrid, or whether it is new. Borrowing from statistics, if the number of specimens is small, the degree of difference must be large in order for it to be meaningful. If you have many specimens, then even a minute difference may turn out to indicate a new species. As an example consider *Neopilina galatheae* Lemche, 1957, which was described from very few specimens. However, it was so radically different from anything else that a new genus and new species were erected for it and the creature was assigned to the class Monoplacophora, then believed to be extinct since the Devonian Era (about 400 million years ago).

10.3.5 What is a species? This question is clearly related to how to recognize a species, because you should know what you try to find and how to recognize it. Yet, this question is one of the most difficult in biology. There are a number of different species concepts. Each of them has clear applications in some specific cases, and clearly fails under other circumstances. It is advantageous to be aware of the differences and to be able to apply them to particular cases. Overviews can be found in any textbook on evolution (e.g., Smith 1994, Ridley 1996, Futuyma 1998, Freeman and Herron 2001) and a recent summary can be found in Wheeler and Meier (2000).

Biological species concept: A species is composed of actually or potentially interbreeding individuals. Individuals that do not do so belong to two different species. This species concept applies beautifully to most mammals and birds, but in most cases it is difficult to apply for invertebrates, because reproductive barriers are usually unknown. Other complications arise when fertile hybrids between two "species" are formed, and in asexual organisms.

Morphological species concept: A species is composed of similar looking individuals. For paleontology, this is the only species concept that applies, and can be used for asexual organisms. Problems arise in cases of sexual dimorphism, and if a lineage changes continuously without branching off daughter species. In cases in which a metamorphosis is encountered during development (e.g., tadpoles and frogs, caterpillars and butterflies), the morphological definition can be segregated into stage specific units.

Phylogenetic species concept: A species is the least inclusive monophyletic group (see Chapter 11). This definition is derived from modern phylogenetic thinking but requires a phylogeny as a starting point. However, isolated populations of one biological species can be regarded as distinct phylogenetic species. Furthermore, for asexual organisms each lineage would become its own separate species.

Evolutionary species concept: A species is a discrete evolutionary lineage. Although closely related to the Phylogenetic Species Concept, the Evolutionary Species Concept has a temporal component to it. It is difficult to apply to asexual organisms, and hybridization in which evolutionary lineages intersect poses problems.

Cohesion species concept: A species is a group of individuals that self-regulates its membership (cohesion). The problems of occasional hybridizations and asexual lineages are accommodated here. However, it is open how one measures cohesion.

Chronospecies concept: Specimens that lived several million years ago cannot interbreed with contemporary individuals; therefore, there is some temporal barrier to a species. This concept allows subdividing lineages that change continuously without branching off daughter species.

Subspecies: Subspecies are usually defined as a geographically isolated form with a constant, unique character, but may potentially interbreed with the other population. The definition of a subspecies is even more vague than that of species; so many modern practicing taxonomists do not deal with subspecies.

From the above the question, "what is a species?," arises whether it is better to excessively subdivide (split) or to excessively unite (lump). Rosenberg (1996) argued that it is easier to lump after the fact, than to split after the fact. It is easier to unite the distributions of four subspecies, than to divide the distribution data of a single species into four subareas of the subspecies. Yet, if there is good reason to do away with excessive names that only label endpoints of a continuum of variation, then one should not shy away from lumping. Splitting has the undesirable property of hiding natural variation.

10.4 ANATOMY OF A SPECIES OR GENUS DESCRIPTION

Whenever the word species is used here, it also applies to subspecies. Whenever the word genus is encountered, it also applies to the subgenus, usually applies to tribes, and may serve as a guideline for all higher categories.

10.4.1 Title and abstract. The title should be concise yet indicate specifically what the contribution addresses. The title "Description of a new species" is too vague, so should be modified to "Description of a new species of *Conus* (Mollusca: Gastropoda: Neogastropoda) from the Philippines". There is some debate whether the new species name can, should, or should not be included in the title. Some journals publish their table of contents with the titles of the articles ahead of the actual issue, which may have nomenclatural consequences. However, the inclusion of the species name can be helpful to subsequent authors engaging in revisions. If you are unsure, consult with the editor of the journal.

The abstract contains a summary of the results of your paper. Only the main points are mentioned, and any more detailed analysis is omitted. The abstract is a self-contained unit of the paper, which means that on first mention of a species the author and date must be mentioned. Usually the abstract does not contain literature citations. Consider that the abstracts of many journals are available in electronic form that can be searched electronically. Hence, use widely known but specific terms such as "new species," "distribution," "anatomy," "habitat," the name of the family, and a common name for the group of mollusk in order to facilitate searches by future workers. Often some keywords may be added to the abstracts, in other cases make sure that all the keywords are included in the abstract proper.

10.4.2 Introduction. The introduction gives some background on the type of organism and mentions some of the main works published on the group at large. What kind of an organism is it? What is its classification? Describe its general life history. List monographs and recent additions. List faunal treatments from that geographical area. In some cases, these points may not be applicable. Often the reason for writing the present contribution is stated near the end of the introduction: new material became available or a long-standing confusion about a taxon needs to be resolved.

10.4.3 Material and methods. All material investigated should be listed. In order to avoid redundancy, type material (see Section 10.4.6 below) and specimens discussed in the treatment of the taxon are excluded, though. The list of institutional acronyms is given in this section. Any activities involving the specimens are described here: specimens were collected by hand, snorkeling, SCUBA, dredging, donated by, etc.; specimens were preserved in ethanol, formalin (indicate percentage and time), dried, etc.; specimens were cleaned with brush, sodium hydroxide, etc.; the animal was extracted by crushing the shell, dissolving the shell in; statistical analysis was performed with; standard terminology of (author, date) is followed here. The purpose of this section is to give other workers the details needed to be able to repeat your study.

10.4.4 Systematics. This is the core part of the manuscript. It starts out with a classification of the organism:

Order Vetigastropoda Salvini-Plawén, 1980
Family Haliotidae Rafinesque, 1815
Genus *Haliotis* Linnaeus, 1758

Some remarks as to the placement of one of the taxa or some recent taxonomic decisions or changes may be noted in a subsection called *Remarks*. The new taxon is clearly labeled as such by adding "new species," "new genus," "*gen. nov.*"and "*sp. nov.*," or an equivalent term after the taxon. Any illustrations accompanying the description in this work are given in parentheses. If new taxa at more than one rank are given (genus and species), the higher rank is described first.

10.4.5 New genus. The description is written in telegraphic style, i.e., it excludes all verbs and articles: "the shell is round" becomes "shell round." Each sentence should start with the feature being described: "columella with folds," not "folds on columella." Follow some logical sequence in the description (top to bottom, outside to inside), and mention the primary feature before secondary ones. If there are diagnostic characters you may highlight them, but you may also elect to write a separate section called *Diagnosis*. The description contains all observable features, whereas the diagnosis contains only those that are unique for the taxon in question.

Type species: For a description of a new genus, a type species should be designated explicitly by the original author. If the type species was already described, it is advisable to cite the publication of the original description for the designated type species, as well as the location of the type specimens. All other species (new or already described) that are considered to belong in that taxon should be listed.

Etymology: The derivation of the name should be indicated, including its gender. There are a number of conventions on how to construct a genus name. Novice authors should consult with experienced ones on potential names to avoid problems. A genus name must be unique in the animal kingdom. Hence, careful consideration should be given when selecting a genus name. Taxonomic resources such as Neave (1939-1940, 1950), Edwards and Hopwood (1966), Edwards and Vevers (1975), Edwards and Tobias (1993), Edwards *et al.* (1996), Sherborn (1902, 1922, 1932), the *Zoological Record*, and the Internet should be consulted.

Biology: Any information on the biology, ecology, or distribution may be indicated in one section or may be further subdivided into sections of their own depending on the amount of information available.

Comparison: This section is also called *Differential Diagnosis*, or is subsumed under *Remarks*. It compares the new genus to related and similar other genera and shows how they are different. This is possibly one of the most useful sections particularly for the non-specialist who needs to be shown how the new genus can be recognized.

Remarks: Any piece of information that does not fit in the above sections can be placed under remarks.

10.4.6 New species. The description and diagnosis are comparable to that of a new genus. If the species has been discussed previously, either under another name, as a *nomen nudum* (see below), or as an unidentified species, these instances should be cited.

Type specimens: Type specimens must be designated. A number of different categories of type specimens can be encountered. The *holotype* is a single specimen given preferential nomenclatural standing over other specimens. *Paratypes* are specimens that belong to the new species but are of lower standing. In theory, multiple specimens of equal standing (*syntypes*) can be designated, but that practice is strongly opposed because such specimens are often a source of nomenclatural problems later on (if it is found that they represent multiple species).

Any material that was inspected but is not considered type material must be specifically excluded from the type material. It is expected that the holotype and at least some of the paratype material will be found in or donated to a public collection. Most editors go beyond the letter of the ICZN code and demand deposition of at least the holotype in a public collection.

Etymology: The derivation of the name should be indicated. The species epithet needs to be unique within that genus. Be mindful, that the genus assignment may change; hence, species names found in closely related genera should be avoided (see *homonymy* below). One of the most common derivations is in honor of a person (*Haliotis fatui* Geiger, 1999: for John Fatu), which is indicated with the Latin genitive suffix in the proper gender (-i for masculine singular, -ae for feminine singular, -orum for masculine or mixed gender plural, -arum for feminine plural). The origin from a particular place is indicated with the suffix –ensis [*Conus bengalensis* (Okutani, 1968) from the Bay of Bengal]. A descriptive adjective is also often encountered, such as *Haliotis rubra* Leach, 1814, for the Australian red (Latin = ruber [m, n], rubra [f]) abalone. Such adjectives need to be properly inflected to the gender of the genus. Furthermore, a species epithet can be any noun (noun in apposition): *Strombus goliath* Schroeter, 1805. For the derivation of species epithets, Brown (1956), Borror (1960), and Werner (1972) are good sources. A number of Latin and Greek dictionaries, and a Latin grammar book (occasionally found as appendices in dictionaries) should also be at hand. Good sources for Latin (and Greek) dictionaries are second hand and seminary bookstores.

Type locality: The location where the holotype was found is the type locality. It should be as specific as possible, including coordinates.

Other sections: The sections on *Biology, Comparisons*, and *Remarks* are structured as detailed under *New Genus* above.

10.5 REVISIONS

A taxonomic revision serves the purpose of summarizing all previous taxonomic accounts and providing some new insights. A revision may tackle a single species, a genus, or an entire family. Preparing a good revision is a significant task that can take many years. It comprises thoroughly and meticulously searching the literature, inves-tigating as much material as possible, and making a particular effort to see all type material. Many early taxonomic treatments were written in languages other then English. A varied selection of dictionaries should be kept at hand. For pre-Linnean texts a Medieval Latin dictionary (e.g., Latham 1980) can be useful. Today there are some translation computer programs and the Babelfish website <babelfish.altavista.com/translate.dyn> that produce reasonable first steps and can reduce the baseline work significantly; however, the scientific language and the peculiar use of certain words in scientific contexts can result in incomprehensible text. Latin texts often contain errors, either in terms of grammar or in terms of type setting (e.g., "niti dissimus" should read "nitidissimus": Lichtenstein, 1794: lot 896). A good revision contains the following elements at the (sub)generic and the (sub)specific level.

10.5.1 Genus. The genus name with author and literature citation for the original description is given. The type species and the mode of designation are indicated. Three modes are found: by original designation (OD) in which the original author stated explicitly that a particular species is the type species; by monotypy (M) where the original author described or discussed only a single species in that new genus so that this species is taken by default as the type species; by subsequent designation (SD) where a later author made a decision on which species should be regarded as the type species. In some instances, no type species had ever been designated. The reviser should carefully evaluate the specific case and make a sensible choice, i.e., make a SD. A SD must be clearly labeled as such by "type species here designated."

Species contained in genus: For any revision, a complete listing of all species belonging to that genus should be given. In larger revisions spanning multiple genera, some species may be impossible to assign to a genus. In this case the category "Family - *incertae sedis*" (= of uncertain seat = affinity) should include those taxa. Every species within the taxonomic scope of the work must be assigned. Occasionally it is also helpful to list species (or genera) previously contained in that group, but no longer considered to be so. For instance, in a treatment of the helmet shells (Cassidae) it is worthwhile to note that the genus *Morum* is no longer in Cassidae, and has been transferred to Harpidae.

Description and diagnosis: The genus should be carefully described and the diagnosis given. This is particularly important when no such diagnosis had been made previously.

Comparisons: Comparisons to related and similar (sub)genera should be provided.

Remarks: Any other noteworthy information is to be included.

10.5.2 Species. The species name with author and literature citation for the original description is given. The consideration of type specimen(s) must be carried out with particular care and attention to detail because any statements in association with types will become permanent. The type specimen(s) as inferred from the literature or as found in museum collections are discussed. Discrepancies must be carefully evaluated; omissions in the original text as well as erroneous labeling in museum collections can be a source of error. Consult with the senior curator of the collection for insight into the history of the collection and the particular specimen. The status of each specimen is to be determined. The following categories are encountered.

Holotype: The holotype is a single specimen given preferential standing by the original author. In cases where the holotype and the paratypes of a taxon belong to two different species, the species of the holotype has its name. Hence, the holotype is also called the *name bearing type*, or the *primary type*.

Paratype: Zero to multiple paratypes may accompany the holotype. The original author usually explicitly designates paratypes.

Syntypes: Multiple specimens of equal standing, all together being one name-bearing type series, can be frequently found with older names. The specimens in the syntype series can belong to one or multiple species. It is due to the possibility that a syntype may represent more than one species that a holotype is preferred. Another name for syntype is cotype.

Lectotype: A lectotype is a single specimen selected from a syntype series that is elevated in rank and becomes the name-bearing type. All other specimens, regardless of whether they belong to the species of the lectotype or not, become *paralectotypes*. The selection of a lectotype should be carried out with great care, particularly if the syntype series comprises more than one species. Preference should be given to a specimen figured in the original description or to the one that best fits the written description of the species. Under the new edition of the ICZN-code a lectotype designation must contain "an express statement of taxonomic purpose" (ICZN, 1999: Art. 74.7.3). Currently there is some debate on the meaning of this phrase, but it seems that as little as "lectotype here selected" should suffice to satisfy the condition. It had been included to prevent the problem known as inadvertent designation of a lectotype.

Neotype: In case the holotype or lectotype has been lost, a new specimen can be designated as the name-bearing, primary type; there are no 'paraneotypes'. Such a designation should be carried out only if the identity of the name is controversial (e.g., Geiger 1996). For many taxa, the whereabouts of the primary type is unknown, but it is not necessary to designate a neotype.

Other types: There is a bewildering diversity of type categories. However, it is important to remember that most of these types are not important in taxonomic descriptions. For a comprehensive listing of types see Fernald (1939).

Synonymy: A listing of all synonyms of this species should be given. The listing should include author and date for each taxon, and the citation of each original description. The listing of type material of all synonyms, and their illustration adds further information. Most synonymies stem from the rule of priority and are easily identified from the later date for the taxon. Other cases (pre-Linnean, non-binominal, homonyms, suppressed taxa, and first reviser's principle) should be explained. Some taxa may be unidentifiable or problematic and are listed as *nomina dubia* (doubtful names) or as *nomina inquirenda* (names to be further investigated).

The synonymy should also include a listing of illustrations of the particular species under consideration, under the correct name or as any of its synonyms. Whether the listing should be exhaustive or selective depends on the scope of the revision and how often a species has been illustrated; an exhaustive listing for *Mytilus edulis* may be unnecessary and the 10-20 best and most widely available sources may suffice.

Altered placement: During revision, it often becomes apparent that a species has been misplaced in a genus. It is self-evident from the synonymy listing if the generic assignment has been changed. If an author wants to emphasize the new assignment, the term new combination or "*comb. nov.*" can be added after the author and date. New combinations do not count as nomenclatural acts, hence, the author who first proposed the new combination does not receive special recognition.

Type locality: Usually the type locality is given in the original description. If this is not the case, a locality can be designated in accordance with ICZN article 76. Such new designations should be clearly marked with "here designated." The stratigraphic unit of the type is subsumed under the type locality. The value of subsequent designation of type localities is somewhat controversial; it certainly carries much less weight than lectotype selections.

Other sections: The sections on *Description, Diagnosis, Etymology, Biology, Comparisons,* and *Remarks* are written as discussed under *Genus descriptions* above.

10.6 THE ICZN CODE

In the following, a brief overview of the code is given, in which some of the more recent developments are emphasized. Although many of the specific rules may sound at first rather obscure, the principles at work are straightforward and based on common sense. Authors intending to engage in any nomenclatural activity must read the entire code cover to cover at least once for minimal preparation. The sections below cannot substitute for a full immersion into the matters of the code. The responsibility for acting within the rules of the code lies solely with the author, and cannot be delegated to either reviewers or editors. These points cannot be stressed enough, because nomenclatural acts are destined for eternity.

10.6.1 History. The start of zoological nomenclature is January 1st 1758, the date of publication of Linnaeus' 10th edition of *Systema Naturae*. For over 150 years there were no formal guidelines on how to pursue taxonomic activities, although a certain informal understanding developed in the zoological community (Stricklandian code of 1842), which was superseded by more formal adoptions (*Règles* of 1905). Only in 1961 when the first edition of the *International Code of Zoological Nomenclature* was published were the rules of nomenclature fixed. The fourth edition of the code has come into effect on January 1st 2000. A more thorough historical account was provided by Melville (1995).

10.6.2 Aim (Articles 1-3). The purpose of the code is to provide a unifying framework for nomenclatural affairs in zoology. There are separate codes with somewhat different rules for plants and for bacteria. The main objectives of the code are universality and stability of names, while permitting freedom of taxonomic thought and flexibility if circumstances warrant it.

10.6.3 Principles (Articles 3-6). The principles are surprisingly few. First, it is based on the Linnean binominal system. The term is derived from the Latin for two names: bi-nomen. "Binomial" is a significant misspelling in nomenclature, because it relates to a type of number distribution in statistics. A species is identified with a combination of two names, that of a genus and that of a species. Second, taxa must be based on types (principle of typification). These may either be type taxa or in the case of species-level names, type specimens. Third, the name that was given to a species first is maintained in the future. This is the principle of priority.

10.6.4 Publication (Articles 7-9). A work must be published in order for the nomenclatural act to be available. Traditional publications are books and periodicals, but exclude manuscripts. In today's electronic age, under certain circumstances works on laser disk and CD-ROM can count as publications, however, any content of web pages is excluded. A full discussion of what constitutes a publication is beyond the scope of this brief introduction (e.g., Ph.D. theses).

10.6.5 Availability (Articles 10-20). These provisions detail whether a name is considered available for the purpose of zoological nomenclature. The provision depends on when the particular nomenclatural act was published, for which four periods are defined (pre 1931, 1931-1960, 1961-1999, 2000-present).The specific provisions under these paragraphs are some of the more esoteric ones.

10.6.6 Publication date (Articles 21-22). Because of the principle of priority, the earliest date for which it is credibly determined that the work was published is used. A work that is only dated "1855" has a publication date of December 31st, 1855. However, if a second author cited that work in a different publication that is dated May 21st, 1855, then May 21st is the publication date of the first "1855" publication.

10.6.7 Validity (Articles 23-24). These articles specify under which conditions a nomenclatural act has standing. Essentially, it states the principle of priority. The oldest name (= *senior synonym*) of a species is the valid name, all others are *junior synonyms*. The more common *subjective junior synonym* is found when the same species is described from two different name-bearing type specimens. Rarely, two names are given to the same name-bearing type, resulting in a case of *objective synonymy*. Sometimes two names are given to the same species in the very same work, so that the rule of priority cannot apply. In this case, the person who first publishes this synonymy (usually in the course of a revision) has the privilege of making a sensible choice under the provision of the *first reviser's principle*. It is a common misconception that the one appearing first in the publication (page priority) gets to be the senior synonym. The reviser chooses which taxon retains the name.

Occasionally, two species are inadvertently given the same name (*homonymy*). The senior homonym is valid, whereas the junior homonym usually cannot be used as a valid name. The exception is for newly discovered senior synonyms and senior homonyms, which can be designated *nomina oblita* (forgotten names) that protect the younger, but established names (*nomina protecta*) under certain well-defined conditions (ICZN article 23.9). This ruling resurrected from the second edition of the code serves the purpose of stability.

10.6.8 Formation of names (Articles 25-34). These articles specify that names must be latinized, and how this is achieved. In theory, any word that could be an attempt to emulate a Latin word (e.g., keyboardus) is admissible. Random letter combinations, 'names' containing numbers, and names with diacritic symbols (é, ö, ñ, ç) cannot be used nor hyphenated names. Recommendations on how to transform unconforming spellings are given. For instance, ö becomes either o or oe. The endings for tribes (-ina), subfamilies (-inae), families (-idae), and superfamilies (-oidea) are also standardized.

10.6.9 Family, genus, and species level names (Articles 35-49). The *principle of coordination* states that for a family, genus, or species level name the author and date also apply for super- and sub-families, subgenera, or subspecies. For instance, Rafinesque, in 1815, authored the family Haliotidae; the authorship of Haliotoidea and Haliotinae is also Rafinesque, 1815, although he never mentioned these names. The articles further specify in case of subdivision of a taxon, that the *nominotypical taxon* is that in which the type species of the higher taxon is found. For example, the type species of *Haliotis* is *H. asinina* Linnaeus, 1758. The subgenus *Haliotis* is the subgenus that contains *H. asinina*.

10.6.10 Authorship (Articles 50-51). The author(s) of a work containing a nomenclatural act are considered responsible together and are cited together. An exception is found if part of a work was clearly written by another person. In that case, the writer of the section (e.g., Smith) is listed first connected by the word 'in' in italics, followed by the author(s) (e.g., Miller and Miller) of the overall work: Smith *in* Miller and Miller.

10.6.11 Homonymy (Articles 52-60). Homonymy can occur at every taxonomic level and must be resolved. Single letter differences are sufficient to preclude homonymy in family and genus level names. In species level names, significant single letter differences conveying a different meaning do also prevent homonymy. However, single letter differences that are only spelling variations (Articles 58.1-58.15) are considered homonymous. The assignment of two homonymous species level taxa to two different subgenera in the same genus does not remove them from homonymy. Two types of homonyms have to be considered. A *primary homonymy* is encountered if the original authors described the two species in the same genus. A *secondary homonymy* arises if one taxon is transferred from one genus to another genus already containing a species with the same species epithet. If a senior synonym is also a junior homonym, then a *nomen novum* (new or replacement name) is given to that species, because the name of the junior homonym has already been taken (is *preoccupied*).

10.6.12 Types (Articles 61-76). The designation and recognition of types at every taxonomic level (family to species) is covered. At each level, a type must be designated according to the principle of typification. In case of the absence of types, their subsequent selection is detailed. For family and genus level names, only other names are concerned, whereas for the species group actual specimens are designated as types (holotypes, lectotypes, and neotypes). A name lacking a proper description is considered a *nomen nudum* (naked name) and is not available.

10.6.13 The Commission (Articles 77-90). In some instances, an action not covered by the articles of the code is warranted. Acts requiring the intervention of the "Commission on Zoological Nomenclature" include suppression of certain senior synonyms and homonyms; de-selection of a poorly chosen lectotype; and suppression of a publication. In such a situation, a formal request in the form of a *case* can be brought to the Commission. The application will be published in the *Bulletin of Zoological Nomenclature* and comments from the community at large are collected for the period of one year. Thereafter, the Commission will vote (exercise plenary power) on the case and publish its decision as an *opinion*. Publications that are made unavailable or names that have been suppressed are placed on the appropriate *Official Index*. Names that have been retained, or works that are considered available, are placed on the appropriate *Official List* (Melville and Smith 1987).

10.6.14 Code of Ethics (Appendix A). If a person knows that another person is already working on the description of a new taxon, then the former individual shall not attempt to beat the latter to publication. If a living author publishes (unknowingly) a homonym, the person who discovers the homonymy needs to give the original author at least one year to propose a *nomen novum*. Names should not be of offensive nature. Certain decorum should be maintained when selecting names.

10.7 NOMENCLATURE, TAXONOMY, AND CLASSIFICATION

Nomenclature and taxonomy are often confused and/or used interchangeably. There are some subtle yet significant differences. Nomenclature is only concerned with giving names but does not place these names into a classification system. The latter process, namely placing existing names in an overall hierarchy, is called taxonomy. Giving a new species a name is a nomenclatural act, placing it in a particular genus concerns taxonomy.

The way names are transformed into a classification has been a hotly debated topic in the last decade. Two schools can be identified: traditional, phylogenetic taxonomy and "new" phylogenetic taxonomy (e.g., de Queiroz and Gauthier 1994), the latter also being called "node pointing system" (Nixon and Carpenter 2000: 298). Both have the same goal, namely representing the existing name in a system in accordance with the phylogeny of the group in question and naming monophyletic groups only (see Chapter 11, Cladistics and Molecular Biology). The main argument of the proponents of traditional taxonomy is that the system has worked well for the past 250 years; hence, there is no reason to replace it with an untested system. The proponents of the node pointing system argue that the new system will be more stable (fewer name changes), avoids mandatory ranks of the Linnean binominal system, and uses phylogenetic hypotheses explicitly to define the taxa.

The node pointing system has gathered little attention amongst most practicing taxonomists (the silent majority), but a trickle of publications by a handful of proponents has placed them in the spotlight in the larger arena of science (Pennisi 1996). Recently, Nixon and Carpenter (2000) and Benton (2000) have critically examined the node pointing system and have resoundingly rebuked its claims. Particularly, with respect to stability as denoted by constancy of taxa included, the node pointing system fares far worse than traditional taxonomy. The nuisance of the node pointing system is destined to be but an insignificant ripple in the development of systematics. The various alternative classification and naming schemes that have been proposed during the past decades (e.g., numerical species names, indented classifications, prefix ranking, Phylocode, uninominals: reviewed by Kron 1997) have all vanished. For various reasons they have turned out to be nothing but futile academic exercises of needless innovation. All have blatant deficiencies that are by far more serious than what is considered suboptimal at present. The Linnean system thrives as at its inception.

What is the purpose of a classification? The intention of an artificial classification is not to reproduce a (real) phylogeny (Benton 2000). Its primary function is that of a communication tool and that of a summary statement. It is also clear that ranks cannot be discovered in nature; they are artificial constructs in an artificial classification. Yet this classification should not contradict the real phylogeny of the organisms it classifies. In that sense, a classification is a rough sketch of a phylogeny. The function of ranks is also one of communication, despite the fact that ranks in different groups cannot be compared: a family of mollusks is quite different from a family of bacteria. However, families of mollusks within a molluscan order are meaningful for the purpose of communication. Ranks serve as a tiered retrieval system for information concerning these organisms.

10.8 ACKNOWLEDGMENTS

This chapter benefited from the comments of Jim McLean, Christine Thacker, and the reviewers.

10.9 LITERATURE CITED

Benton, M. J. 2000. Stems, nodes, crown clades, and rank-free lists: Is Linnaeus dead? *Biological Reviews* **75**: 633-648.

Borror, D. J. 1960. *Dictionary of Word Roots and Combining Forms*. Mayfield Publishing Co., Mountain View, California. 134 pp.

Bouchet, P. 1997. Inventorying the molluscan diversity of the world: what is our rate of progress? *Veliger* **40**: 1-11.

Brown, R. W. 1956. *Composition of Scientific Words*. Smithsonian Institution Press, Washington. 882 pp.

Council of Biology Editors. 1994. *Scientific Style and Format. The CBE Manual for Authors, Editors, and Publishers*, 6th Ed. Cambridge University Press, Cambridge. 825 pp.

de Queiroz, K. and J. Gauthier. 1994. Towards a phylogenetic system of biological nomenclature. *Trends in Ecology and Evolution* **9**: 27-31.

Edwards, M. A. and A. T. Hopwood. (eds.) 1966. *Nomenclator Zoologicus*, Vol. VI, 1946-1955. Zoological Society of London, London. 329 pp.

Edwards, M. A. and H. G. Vevers. (eds.) 1975. *Nomenclator Zoologicus*, Vol. VII, 1956-1965. Zoological Society of London, London. 374 pp.

Edwards, M. A. and M. A. Tobias. (eds.) 1993. *Nomenclator Zoologicus*, Vol. VIII, 1966-1977. Zoological Society of London, London. 620 pp.

Edwards, M. A., P. Manyly, and M. A. Tobias. (eds.) 1996. *Nomenclator Zoologicus*, Vol. IX, 1978-1994. Zoological Society of London, London. 747 pp.

Fernald, H. T. 1939. On type nomenclature. *Annals of the Entomological Society of America* **32**: 689-702.

Freeman, S. and J. C. Herron. 2001. *Evolutionary Analysis*, 2nd Ed. Prentice Hall, Upper Saddle River, New Jersey. 704 pp.

Futuyma, D. J. 1998. *Evolutionary Biology*, 3rd Ed. Sinauer Associates, Sunderland, Massachusetts. 810 pp.

Geiger, D. 1996. Haliotids in the Red Sea with neotype designation for *Haliotis unilateralis* Lamarck, 1822 (Gastropoda: Prosobranchia). *Revue Suisse de Zoologie* **103**: 339-354.

ICZN, 1999. *International Code of Zoological Nomenclature*. International Trust of Zoological Nomenclature, London. 306 pp.

Kron, K. A. 1997. Exporing alternative systems of classification. *Aliso* **15**: 105-112.

Latham, R. E. 1980. *Revised Medieval Latin Word-List from British and Irish Sources, with Supplement*. Oxford University Press, Oxford. 535 pp.

Lichtenstein, A. A. H. 1794. *Catalogus Rerum Naturalium Rarissimarum*, Sectio Secunda. Schniebes, Hamburg. 118 pp.

Mayr, E. and P. D. Ashlock. 1991. *Principles of Systematic Zoology*, 2nd Ed. McGraw-Hill, New York. 475 pp.

Melville, R. V. 1995. *Towards Stability in the Names of Animals: A History of the International Commission on Zoological Nomenclature, 1895-1995*. International Trust for Zoological Nomenclature, London. viii + 92 pp.

Melville, R. and J. D. D. Smith, eds. 1987. *Official Lists and Indexes of Names and Works in Zoology*. International Trust for Zoological Nomenclature, London. 366 pp.

Neave, S. A., ed. 1939-1940. *Nomenclator Zoologicus: A List of the Names of Genera and Subgenera in Zoology from the Tenth Edition of Linnaeus, 1758, to the end of 1935*. 4 Volumes. The Zoological Society of London, London. xiv + 3805 pp.

Neave, S. A., ed. 1950. *Nomenclator Zoologicus*, Vol. V, 1936-1945. The Zoological Society of London, London 308 pp.

Nixon, K. C. and J. M. Carpenter. 2000. On the other "phylogenetic systematists." *Cladistics* **16**: 298-318.

Pennisi, E. 1996. Evolutionary and systematic biologists converge. *Science* **273**: 181.

Reid, D. G. 1996. *Systematics and Evolution of Littorina*. The Ray Society, London. 463 pp.

Ridley, M. 1996. *Evolution*, 2nd Ed. Blackwell Science, Cambridge. 719 pp.

Rosenberg, G. 1996. Conchatenations. *American Conchologist* **24**: 10-11.

Sherborn, C. D. 1902. *Index Animalium*. Section 1, 1758-1800. C. J. Clay and Sons, London. 1995 pp.

Sherborn, C. D. 1922. *Index Animalium*. Section 2, 1801-1850. C. J. Clay and Sons, London. 7056 pp.

Sherborn, C. D. 1932. *Index Animalium*. Section 2, 1801-1850. Epilogue and Additions. Cambridge University Press, Cambridge. 1096 pp.

Smith, A. B. 1994. *Systematics and the Fossil Record: Documenting Evolutionary Patterns*. Blackwell Scientific, Oxford. 223 pp.

Werner, F. C. 1972. *Wortelemente Lateinisch-Griechischer Fachausdücke in den biologischen Wissenschaften*. Suhrkamp Taschenbuch 64. 475 pp.

Wheeler, Q. D. and R. Meier. (eds.) 2000. *Species Concepts and Phylogenetic Theory: A Debate*. Columbia University Press, New York. 230 pp.

Winston, J. E. 1999. *Describing Species: Practical Taxonomic Procedure for Biologists*. Columbia University Press, New York. xx + 518 pp.

C. F. Sturm, T. A. Pearce, and A. Valdés. (Eds.) 2006. *The Mollusks: A Guide to Their Study, Collection, and Preservation.*
American Malacological Society.

CHAPTER 11

CLADISTICS AND MOLECULAR TECHNIQUES: A PRIMER

DAVID CAMPBELL

11.1 INTRODUCTION

You have probably heard about new studies using cladistic analyses, DNA data, and other exotic-sounding methods. Three questions arise for the average collector: How does it work? Can I try it myself? Will it affect the classification of my collection? In this chapter, we will explore these types of questions. For further information see Mayr and Ashlock 1991, Wiley *et al.* 1991, Graur and Wen-Hsiung Li 1996, Hillis *et al.* 1996, Page and Holmes 1998, Schuh 2000, Felsenstein 2003, Lydeard and Lindberg 2003, and Hall 2004.

11.2 CLADISTICS - WHAT IS IT?

Cladistic phylogenetic analysis tries to provide a more objective method of reconstructing evolutionary history than the traditional approach of simply looking at the data and deciding what seems plausible. It tries to do this by starting with an explicit definition of the features of the organisms under study (called characters, see Table 11.1) and some method of creating an evolutionary or phylogenetic tree and evaluating it. For each character, you need to determine the character state for each taxon. For example, one character might be the "number of valves in the shell." The character state for *Pecten* or *Berthelinia* would be "two," for *Strombus* or *Neopilina* would be "one," for *Chaetoderma* or nudibranchs would be "zero," and for *Chiton* would be eight.

Mathematical analysis of the data then generates a phylogenetic tree. This tree is called a cladogram if the branch ends line up and their lengths are arbitrary (Figure 11.1) or a phylogram if the branch lengths on the tree are proportional to the amount of evolutionary change occurring on the branch (Figure 11.2). Phylograms are thus more informative but usually not as easy to read. For example, in Figure 11.2, it is difficult to judge the relationships between A, B, and C. However, it is easy to see that there are very few changes differentiating A, B, and C but many that group these three together. In contrast, the relationships among A, B, and C are easy to see in Figure 11.1, but there is no indication about the number of changes supporting each group.

A group of taxa associated by a common ancestor on a phylogenetic tree is called a clade. In contrast,

Table 11.1 Characters.

Character	Definition	Example
Apomorphy	A derived character found in a descendant but not in the ancestor	Torsion in gastropods
Plesiomorphy	A primitive character found in both the ancestor and some of its descendants	A shell in most gastropods
Synapomorphy	A derived character that is found in two or more taxa and their nearest ancestor and not found outside this group	A bivalved shell in bivalves

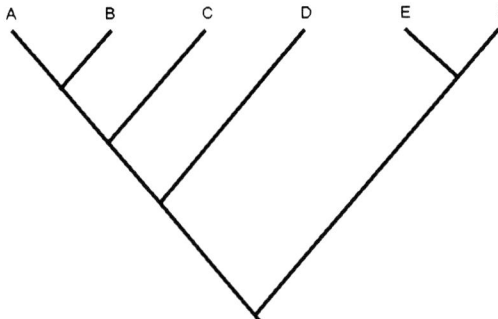

Figure 11.1. A cladogram.

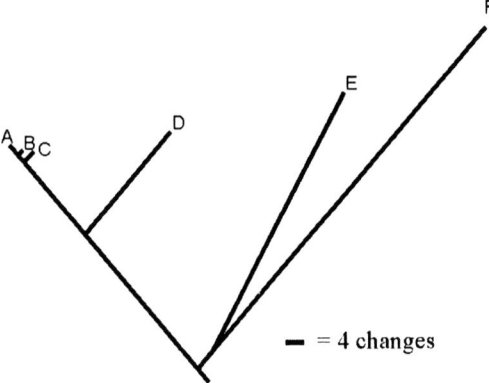

Figure 11.2. A phylogram.

a grade is a group of taxa at a similar evolutionary state, not necessarily closely related. In Figures 11.1 and 11.2, AB, ABC, ABCD, EF, and ABCDEF are all clades. They are monophyletic, including a common ancestor and all its descendants. BCD, on the other hand, is paraphyletic. It includes some members of a monophyletic group but not all (A is missing). AF is polyphyletic (specifically, diphyletic). It includes taxa that evolved separately. Another way of thinking of a polyphyletic clade is to think of it as a group containing members from 2 or more ancestral groups (see Table 11.2).

Almost everyone agrees that polyphyletic taxa should be eliminated from taxonomic use, although some older classifications included known polyphyletic groups. More often, older polyphyletic groups were based on inadequate data. One such example is the Linnaean genus *Patella*, which included several limpet-shaped gastropods now assigned to multiple unrelated genera, including *Patella* (Patellogastropoda), *Siphonaria* (Pulmonata) and *Umbraculum* (Opisthobranchia).

The treatment of paraphyletic taxa is debated. On the one hand, they are not discrete evolutionary units because they omit part of the clade. On the other hand, they may represent a distinct phylogenetic grade and have various shared characteristics. For example, Monoplacophora *sensu lato* is believed to include the ancestors of Bivalvia, Gastropoda, Cephalopoda, Scaphopoda, Rostroconchia, and other classes. Thus, Monoplacophora in the broad sense is not monophyletic and is not strictly equivalent to the other classes. On the other hand, monoplacophorans are all generally similar. Deciding whether a particular cap-shaped early Cambrian fossil is closer to gastropods or to bivalves may not be possible. Thus, Monoplacophora has its usefulness as a concept but also its problems.

11.2.1 The root, the ingroup, and the outgroups. Often, you are not only interested in evolution within your taxon but also how the group is related to other organisms. Note how the base of the tree in Figure 11.1 has an additional branch. This indicates that the studied taxa are related to other organisms at this point in the phylogeny. This branch is known as the root and the tree is rooted. Usually a root is determined by including taxa outside the clade of interest (the ingroup). These taxa, or outgroups, must be carefully chosen to be

Table 11.2 Clades

Clade	Definition	Examples from Figure 11.1
Monophyletic	A group containing an ancestor and all of its descendants	AB, ABC, ABCD, EF, ABCDEF
Paraphyletic	A group containing an ancestor and some, but not all, of its descendants	AC, BC, ABD, ACD, BCD, BCDE
Polyphyletic	A group containing members from 2 or more ancestral groups	AF, ABE, BEF, CEF

sure that they really are outside the clade of interest, but not so distantly related as to provide little useful information. For example, suppose you were studying the evolution of Gastropoda as a whole. Bacteria are obviously not gastropods, so they would be a safe outgroup, whereas bellerophonts may or may not be gastropods and therefore are not a safe outgroup. However, bacteria are so distantly related to gastropods that they provide almost no useful information about the evolutionary patterns. Bivalves, cephalopods, or definite monoplacophorans would be better outgroups for the gastropods. With a known root, you can determine some of the sequence of evolutionary events. The first event in the evolution of the ABCDEF clade was the split between the ancestor of ABCD and the ancestor of EF. D's ancestor split off before C's, which in turn split before A and B split. As Figure 11.1 is a cladogram, we cannot use it to tell how the E-F split compares in time to the splits within ABCD without further information. From Figure 11.2, we can see the relative amount of change on each branch. This suggests that the E-F split was earlier than the splits within ABCD, if we assume a relatively constant rate of change. However, that assumption may be incorrect.

11.2.2 Parsimony. The most popular method of phylogenetic analysis is generally called maximum parsimony, although all methods involve the concept of parsimony. This tries to find an evolutionary tree that minimizes the number of postulated evolutionary changes. For example, Table 11.3 could be part of the data generating the tree in Figure 11.1. Character 2 could be explained by a single change from state 0 to state 1 on the branch leading to ABCD (Figure 11.3). A change from 1 to 0 on the branch leading to EF would also require only one change.

Table 11.3 Sample data.

Species	\multicolumn{7}{c}{Character}						
	1	2	3	4	5	6	7
A	1	1	2	1	1	0	1
B	1	1	1	1	1	0	1
C	1	1	1	1	2	0	1
D	1	1	1	2	3	3	3
E	1	0	3	1	4	1	2
F	1	0	4	3	5	2	4

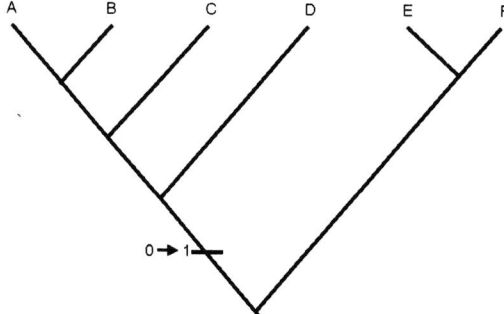

Figure 11.3. A parsimonious hypothesis about change in character 2.

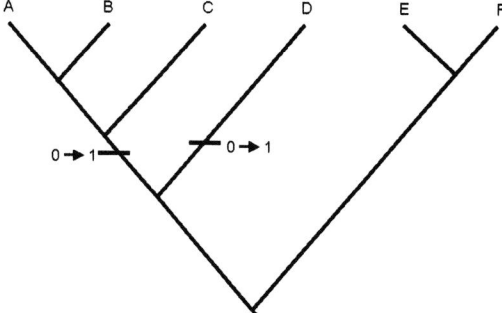

Figure 11.4. A less parsimonious hypothesis about change in character 2.

These options are equally parsimonious. Although the data in Table 11.3 could also be explained by a change from 0 to 1 on the branch leading to D and another change from 0 to 1 on the branch leading to ABC (Figure 11.4), this requires more changes and is thus less parsimonious. Notice that some of the characters are incompatible with others. For example, character 3 suggests that BCD are more similar to each other than any are to any other character species, whereas character 5 suggests that AB are more similar to each other than to any other. In the valve example, if we relied solely on the "number of valves" character, parsimony would group the taxa that have the same number of valves. However, other evidence indicates that these examples are all convergent, and parsimony would give a hypothesis of relationships quite different (and unbelieveable) from an analysis that included more characters (thus we might be fooled or misled by such results). As a result, the choice of data to include in an analysis is very important. Additionally, if you think parts of your data are better or worse, you can put more weight on them in the analyses. Although this is an assumption that must be justified, the common default of giving equal weight to all analyzed data is also an assumption. Maximum parsimony can be used on almost any sort of data, and the principle of minimizing the amount of assumed evolutionary change is intuitively appealing, even though it might not hold up in some circumstances. It is also less computationally intensive than most likelihood methods. As a result, it is probably the most widely used cladistic method. Other analytical techniques may be less sensitive to such problems, but have other weaknesses or difficulties. The choice of species to include is also very important. The valve example selected a small assortment of distantly related species, some of them rather unusual, and relied on only a single character. Thus, both poor sampling of species and inadequate data are to blame.

11.2.3 Distances. For some types of data, you can calculate distances among your taxa. Certain molecular techniques, such as DNA-DNA hybridization, generate distances, as can statistical comparisons of faunas (*e.g.*, similarity indices). In cases such as these, a distance approach may be the only possible way to analyze the data. Additionally, formulae exist for calculating distances from some other data. Because distance calculations typically require modifying and simplifying the data, they generally do not provide as powerful an analysis as those methods that use the unmodified data. In addition, distance methods often have trouble dealing with missing data. However, distance methods are generally much less computationally intensive than parsimony or likelihood methods. They may therefore be the only feasible way to analyze some datasets.

One of the most widespread distance techniques is UPGMA (Unweighted Pair Group Method using Arithmetic averages). This method assumes that the characters have evolved at a constant rate. If they have not (which is often the case), UPGMA can perform very badly. However, it is computationally relatively simple and therefore was one of the first methods implemented on computers. Thus, it is more common in older analyses, though still important for some types of data. Another popular distance technique is neighbor-joining.

Neighbor-joining compensates for variation in the apparent evolutionary rate among taxa, so it is suitable for many more situations than UPGMA. It proceeds by joining the most similar taxa (after compensating for evolutionary rates) until everything is associated into a phylogenetic tree. However, once you have generated this tree, neighbor-joining provides no guidelines for comparing this tree to other trees. You cannot tell whether your tree is better or worse than another simply by relying on neighbor-joining criteria. Therefore, neighbor-joining may be better used to generating preliminary trees to evaluate by another method instead of using it to generate a final tree.

11.2.4 Maximum likelihood. A third category of analyses is maximum likelihood. If you can assign probabilities to the different character-state changes, you can try to find the most probable tree. Assigning probabilities to morphological changes is very difficult if not impossible, but several models and empirical data can be used to produce probability models for molecular sequence data. For example, DNA is composed of the four bases adenosine (A), cytosine (C), guanine (G), and thymine (T). In DNA, A is more similar to G than to T or C, so you might assign a higher probability to A to G mutations than A to T or A to C mutations. Models with greater complexity take into account more considerations about the evolutionary factors and so are probably more realistic. However, they are also more computationally intensive. Complex models also seem more sensitive to discrepancies between model and the real evolutionary pathway than simpler models. For relatively short DNA sequences (a few hundred bases), this sensitivity is probably important. It is less of a problem for long sequences. Maximum likelihood is the most computationally intensive approach and has only recently become feasible for many data sets. However, if it is feasible, its assumptions about the evolutionary pattern are probably the most realistic and thus it is less likely to be misled than parsimony under certain situations.

One type of likelihood analysis gaining in popularity is the Bayesian analysis. It generates a set of evolutionary trees based on probability models and uses a large sample of the most probable trees that it finds to assign probabilities to various clades. It can also be used to give relative probabilities of two evolutionary trees.

11.2.5 Branch-and-bound and heuristic methods.

Both parsimony and likelihood approaches require some technique of generating phylogenetic trees, which are then analyzed to see how parsimonious or likely they are. The simplest approach is an exhaustive search, in which every possible tree is checked. However, there are almost 34.5 million trees possible with 11 taxa (without rooting the tree, which would add more possibilities). Trying to evaluate all the trees for data sets this large or larger is often impractical. Another approach is a branch-and-bound method. This eliminates obviously bad options from consideration, allowing the computer to focus on the promising possibilities. For example, if one tree containing only some of the studied taxa is less parsimonious than the most parsimonious tree found so far for the entire data set, adding more taxa to the bad tree configuration will only make it worse. Thus, all trees based on the unparsimonious partial tree can be eliminated from consideration. Even this approach often takes too long to be feasible. Several approximate, or heuristic, methods are available. These search for an acceptable tree, but cannot guarantee finding the best tree. If there is a tree better than all of the relatively similar trees, but still worse than the overall best tree, heuristic methods may find this relatively good tree rather than the best. Several techniques exist to try to improve the chances of finding the overall best tree. In general, if several different methods all come up with the same tree, then it is probably the best overall. If different methods come up with many different trees, then you have more concerns.

11.2.6 Consensus.

Often, a search will find many equally good trees. To show these data, consensus trees are often used. A strict consensus shows those clades supported by every tree. Majority rule consensus includes clades supported by a majority of the trees. Other consensuses are less widely used.

Several methods can provide a quantitative assessment of trees. Numerous methods and formulae have been proposed, some of which have since been shown to have definite problems. However, a few measures are widely used. As a rule, the statistical significance of these approaches remains unknown.

Consistency indices try to measure the relative proportion of features that support or contradict the resulting tree. Several variations exist in the formulas used to calculate them, but they generally are based on a comparison between the actual tree and theoretical ideals. For example, in a parsimony analysis, each character with two character states (e.g., presence or absence) must change a minimum of one time on the tree. Thus, an analysis based on three presence/absence characters would have a minimum tree length of three, considering the characters separately. However, if the characters do not agree with each other, the minimum tree length found by a parsimony search would be greater than three and so the consistency index would be lower. Consistency indices are affected by the type and amount of data, so you cannot identify a particular value as good for all analyses.

11.2.7 Bootstrapping and jackknifing.

Bootstrapping and jackknifing involve resampling the data set to create a new data set. In bootstrapping, the data set is sampled with replacement, creating a new data set the same size as the original but randomly omitting some characters (or taxa) and including others more than once. Jackknifing randomly omits characters (or taxa) to create a smaller data set. These modified data sets are then analyzed, and the process is repeated many times. If only a few characters supported a clade produced in the initial search, then the random sampling may omit some of those characters and produce a significantly different tree than before. The percentage of replicate analyses that support a given clade is then referred to as the bootstrap or jackknife support. High values of support (depending on the data set, probably a minimum of at least 70% and possibly higher) suggest good results. Another approach is the use of decay indices (= Bremer Support). The decay index for a clade is the difference between the overall best tree and the best tree that does not support the clade. For example, if your most parsimonious tree were 100 steps long, but one clade supported in it did not appear in some trees 101 steps long, then you do not have much confidence in that clade. If, on the other hand, a clade occurs in all trees that are less than 150 steps long, then you have much higher confidence in the clade.

All phylogenetic analyses rely on certain assumptions. Two key assumptions are that the characters are independent and that they evolved according to the model assumed by the analysis. Real data may often violate both of these. For example, the presence of a tooth in the right hinge of a bivalve is not independent of the presence of a socket in the left hinge. Coding those as two independent features would really be entering the same feature twice. Although no one would include such obviously redundant characters, other features may have subtle genetic or evolutionary links. A single mutation in a gene affecting early development may have numerous effects on the morphology of the adult. Similarly, seemingly unrelated features may have associated functions. The strength of hinge teeth in bivalves and the shell microstructure do not have an obvious connection. However, it turns out that a flexible microstructure, such as that of the margins of pearl oysters, enables the shell to seal the edges well even if the two valves do not quite line up. The two valves line up precisely if there are strongly developed hinge teeth, but may not line up exactly if there are no hinge teeth.

11.2.8 Long branch attraction. The example of valve number above points to the most frequent type of violation of the model. If evolution has been rapid relative to the sample, then there is a high probability of multiple mutations that mask the true evolutionary pattern. By examining only a few scattered representatives of modern taxa that diverged from each other no later than the Cambrian, the valve number study above was misled. With more extensive data and additional characters, we would have discovered the broad patterns of valve number and not been misled. However, what if the additional data are just as inappropriate? This produces the problem known as long branch attraction. A DNA sequence has only four possibilities: A, G, T, or C (deletion could be considered a fifth possibility). Thus, two random DNA sequences will have about a 1 in 4 probability of matching at any one point in the sequence. If an organism has evolved so much relative to the others in the study that it is practically randomized, then parsimony analyses will probably mistake the random similarity for evolutionarily significant events. The main effect of this is to artificially group taxa with very long branches together. For example, in Figure 11.2, the branches leading to E and F are both very long, whereas the branch joining them is short. This suggests that long-branch attraction may be a problem. Likewise, an analysis of gastropod molecular phylogeny that included 10 neogastropods, 10 euthyneurans, one patellid limpet, and a bivalve and a chiton as outgroups might group the patellid with the bivalve or the chiton rather than with the other gastropods because the random similarities outweigh the few evolutionary similarities. Adding more DNA data with a similar evolutionary rate would only make the problem worse by increasing the number of random characters. You can solve the problem by adding additional taxa (so that the random characters would no longer outweigh the meaningful ones), by adding data for a more slowly evolving gene or morphological feature, by removing the isolated taxa from the analysis, or by using a different analytical technique. A maximum likelihood analysis may also be misled by such data if its model does not adequately consider the possibility of high rates of mutation; however, maximum likelihood is generally less vulnerable to this problem.

Several computer programs are available to do these sorts of analyses. PAUP* is currently the most popular, but MRBAYES, PHYLIP, and HENNIG are also popular, and many others exist. An extensive listing of such programs is available at <evolution.genetics. washington.edu/phylip/software.html> and includes programs ranging from free to rather expensive.

11.3 HOW TO DO DNA STUDIES

Analysis of DNA sequence data requires collection of appropriate samples, extraction of the

DNA, amplifying the gene of interest, sequencing it, and aligning the sequences before finally performing a phylogenetic analysis. A complete set of equipment for all this would cost over $100,000 to purchase, plus operating expenses. However, it is easy to collect specimens suitable for DNA work in collaboration with someone who has access to the necessary equipment. Other molecular techniques (such as those used for study of protein) differ in detail, but follow similar general patterns.

11.3.1 Collecting. DNA breaks down rapidly after death, so specimens must be kept alive or rapidly preserved. Although it may be possible to obtain DNA from old specimens, including dried tissue or museum material in miscellaneous preservatives, it is best to start with live material. However, if the specimen dies before DNA extraction can begin, the DNA will degrade rapidly. Probably the handiest way to preserve tissues for DNA work is to immerse the specimen in concentrated ethanol.

The ethanol must penetrate the tissues rapidly, so bivalves that can tightly close their shells or operculate gastropods may require cutting open before preservation. In addition, water in the tissues will dilute the ethanol, so use plenty of ethanol. Changing the ethanol after a day or two is also a good idea, as it replaces the ethanol that has been diluted by water from the tissues. Keeping the preserved specimens cold is helpful, but not necessary for the duration of a field trip or for mailing a sample. Another option is freezing. Although extremely low temperatures (below -70°C) are needed for long-term storage, an ordinary freezer will do in the short term. Thawing degrades the DNA, so it must be kept cold. Several other recipes exist for preserving DNA, but these typically require specialized chemicals or supplies. Dried tissue from dead specimens may be usable, but gooey, decaying tissue is probably no good.

For something large, part of the specimen will yield plenty of DNA. Muscle tissue is good. Mantle tissue is not as good because the mucous can interfere with the DNA extraction. However, clipping a small piece of the mantle and putting it directly into ethanol is often the best way to obtain DNA while keeping the specimen alive. The digestive tract is also unsafe for analysis, because it may contain DNA from the last meal as well as DNA of the organism itself.

For preserving DNA, ethanol should probably be at least 80% (160 proof) and preferably 95-100% (190-200 proof). Labeling and transporting the specimens with ethanol can be challenging. Most ink, including that of permanent markers, dissolves in ethanol. Pencil does not dissolve, or you can keep the ethanol off the label. I have also found some zip-closure bags that seem watertight but allow ethanol to leak through, probably because the plastic was somewhat soluble. This has been a problem with cheap bags but not brand-name ones. However, plastic jars are safer, if they seal well. Double bagging is also good. For transport and mailing, replace as much of the ethanol as possible with absorbent cotton wool, and pack the vials in absorbent cotton wool. This should keep the specimens saturated while decreasing both the weight of the sample and the risks from shipping a flammable liquid. Make sure containers and specimens are well padded for mailing; otherwise, they may be crushed, letting the ethanol leak out. Also, check if the shipping company (FedEx, United States Postal Service, etc.) has any regulations about shipping specimens in ethanol.

11.3.2 Processing the Sample. When the specimen gets to the lab, the tissue is ground up and mixed with a solution to break down the tissue and dissolve the DNA. Several chemical steps are required to separate the DNA from the rest of the material making up the organism, such as proteins and lipids. A wide range of techniques is used, with different chemicals used for different situations. The DNA is then amplified using the polymerase chain reaction (PCR). This is an engineered version of bacterial DNA copying. To your DNA sample, you add the DNA-copying enzyme, the bases for making new DNA copies, and an excess of copies of primer sequences. The primers are short DNA sequences (roughly 15-20 bases long) that bind to the beginning and end of the sequence that you want to amplify and provide a starting point for the enzyme. The sample is then heated almost to boiling to separate the DNA strands (Figure 11.5A). It is allowed to cool, but because of the high concentration of primer, the sequence that reattaches is likely to be the primer rather than the original complementary strand (Figure 11.5B). The sample is then heated to a good temperature for the enzyme and it starts copying (Figure 11.5C). By repeating this sequence of heating and cooling about twenty-five times, the number of copies increases exponentially (from 2 to 4 to 8 to 16…); up to a million copies can be made as the copies are copied (Figure 11.5D). The resulting DNA sample, after a purification step, is run through a similar process to generate the sequence.

However, in addition to the standard bases, modified bases are added. For example, in dye sequencing, some of the bases are labeled with a distinctive dye color. Each base has a different color. The result is a mixture of sequences of various lengths ending on A with one color of dye, sequences ending on T with another color, etc. The samples are then run on an electrophoresis gel or through a capillary tube. The gel is a gelatin-like slab, with holes in one end to hold the samples. When an electric current is applied to the gel, the DNA migrates towards the positive terminal. Smaller DNA molecules can move more quickly. Thus, after running the gel for a while, the shortest DNA pieces will have moved the farthest. As a result, the first band on the gel will have the dye for the first base in the sequence. In automated sequencing, a scanner measures the color and sends the data to a computer. (Figure 11.6).

You can look up what sequences are known online. GenBank is the US national database for gene sequences, and the data are shared with the European and Japanese databases. Most journals and book publishers require that you submit sequence data to GenBank before it can be published. You can search for particular taxa, or follow the classification down to more specific levels. Several taxa have a few genes already sequenced, but the first molluscan genome project is only now underway (*Biomphalaria glabrata* (Say, 1818), a planorbid gastropod and schistosomiasis host).

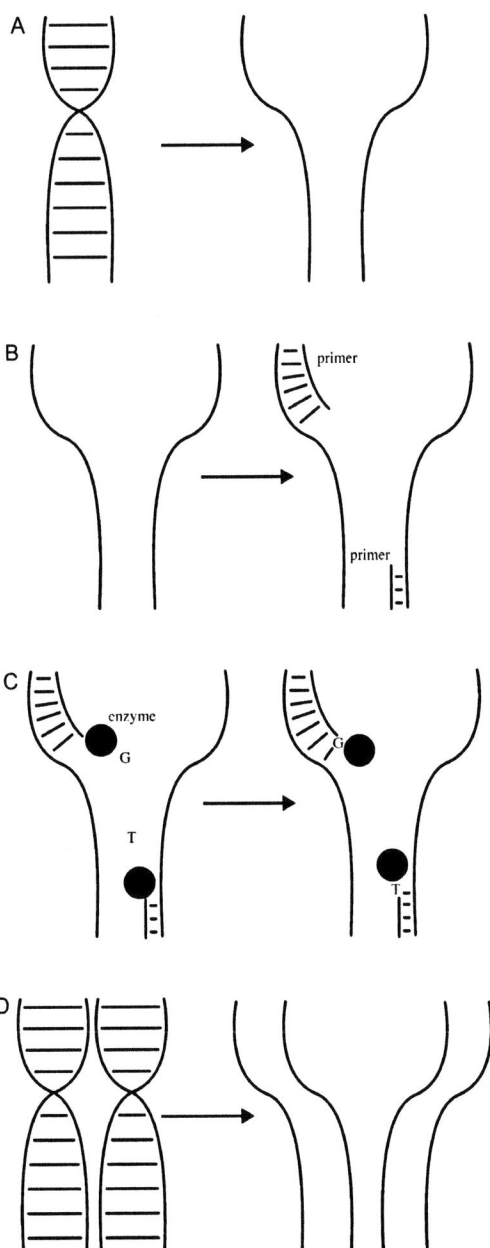

Figure 11.5 DNA amplification by the polymerase chain reaction.
A, The first step in PCR is heating the DNA sample to break the hydrogen bonds between the two strands. B, The second step is cooling the sample to a temperature that allows the primer to bind to the sample DNA. C, The sample is then held at a temperature suitable for the DNA-copying enzyme to work. D, When this is repeated, the copies can be copied as well as the originals.

To perform a cladistic analysis, these sequences from different taxa must be aligned in some fashion. Because there are insertions and deletions in many DNA sequences, some adjustment is necessary to avoid comparing the wrong bases. Several alignment methods exist, with advantages and disadvantages. The problem arises when there are multiple possible alignments for any two sequences. For example, AAATTTG and AAACTTG could be aligned as is, suggesting that a single mutation occurred. However, they could also be aligned by inserting gaps in the sequence, *e.g.*: AAAT-TTG and AAA-CTTG, suggesting insertions or deletions (collectively called indels). If you assume enough indels, you can eliminate mutations; if you assume enough mutations, you can eliminate indels for sequences of the same length. Usually, multiple plausible alignments are possible (especially if many taxa are considered). Computerized alignments typically assign penalties to creating a gap, to mismatches between bases, and to lengthening an existing gap. Depending on the relative magnitude of these penalties, different alignments may be favored. Testing a range of values for the penalties will determine whether the alignment is stable or sensitive to such assumptions. Most workers also look over the resulting alignments, adjusting anything that looks wrong. Once the sequences have been selected and aligned, the aligned sequences are then analyzed using the techniques discussed above.

With enough evolutionary change between taxa, a satisfactory alignment may not be possible. In this case, use of a different part of the gene or a different gene may help. Different genes evolve at different rates and with different evolutionary constraints, so data from multiple genes are better than reliance on a single sequence. In addition, different genes will be better for different studies. For example, the spacer regions are typically highly variable, and might be examined to determine if two populations are conspecific. However, they would probably be so variable as to be useless to try to look at relationships between classes. One widely used variation in evolutionary rate within a gene is the high evolutionary rate of third positions in codons of protein-coding genes. Each amino acid in a protein is coded for by a group of three DNA bases, called a codon. However, in most codons, the third base is at least partially redundant and therefore is freer to mutate than the other two. Thus, many protein-based studies ignore third position bases, unless they focus on low taxonomic levels.

Of the numerous sequences in GenBank, most represent only a few kinds of genes. A few genes of commercial or other research interest have been sequenced, but most sequences are for phylogenetic study. Mitochondrial genes have received extensive study. Mitochondria, organelles within the cell, have their own DNA and typically are inherited directly from the mother. They typically have a moderate to high evolutionary rate and thus are useful for a variety of studies. Popular genes in the mitochondrial genome include cytochrome oxidase I (COI) and the genes for ribosomal 12S and 16S RNA (12S and 16S respectively). Some studies have looked at the order of genes within the mitochondria as well as the sequence of DNA bases. Many mollusks have peculiarities in their mitochondrial genomes, adding to their interest. However, they pose additional problems because some mollusks have independent male and female lineages of inheritance of mitochondria. Evolutionarily, the male or female mitochondrial lineage could die out, to be replaced by the other. Thus, the evolution of the mitochondria might not match the evolution of the organism as a whole. The "doubly uniparental inheritance" of mito-

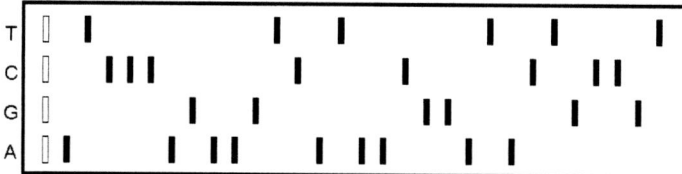

Figure 11.6 Sequencing gel.
A hypothetical sequencing gel using the old four-tube method. The left edge has the holes in the gel where the sample was inserted. The right hand side has the shortest DNA pieces, since they moved the fastest, so the band there indicates the first base in the sequence (TGCCGTCAT...). The current method labels each terminal base with a different color and has everything in one tube, but doesn't make a good black and white picture.

chondrial genes, with male and female lineages, is characteristic of bivalves. A popular group of nuclear genes are the nuclear ribosomal RNA genes, also called rDNA. These include the 18S ribosomal gene and the 28S ribosomal gene, coding for different parts of the ribosome.

11.3.3 What are the results? As conventional classifications already disagree, it is not too surprising that the new analyses have not resolved everything, either. Studies range from the population level to all organisms. Some species have been demoted to subspecies or variety, while other varieties and subspecies have been elevated. In a few cases, previously unsuspected variation has been found using molecular techniques.

The relationships of Mollusca to other phyla have received some elucidation from molecular data, though the picture remains somewhat murky. Mollusca appear most closely related to protostome phyla that lack a regularly molted cuticle. Annelida, Sipunculida, Echiura, and Brachiopoda are among the other members of this group. Relatively few published sequences exist for aplacophorans, and none for monoplacophorans. For the other extant classes, coverage is generally wide, but with key taxonomic gaps, and often genes well documented for one group will be unstudied for other groups. Thus, our understanding of relationships among the classes remains poor. Most morphology-based cladistic analyses group the Conchifera (ancestrally one-shelled mollusks, including Bivalvia) as monophyletic. Within the classes, most traditional taxa appear monophyletic, with some exceptions, but the relationships among taxa often provide surprises. Within Gastropoda, some recent studies suggest that Pulmonata is polyphyletic. However, these results remain relatively untested. In the Bivalvia, Myoida appears polyphyletic. This was suspected already on morphological grounds, and has been supported by several studies, so it seems safe to abandon the order as a polyphyletic assortment of veneroid bivalves adapted for deep burrowing.

Previous phylogenetic ideas often propose a wide range of mutually conflicting ideas. Both DNA and morphological studies can help resolve these issues. In the Gastropoda, these studies support a close relationship between Caenogastropoda and Euthyneura, with "Archaeogastropoda" representing a paraphyletic grade. Within the Bivalvia, Ostreoidea has alternatively been grouped with Pterioida or Pectinoida. DNA strongly favors the Ostreoida-Pterioida grouping while placing Plicatuloidea in Pectinoida.

At the species level, molecular data can help determine whether physical variation is associated with a genetic barrier or just phenotypic plasticity. In some cases, morphologically similar populations prove to have high levels of DNA disparity, suggesting that they represent multiple species. These cryptic species are most commonly discovered in the case of taxa with supposedly very wide ranges. Other morphologically variable populations show no clear differences in their DNA. However, you cannot simply declare a certain level of DNA difference to indicate species-level or subspecies-level features. Some taxa have generally high levels of variation whereas others show very little. Variation within and among populations must be examined to determine if a given taxon can be supported from the molecular data.

Despite all these studies, the results remain relatively preliminary. The constant rapid improvements in methods and computing ability keep any analysis from being the final word, and many important groups remain poorly studied. Thus, these methods will be providing important revisions to molluscan systematics. However, do not change the labels in your collection with every new idea. If a new idea has stood the test of time for 10 or so years, then update your collection. For example, feel free to remove the Architectonicidae from the Mesogastropoda and join them with the Pyramidellidae in the Heterogastropoda, however, leave the Conidae and Turridae alone for a while, until more studies are done.

11.4 LITERATURE CITED

Felsenstein, J. 2003. *Inferring Phylogenies*. Sinauer Associates, Sunderland, Massachusetts. 580 pp.

Graur, D. and Wen-Hsiung Li. 1996. *Fundamentals of Molecular Evolution*. Sinauer Associates, Sunderland, Massachusetts. 443 pp.

Hall, B. G. 2004. *Phylogenetic Trees Made Easy*, 2nd Ed. Sinauer Associates, Sunderland, Massachusetts. 192 pp.

Hillis, D. M., C. Moritz, and B. K. Mable. 1996. *Molecular Systematics*, 2nd Ed. Sinauer Associates, Sunderland, Massachusetts. 655 pp.

Lydeard, C. and D. Lindberg. 2003. *Molecular Systematics and Phylogeography of Mollusks*. Smithsonian Books, Washington D.C. xv + 312 pp.

Mayr, E. and P. D. Ashlock. 1991. *Principles of Systematic Zoology*, 2nd Ed. McGraw-Hill, Inc., New York. xx + 475 pp.

Page, R. D. M. and E. C. Holmes. 1998. *Molecular Evolution. A Phylogenetic Approach*. Blackwell Scientific, Malden, Massachusetts. 352 pp.

Schuh, R. T. 2000. *Biological Systematics: Principles and Applications*. Comstock Publishing Associates, a division of Cornell University Press, Ithaca, New York. ix + 236 pp.

Wiley, E. O., D. R. Brooks, D. Siegel-Causey, and V. A. Funk. 1991. *The Compleat Cladist: A Primer of Phylogenetic Procedures (Special Publication 19)*. University of Kansas Museum of Natural History, Lawrence, Kansas. 158 pp.

CHAPTER 12

ORGANIZATIONS, MEETINGS, AND MALACOLOGY

CHARLES F. STURM

12.1 INTRODUCTION

Organizations serve several functions. Apparent functions include sponsoring meetings and workshops, publishing journals, newsletters and books, and maintaining web sites. They also provide a voice to lobby for interests relating to malacological issues such as beach erosion, pollution, dangers of introduced organisms to native species, and the endangered species acts. Equally important, we are social organisms, and what better way to share our interests? Thus, getting away for a day to a week with other mollusk lovers is just plain fun. When you attend a meeting, get out and mingle. Whether it is a professional function or an amateur get together, mingle. Sometimes, far more is accomplished informally than is accomplished at the formal presentations.

Hatching plans for future collaboration is an important function of these informal liaisons. Another informal function that takes place is looking for a job. Often people will first find out about future job opportunities at meetings. While these activities apply mostly to professionals, amateur meetings are much the same.

Conchologists of America (COA) meetings and local shell club meetings have formal programs. There is also much wheeling and dealing among people trying to swap shells and books, plan future trips, collaborate on other upcoming meetings, etc.

The organizations and meetings that I am going to mention are driven by a voluntary workforce. Few of these organizations have any paid personnel. Officers and organizers are sometimes reimbursed for postage and telephone expenses, but no salaries are provided. To be an officer is a badge of peer recognition. We are acknowledging those that we believe have knowledge and ability in our field and we are asking them to lead us. It is also a commitment from those who serve. Often they have daytime jobs. Whether it is as a college professor, museum curator, or a bricklayer, when people step up to be an officer, organizer, or editor, they are often trying to find time in an already busy schedule to accomplish the task. Thus if things are not just what you would like them to be, do not gripe. Instead, offer your time and expertise to help improve the situation. I doubt you will meet with any rejection of your offer to help.

In the following section, I will be discussing a number of malacological societies. While I will be mainly focusing on societies from the United States, there are many such organizations elsewhere in the world.

12.2 ORGANIZATIONS AND MEETINGS

The main professional malacological organization in the United States is the American Malacological Society (AMS). It was founded in 1931 as the American Malacological Union (AMU), a name some may still be using as the change from Union to Society occurred recently, in 1999. The AMS holds one meeting a year. The location varies across the country. The meeting is generally four to six days in length and often there is a choice of several field trips during or following the meeting. The meeting is generally held in the continental United States, though there have been a few held elsewhere:

Hawaii (1995), Canada (1939, 1960, 1967), Cuba (1938), and Austria (2001, a joint meeting with Unitas Malacologia).

The meeting is generally organized around several themes. These theme sessions have several speakers and constitute a symposium or workshop. There are sessions for open presentations. Here papers of a general and sometimes unrelated nature are presented. There is generally at least one poster session. A poster session is an informal way of presenting information. In a given room, there will be several large boards. On one of these, the presenter will post his or her presentation. Posters are an efficient way to simultaneously display numerous presentations and thus save precious meeting time. The attendees at the meeting will wander through the room and if they see something of interest, they will stop and discuss their ideas or questions with the presenter. If they are not interested in the topic, they will wander on to the next poster.

Poster presentations are a nice way to present preliminary results and obtain the input of others. It can be a way to present a new idea and receive informal comments. It can also be a way of presenting factual data that are not necessarily amenable to a formal presentation. One such example is a presentation that I posted on the composition of the Carnegie Museum's molluscan collection. This was not information that warranted a formal presentation from a lectern and stage, but was ideal for presentation as a poster. People interested in using the collection would stop and ask me if the Carnegie's collection had material that would be useful to them, and how to access the collection. Poster presentations tend to last from two to four hours.

Many people enjoy the poster sessions since they allow for one-to-one discussions with the presenter, they can spend as much time as they would like viewing the poster, and the viewers can choose the topics that are of interest to them. The presenter of a poster reaches fewer people than he or she would in a formal talk. However, the people he or she reaches tend to be interested in the topic and useful dialog is often the outcome.

There is generally an annual auction at the AMS meeting. Members donate books, art objects, journals, etc. These are auctioned at an informal get together that is held in the evening. The proceeds from the auction sponsor student grants. At AMS meetings, unlike some other organizations, shells are not auctioned.

The AMS publishes the *American Malacological Bulletin*, a biannual journal. Often, papers presented at the meetings are published in the journal. Sometimes they are published in a special publication. This book is one such example. Several of the chapters from a workshop at the AMS meeting in 1999 formed the beginnings of this book. Other contributors were then sought to write chapters on topics not presented at the meeting. Lastly, the AMS publishes a newsletter two or three times a year to inform its members of topical issues and it publishes a yearly membership directory. More information regarding AMS can be obtained from their Web site <www.malacological.org>

A second professional organization is the Western Society of Malacologists. Originally formed as the Pacific Division of AMU, the WSM went independent in 1968. The WSM holds a yearly meeting and sometimes it holds a joint meeting with AMS (when the latter meets on the west coast of North America). Similar to AMS, the annual meeting is built around topical symposia and open sessions. The WSM does publish a journal, in the form of the abstracts from its annual meeting and sometimes expanded abstracts and short papers based upon the papers presented. They also hold an auction and the proceeds fund student grants. Visit their Web site at <biology.fullerton.edu/orgs/wsm/> for more information.

The Freshwater Mollusk Conservation Society (FMCS) was organized in 1998 and held its first annual meeting in 1999. It was formed by a coalition of malacologists, biologists, ecologists, etc. from industry, museums, academia, and governmental agencies. They banded together to form an organization whose mission is advocating for the conservation and preservation of freshwater mussels (Unionids) and lately, all freshwater mollusks. The progenitor of this group had a newsletter, the *Triennial Unionid Report*, which has been redesigned as *Elliptio*, the current organization's newsletter. In addition, they will also be the new publisher for the periodical Walkerana. More information can be obtained at their Web site: <www.sari.org/FMCS_General_Information.htm>.

Unitas Malacologica is a European organization. I have decided to include it here because of its involvement with other organizations from the United States. In 1998, Unitas along with AMS and WSM hosted a joint meeting called the World Congress of Malacology. This meeting was held in Washington, D.C. In 2001, AMS and Unitas held a joint meeting in Vienna, Austria. Unitas holds meetings every three years and publishes the abstracts of the proceedings. It also publishes a newsletter. Proceedings of special sessions are usually published as special volumes in various journals or sometimes as books, for example, the *Journal of Conchology* just published the proceedings of the conservation session from this Austrian meeting.

The Conchologists of America is a national organization in the U.S.A. whose membership is primarily amateur malacologists and shell enthusiasts. COA was founded in 1972 (as an offshoot of AMU) and has expanded to become an organization with some 1500 members. The purpose of the organization is to bring together shell enthusiasts and shell clubs for the mutual enjoyment of malacology as well as to promote conservation of molluscan resources. By design, COA and its members tend to be more interested in shells and their study (the strict sense of conchology) as opposed to the study of mollusks in the broader sense of molluscan anatomy, physiology, biochemistry, etc. (malacology). Despite these dichotomies, there is overlap between the membership of COA and organizations such as AMS and WSM.

COA publishes the periodical *American Conchologist*. It is published quarterly and contains a broad range of articles. Some of the articles are travelogues. These are especially helpful if you are thinking of traveling to the location discussed. Along with information on the types of mollusks to be found, you will also learn of the risks and pitfalls of traveling to the location. It also publishes articles that review a given genus or family. Some articles are biographies of prominent shell enthusiasts, professional and amateur alike. There are also book reviews and news items.

COA also holds a yearly meeting. Like other national organizations, the meeting is held in a different city each year. The meeting focuses on making current conchological knowledge accessible to the average person. One of the central features of the COA meeting is the bourse. Here you will find dealers of malacological books and shells from all over the world. There is also an auction. The proceeds from the auction provide funds for the COA Scholarship Program. This program provides funds to college students undertaking research in malacology. Since COA is not a professional scientific organization, original research tends not to be presented at its meeting. Finally, there are often several field trips held before, during, or after the meeting. You will not find an event of this magnitude anywhere else in the United States. For more information about COA, visit their web site at <www.conchologistsofamerica.org/home/>.

In addition to the major meetings mentioned above, there are many shell clubs in the United States and they generally hold meetings on a monthly basis. These meetings tend to be a mix between social and educational activities. Often club members will mill around and socialize before and after the formal presentation. Such presentations may be made by an invited guest or by a member of the club. They may be somewhat formal and technical or very informal. Sometimes a club will host a meeting where members can present interesting shells that they found. Sometimes a member will bring part of his or her collection to put on display. Some clubs have a yearly auction. The money may be used for club expenses, or if the club is big enough, to support a scholarship.

Many clubs have a yearly shell show. The size of these shows often relates to the size of the club. There may be only a handful of exhibits, or there may be dozens. If the show is large enough, shell dealers and booksellers may also have display booths. There are often several awards presented in categories such as best scientific display, most artistic display, best self collected display, shell of the show, best display of the show, etc. Sometimes, awards sponsored by museums or national organizations are handed out. Some examples are the COA Trophy Award, the Academy of Natural Sciences (Philadelphia) Pilsbry Award, and the duPont Award.

Club activities may be varied. In addition to those mentioned above, they might maintain a library, sponsor field trips, put on presentations for schools, youth groups, or other organizations, and publish a monthly newsletter. If a club is sponsored by a museum, its members might help and volunteer time to take care of the museum's mollusk collection. One advantage of clubs is in their smaller size. Being smaller than national organizations, they often act as the incubators where new ideas are initiated, carried out, and tested. Their smallness allows them to act with greater swiftness than what you would see in larger groups.

One example of this is seen in the web sites of some clubs. They may consist of dozens of web pages, cover the marine, freshwater, and terrestrial fauna, and provide links to other sites where one may obtain further information. The sites of some clubs are far more extensive than those of large museums and national organizations. Because bureaucracy does not hamper them, they can modify their web site and experiment with different ideas. They can react to feedback with greater speed than one is accustomed to seeing in a larger organization.

However, their smallness is also one of their greatest dangers. Often the work of running the club falls on the shoulders of a selected few. For a successful club, all members must be willing to participate. The officers cannot be expected to run the club, prepare and carry out all of the field trips, present at all the meetings, and put out the monthly newsletter. You must be proactive in a shell club if it is to be viable. Do not sit back and complain that nothing ever happens. If just three members a month get together and each speaks for 15 minutes on a shell, group of shells, or a collecting site that is of interest to them, you have a monthly meeting with minimal effort on each person's part. The bottom line is, get involved!

While most shell clubs in the United States are found in coastal areas, with Florida having the greatest concentration, there are some inland clubs as well. You can find them in Cleveland, St. Louis, Chicago, and Las Vegas, just to name a few non-coastal sites. For a listing of shell clubs, get a copy of Tom Rice's book (Rice 2003) or go to the COA web site: <www.conchologistsofamerica.org/home/>.

I would like to mention one final group of meetings. These groups have no officers, no dues, no publications, and no membership requirements other than showing up! The original one in this group of meetings and organizations is the Bay Area Malacologists (BAM) held in the San Francisco area and started by Eugene Coan in 1968. This was followed in southern California by the Southern California Unified Malacologists (SCUM) and on the East Coast of the United States by the Mid Atlantic Malacologists (MAM). The only requirement to attend these meetings is to have an interest in mollusks.

Professionals and amateurs, paleontologists and neontologists, undergraduate and graduate students, active collectors and couch potatoes alike show up for these meetings. They all come because of an interest in mollusks. The meetings are held once a year. Someone decides to hold it at their institution and posts a notice about the meeting. Anyone can present at these meetings. The format is flexible. One presentation may deal with someone's trip to the Caribbean, while the next deals with an unusual shell, and is then followed by a short research presentation by a student. To give you an idea of what can happen I will recount one incident from a MAM meeting.

One attendee presented a shell that might have represented a new species or might just have been an extreme variant of the normal range of variability. It was being presented to obtain reactions from the other participants. From here, a discussion regarding what constitutes a species ensued. Most of the discussion dealt with the biological species concept. This was expanded upon with insights from a paleontologically inclined participant who brought up the morphological species concept, which is prevalent in paleontology. Amateurs and professionals joined in the discussion, a true democratic sharing of ideas. Often someone will write a summary of the presentations at these meetings and publish them in a malacological periodical.

Since I have attended only the MAM meetings, I will discuss them in a bit of detail. The MAM meeting is held at the Delaware Museum of Natural History. It draws participants from Washington, D.C., north to New York and as far west as Pittsburgh. The meeting begins with some light refreshments, and then there are about two hours worth of presentations. We then stop for lunch. People break up into small groups and visit local restaurants. We reassemble at the Museum for another hour or two of talks. After the last talk, those who need to use the library or would like to use the collection arrange with the curator for such access. If any summary of the proceedings is written, it is posted to the Conch-L discussion group on the Internet. Some magazines such as *American Conchologist* and *La Conchliglia* may print this report as a news item, but it is not an official publication of the meeting. At the end of the day, we all return home and wait to be notified of the next such meeting one year hence.

12.3 THE INTERNET

While not a club or organization, discussion groups (mailing lists) on the Internet serve some of the same purposes. They are forums where people with like interests can meet, exchange ideas, ask questions, and inform one another of happenings around the world. Discussion lists may be moderated or unmoderated.

One joins the list by sending an e-mail message to a computer at the host site. Generally within 24 hours of joining, you can post messages to the list. When you send a message in, it is then sent to everyone who has subscribed to the list. Others, who want to reply to your message, send their messages back to the list, and so it goes, until the topic is played out.

Some individuals may be very active on such discussion groups. Sometimes people are lurkers. These are people who read and follow the messages posted but rarely, if ever, join in the discussions. These folks are often newcomers who want to learn about mollusks, but feel that they do not yet have much to offer. One of the important aspects of discussion groups is that they are often international in character. People from around the world often sign up and are connected to one another electronically. This allows people that are in out of the way areas to participate in discussions with kindred spirits.

Several discussion groups deal with molluscan related issues. Some are more active than others. Some groups are more professional while others are more popular. Some deal with a particular group of mollusks while others deal with any question that one would like to post.

Sometimes people are dismayed and overwhelmed by the volume of e-mail messages that they receive from some groups. It is important to remember the "delete" function. Many list servers offer a digest service, in which all messages of the day (or other time period) are bundled together into one message. An advantage of digests is fewer separate messages, but a disadvantage is that you do not receive the messages as soon as they are posted.

Just because you attend a large meeting does not mean that you must attend every presentation or read and discuss every poster. You choose what is most useful to you. It should be the same with discussion groups. If a message or series of messages (a thread) are of no interest to you, delete them. Do not waste your time. Move on to the next message or thread. If you hit on something of interest to you, read the message. If you have something to offer to the discussion, join in and post a reply. Over time, people will even seek you out as they learn of your particular interests and areas of expertise. You may eventually get to meet some of the other discussion group participants at national or international meetings. Below are some discussion groups that you might be interested in joining.

The outline for entries is:
(1) LIST NAME

(2) <URL> (Unique Resource Location) for more information about the group.

(3) Description of the discussion group from the groups' web site except for Permit-L. This group does not have a formal web site. The information was taken from the message announcing the formation of the group.

(4) Address used to subscribe. This is the e-mail address used to join the group.

(5) Command used to join the group. This is to be typed in the body of your e-mail message. Be sure your signature file is deactivated. If not, there may be difficulties in the host computer knowing who is trying to join.

(6) Address used to post messages once you are a member of the List. Please note that this address differs from the address used to join the group. If you want to post a message or reply to one, use this address, not the one listed above as address used to join.

12.3.1 MOLLUSCA LIST.
<www.ucmp.berkeley.edu/mologis/mollusca.html>

"It is our intent to provide an informal and rapid response forum for discussions of molluscan evolution, paleontology, taxonomy and natural history. A special emphasis of this list server will be to provide an interface between paleontological and neonotological molluscan workers. We will also post notices of meetings, symposia, literature, software and other electronic happenings that may be of interest to the malacological community."

{address to subscribe} Majordomo@listlink.Berkeley.Edu
subscribe molluscalist [your name here without brackets]
{address to post messages} molluscalist@listlink.berkeley.edu

12.3.2 CONCH-L.
<www.conchologistsofamerica.org/conch-L>

"Are you having trouble finding the online information you need about the hobby of conchology? ...Conch-L will provide a forum for conchologists, indeed, for anyone with an interest in mollusks, to discuss any topic related to conchology."

{address to subscribe} listserv@listserv.uga.edu
subscribe CONCH-L [Your Full Name without brackets]
{address to post messages} CONCH-L@listserv.uga.edu

12.3.3 UNIO.
<my.fit.edu/~rtankers/unio.htm>

"UNIO is an unmoderated Internet listserver focusing on the biology, ecology and evolution of freshwater unionid mussels. The primary objectives of the list are (1) to foster communication and collaboration among scientists, researchers, and students engaged in mussel-related activities and (2) to facilitate the informal discussion of regional and federal research priorities."

{address to subscribe} join-unio@lyris.fit.edu
no message in the body of the subscription e-mail
{address to post messages} unio@lyris.fit.edu

12.3.4 PaleoNet.
<www.nhm.ac.uk/hosted_sites/paleonet/>

"PaleoNet is a system of listservers, www pages, and ftp sites designed to enhance electronic communication among paleontologists. While primarily designed as a resource for paleontological professionals and graduate students, PaleoNet welcomes input and participation from all persons interested in the study of ancient life."

{address to subscribe} PaleoNet-Request@nhm.ac.uk

subscribe PaleoNet <e-mail address>
{address to post messages} PaleoNet@nhm.ac.uk

12.3.5 NHCOLL-L

"The Natural History Collections Listserver, or NHCOLL-L, is a general purpose electronic forum for those with an interest in the care, management, computerization, conservation and use of natural history collections. NHCOLL-L is co-sponsored by the Association for Systematic Collections (ASC) and the Society for the Preservation of Natural History Collections (SPNHC)."

{address to subscribe} LISTPROC@LISTS.YALE.EDU
SUBSCRIBE NHCOLL-L [your name, without brackets]
{address to post messages} NHCOLL-L@LISTS.YALE.EDU

12.3.6 PERMIT-L.

"PERMIT-L is a moderated cross-disciplinary listserv, hosted by the Smithsonian Institution, intended to facilitate discussion and information flow on all issues related to the rapidly changing terrain of biological collecting, permits, access, and import/export transactions."

{address to subscribe} LISTSERV@SIVM.SI.EDU.
Subscribe PERMIT-L [Firstname Surname, without brackets]
{address to post messages} PERMIT-L@SIVM.SI.EDU

12.4 SUMMARY

There are many different meetings, organizations, and venues for conchologists and malacologists to meet and exchange ideas. They are important not only for professionals but also for amateurs. They help develop our professional and social enjoyment of this field. This chapter has exposed you to some of the options available to you. It is now up to you to jump in and explore the different options, embrace one or two of them, and expand your enjoyment of the Phylum Mollusca.

12.5 LITERATURE CITED

Rice, T. 2003. *Tom Rice's A Shellers Directory of Clubs, Books, Periodicals and Dealers,* 26th Ed. Of Sea and Shore Publications, Port Gamble, Washington. 114 pp.

CHAPTER 13

MUSEUMS AND MALACOLOGY

CHARLES F. STURM

13.1 INTRODUCTION

This chapter is composed of two parts. The first part discusses the role of museums of natural history. As you will discover, museums provide several benefits to members of our society. The second part discusses some of the major malacological collections to be found in museums. This part has two limitations. The first is that it focuses only on museums in the United States and Canada. The second relates to which museums were selected for inclusion. For the most part, the institutions selected have collections of regional, national, or international reputation and contain at least 40,000 lots. Thus, many smaller local or regional museums are not addressed. For a more complete listing of museums, refer to the Web site by Cummings, Oleinik, and Slapcinsky at <www.inhs.uiuc.edu/cbd/collections/mollusk_links/museumlist.html>. This site is a compilation of natural history museums found worldwide.

13.2 MUSEUMS AND SOCIETY

Much of this chapter does not represent either radical thinking about museums or original thought. My thoughts have developed over the past 10 years and I was pleasantly surprised to find all of them already in print. Thus, my discussion is more of a summary of the way many people working in museums view the role of their institution in society. As I see it, museums have three major functions: education, maintaining collections, and research.

13.2.1 Education. For most of us, our first exposure to a museum was during a visit to one with our school class or with our family. As young children, we would follow an adult around, or alternately, they would be running after us, as we explore the museum. It was hoped that we gained some insight into the mysteries of nature during our visit. During our childhood, there may have been several such visits. Thus, the first mission of the museum is education.

Museums display items. More importantly, they interpret these items and put them into a context that helps us understand them. Informative exhibits are built around these items. In the past, most of the exhibits were static dioramas that changed very little over time. Today, some museums are embracing a "digital fever". Exhibits are completely computerized and no specimens are exhibited. As we can provide all this whiz-bang technology over the Internet, why travel to the museum? I believe that the future will require us to meld these two different approaches into a new paradigm.

There is a certain satisfaction of seeing an object with your eyes. It is difficult to comprehend the size of some dinosaurs or a whale without standing next to it. A picture on a computer screen of the object with a mannequin next to it for size comparison just does not measure up to the original. Even with the best of technology, it is hard to present accurately the full range of color and texture of an object in an image. Thus, museums are still needed so that we can see actual items.

The addition of digital technology to an exhibit hall opens many new possibilities. You are no longer limited to the information that is presented on a label outside a diorama, or the description in the guidebook. By presenting information through a computer kiosk, you will now be able to select from a menu of topics about the objects that you are viewing. You will be able to select from several video clips to explore different aspects of the exhibit. The computer terminal may even allow you to explore topics not covered by the exhibited objects. The expectation is that the computer display will heighten your engagement and by being an active participant in your learning, you will walk away as a more informed member of society. In addition, digital technology will allow the information in exhibits to be updated more frequently and in an easier manner than has been the case in totally static exhibits.

The second educational aspect of the museum is through classes that are offered at the museum. Sometimes these are conducted by the museum staff and at other times by outside individuals. The variety of what can and is presented in this venue is impressive.

These classes may be a single lecture being presented by a renowned authority or a workshop for teachers. They can be hands-on experiences for children or even field trips. Whatever the activity, the purpose is to build on the educational mission of the institution and present additional information that may or may not be covered in the exhibit halls. These educational activities complement our formal education.

13.2.2 Collections. The next major role of museums is to provide a repository for items. As a friend of mine states it, "they store stuff". This is one of the things that differentiates museums from certain other institutions. Science centers share some of the same educational objectives of museums. However, science centers do not maintain collections, and thus they are fundamentally different from a museum. Zoological parks overlap slightly with museums. They house live specimens for exhibit but they do not maintain collections of specimens. This lack of maintaining collections again differentiates them from museums. Nature centers serve to help us understand the natural world around us. They may have natural history objects on display to help us gain this appreciation, but they do not maintain collections. Again, this differentiates them from museums.

Collections are the heart and soul of a museum. Collections may be large or small, worldwide, regional, or local in scope. They can cover many different groups of natural history objects, or just one. What is important is that a museum has such a collection and maintains it as a "library of natural history items." Just what do I mean by this?

If you enter a library, you pick up a book. You then read what is in the book to gather new information. By referring to several books, you may obtain a broader view of the topic. So it is with objects in a museum. Each object is associated with data. Each object has a story to tell. By examining the object and studying the data, you can learn something about the object. By looking at other objects of the same or similar kinds, you can broaden your picture about some aspect of nature.

By way of example, let us imagine holding a *Mimachlamys* shell in our hand. Our observations, teach us the shape, texture, and color of the shell. The associated data shows us where and when it was collected. We may learn who collected it and other information as well. By examining many such shells, we may realize that while the shapes are all similar, there is a wide range of colors including violet, rose, orange, yellow, and some pale tan specimens. We may even notice that the colors do not represent different geographical areas but can be found throughout the range of the *Mimachlamys*. Therefore, by having multiple specimens in a collection, one can discern patterns of variation and evolution in nature. Often, one museum may not have a large enough collection of specimens and a researcher may consult the collections of several museums. Specimens also document past occurrences of organisms, biogeographical distribution, and sometimes represent undescribed taxa.

The above lets you see the importance of not only the object but also the data associated with it. Without data (e.g., where and when collected), most items in a museum would be mere curios. With data, they represent a portion of the vast tapestry of the natural world. For more on the importance of data associated with a specimen and what data may be important to malacological specimens, see Chapter 14.2-4.

Another aspect of a collection is the help it provides as an identification tool. Members of the public frequently approach museums for help in identifying items. A collection assists in this task. It also provides a basis for identification services provided to governmental agencies, private industry, and researchers. Often museum collections are helpful as many species are not adequately mentioned or illustrated in guidebooks and someone can find material for comparison only in a museum collection.

Museums build their collections by various means. At one time, museums would buy large collections to expand their holdings. This is less common today, as most museums do not have the capital to fund such acquisitions. If the museum's staff is undertaking an active program of fieldwork, the specimens collected will be deposited in the collection. Outside specialists will often deposit specimens from their work with a museum. These specimens are known as voucher specimens and are important for researchers who want to review the specialist's work in the future. Lastly, many museums have depended on donations of specimens from amateurs, and this is especially true in malacological collections (see Chapter 14 for more on this topic).

We see that museums maintain collections. What is important is to realize that they also look at a collection in terms of the future. As stewards of the collections, curators and collection managers are constantly wondering how best to protect the collection entrusted to them. Limiting access to the collection helps preserve it for future generations. How access is limited varies from institution to institution.

13.2.3 Research. The third major mission of the museum is research. It is important for museums to use their collection to expand the bounds of knowledge, and to make them available for others to do so. Depending on the size of a museum, there may be one or more people involved with the collection. A small institution may have only one scientific specialist on staff. A large museum may have several curators, a collection manager, research specialists, and support staff working with just one specific collection. In addition to undertaking a research role, the museum staff communicates what they learn to others. This may be through presentations at a meeting or by publishing the results in scientific books and journals.

Thus, we see that there are three major functions that an institution must embody for us to think of it as a museum. To recapitulate, these functions are education, maintenance of collections and research. For further information on these topics see Schmidt (1958), Colbert (1961), Parr (1962), Amadon (1971), Danks (1991), Hoagland (1994), Watkins (1994), and Pettitt (1997). These references cover how people have viewed natural history museums over the past 50 years.

13.3 UNITED STATES AND CANADIAN MUSEUMS

I will now review some of the institutions that have prominent molluscan collections in the United States and Canada. Some of these institutions may be well known for their exhibits of mollusks, others for their publications. Some institutions may be well known for the famous curators and staff that have been affiliated with them. Some may be well known for the collections that they house.

Selecting the institutions that are presented here was no easy task. I limited this review to museums in North America with collections of 40,000 lots or more. I included one collection that is smaller than this size due to the importance of the collection and the outreach of those in control of the collection to the scientific community and general population. The information presented is taken from a survey that I sent to these institutions in 2000-2001. Of the 22 institutions, all but one of them responded to the survey. The information from the one non-responding institution is taken from Cummings *et al.* (2000).

This tour will start in Washington, D.C. with the National Museum of Natural History, a component museum of the Smithsonian Institution. From here, we will travel north along the Atlantic Coast. We will then proceed westward and then down the Pacific coast of the United States. We will then travel east through the West and Southwest and finish our museum tour in Florida.

Defining several terms will make it easier to understand this section. A "lot" is a collection of one type of organism made at one place at one time. A lot may consist of one specimen or hundreds. Most museums measure their collections in terms of lots and not numbers of specimens. The numbers given below should be taken as estimates and not with certainty. Collections always have uncataloged material coming in and specimens being culled from the collection on the way out. Thus, exact numbers are unlikely, even if the collection is electronically databased. A good assumption is that the numbers are accurate to ±10%.

Type specimens are specimens that were used in establishing a new name. They are especially important when deciding if two different specimens are the same taxon. Most museums segregate type specimens from the regular collection and afford them special treatment. The size of the type collection of a museum is listed if I have obtained this information.

Some malacological collections include only Recent specimens while others also include fossil specimens. The Delaware Museum of Natural History incorporates its small holding of Cenozoic fossils in with its Recent malacological holding while the Natural History Museum of Los Angeles County maintains separate collections for Recent and fossil specimens. For some museums, I have listed if they have a joint collection or if the fossil and Recent material is housed separately. Incorporated in the description of each institution is their Web site address (URL). This will allow you to obtain further information regarding the respective institutions. Further information can be found in Solem (1975) and Cummings *et al.* (2000).

The National Museum of Natural History (of the Smithsonian Institution) is located in Washington, D.C. It has over 900,000 lots in its collection and fossil specimens are housed in a separate department. The Museum was founded in 1866 and the type collection numbers some 13,000+ lots. The collection is worldwide in scope and is 60% marine, 20% terrestrial, and 20% freshwater. The large size of this collection is due in part to the Museum being the repository for other governmental agencies. Some of the historic collections in the NMNH include material from I. Lea, A. A. Gould, W. H. Dall, C. T. Simpson, and P. Bartsch.

The Delaware Museum of Natural History is located in Wilmington, Delaware. It has some 220,000 lots, a type collection of 1217 lots and the breakdown of the collection is as follows: 50% marine, 25% terrestrial, and 25% freshwater. John E. duPont founded the Museum in 1958, and R. Tucker Abbott was appointed the first curator of mollusks in 1969. While the collection is worldwide in scope, it is especially strong in Indo-Pacific marine mollusks and terrestrial Philippine gastropods. The DMNH also houses the duPont Volutidae collection.

The Academy of Natural Science in Philadelphia is the oldest natural history museum in the United States. It has 550,000 lots with a type collection of 18,000 lots. Its collection is approximately 35% marine, 30% terrestrial, 20% freshwater, and 15% fossil. Some of its strengths include marine mollusks from the Indo-Pacific and Western Atlantic Regions, worldwide terrestrial and freshwater mollusks, and fossils from the Atlantic and Gulf Coast regions of the United States. Some of the famous contributors to the Academy include I. Lea, T. A. Conrad, A. Heilprin, T. Say, H. B. Baker, and H. A. Pilsbry.

The American Museum of Natural History is located in New York City. It was founded in 1869. Its collection contains 328,000 lots with a type collection of approximately 1,850 lots. Fossils are housed in a separate department. It is strong in marine taxa from the Indo-Pacific and Western

Atlantic regions. The famous collection built by John Jay is housed in the AMNH.

The Peabody Museum of Yale University is located up the coast in New Haven, Connecticut. The collection consists of 70,000+ lots of dry specimens and another 4,000 lots of fluid preserved specimens. The collection is 80% marine, 10% freshwater, and 10% terrestrial. Fossils are housed in a separate department. Historic holdings include the collections of A. E. Verrill from the later part of the 19th Century, the Woolsey Caribbean Collection, and the Gray Museum Collection (Marine Biological Laboratory) assembled under the direction of M. Carriker in the 1960s.

The Museum of Comparative Zoology of Harvard University is the northeastern most institution before we head inland. This collection in Cambridge, Massachusetts numbers about 620,000 lots with a type collection of 5,000 lots. The collection is 45% marine, 30% terrestrial, and 25% freshwater. Fossils are in a separate department. The collection is particularly strong in Unionidae and other freshwater groups and terrestrial gastropods of North America. This museum, founded in 1860, contains the historic collections of W. Clench and R. D. Turner (Teredinidae).

The Paleontological Research Institution, located in Ithaca, New York, was founded in 1932 by G. D. Harris to house his fossil collection. It has grown to be one of the largest fossil collections in the country and, surprising to some, it has a sizable Recent shell collection. The collection has about 3 million specimens and 30,000 in the type collection. Eighty percent of the collection is fossil specimens. The collection is 60% marine, 15% terrestrial, and 25% freshwater. The collection includes sizable holdings of Cenozoic marine mollusks and the Cornell University Recent mollusk collection is housed here.

The Canadian Museum of Nature in Ottawa, Canada was founded in 1950. The collection was previously at the Geological Survey of Canada. The mollusk collection is 130,000 lots including 30,000 fluid-preserved lots. The breakdown of the collection is 35% marine, 35% freshwater, and 30% terrestrial. The type collection contains 280 type specimens. The museum has significant holdings of Canadian and U.S.A. freshwater unionids and gastropods, and eastern North American terrestrial gastropods. It also contains the largest Canadian marine mollusk collection (including the Arctic Province), dating back to the 1850s.
<www.nature.ca/nature_e.cfm>

The Carnegie Museum of Natural History is located westward across the United States in Pittsburgh, Pennsylvania. The museum was founded in 1895. The collection has 115,000 lots and a type collection of 1,200 lots. The collection is 18% marine, 46% terrestrial, and 36% freshwater. The collections' strengths include freshwater bivalves (Unionidae and Sphaeriidae) and gastropods of North and South America, and terrestrial gastropods of North America. Significant historical collections include the G. H. Clapp Collection of terrestrial gastropods, the V. Sterki collection of Sphaeriidae, and the A. E. Ortmann Unionidae collection.
<www.carnegiemuseums.org/cmnh/>

The Museum of Biological Diversity of the Ohio State University is located in Columbus Ohio. Founded in 1890, the collection has over 100,000 lots. The collection is 10% marine, 20% terrestrial, and 70% freshwater. There are 60 lots in the type collection. The strength of this collection is its unionid holding. It also has a sizable fluid preserved collection. Its bivalve collection is computerized and searchable over the Internet.
<www.biosci.ohio-state.edu/~molluscs/OSUM2>

The Field Museum of Natural History in Chicago was founded in 1893. This collection numbers about 340,000 lots with a type collection of 5,800 lots. The collection is 18% marine, 68% terrestrial, and 14% freshwater. Its major strengths are the terrestrial gastropods of North and South America, Europe, Australia, and the Pacific Islands. It also has strong holdings in the marine taxa of the tropical and subtropical Western Atlantic region.

Historical collections include those of G. D. Gude, W. F. Webb, W. J. Eyerdam, J. Ferriss, A. Solem, and the largest historic collection at the Field Museum is the L. Hubricht collection of terrestrial gastropods.

The University of Michigan Museum of Zoology in Ann Arbor, Michigan began to accumulate its mollusk collection around 1840 and the Museum was formally organized in 1882. The collection numbers about 260,000 lots. The collection is 5% marine, 35% terrestrial, 57% freshwater, and 3% fossils, microscopic slide mounts of snail radulae, and freeze-dried snails. The collection has one of the largest collections of freshwater mollusks in the United States and the Bryant Walker collection of terrestrial gastropods is housed here.

The Milwaukee Public Museum is located in Milwaukee, Wisconsin and was founded in 1883. Its collection numbers 186,000 specimens (18,775 lots) and is 50% marine, 10% terrestrial, and 40% freshwater. A different department houses the paleontological collections. The strengths of this museum are its holdings in Wisconsin mollusks and the historic C. M. Wheatly Collection of freshwater mollusks.

The California Academy of Sciences in San Francisco on the Pacific Coast of the United States has one of the larger collections of mollusks in the northern half of California. Its' collection is roughly 250,000 lots and a wet collection of 40,000 lots [Cummings *et al.* (2000)].

The Santa Barbara Museum of Natural History is located further south in the city of Santa Barbara. This collection is 160,000 lots of mollusks with a type collection of 1,800 lots. The museum was founded in 1959. The collection is especially strong in worldwide marine gastropods, bivalves, and cephalopods as well as western North American Pulmonates.

The Natural History Museum of Los Angeles County is located in Los Angeles. This collection is approximately 450,000 lots and breaks down to 86% marine, 8% terrestrial, and 6% freshwater. Fossils are handled by a separate department. There are approximately 1,700 lots in the type collection. The collection is very strong in marine mollusks from the Pacific coastal regions of North and South America, as well as many families of micro marine gastropods.

The San Diego Museum of Natural History is located in Balboa Park, San Diego. It has a collection of some 95,000 lots of mollusks. Within the collection, 80% are marine, 12% terrestrial, and 8% freshwater. The type collection numbers 1059 type specimens. The marine collection is particularly strong in material from southern California and the Panamic Province. The collections of J. L. Bailey, H. N. Lowe, and Fred Baker are housed here.

The Bernice Pauahi Bishop Museum is located off the mainland in Honolulu, Hawaii. This museum was established in 1889. It includes some 260,000 cataloged lots and an estimated 80,000 uncataloged lots. The type collection numbers some 2000 lots, 25% of them representing marine taxa and the rest representing terrestrial gastropods. The main collection is 30% marine, 60% terrestrial, 5% freshwater, and 5% fossil and concentrates mostly on taxa from the Indo-West Pacific region. The Bishop Museum has the largest collection of land snails from Pacific Islands in the world.

The University of Colorado Museum in Boulder, Colorado was founded in 1902. This collection numbers some 44,000 lots. It breaks down to being 50% marine, 25% terrestrial, and 25% freshwater. The fossil collection is in another department. The type collection contains 273 specimens. The strengths of the collection are the non-marine mollusks of the Great Plains, Great Basin, and Rocky Mountain regions of the United States. The J. Henderson Collection is found here.

The Houston Museum of Natural History is located in Houston, Texas. Its collection contains 51,000 lots and no type collection. The collection is 95% marine and 5% terrestrial. Over 40% of the holdings represent taxa from the Gulf of Mexico. This specialization makes this collection second only to the National Museum of Natural History for richness of Gulf Coast representation.

The Florida Museum of Natural History in Gainesville is one of two major museums in Florida. Founded in 1917, it has some 430,000 lots in its collection. The collection is 25% marine, 50% terrestrial, and 25% freshwater. A department of invertebrate paleontology houses the fossil collection. There is a type collection that contains 740 holotypes. The collections of F. G. Thompson and W. Auffenberg are found here.

The Bailey-Matthews Shell Museum in Sanibel is unique and is an appropriate place to end this trip around the North American mollusk museums. It has the distinction of being the only major museum in North America that deals solely with mollusks. Established in 1995, it is the youngest museum listed here. Despite its youth, it has amassed a collection of approximately 200,000 lots. It breaks down to 85% marine, 4% terrestrial, 4% freshwater, and 7% fossil. There are about 15 lots in the type collection. Though the collection is worldwide in scope, its strength is in the mollusks found in the areas of Florida, the Florida Keys, the Gulf of Mexico, and the Caribbean.

13.4 LITERATURE CITED

Amadon, D. 1971. Natural history museums -Some Trends. *Curator* **14**: 42-49.

Colbert, E. H. 1961. What is a museum? *Curator* **4**: 138-146.

Cummings, K. S., A. Oleinik, and J. H. Slapcinsky. 2000. Systematic Research Collections (Recent and Fossil Mollusca). <www.inhs.uiuc.edu/cbd/main/collections/mollusk_links/museumlist.html>

Danks, H. V. 1991. Museum collections: Fundamental values and modern problems. *Collection Forum* **7**: 95-111.

Hoagland, K. E. 1994. Risks and opportunities for natural history collections: moving towards a unified policy. *Curator* **37**: 129-132.

Parr, A. E. 1962. Museums and museums of natural history. *Curator* **5**: 137-144.

Pettitt, C. 1997. The cultural impact of natural science collections. *In:* J. R. Nudds and C. W. Pettitt, eds., *The Value and Valuation of Natural Science Collections*. The Geological Society, London. Pp. 94-103.

Schmidt, K. P. 1958. The nature of the natural history museum. *Curator* **1**: 20-28.

Solem, A. 1975. The Recent mollusk collection resources of North America. *Veliger* **18**: 222-236.

Watkins, C. A. 1994. Are museums still necessary? *Curator* **37**: 25-35.

C. F. Sturm, T. A. Pearce, and A. Valdés. (Eds.) 2006. The Mollusks: A Guide to Their Study, Collection, and Preservation.
American Malacological Society.

CHAPTER 14

DONATING AMATEUR COLLECTIONS TO MUSEUMS

TIMOTHY A. PEARCE

14.1 INTRODUCTION

Practically everyone collects something. Some people collect matchbook covers or stamps, and many people collect shells. Collecting seems to fill a need in us. We enjoy collecting things, and we enjoy owning the things we have collected. We often put a considerable amount of time and energy into our collections. As with many of the things we enjoy, we often feel a need to justify to others the time and effort we spend on our collections.

Some common justifications of collecting include monetary value, aesthetics, and making a contribution toward the good of society. Some people collect objects as an investment, choosing objects that they think will become more valuable in the future. Some people collect objects for their aesthetic appeal, and indeed, we humans like to surround ourselves with objects of beauty. Some people collect to make a contribution to scientific knowledge, and the idea that they are contributing to the greater good gives many people a great deal of satisfaction. By donating their collections to museums, collectors can make valuable contributions that will be enjoyed and used by others in the future. Specimens in museums are used in educational exhibits and by scientific researchers.

Both museums and donors benefit when collections are donated. Museums benefit when the size, scope, and scientific value of their collections increase. You, as a donor, benefit through tax deductions, by knowing that your lovingly assembled collection will have lasting scientific and educational value, and by gaining immortality. Since most museums keep donor and collector information with specimens indefinitely, donating specimens to museums can assure that the donor's name will live on through the centuries. Occasionally, donors benefit monetarily if a museum purchases a collection, but these days, museums are on tight budgets so purchases are rare.

Amateur collections donated to museums are essential to our ability to conduct scientific research. At the Delaware Museum of Natural History (DMNH), amateur donations account for more than 80% of the mollusk specimens in the collections. Amateurs originally collected 80% (Solem *et al.* 1981) to 85% (Solem 1975) of the mollusks in major institutional collections.

While museums are actively building their collections, they cannot indiscriminately accept every specimen offered. The realities of finite storage space and limited funding for curating specimens means that museums must selectively accept and curate the specimens that are most likely to be needed by future generations. Donations that are perceived as lower priority may sit on shelves for years before being curated. A museum might dispose of specimens that lack locality data or have low scientific potential.

As a collector, you can take steps to make your specimens more valuable to museums, and to increase the speed and likelihood that museums will accept and incorporate your specimens into their collection. Understanding the space and funding constraints that museums face and understanding the ways museums use specimens can help you assure that your specimens have maximum value to museums. Collectors who contact the curatorial staff at a museum can learn the types of specimens the museum desires, and the best preservation and storage practices. Finally, if you can make a financial donation to the museum or volunteer time to curate your donation, you can dramatically increase the speed and likelihood that it will be incorporated into the museum's collection.

14.2 WHAT MAKES A SPECIMEN VALUABLE TO A MUSEUM?

Museums are the libraries where scientists find specimens to study (Allmon 1994). In keeping specimens for future generations, museums make their best guess of what specimens and what preservation methods future scientists will need. Curators want collections that are reference tools for taxonomic, zoogeographic, and ecologic studies, rather than accumulations of specimens that are of little worth other than as curiosities (Emerson and Ross 1965). Who could have predicted in the 1960s that twenty years later researchers would routinely extract DNA from specimens preserved in ethanol? Fortunately, modern researchers can use specimens that museums have kept in ethanol from decades gone by. We can only guess what uses future scientists will have for museum specimens.

Specimens are more valuable to museums (1) if they have associated data such as locality data, (2) if they are within the scope of the museum's mission or existing collections, and (3) if soft parts are preserved along with dry shells.

14.3 HIERARCHY OF USES

Natural history museums put specimens to a variety of uses, including scientific research, exhibits, and education, so specimens that are not useful in one area may be useful in another. Because the primary use of most natural history museum collections is scientific research, museums prefer to accept specimens that will be useful to researchers. When specimens are donated to a museum, they are evaluated and used at the highest possible level in a Use Hierarchy. Here I will describe the seven levels of Use Hierarchy at the Delaware Museum of Natural History. Other museums use some variation of this method in deciding how to use donated specimens.

Because a central theme for scientific research collections is the need for accurate data, in particular locality data, the first question is about the quality or completeness of the locality data associated with the specimens. However, even specimens lacking locality data might be useful to museums at some level in the Use Hierarchy.

14.3.1 Research or Systematic Collection. Specimens can be added to the museum's research or systematic collection if they have good locality data and are within the scope of the museum's mission or existing collections. For example, a donation of a deep-sea limpet with information about where and when it was collected would be within the scope of the DMNH mollusk collection, and would be a valuable addition to the research specimens. On the other hand, they may be of no use to a small college museum that specializes in freshwater mollusks.

14.3.2 Specimens for exchange. Specimens can be used in exchanges with other institutions if they have good locality data but are outside the scope of the museum's collections. Exchanges can be used to obtain specimens that are within the museum's scope. For example, a donation of an Australian earthworm with good locality information would be outside the scope of the DMNH collection because we do not have a worldwide annelid collection. However, we could exchange the earthworm for a mollusk specimen with another museum that does have an annelid collection.

14.3.3 Exhibits. The exhibits department might use specimens lacking locality data if the specimens are particularly good examples or illustrate points well. For example, a particularly large *Tridacna* (giant clam) shell can make a striking exhibit even if its locality and date of collection are unknown.

14.3.4 Education. The education department might use specimens in hands-on programs if they are not needed for exhibits. Children and adults enjoy touching and holding specimens and seem to learn better when they can touch specimens. However, specimens handled by museum visitors eventually become damaged, so we prefer to let visitors handle specimens that are of little use to researchers. Shells of snails and bivalves that are roughly fist-sized tend to be most useful in educational programs.

14.3.5 Specimens for sale. Sometimes, shells are sold in the museum store if the education department cannot use them. We include an informative note with the shells so that they play an educational role, and we explain that these shells lack locality data so are not useful to researchers. Shells for the store must be large enough to appeal to shoppers (usually larger than 2-3 cm), and in relatively good condition (not beach worn).

14.3.6 Crafts. Volunteers might use shells to make crafts if the shells are not appropriate to sell in the store. Crafts might include refrigerator magnets, Christmas tree ornaments, or knick-knacks. The crafts might then be sold in the store.

14.3.7 Specimens to be discarded. We discard the few shells that we cannot use higher in this Use Hierarchy. Shells that we discard are usually badly broken, beach worn, or very small.

14.4 WHAT AND HOW MUCH DATA?

Now that you understand the importance of locality data, you might be asking yourself how much and what kinds of information to record. Many collectors think the species name is one of the most important pieces of information. However, the species identity can be determined or verified later by examining the specimen. In contrast, the locality of collection cannot be determined later by examining the specimen. Consequently, locality information is much more important than species identity. There is no single answer to how much information to record, but some minimum data would include date of collection and collector, locality in sufficient detail that a future worker could find the place again, and ecological habitat information (Solem *et al.* 1981). As a guide, ask yourself how future researchers might use the specimen, and what they might want to know about it.

Let me introduce the concepts of intrinsic versus extrinsic information. Intrinsic information comes from examining the specimen itself. You can determine the length, the color, and usually the species' identity by looking at a specimen. Extrinsic information, on the other hand, is information external to the specimen itself. Locality and date of collection are extrinsic information. In order to find out the elevation or water depth where a specimen was collected, you would have to refer to the notes of the collector. The extrinsic information makes a specimen scientifically valuable.

Because we cannot know exactly what questions future researchers will ask, we do not know exactly what information they will need. I try to record as much information as practical when collecting. We do know that a common use of specimens is the production of identification guides, which include distribution and habitat information. Consequently, recording locality and habitat information will always be a good idea.

How detailed should locality information be? Again, include as much detail as practical. Imprecise locality information such as "Indo-Pacific" would not be very useful to someone preparing a field guide to Mollusks of the Solomon Islands. Within the U.S.A., many localities of non-marine species have been recorded to the level of county or parish. However, in contrast to insects, for which county level information is usually adequate because insects tend to be quite mobile, locality information for mollusks these days should ideally be more detailed than county. Conservation organizations trying to manage and protect endangered species often need to know within one hundred meters where rare populations occur.

How much detail would a researcher need in order to revisit the collecting site? If the species is abundant and widespread, less detail may be adequate, but for a rare species occurring on only one side of a cliff face, a great deal of detail may be necessary for the researcher to find the population again. I try to indicate localities within 100 m. In locality descriptions, refer to features that are likely to be permanent, such as intersections of roadways, rather than a particular building. With good maps available to collectors these days, and with the increasing availability of GPS (global positioning systems), we can expect future specimens coming in to collections to have more detailed and more accurate locality information than specimens of the past.

Besides locality in space, position in time is important to researchers. The year of collection can help researchers understand when species invaded an area or became locally extinct. The month and date of collection can be important to researchers seeking to document life history, for example, do individuals mature at a particular time of year?

Researchers also need to know habitat information. The habitat in which a population of organisms occurs can be an important clue in determining whether a group of similar individuals represents one species or two, and such information is often used in preparing identification guides. The more restricted and limited the habitat is locally, the more precise must be the locality description to allow a future researcher to find the place (Solem *et al.* 1981). Although rarely recorded because it is usually obvious, knowing whether a specimen came from a marine, freshwater, or terrestrial environment can be important to knowing its habitat, or where to look if one wished to find more specimens. Depth or elevation at which a specimen was collected can be important information to someone interested in the creature's life history or distribution. Other aspects of habitat information might include substrate type, and forest or vegetation type.

Behavioral observations can be very useful to researchers. We do not know much about the behavior of most species. Dead specimens in a museum have no active behavior, so behavioral observations on living animals can be crucial to a scientist, for example, when trying to interpret the functions of particular morphological features.

Researchers studying fossils need to know not only the geographical locality, but also the vertical position in the sediment. When collecting fossils, measure up or down from a landmark such as a conspicuous contact or bed, and note the rock formation if possible. Furthermore, note the sediment type, or better yet, collect a small sample of the sediment, so researchers will be able to draw conclusions about the environment in which the specimen was deposited.

In short, consider recording any extrinsic information that a future researcher might find useful. Too much information will never be a reason to refuse a specimen, but too little information might be.

14.5 IS SPECIMEN QUALITY IMPORTANT?

In contrast to shell collectors who desire high quality specimens, scientists may not place much importance on the quality of a specimen. Of course, given a choice, museums and researchers prefer specimens in good shape. However, the information associated with a specimen is often more important than the specimen itself. To a scientist, the specimen will often be simply a way to verify the species that goes with the data. A researcher can usually identify a specimen of high or low quality equally easily. In this case, the specimen quality would have little importance. Similarly, specimen quality would be unimportant to a researcher measuring specimen dimensions, unless the specimen is badly worn or broken.

Both live-collected and dead-collected specimens provide valuable locality information. I include locality records based on both live and dead specimens in my land snail distribution surveys. A live-collected specimen does give more reliable locality information, though, because the specimen was definitely living at the time and place it was collected. In contrast, a dead shell could have been transported by currents, could have been carried by animals such as hermit crabs, fish, or birds, could have fallen from the necklace of a tourist, or could be left over from a population that is now locally extinct.

Sometimes specimen quality does matter to researchers. Studies of the protoconch, shell surface microsculpture, or color would require specimens to be of good quality, not worn or faded. Often juvenile shells must be used to observe protoconch features because even the highest quality adult shell might have a worn protoconch.

So, specimen quality is less important to museums and researchers than it is to shell collectors. However, soft parts of a live-collected specimen, if saved, immensely increase the scientific value of a specimen.

14.6 SOFT PARTS

Researchers can learn a lot from studying mollusk shells, but they can learn much more from studying the soft parts. This additional information can be essential to systematists studying relationships among species. Soft parts provide information on external soft anatomy, internal anatomy (especially genital anatomy, but more and more researchers are studying other organ systems), morphology of the radula, DNA and other molecules, and stomach contents (which provide clues to behavior).

Many collectors preserve soft parts of specimens. Certainly, for species that lack hard parts, such as nudibranchs and octopus, preservation in fluid is the best option. Even for shelled mollusks, having the soft parts preserved with the shell provides much information to researchers.

On the other hand, many collectors take live mollusks but discard the soft parts and keep only the shell. This practice is unfortunate from a scientific perspective, because the soft parts could provide much valuable information to researchers. In an effort to avert this tremendous loss of information, several museums including the Delaware Museum of Natural History and the Carnegie Museum of Natural History accept donations of reasonably intact soft parts that have been removed from their shells under four conditions: (1) the specimens have good locality data, (2) the shell from which the soft parts came is unambiguously identifiable so a researcher could locate the shell that corresponds with particular soft parts, (3) the shell can stay in the collector's collection but must be available for researchers to borrow, and (4) arrangements are made for eventual deposit of the shell in a scientific museum, preferably the one to which the soft parts were donated. Check with the curator of your museum if you are interested in undertaking this course of action.

14.7 REMOVING THE SOFT PARTS FROM A SHELL

Mollusk bodies attach to their shells. People wishing to study the soft parts must remove the bodies from the shell. Many collectors who are interested only in the shell will allow the body to rot within the shell until it can be removed from the shell and discarded. That practice will produce very poor soft parts for research. Of course, the shell could be broken to remove the intact body, but that practice would produce a very poor shell for the collection.

One way to remove the body while keeping the gastropod shell intact is to immerse the mollusk in water heated to 60-65°C (140-150°F). At this temperature the columellar muscle relaxes its hold on the shell so the body can be twisted out of the shell. In my experience, if the shell cools before you twist the body out, the muscle may reattach to the shell, so remove the body while it is still warm.

Once the body is removed from the shell, it can be preserved in alcohol. If you use heat to remove bodies from shells, put a note with the specimen stating what temperature you used. This information will be useful to researchers, because too high a heat might make certain studies difficult.

14.8 PRESERVATIVES FOR SOFT PARTS

There are many different uses for soft parts, and some uses require special preservation methods, so the best method of preservation depends on the uses to which the specimens will be put. The preservative that allows the greatest number of common uses is 80% ethanol. Specimens preserved in 80% ethanol are excellent for gross anatomical studies, and DNA can be extracted from alcohol-preserved specimens using PCR (polymerase chain reaction) techniques, as long as the specimens have never been in formalin.

Other types of alcohol can be used. While unadulterated ethanol is the best, it is possible to get good results with denatured ethanol, although some denaturants might cause hardening of the tissues. Isopropyl alcohol (rubbing alcohol) can be used, although again, it can cause tissues to harden over time, making dissection difficult. Some collectors have successfully used whiskey or vodka for temporary preservation when other preservatives were unavailable. The higher the proof of the alcoholic beverage, the greater is the ethanol concentration. For example, an alcoholic beverage that is 160 proof is 80% ethanol; one that is 80 proof is 40% ethanol.

Other methods of preservation include fixatives such as formalin (a 37-40% aqueous solution of formaldehyde) or Bouin's solution (e.g., for histology and microscopic anatomical studies), glutaraldehyde (e.g., for electron microscopic studies), FAA (formalin, acetic acid, and ethyl alcohol, for chromosome studies), and ultra cold freezing (at temperatures of liquid nitrogen, for recovering molecules such as DNA or enzymes).

For general preservation, I strongly advise against the use of formalin unless there is a specific reason to use it. Some disciplines, especially vertebrate collections, fix specimens for a few days in formalin and then transfer them to ethanol. The formalin helps the body and organs keep their shape by cross-linking proteins, but using formalin almost completely destroys the ability to extract DNA from tissues.

For long-term storage, preservatives should be chemically buffered to prevent the shell from dissolving. Preservatives can become acidic from natural fluids leaching out of the soft tissues. The acidity can dissolve the calcium carbonate from the shells. A number of buffers exist to control pH. An inexpensive buffer used at the Delaware Museum of Natural History is a saturated solution of borax (sodium borate) in the ethanol (less than one gram per liter).

14.9 INCREASING SPEED AND LIKELIHOOD OF DONATIONS BEING INCORPORATED INTO MUSEUMS

Sometimes it takes a long time for specimens donated to a museum to be curated into the collection and not every specimen will become part of the scientific reference collection. The following five bits of advice will help donors who wish to increase the speed and likelihood that their donations will be incorporated into museums.

14.9.1 Establish the fate of the collection ahead of time. The single most important advice I can give is for you to arrange ahead of time for your collection to be donated in case you die or become incapacitated (see Teskey 1974). Tell your next of kin and make clear in your will your intentions about your collection and associated locality data books. Don't burden your loved ones with trying to figure out what to do with the shell collection if you become incapacitated - often when that happens, the locality data books get lost, and the scientific value of the collection plummets.

Every curator has horror stories of receiving neatly curated collections, with numbers carefully marked on the shells, but there is no locality data and no clue about what the numbers on the shells mean. In such cases, the notebooks of locality data probably went to the landfill, their value unrecognized. In other cases, specimens have been sitting in the attic for 30 years, or have been moved several times. Through the jostling, the data slips may no longer be uniquely associated with the specimens, or may have been eaten by silverfish.

14.9.2 Associate data with specimens. Associating data unambiguously with specimens will assure that the specimens will have lasting scientific value. If data slips are lost or mixed with other specimens, if notebooks are lost, or if your code for associating locality data with specimens is undecipherable, then the specimens become worthless to researchers. Unfortunately, these losses happen all too often.

If you have documentation such as collecting permits that show you had permission to collect the specimens, or if you have permits allowing you to possess an otherwise protected or regulated species, include a copy of those permits with the donated specimens. Museums are coming under increased scrutiny to show proper documentation for the specimens in their collections.

14.9.3 Contact a Museum. Contact the curatorial staff at a museum and talk about curatorial practices while you are building your collection. Talk about what kind of supplies the museum uses, the kinds of data and formats they use, what reference materials are helpful. Also, ask about volunteer opportunities. Since every museum has certain peculiarities, contacting an individual museum to learn how they curate specimens can make it even easier for them to incorporate your specimens.

A museum might help you by providing curatorial supplies and alerting you to curation standards. If curators at a museum think that they will eventually receive your collection, they may be willing to supply you with the materials they use, in order to make incorporating your collection easier. Materials might include vials, acid-free specimen boxes, archival label paper, etc.

Many people these days enter information about their collections into a computer database. If you computerize your collection, contact a museum as you begin the process to learn what information they capture, and what formats they use. If you computerize your collection using similar data fields and formats as a museum, incorporating your collection into the museum will be much easier. These days transferring information from one database program to another is relatively easy, but you want to be sure that your locality descriptions, for example, are brief enough to fit into the locality field(s) that the museum uses.

Museums have good reference materials that can help you with geographical locality names, with identifying specimens, and with determining the currently correct species names.

14.9.4 Donate money or time. Two other things that can ensure your collection will be added to a museum collection quickly, well ahead of other collections that may have been donated years ago, are (1) a financial donation to help with the cost of curating, and (2) volunteering to help curate your collection. The financial donation can be made after you die, as long as you have provided for it in your will or by some other mechanism. Volunteering must be done while you are still able to function.

Many collectors who plan to donate their collections do not realize that there are real costs to the museum in terms of materials, salaries, and space to curate and house donated collections. Because of these costs, donations with an accompanying financial contribution gain a higher priority.

I want to emphasize the benefits of volunteering. First, volunteering will speed incorporation of your collection into the museum's research collection. Beyond that, volunteering will allow you to enjoy working with the much larger collection in the museum, much as you enjoyed your own collection. It can feel very rewarding to get all of the museum's specimens of your favorite family identified correctly and in proper taxonomic order. There are always volunteer opportunities at museums, and most museums rely on volunteer help.

14.9.5 No special conditions. A final piece of advice for getting your collection added to a museum collection is not to attach special conditions. The most common special condition I hear from donors is that they want their collection to remain intact. However, museums usually arrange specimens in taxonomic order. Keeping a collection intact would be a bookkeeping difficulty. Imagine a researcher visiting a museum to examine specimens of <insert your favorite group here>. The curator says, "These cabinets contain the specimens you need, oh, and then there may be some specimens in these 30 cabinets of donations that we have to keep intact, so you'll have to look through all of those as well." Most museums would refuse a donation if they had to keep it intact.

Another common request is that the collection be put on public display (for some usually unexpressed period of time). Again, this request presents difficulties. First, putting specimens on display can take months of planning to find display space, prepare labels, and arrange specimens. Secondly, displays at museums try to make a particular point or tell a story, so specimens that best tell the story are typically selected. Sometimes museums have displays such as "A recently donated shell collection," but such displays are competing for display space with other exhibits that tell scientific stories such as "Radula: versatile feeding structure of Mollusks" or "Unionidae: America's disappearing treasure." It is usually best to let the museum decide when and which specimens to put on display.

14.10 SUMMARY

As an amateur, you can make a lasting scientific contribution by donating your natural history collection to a museum. Specimens having the greatest scientific value are those with as much extrinsic information as possible, particularly, those with good locality data, and those with preserved soft parts. You can improve the speed and likelihood that your collections will become incorporated into a museum collection by making arrangements for the donation before becoming incapacitated, by clearly associating the data with the specimens, by contacting a museum regarding curation standards, and by volunteering time or donating financially toward incorporating the specimens into the museum collection. Let your hobby live on for the enjoyment and benefit of others by donating your collection to a museum.

14.11 LITERATURE CITED

Allmon, W. D. 1994. The value of natural history collections. *Curator* **37**: 83-89.

Emerson, W. K. and A. Ross. 1965. Invertebrate collections: trash or treasure? *Curator* **8**: 333-346.

Solem, A. 1975. The Recent mollusk collection resources of North America. *Veliger* **18**: 222-236.

Solem, A., W. K. Emerson, B. Roth, and F. G. Thompson. 1981. Standards for malacological collections. *Curator* **24**: 19-28.

Teskey, M. C. 1974. "I hereby bequeath." *In:* M. K. Jacobson, ed., *How to Study and Collect Shells,* 4th Ed. American Malacological Union, Wrightsville Beach, North Carolina. Pp. 106-107.

CHAPTER 15

FOSSIL MOLLUSKS

DAVID CAMPBELL

15.1 INTRODUCTION

A fossil is evidence of a previously living organism. Generally, scientists use an age of 10000 years or older to classify something as a fossil (Table 15.1). Materials younger than this are called sub-fossils. Fossils may be part of the actual organism, a mold or cast of it, or some other evidence of the organism such as a burrow. Mollusks fossilize easily because of their hard calcareous shell. Mollusks are best preserved in sedimentary rocks, such as limestone, sandstone, and shale, or in sand, silt, or clay.

This chapter is an introduction to paleoconchology; the study of fossil mollusks. In it I will briefly cover some of the main topics dealing with fossil collection and preparation. More information can be found in Shimer and Shrock (1944), Moore *et al.* (1952), MacDonald (1983), Boardman (1987), and Fenton *et al.* (1989). Studying fossil mollusks provides both opportunities and challenges slightly different from work on modern forms. In particular, studying fossil mollusks gives an opportunity to trace the ancestry of modern forms.

15.2 FOSSIL MOLLUSKS

Often, fossil shells are simply very old shells. The colors are generally faded or gone, and there are no smelly bits of animal left to clean out, but otherwise they may look very similar to shells that you find at the beach. In fact, some beaches have a confusing mix of fossil and modern shells. In addition to being less fresh, the fossil shells usually will include species no longer living in the area, probably including some that are extinct.

However, sometimes you will find fossil shells consisting of a different substance than the original shell. This may happen in one of two ways. One mineral can replace another, or the shell may dissolve away and another mineral be deposited in the space, forming a cast of the original shell. Common replacement minerals include calcite, microcrystalline quartz, pyrite, and various calcium phosphate minerals. These may be more resistant to acid than the rock, allowing extraction by acid. Storage concerns are largely the same as for modern taxa (see Chapter 5).

Most mollusks make their shells out of aragonite. Aragonite is a form of calcium carbonate that is thermodynamically unstable under conditions at the earth's surface. Calcite is the stable form of calcium carbonate. For the life of the animal, this is not a problem. In fact, aragonite can have advantages under certain conditions. However, over geologic time, the shell tends to crumble away. (This is similar to the relationship between diamond and graphite. Diamond is more stable at high pressure and so is found in rocks deep inside the earth. Although diamond rings do not spontaneously change into graphite, they can change under the right pressure and temperature conditions.)

A few kinds of mollusks use calcite in their shells, including oysters (Ostreidae), scallops (Pectinidae), wentletraps (Epitoniidae), nerites (Neritidae), and certain extinct groups like the hippuritids (Hippuritidae) and belemnites (Cephalopoda). Other organisms also vary in their use of aragonite versus calcite. Thus, a deposit in which the aragonitic

Table 15.1. The geological time-scale. MYA = Millions of years ago.

MYA	Period or *Epoch*	Events
0.01-0	*Recent or Holocene*	Human-caused changes
1.8-0.01	*Pleistocene*	Extensive glaciers, variable climate
5.3-1.8	*Pliocene*	Panama seaway closes, northern glaciers expand
23.8-5.3	*Miocene*	Mediterranean briefly cut off from oceans and dries up
33.7-23.8	*Oligocene*	Extensive Antarctic glaciation
54.8-33.7	*Eocene*	Mostly warm, global cooling near end, Antarctica-Australia-South America split complete
65-54.8	*Paleocene*	New species evolving after the extinction; extremely warm at end
65		Bolide impact, mass extinction-ammonites and many others die out
144-65	Cretaceous	Warm, high sea level, southern continents splitting up
206-144	Jurassic	Warm, further splitting up of northern continents
248-206	Triassic	Pangea starts to split up, major extinction near the end
248		Massive extinction, cephalopods hit hard
290-248	Permian	All continents assembled into Pangea
354-290	Carboniferous	Cooling, continents coming together
417-354	Devonian	First known non-marine mollusks
443-417	Silurian	Diversification of many groups
490-443	Ordovician	Many mollusk orders appear, first known scaphopods
540-490	Cambrian	Shells become common, most mollusk classes appear
4500-540	Precambrian	First animal fossils near end

shells have dissolved may preserve calcitic shells nicely. If both calcitic and aragonitic shells are preserved, often they have different appearances. For example, in the Miocene to Pleistocene of the United States Atlantic Coastal Plain, aragonitic shells are typically white whereas calcitic shells are brown or gray.

Unlike collecting living marine or freshwater taxa, you can often collect fossils without getting wet (although I fell into a river once, to the amusement of the students on the field trip) and have no smell to deal with while cleaning them.

Most modern groups of mollusks with shells are well represented in the fossil record. Pleurotomariids and *Pholadomya* are examples of groups easily collected as fossils though now rare. Also, because of plate tectonics and climate changes, different faunas lived in different parts of the world, so a tropical fossil fauna may be found in what is today a cold climate. For good diagrams of paleogeography through time, try <www.scotese.com/>

On the other hand, there are a few major kinds of fossil mollusks that are extinct and it is challenging to recognize them even as mollusks. *Kimberella*, known from the late Precambrian of Russia and Australia, is the oldest known fossil attributed to the phylum. There are late Precambrian traces apparently made by mollusks scraping algae with their radulae, much like traces left by modern limpets (Patellogastropoda) and chitons (Polyplacophora). Mollusks diversified rapidly in the Cambrian, producing most of the living major groups and several extinct forms.

Most of the Cambrian forms were very small, often less than 1 cm in maximum dimension. Varied small shelly fossils occur in the lower Cambrian. At least some of these are mollusks, and others may be. Often we are not sure whether a particular shell enclosed most of the organism (as for most snails) or only a small part (like many other snails) or if one animal had multiple shells (like a chiton). Other groups are better known. The hyolithids are sometimes considered an extinct class of mollusks and sometimes a closely related phylum. Their shells are pointed and roughly conical with a lid-like valve over the open end.

Several forms, broadly considered monoplacophorans, seem to include the ancestors of other extant classes. Because of this, Monoplacophora (Figure

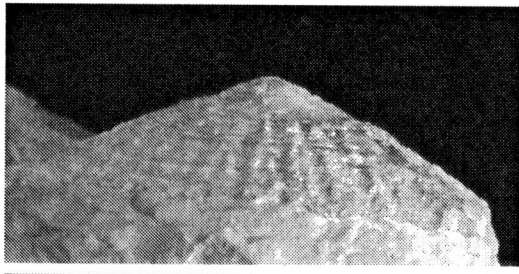
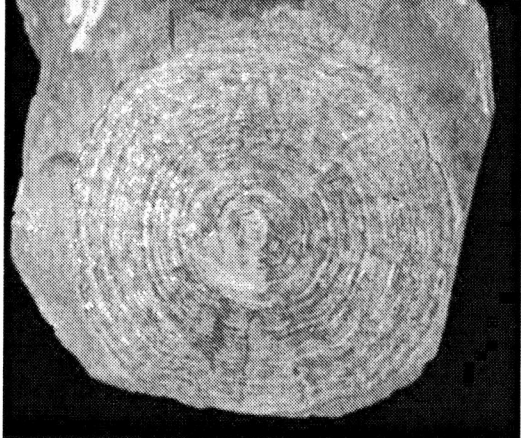

Figure 15.1 Monoplacophora.
Monoplacophora, genus and species unidentified, Cretaceous, Basses Alps, France, 30 mm in diameter, top - lateral view, bottom - apical view, CMNH-P 29682. (CMNH-P = Carnegie Museum of Natural History Section of Invertebrate Paleontology).

15.1) is often split into multiple classes to better distinguish among the major lineages. However, it remains a useful term for a general grade of evolution. Fossil monoplacophorans include, in addition to the limpet-like forms similar to the living monoplacophorans, laterally compressed monoplacophorans ancestral to the bivalves and some tall conical forms that appear close to cephalopods. See Chapter 17 for more information on the Monoplacophora.

Distinguishing gastropods from monoplacophorans can be particularly difficult, as either can appear limpet-like or coiled. Torsion is the key difference, but this may not be obvious from a fossil shell. The Class Paragastropoda was proposed for snail-like coiled shells that were not torted (Linsley and Kier 1984). Other forms, the bellerophonts (Bellerophontoidea, Figure 15.2), are planispirally coiled, a bit like very fat *Planorbis* snails. Whether they were gastropods or monoplacophorans remains unclear (Gubanov and Peel 2000).

Another extinct class was Rostroconchia (Figure 15.3). As adults, these resemble bivalves, but the larval shell is not divided into two. Instead of a hinge like the bivalves, they had solid shell across the dorsal side. Almost all of these groups became extinct by the end of the Paleozoic in the massive Permo-Triassic extinction. Bellerophonts, however, survived up to the end of the Triassic.

The extant classes include a few peculiar extinct forms. The strangest of the fossil bivalves are the rudists (Hippuritoidea, Figure 15.4). The ancestral forms are superficially similar to *Glossus* (Bivalvia). However, the rudists developed an epifaunal lifestyle, either attaching to a hard surface or else sticking up out of the sediment. At least one valve is large and conical, sometimes extensively coiled. The other valve may be similarly formed or cap-shaped. They were abundant in shallow-water Cretaceous limestones, similar to the modern coral reef environment, but became extinct by the end of the Cretaceous. Some reached enormous size, surpassing the modern *Tridacna* (Bivalvia).

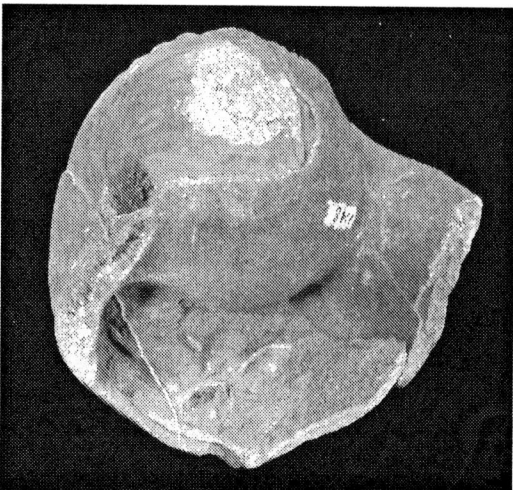

Figure 15.2 Bellerophontoidea.
Aglaoglypta tuberculatus (de Ferussac and d'Orbigny, 1840) (Mollusca: Bellerophontoidea), Middle Devonian, Westphalia, Germany, 72 mm wide, CMNH-P 1148. There is a small *Murchisonia* (Gastropoda) (or related genus) in the matrix of its aperture.

Figure 15.3 Rostroconchia.
?*Bransonia* sp. (Mollusca: Rostroconchia), Lower Mississippian, Humboldt, Iowa, 11 mm, CMNH-P 40061.

Not to be outdone, the cephalopods (Cephalopoda) also had large, strange forms. Unlike the coiled *Nautilus*, many fossil cephalopods had straight shells (Figure 15.5). Some of these were quite large, perhaps even a few meters long. Although coiling makes the shell more compact, various coiled forms may reach a meter in diameter. One particularly successful group of coiled cephalopods was the ammonoids (Ammonoidea, Figure 15.6). In contrast to the simple form of *Nautilus* and other shelled cephalopods, these developed complex wrinkled surfaces to their chambers. A few even uncoiled and took on unusual shapes. The ammonoids evolved rapidly across wide geographic ranges, making them useful index fossils for various ages of rocks. They were quite successful in the mid- to late Paleozoic and the Mesozoic despite two brushes with extinction. The ammonites, last of the ammonoids, finally disappeared at the end of the Cretaceous along with many mollusks of other

Figure 15.4 Hippuritoidea.
Hippurites radiosus Des Moulins, 1826, (Mollusca: Bivalvia: Hippuritoidea) Upper Cretaceous, France, 140 mm in diameter, CMNH-P 39966.

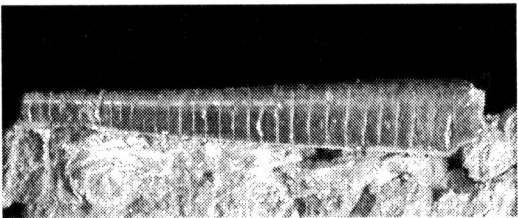

Figure 15.5 Nautiloidea.
Pseudorthoceras knoxense (McChesney, 1860), (Mollusca: Cephalopoda: Nautiloidea) Brush Creek Limestone, Carboniferous (Pennsylvanian), Pennsylvania, 59 mm, CMNH-P 40396.

groups, as well as other organisms ranging from microplankton to giant reptiles.

The belemnites (Belemnoidea, Figure 15.7) were also cephalopods, ancestral to modern squids. The cuttlebone of modern *Sepia* (Figure 20.5) is descended from the aragonitic, chambered portion of the belemnite shell. However, a calcitic rostrum, lost in later squids, covered the tip of the shell. This rostrum is bullet shaped with a conical depression in one end. Isolated belemnite rostra do not look obviously molluscan. Because the rostrum is calcite and the chambered potion aragonite, often only the rostrum is preserved. True belemnites also did not survive past the Cretaceous, but their descendants, the sepiid squids, retained some calcitic tip to their shells into the Eocene. See Chapter 20 for more details on the Cephalopoda.

15.3 COLLECTING

15.3.1 In the field. Collecting fossils may require equipment beyond the standard shell collecting gear. Normal collecting tools for exhuming mollusks from their rock graves are hammers, chisels, heavy sledgehammers, pry bars, shovels, and pickaxes. Rich fossil sites sometimes require only a large bag, patience, and time to collect with your fingers. Sometimes an exquisite specimen can be had by breaking off a large piece of rock and then doing the more detailed extraction at home rather than in the field, so be as generous as possible in collecting the matrix surrounding the specimen.

Additional equipment needed to facilitate your collecting trip are: drinking water, sunscreen, and

Figure 15.6 Ammonoidea.
Parkinsonia parkinsonia (Sowerby, 1821) (Mollusca: Cephalopoda: Ammonoidea), Jurassic, Dorset, England, 110 mm, top - external view, bottom - polished internal view showing chambers, CMNH-P 44343.

mosquito repellent for those long hiking and quarry collecting adventures. When collecting with high rock overhangs or quarries, safety gear such as a first-aid kit and a hard hat are also advisable. Many active mining operations require hard hats, safety goggles, or other protective gear if you are collecting anywhere near the active area.

Fossil collecting involves most of the same hazards as land snail collecting. Watch your step to avoid falling. Be sure you know how to avoid hazardous flora and fauna like poison ivy, fire ants, or hostile natives.

In particular, be sure you obtain proper permission from the landowner before collecting. Not only does this help create a friendly relationship for return trips, but also it lets the landowner inform you of potential hazards. For example, you do not want to be wandering around a quarry when they are about to set off explosives.

Fossils preserve the maximum amount of scientific information when undisturbed in their original location, providing the collector with an opportunity to note its relationship to other fossils or physical structures, the position of the shell, what kinds of fossils are associated, etc. On the other hand, fossils provide no scientific information when they are eroded away, ground into gravel, or permanently paved over. Thus, the decision of how much to collect must balance the need to preserve the site for future scientific study with the risk that the specimen (or entire locality) may be destroyed.

15.3.2 Transporting material home. You will need something in which to carry your finds. Any sturdy box, bucket, or bag will suffice. Padding may be required for delicate specimens. Restaurants, professional painters, and others who buy supplies in five-gallon buckets may be willing to give you the empty buckets or sell them to you for a token amount. For small specimens, empty plastic medicine bottles or plastic film canisters do well. Photo developers often will give away empty film canisters, but be sure to get the lids as well.

Especially for smaller species, it may be more efficient to collect a bucketful of sediment and then examine it more carefully at home with a microscope or magnifying glass. Fragile specimens may also transport better surrounded by the original matrix.

Take careful notes as to where your collection originates by plotting the site on a topographic or geologic map or use a GPS receiver. This may help in identifying your specimens and providing others (scientists) with valuable information for later fossil dating by microfossils. Museum samples without adequate label information have been more precisely located by examination of adhering sediment, but do not count on this as adequate for your specimens.

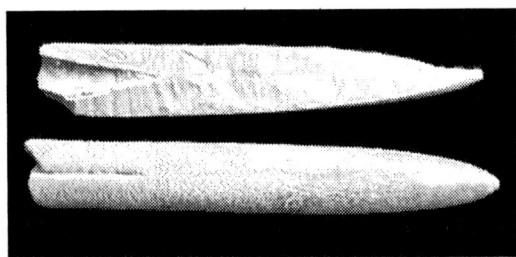

Figure 15.7 Belemnoidea.
Rostrum of *Belemnitella americana* (Morton, 1830) (Mollusca: Cephalopoda: Belemnoidea), Upper Cretaceous, Marlboro, New Jersey, 96 mm, CMNH-P 29818.

Shells can be fragile, so be careful how close you hammer on the rock; otherwise they may break. In particular, if you can see part of a promising specimen sticking out of the matrix, you should excavate well away from the specimen in case it is bigger than you expected. Some of the same tools used for cleaning difficult modern shells may also be useful for both collecting and cleaning fossils, including dental picks, drills, and brushes of varying stiffness.

A sieve can be used for matrix (gravel, sand, and mud) that may contain small shells. When you get a sample of sediment home, you need to sieve and sort it. If the locality has water available, sieving on-site may also be effective. Washing through a variety of sieve sizes and then drying and sorting is often effective. Old toothbrushes and dental tools are good for removing adhering dirt.

For micro-mollusks, nylon stockings provide a cheap alternative to a fine-mesh sieve. An effective way of collecting micro-gastropods is to dry a sample thoroughly after sieving out the larger material. Once it is dry, quickly dump the sediment into a bucket of water and stir it around a bit. The gastropods might have air in them and float. You then pour the water though the stocking. Mud washes through, leaving behind the shells.

15.3.3 Cleaning and preparing fossils. Most fossils have some adhering dirt. Washing and scrubbing will usually take care of most of it. An old toothbrush or a dental pick may help. If the sediment is a bit difficult to remove, a surfactant like soap may help. Ultrasonic cleaning baths marketed for cleaning jewelry may be effective for cleaning small specimens (see Chapter 2.7.10).

When the fossils are in rock, extraction of the shells is much more difficult. Mollusk shells dissolve easily in acid, so most methods of dissolving rock will also dissolve the shell. For softer mudstones repeated wetting and drying, and perhaps some soap, may work. The safest approach is to take a sample you can spare and experiment. Careful work with dental tools, abrasives, very weak acid, or the like may also be worth trying. In some cases, natural weathering may be the best way to extract a specimen, but it is also very slow. A rock saw may be used to cut off excess material.

A good collection of fossil molds takes up a lot of space, as you must collect the rock and not just the shell itself (Figure 15.8). Often the best way to identify and study molds is to make casts. Both calcite and aragonite readily dissolve in acid. Often, the shell will dissolve entirely, leaving a hole in the rock. Even though the shell dissolved away, the mold may preserve a very faithful replica of the original shell. Microscopic examination of the shell sculpture or form of the protoconch may be possible.

A cast of the fossil can be made with liquid latex or silicone. These materials remain flexible after hardening. Silicone usually lasts longer but also is more expensive. Before casting, be sure to clean the mold carefully. Add the casting material slowly and

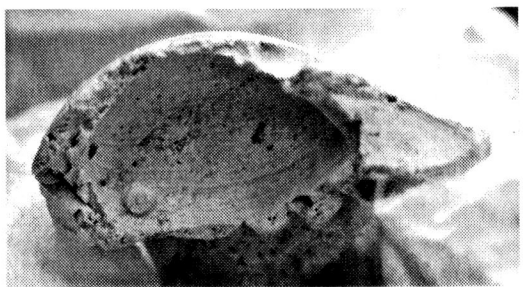

Figure 15.8 Fossil mold.
A mold of a Cenozoic bivalve (Mollusca: Bivalvia: Corbuloidea), Eocene, Catherine Lake, North Carolina. The disc in the lower left shows evidence of boring by a gastropod. From author's collection.

carefully to ensure there are no air bubbles. Allow plenty of time for the cast to harden and then carefully pull it out. A similar technique may work for shells not readily extracted from rock. Some studies have used a mild acid to dissolve the shells and then made casts of the resulting molds.

15.4 LOCALITY AND STRATIGRAPHIC DATA

Exact locality data are even more important for fossils than for modern shells. If you know a modern shell was collected alive, you at least know that it is Recent. A fossil from an unknown locality could be almost any age. The exact location, relative to other features of the exposure, is particularly important. Many localities have more than one layer present. Determining the exact source layer will give the age, environment, and several other pieces of information about your specimen.

If possible, try to look at the exposure from a distance as well as up close. Are there any abrupt or gradual changes in the deposit, such as a shift from sand to silt or clay? Are there layers with pebbles, teeth and bones, and worn shells? Such features indicate a change in the environment of deposition and possibly a significant gap in time. Note the distance above and below the layers from which your specimen comes. In addition, caution is required near the contact between layers.

Remembering that fossils preserve maximum information when still in place, try to get as much information as possible before removing the specimen. Consider the potential for further studies of material in place, balanced against any risk that the specimen or the locality might be lost before further study.

A deep burrower, such as *Panopea*, below the contact might have burrowed in from above. Worn shells near the base of the upper layer may have been reworked from the lower layer and re-deposited. A measuring tape is both an inexpensive and precise tool for detailing the location within an outcropping, though a GPS receiver is easier for measuring longer distances.

15.5 IDENTIFYING FOSSILS

Once you have collected your fossils, the next challenge is identification. Finding appropriate references will depend on the locality and age. In most cases, you can get a general idea of the major group to which a fossil belongs by using references on modern mollusks. Various extinct groups may at first appear to be something besides a mollusk. In general, the younger a sample, the more it appears to resemble a living taxon. Locating detailed information on the fossils of a particular deposit can be difficult. A good source for locating geological information can be found by contacting state and national geological surveys. The survey's exact name varies, so if you have difficulty finding information about your own state's geological survey, try the bureau of mines or the department of natural resources. Many of their publications are still available, though some may be out of print.

Local museums or college geology departments can be helpful and they often have useful libraries. Two general references are Georef and *The Treatise on Invertebrate Paleontology*. Georef is a compilation by the American Geophysical Institute, available by subscription online or on CD, which attempts to provide a bibliographic compilation of all publications on geology. It is accessible from many university libraries.

The *Treatise on Invertebrate Paleontology* is a multi-volume compendium that attempts to describe all known genera. Although some volumes are somewhat out of date, they provide the most extensive reference available at this level. The gastropod volumes have yet to be completed, but most other mollusks are covered to the extent that they were known at the time of publication. The Geological Society of America publishes this series in cooperation with the University of Kansas and it is available at most major university and museum libraries.

Some paleontology references can also be found online. Check the website for the geological survey for the state, province, or country in which you have an interest.

It is very helpful to determine the age of your specimens. This will greatly narrow the options for identification. A geological map (available from geological surveys, often online) of the region will show you what type and age of rocks are present in a given area. Over geologic time, the sea level has gone up and down, and seafloor has been pushed up into mountains, so many areas of land now have marine fossils. In general, in North America, relatively young fossils (Cretaceous and Cenozoic) occur in the coastal plain, the low-lying areas extending from Long Island south along the Atlantic coast to Mexico and up the Mississippi Valley as far as southernmost Illinois. Similar deposits, though more patchy, can be found along the Pacific coast. North of Long Island, there are some glacial deposits with fossil shell. Much of the Great Plains region is underlain by Cretaceous marine rocks. Much of the Appalachians, including much of eastern Canada, and the Midwest, extending south into northern Alabama, have much older rocks, spanning much of the Paleozoic. Almost all the Paleozoic genera and many families or higher categories are extinct. Also, brachiopods were much more common in the Paleozoic and may confuse someone who expects all shells to be mollusks (see Chapter 29.2.9). Paleozoic rocks also occur in western Texas and in much of the mountainous regions of the West. Mesozoic rocks are also widespread in this region. Non-marine mollusk fossils are generally less common, but may occur in continental glacial deposits in the northeast, midwest, and northern plains. There are also extensive Mesozoic non-marine beds in Alberta south to New Mexico and Texas, with fossil mollusks as well as dinosaurs. The Green River Formation of Colorado and Wyoming, famous for its fossil fish, has many early Cenozoic non-marine mollusks. Non-marine shells also turn up in some of the same regions as the marine fossils, from times when the sea level was lower.

Unfortunately, the vagaries of preservation often make fossils more difficult to identify compared to modern shells. Sometimes the preservation pattern can be confusing. For example, an outer shell layer may have been lost before fossilization. Sometimes fossils can be warped or squashed by geologic forces. This may superimpose external and internal features of the shell, especially if the original shell was thin.

15.6 AMATEUR ACTIVITIES AND OPPORTUNITIES

Paleontology is an area where amateurs can readily provide important new scientific data. Amateur paleontologists often provide important collecting data that are new to science.

An amateur collector can undertake several kinds of scientifically important projects on fossil mollusks. Perhaps most importantly, you may discover and report on new localities and specimens. If you live in an area with fossils, any excavation or other event that exposes rock and sediment may unearth something new. By the time a professional paleontologist arrives, the exposure may be largely gone or important features may be lost. Thus, early work on a site may preserve valuable materials and information, especially if done to professional standards.

Conversely, persistent study of a supposedly well-known locality will almost certainly turn up taxa not previously known from the site. Such documentation of a fauna, especially if it includes information on the frequency as well as the presence of different species, provides important scientific information regarding the environment and biogeography. By documenting the variation within a species or a species group, you can enhance our knowledge of the systematics of the group.

Many fossils remain undescribed, so you may get an opportunity to name new species or higher taxa. The equipment needed is the same as for modern shells: a microscope, measuring tools such as calipers, a good library, and some hard work (see Chapter 10).

Your network of friends provides some unique connections that a professional researcher will not have. You may happen to know a landowner, or a friend who happened to notice some shells when out bird watching, or a construction worker. Cultivating a good relationship with the landowner is also crucial to continued access for you and other collectors. Many excellent fossil localities are now off limits to the public because of previous trespassing, disregard for safety regulations, or property damage by irresponsible collectors.

Some laws may apply to fossil collecting, particularly on government-owned land. In general, attention has focused on vertebrate fossils, but collection of invertebrates may be affected as well. An overzealous Canadian customs agent once impounded a small box containing a couple of Paleozoic bivalves that my advisor was returning to a Canadian museum. Apparently identifying the contents as "fossils" sounded impressive.

Finally, if you find something unusual, donating it to a museum will help ensure that the specimen receives appropriate curation and scientific study. Of course, you can study it first yourself.

Ethical considerations apply to the collection of fossil shells as opposed to those from recent times, but for slightly different reasons. Over-collection will not lead to the unnecessary death of any fossil mollusk, however, it can be thought of as stealing from other collectors who may never have the chance to have the enjoyment you did. A single careless collector can totally and permanently deplete a small locality that may be a valuable scientific and educational resource. Collect only what you need for your use and possibly for trade with others. You can gauge what is appropriate to collect based on the quantity of material present.

15.7 LITERATURE CITED

Boardman, R. S., A. H. Cheetham, and A. J. Rowell, eds. 1987. *Fossil Invertebrates*. Blackwell Scientific Publications, Palo Alto, California. xi + 713 pp.

Croucher, R. and A. R. Woolley. 1982. *Fossils, Minerals and Rocks; Collection and Preservation*. British Museum (Natural History) and Cambridge University Press, London. 60 pp.

Cvancara, A. M. 1985. *A Field Manual for the Amateur Geologist; Tools and Activities for Exploring Our Planet*. Prentice Hall, Inc., New York. xii + 257 pp.

Feldmann, R. M., R. E. Chapman, and J. T. Hannibal, eds. 1989. *Paleotechniques. Paleontological Society Special Publication No. 4*. Paleontological Society, Knoxville, Tennessee. iv + 358 pp.

Fenton, C. L. and M. A. Fenton [revised and expanded by P. V. Rich, T. H. Rich, and M. A. Fenton]. 1989. *The Fossil Book, A Record of Prehistoric Life*. Doubleday, New York. 740 pp.

Gubanov, A. P. and J. S. Peel. 2000. Cambrian monoplacophoran molluscs (Class Helcionelloida). *American Malacological Bulletin* **15**: 139-145.

Linsley, R. M. and W. M. Kier. 1984. The Paragastropoda: A proposal for a new class of Paleozoic Mollusca. *Malacologia* **25**: 241-254.

Macdonald, J. R. 1983. *The Fossil Collector's Handbook; A Paleontology Field Guide*. Prentice-Hall, Inc., Englewood Cliffs, New Jersey. xi + 193 pp.

Moore, R. C., C. G. Lalicker, and A. G. Fischer. 1952. *Invertebrate Fossils*. McGraw Hill Book Company, Inc., New York. xiii + 766 pp.

Shimer, H. W. and R. R. Shrock. 1944. *Index Fossils of North America*. The MIT Press, Cambridge, Massachusetts. ix + 837 pp.

Yochelson, E. L. 2000. Concerning the concept of extinct classes of Mollusca: or what may/may not be a class of mollusks. *American Malacological Bulletin* **15**: 195-202.

CHAPTER 16

APLACOPHORA

AMÉLIE H. SCHELTEMA

16.1 INTRODUCTION

A discussion about the marine Aplacophora may seem like an anachronism in a book that is chiefly about shelled mollusks, for they lack a shell (*a-placo-phora*, bearing no shell) and are seldom collected except by great effort and expense. In fact, these vermiform organisms with their shining coats of innumerable calcium carbonate sclerites (scales or needle-like spicules) would not strike anyone at first view as even being mollusks. Nevertheless, upon examination, these creatures prove to be thoroughly molluscan, and they bear biogeographic information about their primary habitat-the deep sea-and are of great interest in studies of the evolutionary relationships of mollusks. A few aplacophorans have been collected just below the low tide mark, so you might be so fortunate as to find one. The small size, usually 3-20 mm, worm shape, and glistening, sclerite-covered body will immediately identify it as an aplacophoran mollusk and not any other invertebrate.

16.2 ORGANIZATION

Within their sclerite-covered, worm-like body, there is the true sign of a mollusk, a radula (lacking in some species). There is a mouth anteriorly and a small mantle cavity at the posterior end. There are further characteristics that separate the two aplacophoran taxa from each other.

The Neomeniomorpha (neomenioids), or Solenogastres (Figures 16.1 and 16.2), have a narrow ventral groove within which lies a foot; it is little more than a nonmuscular, ciliated ridge with which they glide. Above the mouth is a space open to the exterior with numerous sensory papillae. The gut is straight with a combined stomach and digestive gland. The paired hermaphrodite gonad (combined male and female) lies above the gut, and posteriorly there are often ornately shaped copulatory spicules of calcium carbonate. The neomenioids are carnivorous and feed mostly on hydroid and octocoral cnidarians. Their shapes vary from nearly spherical, to elongate with a broad or narrow body; size varies from 1 millimeter to 30 centimeters. One species from the continental shelf off the Falkland Islands is the size and shape of a bagel (but undoubtedly not as tasty).

Chaetodermomorpha (chaetoderms), or Caudofoveata (Figures 16.3 and 16.4), unlike the neomenioids lack both a foot and ventral furrow and the mouth is surrounded by a cuticular oral shield. Much of their internal anatomy is indicated by their body shape. The anterior head end is often set off from the main body by a constriction. The stomach is anterior to the digestive gland and gonad. The terminal mantle cavity is bell-shaped and contains a pair of gills. Chaetoderms are dioecious (males and females are separate) and presumably spawn their eggs and sperm into the water. They are chiefly carnivorous or omnivorous and feed on Foraminifera or other organisms and organic detritus as they burrow through the silt and mud of soft bottoms. Their size varies from 2 mm to more than 12 centimeters, and their elongate shape varies from broad to narrow.

Recent sources in English for a general account of the Aplacophora are Salvini-Plawen (1985a), Scheltema *et al.* (1994), and Scheltema (1998a). Descriptions of the aplacophoran fauna off southern

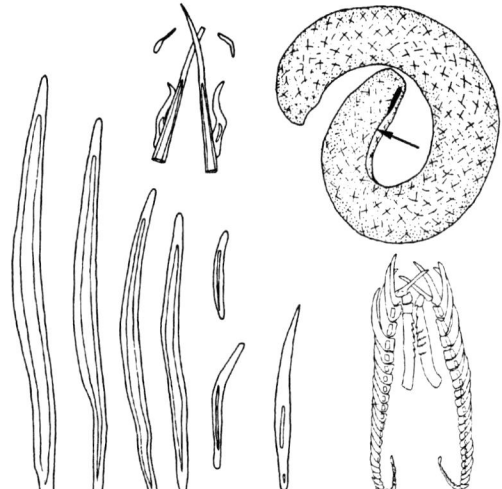

Figure 16.1 Neomeniomorpha (Solenogastres).
Upper right, *Simrothiella abysseuropaea* Salvini-Plawen, 2004, 10 mm long, from off Bergen, Norway; the arrow points to the ventral line of the foot groove. The crossed copulatory spicules on left above are 1.2 mm long, and body sclerites on left below are up to nearly 0.3 mm long. The radula on the lower right has two teeth per row; it is 2 mm in total length. [From Scheltema and Schander (2000), as *S. margaritacea*].

California can be found in Scheltema (1998b). A natural history has been published on one colorful, large tropical species (Scheltema and Jebb 1994), and observations of other living aplacophorans can be found in Salvini-Plawen (1968a, b). An aplacophoran web page is under construction (Woods Hole Oceanographic Institution 2004).

16.3 ECOLOGY

The first recorded aplacophoran, *Chaetoderma nitidulum* Lovén, 1844, was collected at 38 meters depth by dredge off the west coast of Sweden. It was not until 1875 that a second species, *Neomenia carinata* Tullberg, 1875, was described, also from Scandinavian waters. Since that time, more than 235 neomenioid and 130 chaetoderm species have been described from near- and offshore localities in all parts of the marine world from a few to more than 9,000 m depth. The list grows yearly.

The aplacophoran mollusks still remain a largely unknown group. More than 60% of the planet lies

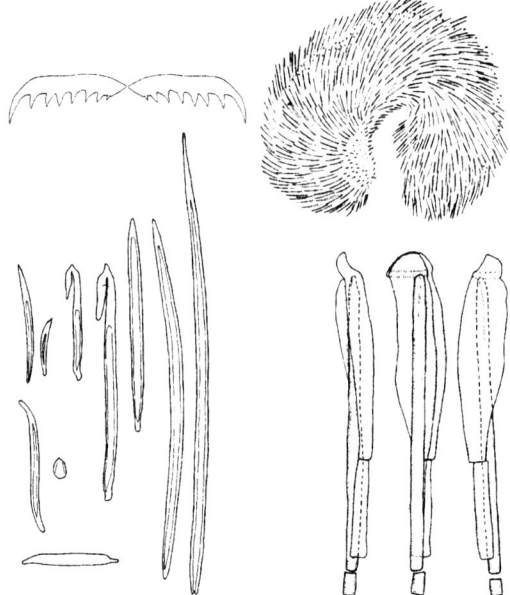

Figure 16.2 Neomeniomorpha (Solenogastres).
Upper right, a species of perhaps *Eleutheromenia* from a depth of 140 m in Bass Strait, Australia, 3 mm long; at upper left, a radula with one of 23 rows with two teeth 0.14 mm from left to right. At lower left, the longest of the sclerites from the body is 0.6 mm long, and the copulatory spicules to the right are nearly a millimeter long, or one-third the length of the animal. [From Scheltema and Schander (2000)].

beneath the sea surface at depths below 200 m, and the benthos, or bottom-dwelling fauna, of this vast area has scarcely been sampled. Aplacophorans of both taxa have been collected in greatest numbers and diversity from depths less than 3,500 m. Some species have very broad distributions (see Figure 16.4), and many genera are found worldwide. In particular localities, some species are the numerically dominant organism among the macrofaunal benthos (collected by sieving, too small to pick out by hand). Species have been found on and within level bottom sea-floor muds, on hydroids and octocorals around which neomenioids wrap themselves, on sea mounts (Scheltema 2001), in the Arctic and Antarctic polar regions (Salvini-Plawen 1978, and unpublished observations), from hydrothermal vents (Scheltema 2000), from whale bones (unpublished observations), and from oceanic trenches 7,000 to more than 9,000 m deep (Salvini-Plawen 1978, Scheltema 1985, Ivanov

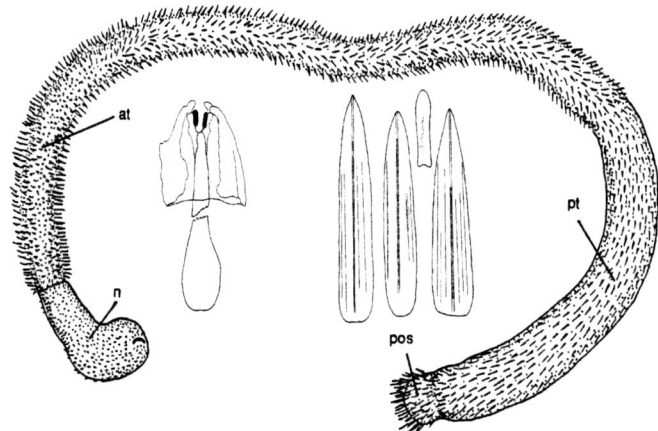

Figure 16.3 Chaetodermomorpha (Caudofoveata).
Chaetoderma elegans Scheltema, 1995 from off the coast of southern California between 50 and 1,800 m depth. Length of specimen 32 mm, with neck (n), anterior trunk containing stomach (at), posterior trunk with digestive gland and gonad (pt), and posterium with mantle cavity (pos). The radula, shown at left center, is 0.2 mm long; it comprises a cone and two denticles shown in black, typical for the family Chaetodermidae. Body sclerites are from various parts of the body and are up to 0.23 mm long. [From Scheltema (1998b)].

1996). Small aplacophorans 3 mm or less may be collected interstitially, living in the spaces between sand grains or broken shell hash (Morse 1979, Salvini-Plawen 1985b).

16.4 COLLECTION TECHNIQUES

As can well be imagined, most aplacophorans are now being taken by dredges, trawls, grabs, and cores from oceanographic research vessels (see Chapter 3, Remote Bottom Sampling). Collections are also made from submarines (Research Submersible Vessels, or RSVs), and readers may have seen the wonderful submarine dives on hydrothermal vent communities on their televisions. One can also dredge in relatively shallow waters from small ships, such as fishing vessels, if they are equipped to put over the side and retrieve a dredge or grab.

A quantity of mud is brought up, which then is sieved through 0.5 or 1.0 mm screens, preferably while still aboard the vessel using gentle flotation in seawater. If you have captured aplacophorans, they can be sorted from the rest of the organisms back on land. Techniques for collecting interstitial aplacophorans may be found in Morse and Scheltema (1988).

Diving is another way that aplacophorans have been collected recently. Divers have discovered several species on a variety of surfaces: on rocks, amongst turtle grass rhizomes, upon alcyonarian soft corals, and interstitially.

Aplacophorans from shallow depths can be kept alive and brought back in seawater; however, deep-water forms come from water of only 1-2°C and should be preserved as soon as possible. The rise in temperature, rather than decreased pressure, will kill the organisms quickly. Buffered formalin is a good general preservative, but preservation over time should be in buffered alcohol in order to preserve the sclerites. Borax is a good buffering agent; 1 teaspoon (10 g) to a pint (500 cm^3) of liquid gives a saturated solution.

16.5 EXAMINATION

Should you find an aplacophoran, you will do well to examine it under a dissecting microscope or, if unavailable, through a hand lens. Look first for a tell-tale ventral line that if present indicates you have a neomenioid (Figure 16.1). A cuticular oral shield, discrete differences in width along the body, or a tail-like posterior end indicates a chaetoderm (Figures 16.3 and 16.4). The beautiful sclerites, of all manner of shapes and sizes according to species (Figures 16.1-16.4), can be removed by needle into a drop of glycerin in a depression slide or, alternatively, onto a flat slide, dried, and permanently mounted with a commercially available histological mounting medium.

Good slide preparations of the sclerites are of great importance for identification, just as are the shells of other mollusks. In order to see the radula, you will need to cut off the anterior end and remove the tissue in household bleach (5% sodium hypochlo-

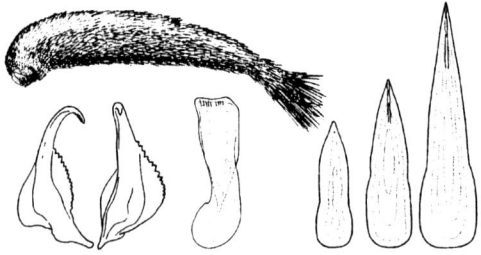

Figure 16.4 Chaetodermomorpha (Caudofoveata).
A common species, *Prochaetoderma yongei* Scheltema, 1985, found in the Atlantic Ocean at depths of 800 to 2,100 m from north of Cape Hatteras to Iceland and the entire length of the eastern Atlantic. Individuals are less than 3 mm long. There are two radular teeth per row, lower left, less than 0.1 mm in greatest dimension, and a pair of jaws up to 0.4 mm long, middle of lower row. Sclerites on right are up to 0.2 mm long. [From Scheltema (1985)].

rite). The radula is best kept wet by placing it in a drop of glycerin. Further descriptions on handling these animals may be found in Salvini-Plawen (1975), Scheltema (1998b), Scheltema and Ivanov (2000, 2004), Scheltema and Schander (2000), and Mizzaro-Wimmer and Salvini-Plawen (2001). You may wish to turn to an expert for identification, and you might discover that you have collected a new species!

16.6 LITERATURE CITED

Ivanov, D. L. 1996. *Chevroderma hadalis*, a new species of Prochaetodermatidae (Caudofoveata, Aplacophora) from the North-West Pacific. *Ruthenica* **6**: 83-84.

Mizzaro-Wimmer, M. and L. v. Salvini-Plawen. 2001. *Praktische Malakologie (Practical Malacology)*. Springer-Verlag, Wien. 187 pp.

Morse, M. P. 1979. *Meiomenia swedmarki* gen. et. sp. n., a new interstitial solenogaster from Washington, USA. *Zoologica Scripta* **8**: 249-253.

Morse, M. P. and A. H. Scheltema. 1988. Aplacophora. In: R. P. Higgins and H. Thiel, eds., *Introduction to the Study of Meiofauna*. Smithsonian Institution Press, Washington, D.C. Pp. 447-450.

Salvini-Plawen, L. v. 1968a. Über Lebendbeobachtjungen an Caudofoveata (Mollusca, Aculifera). *Sarsia* **31**: 105-126.

Salvini-Plawen, L. v. 1968b. Über einige Beobachtungen an Solenogastres (Mollusca, Aculifera). *Sarsia* **31**: 131-142.

Salvini-Plawen, L. v. 1975. Mollusca: Caudofoveata. *Marine Invertebrates of Scandinavia* **4**: 1-55.

Salvini-Plawen, L. v. 1978. Antarktische und subantarktische Solenogastres (eine Monographie: 1898-1974). *Zoologica (Stuttgart)* **44**: 1-315.

Salvini-Plawen, L. v. 1985a. Early evolution and the primitive groups. In: E. R. Trueman and M. R. Clarke, eds., *The Mollusca*, Vol. 10. Evolution. Academic Press, Inc., Orlando, Florida. Pp. 59-150.

Salvini-Plawen, L. v. 1985b. New interstitial Solenogastres (Mollusca). *Stygologia* **1**: 101-108.

Scheltema, A. H. 1985. The aplacophoran family Prochaetodermatidae in the North American Basin, including *Chevroderma* n.g. and *Spathoderma* n.g. (Mollusca: Chaetodermomorpha). *Biological Bulletin* **169**: 484-529.

Scheltema, A. H. 1998a. Class Aplacophora. In: P. L. Beesley, G. J. B. Ross, and A. Wells, eds., *Mollusca: the Southern Synthesis. Fauna of Australia*, Vol. 5, Part A. CSIRO Publishing, Melbourne. Pp. 145-157.

Scheltema, A. H. 1998b. Aplacophora. In: P. V. Scott and J. A. Blake, eds., *Taxonomic Atlas of the Benthic Fauna of the Santa Maria Basin and the Western Santa Barbara Channel*, Vol. 8. The Mollusca. Part I. Santa Barbara Museum of Natural History, Santa Barbara, California. Pp. 3-47.

Scheltema, A. H. 2000. Two new hydrothermal vent species, *Helicoradomenia bisquama* and *Helicoradomenia acredema* from the eastern Pacific Ocean (Mollusca, Aplacophora). *Argonauta* **14**: 15-25.

Scheltema, A. H. 2001. Neomenioid aplacophorans are numerically dominant on two East Pacific seamounts. In: L. Salvini-Plawen *et al.* eds., *Abstracts of the World Congress of Malacology 2001*. Unitas Malacologica, Vienna. P. 312.

Scheltema, A. H. and D. L. Ivanov. 2000. Prochaetodermatidae of the eastern Atlantic Ocean and Mediterranean Sea (Mollusca: Aplacophora). *Journal of Molluscan Studies* **66**: 313-362.

Scheltema, A. H. and D. L. Ivanov. 2004 Use of birefringence to characterize Aplacophora sclerites. *The Veliger* **47**: 153-156.

Scheltema, A. H. and M. Jebb. 1994. Natural history of a solenogaster mollusc from Papua New Guinea, *Epimenia australis* (Thiele) (Aplacophora, Neomeniomorpha). *Journal of Natural History* **28**: 1297-1318.

Scheltema, A. H. and C. Schander. 2000. Discrimination and phylogeny of Solenogaster species through the morphology of hard parts (Mollusca, Aplacophora, Neomeniomorpha). *Biological Bulletin* **198**: 121-151.

Scheltema, A. H., M. Tscherkassky, and A. M. Kuzirian. 1994. Aplacophora. In: F. W. Harrison and A. J. Kohn, eds., *Microscopic Anatomy of Invertebrates*, Vol. 5. Mollusca I. Wiley-Liss, New York. Pp. 13-54.

Woods Hole Oceanographic Institution. 2004. *The Taxonomy of the Aplacophora (Chaetodermomorpha or Caudofoveata and Neomeniomorpha or Solenogastres), Sclerite-Bearing Deep-Sea Mollusks*. <www.whoi.edu/science/B/aplacophora/>

CHAPTER 17

MONOPLACOPHORA

CLEMENT L. COUNTS, III

17.1 INTRODUCTION

The Monoplacophora are univalved, limpet shaped, untorted, mollusks with pseudometamerism of repeated organs and muscles. Though Abbott (1986) put them in his popular field guide, it is highly unlikely that you will ever encounter the shell of a living species of the molluscan class Monoplacophora. There is an even smaller chance of finding an entire living specimen. This is because living members of the class are found, thus far, only in the deep-sea and have been collected using equipment and facilities whose cost is beyond the financial wherewithal of all but the wealthiest of shell enthusiasts and the majority of professional biologists. This does not mean to say that you will never find examples of monoplacophorans from the deep-sea. In fact, the Monoplacophora can be found in three depth zones: the abyssal zone, the continental shelf, and the continental slope. The shallowest depth from which a monoplacophoran has been collected is 180 m. Warén and Gofas (1996) have reported specimens of *Veleropilina reticulata* (Segueanza, 1876) from the Tyrrhenian Sea off Italy's eastern coast that were in the hands of amateur collectors. This is clearly an unusual exception to the rule.

There is, however, a much better chance that fossilized specimens may be encountered as many species of monoplacophorans have been found in strata dating from the late Cambrian (about 500 million years old) to the Devonian (about 320 million years old) (Shimer and Shrock 1944, Knight and Yochelson 1960, Yonge 1960). Specimens have been taken from these strata the world-over and, until the 1950s, monoplacophorans were thought to be an extinct group of mollusks. Since the monoplacophorans were recognized as the fossil remains of limpet-like Gastropoda (e.g., *Acmaea euglypta* Dautzenberg and Fischer, 1897) by earlier malacologists, there was no need for something like the Class Monoplacophora.

In The Treatise on Invertebrate Paleontology, C. M. Yonge (1960) recorded that the Class Monoplacophora was erected because the anatomy of modern representatives, as reflected in fossilized remains, were distinctive enough to warrant their being recognized as a separate group within the Mollusca. What Yonge was talking about is the pairing of the ctenidia, muscles, and other internal organs seen in the (then) newly discovered living representatives. These pairings, particularly the 8 muscle scars of fossil species, led Wenz (1940) to speculate that the Monoplacophora were chitons (Polyplacophora) whose eight shell plates had fused and later to erect the Class Monoplacophora (Wenz in Knight 1952).

Until the discovery of a living species of Monoplacophora, *Neopilina galatheae* Lemche, 1957, all members of the class were believed to be extinct. Many zoologists were very excited about the discovery of living members of the class because some thought they may represent, or at least tell us something about, the oft-described archetypical ancestral mollusk. While speculating on the ancestral mollusk's anatomy is an interesting intellectual exercise, we probably do not know enough about exactly where the Mollusca came from to relate a newly-found species, even one as interesting as *Neopilina galatheae*, to the grand evolutionary development of the Phylum Mollusca.

Monoplacophorans have been popularly described as living fossils (Siekman 1987) but such an attribution may not be entirely correct. Monoplacophoran mollusks were represented in shallow waters of the early Paleozoic by a wealth of limpet-like and planispirally coiled forms. Shallow water marine environments can be physically more variable, and therefore more physiologically demanding, than those of deeper waters (Nybakken 2001). While the extinct species of monoplacophorans were found in strata formed by shallow waters, the first living specimens of the class seen by humans came from a deep-sea trench and this supported a long-held view that presumably extinct marine fauna might be found in the refugia of the deep-sea. The trenches represent an essentially unchanging environment: cold and dark with little variation in salinity and pressure. It is this environmental sameness over vast stretches of time that has led some to search for the living fossils of other extinct organisms, such as Trilobita, in the deep-sea on the chance that they too survive out of the reach of the vicissitudes of shallow water life.

While this is a hope that has not always come true for most extinct animals, living monoplacophorans have thus far all been taken from deep-sea environments of the Atlantic and Pacific oceans and the Mediterranean Sea. However, living monoplacophorans should not be thought of as living fossils but rather as a highly modified and specialized offshoot of the molluscan evolutionary tree (Schaefer and Haszprunar 1997).

17.2 BIOLOGY

The class Monoplacophora is distinguished by the presence of repeated internal and external structures. There are: four to eight pairs of foot retractor muscles; as many as six kidneys (nephridia) paired with nerve ganglia and three to six pairs of gills. The fossil specimens had muscle scars of paired retractor muscles but the presence of other paired structures was unknown until the living species were collected.

The biology of the Monoplacophora is very much a work in progress. The difficulty and expense of observing living specimens in their natural habitats seems to work against our direct study of these mollusks (Lowenstam 1978). While we have made great strides in determining the anatomical variation within the class, only now are we beginning to understand the biological processes and life histories of Monoplacophora.

17.2.1 Shell. The shell is a shallow, somewhat spoon-shaped dorsal structure, which at first glance looks like a limpet. There is an anteriorly directed apex and many are thin and fragile (Warén and Gofas 1996) (Figure 17.1). On the inner surface of the shell can be found the pairs of muscle scars, the number of which can be diagnostic for the species. Structurally, the shell of *Neopilina galatheae* was reported to consist of three shell layers: an outer periostracum, a thick underlying prismatic layer, and an inner lamellate nacreous layer (e.g., Hyman 1967). The sizes of monoplacophoran shells range from 3 mm in diameter [*Rokopella oligotropha* (Rokop 1972)] to *Neopilina galathea* at 37 mm. The surface sculpture may be coarse concentric ribs (genus *Veleropilina*), radial threads, reticulate ridges, concentric threads, or concentric undulations. There is evidence of an embryonic shell, called an apical cap in some works, which may be pitted. One specimen of *Neopilina galatheae* demonstrated a spiral larval shell, which was noted as being usually associated with torsion.

17.2.2 Locomotion. Aside from the shell, the foot is the largest anatomical structure of the Monoplacophora. *Neopilina galatheae* has a weakly muscled foot that is circular and Lemche (1957) thought this meant the animals lay on their back (on their shells) waiting for food to pass by. Warén and Gofas (1996), in their beautiful comparative study of several species of living monoplacophorans, noted that the size of the foot in relation to the size of the shell was variable and could depend on how and when observations of the foot are made. They found intraspecific variation depending upon whether the measurement was made while the animal was alive, preserved, or critical point dried for observation with a scanning electron microscope. They found that the foot of living monoplacophorans was usually twice as large as that of those subjected to critical point drying.

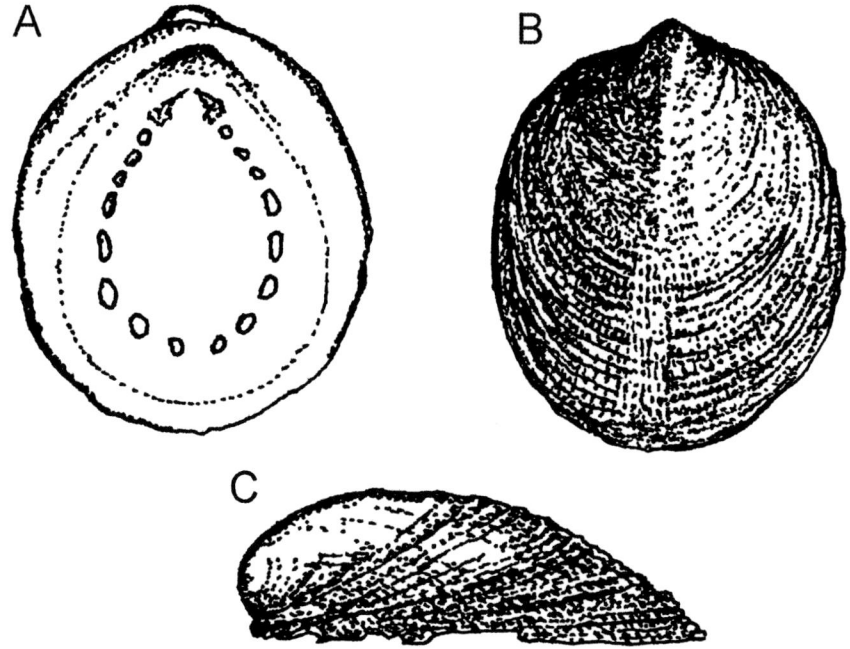

Figure 17.1 The Monoplacophoran Shell.
A. Interior view of the shell of *Neopilina galatheae* showing the eight pairs of foot retractor muscles characteristic of the species (after Lemche and Wingstrand 1959). B. Dorsal view of *Veleropilina reticulata* (after Warén and Gofas, 1996). C. Left lateral view of the shell of *V. reticulata* (after Warén and Gofas, 1996).

17.2.3 Digestion and diet. Briefly, the digestive tract of the monoplacophorans has an anterior mouth that is surrounded by fleshy palps. There is little in the way of head development (cephalization) in members of this class but the mouth is located at this head and in front of the foot. Just in front of, and behind the mouth is a fold of tissue (a preoral and postoral fold called the velum). The front fold becomes a ciliated palp or tentacle on either side of the mouth and the hind fold becomes a second pair of tentacles (Rupert and Barnes 1994).

The mouth contains a radula. The appearance of the radula is frequently used as a characteristic by which evolutionary relationships can be determined. The stomach of monoplacophorans is cone-shaped and contains a style sac in which the crystalline style is found but there is no gastric shield. The intestine is reported to be greatly coiled (usually between 4 and 6 loops), which is a characteristic of an unspecialized herbivore or deposit feeder (Morton and Yonge 1964). Hyman (1967) noted that the digestive system of monoplacophorans did not differ in any significant way from that of other mollusks. The digestive tract ends in a posterior anus.

Monoplacophorans are believed to be deposit feeders as the fleshy structures surrounding the mouth may function in gathering food (Allen 1983). Some of the reported foods of various monoplacophorans include protozoans, radiolarians, diatoms, foraminiferans, and sponges.

17.2.4 Gas exchange and water balance. Five pairs of gills are located along the foot in *Neopilina galatheae* and six pairs are found in members of the genus *Vema*. The number of gill pairs that is characteristic of Monoplacophora in a particular genus or species can change with new systematic arrangements and does not seem to be a stable characteristic of higher classification. Water flow over the gills is hypothesized to be from front-to-back with the water current exiting in a stream behind the anus.

17.2.5 Reproduction and development. The gonads of *Neopilina galatheae* are paired ventral structures. The sexes are separate and there are two pairs of ovaries and two pairs of testes connected by ducts to the third and fourth renal organs. Parental care of the young has been reported in *Micropilina arntzi* Warén and Hain, 1992, where they were reported to be brooded in the distal oviduct and the pallial groove and were released at a size of about 300 µm.

17.2.6 Substratum. The ecology of the Monoplacophora was first described by Menzies *et al.* (1959). Just by looking at the foot (broad) and shell (low with little hydrodynamic resistance) of a monoplacophoran, one would think that their usual habitat is something like the high-energy rocky intertidal area. Actually, *Neopilina galatheae* has been reported to occur on soft, dark clay or rocky substrata. The shell and foot shape may be an adaptation to current conditions or a relic of earlier habitats. Many species have been found near or on manganese or iron nodules. Others have been found amongst calcareous coral rubble, basalt, and muddy or rocky bottoms.

17.3 ZOOGEOGRAPHIC DISTRIBUTION

17.3.1 Fossil species. Fossil monoplacophorans are the species of monoplacophorans most likely to be encountered by collectors. There have been numerous fossil species described and several summaries of these have been produced (Shimer and Shrock 1944, Knight and Yochelson 1960, Strusz 1996). Paleontologists have studied and described many species of monoplacophorans. Some, after an in-depth morphological analysis, have suggested the Class Monoplacophora should be divided into Class Tergomya and Class Helcionelloida (e.g., Peel 1991). Additionally, it has been hypothesized that one genus *Scenella*, placed in the Class Helcionelloida, is in fact a chondrophorine or floating coelenterate (Yochelson and Gil Cid 1984, Briggs *et al.* 1994).

17.3.2 Living species. Hyman (1967) stated there were four known species of Monoplacophora. Keen (1971) listed three species of *Neopilina* from tropical eastern Pacific waters and two of them she placed in the subgenus *Neopilina* (defined by having five pairs of gills). In the last edition of American Seashells published by Abbott (1974), no species of monoplacophorans were discussed at all. Batten (1984) provided a systematic list of eight species distributed equally among the genera *Neopilina* and *Vema*. Rupert and Barnes (1994) reported eleven species in three genera: *Neopilina*, *Vema*, and *Monoplacophorus*. Warén and Gofas (1996) reported eighteen known living species of Monoplacophora in eight genera. They based their determination on a number of factors including radular tooth shape, number of gills, muscle scars, shell shape and sculpture, and the position of the apex. A brief synopsis of the distribution of species within the class is given below in Table 17.1.

17.4 COLLECTING AND STORAGE TECHNIQUES

17.4.1 Recent Monoplacophora. Warén and Gofas (1996) noted that small mollusk shells can be damaged by exposure to ethanol due to the formation of complex water-soluble ions with the alcohol. This will cause decalcification and perhaps recrystallization of the shell's calcium resulting in total destruction of the shell in a few months' time. They went on to note that higher concentration of ethanol (90%) may slow this chemical process but would make soft tissues more difficult to examine histologically. Switching to another type of alcohol (methanol, isopropanol, etc.) would have a faster chemical reaction than does ethanol. Their suggestion is to store the shells dry and preserve the soft tissues in 70% ethanol. If the use of transmission electron microscopy (TEM) is contemplated, consider the techniques of Haszprunar *et al.* (1995), who used preservation of tissues in the field in 2.5% gluteraldehyde that had been buffered in 0.1 M cacodylate buffer.

If I may be permitted a personal plea, remember that living monoplacophora are rare entities. If you should collect one or more outside of a scientific endeavor, please consider donating them to a museum or scientist that is working on this group.

Table 17.1. Species of Monoplacophora.

Taxon	Locality	Depth	Remarks
Neopilina			
N. galatheae Lemche, 1957	Pacific Ocean off of Costa Rica and Mexico	2,739-3,570 m	The first living monoplacophoran discovered.
N. bruuni Menzies, 1968	Peru Trench	4,825 m	This taxon is known from a single specimen.
Laevipilina			
L. rolani Warén and Bouchet, 1990	Atlantic Ocean off the coast of Spain	840 m	A single specimen was found on manganase nodule.
L. antarctica Warén and Hain, 1992	Weddell Sea	210-650 m	This taxon is known to have bacterial epidermal symbionts.
L. hyalina (McLean, 1979)	Santa Rosa - Cortes Ridge	229-388 m	This taxon is known from three localities.
Laevipilina cachuchensis Urgorri, García-Álvarez, and Luque, 2005	off north Iberian Peninsula	580-600 m	Known from two live collected specimens.
Rokopella			
R. oligotropha (Rokop, 1972)	680 miles north of Hawaii	6,065-6,079 m	
R. euglypta (Dautzenberg and Fischer, 1897)	Mid-Atlantic Ridge and seamounts from 34-38°N latitude	1,200-1,600 m	Originally described as an *Acmaea* in 1897.
R. brummeri Goud and Gittenberger, 1993	east of Mid-Atlantic Ridge	2,162 m	
Veleropilina			
V. veleronis (Menzies and Layton, 1963)	near Cedros Island, Mexico	2,730-2,769 m	
V. zografi (Dautzenberg and Fischer, 1896)	Mid-Atlantic Ridge and seamounts from 30-38°N latitude	600-1,400 m	The first living monoplacophoran collected in the Atlantic Ocean, 1896.
V. reticulata (Seguenza, 1876)	Mediterranean and Tyrrhenian Seas	180-600 m	Collected from Pliocene-Pleistocene deposits; living specimens possible but not yet found.
Micropilina			
M. minuta Warén, 1989	Wyville-Thomson Ridge off Scotland		Was reported from Pleistocene sites in southern Italy.
M. rakiura Marshall, 1998	New Zealand		
M. arntzi Warén and Hain, 1992	Weddell Sea	210-650 m	The smallest monoplacophoran in the Antarctic Ocean.
M. tangaroa Marshall, 1990	New Zealand		
Vema			
V. levinae Warén and Gofas, 1996	12°56'N, 103°29'W	1,058 m	This taxon was described from fragmentary materials that consisted of radula, pieces of shell and soft parts.
V. ewingi (Clarke and Menzies, 1959)	Peru-Chile Trench	5,821 m	This taxon was found at nine locations, all within the Peru-Chile Trench.
Adenopilina			
A. adenensis (Tebble, 1967)	off Oman, Gulf of Aden	3,000-3,950 m	This taxon was described from a single specimen.

These organisms are so rare that you are encouraged not to keep them in a private collection but instead deposit them in a public collection where they are available to the research community. Your satisfaction will be in making a lasting and significant contribution to science.

17.4.2 Fossil Monoplacophora. Fossil materials should be appropriately identified as to the geologic strata from which the specimens were taken, and detailed locality data. As anyone who has looked at fossils available through commercial dealers can attest, the locality data usually accompanying their specimens is uncommonly poor. What you need to record is geographic coordinates, state and county of occurrence, strata, depth, geologic age, nearest city, town or other identifiable landmark, nearest road, date of collection, collector(s), fossil floral and faunal associations, and any other data that will increase the scientific value of the materials.

17.5 LITERATURE CITED

Abbott, R. T. 1974. *American Seashells*, 2nd Ed. Van Nostrand Reinhold, New York. 663 pp.

Abbott, R. T. 1986. *Seashells of North America: A Guide to Identification*. Golden Press, New York. 280 pp.

Allen, J. A. 1983. The ecology of deep-sea molluscs. *In:* W. D. Russell-Hunter, ed., *The Mollusca*, Vol. 6. Ecology. Academic Press, Orlando, Florida. Pp. 29-75.

Batten, R. L. 1984. *Neopilina*, *Neomphalus*, and *Neritopsis*, living fossil molluscs. In: N. Eldridge and S. M. Stanley, eds., *Living Fossils*. Springer Verlag, New York. Pp. 218-224.

Briggs, D. E. G., D. H. Erwin, and F. J. Collier. 1994. *The Fossils of the Burgess Shale*. Smithsonian Institution Press, Washington, D.C. xvii + 238 pp.

Haszprunar, G., K. Schaefer, A. Warén, and S. Hain. 1995. Bacterial symbionts in the epidermis of an Antarctic neopilinid limpet (Mollusca, Monoplacophora). *Philosophical Transactions of the Royal Society of London* B **347**: 181-185.

Hyman, L. H. 1967. *The Invertebrates*. Vol. VI: Mollusca I. McGraw-Hill Book Company, New York. vii + 792 pp.

Keen, A. M. 1971. *Sea Shells of Tropical West America: Marine Mollusks from Baja California to Peru*, 2nd Ed. Stanford University, Stanford, California. 1064 pp.

Knight, J. B. 1952. Primitive fossil gastropods and their bearing on gastropod classification. *Smithsonian Miscellaneous Collections* **117**: 1056.

Knight, J. B. and E. L. Yochelson. 1960. Monoplacophora. *In:* R. C. Moore, ed., *Treatise on Invertebrate Paleontology*, Part I: Mollusca 1. Geological Society of America and University of Kansas Press, Lawrence, Kansas. Pp. 177-184.

Lemche, H. 1957. A new living deep-sea mollusc of the Cambro-Devonian class Monoplacophora. *Nature* **179**: 413-416.

Lemche, H. and K. G. Wingstrand. 1959. The anatomy of *Neopilina galatheae* Lemche, 1957 (Mollusca: Tryblidiacea). *Galathea Reports* **3**: 9-72.

Lowenstam, H. A. 1978. Recovery, behavior, and evolutionary implications of live Monoplacophora. *Nature* **213**: 231-232.

Menzies, R. J., M. Ewing, J. L. Worzel, and A. H. Clarke. 1959. Ecology of Recent Monoplacophora. *Oikos* **10**: 168-182.

Morton, J. E. and C. M. Yonge. 1964. Classification and structure of the Mollusca. *In:* K. M. Wilbur and C. M. Yonge, eds., *Physiology of the Mollusca*, Vol. I. Academic Press, New York. Pp. 1-58.

Nybakken, J. W. 2001. *Marine Biology: An Ecological Approach*. Benjamin Cummings, San Francisco. xi + 516 pp.

Peel, J. S. 1991. The classes Tergomya and Helcionelloida, and early molluscan evolution. *Bulletin Gronlands Geologiske Undersokelse* **161**: 11-65.

Rupert, E. E. and R. L. Barnes. 1994. *Invertebrate Zoology*, 6th Ed. Saunders College Publishing, Fort Worth, Texas. xii + 1056 pp.

Schaefer, K. and G. Haszprunar. 1997. Anatomy of *Laevipilina antarctica*, a monoplacophoran limpet (Mollusca) from Antarctic waters. *Acta Zoologica Stockholm* **77**: 295-314.

Shimer, H. W. and R. R. Shrock. 1944. *Index Fossils of North America*. The M.I.T. Press, Cambridge, Massachusetts. ix + 837 pp.

Siekman, L. 1987. *Neopilina*, a living fossil. *Hawaiian Shell News* **35**: 7, 8.

Strusz, D. L. 1996. CPC Catalogues: 6. Catalogue of type, figured and cited specimens in the Commonwealth Paleontological Collection. Palaeozoic Mollusca and Hyolitha. *Australian Geological Survey Organisation Record* **49**: iii + 130 pp.

Warén, A. and S. Gofas. 1996. A new species of Monoplacophora, redescription of the genera *Veleropilina* and *Rokopella*, and new information on three species of this class. *Zoologica Scripta* **25**: 215-232.

Wenz, W. 1940. Ursprung und frühe Stammesgeschichte der Gastropoden. *Archiv für Molluskenkunde* **72**: 1-110.

Yochelson, E. L. and M. D. Gil Cid. 1984. Reevaluation of the systematic position of *Scenella*. *Lethaia* **17**: 331-340.

Yonge, C. M. 1960. General characteristics of Mollusca. *In:* R. C. Moore, ed., *Treatise on Invertebrate Paleontology*, Part I: Mollusca 1. Geological Society of America and University of Kansas Press, Lawrence, Kansas. Pp. 13-136.

CHAPTER 18

POLYPLACOPHORA

ENRICO SCHWABE
ANDREAS WANNINGER

18.1. INTRODUCTION

The Polyplacophora, commonly known as chitons, are a group of morphologically conservative, exclusively marine mollusks. The interest in this group has grown, probably due to the monograph series of Kaas and Van Belle (1985a-1994), a contribution that together with their species catalogue (Kaas and Van Belle 1998) allows a rather easy identification of the most common taxa. Nevertheless, identification down to the species level is difficult, since examination of features like radula morphology, ctenidium arrangement, and girdle elements are required.

General information on the anatomy and morphology of chitons is provided by Kaas and Van Belle (1985a) and Eernisse and Reynolds (1994). This chapter summarizes the most important aspects of polyplacophoran biology, life history, anatomy, and ontogeny.

18.2. GROSS MORPHOLOGY

Although a very old group, with the first representatives occurring as early as in the Cambrium (Yates *et al.* 1992), the external anatomy of the chitons has not changed significantly over time. Chitons are unsegmented (in the sense of an annelid-like segmentation), bilateral symmetrical, dorso-ventrally flattened animals with eight, usually overlapping, plates that may be reduced (e.g., *Amicula*) (Figure 18.1F). The body shape ranges from broad oval (e.g., *Placiphorella*, Figure 18.1A) to wormlike (e.g., *Cryptoplax*, Figure 18.1C). Chitons are able to roll up ventrally.

The broad, fleshy foot is situated ventrally and is surrounded by the mantle cavity. Chitons lack a true head, their cephalic region is eyeless and without tentacles. The anus is situated posteriorly, from where rows of ctenidia extend laterally on both sides of the foot in an anterior direction (Figure 18.2B), either without a space between the anus and the last ctenidium (in the order Lepidopleurida) or with a small space between the ctenidia and the anus (in the order Chitonida).

What is remarkable is the rather broad, fleshy perinotum (girdle), which in most cases is dorsally densely beset with different kinds of protective (scales, bristles, hairs, needles, corpuscles) and sensory organs. Ventrally, the cuticle is normally covered with scales but may rarely be naked (*Ferreiraella*).

Chitons vary in length from a few millimeters to about 15 cm. The world's largest species, however, is the North Pacific *Cryptochiton stelleri* (von Middendorff, 1847), which may attain a length of more than 30 cm (Figure 18.2A).

18.2.1 The plates (valves). The plates are four-layered. Uppermost, a thin aragonite layer (properiostracum) is found, which differs from the conchiferan periostracum. Usually, the second layer (tegmentum) is of the highest taxonomic relevance. It is often colorful and mostly intricately sculptured. It consists of inorganic as well as of organic material and is penetrated by numerous branching canals, in which the nerves of the aesthetes (sensory organs) are situated. The latter have photosensory

The third layer, also of aragonitic origin, extends ventrally under the tegmentum in anterior and lateral direction. In the anterior regions of the intermediate valves and the tail valve, the so-called articulamentum builds two thin processes, the apophyses, which are overlapped by the preceding valve. Laterally, they form slitted or unslitted insertion plates, which connect the valves to the girdle. The nerves that innervate the aesthetes run within the pores of the radially arranged slit rays. The fourth plate-layer (myostracum) is very thin and forms the attachment site of the dorso-ventral musculature.

For private collections, abnormal plates are often of significant interest. An extensive list of all abnormal chitons reported so far is found in Dell'Angelo and Tursi (1990). Four cases of abnormality can be distinguished: hypomerism - less than the regular eight plates; hypermerism - more than the regular eight plates; coalescence - fusion of two adjacent plates; splitting - the division of a plate into several parts (Schwabe 2001).

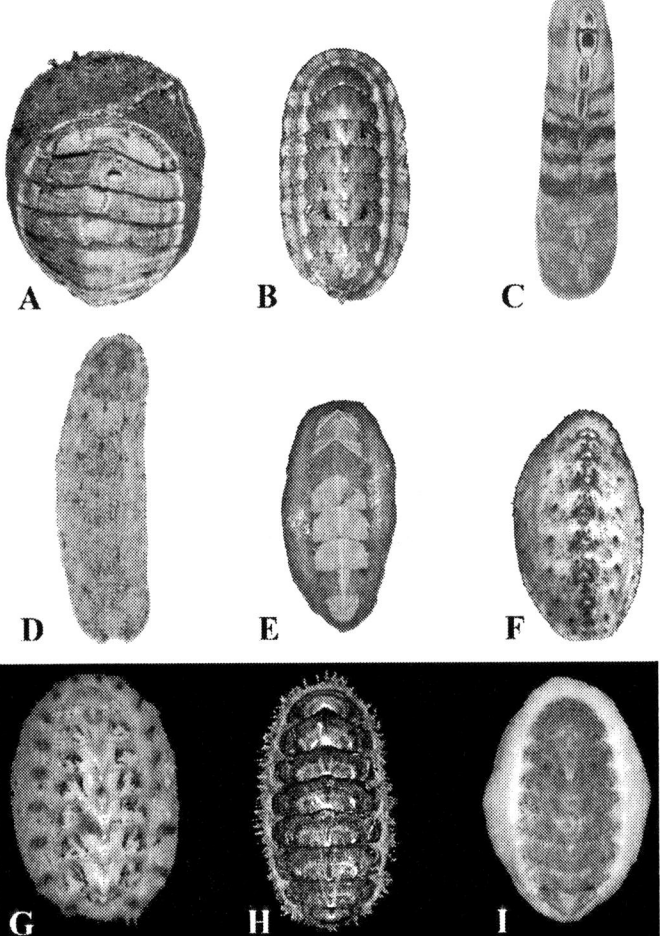

Figure 18.1 Polyplacophora.
A. *Placiphorella atlantica* (Verrill and S. I. Smith in Verrill, 1882), 14.8 mm, North Atlantic (coll. Zoologisches Museum Hamburg); B. *Chiton (Tegulaplax) hululensis* (E. A. Smith, 1903), 23.2 mm, Egypt (coll. E. Schwabe); C. *Cryptoplax larvaeformis* (de Blainville MS, Burrow, 1815), appr. 60 mm, Indonesia (coll. H. Strack); D. *Schizochiton incisus* (Sowerby, 1841), appr. 50 mm, Indonesia (coll. H. Strack); E. *Nuttallochiton mirandus* (E. A. Smith MS, Thiele, 1906), appr. 40 mm, Antarctica (coll. Zoologische Staatssammlung Muenchen); F. *Amicula vestita* (Broderip and Sowerby, 1829), 44 mm, East Sibirian Sea (coll. E. Schwabe); G. *Acanthochitona leopoldi* (Leloup, 1933), 5.4 mm, Indonesia (coll. E. Schwabe); H. *Chaetopleura peruviana* (Lamarck, 1819), 50 mm, Chile (coll. E. Schwabe); I. *Tonicia (Lucilina) sowerbyi* Nierstrasz, 1905, 9.8 mm, Indonesia (coll. E. Schwabe).

and probably additional mechano- and chemosensory functions. The aesthetes may bear large lenses (e.g., *Tonicia*) or they may be black pigmented (e.g., *Callochiton*). The tegmentum layer is divided into distinct areas as shown in Figures 18.2 D-E.

18.2.2 The perinotum (girdle). A thick cuticle covers the fleshy perinotum. Dorsally, it may bear different kinds of elements such as imbricating calcareous scales (*Ischnochiton*), aragonitic spines (*Acanthopleura*), corneous bristles (*Mopalia*), small calcareous corpuscles (*Lepidochitona*), as well as sensory organs. Ventrally, the girdle is paved by radiating rows of more or less rectangular scales, except in the deep-water genus *Ferreiraella*, where it is naked. The size of the perinotum varies considerably and may be very small as in members of Leptochitonidae, or extremely anteriorly extended as in members of *Placiphorella*. It may also cover almost the whole animal, as in *Cryptochiton stelleri*. There are several species with a notched terminal girdle, e.g., the members of the Indo-Pacific family Schizochitonidae. For further information on girdle morphologies see Fischer *et al*. (1988), Leise (1988), and Eernisse and Reynolds (1994).

18.2.3. The ctenidia (gills). Although the term gill is commonly used, this refers to a respiratory organ in a strict sense. The basal molluscan ctenidium, however, is a ventilation organ, which may acquire a secondary function for gas exchange (e.g., in higher gastropods). Since chitons usually show the primary condition, the term "ctenidium" is used herein.

The ctenidia are situated in the mantle cavity on both sides of the foot (Figure 18.2 B). They may reach from the anus towards the head region (holobranchial) or may be restricted to the posterior region (merobranchial). In members of the suborder Lepidopleurina only, the ctenidia are in direct contact with the anus. If the largest ctenidium (or the group of largest ctenidia) is the most posterior one, it is usually referred to as the abanal type, in contrast to the adanal type, which is defined by the occurrence of smaller ctenidia between the large ctenidia and the anus (for a detailed definition, see Sirenko 1993). Recent data on ctenidium morphology and function are found in Russell-Hunter (1988), Fischer *et al.* (1990), Sirenko (1993), and Lundin and Schander (2001).

18.3. ANATOMY

It would exceed the framework of this overview to provide a detailed review of the whole polyplacophoran anatomy. Useful information can be found in Plate (1897, 1899), Fischer-Piette and

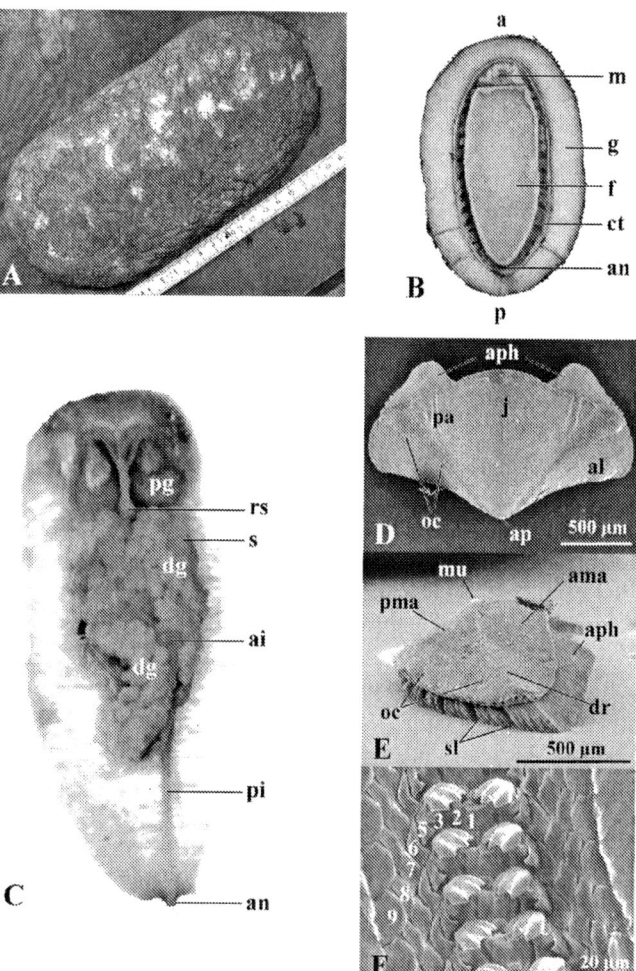

Figure 18.2 Polyplacophoran anatomy.
A. *Cryptochiton stelleri* (von Middendorff, 1847), 224 mm, North-East Pacific (coll. Zoologische Staatssammlung Muenchen); B. *Liolophura japonica* (Lischke, 1873), 36 mm, Japan (coll. E. Schwabe); C. *Craspedochiton* sp., 5.8 mm (length of the illustrated digestive tract), Thailand (coll. Phuket Marine Biological Center Reference Collection); D. *Tonicia indica* Leloup, 1981, second valve - dorsal view, Mascarene Islands (coll. National Museum and Galleries of Wales - NMGW); E. as fig. D, tail valve - lateral view; F. *Callochiton deshayesi* Thiele, 1909, radula, Mascarene Islands (coll. NMGW). Abbreviations: a - anterior; ai - anterior intestine; al - area lateralis; ama - antemucronal area; an - anus; ap - apex (beak); aph - apophyses; ct - ctenidia; dg - digestive gland ; dr - diagonal ridge; f - foot; g - girdle (perinotum); j - jugum (jugal area); m - mouth; mu - mucro; oc - ocelli (lenses); p - posterior; pa - pleural area (pa + j = central area); pg - pharyngeal gland; pi - posterior intestine; pma - postmucronal area; rs - radula sac; s - stomach; sl - slit; 1 - central tooth; 2 - first lateral tooth; 3 - second or major lateral tooth; (4) - first uncinal tooth (covered by the second uncinal tooth in this view); 5 - second uncinal tooth; 6 - third or spatulate uncinal tooth; 7 - first marginal tooth; 8 - second marginal tooth; 9 - third marginal tooth.

Franc (1960), Yonge (1960), and Eernisse and Reynolds (1994). The latter includes references to many other works.

18.3.1. The radula. Chitons are grazers that use their radulae to scrape the substrate. The moderately large chiton radula can reach a length of at least one third of the body length. Approximately 25 to 150 rows of teeth are situated on a thin radula membrane, usually with 17 teeth per row (in members of the genera *Micichiton, Nanichiton*, and *Juvenichiton* the number is reduced to 11 to 13 teeth per row). Eernisse and Kerth (1988) have shown that tooth rows, in recently metamorphosed juveniles, lack several teeth. The single tooth row is symmetric and half a row consists of the central tooth, the first lateral tooth, the major lateral tooth, two uncinal teeth, a spatulate uncinal tooth, and three marginal teeth (Figure 18.2 F). The shapes of the three centrally situated teeth are major distinctive characters. The most dominant of them, the major lateral tooth, bears a large magnetite-capped blade, which is hardened and allows for scraping on hard substrates. The spatulate uncinal tooth seems to be functionally linked to the major lateral tooth, since in most cases its shape matches a posterior depression in the blade. An outstanding reference list of papers dealing with radula formation can be found in Eernisse and Reynolds (1994). Additional data are provided by Sirenko (1974), Evans *et al.* (1994), Matsukuma and Tsubaki (1995), Evans and Alvarez (1999), Brooker and Macey (2001), and Saito (2004).

18.3.2 The digestive system. The mouth lies ventrally in the center of the cephalic region and leads to the buccal tube (cavity) of the mouth, which posteriorly forms blind subradular and radular sacs ending just beneath the paired subradular organ, which serves as a taste organ. Where the mouth cavity passes into the gullet, it is equipped with mucus secreting salivary glands. The strong, muscular-walled gullet is strengthened with chitin and forms a firm underlayer for the radula. From the sac-shaped radular sheath, which extends ventrally from the digestive tract as far as the third or fourth valve, the radula emerges distally into the pharynx. A very complicated muscular system permits the radula to move back and forth across a pair of connective tissue bars (bolsters). The pharyngeal glands (sugar glands) are situated laterally to the pharynx. The stomach is a rather large sac, shaped by the surrounding lobes of the mid-gut and the bilobed digestive gland. The smaller right lobe extends anteriorly over and around the anterior part of the stomach, while the larger left lobe is situated above the posterior end of the stomach. The two lobes open separately into the stomach just before it passes into the looped intestine. The intestine, about four or five times the length of the animal, is separated by a sphincter into an anterior and a posterior portion. Contraction of the anterior sphincter musculature isolates a portion of the food string in the posterior part. It rotates the material into a firm fecal pellet. The posterior intestine is extensively coiled and lined with a ciliated, mucus-producing epithelium. It leads to the rectum, a short ciliated tube that passes through the body musculature, and opens to the exterior through the anus (Figure 18.2 C). For detailed illustrations and comments on the digestive tract and other organs, see Plate (1897, 1899).

In general, the diet of chitons consists of diatoms, detritus, and fleshy and encrusting algae. However, several specialist feeders are known: species of *Ferreiraella* feed on sunken wood, probably using bacteria or fungi to digest the cellulose (see Sirenko 2001); carnivorous members of *Placiphorella*, *Craspedochiton*, and *Loricella* hunt their prey (mostly small crustaceans) by a rapid clamping movement of their precephalic tentacles or anterior mantle edges (McLean 1962, Ludbrook and Gowlett-Holmes 1989). Other species are associated with hydrothermal vents (Saito and Okutani 1990), or exclusively feed on sponges (e.g., some *Notoplax* species) or sea grass (*Stenochiton*).

18.3.3 Nervous system and sensory organs. The following is a brief description of the neuronal anatomy and a short overview of the sensory organs that occur in adult chitons, together with the most important references for more detailed descriptions.

The amphi-tetraneuran, cordlike nervous system of chitons consists of paired lateral and pedal nerve cords. They are interconnected by lateropedal commissures and, between the pedal nerve cords, by pedal commissures. Each ctenidium is joined by two nerves (associated with venous and arterial blood sinuses), which arise from the pedal nerve cord (Fischer et al. 1990). A wide cerebro-buccal ring is found in the cephalic region of the animal. Dorsal to this, a nerve ring is situated, which forms a pair of esophageal ganglia (which are interconnected by the dorsal esophageal commissure) and a more posteriorly situated supraradular ganglion. This nerve ring is connected by buccal connectives with the cerebro-buccal ring. The latter also innervates a pair of subradular ganglia.

The best-known polyplacophoran sensory organs are the aesthetes, which are embedded in the tegmentum and are innervated by very fine branches of nerve endings. Moseley (1885) was the first to discover this organ system. He was able to separate large macroaesthetes from the numerous microaesthetes, based on the size of the pores in the shell surface. Aesthetes show significant ultrastructural differences from species to species. They were examined by several workers (Blumrich 1891, Boyle 1974, Fischer 1978, 1979, 1988, Baxter et al. 1990, Sturrock and Baxter 1995, Reindl et al. 1997) and a summary of their morphology, histology, and function is found in Eernisse and Reynolds (1994). Many aesthetes serve as photosensory organs but several additional functions such as mechano- or chemoreception are nowadays discussed.

Light microscopic examinations of the valve surface show different kinds of macroaesthetes. The pores may be even with the valve surface (*Tonicella, Boreochiton*), elevated (*Nierstraszella, Ferreiraella*), or may be situated on granules (*Hanleya, Lepidopleurus, Leptochiton*) and they may contain black pigmentation, the so-called "black shell eyes" (*Callochiton*), or lenses (*Schizochiton, Tonicia, Onithochiton*) (Figure 18.2 D-E). Microaesthetes can be arranged around the macroaesthetes (*Hanleya, Acanthochitona*), around the base of granules (partly in *Leptochiton*), or longitudinally in fine grooves (e.g., *Tonicella, Callochiton*). An unnamed sensory organ embedded in the valves of *Cryptoplax mystica* Iredale and Hull, 1925 (see Currie 1992), with a similar structure to other aesthetes might serve as a mechanoreceptor or as a balance organ.

The perinotum also bears numerous sensory organs, which might function as mechanoreceptors or photoreceptors (Fischer et al. 1980, Fischer et al. 1988, Leise 1988). These have been shown to have a structure similar to aesthetes and are probably homologous. Sensory organs are also found in the mantle cavity. Ischnochitonids and acanthochitonids bear a posterior osphradial organ. In contrast, lepidopleurids show branchial and lateral sense organs that do not appear to be homologous to these osphradia. The lateral sense organs are situated in the outer wall of the mantle cavity, distal to the ctenidia. They are a series of small patches innervated by the lateral nerve cord. The branchial sense organs, innervated by the ctenidial nerves, are situated in the inhalent chamber.

The function of the osphradia is probably a chemosensory one (Plate 1901). Haszprunar (1987), who showed that the sense organs in the mantle cavity vary positionally and ultrastructurally within the Polyplacophora, proposed a reproductive role involving synchronization of male and female spawning. Kamardin (1989) suggested that the osphradium is needed for chiton homing.

18.3.4 The body musculature. Adult Polyplacophora show numerous complicated muscle systems, which have been described in detail by Wingstrand (1985) and are summarized in Haszprunar and Wanninger (2000). The most striking muscular elements are the eight paired sets of dorso-ventral muscle units, which insert at the shell plates and each consists of various distinct subunits of mainly obliquely running muscle bundles. As is diagnostic for Mollusca, the shell plate muscles intercross ventrally, thus forming a muscular basket in which the soft parts of the animal are embedded (see Wingstrand 1985: figure 26). A polyplacophoran apomorphy is the dorsally situated musculus rectus, which runs underneath the shell plates in anterior-posterior direction. Laterally, the body is engulfed by the circular enrolling muscle, which allows chitons to coil up ventrally if separated from the substratum. Additional muscle systems include the complicated buccal apparatus, transverse muscle cushions underneath each shell plate, and the massive musculature of the foot, mantle, and girdle.

18.4 REPRODUCTION, LIFE HISTORY, AND ORGANOGENESIS

Most polyplacophoran species have separate sexes and are free spawners with external fertilization. This is regarded as basal for the class (as it is for the entire Mollusca), but exceptions do occur. Accordingly, brooding is observed in a few genera (a summary of all brooding species is given in Strack 1987) such as *Lepidochitona* (see Eernisse 1984, 1986) or *Onithochiton* (see Creese 1986), and even ovovivipary [*Calloplax vivipara* (Plate, 1899)] and hermaphroditism [*Lepidochitona caverna* and *L. fernaldi* (Eernisse 1986)] have been described. The life cycle of basal chitons is summarized in Figure 18.3.

The hulls of mature eggs often show a characteristic, species-specific ornamentation (Figure 18.3 A) and are usually red or green in color. Cleavage is spiral, and a free-swimming, non-feeding (i.e., lecithotrophic) trochophore-like larva hatches. It shows a distinct prototroch, an apical ciliary tuft, and an apical organ (i.e., a larval apical sensory system including the cells of the apical tuft) (Figure 18.3 B). The planktonic stage can last from a few hours to more than a week. Towards metamorphic competence, the body elongates and the anlagen of the first seven shell plates form along with the first girdle spicules (Figure 18.3 C, D; see also Wanninger and Haszprunar 2002). Early juveniles still lack the eighth valve, which can take several months to develop. Recent gene expression pattern analyses indicate that the homeobox gene *engrailed* is involved in polyplacophoran plate formation, as it is in the morphogenesis of bivalve, gastropod, and scaphopod embryonic but not adult shells (Moshel *et al.* 1998, Jacobs *et al.* 2000, Wanninger and Haszprunar 2001). These data, however, appear not yet sufficient to answer fully the question regarding the homology of chiton valves and conchiferan shells.

One pair of protonephridia form the larval excretory system and they persist in the juvenile stage (personal observation). However, it is still unknown whether they are resorbed or remodeled eventually to form the adult kidney in later stages. Chiton larvae develop one pair of photosensory organs (ocelli) just behind the prototroch, which are histologically distinct from the aesthetes of the adults (Fischer 1978, 1980). They are innervated by the pedal nerve cord (Eernisse and Reynolds 1994), are carried over into early juvenile stages, and are eventually lost.

In addition, a unique larval sensory system comprising so-called "ampullary cells" and mainly

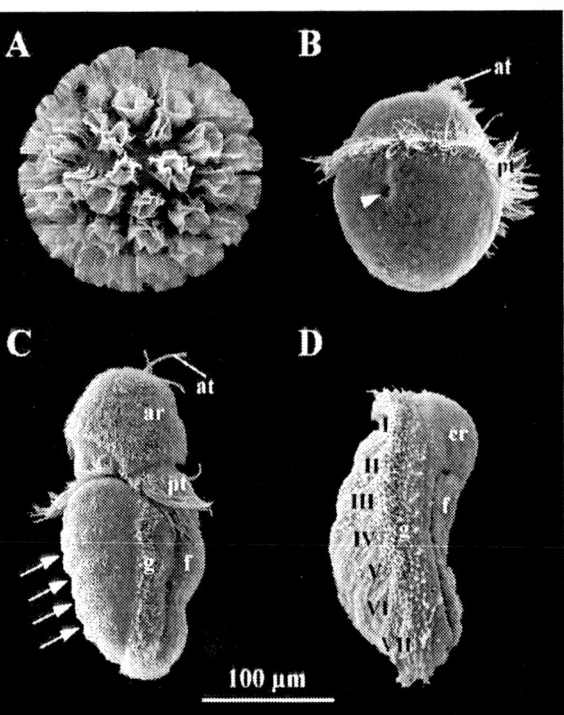

Figure 18.3 Polyplacophoran ontogeny (*Mopalia muscosa*).
A. Recently spawned mature egg with richly ornamented hull (image kindly provided by S. Friedrich, Muenchen). B. Early trochophore larva immediately after hatching with distinct apical tuft (at), prototroch (pt), and blastopore (arrowhead). C. Metamorphic competent larva with prominent apical region (ar) and anlagen of the adult shell plates (arrows), girdle (g), and foot (f). D. Early juvenile showing the first seven shell plates (I-VII) and the girdle (g) with spines. The larval apical region including the apical tuft has been reduced and is replaced by the adult cephalic region (cr).

FMRF-amide-positive neurons is expressed, which is reduced during metamorphosis (Haszprunar et al. 2002, Voronezhskaya et al. 2002). The larval prototroch is underlain by a serotonergic nerve net, which disintegrates once the juvenile body plan is established (Friedrich et al. 2002).

Compared to representatives of other molluscan classes, metamorphosis appears as a very gradual process in chitons and follows initial settlement of the competent larva. Subsequently, the prototroch and all larval neuronal components are lost, the dorso-ventral axis of the animal flattens, and the juvenile starts its creeping life stage (see Friedrich et al. 2002, Wanninger and Haszprunar 2002).

The question whether polyplacophorans may represent an annelid-like segmented body plan has been widely debated due to the "eight-metameric" arrangement of adult valves and muscle units (see above). Ontogenetic data on muscle development, however, show that this is clearly a secondary condition and that the adult muscle bauplan is established via an initial stage of numerous serially repeated dorso-ventral muscle bundles, which resembles the situation found in adult neomeniomorphs (solenogastres) or other worm-shaped, non-segmented taxa such as flatworms or nemertines. Moreover, an anterior muscle grid is found in the chiton larva, which is lost at metamorphosis and may be regarded as an ontogenetic relic of an ancestral worm-like body wall musculature (Wanninger and Haszprunar 2002). Recent works on polyplacophoran neurogenesis likewise argue against an annelid-like segmented body plan of chitons, since the formation of the pedal commissures occurs random-like and does not follow a strict anterior-posterior pattern as in the segmented annelids (Friedrich et al. 2002, Voronezhskaya et al. 2002).

18.5 HABITAT

Chitons are distributed worldwide in all seas (highest diversity is in the warm temperate zones of the world) with a bathymetrical range from the splash zone (e.g., *Acanthopleura*) down to the hadal depths (e.g., *Ferreiraella*). They can be associated with hydrothermal vents (*Thermochiton*) and are also found in Arctic and Antarctic waters (e.g., *Leptochiton*). Chitons are euryhaline showing a certain tolerance regarding fluctuations in the salinity of the seawater. Thus, polyplacophorans are found in the Baltic Sea [Bornholm Island - 15 PSU (practical salinity units)] as well as in the Mediterranean (36 PSU) (Sirenko 1998).

Chitons generally live on all kinds of hard substrata, but members of the genera *Pseudotonicia* or *Bassethullia* are known to inhabit coarse sand as well. Members of the genus *Acanthopleura* are known to stay close to their home places and field studies have shown that the animals use (more or less exactly) their paths regularly. This so-called "homing" was investigated by several researchers (Lyman 1975, Kamardin 1989, Hulings 1991, Yoshioka and Fujitani 2001). Since many species of Polyplacophora are photonegative, they are often found underneath rocks and in crevices.

A remarkable habitat is that of the Northwest-American species *Lepidochitona caverna*. Gómez (1975) demonstrated that this species [misidentified as *L. dentiens* (Gould, 1846); see Eernisse 1986: 15] searches actively for *Nuttallina californica* (Reeve, 1847) to crawl under its girdle if other chitons or the gastropod *Acmaea* occurs nearby. In addition, chitons are known to be hosts for several parasites. Baxter et al. (1989), for example, reported a high percentage of castration in the European *Lepidochitona cinerea* (Linnaeus, 1767) if they were infected by the sporozoan *Minchina*. Franz and Bullock (1990) reported parasitic copepods in Central-American species while Avdeev and Sirenko (1991) found a copepod in the North Pacific *Tonicella submarmorea* (Middendorff, 1846).

18.6 COLLECTION AND PREPARATION

The German North Sea population of *Lepidochitona cinerea* and Polyplacophora along the coast of California are currently protected taxa. No chiton species is currently considered to be endangered. This is probably only a result of our limited knowledge of the distribution of the different species. More data are needed for a final statement about whether or not a certain species is endangered.

Chitons should not be kept together in a collecting container with large gastropods, crustaceans, or pieces of hard bottom substratum, since these may mechanically damage them. A description for the preparation of collecting containers is found in Geiger (1997) (see Figure 24.4). Besides suitable boxes, a sharp-pointed knife is required. When a chiton is discovered, the knife should be placed immediately under the somewhat uplifted girdle to avoid the adhesion of the foot, which would make its removal impossible. Injury to the soft parts of the animal must be avoided.

Chitons on smooth surfaces may easily be removed by pushing the animal anteriorwards using the thumb. Chitons of equal sizes should be placed in the same box; if animals of different sizes are put together, the effect will be that the smaller animals will curl into the larger ones and may thus not be prepared in a flat state. If possible, the boxes should be kept cool and the seawater should be changed frequently.

The killing and subsequent storage of the animals depends on their later use. Chitons caught for private collections should be dried and preserved flat. Placement in boiling water for a few seconds (if the specimen is not needed for histological or molecular studies) should kill the animal. To avoid curling of the animal, the specimen should be placed on a wooden stick of suitable size and fixed with a soft bandage. It may be placed in fresh water and drowned, but the cell structure will be destroyed. This makes specimens useless for histological examinations. Moreover, the animal will die slowly. This technique is suitable for fresh and agile specimens only.

For molecular analyses, it is necessary to kill and store the animal in highly concentrated ethanol (preferably at least 95%). The specimens should not be pre-fixed in formalin as this may result in degradation of the DNA material. If in this case a flat-preserved specimen is desired, it should be anesthetized with 7% magnesium chloride (see Chapter 2.5).

If the anatomy of the animals is to be studied later and histological examinations are planned, specimens can be preserved in ethanol or, preferably, in 4% formalin. If ethanol is preferred and no molecular analyses are planned, a concentration of 80% is sufficient. A higher concentration hardens the body, which makes histological sectioning more difficult. For histological investigations based on semi-thin sections, phosphate-buffered paraformaldehyde or, for better tissue preservation, cacodylate-buffered glutaraldehyde in combination with osmium-tetroxide postfixation are commonly used (for detailed protocols, see Wanninger et al. 1999). It should be noted, however, that delicate girdle elements such as minute spicules may dissolve even if the specimens are stored in the respective buffer solution after fixation.

Most amateurs will prefer a dried sample; the soft parts can be removed so that only the plates and the girdle remain. A preparation using a mix of ethanol and glycerin (1:1) is useful but the animal has to be kept therein for at least two weeks, depending on the size of the specimen. Afterwards, the specimen has to be dried for at least the time it has spent in this solution. Comprehensive instructions on this topic can be found in Berry (1966), Kaas and Van Belle (1985a), on the Internet <www.worldwide-conchology.com/PreservingChitons.html>, and in Chapter 2.4.12.

18.7 POLYPLACOPHORAN PHYLOGENY

The systematic position of polyplacophoran taxa higher than at species level has been widely debated for many years. Traditional studies, dealing with external morphological characters alone, provide a different phylogenetic system of Polyplacophora than analyses using molecular techniques (Okusu et al. 2003). For the future, it is expected that analyses combining all available characters should provide more accurate phylogenetic trees.

The following is a brief overview of genera reported from the North and Central American regions. To determine species from this and other parts of the world, the monograph series of Kaas and Van Belle (1985a-1994) should be consulted. Slieker (2000) is another useful reference.

Ferreiraellidae Dell'Angelo and Palazzi, 1991
Ferreiraella Sirenko, 1988

Leptochitonidae Dall, 1889
    Leptochiton Gray, 1847
    Hanleyella Sirenko, 1973

Protochitonidae Ashby, 1925
    Oldroydia Dall, 1894
    Deshayesiella Carpenter MS, Dall, 1879

Ischnochitonidae Dall, 1889
    Ischnochiton Gray, 1847
    Stenosemus von Middendorff, 1847
    Stenoplax Carpenter MS, Dall, 1879
    Lepidozona Pilsbry, 1892

Callistoplacidae Pilsbry, 1893
    Ischnoplax Carpenter MS, Dall, 1879
    Callistochiton Carpenter MS, Dall, 1879
    Callistoplax Carpenter MS, Dall, 1882
    Ceratozona Dall, 1882
    Calloplax Thiele, 1909

Chaetopleuridae Plate, 1899
    Chaetopleura Shuttleworth, 1853

Chitonidae Rafinesque, 1815
    Chiton Linnaeus, 1758
    Acanthopleura Guilding, 1829
    Tonicia Gray, 1847

Tonicellidae Simroth, 1894
    Lepidochitona Gray, 1821
    Nuttallina Carpenter MS, Dall, 1871
    Tonicella Carpenter, 1873
    Boreochiton G. O. Sars, 1878
    Juvenichiton Sirenko, 1975
    Micichiton Sirenko, 1975

Schizoplacidae Bergenhayn, 1955
    Schizoplax Dall, 1878

Mopaliidae Dall, 1889
    Mopalia Gray, 1847
    Amicula Gray, 1847
    Placiphorella Carpenter MS, Dall, 1879
    Katharina Gray, 1847

Hanleyidae Bergenhayn, 1955
    Hanleya Gray, 1857

Acanthochitonidae Pilsbry, 1893
    Choneplax Carpenter MS, Dall, 1882
    Cryptoconchus de Blainville MS, Burrow, 1815
    Acanthochitona Gray, 1821
    Cryptochiton von Middendorff, 1847

18.8 SELECTED WEB RESOURCES

Chitons of Puerto Rico <cuhwww.upr.clu.edu/~cgarcia/quitones>

General information on chitons and other Mollusca <www.worldwideconchology.com/MainFrames.htm>

Chitons of Bali, Indonesia <www.worldwideconchology.com/BaliChitons.html>

Short notes and illustrations on worldwide chitons, arranged by geographical areas <home.inreach.com/burghart/>

Links to chiton illustrations on the Internet, arranged systematically <biology.fullerton.edu/people/faculty/doug-eernisse/chitons/index.html>

Chitons of Alaska <www.jaxshells.org/010603.htm>

18.9 LITERATURE CITED

Avdeev, G. V. and B. I. Sirenko. 1991. Chitonophilidae fam. n., a new family of parasitic copepods from the chitons of the North-Western Pacific. *Parasitologija* **25**: 370-374.

Baxter, J. M., A. N. Hodgson, and M. G. Sturrock. 1989. Variations in infestation rates of *Lepidochitona cinereus* (Polyplacophora) by *Minchinia chitonis* (Sporozoa) in twelve populations in Scotland and Northern Ireland. *Marine Biology* **102**: 107-117.

Baxter, J. M., M. G. Sturrock, and A. M. Jones. 1990. The structure of the intrapigmented aesthetes and the properiostracum layer in *Callochiton achatinus* (Mollusca: Polyplacophora). *Journal of Zoology, London* **22**: 447-468.

Berry, S. S. 1966. Chitons, their collection and preservation. *In:* R. T. Abbott, M. K. Jacobson, and M. C. Teskey, eds., *How to Collect Shells*, 3rd Ed. American Malacological Union, Marinette, Wisconsin. Pp. 43-48.

Blumrich, J. 1891. Das Integument der Chitonen. *Zeitschrift für wissenschaftliche Zoologie* **52**: 404-476, pls. 23-29.

Boyle, P. R. 1974. The aesthetes of Chitons. II. Fine structure in *Lepidochitona cinereus* (L.). *Cell and Tissue Research* **153**: 383-398.

Brooker, L. R. and D. J. Macey. 2001. Biomineralization in the radula teeth of the chiton genus *Acanthopleura* (Mollusca: Polyplacophora) and its significance to systematics within this genus. *In: 4th International*

Workshop of Malacology "Systematics, Phylogeny and Biology of Polyplacophora." Italian Malacological Society, Menfi, Italy. P. 13 [Abstract].

Creese, R. G. 1986. Brooding behavior and larval development in the New Zealand chiton, *Onithochiton neglectus* de Rochebrune (Mollusca: Polyplacophora). *New Zealand Journal of Zoology* **13**: 83-91.

Currie, D. R. 1992. Photoreceptor or statocyst? The ultrastructure and function of a unique sensory organ embedded in the shell valves of *Cryptoplax mystica* Iredale and Hull, 1925 (Mollusca: Polyplacophora*). Journal of the Malacological Society of Australia* **13**: 15-25.

Dell'Angelo, B. and A. Tursi. 1990. Abnormalities in chiton shell-plates. *Oebalia* n.s. **14**: 111-138.

Eernisse, D. J. 1984. *Lepidochitona Gray, 1821 (Mollusca: Polyplacophora) from the Pacific Coast of the United States: Systematics and Reproduction.* Ph.D. Dissertation. University of California, Santa Cruz. 358 pp.

Eernisse, D. J. 1986. The genus *Lepidochitona* Gray, 1821 (Mollusca: Polyplacophora) in the northeastern Pacific Ocean (Oregonian and Californian provinces). *Zoologische Verhandelingen, Leiden* **228**: 1-52.

Eernisse, D. J. and K. Kerth. 1988. The initial stages of radular development in chitons (Mollusca: Polyplacophora). *Malacologia* **28**: 95-103.

Eernisse, D. J. and P. D. Reynolds. 1994. Polyplacophora. *In:* F. W. Harrison, ed., *Microscopic Anatomy of Invertebrates*, Vol. 5: Mollusca I. Wiley-Liss, New York. Pp. ix- xiv, 1-390.

Evans, L. A. and R. Alvarez. 1999. Characterization of the calcium biomineral in the radular teeth of *Chiton polliserpentis*. *Journal of Biological Inorganic Chemistry* **4**: 166-170.

Evans, L. A., D. J. Macey, and J. Webb. 1994. Matrix heterogeneity in the radular teeth of the chiton *Acanthopleura hirtosa*. *Acta Zoologica* **75**: 75-79.

Fischer, F. P. 1978. *Untersuchungen an den Ästheten dreier Polyplacophoren-Arten.* Inaugural-Dissertation zur Erlangung des Doktorgrades des Fachbereiches Biologie der Ludwig-Maximilians-Universität, München. 118 pp.

Fischer, F. P. 1979. Die Ästheten von *Acanthochiton fascicularis* (Mollusca, Polyplacophora). *Zoomorphologie* **92**: 95-106.

Fischer, F. P. 1980. Fine structure of the larval eye of *Lepidochitona cinerea* L. (Mollusca, Polyplacophora). *Spixiana* **3**: 53-57.

Fischer, F. P. 1988. The ultrastructure of the aesthetes in *Lepidopleurus cajetanus* (Polyplacophora: Lepidopleurina). *American Malacological Bulletin* **6**: 153-159.

Fischer, F. P., M. Alger, D. Cieslar, and H. U. Krafczyk. 1990. The chiton gill: ultrastructure in *Chiton olivaceus* (Mollusca: Polyplacophora*). Journal of Morphology* **204**: 75-87.

Fischer, F. P., W. Maile, and M. Renner. 1980. Die Mantelpapillen und Stacheln von *Acanthochiton fascicularis* L. (Mollusca, Polyplacophora). *Zoomorphologie* **94**: 121-131.

Fischer, F., B. Eisensamer, C. Miltz, and I. Singer. 1988. Sense organs in the girdle of *Chiton olivaceus* (Mollusca: Polyplacophora). *American Malacological Bulletin* **6**: 131-139.

Fischer-Piette, E. and A. Franc. 1960. Classe des Polycophores. *In:* P.-P. Grassé, ed., *Traité de Zoologie. Anatomie, Systématique, Biologie*, Tome V. Masson, Paris. Pp. 1053-2219.

Franz, C. J. and R. C. Bullock. 1990. *Ischnochitonika lasalliana*, new genus, new species (Copepoda), a parasite of tropical Western Atlantic chitons (Polyplacophora: Ischnochitonidae). *Journal of Crustacean Biology* **10**: 544-549.

Friedrich, S., A. Wanninger, M. Brückner, and G. Haszprunar. 2002. Neurogenesis in the mossy chiton, *Mopalia muscosa* (Gould) (Polyplacophora): Evidence against molluscan metamerism. *Journal of Morphology* **253**: 109-117.

Geiger, D. L. 1997. SACOS: An inexpensive and robust underwater sampling device for SCUBA divers. *The Festivus* **29**: 32-34.

Gómez, R. L. 1975. An association between *Nuttallina californica* and *Cyanoplax hartwegii*, two west coast polyplacophorans (chitons). *The Veliger* **18** (Suppl.): 28-29.

Haszprunar, G. 1987. The fine morphology of the osphradial sense organs of the Mollusca. III. Placophora and Bivalvia. *Philosophical Transactions of the Royal Society of London* (B) **315**: 37-61.

Haszprunar, G. and A. Wanninger. 2000. Molluscan muscle systems in development and evolution. *Journal of Zoological Systematics and Evolutionary Research* **38**: 157-163.

Haszprunar, G., S. Friedrich, A. Wanninger, and B. Ruthensteiner. 2002. Fine structure and immunocytochemistry of a new chemosensory system in the chiton larva (Mollusca: Polyplacophora). *Journal of Morphology* **251**: 210-218.

Hulings, N. C. 1991. Activity patterns and homing of *Acanthopleura gemmata* (Blainville, 1825) (Mollusca: Polyplacophora) in the rocky intertidal of the Jordan Gulf of Aquaba. *The Nautilus* **105**: 16-25.

Jacobs, D. K., C. G. Wray, C. J. Wedeen, R. Kostriken, R. DeSalle, J. Staton, R. D. Gates, and D. L. Lindberg. 2000 Molluscan engrailed expression, serial organization, and shell evolution. *Evolution and Development* **2**: 340-347.

Kaas, P. and R. A. Van Belle. 1985a. *Monograph of Living Chitons (Mollusca: Polyplacophora) 1, Order Neoloricata: Lepidopleurina.* Brill/Backhuys, Leiden. 240 pp.

Kaas, P. and R. A. Van Belle. 1985b. *Monograph of living chitons. (Mollusca: Polyplacophora) 2, Suborder*

Ischnochitonina, Ischnochitonidae: Schizoplacinae, Callochitoninae and Lepidochitoninae. Brill/Backhuys, Leiden. 198 pp.

Kaas, P. and R. A. Van Belle. 1987. *Monograph of Living Chitons (Mollusca: Polyplacophora) 3, Ischnochitonidae: Chaetopleurinae, Ischnochitoninae (pars), additions to vols 1 and 2.* Brill/Backhuys, Leiden. 302 pp.

Kaas, P. and R. A. Van Belle. 1990. *Monograph of Living Chitons (Mollusca: Polyplacophora). 4, Suborder Ischnochitonina: Ischnochitonidae: Ischnochitoninae (continued). Additions to vols 1, 2 and 3.* Brill/Backhuys, Leiden. 298 pp.

Kaas, P. and R. A. Van Belle. 1994. *Monograph of Living Chitons (Mollusca: Polyplacophora). 5, Suborder Ischnochitonina: Ischnochitonidae: Ischnochitoninae (concluded). Callistoplacinae; Mopaliidae; Additions to Volumes 1-4.* Brill/Backhuys, Leiden. 402 pp.

Kaas, P. and R. A. Van Belle. 1998. *Catalogue of Living Chitons (Mollusca, Polyplacophora).* 2nd revised Ed. Backhuys, Leiden. 204 pp.

Kamardin, N. N. 1989. Study of homing in *Acanthopleura gemmata* (Polyplacophora, Mollusca). *Vestnik Leningradskogo Universiteta Biologiya* 3: 58-62.

Leise, E. M. 1988. Sensory organs in the hairy girdles of some mopaliid chitons. *American Malacological Bulletin* 6: 141-151.

Ludbrook, N. H. and K. L. Gowlett-Holmes. 1989. Chitons, gastropods, and bivalves. *In:* S. A. Sheperd and I. M. Thomas, eds., *Marine Invertebrates of South Australia*, Part II. South Australia Government Printing Division, Adelaide. Pp. 504-724.

Lundin, K. and C. Schander. 2001. Ciliary ultrastructure of polyplacophorans (Mollusca, Amphineura, Polyplacophora). *Journal of Submicroscopic Cytology and Pathology* 33: 93-98.

Lyman, B. W. 1975. Activity patterns of the chiton *Cyanoplax hartwegii* (Mollusca: Polyplacophora). *The Veliger* 18 (Suppl.): 63-69.

Matsukuma, A. and Y. Tsubaki. 1995. Radular morphology and feeding tracks of *Liolophura japonica* (Mollusca. Polyplacophora). *Science Report, Department of Earth and Planetary Sciences Kyushu University* 19: 81-92.

McLean, J. H. 1962. Feeding behaviour of the chiton *Placiphorella. Proceedings of the Malacological Society of London* 35: 23-26.

Moseley, H. N. 1885. On the presence of eyes in the shells of certain Chitonidae, and on the structure of these organs. *Quarterly Journal of Microscopical Science* 25: 37-60.

Moshel, S. M., M. Levine, and J. R. Collier. 1998. Shell differentiation and engrailed expression in the *Ilyanassa* embryo. *Development, Genes and Evolution* 208: 135-141.

Plate, L. H. 1897. Die Anatomie und Phylogenie der Chitonen. Fauna Chilensis1 (1). *Zoologisches Jahrbuch (Syst.)* 1 (Suppl. 4): 1-243, pls. 1-12.

Plate, L. H. 1899. Die Anatomie und Phylogenie der Chitonen. Fauna Chilensis2 (1). *Zoologisches Jahrbuch (Syst.)* 2 (Suppl. 5): 15-216, pls. 2-11.

Plate, L. H. 1901. Die Anatomie und Phylogenie der Chitonen. Fauna Chilensis 2 (2). *Zoologisches Jahrbuch (Syst.)* 3 (Suppl. 5): 281-600, pls. 12-16.

Okusu, A., E. Schwabe, D. J. Eernisse, and G. Giribet. 2003. Towards a phylogeny of chitons (Mollusca, Polyplacophora) based on combined analysis of five molecular loci. *Organisms Diversity and Evolution* 3: 281-302.

Reindl, S., W. Salvenmoser, and G. Haszprunar. 1997. Fine structural and immunocytochemical studies on the eyeless aesthetes of *Leptochiton algesirensis*, with comparison to *Leptochiton cancellatus* (Mollusca, Polyplacophora). *Journal of Submicroscopic Cytology and Pathology* 29: 135-151.

Russell-Hunter, W. D. 1988. The gills of chitons (Polyplacophora) and their significance in molluscan phylogeny. *American Malacological Bulletin* 6: 69-78.

Saito, H. 2004. Phylogenetic significance of the radula in chitons, with special reference to the Cryptoplacoidea (Mollusca: Polyplacophora). *Bollettino Malacologico*, Suppl. 5 [2003]: 83-104.

Saito, H. and T. Okutani. 1990. Two new chitons (Mollusca: Polyplacophora) from a hydrothermal vent site of the Iheya Small Ridge, Okinawa Trough, East China Sea. *Venus* 49: 165-179.

Schwabe, E. 2001. Abnormality of shell-plates in *Chiton cumingsii* Frembly, 1827 (Mollusca: Polyplacophora: Chitonidae). *Bollettino Malacologico* 37: 5-6.

Sirenko, B. I. 1974. Development and evolution of radula in Polyplacophora. *Zoologiceskij Zurnal* 53: 1133-1139. [In Russian, sometimes transliterated as Zoologicheskii Zhurnal].

Sirenko, B. I. 1993. Revision of the system of the order Chitonida (Mollusca: Polyplacophora) on the basis of correlation between the type of gills arrangement and the shape of the chorion processes. *Ruthenica* 3: 93-117.

Sirenko, B. I. 1998. Relict settlement of the chiton *Lepidochitona cinerea* (Mollusca, Polyplacophora) in northern Norway. *Archive of Fishery and Marine Research* 46: 139-149.

Sirenko, B. I. 2001. Composition, origin and development of several groups of chitons which live and feed on sunken wood. *In: 4th International Workshop of Malacology "Systematics, Phylogeny and Biology of Polyplacophora."* Italian Malacological Society, Menfi, Italy. P. 8 [Abstract].

Slieker, F. J. A. 2000. *Chitons of the World: An illustrated Synopsis of Recent Polyplacophora.* Mostra Mondiale Malacologia, Cupra Marittima, Italy. vi + 154 pp, 50 pls.

Strack, H. L. 1987. The Polyplacophora of Gran Canaria, including a worldwide survey of the brooding species. *Iberus* 7: 179-187.

Sturrock, M. G. and J. M. Baxter. 1995. The fine structure of the pigment body complex in the intrapigmented aesthetes of *Callochiton achatinus* (Mollusca: Polyplacophora). *Journal of Zoology* **235**: 127-141.

Voronezhskaya, E. E., S. A. Tyurin, and L. P. Nezlin. 2002. Neuronal development in larval chiton *Ischnochiton hakodadensis* (Mollusca: Polyplacophora). *The Journal of Comparative Neurology* **444**: 25-38.

Wanninger, A., B. Ruthensteiner, S. Lobenwein, W. Salvenmoser, W. J. A. G. Dictus, and G. Haszprunar. 1999. Development of the musculature in the limpet *Patella* (Mollusca, Patellogastropoda). *Development, Genes and Evolution* **209**: 226-238.

Wanninger, A. and G. Haszprunar. 2001. The expression of an engrailed protein during embryonic shell formation of the tusk-shell, *Antalis entalis* (Mollusca, Scaphopoda). *Evolution and Development* **3**: 312-321.

Wanninger, A. and G. Haszprunar. 2002. Chiton myogenesis: perspectives for the development and evolution of larval and adult muscle systems in molluscs. *Journal of Morphology* **251**: 103-113.

Wingstrand, K. G. 1985. On the anatomy and relationships of Recent Monoplacophora. *Galathea Report* **16**: 7-94.

Yates, A. M., K. L. Gowlett-Holmes, and B. J. McHenry. 1992. *Triplicatella disdoma* Conway Morris, 1990, reinterpreted as the earliest known Polyplacophoran, abstracted. *Journal of the Malacological Society of Australia* **13**: 71.

Yonge, C. M. 1960. General characters of Mollusca. In: R. C. Moore, ed., *Treatise on Invertebrate Paleontology*, Part I, Mollusca 1. Geological Society of America, University of Kansas Press, Lawrence, Kansas. Pp. 3-36.

Yoshioka, E. and E. Fujitani. 2001. Activity patterns and homing behaviors of *Acanthopleura gemmata* and *A. tenuispinosa* chitons. *In: 4th International Workshop of Malacology "Systematics, Phylogeny and Biology of Polyplacophora."* Italian Malacological Society, Menfi, Italy. P. 30 [Abstract].

CHAPTER 19

SCAPHOPODA: THE TUSK SHELLS

PATRICK D. REYNOLDS

19.1 INTRODUCTION

Members of the Class Scaphopoda are commonly known as tusk shells for the characteristic shape of their shell (Figure 19.1, Figure 31.1). They are entirely infaunal, marine mollusks with a global distribution, and are found from the intertidal to 7,000 m depths, burrowing into soft sediments. There are 14 families in the class, divided between two orders, Dentaliida (Palmer 1974) and Gadilida (Starobogatov 1974). The Class has about 520 extant species, currently placed in 46 genera; another 15 genera and approximately 500 species are known only from the fossil record. The supra-specific classification of this class is presented in Steiner and Kabat (2001).

The biology of the scaphopods is poorly known compared to that of other molluscan classes, although there is a rich 19th Century literature on basic anatomy, development, and behavior, largely restricted to two northeast Atlantic species, *Antalis entalis* (Linnaeus, 1758) and *A. vulgaris* (Da Costa, 1778) [e.g., Deshayes 1825, Lacaze-Duthiers 1856–1857]. These species have also been used extensively as model organisms in experimental embryology (for review, see Reverberi 1971, Reynolds 2002, and references therein).

Although worldwide in distribution, in most instances scaphopods are relatively minor components of benthic communities, and many species are found only in deeper waters; they are therefore often difficult to obtain for comparative study.

Apart from placement within the subphylum Conchifera (the univalve- and bivalve-shelled mollusks), the position of the Scaphopoda within the phylum Mollusca has been uncertain; the Cephalopoda, Gastropoda, Bivalvia, and the extinct bivalved Rostroconchia (both Conocardioida and Ribeirioida subclasses) have all been argued to be sister taxa, although recent evidence is currently

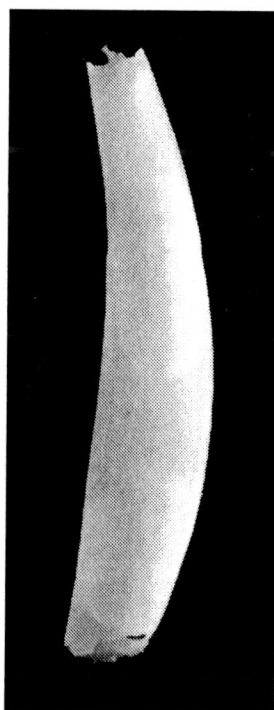

Figure 19.1 *Gadila aberrans.*
Individual of *Gadila aberrans* (Whiteaves, 1887); anterior is to the bottom, ventral to the right. Narrowing of the anterior aperture is characteristic of the Subfamily Gadilinae (Family Gadilidae, Order Gadilida); other scaphopods are widest at the anterior aperture. Note notched apex. Patrick D. Reynolds©.

indicating the cephalopods or gastropods as the most likely closest molluscan relatives (reviewed in Reynolds 2002, Steiner and Reynolds 2003).

19.2 BIOLOGY

Comprehensive treatments of scaphopod biology, and extensive compilations of literature sources, can be found in Lacaze-Duthiers (1856-1857), Fischer-Piette and Franc (1968), Shimek and Steiner (1997), and Reynolds (2002).

19.2.1 Shell. The tusk-shaped shell of scaphopods is a hollow, curved, conical tube open at both ends; for members of the family Gadilidae, the anterior aperture narrows somewhat (Figure 19.1). Shell sculpture is absent or consists of longitudinal ribs; a few genera (e.g., *Omniglypta*) have annular rings. The shell consists of 2-4 prismatic and crossed-lamellar aragonitic layers; a periostracum is absent or worn away. The posterior end or apex of the shell, often notched (Figure 19.1), is periodically cast off or decollated, thereby widening the aperture and accommodating the passage of respiratory currents, gametes, and wastes between the mantle cavity and external environment (Reynolds, 1992).

19.2.2 Mantle cavity. While molluscan morphology is famously disparate among the classes, the scaphopods are still notable for their loss or reduction of typical molluscan organ systems. The absence of an osphradium (a unique molluscan sense organ), ctenidia (gills) and associated auricles, and reduction of the ventricle, are striking characteristics of the class, coincident with a restriction in mantle cavity volume along the elongated functional anterior-posterior axis (also considered the anatomical ventro-dorsal axis) (Figure 19.2). A hindrance to understanding scaphopod anatomy is the relative lack of information on scaphopod organogenesis and larval biology.

The anterior mantle margin is primarily responsible for secreting the scaphopod shell, although additional shell material is added by the general exterior epithelium of the mantle and body wall. The posterior mantle consists of a sleeve of mantle tissue, open along the ventral side, called the pavilion. At

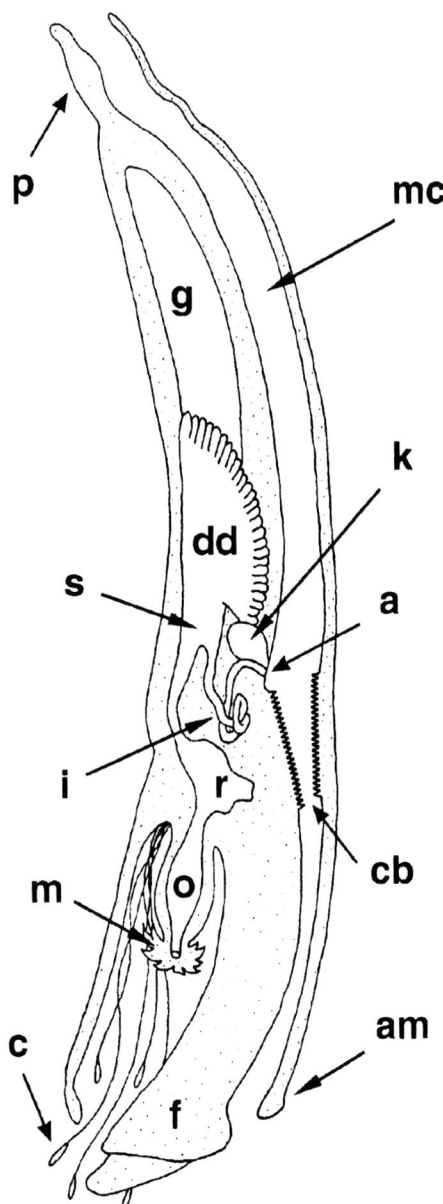

Figure 19.2 Scaphopod body plan
Diagram of the body plan of a dentaliid scaphopod (Order Dentaliida) with the shell removed; same orientation as in Figure 19.1 (*a*, opening of rectum; *am*, anterior mantle margin; *c*, captacula; *cb*, ciliated ridges of the mantle cavity; *dd*, digestive diverticula; *f*, foot; *g*, gonad; *i*, intestine; *k*, kidneys; *m*, mouth and labial palps; *mc*, mantle cavity; *o*, oesophagus; *p*, posterior mantle or pavilion; *r*, radula; *s*, stomach). Redrawn with permission from Pelseneer (1906: figure 182). A. & C. Black[c].

its base is a horizontal (in Dentaliida) or vertical (in Gadilida) slit-like valve opening to the mantle cavity. The interior surface of the pavilion has many sensory receptors, typical of inhalant siphons in other mollusks; its edge and outer surface function in decollation of the posterior shell, and secretion of a secondary shell pipe in some species.

Within the mantle cavity, the anus is borne on a pulsatile anal bulb; arranged on either side of the anal bulb are paired slit-shaped epithelial openings to the hemocoel and paired excretory pores. Anterior to the anal bulb is a series of transverse ciliated bands, ranging in number from one to more than 30 depending largely on body size, that circulates water through the mantle cavity from the posterior opening to the anterior one (Figure 19.2). The mantle and body wall beneath the ciliated bands is highly vascularized, and the bands may be a specialized site of gas exchange. Water is also expelled from the mantle cavity through the posterior opening by withdrawal of the foot; this is the route by which feces and gametes typically exit.

19.2.3 Foot. A muscular, cylindrical foot (Figure 19.2) extends from the anterior aperture and functions in burrowing, but also has a role in feeding. The muscular organization of the foot differs markedly between orders (Steiner 1992). In members of the Dentaliida, it contracts and may fold upon itself within the anterior mantle cavity upon withdrawal into the shell, whereas the foot in Gadilida withdraws by introversion. The end of the scaphopod foot has lobes or a papillate disk to anchor the extended foot into the sediment for burrowing, during which, in many species, the concave (dorsal) surface of the shell remains uppermost. The animal may remain close enough to the sediment surface for the apical shell to protrude slightly from the sediment surface, particularly in dentaliids, although this has been over-generalized through text illustrations and artifact of lab observation; in contrast, some gadilids have been reported to burrow up to 40 cm at rates of 1 cm/ sec (Shimek 1989, 1990).

19.2.4 Feeding. Scaphopods have a microomnivorous to microcarnivorous diet that for many species consists predominantly of foraminiferans (Shimek 1990). The foot is responsible for creating the space in which the feeding tentacles, or captacula (Figure 19.2), can operate. The captacula are ciliated, tentaculate organs that have a bulbous adhesive tip and both circular and longitudinal muscle (Shimek 1988, Byrum and Ruppert 1994). They move over and among sediment grains by ciliary gliding, selecting and transporting food items to the labial palps and thence, ingestion. Scaphopods have been described as ciliary deposit feeders, carnivores, or microphagous burrowers.

After storage in the paired buccal pouches, testate prey is crunched by the relatively large, mineralized radula. The stomach has one or two openings to the digestive gland, and the coiled intestine of this effectively U-shaped gut opens to the mantle cavity through a pulsatile anal bulb located about midway along the mantle cavity or body length (Figure 19.2).

19.2.5 Circulation and excretion. The circulatory system in scaphopods is characterized by the loss of auricles (associated with the loss of ctenidia) and a modified ventricle. This "heart" was historically considered lost, or reduced to a thin dorsal fold in the pericardium, but more recent study supports the strongly beating perianal sinus (largely constituting the anal bulb) as the ventricle homologue (Reynolds 1990a). The circulatory system otherwise consists of an open series of vessels and blood sinuses. The excretory system follows the molluscan ground plan: where the pericardial epithelium overlies the heart (i.e., perianal sinus), it is modified to include podocytes, cells that typically function in ultrafiltration of blood, the ultrafiltrate passing from the heart into the pericardial cavity. Reno-pericardial canals provide a pathway for this primary urine to at least one nephridium, where secretion and reabsorption modify the urine before release to the mantle cavity and exterior (for review see Reynolds 1990b, 2002, and references therein).

19.2.6 Nervous system and sensation. The nervous system of scaphopods consists of a basic conchiferan tetraneury with major cerebral, pedal, and pleural ganglia, and smaller visceral, buccal, subradular, and pavilion ganglia. Major connectives are a pair of fused cerebro- and pleuropedal nerves, and the visceral nerves that connect pleural, visceral, and pavilion ganglia. Ciliated sensory receptors have been described from the anterior mantle edge, captacula, and posterior mantle edge; gustation and orientation are mediated by a sub-radular organ and paired statocysts, respectively.

19.2.7 Reproduction and development. Sexes are separate, although hermaphroditic individuals have been occasionally found in one or two species. The single gonad is an infolded organ occupying the apex and upper half of the shell (Figure 19.2). Sperm is typically bullet-shaped, the unmodified aquasperm type. The egg is relatively large, with obvious polar-lobe extrusion following fertilization; these characteristics attracted experimental embryologists to employ scaphopods as a model system, particularly for the localization of morphogenetic factors, from 1899 through the early 1980's (e.g., Wilson 1904, Verdonk and van den Biggelaar 1983). Gametes are released in masses to the mantle cavity through the right kidney and to the exterior through the posterior aperture or, in some species, single eggs are released through the anterior aperture. Little is known about the reproductive ecology of scaphopods.

Typical molluscan-pattern spiral cleavage produces a coeloblastula that invaginates to produce a gastrula with persistant blastopore. Larval development is well known; an early unshelled lecithotrophic stenocalymma or trochophore leads to a single shell-field (rather than bivalved), veliger-type larva. This embryonic shell forms a bulbous, saddle-shaped, and eventually tubular larval shell, which is later lost from the posterior of the adult shell (see Reynolds 2002, and references therein, for review).

19.3 ECOLOGY

19.3.1 Distribution. Scaphopods are found burrowing into soft marine sediments, ranging from fine mud to gravel, and have a global distribution. Identification guides to the Scaphopoda are naturally organized by region, and provide much distributional information. Sources that are particularly useful in trying to identify these organisms include: for specimens from the Atlantic Ocean, Henderson (1920) and Scarabino (1979, 1986); the Indian and Pacific Oceans, Shimek (1989, 1997), Scarabino (1995), and Shimek and Moreno (1996); from waters near Australia, Lamprell and Healy (1998); near New Zealand, Dell (1957); near Japan, Habe (1964); the waters around Antarctica are covered by Dell (1990), and the Magellanic Province by Steiner and Linse (2000).

Latitude: As has been noted in several marine and terrestrial taxa, there is evidence for a latitudinal gradient in scaphopod diversity in each hemisphere of the Atlantic and Pacific oceans (Reynolds 2002). In the Atlantic, diversity peaks near 20°N, and in the Pacific diversity peaks near the equator. The gradients are generally gradual, diversity diminishing toward the poles. The underlying causative factor for these gradients is uncertain; species-area relationships, recent geologic history of oceanic basins, and sea-surface temperatures are all possibilities. There is evidence that decreasing latitudinal range size with decreasing latitude (Rapoport's Rule) may explain the latitudinal gradients, at least in part (Reynolds 2002).

Depth: Scaphopods have a considerable bathymetric range in the world's oceans. They have been recorded from the low intertidal in France and Australia to depths of 6900-7000 m in the Sunda Trench (Knudsen 1964). Bathymetric ranges of individual species have been recorded as narrow as 100 m to as broad as thousands of meters (McFadien 1973, Lamprell and Healy 1998). It has been generally considered that scaphopods are relatively diverse in the deep sea. However, studies examining scaphopods in the Indo-Pacific and world-wide (Scarabino 1979, 1995, Reynolds 2002) have found generally fewer scaphopods with increasing depth, but with diversity peaks in the 500-800 m and 1200-2000 m ranges, the latter associated with gadilids. The emerging patterns point to distinct radiation patterns of Dentaliida and Gadilida in the deep-sea. The reasons for variation in scaphopod diversity with depth are several, but substrate, sediment characteristics, and community species richness are thought to be more important factors in scaphopod distribution patterns than depth alone.

Stratigraphic: The earliest reports of scaphopods in the geologic record, from the Ordovician [505-440 MYA (million years ago)] and Devonian (410-360 MYA), have been the source of some discussion (e.g., see Yochelson 1999); the earliest undisputed records date from the Mississippian Carboniferous (360-325 MYA). Dentaliida precedes the Gadilida in the fossil record, a disputed species of the latter appearing in the Permian (286-245 MYA), and an unquestioned record as late as the Paleocene Epoch of the Tertiary (65-54.8 MYA). Scaphopods are commonly members of paleocommunities but are invariably minor components of those communities, both in terms of taxonomic diversity and numerically. There have been a few attempts to trace the stratigraphic range of species and genera, to represent stratigraphic or temporal species diversity (e.g., Reynolds 2002), which indicate an extinction/ diversification pattern that reflects temporal patterns seen in other molluscan taxa.

19.3.2 Interspecific associations. Reports of associations and interactions between scaphopods and other organisms are scattered in the literature, often as short notes or anecdotal reports. Except as noted below, a more detailed review of these reports can be found in Reynolds (2002).

Commensalism: Symbiotic associations with scaphopods include common marine invertebrate commensals such as trichodine ciliates, attached within the mantle cavity. Symbiotic bacteria have also been found in the mantle cavity, associated with the ciliated bands, and have also been found, presumably in a commensal relationship, bound to the exterior surface of the egg at the vegetal pole of some dentaliids. Small canals in the outer layers of the shell in some species have been attributed to a boring alga of the genus *Hyella* (Chamaesiphoneae, Pleurocapsales, Pleurocapsaceae).

Mutualism: A mutualistic relationship has been surmised for the oft-noted phenomenon of anemones, corals, and barnacles attached to the exterior of living scaphopod shells. At least two or possibly three species of such anemones have been identified; the best-studied and most frequently encountered relationship is between *Fissidentalium actinophorum* Shimek, 1997, from ~4000 m off southern California, and the anemone *Anthosactis nomados* White, Wakefield, and Fautin, 1999, which lives exclusively attached to the shell of living scaphopods.

While attached anemones and barnacles are usually centered on the concave surface of the scaphopod shell, the solitary coral, *Heterocyathus japonicus* (Verrill, 1866), frequently attaches to the apical end of a living individual of *Fissidentalium vernedei* (Sowerby, 1860). It thereby maximizes its height and proximity to the sediment surface while the scaphopod burrows.

Parasitism. Reports of parasitism in scaphopods are only several and brief. They predictably include ciliates, platyhelminth cercaria (*Ptychogonimus* sp.: Ptychogonimidae, Hemiuroidea) in the blood, adult flatworms in the gonad, and sporozoans [*Minchinia dentali* (Arvy, 1949): Haplosporea] in muscle and connective tissue (Desportes and Nashed 1983).

Predation: Scaphopods near the sediment surface are likely taken by a variety of demersal or bottom-feeding fish, and anecdotal reports of shells having been found in stomachs of ratfish and trumpeter fish support this. Predation by naticid gastropods is easily recognizable by the characteristic beveled borehole (Dávid 1993), usually found in the mid region or posterior 1/3 of the shell. Bored scaphopod shells are commonly reported from both surface sediments and the fossil record, where predation rates of over 50% have been reported (Yochelson *et al.* 1983).

Shell squatting: Sipunculans and hermit crabs often take up residence in empty scaphopod shells. While individuals of several sipunculan species have been found in a wide variety of tusk shells, *Phascolion strombii* (Montagu, 1804) is the more usual occupant found. Hermit crabs have a variety of interesting adaptations to scaphopod shells; members of *Pagurus imafukui* McLaughlin and Konishi, 1994 have an asymmetrical abdomen, as expected in occupants of helically spired (i.e., gastropod) shells, but nonetheless have a pronounced preference and special uropod modification for shells of the scaphopods *Gadilina insolita* (E. A. Smith, 1894) and *Striodentalium rhabdotum* (Pilsbry, 1905). In contrast, pagurid hermit crabs of the Family Pylochelidae (Decapoda, Coenobitoidea) are found in a variety of straight tubular cavities (e.g., decaying wood, bamboo shoots, stone fragments), and have a symmetrical abdomen; individuals of *Bathycheles incisus* (Forest, 1987) are often also found in shells of *Fissidentalium magnificum* (E. A. Smith, 1896). Members of *Pomatocheles jeffreyseii* Miers, 1879 exclusively choose scaphopod shells; one cheliped is shaped to serve as an operculum for the anterior scaphopod shell aperture.

19.4 SAMPLING AND CURATION

19.4.1 Collection. Living in soft, almost exclusively subtidal, sediments, scaphopods can be collected by SCUBA, but as individuals of most species are small and often burrow some distance beneath the sediment surface, mud dredging is the most effective and practical approach. Effective dredge design and duration of tow can vary widely depending on density and size of the scaphopods, but in the vast majority of situations a considerable volume of mud must be brought onboard and sieved with hand-held hoses, preferably with some control of stream pressure. Mesh size can vary depending on volume and consistency of mud, size of scaphopods, and ability to control the sieving rate so that scaphopods can be retrieved (with soft forceps or fingers) during the process; they are usually readily seen as the sediment is gradually washed away and can be plucked from the mud, although the shells are sometimes quite glossy and slippery.

Scaphopods are generally hardy animals with strong shells, into which they withdraw rapidly, and they can withstand fairly rough treatment during dredging and sieving operations. As scaphopods are often in low density, a single dredge can provide hours of enjoyment through careful sieving and picking, often lasting well after the ship has returned to port and the rest of the crew gone home. This is definitely a case of many hands at least lightening the work.

19.4.2 Maintenance. Scaphopods should be kept in seawater, preferably sediment as well, and kept at their ambient seawater temperature. So, while sorting through sediment, collected cold-water scaphopods should be kept in seawater in a thermos flask or a cooler. Several small, wide-mouth jars or nalgene bottles, stored in a cooler, work best. I like to keep the scaphopods horizontal in the containers so that they can extend their foot. Depending on the length of time needed to keep them alive (volume of mud to sort, complexity of specimen preparation, distance at sea, etc.) I often give them a little sediment to burrow into or feed upon as well. Members of some species, such as *Rhabdus rectius* (Carpenter, 1865) [NE Pacific], will continuously try to burrow with the foot, and in an effort to minimize stress on the animal a couple of centimeters of fine sediment will keep them happier. Care should be taken, however, that sediment does not become compacted by engine vibration during transport.

If scaphopods are to be maintained in the lab for any length of time, the same recommendations apply. Most scaphopods will live in plain cool seawater for a couple of days, but in my experience sufficient sediment for horizontal burrowing will enhance this to a few or several days; sufficient sediment from their habitat for more vertical burrowing, and thus adequate diet, will maintain them for several weeks or months. In the latter case, containers that allow aeration of the sediment are necessary; plastic beakers of sediment, with the sides replaced by fine mesh and the whole immersed in circulating seawater, have worked well. Sediment that has been sieved to remove anything larger than 1 mm provides an ideal environment for mud-dwelling species. Poon (1987) has also successfully used individual-sized aquaria for behavioral observations.

Some things to look out for when maintaining scaphopods in the lab include ensuring they are cleared of sediment upon arrival back from the dredging trip. Scaphopods typical withdraw deeply into their shell when disturbed, not uncommonly to the posterior $3/5$ or $1/2$ of the shell. Sometimes, during the sieving and sorting operations, sediment gets packed into the wider anterior part, and if left alone I have found some scaphopods unable to clear the aperture and they die within. So, I make sure the aperture is clear before placing them in an aquarium; a stream of seawater from a water bottle or blunt forceps tip does the trick. Spawning and raising larvae are beyond the scope of this review, but the approaches are similar to those for optimizing spawning and larvae-rearing conditions for many species of marine mollusks.

19.4.3 Preservation and storage. In museum collections, scaphopod shells are usually kept dry; often the soft tissues are simply left within, useless for dissection or histological work, although the radula can sometimes be extracted and examined. Presumably, DNA extractions are theoretically possible on such dried material. Obviously, wet preservation preserves more of the animal, but this can be tricky. Immersing the animal into formalin will elicit withdrawal, and diffusion and penetration of the fixative becomes extremely limited; the anterior aperture is blocked by the thick muscular mantle, and the posterior aperture is insufficiently large. In museum wet collections, I have often found a well-preserved foot and the rest of the soft tissue completely decomposed.

Relaxing the animal can enhance fixation and preservation. However, even with relaxation, the long narrow form of the scaphopod shell limits the diffusion of the fixative. The only way I have found to ensure fixation is to crack the shell 2/3 of the way back from the anterior aperture. Therefore, if the animal withdraws, fixative can still penetrate into the mid-region of the body. While some scaphopod shells are rather small, thick, and glossy, and hard to break, I have usually found I can make a reasonably clean break in the shell by cracking it gently like a small twig with my hands, or by applying pressure with the point of a forceps' tine. I gently pull the two pieces apart to expose a few mm of tissue in the mid-region. If the larger anterior shell piece slides off the animal easily, without damaging the soft tissue (one smooth, firm, continuous, movement works best), I remove it entirely and just store the pieces together in the same vial. I can then still use the shell pieces later for identification, scanning electron micrographs, etc. In some very difficult circumstances, I sacrifice the shell of a couple of specimens entirely by using forceps to gradually break it up, proceeding from the anterior aperture, so that I am sure to have well-preserved soft tissue. In all cases, one can be guided by maximizing the penetration of the fixative.

There is no pigmentation in the soft tissues of especial taxonomic significance, so transfer to long-term storage in ethanol is not a problem in that respect. Some species do have pigmented shells, and either dry or wet preservation seems to preserve the coloration successfully.

19.5 MISCELLANEOUS

Of interest to include here is the role of scaphopod shells in native North American culture, particularly in the Pacific Northwest. Scaphopod shells, mostly *Antalis pretiosum* (Sowerby, 1860) (Figure 31.1), were used as currency, apparently as far as central Canada and central California (Clark 1963). The method of capture was an enigmatic broom-like apparatus, recreated and tested by Nuytten (1993); the unusually shallow sites where this could be accomplished were presumably limited, supply controlled, and so the shells' use as currency viable. Importation of shells of *Antalis* species (*Antalis entalis*) by eastern white traders undermined the currency, but the shells were also used as adornment and seemingly a demonstration of wealth or position. Examples of the latter can be seen in several museums, including The Field Museum (Chicago), the Museum of Anthropology (University of British Columbia, Vancouver), the Royal British Columbia Museum (Victoria, British Columbia), and in several Edward Curtis photographs. Tusk-shell earrings can often be purchased in museum shops.

19.6 ACKNOWLEDGMENTS

My sincere thanks to the Editor, Charles F. Sturm, for his work on this contribution and volume; to two anonymous reviewers for suggested improvements to the manuscript; and to the Reference Staff of the Burke Library, Hamilton College, for their excellence and continuing help. I am especially grateful to the late Prof. Brendan F. Keegan, of The Martin Ryan Marine Science Institute, National University of Ireland, Galway, not only for hosting a sabbatical leave during which this chapter was written, but also for sparking my interest in mollusks and invertebrate biology some years ago.

19.7 LITERATURE CITED

Byrum, C. A. and E. E. Ruppert. 1994. The ultrastructure and functional morphology of a captaculum in *Graptacme calamus* (Mollusca, Scaphopoda). *Acta Zoologica* **75**: 37-46.

Clark, R. B. 1963. The economics of *Dentalium. Veliger* **6**: 9-19.

Dávid, Å. 1993. Trace fossils on molluscs from the Molluscan Clay (late Oligocene, Egerian) - a comparison between two localities (Wind Brickyard, Eger, and Nyárjas Hill, Novaj, NE Hungary). *Scripta Geologica, Special Issue* **2**: 75-82.

Dell, R. K. 1957. A revision of the Recent scaphopod Mollusca of New Zealand. *Transactions of the Royal Society of New Zealand* **84**: 561-576.

Dell, R. K. 1990. Antarctic Mollusca, with special reference to the fauna of the Ross Sea. *Bulletin of the Royal Society of New Zealand* **27**: 1-297.

Deshayes, G. P. 1825. Anatomie et monographie du genre Dentale. *Mémoires de la Society d'histoire naturelle de Paris* **2**: 321-378.

Desportes, I. and N. Nashed. 1983. Ultrastructure of sporulation in *Minchinia dentali* (Arvy), an haplosporean parasite of *Dentalium entale* (Scaphopoda, Mollusca); taxonomic implications. *Protistologica* **19**: 435-460.

Fischer-Piette, E. and A. Franc. 1968. Classe des Scaphopodes. Scaphopoda (Bronn 1862). *In:* P.-P. Grassé, ed., *Traité de Zoologie*, Tome 5(3), Mollusques, Gastéropodes et Scaphopodes. Masson et Cie, Paris. Pp. 987-1017.

Habe, T. 1964. *Fauna Japonica. Scaphopoda (Mollusca)*. Biogeographical Society of Japan, Tokyo. 59 pp.

Henderson, J. B. 1920. A monograph of the East American scaphopod mollusks. *Smithsonian Institution, U.S. National Museum Bulletin* **111**: 1-177.

Knudsen, J. 1964. Scaphopoda and Gastropoda from depths exceeding 6,000 metres. *Galathea Report* **7**: 125-136.

Lacaze-Duthiers, H. 1856-1857. Histoire de l'organisation et du développement du Dentale. *Annales des Sciences Naturelles, Quatrième Serie* **6**: 225-281, pls. 8-10; 319-385, pls. 11-13; **7**: 5-51, pls. 2-4; 171-255, pls. 5-9; **8**: 18-44.

Lamprell, K. L. and J. M. Healy. 1998. A revision of the Scaphopoda from Australian waters (Mollusca). *Records of the Australian Museum, Supplement* **24**: 1-189.

McFadien, M. S. 1973. Zoogeography and ecology of seven species of Scaphopoda. *Veliger* **15**: 340-347.

Nuytten, P. 1993. Money From the Sea. *National Geographic* January 1993: 108-117.

Pelseneer, P. 1906. Mollusca. *In:* E. R. Lankester, ed., *A Treatise on Zoology*, Part V. A. & C. Black, London. Pp. 1-355.

Poon, P. A. 1987. The diet and feeding behavior of *Cadulus tolmiei* Dall, 1898 (Scaphopoda: Siphonodentalioida). *Nautilus* **101**: 88-92.

Reverberi, G. 1971. *Dentalium. In:* G. Reverberi, ed., *Experimental Embryology of Marine and Freshwater Invertebrates*. North-Holland, Amsterdam. Pp. 248-264.

Reynolds, P. D. 1990a. Functional morphology of the perianal sinus and pericardium of *Dentalium rectius* (Mollusca: Scaphopoda) with a reinterpretation of the scaphopod heart. *American Malacological Bulletin* **7**: 137-149.

Reynolds, P. D. 1990b. Fine structure of the kidney and characterization of secretory products in *Dentalium rectius* (Mollusca, Scaphopoda). *Zoomorphology* **110**: 53-62.

Reynolds, P. D. 1992. Mantle-mediated shell decollation increases posterior aperture size in *Dentalium rectius* (Scaphopoda: Dentaliida). *Veliger* **35**: 26-35.

Reynolds, P. D. 2002. The Scaphopoda. *Advances in Marine Biology* **42**: 137-236.

Scarabino, V. 1979. *Les Scaphopodes Bathyaux de l'Atlantique Occidental (Systematique, Distribution, Adaptations). Nouvelle Classification pour l'Ensemble de la Classe*. Ph.D. Dissertation, Université d'Aix Marseille, France. 154 pp.

Scarabino, V. 1986. Systematics of Scaphopoda (Mollusca), 1. Three new bathyal and abyssal taxa of the Order Gadilida from south and north Atlantic Ocean. *Comunicaciones Zoologicas del Museo Nacional de Historia Natural de Montevideo* **11**: 1-15.

Scarabino, V. 1995. Scaphopoda of the tropical Pacific and Indian Oceans, with description of 3 new genera and 42 new species. *Mémoires du Muséum National d'Histoire Naturelle* **167**: 189-379.

Shimek, R.L. 1988. The functional morphology of scaphopod captacula. *Veliger* **30**: 213-221.

Shimek, R. L. 1989. Shell morphometrics and systematics: a revision of the slender, shallow-water *Cadulus* of the Northeastern Pacific (Scaphopoda: Gadilida). *Veliger* **32**: 233-246.

Shimek, R. L. 1990. Diet and habitat utilization in a northeastern Pacific Ocean scaphopod assemblage. *American Malacological Bulletin* **7**: 147-169.

Shimek, R. L. 1997. A new species of eastern Pacific *Fissidentalium* (Mollusca: Scaphopoda) with a symbiotic sea anemone. *Veliger* **40**: 178-191.

Shimek, R. L. and G. Moreno 1996. A new species of eastern Pacific *Fissidentalium* (Mollusca: Scaphopoda). *Veliger* **39**: 71-82.

Shimek, R. L. and G. Steiner. 1997. Scaphopoda. *In:* F. W. Harrison and A. J. Kohn, eds., *Microscopic Anatomy of Invertebrates*, Vol. 6B, Mollusca II. Wiley-Liss, New York. Pp. 719-781.

Steiner, G. 1992. The organisation of the pedal musculature and its connection to the dorsoventral musculature in Scaphopoda. *Journal of Molluscan Studies* **58**: 181-197.

Steiner, G. and A. R. Kabat. 2001. Catalogue of supraspecific taxa of Scaphopoda (Mollusca). *Zoosystema* **23**: 433-460.

Steiner, G. and K. Linse. 2000. Systematics and distribution of the Scaphopoda (Mollusca) in Beagle Channel (Chile). *Mitteilungen Hamburgisches Zoologische Museum und Institüt* **97**: 13-30.

Steiner, G. and P. D. Reynolds. 2003. Molecular systematics of the Scaphopoda. *In:* C. Lydeard and D. R. Lindberg, eds., *Molecular Systematics and Phylogeography of Mollusks*. Smithsonian Institution Press, Washington, D.C. Pp. 123–139.

Verdonk, N. H. and J. A. M. van den Biggelaar. 1983. Early development and the formation of the germ layers. *In:* N. H. Verdonk, J. A. M. van den Biggelaar, and A. S. Tompa, eds., *The Mollusca*, Vol. 3, Development. Academic Press, New York. Pp. 91-122.

Wilson, E. B. 1904. Experimental studies on germinal localization. *Journal of Experimental Zoology* **1**: 1-72.

Yochelson, E. L. 1999. Scaphopoda. *In:* E. Savazzi, ed., *Functional Morphology of the Invertebrate Skeleton*. John Wiley and Sons, Chichester, UK. Pp. 363-367

Yochelson, E. L., D. Dockery, and H. Wolf. 1983. Predation on sub-holocene scaphopod mollusks from southern Louisiana. *Professional Papers, U.S. Geological Survey* **1282**: 1-13.

CHAPTER 20

CEPHALOPODA

FRANK E. ANDERSON

20.1 INTRODUCTION

At first glance, cephalopods seem to have little in common with snails and clams. Unlike other mollusks, most cephalopods are dynamic animals that possess complex behavioral repertoires and a unique suite of characteristics, including muscular appendages that can be used to seize prey and manipulate objects, a parrot-like beak, and seemingly shell-free bodies. Despite external appearances, however, cephalopods exhibit many molluscan anatomical characteristics. Most have a shell of some sort (internal in most cephalopods), a muscular mantle and a radula (a small file-like ribbon used for feeding). The chambered nautilus, which has the external calcified shell typical of most mollusks as well as the appendages and beak seen in other cephalopods, provides compelling evidence that cephalopods are just highly specialized mollusks.

Cephalopods are exclusively marine organisms, and can be found in nearly all seas. In many regions, cephalopods constitute an important intermediate link in marine food webs, feeding on smaller invertebrates and fish and serving as prey for fish, birds and marine mammals. Cephalopods are important organisms for humans as well. Many cephalopod species are important objects of artisanal and commercial fisheries in many parts of the world, and giant axons from cephalopods have been valuable model systems for neurophysiological research. Cephalopods have sparked a lifelong interest in marine biology and malacology in many people, and there is a great deal left for us to learn about these astounding, enigmatic animals.

20.2 TAXONOMY, BIOLOGY, ECOLOGY, AND BEHAVIOR

20.2.1 Basic biology. Cephalopods can be found throughout the world ocean, from surface waters to abyssal depths in tropical, temperate, and polar regions. Some cephalopods, such as squids (Figure 20.1), nautiluses (Figure 20.2), and some octopods, live their entire lives in the water column, spending little if any time on the bottom. Others, including most incirrate (finless) octopods and cuttles, are primarily benthic (bottom-living) or semi-benthic animals (Figure 20.3).

Cephalopods are almost exclusively predatory. Cephalopods as a group are known to consume many types of prey, including crustaceans (e.g., crabs, shrimps, and lobsters), fish, bivalves, gastropods, other cephalopods (cannibalism has been documented for several cephalopod species) and,

Figure 20.1 Teuthida.
Sepioteuthis australis Quoy and Gaimard, 1832, Australia, 1832. Daniel Geiger Stock Photo 7109-1 [1].

Figure 20.2 Nautiloidea.
Shell of *Nautilus pompilius* Linnaeus, 1758, CMNH 65680 (CMNH = Carnegie Museum of Natural History Catalog Number).

more rarely, other invertebrates such as annelids and chaetognaths (Mangold 1983).

In general, cephalopods are relatively short-lived, rapidly growing animals. It was once believed that nearly all cephalopod species (except *Nautilus*) were semelparous, producing all of their offspring in one burst of reproduction at the end of the life cycle. More recent research, summarized in Rocha *et al.* (2001), has shown that cephalopod reproductive strategies are more variable than previously thought. Reproductive behaviors vary widely across Cephalopoda. Elaborate color displays (see below) play an important role in mating, male-male interactions, and mate guarding in some cephalopods.

20.2.2 Behavior. The intelligence and stunningly rich behavioral repertoires of many cephalopods are well known, and have been the subject of several nature shows on television. The intellectual feats of octopuses in particular have been the subject of many articles, books, and apocryphal stories. Many cephalopods have amazing capacities for learning. As Hanlon and Messenger (1996) have noted, cephalopod intelligence seems to be on par with that of the fishes, birds, and mammals with which they interact, and their intellectual abilities are only truly remarkable when one remembers that cephalopods are mollusks.

The ability of many cephalopods to rapidly alter their body patterns and skin texture to camouflage themselves from predators or to send signals to conspecifics makes them unique among animals. Most cephalopods are covered with chromatophores, small organs in the skin that contain sacs filled with different pigments. The edges of the pigment sac are connected to muscles. When these muscles are relaxed, the pigment sac is small and almost invisible to the naked eye. When the muscles contract, the sac is stretched into a broad disc. Chromatophores are under the direct control of the brain, allowing the animal to change color or flash patterns (such as stripes or spots) in a fraction of a second. Other skin organs, including reflecting cells like leucophores and iridophores, allow some cephalopods to reflect and diffract ambient light. Chromatophores and reflecting cells are critical components in crypsis - the ability of many cephalopods to blend into their surroundings as a predation avoidance mechanism - and for intraspecific interactions like mating.

As soft-bodied animals, cephalopods are favored prey items for many animals, and a number of sophisticated anti-predator abilities and behaviors have evolved in response to predation pressure. For example, many octopods, cuttles, and some squids can alter both the color and the texture of the epidermis, allowing them to match closely the appearance of their surroundings. This is particularly important in crypsis, enabling an octopus to adopt the look and texture of, for example, an algae-covered rock. The ink sac constitutes another key aspect of predation avoidance for many cephalopods. When threatened, most cephalopods can jet away rapidly while releasing ink. The ink can be released as a diffuse cloud

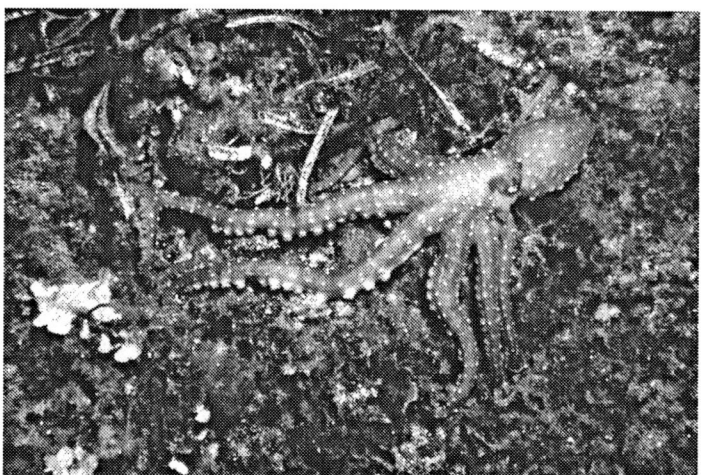

Figure 20.3 Octapoda.
Octopus macropus Risso, 1826, Malta, Daniel Geiger Stock Photo 755 [1].

(somewhat like a smokescreen) or as a pseudomorph - a solid-looking, squid-sized blob made of ink and mucus. The pseudomorph may give a predator a false target to strike while the cephalopod escapes. The ink itself contains various agents, including L-DOPA and dopamine, that seem to affect predator chemosensory organs and may act as an alarm substance.

Finally, many cephalopods (including vampires, several squids and sepiolids, and some octopods) have bioluminescent (light-generating) organs called photophores. Bioluminescence can be either intrinsic (i.e., produced by the animal itself) or bacterial, in which bioluminescent bacteria are housed in crypts within the photophores. Like chromatophores, photophores are mainly used for two purposes: predation avoidance and intraspecific signaling. Many cephalopods use photophores to counterilluminate their ventral surfaces, masking them from predators swimming below. Some deep-water squids (e.g., *Selenoteuthis*) have sets of photophores that emit light of different wavelengths. These photophores may play an important role in intraspecific signaling. Some cephalopods (including *Heteroteuthis*) even eject clouds of bioluminescent secretions in an act known as fire shooting.

Detailed behavioral studies have only been performed on a handful of cephalopod species (primarily octopods, cuttles, and shallow-water or reef-dwelling squids), and much remains to be learned. Hanlon and Messenger (1996) is an excellent overview of our current understanding of cephalopod behavioral diversity, complete with color diagrams, photographs, and an extensive bibliography.

20.2.3 Taxonomic overview.

Cephalopoda is traditionally recognized as a class within the phylum Mollusca. Cephalopod taxonomy and phylogeny remain somewhat confused, but several major groups are recognized by most cephalopod researchers (Table 20.1). Cephalopoda comprises several living and extinct subgroups: Ammonoidea (now extinct, Figure 15.6), Nautiloidea (with very few extant representatives) and Coleoidea, comprising two groups: Belemnoidea (extinct, Figure 15.7) and Neocoleoidea (all extant cephalopods except *Nautilus*). Nautiloids are the only living cephalopods with external shells. All other living cephalopods (neocoleoids) either completely lack a shell, or have internal calcified or chitinous "shells."

Within Neocoleoidea, several major groups are typically recognized, sometimes under slightly different

Table 20.1. An indented list of major cephalopod taxa. [†] = extinct.

Cephalopoda
  Ammonoidea[†] - ammonites (Figure 15.6)
  Nautiloidea - nautiluses (Figure 20.2)
  Coleoidea
    Belemnoidea[†] - belemnites (Figure 15.7)
    Neocoleoidea
      Octopoda - octopuses (Figure 20.3, 4)
        Cirrata - finned octopuses
        Incirrata - finless octopuses
      Sepiida - cuttles (Figure 20.5)
      Sepiolida - "stubby squids"
      Spirulida - *Spirula spirula* (Figure 20.6)
      Teuthida (Teutheoidea) - "squids" (Figure 20.1)
        Myopsina (Myopsida)
        Oegopsina (Oegsopsida)
      Vampyromorpha - vampires (or vampire squids)

Figure 20.4 Octopoda.
The egg case of the paper nautilus *Argonauta nodosa* Solander, 1786 (Argonautidae), a pelagic octopus from the Pacific Ocean, CMNH 63.31.

names or taxonomic ranks depending on who constructed the classification. These groups are Octopoda (the octopods, Figure 20.3, 20.4), Sepiida (the cuttles or cuttlefish, Figure 20.5), Sepiolida (the stubby squids), Spirulida (containing only the enigmatic *Spirula spirula* Linnaeus, 1758, Figure 20.6), Teuthida or Teutheoidea (the squids, Figure 20.1), and Vampyromorpha (the vampires or vampire squids). Members of Sepiida, Sepiolida, Spirulida and Teuthida possess eight sucker-laden arms and two long ventrolateral tentacles (although the tentacles are absent from some adult teuthids), and these groups are sometimes grouped together in Decapodiformes. Teuthida comprises two subgroups, Myopsina (Myopsida) and Oegopsina (Oegopsida); the loliginid squids are perhaps the most familiar myopsid cephalopods. Members of Octopoda and Vampyromorpha lack ventrolateral tentacles, and together they constitute Octopodiformes. Octopoda comprises two subgroups: Cirrata and Incirrata. Cirrate octopods possess a single pair of fins and a row of cirri flanking each side of the single row of suckers on each arm, while incirrate octopods lack fins and cirri.

Sepiida and Sepiolida: True cuttles (Sepiida, Figure 20.5), which possess an internal chambered shell, are not present in the Western Hemisphere. The Caribbean reef squid (*Sepioteuthis sepioidea* Blainville, 1823) is somewhat cuttle-like in appearance, but is actually a loliginid squid (a teuthid). The stubby squids (Sepiolida) are generally rather small animals with stubby, rounded bodies and kidney-shaped fins.

Teuthida: Teuthida (Figure 20.1) is a heterogeneous group that includes several familiar shallow-water squids like *Doryteuthis opalescens* (Berry, 1911) (the California market squid), muscular species from the open ocean [which are sometimes very large, like *Moroteuthis robusta* (Dall *in* Verrill, 1876)], and bizarre (often gelatinous) forms typically found further offshore, like members of Cranchiidae and Histioteuthidae.

Octopoda: Most octopods (Figure 20.3) that you are likely to encounter will be incirrates, and will be primarily benthic animals. Some incirrate octopods (such as argonauts, also known as paper nautiluses, Figure 20.4) and some cirrates are primarily pelagic organisms, spending most or all of their time floating or swimming in the water column.

Vampyromorpha: The Vampyromorpha (vampires or vampire squids) comprises only one species -

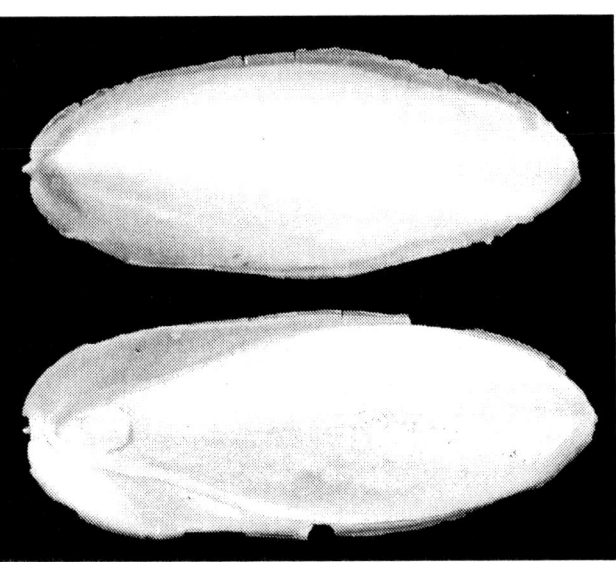

Figure 20.5 Sepiida.
Cuttlebone of *Sepia officinalis* Linnaeus, 1758, 37°40'56.8"N, 27°5'19.6"E, Mediterranean Sea, CMNH 70294.

Figure 20.6 Spirulida.
Spirula spirula (Linnaeus, 1758), northeast coast, Puerto Rico, CMNH 65676.

Vampyroteuthis infernalis Chum, 1903 (the vampire squid from Hell). Vampires are small (11-13 cm mantle length) finned cephalopods found around the world, typically at depths of 700-1,500 meters (although they may be occasionally found much closer to the surface). They have eight arms and two dorsolateral filaments that can be withdrawn into pockets in the web. Little was known about vampire behavior until recently, when several live specimens were encountered by remotely operated vehicles.

Nautiloidea: Living nautiluses are known only from the Indian Ocean and western Pacific (Figure 20.2).

Recent taxonomic research has focused on reconstructing evolutionary relationships among cephalopods and on investigating genetic diversity within species. An evolutionary family tree (a phylogeny) for cephalopods could serve as a vital foundation both for studies of cephalopod anatomical evolution and for a hierarchical classification. Unfortunately, progress toward this goal has been slow, and much work remains to be done.

20.3 COLLECTING TECHNIQUES

Different cephalopods require different collection techniques. The pelagic cephalopods (squids and some octopods) will usually require appropriate fishing gear (such as jigs, traps, or nets of some type) and a boat. Shallow-water benthic octopuses can be hand-collected while snorkeling or scuba diving, or through the use of traps.

20.3.1 Teuthids. Perhaps the easiest way to collect squid for teaching purposes would be to visit the local supermarket, where fresh or frozen squid are usually readily available. Along the coasts, fresh squid can often be obtained at local fish markets or landings, particularly when loliginid squid are spawning. Other species of squid, such as *Doscidicus gigas* (d'Orbigny, 1835) (the jumbo or Humboldt squid), may also occasionally be available in fish markets in some areas.

Of course, many different types of gear can be used to capture squid. Some capture methods include trawling at various depths, jigging (automated or by hand), purse-seining, dip-netting, trapping, and spearing. Bright lights may be used to attract positively phototactic squid (such as the California market squid) to the surface, where they are usually captured by jigging or purse-seining. Larger squids, such as *Moroteuthis robusta* (the north Pacific giant squid) and *Architeuthis*, will generally only be rarely available as bycatch from commercial fisheries, or through rare strandings of dead or moribund individuals.

20.3.2 Sepiolids. Most species of sepiolid squid are bottom dwellers, although some common forms (e.g., *Heteroteuthis* and *Stoloteuthis*) are pelagic. *Euprymna scolopes* Berry, 1913 can be found (often buried in the sand with only their eyes showing) in shallow tidal flats in Hawaii, where individuals can be collected by hand. Other sepiolid species (members of *Rossia* and *Semirossia*) are found in colder waters and at greater depths. These species can be collected in bottom trawls or sometimes by hand by scuba divers. Pelagic sepiolids and *Spirula spriula* may be encountered in midwater trawls.

20.3.3 Octopuses. Several methods for collecting octopuses have been described. The easiest method is simply to collect individuals by hand while tidepooling, snorkeling, or scuba diving. Shallow-water octopuses can sometimes be found in tide pools at low tide, hiding under rocks or in crevasses. Note that even small octopuses possess beaks, and some can inflict painful bites, so care should be taken if octopuses are collected by hand. Octopuses are collected for food in many parts of the world through the use of submerged pots or traps, often attached to longlines, which are deployed overnight and recovered the following morning or several days later.

For more information regarding fishing techniques and gear used to capture cephalopods (including traps), the FAO (Food and Agricultural Organization of the United Nations) website is a good resource (see Section 20.8).

20.4 AQUARIUM CULTURE

In general, I do not recommend that amateurs attempt to maintain cephalopods in closed-water aquarium settings. Cephalopods are notoriously difficult to maintain in aquarium settings, for several reasons:

1. Most cephalopods are extremely active animals and have high metabolic demands for oxygen. Thus, they are highly sensitive to changes in oxygen concentration.
2. In addition to a requirement for well-oxygenated water, many cephalopods are sensitive to changes in water chemistry, and salinity levels in particular - very few species can be found in brackish (low salinity) water. Nitrogenous waste accumulation in aquaria can also adversely affect cephalopods. There is also some evidence that octopuses are sensitive to slight changes in trace metal concentrations. Finally, cephalopods may release their ink when disturbed, producing an additional problem for aquarists.
3. Some cephalopods - for example, many squids - live in groups and are constantly on the move. Animals like these often bump into the sides and ends of conventional rectangular aquaria, damaging their skin in the process.
4. As predators, cephalopods prefer a steady diet of living prey items, although they can be trained to take other food items (Toll and Strain 1988).
5. Octopuses (the most likely candidates for aquaria) are excellent escape artists, and many an octopus has gone exploring "over the side" of an improperly covered aquarium, particularly if they have not been well-fed (see #4).
6. Finally, cephalopods (even large ones, like the giant Pacific octopus *Enteroctopus dofleini*) have relatively short life spans, ranging from a few months to a few years at most.

In short, cephalopods are extremely sensitive, active animals that tend to be picky eaters - not the best (or easiest) choice for most amateur aquarists. In spite of all these difficulties, some aquarists and researchers have had impressive success maintaining various cephalopods in aquaria. Researchers at the University of Texas Medical Branch in Galveston culture several cephalopod species for biomedical use, and public aquaria often maintain various species for display. Many amateur aquarists successfully maintain octopuses in their homes. The Cephalopod Page website, developed by Dr. James Wood, is a great resource for further information on aquarium culture of cephalopods, particularly octopuses (see Section 20.8).

20.5 PREPARATION

20.5.1 Documentation. Several aspects of the collection, fixation, and preservation process are vital for insuring that a specimen will be useful for future research. One important step occurs after the capture of specimens, but prior to fixation. Many living cephalopods have important chromatic and textural patterns on their skin that can be lost (or irrevocably altered) during the process of preservation. If possible, you should take color photographs of the living animal prior to preservation, and simple sketches or descriptions of major color or papilla patterns can also be very helpful.

It is also important to keep detailed records of the collection process itself (including locality information - preferably with GPS coordinates - collection depth, date and time of collection, collector name, and collection method) and preservation of this record along with the collected specimen. A preserved octopus with tightly curled arms may be a bit difficult to study, but an octopus whose collection locality was not recorded may be completely useless for many research projects. If at all possible, collection information should be written in graphite pencil or alcohol-proof ink (a laser printer can also be used) on a small label made from 100% rag-content archival paper.

20.5.2 Preservation for genetic research. Preserved specimens can serve many purposes. Obviously, preserved cephalopods can be used for external and internal anatomical study in either teaching or research settings. They can also be used for genetic studies, including evolutionary or taxonomic studies as well as population genetic studies (for example, stock assessment for fisheries species). I strongly recommend that cephalopod collectors consider preserving animals for both anatomical and genetic study, even if they do not have immediate access to a genetics laboratory.

Unfortunately, optimal preservation techniques differ for animals that are intended for anatomical and genetic research. Initial fixation in a formaldehyde solution (described below) is ideal for most anatomical studies, but is not appropriate for genetic work. Fortunately, most cephalopods are large enough that small tissue samples can be taken from the animal and preserved for genetic work, while the rest of the animal can be fixed and preserved. Ethanol (80-100%) is typically the preservative of choice for genetic work, but many other preservatives are also suitable, including silica gel and a DMSO/EDTA/NaCl solution (Seutin *et al.* 1991). For RNA work, tissue must be frozen at ultracold temperatures or placed in an appropriate preservative such as RNALater (a preservative developed by Ambion for RNA work; <www.ambion.com>).

Tissue samples for use in genetic research can be taken from any part of the cephalopod, but I recommend that thin tissue samples be taken from part of the fin (squids or cuttles) or from the margin of the mantle near the head. Gill tissue is also very good for DNA analysis, but important anatomical measurements are often recorded from the gills. Similarly, arm or tentacle tips are also suitable for genetic work, but arm sucker counts, hectocotylus structure and tentacle sucker numbers and shapes are valuable anatomical characteristics that can be damaged or destroyed if tissue samples are taken from these regions.

You should try to collect at least 25 mg of tissue for genetic analysis. Thin samples (~3 mm or less in thickness, if possible) are preferable, as they allow for rapid penetration of the preservative. If you take thicker samples, you should score them with a sterile razor blade or slice them into smaller, thinner pieces prior to immersion in the preservative. Larger samples are preferable, particularly if the tissue is going to a museum collection for use by multiple geneticists, but the potential use of the specimen for morphological study (and, perhaps, aesthetic considerations) must be kept in mind. Retaining the entire mantle of a large squid for PCR-based DNA work is overkill.

20.5.3 Fixation and preservation for morphological study. Fortunately for collectors, most cephalopods do not have an external shell into which they can retreat. This makes fixation of cephalopods relatively easy. Several fixatives and preservatives have been used by cephalopod researchers to prepare animals for archival storage. The most common fixative is a buffered 10% formalin in sea water solution (e.g., 100 ml of 37-40% formaldehyde stock solution plus 900 mL sea water, buffered to pH 7.0 with a buffering agent; see below). 70% ethanol or 50% isopropanol are common preservatives. Details of recommended fixation and preservation protocols for different cephalopods are listed in the following section.

20.6 FIXATION AND PRESERVATION PROTOCOLS

Fixation and preservation techniques vary somewhat for different groups of cephalopods, and there is some debate regarding which set of techniques are optimal for particular purposes. Typically, however, efforts are made to limit arm curling and other types of physical deformation of the specimen. Much of what follows has been summarized from various publications, primarily Roper and Sweeney (1983), and from conversations with cephalopod researchers around the world, most notably Nancy Voss (University of Miami), Steve O'Shea (Auckland University of Technology, New Zealand), Janet Voight (Field Museum of Natural History, Chicago), Crissie Huffard (University of

California, Berkeley) and Ian Gleadall (Tohoku Bunka Gakuen University, Japan).

20.6.1 Narcotization. If you have collected a live cephalopod that you would like to keep for future study, you should anesthetize it prior to fixation and preservation. During abrupt fixation, a cephalopod will usually strongly contract its arms. This can keep fixative and preservative from reaching all tissues, and (even if fixative does reach all tissues before decay sets in) it can make morphological studies of the specimen extremely difficult. Several methods have been described for narcotizing cephalopods. Slowly decreasing the salinity of the seawater surrounding the specimen by adding freshwater effectively narcotizes many cephalopods, as does lowering the temperature of the water or slowly adding ethanol to the sea water to a final concentration of about 1% (Roper and Sweeney 1983). You should test the response of the specimen to make certain it is narcotized prior to transferring it to the preservation medium.

20.6.2 Octopuses and small squids. The following protocol is recommended by Dr. Nancy Voss (pers. comm.), with some additional suggestions from Crissie Huffard (pers. comm.), and is appropriate for both octopuses and small squids. Live specimens are killed by placing them in fresh water or a very dilute alcohol solution. If the animal is killed in fresh water, the arms will curl less but will still curl when the animal is fixed in formalin (see below). Relaxation in a dilute alcohol solution results in more initial arm curling than relaxation in fresh water, but seems to result in less curling upon fixation. While the specimen is relaxing, the arms can often be massaged by hand to further reduce curling. You can perform this massaging technique several times, until the specimen's arms no longer curl when placed back in fresh water (if you try this technique, you should wear gloves and take care to avoid stretching the arms to an unnatural length).

The animal is then fixed in a 10% formalin in sea water solution (buffered with sodium bicarbonate, sodium borate, or calcium carbonate) for a week, although shorter fixation times may be suitable for very small specimens. With muscular species, injection of formalin or opening of the mantle cavity may be required to ensure fixation of the internal organs. Arm curling will continue during fixation, but this problem can be ameliorated in a couple of ways. Once the arms have been initially straightened during relaxation, you can roll the specimen up in paper to keep the arms straight during fixation. If this method is used, a small cone of paper should be placed between the arms to keep them slightly spread, so that the suckers around the mouth can be studied more easily after preservation. An alternative (and more time-consuming) approach involves holding the octopus by the arm tips and dangling it into a wide-mouthed jar containing 10% formalin until the arms no longer curl when placed in the solution. Small octopuses should remain in the formalin solution for at least twelve hours, while larger specimens should be fixed for several days to a week. After formalin fixation, the octopus is thoroughly rinsed with fresh water and preserved in 70% ethanol.

20.6.3 Large squids. Specimens of large species such as *Architeuthis* are rarely encountered in American waters. If you do happen to obtain a giant squid carcass, it could be of great scientific value (and, if it is properly cared for, it may end up on display!). The following guidelines - developed by Dr. Steve O'Shea (pers. comm.), a researcher who studies giant cephalopods - will help you preserve the animal appropriately for research or public display.

General preparation: If a specimen of a giant species is obtained as bycatch from a commercial fishing vessel, it will often be frozen at sea. Prior to fixation, large frozen specimens must be thawed. Bathing the specimen in a constant gentle shower of cold water (with a gently running hose placed in the mantle cavity) will greatly accelerate thawing. Do not attempt to disentangle arms or pull the specimen apart to accelerate thawing, as this will damage the specimen. In general, you should avoid handling the specimen - hand and finger impressions left on the mantle during thawing can be permanent.

If you think the specimen is suitable for public display, you should avoid performing a ventral midline cut of the mantle (a standard method of dissection for smaller squid). This will expose many of the internal organs, but it will destroy the value of the specimen for display. A series of small, careful cuts can yield as much scientific information as a ventral midline incision. The digestive caecum can be removed through a cut of ~20 cm length, and a smaller incision can be made in the posterior-ventral portion of the mantle through which ovarian tissue can be removed. The penis will protrude through the mantle or funnel, thus sexing and determining male reproductive status will not require a ventral mantle incision. The female nidamental glands can be examined by a mid-ventral incision of ~30 cm length about one quarter the distance up from the anterior mantle margin. Extensive dissection of the digestive system of giant squids to examine their contents is unnecessarily destructive, as the esophagus and intestine are usually empty (however, dissection of the crop of large octopods can be informative). Because many institutions are desperate to receive intact giant cephalopod specimens, only damaged specimens of well-known giant squid and octopus species should be dissected.

Many species of large squid are ammoniacal - the mantle, arms and head are filled with vacuoles containing ammonium chloride. The pH of formalin solutions fluctuates widely when fixing ammoniacal squid, so determining whether or not a squid is ammoniacal is important for preservation. To determine whether squid tissue is ammoniacal, cut ~1 cm^3 cubes from the animal, place them into a narrow-necked glass container, add several crystals of potassium hydroxide and then heat the container. A quick sniff will tell you whether any ammonia vapour is present. If the squid is ammoniacal:

1. Remove representative sucker rings and hooks from the arms and tentacle clubs prior to fixation. The sucker rings of *Architeuthis* have a white carbonate deposit that is lost during formalin fixation; they can only be preserved in buffered ethanol or isopropyl alcohol. Don't forget to obtain tissue samples for genetic analysis!
2. If possible, remove the statoliths with two clean cuts.
3. Inject the squid with 10% buffered formalin solution using a large syringe (150 cm^3 or greater), with long screw-on needle (~150 mm length). This introduces formalin into the tissues faster than immersion. Do not exert too much pressure on the syringe plunger: Clots of formalin solution can form in the tissues. Also, if the needle is blocked, it can detach from the syringe or the plunger can break, resulting in a messy explosion of formalin solution.
4. Slowly inject the following volumes of 10% buffered formalin solution:
   - Mantle (dorsal): ~2 l injected mid-dorsally, deep into the mantle, to reach the digestive gland; about 1 liter injected in the posterior-most portion of the mantle (~ mid-fin length) to reach the gonadal tissues. Mantle (ventral): ~1 l into both sides of the ventral mid-line, in the posterior third of the mantle to reach renal and cardiac tissue; and ~1 l on the specimen's left side in the posterior quarter of the mantle to reach the digestive caecum.
   - Head: dorsal and ventral mantle and head injections are required; ~1 l of formalin solution needs to be injected into the cranium to fix esophageal, nervous and buccal tissue.
   - Arms: at intervals of ~30 cm from the arm base, down each arm, inject ~300 ml of formalin solution, progressively decreasing the volume as you move toward the arm tips.
5. If the tentacles are attached to the squid, sever them at their bases and fix them separately in a buffered formalin solution in an appropriately sized bin (usually they can be coiled within a 0.75 x 0.5 m length/width and 0.5 m depth fixing bin). They are fixed separately to minimize deterioration (they thaw more rapidly than the rest of the carcass) and discoloration.

Fixing tank and formalin solution preparation:
A stainless steel tank 2-3 m long, 1-1.5 m wide and 0.75 m deep, with a close-fitting stainless steel lid, is ideal for fixation of very large squid. For a 200 kg squid a 4:1 formalin solution/squid ratio is recommended. Pre-mark levels inside on the tank wall to indicate volume; add formaldehyde and then salt water to make a 10% solution, agitate, take a pH measurement and buffer (to fix a 200

kg *Architeuthis*, plan to use 10-20 kg of a buffering agent such as sodium bicarbonate). Formalin respirator masks are essential, as are long gloves with elastic cuffs.

Transporting thawed specimens: Giant cephalopods require support from below when they are moved. A stretcher made of two poles and a doubled-over, thick plastic tarpaulin is ideal for transporting squid (a canvas tarpaulin is too abrasive). Slide the specimen onto the tarpaulin mantle-first, then tie up the tarpaulin ends to prevent it from sliding off the stretcher. Should the specimen start to slide off the stretcher do not grab the head or mantle. Lower the stretcher to the ground and reposition the squid.

Once over the fixing tank, the two leading pole-holders need to step into the tank (tall boots or waders are recommended). The two anterior ends of the poles are then lowered and the squid is allowed to slide into the formalin bath in a controlled fashion (other methods can result in an uncontrolled splash of formalin solution, in addition to possibly damaging the specimen). Slightly inflated wine-cask bladders can be gently inserted inside the mantle (with care taken not to damage the gills) to ensure that formalin reaches inner tissues. These should be inserted only after the animal has been immersed in formalin solution.

Monitoring the pH: The formalin solution in which the squid is immersed needs to be regularly monitored for the first week, but particularly during the first 72 hours (this is also true for smaller cephalopods). The pH can decrease to 3 or 4 within the first few hours; if left unchecked for 24 hours, all the sucker rings may be destroyed. To prevent acidification, the buffering agent must be added to the solution, either in powder or slurry form, and the entire solution then manually agitated. Measure pH after the solution has been thoroughly agitated. For best results, the formalin solution needs to be checked every 3 hours for the first 24 hours, every 6 hours for the next 24 hours, and every 12 hours on the 3rd day. After the 3rd day, the specimen needs to be checked only once a day. Once the pH stabilizes you can check it less frequently.

When adding buffering agent, do not allow it to settle on cephalopod tissues, as it will corrode them. By gently gripping the anterior margin of the mantle, raising and lowering it, fresh formalin solution can be circulated throughout the mantle; this exercise should be repeated during every pH check. Similarly, mucous deposits around the buccal membrane and arm bases need to be wiped away, and the arm crown periodically agitated to ensure fresh formalin solution is exposed to the inside of the arms. By the end of the first week, the solution is a revolting yellow to red-brown. By this point, pH should be monitored with a digital meter, as litmus paper and colored solutions no longer give interpretable results.

If the specimen is destined for display, the discolored formalin solution should be replaced with fresh solution so as not to unduly discolor the squid; if not then no damage is done to it by leaving it in.

Preservation: Large specimens must remain in the formalin solution for at least a month in order for it to fix thoroughly. When a month has passed, the specimen needs to be thoroughly soaked and the mantle cavity gently flushed with water to remove residual formalin and miscellaneous grunge. Three water changes are recommended. The specimen can be transferred to a preservative, either 40% isopropyl alcohol or 70% ethanol.

20.7 IDENTIFICATION

Many recognized cephalopod species, particularly in the Indian Ocean and Western Pacific, are now known to be complexes of cryptic species (species that are anatomically identical or nearly so) based on genetic studies (Yeatman and Benzie 1994, Izuka *et al.* 1996). Identification of species in many tropical regions can be difficult, because many similar-looking species have overlapping geographic distributions. Also, for some cephalopods, sex-specific characteristics (particularly the hectocotylus, a structure found only in males) are important for identification, making determinations difficult if only immature specimens or females are collected.

The living North American shallow-water cephalopod fauna is not very diverse, and species-level identification is generally not difficult. Many regional marine invertebrate keys, including Light (1975) for central California, Hoover (1998) for Hawaii, and Ruppert and Fox (1988) and Humann (1996) for the southeastern U.S.A. and the Caribbean, include pictures, drawings, or descriptions of commonly encountered shallow-water cephalopods. The only guides to the global cephalopod fauna that are currently available are Roper et al. (1984) and Norman (2000). The former is a guide to cephalopod species of value to fisheries that is available for download through CephBase and the FAO Fisheries website (a new FAO guide is due to be published soon). Another important and widely used reference is Nesis (1987), but it is currently out of print.

20.8 WEB RESOURCES

- CephBase: <www.cephbase.utmb.edu/>
- Tree of Life:<tolweb.org/tree?group=Cephalopoda>
- Cephalopods at the National Museum of Natural History: <www.mnh.si.edu/cephs/>
- The Cephalopod Page: <is.dal.ca/~ceph/TCP/>
- FAO Fisheries: <www.fao.org/fi/>
- FAO Species Identification Field Guide for Fishery Purposes (Cephalopods): <www.fao.org/docrep/t0726e/t0726e0c.htm#cephalopods>

20.9 SUMMARY

Cephalopods are incredible animals, and we still have a great deal to learn about their biology, ecology, behavior, and evolution. Careful fixation and preservation of cephalopod specimens, coupled with detailed documentation of collection information and color patterns of the living animal, are critical for future research.

20.10 ACKNOWLEDGEMENTS

Figures 20.1 and 20.3 were taken by Daniel Geiger. They are copyrighted and used with permission. The remaining figures are courtesy of the Carnegie Museum of Natural History. Images were taken by C. Sturm.

20.11 LITERATURE CITED

Hanlon, R. T. and J. B. Messenger. 1996. *Cephalopod Behaviour*. Cambridge University Press, New York. 232 pp.

Hoover, J. P. 1998. *Hawai'i's Sea Creatures: A Guide to Hawai'i's Marine Invertebrates*. Mutual Publishing, Honolulu, Hawaii. 366 pp.

Humann, P. and N. DeLoach. 1996. *Reef Creature Identification: Florida, Caribbean, Bahamas*. New World Publications, Jacksonville, Florida. 320 pp.

Izuka, T., S. Segawa, and T. Okutani. 1996. Biochemical study of the population heterogeneity and distribution of the oval squid *Sepioteuthis lessoniana* complex in southwestern Japan. *American Malacological Bulletin* **12**: 129-135.

Light, S. F., R. I. Smith, and J. T. Carlton. 1975. *Light's Manual: Intertidal Invertebrates of the Central California Coast*, 3rd Ed. University of California Press, Berkeley, California. 716 pp.

Mangold, K. 1983. Food, feeding and growth in cephalopods. *Memoirs of the National Museum of Victoria* **44**: 81-107.

Nesis, K. N. 1987. *Cephalopods of the World*. T. F. H. Publications, Neptune City, New Jersey. 351 pp.

Norman, M. D. 2000. *Cephalopods, A World Guide: Pacific Ocean, Indian Ocean, Red Sea, Atlantic Ocean, Caribbean, Arctic, Antarctic*. ConchBooks, Hackenheim, Germany. 318 pp.

Rocha, F., A. Guerra, and A. F. González. 2001. A review of reproductive strategies in cephalopods. *Biological Reviews of the Cambridge Philosophical Society* **76**: 291-304.

Roper, C. F. E. and M. J. Sweeney. 1983. Techniques for fixation, preservation and curation of cephalopods. *Memoirs of the National Museum Victoria* **44**: 29-47.

Roper, C. F. E., M. J. Sweeney, and C. E. Nauen. 1984. *Cephalopods of the World: An Annotated and Illustrated Catalogue of Species of Interest to Fisheries*. United Nations Development Programme, Food and Agricultural Organization of the United Nations, New York. 277 p

Ruppert, E. E. and R. S. Fox. 1988. *Seashore Animals of the Southeast: A Guide to Common Shallow-Water Invertebrates of the Southeastern Atlantic Coast*. University of South Carolina Press, Columbia, South Carolina. 429 pp.

Seutin, G., B. N. White, and P. T. Boag. 1991. Preservation of avian blood and tissue samples for DNA analyses. *Canadian Journal of Zoology* **68**: 82-90.

Toll, R. B. and C. H. Strain. 1988. Freshwater and terrestrial food organisms as an alternative diet for laboratory culture of cephalopods. *Malacologia* **29**: 195-200.

Yeatman, J. and J. A. H. Benzie. 1994. Genetic structure and distribution of *Photololigo* spp. in Australia. *Marine Biology* **118**: 79-87.

CHAPTER 21

FRESHWATER GASTROPODA

ROBERT T. DILLON, Jr.

21.1 INTRODUCTION

Gastropods are a common and conspicuous element of the freshwater biota throughout most of North America. They are the dominant grazers of algae and aquatic plants in many lakes and streams, and can play a vital role in the processing of detritus and decaying organic matter. They are themselves consumed by a host of invertebrate predators, parasites, fish, waterfowl, and other creatures great and small. An appreciation of freshwater gastropods cannot help but lead to an appreciation of freshwater ecosystems as a whole (Russell-Hunter 1978, Aldridge 1983, McMahon 1983, Dillon 2000).

21.2 BIOLOGY AND ECOLOGY

The most striking attribute of the North American freshwater gastropod fauna is its biological diversity. The snails presently dwelling in our lakes and streams, although perhaps sharing some superficial similarities, have their origins in 6-8 separate invasions from the sea. Co-occurring gastropod populations may differ strikingly in anatomy, life history, habitat, food, and ecological requirement. The first distinction to be made among freshwater snails is between the pulmonates and the prosobranchs.

The prosobranchs (Prosobranchia, see Figures 21.1 A-E) are a polyglot group retaining the ancestral gilled condition. They bear relatively heavy shells and an operculum. They are generally slow growing, require at least a year to mature, and live for several years. Sexes are separate in most cases.

The largest-bodied freshwater gastropods (adults usually much greater than 2 cm in shell length) belong to the related families Viviparidae and Ampullariidae. The former family, including the common genera *Viviparus* and *Campeloma*, among others, is distinguished by bearing live young, sometimes parthenogenically. (Eggs are actually held until they hatch internally, so the term "ovoviviparous" is more descriptive.) Viviparids have the ability to filter feed, in addition to the more usual grazing and scavenging habit. The Ampullariidae, tropical or sub-tropical in distribution, includes *Pomacea*, which lays its large pink egg mass above the water, and *Marisa*, which attaches large gelatinous egg masses to subsurface vegetation. Ampullariids have famous appetites for aquatic vegetation. The only ampullariid native to the U.S.A. is the Florida apple snail, *Pomacea paludosa* (Say, 1829), although other ampullariids have been introduced through the aquarium trade.

At the other end of the prosobranch spectrum, the related families Hydrobiidae and Pomatiopsidae are among our smallest freshwater snails, with shell lengths typically less than 5 mm as adults. The former family (*Amnicola, Fontigens, Somatogyrus,* and many other genera) includes diverse inhabitants of clean waters across North America, many species being specially adapted to springs. The latter family, represented by only a few species here, are often amphibious, being found on mud above the water level.

The Pleuroceridae (including such genera as *Pleurocera, Anculosa* [*Leptoxis*], and *Goniobasis* [*Elimia*]) [here and below, when a name follows in brackets it

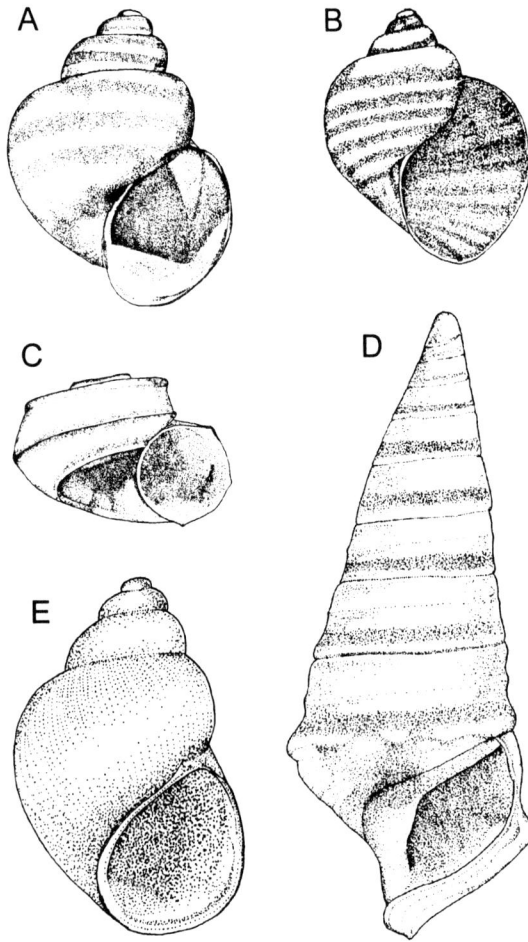

Figure 21.1 Prosobranchs.
A. *Viviparus viviparus* (Linnaeus, 1758). Fox Ferry Point, Potomac River, MD. CMNH 18319. h. = 2.75 cm. B. *Pomacea paludosa* (Say, 1829). Lake Oklawaka, Putnam County, FL. CMNH 63653. h. = 5 cm. C. *Valvata tricarinata* (Say, 1817). Ohio. CMNH 62.7046. w. = 6.3 mm. D. *Pleurocera nobilis* (Lea, 1845). Tennessee River, Florence, AL. CMNH 62.23401. h. = 4.6 cm. E. *Bithynia tentaculata* (Linnaeus, 1758). Ohio Canal, Clinton, OH. CMNH 62.25138. h. = 1 cm. CMNH = Carnegie Museum of Natural History, h. - height, w. - width.

is the one preferred by Turgeon *et al.* (1998)] bear moderately sized shells, perhaps 1- 2 cm as adults. They reach great abundance and diversity in clean, well-oxygenated waters, especially of the southeastern U.S.A. The parthenogenic Thiaridae is a related family common in the tropics and in aquarium shops. Two species, *Melanoides tuberculata* (Müller, 1774) and *Thiara granifera* (Lamarck, 1822) [*Tarebia granifera*], have been introduced to Florida, Texas, and scattered streams elsewhere.

The Valvatidae is a small family of freshwater gastropods with adults generally much less than 5 mm shell length. They are more northerly in their distribution, and are especially found in the deeper waters of lakes. Another noteworthy element of the northern fauna is *Bithynia tentaculata* (Linnaeus, 1758), of the family Bithyniidae, introduced from Europe. Bearing a shell about 1 cm long and a calcareous operculum, *Bithynia* has the ability to filter feed as well as graze.

Snails of the other major group of freshwater gastropods, the pulmonates (Pulmonata, see Figures 21.2 A-G), have lost their gills and now gather oxygen across the simple inner surface of their mantle. The freshwater pulmonates belong to the Order Basommatophora, so named because their eyes are located at the base of their tentacles. This distinguishes them from the more familiar land snails, the Stylommatophora, with eyes at tentacle tips.

Most freshwater pulmonates carry an air bubble under their shell, which they replenish occasionally at the surface, and which serves to adjust their buoyancy. This allows typical pulmonates to inhabit calm, warm, and even stagnant water where dissolved oxygen concentrations may be quite low. It should be noted, however, that some pulmonates (especially limpets and smaller or cold-water species) do not carry air bubbles, and rely on diffusion of oxygen from the water directly into their body tissues.

Pulmonate snails are lightly-shelled and do not bear opercula. They grow quickly, as a general rule, some populations passing multiple generations in a single growing season. They are reproductively hermaphroditic, with the capability of self-fertilization.

The freshwater pulmonates reach their greatest diversity in more northerly latitudes. There are four major families in North America. As a generality, the snails of the family Lymnaeidae bear slender,

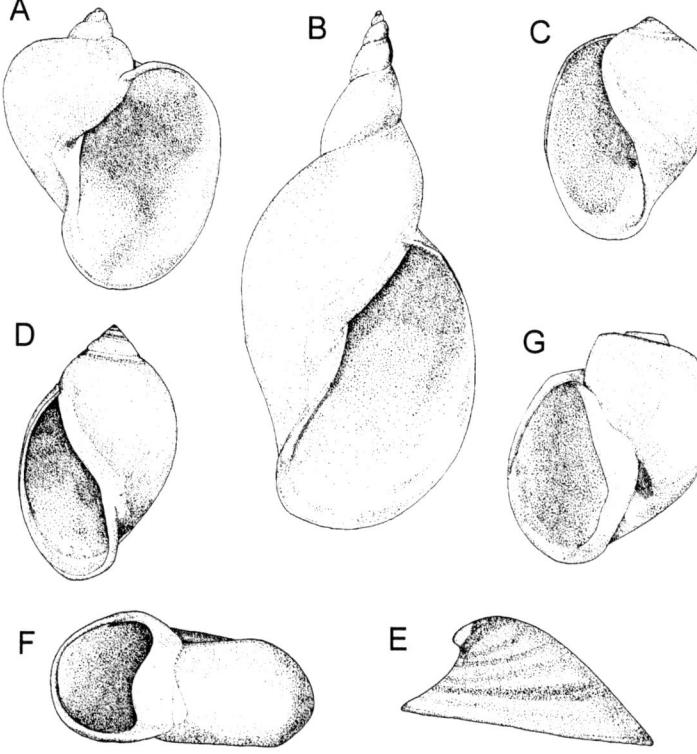

Figure 21.2 Pulmonates.
A. *Lymnaea auricularia* (Linnaeus, 1758) [*Radix auricularia*]. Switzerland (also introduced in North America). CMNH 62.25691. h. = 2.4 cm. B. *Lymnaea stagnalis* (Linnaeus, 1758). Douglas Lake, MI. CMNH 62.32741. h. = 4.4 cm. C. *Physa ancillaria* Say, 1825 [*Physella ancilaria*]. Lake Chautauqua, Chautauqua County, NY. CMNH 62.7495. h. = 1.7 cm. D. *Physa acuta* Draparnaud, 1805 [*Physella acuta*]. Lake Erie, Ottawa County, OH. CMNH 62.32834. h. = 2.2 cm. E. *Ancylus fluviatilis* (Müller, 1774). Long Park, England. CMNH 63654. w. = 1.1 cm. F. *Helisoma campanulata* (Say, 1821) [*Planorbella campanulata*]. Pittsburgh, PA. CMNH 62.33654. w. = 1.9 cm. G. *Helisoma scalaris* (Jay, 1839) [*Panorbella scalaris*]. Palm Beach County, FL. CMNH 47174. h. = 1.6 cm. CMNH = Carnegie Museum of Natural History, h. - height, w. - width.

whether they are national, state, or local. There are 20 endangered or threatened species of freshwater gastropods on the federal list at the present writing, including 9 hydrobiids and 5 pleurocerids, and most states also list freshwater gastropod species among their taxa of special concern. Many additional species of freshwater snails deserve protection. The nonprofit conservation organization, NatureServe, maintains a database listing the conservation status of all American freshwater gastropod species (see Section 21.9 for web address). More information can be found in Chapter 30.

21.4 COLLECTING TECHNIQUES

Dress with the weather in mind. Plan to be challenged by briars, poison ivy, and biting insects on the way to being wet and muddy. Hip boots or chest waders will be required for the mountains in March; shorts and canvas wading shoes are suitable for the swamps in July. Carry with you an assortment of unbreakable containers for specimens, perhaps in a bucket, canvas bag, or knapsack, or in the pockets of a vest. Whirl-pak bags (made of thick plastic with leak proof closures) can be very handy. A scientific collector or serious amateur will always carry at least a couple small vials (4 dram, 15 ml) in his pockets for limpets, hydrobiids, tiny planorbids, and other small snails easily lost in bottles with larger pleurocerids and pulmonates.

right-handed shells of typical appearance, the shells of the Physidae are left-handed, those of the Planorbidae are planispiral (coiled like a watch spring), and those of the Ancylidae are limpet-shaped.

21.3 CONSERVATION

Before embarking on a collecting trip for any element of the biota, it is the responsibility of all good stewards of the environment to become familiar with conservation concerns. The taking of any animal or plant life is generally prohibited in all parks,

Since many elements of the freshwater gastropod fauna are minute, the well-prepared collector will keep a set of fine forceps hanging from a pocket flap or dangling on a string around his neck. A knife or scalpel may be preferable to forceps for collecting limpets, but perhaps more difficult to carry. A sturdy, long-handed net or dipper will be an asset, and/or a sieve, tea strainer, or similar device. Benthic sampling nets may be purchased from biological supply companies in a variety of styles. The mesh need not be fine; 1 mm will capture even newly-hatched gastropods, and should be protected with a canvas or muslin bottom or shroud. A "kick net" with a rectangular or triangular opening is better for rocky bottoms, and a dip net with a more conventional round opening is probably better for vegetation. D-frame nets combine the benefits of both types.

A successful search for freshwater snails begins with an inventory of available habitat types. Rivers should be surveyed in both riffle and pool; lakes should be surveyed both in quiet, protected bays and on exposed shores. Even ephemeral pools and dune ponds can host their own molluscan faunas. The entire range of substrate types should be sampled, including mud, sand, and rock, as should the entire macrophyte flora, both floating and attached. Consider collecting from a boat.

Upon arrival at the collecting site, your first task is to observe, in a manner as unobtrusive as possible. In some environments, especially those characterized by hard water, a preliminary search for relic shells in outwashed fines and sediment scour at lake or river's edge may yield valuable clues regarding the gastropod species to be expected alive. Do not disturb the silt by entering the water, at least initially. It is best to kneel patiently at the water's edge for a while, allowing your eyes to adjust to the scene, alert for small movements and trails in the mud. Some lymnaeids and pomatiopsids are quite amphibious, often being found on the dry tops of floating plants, or muddy flats some distance from the water's edge. Run your fingers through loose sediment for viviparids.

Enter the water only after sampling snails from all visible surfaces. Lift rocks, pull macrophytes, and inspect all surfaces carefully. Examine floating vegetation and debris. Smaller snails can often be conveniently collected by shaking vegetation in a bucket of clean water. My colleague, Amy Wethington (pers. comm.), reports that she is occasionally alerted to the presence of pulmonate snails on lifted debris by the tiny sucking noise they make as they withdraw into their shells.

Plastic bags and floating garbage of many sorts often seem to attract pulmonates. Remove all such materials from the water, inspect them, and transport them to the nearest trash receptacle.

A truly complete survey for freshwater snails will conclude with a number of passes using a stout, long-handled net. The net should be run through both the bottom sediments and any aquatic vegetation in turn, and its contents examined carefully for small gastropods such as the hydrobiids, limpets, and small planorbids. Older sources describe the "Walker dipper" as an alternative to the standard dip net. The following passage from Baker (1942) describes how to construct a Walker dipper. "Thanks to one of our oldest members, Dr. Bryant Walker, now gone from among us, who lived in Detroit, Michigan, a useful dipper was invented which is fittingly called a Walker dipper or dredge. It is so efficient that usually no other implement is necessary. This dipper is about 6 inches in diameter at the top and 5 inches in diameter at the bottom, with a depth of about 3 inches. The bottom is covered with copper wire screening of a sufficiently large mesh to allow the mud and water to run through and retain the shells. It is fastened to a handle 5 or 6 feet in length. A broom handle often makes a good handle. The dipper should be made of copper to prevent rusting. If copper is not available then the sides of the dipper may be made of tinned iron and the bottom may be of copper. A dipper of this kind has been in use by the writer for several years and shows no indication of wearing out" (Figure 2.1 D).

Campeloma can be collected by baiting. Simply tie fish or carrion, partially buried, to a tall stake and sample the surrounding substrate with a net or screen at intervals of several days. The technique seems to work best in shallow, flowing waters with loose sand or silty bottom.

The most convenient method of bringing your samples home is to preserve them immediately with alcohol, together with a field label. If on the other hand it is your intent to relax your specimens before preservation, or to keep your animals alive and healthy, transportation becomes somewhat more challenging. Bring a thermometer and try to maintain the temperature in the transport vessel as close as possible to that prevailing in the natural environment. Small thermos jugs are ideal to transport freshwater living snails, although this can be impractical if a large number of sites are to be visited in a single trip. A low-cost option is to accumulate samples in sturdy plastic bags or milk jugs inside a single large cooler. The tops can be cut off the jugs to remove the animals upon return from the field. Some collectors prefer to transport living snails in wet vegetation. Be sure to field-label containers of living snails on the outside - live snails may consume any slips of paper dropped among them.

Living snails will need transfer to suitable holding facilities promptly upon arrival at the home or laboratory. Ideally, aquaria should be established and conditioned before departure on a live collecting trip. Alternatively, you can transport carboys of water from the field and set up fresh aquaria on your return.

Take good notes for each collection. Record the locality as specifically as possible, ideally on site, completing as many of the first 14 fields of Table 21.1 as possible. Habitat notes and environmental observations are often useful. Upon return from the field, the safest and most systematic approach is to assign each lot a catalog number and to record data in a hardbound journal and/or an electronic database. An Excel spreadsheet formatted in the template of Table 21.1 is available for download from the FWGNA site.

21.5 PREPARATION AND STORAGE

The vast majority of all freshwater mollusk collections, whether they are in national museums or in private cabinets, are of shell. As most freshwater gastropods are not large of body, the preservation of their shells is best accomplished by drying, ideally

Table 21.1 Database fields in use by the Freshwater Gastropods of North America project.

1. Hydrological Unit Code (U.S. Geological Survey system)
2. Site number (catalog number)
3. Date (mm/dd/yyyy)
4. County (record two if on county line)
5. Project (or funding source, if any)
6. Water body name
7. Common location (e.g., "2 km W of Dumpton.")
8. Road No. (route number at bridge or access point)
9. River Basin
10. State (record two if on State line)
11. Latitude (decimal degrees)
12. Longitude (decimal degrees)
13. Collector's name(s)
14. Scientific Name (genus and species)

in some out of the way place where the odor will not become a problem. It can be desirable to take specimens through one or two changes of alcohol over several days before drying them. This seems to mitigate the odor, lowering the intensity although perhaps prolonging the duration. Another approach is to freeze specimens in a container of water overnight. They generally die in an extended condition, and can be thawed and cleaned with forceps.

It may be necessary to clean the largest specimens, especially the ampullariids and viviparids, by boiling. The animals should be placed in a pot of cool water and warmed to boiling gradually. The meat can then be hooked out with a pin, and the operculum saved in the aperture with a plug of cotton.

The dry shells of freshwater gastropods are often small and fragile. For this reason, they are more commonly stored in enclosed containers than most mollusks. Clear glass shell vials plugged with polyester are best, or clear plastic snap-top boxes, either of which may be purchased at specialized biological supply companies. Clear plastic pill bottles can be purchased at the pharmacy. Labels with data should be included with each lot of shells, as described in Chapter 14.4.

From a scientific standpoint, the preservation of soft part anatomy in freshwater gastropod collections can be very important. The application of

DNA methods is also becoming more widespread with each passing year. Thus increasingly we find scientific collections stored in alcohol, 80% ethanol being the recommended standard. Ethanol concentrations of 90% or higher are favored by researchers planning DNA studies, although such high quality reagents are not readily available to the general public. The "rubbing" alcohol stocked by ordinary pharmacies is often 70% ethanol "denatured" with acetone or similar organic solvents. Other rubbing alcohol formulations, such as 70% or 90% isopropanol, can be used to preserve specimens but are not ideal. Because the upper regions of gastropod anatomy are especially liable to decomposition, it is a good idea carefully to crack the shells of a few individuals before placing them in alcohol.

Formaldehyde, which can be hazardous, is not recommended for general use. Some workers recommend brief fixation (no more than a few hours) in 10% formalin before preservation in alcohol, although any contact with formalin will render tissue unsuitable for future DNA studies. Refer to Chapter 5 for information regarding vials and jars appropriate for storing wet collections.

Some researchers prefer to make anatomical observations on specimens that have been relaxed before preservation. Menthol crystals, available from your pharmacist, are among the most convenient of the variety of chemicals used for this purpose. Other anesthetics, such as chloretone, nembutal, or chloral hydrate are more difficult to obtain. Simply transfer animals to be relaxed into a shallow vessel of water (perhaps 1 cm), float a large menthol crystal (or several small ones) on the surface, and leave them cool and undisturbed. Periods of 12-24 hours are typically required for complete relaxation, but decomposition can follow shortly thereafter. Specimens should be probed periodically (a touch to the tentacle will suffice) and transferred to alcohol promptly after death. Menthol crystals can be dried and reused (see Chapter 2.5 for more on relaxing or narcotizing).

21.6 IDENTIFICATION

The identification of freshwater gastropods presents a greater challenge than one encounters with marine or even terrestrial species, at least in North America. The most comprehensive guide available at present is J.B. Burch's *North American Freshwater Snails* (Burch 1989). The work was originally published by the U.S. Environmental Protection Agency, and re-published in the journal Walkerana (Burch and Tottenham 1980, Burch 1982, 1988). Burch's work includes illustrations, historic ranges, synonyms, and a dichotomous key to the species level for most taxa. Other (shorter) references useful at the national level include the keys of Pennak (1989) and Brown (1991).

In addition to the above, there have been a fair number of regional surveys, species lists, and systematic reviews of taxa helpful in special situations. Especially notable are the works for the following regions: Canada (Clarke 1973, 1981), Colorado (Wu 1989), Connecticut (Jokinen 1983), Florida (Thompson 1984), Missouri (Wu *et al.* 1997), New York (Jokinen 1992), and North Dakota (Cvancara 1983). F.C. Baker authored a large and comprehensive monograph on the Lymnaeidae (1911), and his similarly ambitious work on the Planorbidae was published posthumously in an incomplete form (1945). Both of these works were rendered somewhat obsolete by the global-scale monographs of Hubendick (1951, 1955). The Physidae have recently been monographed by Wethington (2003). For information on aids to identifying non-North American taxa see Chapter 9.2.3 and 9.2.6.

The Freshwater Gastropods of North America (FWGNA) project is a long-term, collaborative effort to survey, map, and monograph the entire continental fauna north of Mexico. It is anticipated that both conventional print and electronic resources will be developed to facilitate the identification and conservation of these remarkable animals. A guide to the freshwater gastropods of South Carolina is on line now, with plans to extend throughout all southern Atlantic drainages in the near future. A complete list of all reference materials useful for the identification of North American freshwater gastropods published since 1900 can be found on the FWGNA website (see Section 21.9 for the website address), as well as links to a small but growing number of online resources and databases relevant to freshwater snails.

21.7 AQUARIUM CULTURE

Freshwater snails make interesting pets. Most snail species seem to adapt well to life in standard aquarium conditions, and a growing number of varieties are sold specifically to the hobbyist. Such casual interest in snails as may be displayed by the typical customer in a hobby shop almost certainly derives from the search for additions to aquaria featuring fish. But culturing fish and snails together can be detrimental to the former, and is never good for the latter. Most fish will eat snails, especially the smaller and more fragile pulmonates, and some tropical fishes may require heated waters, never necessary or even desirable for gastropod culture. On the other hand, given the right conditions the populations of some pulmonates can rapidly increase in a fish tank, generating a great deal of toxic ammonia and devouring expensive ornamental plants. It is best to raise freshwater snails by themselves.

The following is a brief review of the freshwater snails commonly available in pet shops, and a bit about culturing them. Have fun with these, but please do not release them into the wild. Exotic gastropod species can multiply in great numbers, and some are documented pests. But even though the adverse consequences of releasing aquarium species to the environment may be less than obvious in many cases, most of us simply prefer that natural communities remain undisturbed.

21.7.1 *Pomacea*. The most popular gastropod pets today are the large and gaudy ampullariids, generally labeled "apple snails" in the aquarium shops, but sometimes also called "mystery snails" (Perera and Walls 1996). Twenty years ago, the most common species was the North American native *P. paludosa*. More recently, the South American *P. bridgesi* (Reeve, 1856) has achieved widespread popularity by virtue of the marvelous color varieties available. Clever and enterprising breeders have brought to market diverse colors of shell and body bearing such names as "golden," "ivory," "blue," and "tuxedo." Also more recently available in pet shops is the "Giant Peruvian" or "Inca" snail, *Pomacea maculata* (Perry, 1810).

Marisa cornuarietis (Linnaeus, 1758) is a planispiral ampullariid native to South and Central America, now often sold in aquarium shops as a "giant Colombian ramshorn." There are both banded and unbanded forms - the unbanded sometimes called a "golden ramshorn".

Macrophytic vegetation comprises the ordinary diet of ampullariids such as *Pomacea* and *Marisa,* as well as occasional small invertebrates, including other snails. Do not attempt to culture ampullariids with aquatic plants or smaller snails about which you care. They seem to grow well on a diet of lettuce, especially Romaine. (Iceberg lettuce tends to cloud the water.) See the bibliography and web resources for special references, both print and electronic, on the apple snails.

21.7.2 *Bellamya* (or *Cipangopaludina)*. This Asian viviparid is widely marketed in the U.S.A. as a "Japanese" or "Chinese trap-door snail," for use primarily in outdoor water gardens. *Bellamya* may also sometimes be called a "mystery snail," although this name tends to confuse them with *Pomacea*. By virtue of the trap-door snail's ability to filter-feed, and its benign relationship with aquatic vegetation, nurseries selling water lilies and other aquatic plants often promote these gastropods to clean the pond water. Like almost all other freshwater snails, however, they will probably do well in the aquarium provisioned with ordinary fish food.

21.7.3 *Melanoides tuberculata*. *Melanoides tuberculata* is an old world thiarid now ubiquitous through tropical and subtropical regions worldwide. It is marketed in aquarium shops as a "Malayan Needle Point." *Melanoides* is among the hardiest of the prosobranchs, and by virtue of its parthenogenic mode of reproduction, one of the most easily cultured in standard aquarium conditions.

21.7.4 *Helisoma trivolvis* [*Planorbella trivolvis*]. *Helisoma trivolvis* (Say, 1817) is a North American native that has long standing in the aquarium trade. It is usually just called a "ramshorn snail." Albinos actually look red, since absence of body pigmentation uncovers their hemoglobin content. (The snails then sell for a premium as red ramshorns). *Helisoma* [*Planorbella*] enjoys lettuce, in addition to fish food, and seems indifferent to aeration. Given an occasional feeding, and a rare water change, it will thrive.

21.7.5 Physa [Physella]. *Physa* most often enters the hobbyist's aquarium as a contaminant on water plants, although they make active and interesting pets. It has recently been shown that the common and widespread North American species, *P. heterostropha* (Say, 1817), *P. integra* (Haldeman, 1841), and *P. virgata* (Gould, 1855), are all synonymous with the old world *P. acuta* Draparnaud, 1805, making *P. acuta* the world's most cosmopolitan freshwater gastropod. These are the cockroaches of malacology, thriving in all conditions of food and culture, and quick to reproduce. Their weak mouthparts make them less dangerous to aquatic vegetation than *Helisoma*, for example, and more dependent on a diet of algae and/or fish food.

21.8 SUMMARY

The North American freshwater snails do not tend to grow as large or as colorful as most groups of mollusks, and consequently do not often attract the attention of hobbyists. Yet they are widespread, easily collected, and adapt easily to the home aquarium. The diversity of freshwater gastropods, and the variety of environments they inhabit, can yield great intellectual rewards to the malacologist, amateur or professional, with the dedication to pursue them.

21.9 WEB RESOURCES

Nets and Freshwater Sampling Gear:
Freshwater Gastropods of North America Project:
  <www.cofc.edu/~dillonr/fwgnahome.htm>
Freshwater Gastropods of South Carolina:
  <www.cofc.edu/%7Edillonr/FWGSC/>
F. G. Thompson (1984) Freshwater Snails of Florida:
  <www.flmnh.ufl.edu/natsci/malacology/fl-snail/snails1.htm>
A. M. Cvancara (1983) Aquatic Mollusks of North Dakota:
  <www.npwrc.usgs.gov/resource/distr/invert/mollusks/mollusks.htm>
U.S. Fish & Wildlife Service Threatened and Endangered Species System:
  < ecos.fws.gov/tess_public/TESSWebpage>
NatureServe Online Encyclopedia of Life
  < www.natureserve.org/explorer/>
Stijn Ghesquiere's Apple Snail website:
  <www.dds.nl/~snc/>

21.10 LITERATURE CITED

Aldridge, D. W. 1983. Physiological ecology of freshwater prosobranchs. *In:* W. D. Russell-Hunter, ed., *The Mollusca*, Vol. 6. Academic Press, Orlando, Florida. Pp. 329-358.

Baker, F. C. 1911. *The Lymnaeidae of North and Middle America, Recent and Fossil*. Special Publication, No. 3. Chicago Academy of Natural Sciences, Chicago. 539 pp.

Baker, F. C. 1945. *The Molluscan Family Planorbidae*. University of Illinois Press, Urbana, Illinois. 530 pp.

Baker, F. C. 1942. Collecting and preserving fresh water snails. *The American Malacological Union, Eleventh Annual Report* 11: 5-9.

Brown, K. M. 1991. Gastropoda. In: J. H. Thorp and A. P. Covich, eds., *Ecology and Classification of North American Freshwater Invertebrates*. Academic Press, New York. Pp. 285-314.

Burch, J. B. 1982. North American freshwater snails: Identification keys, generic synonymy, supplemental notes, glossary, references, index. *Walkerana* 4: 1-365.

Burch, J. B. 1988. North American freshwater snails: introduction, systematics, nomenclature, identification, morphology, habitats, distribution. *Walkerana* 2: 1-80.

Burch, J. B. 1989. *North American Freshwater Snails*. Malacological Publications, Hamburg, Michigan. 365 pp.

Burch, J. B. and J. L. Tottenham. 1980. North American freshwater snails: Species list, ranges, and illustrations. *Walkerana* 3: 1-215.

Clarke, A. H. 1973. The freshwater molluscs of the Canadian Interior Basin. *Malacologia* 13: 1-509.

Clarke, A. H. 1981. *The Freshwater Mollusks of Canada*. The National Museums of Canada, Ottawa. 446 pp.

Cvancara, A. M. 1983. *Aquatic Mollusks of North Dakota*. Report of Investigation, Vol. 78. North Dakota Geologic Survey. 142 pp.

Dillon, R. T. 2000. *The Ecology of Freshwater Molluscs*. Cambridge University Press, Cambridge. 509 pp.

Hubendick, B. 1951. Recent Lymnaeidae. Their variation, morphology, taxonomy, nomenclature, and distribution. *Kungl Svenska Vetenskapasakademiens Handlingar* **3**: 1-223.

Hubendick, B. 1955. Phylogeny in the Planorbidae. *Transactions of the Zoological Society of London* **28**: 453-542.

Jokinen, E. H. 1983. The freshwater snails of Connecticut. *Connecticut Geological and Natural History Survey Bulletin* **109**: 1-83.

Jokinen, E. H. 1992. The freshwater snails (Mollusca: Gastropoda) of New York State. *New York State Museum Bulletin* **482**: 1-112.

McMahon, R. F. 1983. Physiological ecology of freshwater pulmonates. *In:* W. D. Russell-Hunter, ed., *The Mollusca*, Vol. 6, Ecology. Academic Press, New York. Pp. 359-430.

Pennak, R. W. 1989. *Fresh-water Invertebrates of the United States*, 3rd Edition. John Wiley and Sons, New York. 628 pp.

Perera, G. and J. G. Walls. 1996. *Apple Snails in the Aquarium*. T. F. H. Publications, Neptune City, New Jersey. 121 pp.

Russell-Hunter, W. D. 1978. Ecology of freshwater pulmonates. *In:* V. Fretter and J. Peake, eds., *Pulmonates*, Vol 2A. Academic Press, New York. Pp. 336-83.

Strayer, D. L. 1990. Freshwater Mollusca. *In:* B. L. Peckarsky, P. R. Fraissinet, M. A. Penton, and D. J. Conklin, eds., *Freshwater Macroinvertebrates of Northeastern North America*. Cornell University Press, Ithaca, New York, Pp. 33-372.

Thompson, F. G. 1984. *The Freshwater Snails of Florida, A Manual for Identification*. University of Florida Press, Gainesville, Florida. 94 pp.

Turgeon, D. D., J. F. Quinn, Jr., A. E. Bogan, E. V. Coan, F. G. Hochberg, W. G. Lyons, P. M. Mikkelsen, R. J. Neves, C. F. E. Roper, G. Rosenberg, B. Roth, A. Scheltema, F. G. Thompson, M. Vecchione, and J. D. Williams. 1998. *Common and Scientific Names of Aquatic Invertebrates from the United States and Canada. Mollusks*, 2nd Ed. American Fisheries Society Special Publication 26. 526 pp.

Wethington, A. R. 2003. *Phylogeny, Taxonomy, and Evolution of Reproductive Isolation in* Physa *(Pulmonata: Physidae)* Ph.D. Dissertation, University of Alabama, Tuscaloosa, Alabama. 119 p.

Wu, Shi-Kuei. 1989. *Colorado Freshwater Mollusks*. Natural History Inventory of Colorado, Vol. 11. University of Colorado Museum, Boulder, Colorado. 117 pp.

Wu, Shi-Kuei., R. D. Oesch, and M. E. Gordon. 1997. *Missouri Aquatic Snails*. Missouri Department of Conservation, Jefferson City, Missouri. 97 pp.

CHAPTER 22

TERRESTRIAL GASTROPODA

TIMOTHY A. PEARCE
AYDIN ÖRSTAN

22.1 INTRODUCTION

Collecting land snails can range from a pastime to a serious scientific pursuit resulting in significant contributions to scientific knowledge. This chapter summarizes information about where and how to collect land snails. We provide information about their macro- and microhabitat needs that will help you locate good collecting places. We discuss field collecting methods and equipment, as well as methods for separating snails from leaf litter or other material brought back from the field. We remind you that record keeping is important in the field. We will discuss methods for preserving both shells and soft parts of specimens, and literature that should help you to identify your finds.

We encourage you to go beyond using this chapter, and contact workers at museums and other land snail collectors for hints or assistance. They might share with you information on techniques and local collecting spots, and they might be pleasant field companions on a collecting trip.

22.2 BIOLOGY OF LAND SNAILS

Land snails represent multiple invasions of land from marine snail ancestors. The Pulmonata are the most successful group of land snails in numbers of species and in diversity of habitats. In contrast, the operculate snail groups that invaded land are mostly confined to the moist tropics (Solem 1974). The systematics of operculate land snails is currently undergoing revision (Ponder and Lindberg 1997, Barker 2001); results to date confirm the idea that operculate snails invaded land multiple times, but we do not yet know how many times. Pulmonata is apparently a monophyletic group, but the operculate group of snails formerly known as "Prosobranchia" is clearly not monophyletic, and the group of land snails derived from them is polyphyletic. Here we refer to these non-pulmonate land snails as operculate land snails.

Nearly all the operculate land snails possess an operculum, or door, for closing the shell when the animal withdraws. Shell shapes and surface sculptures vary; Figures 22.1E and G show operculate snails with particularly ornate surface sculpture. Operculate snails have one pair of tentacles, with eyes at the bases of the tentacles, and always have a coiled shell into which they can retract (i.e., no slugs). There are separate male and female individuals. Although derived from marine ancestors, the land operculate snails have lost their gills; the neck is not fastened to the mantle anteriorly so the entire mantle cavity is open to the flow of air and the head is prolonged into a proboscis, distinctly separated from the foot (Solem 1974). In temperate areas, fewer than 1% of land snails are operculate, while in the American tropics, about 50% of the species are operculate.

In contrast to the operculate snails, pulmonate land snails never possess an operculum, the head is not prolonged into a proboscis, and instead of an open mantle cavity, the mantle collar is fused to the neck of the snail, so only the pneumostome (breathing hole) connects the mantle cavity to the outside world (Solem 1974). The shells are nearly

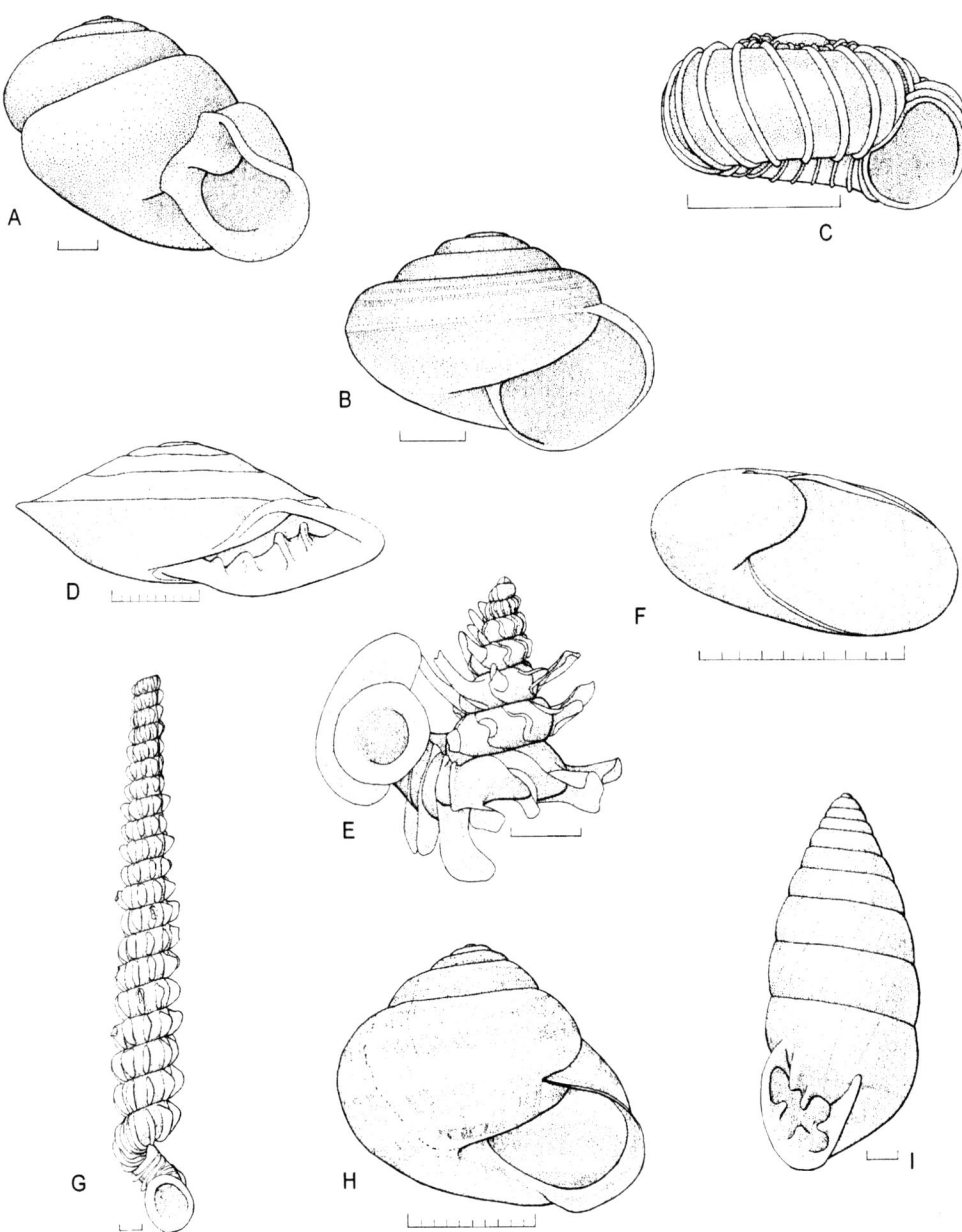

Figure 22.1 Diversity of form in land snails.
A. *Haploptychius andamanicus* (Benson, 1860), CM 62.13336, Andaman Islands, showing skewed coiling axis. B. *Sitala aliciae* Emberton and Pearce, 2000, Madagascar. C. *Planogyra asteriscus* (E. S. Morse, 1857), CM 64976, Isle d'Orleans, Quebec, Canada, showing periostracal ribs. D. *Pleurodonte peracutissimus* (C. B. Adams, 1845), CM 62.2712, Jamaica, carinate shell. E. *Opisthostoma mirabile* E. A. Smith, 1893, CM 65201, Borneo, dextral shell appears to coil sinistrally because coiling direction reverses near end of growth. F. *Vitrinizonites latissimus* (J. Lewis, 1875), CM 65241, Great Smoky Mountains, Tennessee, semislug shell with large aperture. G. *Gongylostoma gemmata* (Pilsbry, 1927), CM 65202, Guane, Pinar del Rio Province, Cuba, narrow with hollow tubercles. H. *Cepaea nemoralis* (Linnaeus, 1758), CM 62.1230, England. I. *Fauxulus capensis* Küster, 1841, CM 62.20188, Cape of Good Hope, South Africa, sinistral. Tickmarks on scale bars are in millimeters. CM = Carnegie Museum. Illustrated by Jessica P. Domitrovic

all dextrally coiling, but some sinistral species are known (Figure 22.1I). Shells may be relatively simply coiled throughout growth (Figure 22.1B), while other species develop a reflected or thickened lip (Figures 22.1A, D, H, I) or apertural barriers (Figures 22.1A, D, I), at the end of growth. Some species have a carinate periphery (Figure 22.1D), and some species have a twisted axis of coiling (Figure 22.1A). The surface sculpture is often smooth or with growth ridges, some species have hairs, and others have periostracal processes such as ribs (Figure 22.1C). Pulmonates are hermaphrodites, meaning that one individual is both male and female; in contrast to the separate sexes in the operculate land snails. Pulmonates comprise three groups: Stylommatophora, Basommatophora, and Systellommatophora. All three of these groups include terrestrial species, although the majority of terrestrial gastropods are Stylommatophora. The Stylommatophora nearly always have two pairs of tentacles that can be retracted into the head through inversion, with eyes situated on the tips of the upper (posterior) pair of tentacles. The few terrestrial species of Basommatophora apparently evolved from freshwater pulmonates, and have two tentacles with eyespots at the bases of the tentacles. The Systellommatophora are mostly tropical slugs having two pairs of tentacles with eyes on the tips of the upper tentacles, but unlike the Stylommatophora, the eyes are contractile, and cannot be inverted (retracted) into the head.

Slugs evolved from snails by reducing and internalizing the shell. In most slugs, the shell is reduced to a flat plate (e.g., Limacidae, *Ariolimax* - banana slugs), a few calcareous granules (e.g., *Arion*), or may be completely absent (e.g., Philomycidae, Veronicellidae). Semi-slugs are between snails and slugs; they have an external shell (Figure 22.1F), but the shell is too small for the animal to withdraw into. There are more species of semi-slugs than there are of slugs (Solem 1974). Most semi-slugs live in tropical areas. A snail has a muscular foot and the internal organs are within the shell. In a semi-slug, the internal organs are in a hump on the slug's back and the foot is usually muscular but it might be partly hollowed to accommodate the internal organs. In a slug, the internal organs are in a hollow cavity occupying most of the foot. Slugs evolved independently from snails at least 10 times and semi-slugs evolved from snails at least 25 additional times (Pearce, unpublished). Consequently, slugs excluding snails are not a natural group (they are polyphyletic, not monophyletic).

22.3 WHERE TO FIND LAND SNAILS

Terrestrial gastropods can be found in moist woodland and arid regions. In this section, we will discuss some of the environmental factors to consider when searching for land snails.

22.3.1 General considerations. While many large species of land snails exist, the vast majority of land snail species are tiny, less than one centimeter in greatest dimension, some being only one millimeter, so more than casual searching is required to find the small ones. Because the casual collector more easily finds larger land snails, a greater proportion of specimens in museums are the larger species, and the larger species are better known. Consequently, the potential for making important new discoveries is greater if you concentrate on collecting tiny land snails, e.g., most of the new species being described these days are the smaller species.

Land snails occur in practically any terrestrial habitat that has some source of moisture and is ice-free for at least a few weeks of the year. Most land snail species occur in forests that retain moisture even during dry periods, but some species, such as *Vallonia* spp. and *Cochlicopa* spp. occur in meadows and fields. Other species occur in seasonally hot and arid regions, for example, *Cerion* spp. on Caribbean Islands, *Holospira* spp. in the southwestern USA and Mexico, and *Albinaria* spp. in coastal Turkey and Greece. While *Sphincterochila boissieri* (Charpentier, 1847) lives in the deserts of the Middle East where sometimes a year may pass between rains (Schmidt-Nielsen *et al.* 1971), *Truncatella* spp. live under the rocks and piles of dead sea weed on marine beaches where they are frequently covered by the waves. Land snails have been recorded above the tree line and in tundra areas toward the Poles. Land snails also occur in urban areas such as roadsides, gardens, greenhouses, and probably in your backyard. Identifiable shell fragments of land snails may be found even in pellets of predatory birds that eat snails, such as owls (Mienis 1971).

While experience is very important in knowing where to look for snails, understanding the ecological requirements of snails can allow useful and usually accurate predictions about which sites will contain the most numerous snails. When looking for snails, think like a snail. Choose places to look that have hospitable conditions throughout the year because snails do not migrate great distances. Although understanding their ecological requirements can be helpful for finding snails, chance can also be a factor in locating them. The relatively small home ranges of snails coupled with the tendency of some species to aggregate, especially before they become dormant, might explain why one log on the forest floor may harbor numerous specimens, whereas another log 10 meters away may have none.

One of the most important needs of land snails is moisture (Riddle 1983) because land snails are like leaking bags of water trying to survive on land. Generally, land snails have mechanical or behavioral strategies for dealing with temporary periods of dryness lasting several weeks to months (or more than a year for desert species). Another requirement for snails is a source of calcium for making shells, although slugs need less calcium. Areas of limestone (calcium carbonate) are famous among land snail collectors for having greater abundances and diversities of snails.

22.3.2 Macrohabitat requirements. The amount of moisture, altitude, topography, rock type, soil composition, and tree species are some of the interdependent factors that influence the distributions of land snails in complex ways (Coney *et al.* 1982). In addition, some snail species are tolerant of a wider range of macrohabitat characteristics than are other species. Therefore, it is not practical to specify definite conditions that would satisfy the requirements of all species. However, we can make some useful generalizations. In general, forested areas tend to be moister than cleared fields, and areas with leaf litter in well-shaded deciduous forests are good places to find snails. Some species, such

as *Vertigo gouldi* (A. Binney, 1843) might prefer conifer-dominated forests (Kralka 1986). Primary forest, especially in tropical areas, will usually have richer snail faunas than secondary forest. Clench (1974) suggested looking above the flood line when searching for snails along streams and lakes because land snails generally cannot tolerate immersion. On the other hand, land snail shells can often be found near rivers in the drift debris deposited by floods (see below).

Slope and aspect (orientation to the sun) of the land can influence moisture. Gentler slopes probably drain more slowly, and land sloping away from the sun (e.g., north facing slopes in the Northern Hemisphere) may retain moisture longer during dry periods. More exposed topography such as hill and ridge tops are probably drier than slopes or valleys and may be poorer sites for locating snails (Emberton *et al.* 1996); however, some *Ashmunella* spp. in the southwestern U.S.A. live exclusively on treeless rocky slopes (Pilsbry 1940). Cain (1983) speculated that aspect might be especially important to snails at the climatic edges of their ranges. Although these topographical features may be less important influences on snail distribution than others such as geography, climate, forest type, and rock type, considering topography might help in locating denser populations of snails.

Abundance and diversity of snails are likely to be lower in deserts. Because arid regions typically have high temperatures with extreme daily and monthly temperature ranges, low and infrequent rainfall, low humidity, and many sunny days with high light intensity, mollusks of arid regions consist mainly of forms with a wide tolerance for temperature, moisture, and sunshine. Snails can be found in the desert regions of the American southwest, in eastern and southeastern California, Arizona, New Mexico, northern Mexico, and the western parts of Texas (Gregg 1974).

Open meadows and pasturelands, including those with forest cover, are usually poor for snails unless there are many logs (Clench 1974), but one can find a few species near grass roots (Anonymous 1929). Trampling may influence the low abundance of snails in pastures (Chappell *et al.* 1971). When humans clear forests for agriculture, snail diversity and abundance decrease, and species composition changes (Evans 1972). On the other hand, meadows are not always poor for land snails. Diversity and abundance of snails in meadows was found to be greater than that in forests on the Kuril Islands in far eastern Russia, where the climate is so rainy that the meadows are almost constantly moist (Pearce 1997).

Areas of limestone are particularly good for land snails, both in abundance and diversity. The reasons for this pattern have not been fully explored, but limestone may be good for snails because it provides abundant calcium, or because it often erodes into deep cracks, providing a refuge for the snails (Burch and Pearce 1990). Limestone ridges are usually rich in snails, especially if there is ample shade and much moss and dead leaves at the base of ledges (Anonymous 1929, Clench 1974, Nekola and Smith 1999). LaRocque (1974a) noted that *Gastrocopta, Hawaiia,* and *Zonitoides* live in large numbers in limestone quarries, under loose blocks of rock, and other species can be abundant in the soil between limestone ledges.

Many species of *Eremarionta, Sonorella, Sonorelix, Helminthoglypta, Ashmunella,* and *Radiocentrum* can be found on rocky hillsides, particularly in rockslides, and sometimes in dry weather, under or among the roots of yucca and agave plants. *Holospira* is found on hillsides where limestone is present, at some times of the year on rock surfaces, but in dry, hot weather, it is found beneath rocks or desert vegetation (Gregg 1974).

22.3.3 Microhabitats. Within a larger habitat, snail abundance varies with microhabitat. Different species live in different microhabitats. Therefore, your choice of microhabitat will influence which species and how many of them you will find. Generally, look for areas that are likely to have a supply of moisture throughout the year, for example, in deep piles of leaf litter, in depressions, under logs and on the undersides of logs, in cracks among rocks, and among moisture-loving plants such as ferns. If the log is rotten enough, break it apart and you might find slugs in the outer few centimeters of the rotten wood (Anonymous 1929).

Native snails in North America tend to be most abundant in leaf litter, under and in rotting logs, and around the base of stones, while introduced species tend to be more urban and may be found under old boards and bricks, and beneath damp litter in towns and cities (Burch and Pearce 1990). Small species of snails spend most of their time within leaf litter (Boag 1985). While snails are most abundant in the top 5 cm of leaf litter, they can be found as deep as 20 cm in the soil (Locasciulli and Boag 1987). Old brush piles might have many snails, but branches of fresh, green brush do not have many snails (Anonymous 1929). Under or near decaying logs and fallen trees is a good place to find snails, probably because logs retain moisture during drier periods. Many small species, for example, *Strobilops* spp., and the juveniles of larger species may be found in the powdered wood that accumulates on rotting trunks. Look for snails in shaded areas of ravines having ample ground moisture. Snails and slugs can be found under the bark of standing and fallen trees, and snails can be found in the crevices of the bark of some living trees and shrubs. Rock outcrops on wooded hillsides that are surrounded by leaf litter are usually good places to find snails.

Considering these needs of snails, we can expect low abundance and diversity of snails in managed forests from which fallen logs have been removed, or in recently forested areas having only sparse leaf litter and few rotting logs. Likewise, pastureland recently converted to forest may completely lack snails, especially large species, if the area lacked snails before trees were planted and snails have not had a chance to colonize.

Some species occur on plants, in trees, and on their epiphytes. Tree-dwelling snails are found more often in tropical areas, but also occur in temperate areas. Smaller species in temperate areas, including Succineidae, climb on plants including plantain (Anonymous 1929), grasses and sedges, and other herbs (Figure 22.2). Some noteworthy tree-dwelling species occurring in North America include *Gastrocopta corticaria* (Say, 1816) of northern

North America, some of the *Vertigo* and *Columella* species, and *Liguus fasciatus* (Müller, 1774) of southern Florida. In addition, many forest snails and slugs, including *Neohelix albolabris* (Say, 1817), *Mesodon thyroidus* (Say, 1816), *Philomycus* spp. and *Anguispira* spp., climb on live or dead trees on warm rainy days, especially in the evenings (Ingram 1941).

Some species are characteristic of open areas such as grasslands, especially grasslands on limestone. For example, many *Gastrocopta, Cochlicopa, Vallonia,* and *Pupilla* species occur in grassland. Their numbers tend to be greater in old shell middens of mussel or oyster shells, probably because of the calcium source.

Many people have reported moss as a good place to find snails, although we have generally had poor luck finding snails in moss. LaRocque (1974a) found *Striatura, Vertigo, Zonitoides,* and *Planogyra* from sifting dried moss from the area around Ottawa, Canada. He suggested collecting moss from the edges of swamps, and from shade under trees or at the bases of cliffs. On the other hand, few snails can be gotten from moss taken from areas flooded in the spring or from more exposed situations.

Besides areas having moisture, look for areas having refuges that might protect snails from predators as well as from desiccation. For example, slugs are more common in gardens having objects lying on the ground, such as loose rocks or boards under which they can hide, and in less cultivated fields that have air spaces in the soil. Cultivation decreases the air spaces in the soil and can help to decrease slug populations (South 1992, Henderson 1996).

During dry and hot or cold periods, land snails aestivate by withdrawing into their shells and becoming dormant. During aestivation, they may adhere to a leaf or rock and secrete one or more epiphragms, or membranous partitions between the animal and the aperture. Some snails aestivate on stems of plants. Other species aestivate in deep cracks that provide some protection from desiccation, such as deep in

Figure 22.2 *Oxyloma retusa* **(Lea, 1834) on** *Typha* **sp. (cattail)**.

a rock pile. During low temperatures of winter, snails crawl into the lowermost levels of the leaf litter and sometimes burrow into the soil (Roscoe 1974). Therefore, during seasonal extremes of heat, cold, or dryness, when snails are usually inactive, it will be difficult to collect those species that hide in deep cracks, in rotting logs, under rocks, and in the soil.

Knowledge of microhabitats in the desert environment is useful toward success in collecting desert snails (Roscoe 1974). In many instances, snails survive in deserts by living in rockslides (Gregg 1974) because rockslides provide deep cracks for refuges. If you are lucky enough to be collecting snails in arid areas during rainy weather, you should have a much easier time finding live ones (Hochberg *et al.* 1987); otherwise you will probably need to move hundreds of kilograms of rocks to find the aestivating snails.

In addition to looking in favorable places, we encourage you to spend a little time looking in places you do not expect to find many snails. Otherwise, you might miss species of unusual habitats. For example, some species in the U.S.A. have been found only in caves (Hubricht 1962) (if you are not experienced in exploring caves, seek help from experienced spelunkers, because entering a cave is potentially dangerous) and on rare occasions, land snails can be found in *Sphagnum* bogs.

If you are looking for particular species of land snails, check for information on habitats in Pilsbry (1939-1948) and Hubricht (1985) for North America, and Kerney and Cameron (1979) or Kerney *et al.* (1983) for Europe.

22.4 FIELD METHODS AND EQUIPMENT

22.4.1 Methods, general considerations. The type of sampling you do will depend on your purpose. For example, in surveying which snails occur in an area, you would like to discover all the species there, and it will not matter whether they are alive or dead. Similarly, for a diversity study, you need to know all the species and you probably need to know the relative abundances of the species. In such scientific sampling that seeks to quantify snails per surface area or per volume of material, it is more important to get a sample of the material containing the shells than it is to pick up individual shells on the surface (LaRocque 1974b). This consideration holds for both modern and fossil shells.

If you are studying intraspecific variation among different populations of one species or interspecific variation among closely related species, you may need to collect large numbers of shells from different locations. Consult Boycott (1928) for invaluable guidelines on how to carry out such collections. In general, shells for a variation study must be collected without bias for the conchological characters visible to the collector, such as dimensions, shape, color, etc. The easiest way to obtain an unbiased collection is to take every adult shell one finds in a given location (Boycott 1928). If the available shells are too numerous, one may take every adult shell until a pre-set limit, say 100, is reached.

Any of the collection and survey methods discussed below, and others we have not discussed, may be used to collect and study the land snails of a location. Several authors recommend a combination of visual search and litter sampling (Lee 1993, Emberton *et al.* 1996). Menez (2001) evaluated the amount of effort necessary to sample Mediterranean sites adequately and concluded that 2.5 hours of visual search plus processing 4.2 liters of substrate were sufficient.

However, species compositions and relative abundances obtained may differ depending on the method used. Runham and Hunter (1970) found that trapping and soil sampling methods in the same area resulted in different proportions of slug species. Oggier *et al.* (1998) compared the results obtained with three of the methods discussed here (mark-release-recapture, cardboard trapping, and soil sampling) and found that soil sampling yielded the most species. However, the soil sampling and mark-release-recapture procedures were more labor intensive than cardboard trapping, and the three methods yielded different proportions of species. They concluded that soil sampling would be most reliable for obtaining complete species lists in small areas, whereas cardboard trapping would be suitable for examining populations of selected species over larger areas.

While collecting shells haphazardly over an indefinite area can result in an interesting assemblage of species for a personal shell collection, systematic surveying can have more scientific and lasting value. Systematic surveys of selected areas ranging in size from a park (Cowie *et al.* 1994, Örstan 1999a) to a peninsula (Pearce and Italia 2002) or an island (Cameron 1986, Pearce 1994, Kerney 1999, Bieler and Slapcinsky 2000) can provide very useful results. Large areas can be surveyed systematically by dividing the area into smaller units, for example, from 1x1 to 10x10 km squares, and taking samples from each unit area. Snail colonies may be very small, and samples from even several meters away might harbor entirely different species (Anonymous 1929). Harry (1998: 8) found that colonies of *Carychium* in Michigan were only a couple of meters in diameter. Alternately, instead of considering every species, a study might concentrate on one species or a genus. For example, Welter-Schultes (1998) recorded every *Albinaria* species in 1 x 1 km UTM squares over about 6,000 km² on the island of Crete.

22.4.2 Visual search. You can find an adequate number of snails, especially larger ones, in any type of suitable habitat without any special equipment simply by looking in the appropriate places. Looking under logs or in leaf litter in moist depressions will usually reveal snails. Be sure to replace logs and rocks in their original positions when you have finished so remaining creatures can continue to survive. In some places such as marshes and mangroves, having nearly constant water, snails and empty shells are usually in the open or under accumulated debris. Similarly, snails are usually easy to find in seasonally hot, arid, limestone areas because most species become dormant in dry soil, in rock crevices, or simply attached to a rock surface. Since leaf litter is scarce or absent from such places, you can easily find empty shells accumulated on the soil surface around limestone rocks.

Many snails and slugs have diurnal patterns of activity on the surface of the ground during early morning and evening hours, so you might find more individuals during these times. An increase in evening activity is probably related to an increase in relative humidity that occurs when the temperature drops (Dainton 1954), although a circadian rhythm may also be involved (Blanc and Allemand 1993). However, during rainy periods with high humidity, land snails may often be found actively crawling regardless of the time of day (Roscoe 1974). For species occurring in arid areas, you will have more luck finding living snails if you wait for rainy weather (Hochberg *et al.* 1987).

Since snails are more active during moist periods, you may be able to induce snails to activity by adding water to an area where snails are suspected to be hiding. We have had success collecting living *Mesodon thyroidus* and *Neohelix albolabris* during dry weather from a mixed hardwood forest in Northern Michigan, using a watering can to sprinkle about 50 liters of water over about 600 m² of forest floor. We captured the snails when they emerged from hiding beneath the leaf litter.

Because snails tend to emerge from their hiding places at night, and be active during warm weather, you can collect them using a portable light. Our best collecting by this method has been in the first few hours after sunset, especially after an afternoon rain.

When searching for snails, be on the lookout for some snail shells that are typically covered with soil particles. *Gastrocopta* spp., *Catinella vermeta* (Say, 1829), and shells of some other species are somewhat camouflaged by the soil particles that stick to the outside of their shells.

For studies needing to be quantitative, you can get a measure of catch per unit effort by recording the number of snails gathered over how much time you searched. Searching for the same amount of time at a number of sites will allow you to make direct comparisons of the relative snail abundances among the sites. If necessary for your study, try to control other factors that might influence your success rate, such as weather and time of day or year. Because a person's ability to find snails improves over time, especially at first, be sure to train new assistants before including the results of their timed searches.

22.4.3 Leaf litter and soil sampling. Collecting leaf litter and removing the snails from it at home or in the laboratory has advantages and disadvantages. The main advantage is the good recovery of specimens less than 3 mm, which are almost completely overlooked in the field (Lee 1993, Emberton *et al.* 1996). Another advantage is that more time may be spent in the field collecting additional samples since processing will occur later.

Also, leaf litter and soil sampling allows quantitative sampling, either by area of the substrate sampled, or by volume of the material, and often recovers numerous examples of the species (Emberton *et al.* 1996). Lee (1993) found 28 species of land snails from a 2 liter soil sample from the Smoky Mountains, U.S.A. Sampling by area of substrate allows population estimates in terms of area. Sampling by litter volume allows relative population comparisons and it tends to be faster to gather a certain volume of leaf litter from promising-looking spots than to take all the litter from 1 m^2 or other size area. Litter sampling in Cameroon, Africa, revealed 97 species of land snails within one km^2 (de Winter and Gittenberger 1998).

The disadvantages of leaf litter sampling include transportation of bulky and heavy litter samples, labor-intensive separation of snails from litter, and the death of live specimens unless samples are processed quickly. The live snails in leaf litter samples from temperate or boreal areas can be stored for several months in a refrigerator (about 4°C). However, tropical species generally do not survive well in the refrigerator (C. Coney, pers. comm.).

For fossil mollusks, careful sampling from a measured section can provide detailed information about changes in the mollusk fauna over time (LaRocque 1966: 7, 1974b). Study of shells from midden deposits can reveal important information about past climates and how ancient humans changed the environment (Evans 1972). For fossil deposits including marl, loess, silts, and peaty material, find a place where a road cut or a river has exposed the deposit and take out a series of layers of the material. The thickness of layers is up to you; LaRocque (1974b) recommended sampling in layers 5 cm thick. Keep each layer in a separate plastic bag or other container and label it clearly, preferably with the distance up or down the section from some reference point, if possible. Also, note the nature and appearance of the deposit in each layer, to help with reconstructing the environment when the snails were alive.

22.4.4 Transporting soil. Transporting soil and leaf litter has potential to move harmful pests or diseases to new places where they can attack native organisms. Use good judgment when disposing of soil or leaf litter after you have finished picking out the snails. If the soil is from your local area (up to several hundred km away), it is probably safe to dump it in your yard or in the trash, because organisms in the soil have probably already had the opportunity to disperse to your area. However, if your sample is from across a mountain range or other barrier to dispersal, please consider sterilizing the soil before you dispose of it. Three methods for sterilizing soil are cooking at 121°C (250°F) for 2 hours, burning it, and soaking it in an oxidizing agent such as household bleach (5% sodium hypochlorite) diluted 1:4 with water, for 15 minutes.

One reason the agricultural inspectors are in airports and at international borders is to keep invasive pests, such as microorganisms in soil, from crossing international borders. If you plan to transport soil or leaf litter across an international boundary, consult the U.S. Department of Agriculture (or equivalent regulatory body) regarding permits and requirements. The U.S. Department of Agriculture currently regulates import of soil from all foreign sources, from Hawaii, Guam, Puerto Rico, the U.S. Virgin Islands, and from certain other parts of the U.S.A.

The only two methods currently approved by the U.S. Department of Agriculture for treating soil is dry heat at 121°C (250°F) for at least 2 hours, or steam heat at 121°C (250°F) for 30 minutes with 15 pounds per square inch pressure (103 kilo Pascals) (USDA 2001). Some malacologists have doused soil samples with ethanol and the agriculture inspection people let the samples into the U.S.A., but ethanol will not always satisfy agriculture inspectors. Heat treatment is not likely to harm shells, but will certainly kill any live snails in the sample, and might make recovery of certain biochemicals difficult, so be sure to note on the specimen label any heat treatment of the specimens, to inform future researchers. If you need to transport untreated soil into the U.S.A., see USDA (2001) for procedures.

22.4.5 Stream drift. Sometimes land snails may be found in stream drift accumulated around an obstruction such as a log, root, or stream bank (Gregg 1974, LaRocque 1974a). Shell material in stream drift has been concentrated for you in a natural process; empty shells float along with sticks and leaves, while soil and rocks sink and are removed from the drift. Flash floods in the desert can carry much organic debris, including snail shells, especially tiny species (Gregg 1974). Sieving and picking the material often recovers many smaller specimens.

The major disadvantage of collecting from stream drift is that one cannot always be sure where the shells originated. Some shells might have traveled far from their original location, and might include fossils. For example, shells from drift in Arizona are often a mixture, difficult to sort, of Recent specimens and late Cenozoic fossils. Therefore, stream drift specimens are unreliable for determining Recent distributions (Bequaert and Miller 1973: xiv). Also, as Boycott (1928) pointed out, shells accumulated by floods (or by the wind) may have been sorted by size and thus, would be unsuitable for analysis of variation. Therefore, be sure to indicate clearly in your records that the specimens are from stream drift.

22.4.6 Trapping. Trapping can be a successful method for collecting snails, including slugs, especially when large numbers of certain species are desired. Trapping methods probably do not capture all the species equally, so do not rely on trapping alone to determine the diversity or relative abundance of mollusks in an area.

For trapping small land snails, such as *Cochlicopa lubrica* (Müller, 1774), in an open pasture, Krull and Mapes (1974) used a wet gunnysack folded 3-6 times and covered with two or more layers of rocks, not heavy enough to press the entire sack to the ground. This arrangement provided air circulation, protected snails from the sun's heat, and provided a cool, shaded, moist area. They checked the traps 2-4 times per week, and found up to 26 specimens per trap.

Bait can be useful in trapping. Many gardeners know that a pan of beer will attract slugs, which crawl into the pan and drown. Apparently, the slugs are attracted to the smell of the hops in the beer. Snail and slug poisons, such as metaldehyde, are usually mixed with a grain product, such as bran, and the grain is the attractive element of the bait. Attractants such as these can be used in traps to attract a variety of land mollusks.

Cardboard trapping is another commonly used method to capture snails (e.g., Oggier *et al.* 1998). Sheets of wet cardboard (or dry cardboard after a rain) are placed on the ground in the woods or in a meadow. The moisture remaining under the cardboard attracts snails. After a day or so, the undersides of the cardboards can be inspected for snails. Instead of cardboard, one may use wet sacks or wooden boards.

22.4.7 Vacuuming, sweeping, and beating. Sampling snails from dense grasslands can be challenging because the roots of the turf are tightly matted. One could pick the turf apart and sieve it, but a simpler method is to use a garden leaf blower in reverse, fitted with a 0.5 mm mesh screen (Ian Killeen, pers. comm.). The intake can be passed closely over the surface of the turf and the suction will pull the snails against the screen. Be sure to invert the apparatus before turning off the blower, and any snails present should be on the screen with little debris. For using this method in an area with looser debris, you might try fitting a coarser pre-screen (e.g., 4-6 mm) over the opening to exclude larger debris.

For sampling small snails from tree trunks, use a brush such as a large paintbrush to brush snails from the bark onto a surface held against the tree trunk (Ian Killeen, pers. comm.). Beating vegetation onto a cloth, plastic sheet, or inverted umbrella can recover species that climb or live on grass, sedges, bushes, and trees. For example, *Columella* and some Succineidae climb grass and herbs.

22.4.8 Mark-release-capture method. Mark-release-capture is a method that ecologists commonly use to estimate animal population sizes or determine snail dispersal rates (e.g., Baur 1988, Schilthuizen and Lombaerts 1994). It may be used to study the members of one species or all the species in an area. Since the method does not require the killing of snails, it is especially suitable for estimating population sizes of endangered species.

In the mark-release-capture method, designated plots are searched for snails. All snails found, or members of designated species, are counted, marked in some way, and released in the same plot. After a suitable time, the plots are searched again, and the snails found are counted and checked for marks. The researcher then determines the proportion of marked to unmarked specimens to estimate the actual population size of the species (population estimate = # marked time 1 x total caught time 2 / # having marks time 2). Marks can be made using permanent but non-toxic paint. If it is desirable to mark snails individually, numbers can be applied with ink (such as India ink), and painting over the mark with clear fingernail polish can make it more permanent. For marking dark shells, one can apply white paper correction fluid or white paint to the shell, write on it with India ink after it dries, and paint over the mark with clear fingernail polish. Other methods for marking snails include gluing tags to the shell, filing notches in certain positions around the lip, or engraving markings if the shells are thick enough (see Chapter 2.6 for additional details).

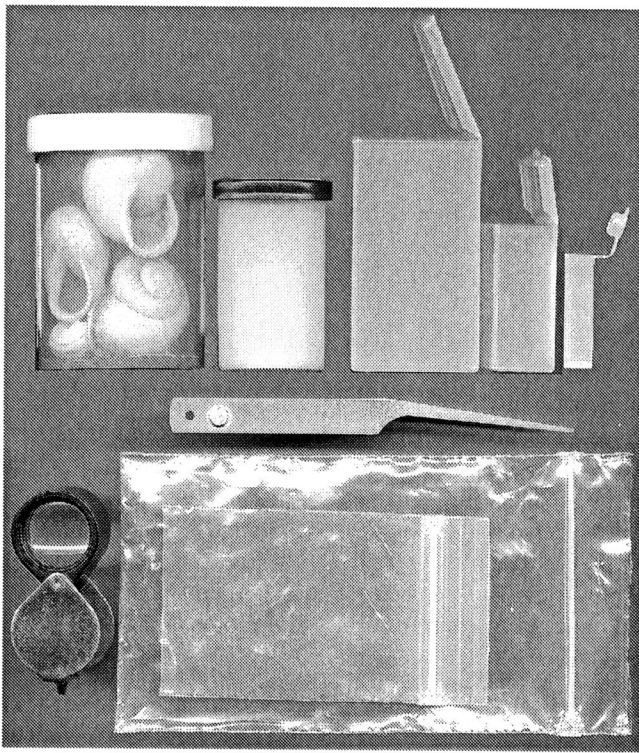

Figure 22.3 Miscellaneous collecting equipment.

To mark slugs, they can be given colored food, for example, agar containing the dye neutral red, or food containing a radioactive marker such as lettuce with radioactive phosphorous (Runham and Hunter 1970). Richter (1976) summarized some slug-marking techniques, and described a new method for marking slugs individually using freeze brands cooled in liquid nitrogen, then applied to the slug for 1 to 5 seconds. Marked slugs are released into the area where collected and allowed to mingle with unmarked slugs.

Other methods can be useful for tracking snails and recording their home or activity ranges. Such methods include radio-tagging (Vail 1979, Auffenberg 1982), and spool-and-line (Pearce 1990).

22.4.9 Containers, shipping, and other equipment. Shells can be collected in any type of suitable container (Figure 22.3). Plastic film canisters or similar plastic containers (for example, medicine bottles with straight necks) are ideal especially for short local trips when only a few shells will be collected. If you will be collecting large numbers of shells or going on an overseas expedition, consider using re-closeable plastic bags. An advantage is that empty bags take very little storage space compared to rigid containers. In addition, shells placed in such bags do not rattle around. However, plastic bags do not offer protection from crushing, so filled bags should be kept in a sturdy container. Avoid putting heavy and fragile shells together in the same container.

It is best to keep very small shells in small vials separate from larger shells; otherwise, they could crawl into (if they are alive) or become stuck in larger shells. Also, keep carnivorous species (e.g., *Euglandina* spp., *Haplotrema* spp., *Oxychilus* spp., at least some *Glyphyalinia* spp.) separate from other snails or you will get home with many empty shells. Scientific supply companies sell plastic vials in various sizes. Those with attached lids are useful for fieldwork since one can open and close them with one hand and not risk losing the cap.

Live and active (not dormant) snails can survive in dry containers for a few days as long as they can receive fresh air and they are not subjected to high temperatures (as in a closed automobile on a hot day). Slugs are especially vulnerable to high temperatures. One way to assure the survival of live snails is to put them into cloth bags with moist leaves (Clench 1974) or crumpled wet newspaper, or into a sturdy container with holes in the lid (or some other arrangement that is gas permeable). Hubricht (1951) recommended wrapping the collecting container in a wet towel and placing it where air can get to it; this evaporative cooling will help to prevent death of specimens in warm weather. Aestivating snails collected during dry periods in arid areas can survive dry for several months in containers that permit air circulation if they are not subjected to temperature extremes or long periods of extremely low humidity.

Land snails can be shipped alive as long as they are shipped in a dry condition; use of paper in packing will prevent excess moisture from developing. If you are shipping snails across international borders, be sure to check with the appropriate governmental organizations regarding permits. Leaf litter and soil may be collected in plastic bags, buckets with lids, or if you want the samples to dry out, use cloth or paper bags.

Besides containers for shells, depending on your needs, you will need to carry other equipment with you. First, a serious collector should never go on a collecting trip without a field notebook, and a pen and/or pencil. We discuss record keeping in detail below. A GPS receiver, a compass, and topographic maps are useful for documenting your collecting localities, relocating stations, and to avoid getting lost. These tools can be especially helpful for determining the coordinates and orientations of collection stations in rural countrysides, extensive grassy plains, or on a mountain. You might want to consider a small scoop or a large spoon to help search through the leaf litter and soil, and to pick up shells. Likewise, a pair of lightweight forceps could be useful for picking up very small shells. In addition, a good magnifying lens, preferably a 7x or 10x Hastings triplet, is useful for identifying very small species in the field. You may consider carrying a camera to take pictures of your collecting stations (see below, and Chapters 6 and 7). Also, if your method requires it, be sure to carry a bottle of preservative with you in the field (the next section discusses preservatives for processing specimens). Finally, carry a first aid kit with you, especially during long expeditions away from civilization.

22.4.10 Other considerations. Here are a few additional items that you might need to consider.

Collecting live specimens: In parts of the world, land snails have decreased in numbers and some species have become extinct (Seddon 1998, Lydeard *et al.* 2004). While habitat loss seems to be the main reason behind the decline of mollusks, in some areas introduced species including non-native snails are responsible for the loss of native snail populations (Hadfield 1986, Murray *et al.* 1988, Cowie 1992). Collectors can help snail populations recover by not collecting live specimens of endangered species, and by minimizing the collecting of live specimens, particularly juveniles, of native species especially if they are known to be rare or if plenty of empty shells of the same species are available. If live specimens are necessary for dissections, collect only a few specimens. If large numbers of live snails are needed for a study, consider raising a pair in captivity (see Chapter 23 for techniques on rearing terrestrial gastropods). Their offspring can then be used for dissections or other studies.

Collectors should be careful not to release live snails into areas where they have not been recorded before. Introduced snails may become agricultural pests or compete with native snails (Rollo 1983a, 1983b). In some cases, exterminating infestations of introduced snails has required great costs (Simberloff 1996), but often eradicating pest snails is too difficult or expensive, so the problem persists. Some invasive snails that have been distributed by humans over many continents, for example, *Rumi-*

na decollata (Linnaeus, 1758), (Selander and Kaufman 1973), can reproduce without mating. Hence, a single introduced snail can potentially start a new population. The carnivorous snail, *Euglandina rosea* (Férussac, 1818), native to the southeastern U.S.A., has been intentionally distributed to many Pacific Islands in ill-fated attempts to control the introduced giant African snail, *Achatina fulica* (Bowdich, 1822). However, *E. rosea* has caused extinctions of many local endemic species, while being ineffective at controlling *A. fulica* (Clarke *et al.* 1984).

Figure 22.4 *Helix lucorum* **Linnaeus, 1758 with epiphragms on Burgazada, Turkey**. Arrows point to epiphragms.

In light of these arguments, whether you are looking for or avoiding live snails, you may frequently need to determine on the spot if a shell you have just found is empty or has a snail in it. But as easy as it may sound, sometimes one cannot be sure if a shell is empty or not. Here are some tips to help you to make a determination.

1. Many snails that have been dormant (i.e., aestivating) for more than a few days completely or partially seal the apertures of their shells with their mucus, forming what is called an epiphragm. In some forest snails, the epiphragm may simply be a thin membrane-like layer of mucus, while during dry seasons, snails that live in arid regions may form an epiphragm having calcium carbonate in it and being almost as hard as the shell itself (Figure 22.4). Thus, the presence of a more or less intact epiphragm at the aperture of a shell usually indicates that there is a snail inside.

2. Snails frequently withdraw deeply into their shells, so their bodies may not be visible from the aperture. Hold a shell suspected to contain a live snail against the bright sky or other light. You should be able to see whether the shell contains the darker body of a snail. This method works best with juvenile shells, which are usually thinner and more translucent than adults are. To verify whether the dark part represents a live snail, as opposed to soil or a dead body, check that the outermost part is perpendicular to the whorls (minimizing surface area). You might even be lucky enough to see the snail's heart beating. If shells are opaque and you are unable to see whether a body is present, you might be able to separate live and dead shells by the heavier weight of living snails. Also, the presence of soil within the body whorl of a shell usually indicates that the shell is empty.

3. The outer surface of most snail shells is an organic layer called the periostracum, which is often some shade of brown. The periostracum of empty shells disintegrates after a while and the shell itself, especially in locations exposed to the sun, bleaches white. Such weathered shells are usually empty.

4. Bring the shells suspected of having live snails in them indoors and wrap them in a wet paper towel or place them in a container with wet

paper towels for up to a day or so. Periodically check to see if any snails have become active. We especially recommend that you do this with the shells you have collected in a foreign country before leaving that country and leave behind any live snails you may find.

If you are collecting live snails for their soft parts, learn to recognize any carnivorous species that you are likely to encounter. Some larger carnivorous snails include *Haplotrema* spp. in the U.S.A., *Euglandina rosea* in Florida (and introduced to many tropical islands), and *Rumina decollata* in Mediterranean areas (also introduced in California, Texas and Florida). Some smaller carnivorous snails include *Glyphyalinia indentata* (Say, 1823) in eastern U.S.A., *Oxychilus* spp. in Europe (and introduced into U.S.A.), and *Aegopinella* in Europe. Keep live carnivorous species separate from the rest of the live snails or they may consume your other snails before you get home.

Because slugs have either no shell (Philomycidae and Veronicellidae), or a shell that is not usually identifiable to species, or even to family, they need to be collected alive or dead no more than a few hours. Furthermore, positive identification of many slug species relies on examining internal genital anatomy, so you want to collect mature slugs. LaRocque (1974c) found the best slug collecting in rainy weather. For example, *Philomycus*, which are normally very secretive, can be found crawling on trees, tree trunks (even 4 m above the ground), and even on bare rock during rainy weather (Hubricht 1951, pers. obs.).

Unusual Specimens: Even broken shells or shells with holes or repairs and otherwise abnormal looking shells should be collected as they may be used to study the predators, parasites, and diseases of the snails. In some seasonally arid regions, for example, along the Mediterranean coasts of Greece and Turkey, snail shells with small uniformly shaped holes are common (Örstan 1999b). Some of these holes are made by the predatory larvae of drilid beetles (Coleoptera: Drilidae), but very little is known about the interactions of these snails with their predators and parasites.

Occasionally you may find shells with insect cocoons, puparia, or insect parts in them. In the U.S.A., the larvae of some flies, for example, sciomyzids (Diptera: Sciomyzidae), are parasitoids of land snails (Berg and Knutson 1978). If shells with intact insect cocoons or puparia in them are collected, they should be kept individually in transparent containers and kept under observation. Any emergent insects should be properly preserved so that they can be identified.

You may also find reverse-coiled shells, that is, sinistrally coiled shells of a normally dextral species, or vice versa. Rarity of such shells makes them valuable not only for collectors but also for scientists (e.g., Gould and Young 1985, Örstan and Welter-Schultes 2002).

Conservation and collection permits: When collecting live individuals, take only the specimens you need, leaving some living individuals to grow and reproduce for the future. Collect responsibly and avoid collecting endangered snails, including their empty shells unless you have an appropriate permit. In the U.S.A., it is illegal to take even the empty shells of endangered snail species such as *Liguus* tree snails (e.g., see the Web site <www.jaxshells.org/guide.htm>) without proper permits. This regulation exists because a law enforcement agent cannot always determine if the shell of an endangered snail was found empty or obtained by killing the snail. Therefore, before starting to collect in an area, determine if there are any endangered snail species in that area and learn what they look like. If necessary, carry their pictures to the field. In the U.S.A., some states have endangered species lists separate from those of the Federal Fish and Wildlife Service. Lists of the endangered species may be obtained from the U.S. Fish and Wildlife Service and from appropriate state agencies.

Whenever necessary, obtain a permit before starting to collect in an area. Do not enter private property without the permission of the owner. In the U.S.A., public areas may belong to independent organizations (for example, the Nature Conservancy), local governments (for example, county parks), states, or the Federal Government. In most cases, all you may have to do to obtain a permit is to contact the park manager. But, it may be more difficult to obtain a permit to collect in federally owned areas. In general, it will be easier to obtain a permit if your interests are scientific and not commercial. In some foreign countries, you may not need collection permits to collect in public areas, but find out beforehand. Retain all written permits with your collection, because if you will ever donate all or parts of your collection to a museum, you may have to demonstrate that the specimens were obtained legally.

Also, you are required by law to declare at the U.S. Customs snail shells you may be bringing into the U.S.A. from other countries. Otherwise, if your luggage is searched and your specimens found, you may have to pay a fine, but worst of all, return home empty handed.

Avoid danger and be prepared for emergencies: If your collection trip is going to take you away from residential areas, roads, park offices or other people, be cautious and be prepared. If you are going on a long trip, chart your course beforehand and leave a copy of your plans with a responsible person who is staying behind. Learn how to use and carry a map, a compass, a GPS receiver, and a cellular phone. Carry a first-aid kit and learn how to treat simple injuries, stop bleeding, and splint broken bones. Learn to recognize and beware of poisonous plants, ticks, snakes, scorpions, and other potentially dangerous animals. Do not go collecting in the woods during a thunderstorm. Do not enter unfamiliar woods (or other areas) at night, because limited visibility, even with a strong flashlight, makes it easy to become disoriented. If you are collecting in hot and sunny weather, carry more than enough water and avoid sunburns. If possible, avoid collecting during hunting seasons in areas where hunting is permitted; otherwise, use caution and wear an orange hat or vest.

22.5 RECORD KEEPING IN THE FIELD

You should note locality information in sufficient detail that a researcher in the future could find the same locality. This way a locality can be assessed for any eventual change. Field observations that cannot be repeated or verified are of little value (Bequaert and Miller 1973: 5).

The shells of most land snail species exhibit extensive intraspecific variation (Goodfriend 1986). While such variation can be a nuisance when identifying species, it is also a source of invaluable information for those who are studying the ecology and evolution of snails. In many cases, the exact causes of this variation are not known, but both genetic and environmental factors appear to be important. It is not unusual for neighboring colonies of a species, sometimes within a few hundred meters of each other, to differ significantly in shell morphology (for example, see Boycott 1920, Wolda 1969, Baur 1988). Obviously, study of such variation will be possible and meaningful only if collections from different locations or habitats, and even microhabitats, are kept separate and properly labeled; therefore, it is extremely important to keep adequate notes in the field and to label specimens properly before leaving a station. Since there cannot be general rules applicable to every species, every location, and every condition, and it is not practical to keep and label every shell individually in the field, the collector must learn to recognize meaningful differences in shells while collecting them and use proper judgment to determine when to keep shells together and when to separate them.

The important point is to take a notebook and something to write with on collecting trips. Medium-sized, spiral-bound artist's sketch books with thick unlined acid-free pages are good because they withstand occasional rain and mud better than does thinner paper. Leave spaces between entries or use only one side of each page so that additional notes may be entered later (for example, species lists or more information on the location). Use a pen, preferably with waterproof permanent ink, but always carry a pencil (pens are known to stop writing at temperature extremes or in the rain). If you are collecting in the rain, you may want to carry a notebook with a special coating on the pages, such as Rite-In-The-Rain® paper, that accepts a pencil even when wet.

If you prefer to scribble temporary notes in the field, you should transcribe them that evening or soon afterward. A scribble that means something to you one day might be incomprehensible weeks or months later.

Be sure to write labels with pencil or permanent ink. Note that the ink from most ballpoint pens dissolves in alcohol! Bottles of specimens with blank labels are basically useless scientifically. India ink is a good waterproof and alcohol-proof ink. Pigma® brand felt tip pens or similar pens with archival ink are convenient pens with alcohol-proof ink.

Most of the following information should be written down in the field, while collecting or before leaving the location. Do not rely on memory!

1. Directions to and description of location, GPS coordinates, landmarks, range of the area where shells were found (important if introduction of snails is a possibility), etc. Give as much locality detail as possible so a future collector could revisit your site, if not exactly, at least within 100 m.
2. Characteristics of macrohabitat (trees, rocks, soil type, ground cover, proximity to residential areas, ruins, water bodies, etc.)
3. Characteristics of microhabitat. Exactly where were the snails? Were they in leaf litter, under logs, on plants (say what kinds)? Describe the level of insolation, the slope, etc.
4. Physical measurements and other conditions at the time of collecting (temperature, rain, wind, etc.)
5. If a snail was found alive, was it dormant or active? If active, what was it doing? If feeding, what was it feeding on?
6. Abundance of shells in general and of individual species.

In addition to carrying a notebook, you can place a small piece of paper in each of your containers before a field trip. Before leaving a collection station, write the date and an identification code for that station on the paper in each container used there (matching the identification code in your field notebook). One advantage of this method is that seeing a piece of paper in a container will remind you to record the necessary information. However, do not place paper records in containers with live snails because most snails will eat paper. Station codes may be written in permanent ink on the outsides of containers of live snails. Alternately, you can place the container and a separate paper record in a plastic bag. A snail-proof method is to engrave the station codes with a slightly pointed tool on small squares of aluminum. These can be cut out of thin aluminum containers.

Besides taking detailed notes, consider taking photographs of your collection stations. Not only can these photographs later help you locate the same station, they will also be a permanent record of the macro- and microhabitat.

22.6 PROCESSING AND STORING SAMPLES

22.6.1 Laboratory recovery of snails from leaf litter and soil. There are several methods of separating snails from leaf litter, the most common ones being picking, floating, and sieving.

Picking and floating: If you have a small amount of litter or soil, less than a few hundred milliliters, it may not be worth sieving it. Place the sample in a wide tray and under a bright lamp, pick through it with a small spatula or a similar tool. If the background is light brown or grey, it will be easier to see most shells. To make sure you have found everything, go through the sample a few times. If you have never done this before, first familiarize yourself with the shapes and sizes of the shells you expect to find by looking at their pictures. For example, the shells of *Vertigo* spp., only a few mm long and brown, are usually difficult to notice against the litter and soil unless your eyes are used to recognizing them.

Another method that works well with small samples (less than a few hundred milliliters) is floatation. Place the litter or soil sample in a wide tray, and cover it with tap water. Gently stir and break any soil clumps that may be present. Then under a bright lamp, examine the surface of the water for shells. This method works because air trapped in empty shells makes them float. One disadvantage of this method is that live snails will sink and escape detection. Therefore, before floating a sample, first pick through it dry and remove the live snails, or save the sinking portion and examine it separately for live snails. Also, if a large portion of a sample consists of wood or plant fragments, these will also float and cover the surface of the water, making it difficult to see the floating shells.

Sieving: Sieving is helpful because: (1) it is easier to separate snails from non-snails if all of the items are roughly the same size, and (2) sieving can remove particles that are smaller than you wish to examine (either because that size contains no snails, or because you are not interested in recovering shells that small). By using several sizes of screens, for example in a nested series of soil sieves, you can transform a single pile of litter into several well-sorted piles of similar-sized particles. You still must examine all the material, but you are more likely to find minute snails if you are sorting them from minute non-snails, than if you were sorting them from the whole range of particle sizes.

What size of sieves to use? Inexpensive screens for use in the field can be made from mesh sizes that are easily obtained, for example in the U.S.A., 6 mm and 1.2 mm window screen (1/4 inch and 1/16 inch) are readily available. Material retained by each sieve fraction can be sorted on any conveniently transportable flat surface such as a Manila file folder. In the laboratory, more expensive, bulkier, or heavier sieves can be used, for example, soil sieves are available in a variety of sizes from many science and field supply companies. Convenient for processing several liters of leaf litter are sieves 20 cm diameter and 5 cm high that have mesh openings of 8, 4, 2, 1, 0.7, and 0.5 mm. The material can be sorted on a manila file folder or on a light-colored no-pattern cafeteria tray.

What is the smallest screen size to use? In samples from North America, the 0.5 to 0.7-mm fraction usually contains only juvenile snail specimens with an occasional adult of *Carychium* or narrow *Gastrocopta* species. In Madagascar, the only adult specimens of *Gulella minuscula* Emberton and Pearce, 2000 were found in the less than 0.8 mm fraction (Emberton *et al.* 1996, as Streptaxid sp. 15). Still smaller screens would be needed in other parts of the world, e.g., where tropical diplommatinids less than 0.5 mm in maximum size occur (Zilch 1959-1960). Furthermore, if all individuals including juveniles must be recovered (e.g., for computing Shannon diversity, or for studies of growth series), still finer screens may be necessary to recover juveniles of very minute species.

To prevent crushing when picking up shells, especially the smaller shells, you can use lightweight forceps to pick them up (such forceps, sometimes called larval forceps, are sold by entomological supply companies, for example, BioQuip Products, on the Internet at <www.bioquip.com/>). Smaller snails can also be picked up with a fine paintbrush moistened as needed (LaRocque 1974b). Shells can also be aspirated from samples on the sorting tray using an aspirator like those used by entomologists to collect small insects, by modifying the intake tube to have a smaller opening. Usually soil and leaf particles will be aspirated along with the shell, but they can be separated later. An advantage of aspirating is the shells will not be crushed, and they will remain dry.

Be sensible with disposal of soil that you have transported some distance (see Section 22.4.4).

For fossil mollusks, some additional processing is usually necessary to disaggregate shells from the substrate. Often soaking in water overnight will be sufficient to disaggregate samples, sometimes boiling will agitate the sample enough to separate shells from soil, or adding detergents can help with disaggregating (LaRocque 1974b). You can sieve the samples while they are wet, washing carefully with water, to wash away much of the fine unwanted substrate. Soaking in kerosene can help disaggregate very stubborn samples.

Laboratory recovery of slugs from leaf litter and soil: Runham and Hunter (1970: 116-123) discussed methods for recovering slugs from quantitative samples from the field. The sample must be large enough to contain enough specimens to assess the population adequately, and they recommended several replicate samples to avoid drawing biased conclusions from sampling one very dense or very sparse part of the population. In one of their studies, they took samples that were a cube, 30 x 30 x 30 cm on a side. One method for recovering slugs from bulk samples taken in the field involves washing the soil through nested sieves (0.85, 0.25, and 0.085 cm mesh), then dipping and agitating the sieves into magnesium sulfate ($MgSO_4$) solution (at least 1.17 mg $MgSO_4$/ml) to float out the slugs. The method worked well except that recovered slugs were not always in good condition, some small slugs less than 12.5 mg and the more fragile eggs (e.g., of *Arion hortensis* Férussac, 1819) were destroyed by the water jet in the initial washing, and immersion of slugs in magnesium sulfate for periods over an hour caused slugs to lose up to one third their weight. This weight loss is a disadvantage in studies in which slug weight is used as a surrogate measure of age.

A simpler method of recovering slugs from soil or turf samples relies on behavior of slugs moving to stay out of rising water. Runham and Hunter (1970) placed intact samples into covered buckets and slowly added water over 3 or 4 days. Slugs were picked off as they crawled up. In a modification of this method for soils that crumble, they placed the sample into a bowl having holes in the bottom, and immersed the bowl into water, raising the water level gradually over 4 or 5 days. By these methods, they recovered about 90% of the slugs in the sample, and the slugs were in good condition. However, they did not recover eggs. They mentioned that hand-picking slugs, as with sieving, can be successful from drier soils, but recognizing dried slugs in very dry samples can be difficult, especially if slugs have soil and leaf particles adhering.

22.6.2 Cleaning and preserving empty shells. If a shell with a dead snail in it is not promptly cleaned, the odor of the rotting snail can dim the enthusiasm of even a seasoned collector (which is another reason for not taking live snails). Several methods exist for removing a dead snail from its shell. If you don't need to keep the body, you can boil shells 6-12 mm diameter for 30 seconds, and shells up to 4 cm or more for 1 minute (Clench 1974), then extract the body with a hooked safety pin or a bent wire. Alternately, you can place the shell outdoors in a container that would permit the entry of insects such as ants. The insects will eventually consume the body, leaving a clean shell, but this method can be time consuming. Another method is to treat the specimen with dilute bleach or hydrogen peroxide to dissolve the body. However, these chemicals may also destroy the periostracum and alter the appearance of the shell. Therefore, use them only after you have determined that the shell will not be affected significantly.

You can boil small shells briefly, and then remove the body with a strong force of water. A fine jet of water can be achieved using an ear syringe, or for very tiny snails, by fashioning a capillary tube from glass tubing and attaching a rubber bulb (Clench 1974). Alternately, you can simply dry the very small specimens, for example in a desiccator, and then store them dry. The smell of very small dried shells should not be noticeable.

Some shells require special handling to prevent loss of color. For example, if you boil *Liguus* shells, contact with steam (not the heat itself) causes the green lines to become bronzy or dirty gray. To avoid fading, heat the snails in an oven at 150°C (300°F) for 5-7 minutes, and then pull the bodies out. Avoid too much dry heat or the pink tints will fade. You can also clean the shells by freezing them, then pulling out the partly thawed bodies (Pilsbry 1946: 52).

22.6.3 Preserving soft parts. While the shells of snails can be preserved dry, mollusks without shells, such as slugs, must be preserved wet (although Crowell 1973, has developed a freeze drying method for preserving slugs dry so they can be mounted on insect pins). Furthermore, even for shelled mollusks, it has long been known that it is desirable to preserve soft parts with the shell (e.g., Anonymous 1837: 153). Because soft parts provide so much valuable information to researchers, we encourage collectors to preserve live-collected snails in a way that will preserve the soft parts. Soft parts allow study of external and internal anatomy, as well as biochemical studies. The preservative will depend on your intended use for the specimens. The best general method to preserve live-collected land snails is to relax them (especially, if they are to be dissected) and then preserve them in 80% ethanol.

Several methods exist for relaxing land snails. The simplest method is to drown them overnight or up to a day in a small watertight container filled with water and few or no air bubbles. It is easier to exclude air bubbles if you place the snail container inside a bucket of water, let all the air bubbles escape, and then cap it. When the specimens do not respond to a pinch from forceps, they can go directly into 80% ethanol. The drowning time in water will be shorter in warmer water (increasing the snail's metabolism to use up oxygen faster) but use care to prevent overheating (Hubricht 1951). The drowning process will be faster if the water is boiled ahead of time to remove oxygen, and then cooled.

Drowning in water may not be very successful for relaxing aquatic or semi-aquatic species, or with some terrestrial species. Various chemicals can be added to the water for drowning aquatic species, or to speed up the drowning of terrestrial forms. Menthol crystals are easy to obtain and use because menthol is not a regulated drug; add one or two menthol crystals to the drowning water. Hubricht (1951) recommended drowning slugs in a chloretone solution (prepared by diluting a saturated solution to 5 to 10%) rather than in plain water, because he indicated that in water slugs struggle and produce much mucus, which obscures color patterns. According to Hubricht, it is not necessary to fill the jar completely, because relaxing requires but a few minutes, and killing in the solution takes 3 to 10 hours. After drowning, put the specimens in preservative. The slugs will be preserved life size with clear color patterns, and without fermentation of stomach contents. Chloretone may be regulated as an addictive drug, so it might be difficult to obtain. Another method for relaxing slugs is to place them in an approximately 5% ethanol solution (Webb 1950), again without necessarily filling the jar. In this solution, most slugs are anesthetized in an extended state within about an hour, after which they should be transferred to 95% ethanol for several hours, and then stored in 80% ethanol (For more information on relaxing mollusks in general see Chapter 2.5).

Use at least five volumes of alcohol for each volume of body tissue to avoid too much dilution of preservative by body fluids. For long term storage in liquids (several months or more) be sure to buffer the solution against acidity so the shells do not dissolve. One substance that is commonly used as a buffer is sodium borate (Borax). Adding a teaspoon (10 g) to a liter of ethanol will provide an adequate buffering capacity.

Use other methods of preserving specimens for special uses. For example, because drowned specimens seem to be incompatible with DNA studies (Schander and Hagnell 2003), either preserve specimens intended for DNA studies directly in ethanol, or snip off a piece of tissue (edge of the foot) and preserve the tissue directly in 95% ethanol before drowning and preserving the rest of the specimen. Other preservatives include FAA (1:1:1 10% formalin, acetic acid, and ethyl alcohol) for cytology or study of chromosomes (preserves specimens for about three months), and formalin or glutaraldehyde for histology or cell microscopic studies.

After a snail has drowned, it is possible to remove it from its shell without destroying either the shell or body, by heating the snail in water to 65°C (about 150°F) at which point the columellar muscle will loosen its grip on the columella, and the body can be twisted out of the shell, using a pair of fine forceps for small specimens. Because sometimes a snail's body may shrink when heated, be sure to note on the label that you used heat for removing the body. If you separate the body and shell, be sure to label both the body and the shell so they can be reunited unambiguously in the future if a researcher needs to do so. You can write a number with permanent ink on the shell, and write the same number and other information on alcohol-proof paper. You can keep the body alone in a vial, or if you wish to put several specimens together, you can attach the label to the body with a needle and thread, for example, sewing through a section of the foot.

During hot weather, one may occasionally chance upon slugs dried up on a hot sidewalk or a driveway. Such specimens may be rehydrated by placing them in an approximately 0.5% aqueous solution of trisodium phosphate (available in hardware stores) for a few hours (Van Cleave and Ross 1947). Rehydrated specimens should then be soaked in water to remove the trisodium phosphate, which is insoluble in ethanol, and then placed in 80% ethanol. Trisodium phosphate may also be used to rehydrate alcoholic specimens that have dried out, but dry specimens that have been stored in alcohol should first be soaked in water to remove ethanol (see Chapter 5.9 for some other methods of reconstituting dried tissue).

If you separate the body from the shell but want to keep only the shell, certain museums (e.g., the Delaware Museum of Natural History; the Carnegie Museum of Natural History) are willing to accept just bodies of specimens under the following conditions. (1) You must be willing to let researchers borrow the shell for study, (2) the shell and body must be marked so it is unambiguous which shell goes with which body, and (3) you must make provision for the shell eventually to go to a museum (preferably the museum housing the body). By donating the soft parts, you will make a contribution to science (see Chapter 14.6 to 14.8).

22.6.4 Vials and closures. For preserving soft parts, use containers with waterproof closures. Glass containers are preferred, because they are inert towards common preservatives and allow the contents to be seen. Various plastic containers can be used for shorter times of storage. Polyethylene and polypropylene are not as transparent as glass, but they will probably be stable for many years. Polystyrene, as in pill bottles, although transparent, can become brittle and crack after one to several years storing alcohol. Furthermore, the plastic lids on snap cap vials are not sufficient for containing alcohol more than several months.

For storing larger specimens, you can use glass food storage jars with glass lids (not metal, which will corrode after one to several decades) and rubber gaskets. Other small to large specimens can be stored in glass bottles with lids having some sort of plastic seal; do not use lids with paper seals. Contact a local museum to learn what they use. If you plan eventually to donate your specimens to that museum, they might be willing to give you good containers for storing your specimens (see Chapter 5.5 for more on this topic).

22.6.5 Labeling and keeping good records. Detailed locality and habitat data are often as important to researchers as the specimen itself. The more extrinsic information you can record (see Chapter 14.4), the more useful the specimen will be to a researcher studying the distribution or habitat of the species. Because of the importance of locality data, collectors on extended field trips sometimes make a carbon copy or photocopy of their field notes, with field numbers, detailed locality, and habitat information, to send home periodically in case the field book is lost or destroyed (Clench 1974).

Other information that is important to record includes information on how you handled the specimens. Did you heat (how hot) the specimen to remove the body? What preservative(s) have you used, for example, has the specimen ever been in formalin? Information like this is important to researchers looking for specimens for particular uses. It is nearly impossible to obtain DNA from specimens that have been in formalin.

It is important to associate specimens unambiguously with the collecting information. For labeling larger shells, you can write a unique number directly on the shell. For smaller specimens, be sure the container is labeled with a unique field number. You can write numbers on the outside of the vial, or put a label inside the container, or both. Use care if you put labels in vials with living specimens, because many snails eat paper. Also, be sure to write labels with pencil or permanent ink; alcohol will dissolve ink from most ballpoint pens, so do not use ballpoint pen for writing labels.

22.6.6 Storing and display. Now your specimens are ready for study, displaying in your personal collection, or donating to a museum where they can be studied by current and future researchers. Research collections are the libraries where researchers find the specimens they need for their studies. Specimens in museums may be used in a variety of ways, for example, researchers studying the systematics of a group of species need many specimens to assess how variable the specimens are within a species versus among species, allowing them to draw conclusions about how species are related. Researchers rely on specimens with good habitat and locality data in assembling field guides and distribution maps. Along these lines, keep in mind Boycott's (1928) advice that to retain their value for future researchers, lots containing large numbers of randomly collected shells (see above) should not be broken into smaller sets to distribute to museums or individuals or combined with lots from different locations.

Museums strive to keep specimens in good shape for hundreds or thousands of years. Consequently, museums try to avoid environmental conditions that can degrade specimens over time. For example, light can fade colors and fluctuations in temperature and humidity can degrade specimens over time. In particular, fluctuating humidity in acid conditions (such as oak-wood cabinets, and non-archival paper labels or paper boxes) can lead to Bynesian Decay (see Chapter 5.2 for more details). You might consider these environmental conditions when storing your specimens.

22.7 HOW TO IDENTIFY LAND SNAILS

A serious collector needs access to certain publications that will help identify the snails of interest. We recommend that collectors planning to collect in new areas should familiarize themselves beforehand with the species they are likely to encounter. This familiarity will aid in identifying most species on the spot, at least to genus, and in recognizing rare, endangered, introduced, and perhaps an occasional undescribed species.

A complete list of identification guides for the world's land snails is beyond the scope of this chapter (see Chapter 9.2.3). We will instead concentrate on the identification aids for the U.S.A. Pilsbry's four-part monograph (1939-1948), although now more than 50 years old, remains the most complete and essential work for the North American land snails. As of February 2005, some of Pilsbry's monograph was still available from the Academy of Natural Sciences in Philadelphia <www.acnatsci.org/library/scipubs/index.html>. Burch (1962) is a pictured key version of Pilsbry's monograph for the eastern North American land snails; unfortunately, it is out of print, but you can probably find a copy in a library.

Many new species have been described in the U.S.A. since the publication of Pilsbry's monograph. Unfortunately, these have not yet been compiled in a work comparable to that of Pilsbry's. In many cases, it may be necessary to consult the original publications, which are too numerous to cite here individually. Two works that can be consulted for references to new species descriptions and taxonomic changes from 1948-1985 are Miller *et al.* (1984) and Hubricht (1985). Furthermore, Hubricht's county-based distribution maps will be useful in identifying the native eastern land snail species by suggesting which species are likely to be found in an area. However, be careful using geography to identify species, because many of the maps are known to be incomplete. Hubricht also gave brief habitat information for all the species, which can be helpful when searching for certain species. For some of the polygyrid genera, we suggest that you consult Emberton's (1988, 1991) revisions. Burch and Pearce (1990) gave an illustrated and updated key to the genera of land snails in North America. For land slugs of northeastern U.S.A., Chichester and Getz (1973) can be helpful. Most of our introduced slugs and some of the shelled snails are from Europe, so Kerney and Cameron's (1979) guide to European land mollusks should be useful. Land mollusk taxonomy is constantly being updated. A good reference to keep current on taxonomy of North American land snails is that by Turgeon *et al.* (1998), who list all known mollusk species in North America, and an updated edition is produced every ten years.

Regardless of where you collected your specimens and what identification keys you are using, the following general process will help you identify your snails. First, prepare a list from literature or museum records of snail species that have been recorded in the area where you got your specimens. Also, include in this list the species that have been recorded in neighboring areas and which, judging from habitat requirements, you think might occur in your particular area. For example, for the eastern U.S.A, using Hubricht (1985), you might compile the snails recorded in the county where your specimens are from, then add any different species from the surrounding counties. Pilsbry (1939-1948) can be used for compiling distribution records for the western U.S.A. Then, using the available keys and pictures for the species on your list, try to identify your specimens as best as you can. Some species that have unique characteristics will be easy to identify, while others will require more work.

The next step in getting or confirming identifications is to contact other collectors, specialists, and museums. If you have a digital camera or a scanner, you might get quick help with identifications by e-mailing images of your shells, preferably viewed from the top, the bottom, and the apertural side, to specialists or to members of a mollusk list-server and asking for their opinions (see Chapter 12.3 for a list of such groups). When posting or sending pictures, always include the dimensions of the shells and location information.

Finally, you might be able to identify your specimens by comparing them with the already identified shells in museum collections. Most of the shell collections in museums are kept in areas that are not normally open to the general public. Therefore, before taking your shells to a museum, contact the curator of the mollusk collection and explain your needs. You may then be able to set up an appointment with the curator to use the collections of the museum.

A stereomicroscope is essential to examine and identify very small shells and to carry out dissections. For example, the minute North American species, *Striatura meridionalis* (Pilsbry and Ferriss 1906) is distinguished from the similar *S. milium* (E. S. Morse, 1859) by the very fine striae on its protoconch visible at magnifications near 40X. Dissection is necessary to distinguish among some species (for example, succineids) that have identical or variable shells whose properties overlap those of shells of other species. Large snails are easier to dissect. Kerney and Cameron (1979) gave good general guidelines for dissecting snails. Do not attempt to dissect small or rare snails until you have gained experience with larger ones.

22.8 LITERATURE CITED

Anonymous. 1837. Directions for preparing specimens of natural history. *Transactions of the Maryland Academy of Sciences and Literature* **1**: 148-156.

Anonymous. 1929. Directions for collecting land snails. *Carnegie Museum, Invertebrate Zoology Leaflet* **1**: 7 pp. [unnumbered]

Auffenberg, K. 1982. Bio-electric techniques for the study of molluscan activity. *Malacological Review* **15**: 137-138.

Barker, G. M. 2001. Gastropods on land: phylogeny, diversity and adaptive morphology. In: G. M. Barker, ed., *The Biology of Terrestrial Molluscs*. CABI Publishing, New York. Pp. 1-146.

Baur, B. 1988. Microgeographical variation in shell size of the land snail *Chondrina clienta*. *Biological Journal of the Linnean Society* **35**: 247-259.

Berg, C. O. and L. Knutson. 1978. Biology and systematics of the Sciomyzidae. *Annual Review of Entomology* **23**: 239-258.

Bequaert, J. C. and W. B. Miller. 1973. *The Mollusks of the Arid Southwest, with an Arizona Checklist*. University of Arizona Press, Tucson, Arizona. xvi + 271 pp.

Bieler, R. and J. Slapcinsky. 2000. A case study for the development of an island fauna: Recent terrestrial mollusks of Bermuda. *Nemouria, Occasional Papers of the Delaware Museum of Natural History* **44**: 1-99.

Blanc, A. and R. Allemand. 1993. L'escargot Turc *Helix lucorum* L. (Gasteropoda Helicidae), espece acclimatee dans l'agglomeration lyonnaise: Comparaison du rythme d'activite avec celui de deux especes voisines autochtones. [The Turkish snail *Helix lucorum* L. (Gastropoda Helicidae), a species acclimatized to the region of Lyon (France): circadian rhythm of activity and comparison with two indigenous species]. *Bulletin de la Societe Zoologique de France* **118**: 203-209.

Boag, D. A. 1985. Microdistribution of three genera of small terrestrial snails (Stylommatophora: Pulmonata). *Canadian Journal of Zoology* **63**: 1089-1095.

Boycott, A. E. 1920. On the size variation of *Clausilia bidentata* and *Ena obscura* within a "locality".

Proceedings of the Malacological Society, London **14**: 34-42.
Boycott, A. E. 1928. Conchometry. *Proceedings of the Malacological Society, London* **18**: 8-31.
Burch, J. B. 1962. *How to Know the Eastern Land Snails.* William C. Brown Co., Dubuque, Iowa. 214 pp.
Burch, J. B. and T. A. Pearce. 1990. Terrestrial Gastropoda. *In:* D. L. Dindall, ed., *Soil Biology Guide.* John Wiley and Sons, New York. Pp. 201-309.
Cain, A. J. 1983. Ecology and ecogenetics of terrestrial molluscan populations. *In:* W. D. Russell-Hunter, ed., *The Mollusca*, Vol. 6, Ecology. Academic Press, New York. Pp. 597-647.
Cameron, R. A. D. 1986. Environment and diversities of forest snail faunas from coastal British Columbia. *Malacologia* **27**: 341-355.
Chappell, H. G., J. F. Ainsworth, R. A. D. Cameron, and M. Redfern. 1971. The effect of trampling on a chalk grassland ecosystem. *Journal of Applied Ecology* **8**: 869-882.
Chichester, L. F. and L. L. Getz. 1973. The terrestrial slugs of northeastern North America. *Sterkiana* **31**: 11-42.
Clarke, B., J. Murray, and M. S. Johnson. 1984. The extinction of endemic species by a program of biological control. *Pacific Science* **38**: 97-104.
Clench, W. J. 1974. Land shell collecting. In: M. K. Jacobson, ed., *How to Study and Collect Shells*, 4th Ed. American Malacological Union, Wrightsville Beach, North Carolina. Pp. 67-68.
Coney, C. C., W. A. Tarpley, J. C. Warden, and J. W. Nagel. 1982. Ecological studies of land snails in the Hiwassee River basin of Tennessee, U.S.A. *Malacological Review* **15**: 69-106.
Cowie, R. H. 1992. Evolution and extinction of Partulidae, endemic Pacific island land snails. *Philosophical Transactions of the Royal Society of London* (B) **335**: 167-191.
Cowie, R. H., G. M. Nishida, Y. Basset, and S. M. Gon, III. 1994. Patterns of land snail distribution in a montane habitat on the island of Hawaii. *Malacologia* **36**: 155-169.
Crowell, H. H. 1973. Preserving terrestrial slugs by freeze-drying. *Veliger* **15**: 254-256.
Dainton, B. H. 1954. The activity of slugs: I. The induction of activity by changing temperatures. *Journal of Experimental Biology* **31**: 165-187.
de Winter, A. J. and E. Gittenberger. 1998. The land snail fauna of a square kilometer patch of rainforest in southwestern Cameroon: High species richness, low abundance and seasonal fluctuations. *Malacologia* **40**: 231-250.
Emberton, K. C. 1988. The genitalic, allozymic, and conchological evolution of the eastern North American Triodopsinae (Gastropoda: Pulmonata: Polygyridae). *Malacologia* **28**: 159-273.
Emberton, K. C. 1991. The genitalic, allozymic and conchological evolution of the tribe Mesodontini (Pulmonata: Stylommatophora: Polygyridae). *Malacologia* **33**: 71-178.
Emberton, K. C., T. A. Pearce, and R. Randalana. 1996. Quantitatively sampling land-snail species richness in Madagascan rainforests. *Malacologia* **38**: 203-212.
Evans, J. G. 1972. *Land Snails in Archaeology, with Special Reference to the British Isles.* Seminar Press, London. xii + 436 pp.
Goodfriend, G. A. 1986. Variation in land-snail shell form and size and its causes: a review. *Systematic Zoology* **35**: 204-223.
Gould, S. J. and N. D. Young. 1985. The consequences of being different: sinistrally coiling in *Cerion*. *Evolution* **39**: 1364-1379.
Gregg, W. O. 1974. Finding snails in the desert. *In:* M. K. Jacobson, ed., *How to Study and Collect Shells*, 4th Ed. American Malacological Union, Wrightsville Beach, North Carolina. Pp. 76-77.
Hadfield, M. G. 1986. Extinction in Hawaiian achatinelline snails. *Malacologia* **27**: 67-81.
Harry, H. W. 1998. *Carychium exiguum* (Say) of Lower Michigan; morphology, ecology, variation and life history (Gastropoda, Pulmonata). *Walkerana* **9**: 1-104.
Henderson, I. 1996. Slug and snail pests in agriculture. *British Crop Protection Council, Symposium Proceedings* **66**: 1-450.
Hochberg, F. G., Jr., B. Roth, and W. B. Miller. 1987. Rediscovery of *Radiocentrum avalonense* (Hemphill in Pilsbry, 1905) (Gastropoda: Pulmonata). *Bulletin of the Southern California Academy of Sciences* **86**: 1-12.
Hubricht, L. 1951. Preservation of slugs. *Nautilus* **64**: 90-91.
Hubricht, L. 1962. New species of *Helicodiscus* from the Eastern United States. *Nautilus* **75**: 102-107.
Hubricht, L. 1985. The distributions of the native land mollusks of the eastern United States. *Fieldiana, Zoology* (n.s.) **24**: i-viii, 1-191.
Ingram, W. M. 1941. Habits of land Mollusca at Rensselaerville, Albany County, New York. *American Midland Naturalist* **25**: 644-651.
Kerney, M. P. 1999. *Atlas of the Land and Freshwater Molluscs of Britain and Ireland.* Harley Books, Colchester, England. 261 pp.
Kerney, M. P. and R. A. D. Cameron. 1979. *A Field Guide to the Land Snails of Britain and Northwest Europe.* Collins, London. 288 pp.
Kerney, M. P., R. A. D. Cameron, and J. H. Jungbluth. 1983. *Die Landschnecken Nord- und Mitteleuropas.* Verlag Paul Parey, Hamburg. 384 pp.
Kralka, R. A. 1986. Population characteristics of terrestrial gastropods in boreal forest habitats. *American Midland Naturalist* **115**: 156-164.
Krull, W. H. and C. R. Mapes. 1974. Trapping small land snails. *In:* M. K. Jacobson, ed., *How to Study*

and Collect Shells, 4th Ed. American Malacological Union, Wrightsville Beach, North Carolina. P. 69.

LaRocque, A. 1966. Pleistocene Mollusca of Ohio. *Ohio Geological Survey Bulletin* **62**: i-iii, 1-111.

LaRocque, A. 1974a. Shells from moss, shells from dead leaves, shells from stream drift, quarries and ledges. *In:* M. K. Jacobson, ed., *How to Study and Collect Shells*, 4th Ed. American Malacological Union, Wrightsville Beach, North Carolina. Pp. 70-71.

LaRocque, A. 1974b. Non-marine Pleistocene Mollusca. *In:* M. K. Jacobson, ed., *How to Study and Collect Shells*, 4th Ed. American Malacological Union, Wrightsville Beach, North Carolina. Pp. 71-73.

LaRocque, A. 1974c. When to collect slugs. *In:* M. K. Jacobson, ed., *How to Study and Collect Shells*, 4th Ed. American Malacological Union, Wrightsville Beach, North Carolina. Pp. 78-79.

Lee, H. G. 1993. Toward an improved strategy for land-snail collecting. *American Conchologist* **21**: 12.

Locasciulli, O. and D. A. Boag. 1987. Microdistribution of terrestrial snails (Stylommatophora) in forest litter. *Canadian Field-Naturalist* **101**: 76-81.

Lydeard, C., R. H. Cowie, W. F. Ponder, A. E. Bogan, P. Bouchet, S. A. Clark, K. S. Cummings, T. J. Frest, O. Gargominy, D. G. Herbert, R. Hershler, K. E. Perez, B. Roth, M. Seddon, E. E. Strong, and F. G. Thompson. 2004. The global decline of nonmarine mollusks. *BioScience* **54**: 321-330.

Menez, A. 2001. Assessment of land snail sampling efficacy in three Mediterranean habitat types. *Journal of Conchology* **37**: 171-175.

Mienis, H. K. 1971. *Theba pisana* in pellets of an Israelian owl. *Basteria* **35**: 73-75.

Miller, W. B., R. L. Reeder, N. Babrakzai, and H. L. Fairbanks. 1984. List of new and revised Recent taxa in the North American terrestrial Mollusca (north of Mexico) published since 19 March 1948. Part 1. *Tryonia* **11**: i, 1-14.

Murray, J., E. Murray, M. S. Johnson, and B. Clarke. 1988. The extinction of *Partula* on Moorea. *Pacific Science* **42**: 150-153.

Nekola, J. C. and T. M. Smith. 1999. Terrestrial gastropod richness patterns in Wisconsin carbonate cliff communities. *Malacologia* **41**: 253-269.

Oggier, P., S. Zschokke, and B. Baur. 1998. A comparison of three methods for assessing the gastropod community in dry grasslands. *Pedobiologia* **42**: 348-357.

Örstan, A. 1999a. Land snails of Black Hill Regional Park, Montgomery County, Maryland. *Maryland Naturalist* **43**: 20-24.

Örstan, A. 1999b. Drill holes in land snail shells from western Turkey. *Schriften zur Malakozoologie* **13**: 31-36.

Örstan, A. and F. Welter-Schultes. 2002. A dextral specimen of *Albinaria cretensis* (Pulmonata: Clausiliidae). *Triton* **5**: 25-28.

Pearce, T. A. 1990. Spool and line technique for tracing field movements of terrestrial snails. *Walkerana* **4**: 307-316

Pearce, T. A. 1994. Terrestrial gastropods of Mackinac Island, Michigan, U.S.A. *Walkerana* **7**: 47-53.

Pearce, T. A. 1997. Land snail ecology on northern Kuril Islands, Far Eastern Russia: habitat versus isolation. Abstract. *Annual Report of the Western Society of Malacologists* **30/31**: 53

Pearce, T. A. and A. S. Italia. 2002. Land snails and slugs in Delaware, U.S.A.: Systematic survey reveals new distribution records. *Annual Report of the Western Society of Malacologists* **33**: 26.

Pilsbry, H. A. 1939-1948. Land Mollusca of North America (North of Mexico). *Academy of Natural Sciences, Philadelphia, Monograph* **3**: Vol. 1, Part 1 (1939): i-xvii, 1-573, i-ix, Vol. 1. Part 2 (1940): i-vi, 575-994, i-ix, Vol. 2, Part 1(1946): i-vi, 1-520, Vol 2, Part 2 (1948): i-xlvii, 521-1113.

Ponder, W. F. and D. R. Lindberg. 1997. Towards a phylogeny of gastropod molluscs: An analysis using morphological characters. *Zoological Journal of the Linnean Society* **119**: 83-265.

Richter, K. O. 1976. A method for individually marking slugs. *Journal of Molluscan Studies* **42**: 146-151.

Riddle, W. A. 1983. Physiological ecology of land snails and slugs. *In:* W. D. Russell-Hunter, ed., *The Mollusca*, Vol. 6, Ecology. Academic Press, New York. Pp 431-461.

Rollo, C. D. 1983a. Consequences of competition on the reproduction and mortality of three species of terrestrial slugs. *Researches in Population Ecology* **25**: 20-43.

Rollo, C. D. 1983b. Consequences of competition on the time budgets, growth and distribution of three species of terrestrial slugs. *Researches in Population Ecology* **25**: 44-68.

Roscoe, E. J. 1974. Collecting mollusks in desert regions. *In:* M. K. Jacobson, ed., *How to Study and Collect Shells*, 4th Ed. American Malacological Union, Wrightsville Beach, North Carolina. Pp. 73-76.

Runham, N. W. and P. J. Hunter. 1970. *Terrestrial Slugs*. Hutchinson University Library, London. 184 pp.

Schander, C. and J. Hagnell. 2003. Death by drowning degrades DNA. *Journal of Molluscan Studies* **69**: 387-388.

Schilthuizen, M. and M. Lombaerts. 1994. Population structure and levels of gene flow in the Mediterranean land snail *Albinaria corrugata* (Pulmonata: Clausiliidae). *Evolution* **48**: 577-586.

Schmidt-Nielsen, K., C. R. Taylor, and A. Shkolnik. 1971. Desert snails: Problems of heat, water and food. *Journal of Experimental Biology* **55**: 385-398.

Seddon, M. B. 1998. Red listing for molluscs - A tool for conservation? *Journal of Conchology Special Publication* **2**: 27-44.

Selander, R. K. and D. W. Kaufman. 1973. Self-fertilization and genetic population structure in a colonizing land snail. *Proceedings of the National Academy of Sciences* **70**: 1186-1190.

Simberloff, D. 1996. Impacts of introduced species in the United States. *Consequences* **2**(2) [available at <www.gcrio.org/CONSEQUENCES/vol2no2/article2.html>]

Solem, A. 1974. *The Shell Makers, Introducing Mollusks*. John Wiley and Sons, New York. xiv + 289 pp.

South, A. 1992. *Terrestrial Slugs: Biology, Ecology and Control*. Chapman and Hall, London. x + 428 pp.

Turgeon, D. D., J. F. Quinn, Jr., A. E. Bogan, E. V. Coan, F. G. Hochberg, W. G. Lyons, P. M. Mikkelsen, R. J. Neves, C. F. E. Roper, G. Rosenberg, B. Roth, A. Scheltema, F. G. Thompson, M. Vecchione, and J. D. Williams. 1998. *Common and Scientific Names of Aquatic Invertebrates from the United States and Canada. Mollusks*, 2nd Ed. American Fisheries Society Special Publication 26. 526 pp.

USDA. 2001. How to import foreign soil and how to move soil within the United States. *U.S. Department of Agriculture, Circular Q-330.300-2, Soil (01/2001)*. U.S. Department of Agriculture, Animal and Plant Health Inspection Service, Plant Protection and Quarantine. 4700 River Road, Unit 133, Riverdale, Maryland 20737-1228, U.S.A. [available at <www.aphis.usda.gov/ppq/permits/soil/circul.pdf>].

Vail, V. A. 1979. Going collecting? Bring your radio. *Hawaiian Shell News* **27**: 3.

Van Cleave, H. J. and J. A. Ross. 1947. A method for reclaiming dried zoological specimens. *Science* **105**: 318.

Webb, G. R. 1950. New and neglected philomycids and the genus *Eumelus* Rafinesque (Mollusca, Gastropoda, Pulmonata). *Transactions of the American Microscopical Society* **69**: 54-60.

Welter-Schultes, F. W. 1998. *Albinaria* in central and eastern Crete: Distribution map of the species (Pulmonata: Clausiliidae). *Journal of Molluscan Studies* **64**: 275-279.

Wolda, H. 1969. Fine distribution of morph frequencies in the snail *Cepaea nemoralis* near Groningen. *Journal of Animal Ecology* **38**: 305-327.

Zilch, A. 1959-1960. Euthyneura. *In:* W. Wenz, ed., *Gastropoda*, Vol. 2. Gebrüder Borntraeger, Berlin. Pp. i-xii + 1-834.

CHAPTER 23

REARING TERRESTRIAL GASTROPODA

AYDIN ÖRSTAN

23.1 INTRODUCTION

Many species of land snails are relatively easy to keep in captivity; their requirements are simple, they do not take up much space nor do they require constant care. One may keep land snails in captivity for any length of time depending on one's intentions and resources. Short-term captivity lasting a few days may allow one to perform certain activities easily that may be difficult to carry out in the field. For example, photography of live snails, examinations of external anatomy, observations of snail behavior in response to external stimuli, and determination of food preferences are the types of work that may be performed with captive snails in short periods.

Some projects may require captive breeding and maintenance of large numbers of snails for many years. For example, one may want to know if two species will hybridize, or how various conditions will influence the phenotypic characteristics of shells. Depending upon how quickly the captive snails reproduce and their offspring grow, such studies may take several years to complete. One could even collect rare or endangered snail species (with proper permits), captive-breed them, and release their offspring to the wild.

At first glance, the apparent diversity of methods in the literature may give the impression that everyone who has successfully raised land snails has a different protocol. True, there is no ideal container for keeping snails, or a single perfect food that will satisfy all species. However, everyone who intends to establish a healthy colony must satisfy the basic requirements of snails, but the details of how one accomplishes this task will depend on one's intentions, resources, ingenuity, and the species of snails.

23.2 A SURVEY OF THE LITERATURE

For some of the common North American snails, there is plenty of (but, probably not enough) published information dealing with captive rearing, but for most of the species that never have been kept in captivity, there is almost no information. Often, one may find information on how to raise a particular species in a paper dealing primarily with some other aspect of that species, for example, its genetics. Therefore, before bringing snails indoors, it is advisable to conduct a careful literature search to find out if anything on the captive breeding of that particular species has been published. Keep in mind that the data gathered while raising an infrequently studied species, if done and presented properly, could become a useful contribution to the scientific literature.

The following is an annotated list of some of the publications dealing with raising land snails. Krull (1937) gave directions for establishing a terrarium suitable for large species, such as *Mesodon thyroidus* (Say, 1816). Archer (1937) gave detailed instructions for converting an aquarium or a flowerpot into a snail terrarium. Carmichael (1937) presented methods for rearing slugs, including handling instructions for their eggs and juveniles. Kingston (1966) raised 18 species of North American snails, including slugs, and gave practical advice for maintaining these snails and their eggs. Grimm (1974) discussed and offered solutions for the various problems encountered when raising snails. Gray *et al.* (1985) provided information on the captive rearing of several species of North American snails, including slugs. Cowie and Cain (1983) presented information especially useful for the captive maintenance of European land snails.

23.3 SHORT TERM MAINTENANCE OF LAND SNAILS

Snails collected dormant during a dry or a cold season may take several hours to become fully active after having been wetted down or warmed up. Therefore, if time is limited, it is best to start a project with snails that are already active in the field. If an active snail is brought indoors soon after collection and placed in a suitable container, for example, a tray, and left undisturbed, it will soon start exploring its new environment.

Occasionally, the tray and the snail should be gently sprayed with water to prevent the snail from getting too dry and thus withdrawing into its shell. Because slugs lack a protective shell, when they get too dry, they stop moving and eventually shrivel and die at that spot. To prevent slugs from getting dry either keep them in a humid environment or occasionally spray them with water or keep them on a moist substrate, like wet paper towel. Incandescent lights should be placed a safe distance away to prevent the snails from overheating; cooler fluorescent lamps are a better choice for indoor observations.

If you intend to keep the snails for only a few hours, you may not need to feed them. Nevertheless, a piece of carrot or a flake of fish food may naturally immobilize a hungry snail thereby allowing it to be photographed easily or examined under a stereomicroscope. If the intent is to keep the snails for a few days, place them in a secure container lined with moist paper towel or damp soil (a few centimeters thick) and provide them with food (see below). At the end of the work, make every effort to return the snails to the wild, most preferably to the same location where they were collected.

23.4 REARING NORTH AMERICAN WOODLAND SNAILS

Some species of land snails that normally reproduce by outcrossing can also produce offspring without mating, for example *Neohelix albolabris* (Say, 1817) (McCracken and Brussard 1980). However, it is not known if a robust clone can be established from such offspring. Therefore, to build a long-term colony it may be best to start with several individuals to avoid the appearance of phenotypic abnormalities in the offspring produced from the mating of closely related individuals.

23.4.1 Containers. Many different types of containers may be successfully used to keep snails. Plastic or glass containers with clear lids or sides are ideal, because snails can be observed without disturbing them. Fish aquaria may be adapted for snails (Archer 1937). The lid of the container should be tight enough to prevent the snails from escaping, but not airtight. Large snails, such as *Neohelix*, can easily push open loosely closed lightweight lids, and yet if too many of them are kept in airtight containers, they may suffocate. Small ventilation holes may be punched in the lids of containers or netting may be used as a cover. Obviously, such ventilation openings must be small enough to prevent the escape of snails, especially the juveniles. Fit the size of the container to the size and the number of snails; if you put a one-millimeter snail in an aquarium full of soil, you may never be able to find the snail again. In addition, for some species, overcrowding may negatively influence growth and reproduction (Cowie and Cain 1983, Pearce 1997). Small snails may be kept in petri dishes (Mapes and Krull 1951), and the petri dishes then may be placed in larger plastic containers to retain moisture. Trial and error will help you develop a system that will produce the best results.

23.4.2 Substrate. Excess moisture can kill snails. This is especially true for the foul liquid that accumulates when plant material with a high water content, such as lettuce leaves, rots. Therefore, if the intent is to keep snails for longer than a day or two, use a substrate in their containers to absorb excess liquids. Paper towels may be used for this purpose, but after a while the paper itself starts to decompose or develop mold. Soil and leaf litter from the snails' habitat is a natural alternative. Soil absorbs and retains moisture. If the container begins to dry out, some species will bury themselves in the soil, in an attempt to slow their water loss. Many species bury their eggs in soil. Moreover, snails eat decomposing leaves and damp soil, which thereby serve as a backup food source for when one is away or too busy to tend to the snails.

Forest soil, including leaf litter, often contains mites and nematodes that are potentially harmful to captive snails (see the section on snail health below, Chapter 23.5). Heating soil and litter in shallow pans in an oven at about 50°C (120°F) until dry can kill mites and nematodes. Soil can be sterilized by heating at 250°F for 2 hours (USDA 2001). Soil may be heated at higher temperatures, but beware that dry leaves may start to burn at temperatures above 230°C (450°F). Alternately, to avoid the trouble of transporting and heating large quantities of forest soil (and needlessly destroying many inhabitants of the soil in the process), consider using store-bought soil. The best general-purpose substrate for use in snail containers is commercial potting soil manufactured by composting of wood. Composting naturally takes place at raised temperatures, killing many potentially nuisance organisms.

In time, the soil and the sides of a container are covered with snail mucus, to which snail excreta and other debris adhere. Occasionally replace the soil and clean the containers under tap water. Snail mucus may be removed with household bleach (Pearce 1997).

23.4.3 Food.
Table 23.1 lists food items fed to several species of native and introduced North American land snails, including some that do not normally live in the woods. Grimm (1974) recommended additional foods to supplement snails' diets, especially during reproductive periods: potatoes, turnips, radishes, wheat germ, porridge, boiled egg yolk, and bananas. Dry oats have always been recommended for snails (Krull 1937). However, uneaten oats should be promptly removed, because fungi quickly grow on moist oats. Aquarium fish food is also accepted by many species of snails (Table 23.1). Cowie and Cain (1983) fed their snails a mixture of dry skim milk, breakfast cereal, and calcium carbonate. Various fruits, including grapes and apples, may also be fed to snails.

Carnivorous species, for example, *Haplotrema concavum* (Say, 1821), should be given live prey of appropriate sizes (Table 23.1), although some carnivores, for example, *Rumina decollata* (Linnaeus, 1758) can be raised on lettuce only (Table 23.1). Snails need calcium to make their shells and the shells of their eggs. In the wild, they get calcium from their food, the shells of other snails, and calcium containing rocks and soil. Although captive snails get some calcium from the foods they eat (even lettuce has minute quantities of calcium), their diet should be supplemented. Calcium carbonate may be supplied in the form of limestone or marble rocks, cuttlefish bones, empty mollusk shells, eggshells, or any other suitable source. A favorite is powdered limestone, which is cheaply available in gardening supply stores and easily applied from a saltshaker. Powdered limestone will also help prevent the pH of the soil from getting too acidic.

Start out by placing small quantities of food in the snails' containers. If it is quickly eaten, increase the amount. Too much food may spoil before it is eaten and create an unhealthy condition. Place the snails directly on the fresh food. If they have become dormant since last checked, spray them with water, which may quickly revive them. Each time the snails are fed, sprinkle some limestone powder on the food and soil. Excess limestone is not harmful.

As many field collectors have learned, land snails will eat paper. This should not be surprising since paper is mostly cellulose, a main ingredient of the natural diets of many species. Grimm (1974) suggested feeding snails various types of paper products (cardboard, newspaper, etc.). However, since the composition of most inks are a trade secret, you might prefer using paper without printing on it. This removes any potential risk to the snails from the ink.

Table 23.1. Foods for rearing some North American land snails: in addition to a source of calcium.

Species	Food
Native species	
Forest snails	leaf litter, lettuce, cabbage, oatmeal, bran, leaves of *Ailanthus glandulosa* (Archer 1937, Kingston 1966, Grimm 1974)
Neohelix spp.	lettuce, carrots, cucumbers, mushrooms, spinach, oats, fish food (Vail 1978, McCracken and Brussard 1980, Emberton 1994, Örstan, unpublished)
Mesodon spp.	lettuce, oats, fish food (Emberton 1994, Pearce 1997)
Triodopsis spp.	lettuce, carrots, fish food (Gray *et al*. 1985, Örstan, unpublished)
Deroceras spp.	lettuce, carrots, leached maple leaves, oatmeal, leaves of *Ailanthus* sp. (Kingston 1966, Gray *et al*. 1985)
Discus cronkhitei (Newcomb, 1865)	lettuce (Gray *et al*. 1985)
Vallonia spp.	leaf litter (Gray *et al*. 1985)
Zonitoides arboreus (Say, 1816)	leaf litter (Gray *et al*. 1985)
Anguispira spp.	lettuce, oatmeal, dried maple leaves, fish food, carrots (Elwell and Ulmer 1971, Pearce 1997, Örstan, unpublished)
Haplotrema concavum (Say, 1821)	*Succinea ovalis* Say, 1817, *Zonitoides arboreus* (Say, 1816), *Discus catskillensis* (Pilsbry, 1896), *Ventridens suppressus* (Say, 1829), *Allopeas gracilis* (Hutton, 1834), *Anguispira* spp. (including eggs), young of *Mesodon thyroidus* (Say, 1816), *Neohelix albolabris* (Say, 1817), and *Helix aspersa* Müller, 1774 (Pearce and Gaertner 1996, Atkinson and Balaban 1997, Örstan, unpublished)
Cochlicopa lubrica (Müller, 1774)	leaves of *Ailanthus* (Mapes and Krull 1951)
Introduced species	
Limax flavus Linnaeus, 1758	bread, potatoes, sweet potatoes, various leafy vegetables, turnips, milk (Carmichael 1937)
Rumina spp.	lettuce, oats, fish food, carrots, beets, *Helix aspersa* Müller, 1774 (Batts 1957, Dundee 1970, Fisher 1974, Örstan, unpublished)
Oxychilus draparnaudi (Beck, 1837)	*Stenotrema fraternum* (Say, 1824), *Webbhelix multilineata* (Say, 1821), *Anguispira alternata* (Say, 1816), cucumbers (Frest and Rhodes 1982)
Allopeas gracilis (Hutton, 1834)	lettuce, fish food (Örstan, unpublished)
Edible species (*Helix*, *Cepaea*, *Otala*, etc.)	vegetables, fruits, cereals, legumes, garden plants, skim milk powder (Cowie and Cain 1983, Thompson and Cheney 1996)

23.4.4 Moisture, temperature, and light. Dry soil (or other suitable substrates) placed in a snail container should be wetted by spraying it with water. To minimize the growth of fungus and potentially harmful mites and nematodes (see below), avoid continuous high humidity inside the containers. Shelled snails will usually survive if their containers are dried out for brief periods. Slugs however will die if their containers dry out for even a brief period (Kingston 1966).

For snails to remain active year round, keep them within an acceptable temperature range. The limits of this range may be species specific. Most U.S. forest species will remain active within 10-25°C (50-80°F). If kept for a long period at temperatures below 10°C, snails are likely to become dormant until the temperature rises. Also, at low temperatures, the development of eggs will be delayed. However, exactly what a given snail will do under certain conditions is difficult to predict, since many external physical conditions and individual genotypic variations will influence behavior and physiology. In the woods in Maryland on a rainy January day when the temperature was slightly above freezing, I found under the leaf litter five live, but dormant *Neohelix albolabris* with apertures sealed with dried mucus. Nearby was one *Mesodon thyroidus* sticking to the underside of a leaf with its foot partially out of its shell. Some forest species survive freezing (Riddle and Miller 1988) and some such as *Anguispira alternata* (Say, 1816) (Elwell and Ulmer 1971) may need a cold-induced dormant period before reproducing. Familiarization with published literature will be useful in maintaining the proper conditions for the good of the snails.

Elevated temperatures are more likely to kill snails than are low temperatures (Grimm 1974). Although in areas with long, dry summers (for example, in the Mediterranean countries) many species aestivate (and survive) in the sun for many months, the elevated temperature inside a container of active snails on a window sill exposed to the sun on a summer day is most likely to kill the snails. Thus, the maxim is that cool is better than hot. In addition, the intensity and duration of light exposure may influence snails' activity and reproduction. The easiest way to regulate the duration of light exposure (photoperiod) is to keep the snail containers away from direct sunlight and on or near a north-facing windowsill.

23.4.5 Maintenance of eggs and juveniles. It is best to remove the juveniles and the eggs from the containers of adults and keep them in separate containers (Carmichael 1937, Kingston 1966, Vail 1978, McCracken and Brussard 1980). Otherwise, the small juveniles may be inadvertently thrown away when removing old food or replacing the substrate. Occasionally adults may consume the eggs (Grimm 1974). Young hatchlings are also known to eat conspecific eggs (Vail 1978, Atkinson and Balaban 1997, Desbuquois *et al.* 2000). However, it would be difficult and unnecessary to attempt to prevent this behavior. You may find the eggs, usually buried in soil in groups, by gently mixing the soil with a small spatula.

Eggs may be incubated in damp soil, or on damp paper towels in petri dishes, or in other small containers. Avoid getting the eggs too dry or too wet for prolonged periods. Empty transparent pill bottles make good incubators for eggs. Eggs placed on damp pieces of paper towel (or several layers of toilet paper) and sealed inside pill bottles may not require further care until they hatch. For accurate control of temperature, bottles with eggs may be partially immersed in temperature-controlled water baths. Keep the newly hatched snails in small containers where they may be easily found. Transfer them to larger containers as they grow.

23.5 SNAIL HEALTH

If snails are kept under proper physical conditions and given adequate food, most of them will be, as far as a human observer can tell, healthy. Occasionally, snail containers may be infested with nematodes, mites, or protozoans (Arias and Crowell 1963, Kingston 1966, Grimm 1974, Gray *et al.* 1985). In the wild, the numbers of such organisms are controlled by various environmental factors, which may be modified in their favor in an artificial setting, resulting in their populations increasing dramatically. These organisms are usually smaller than a millimeter and the examination of the soil from a snail container under a stereomicroscope or with a magnifying glass may be necessary to detect them.

There are several species of mites (Fain 2004) and nematodes (Morand *et al.* 2004) that are parasitic in land snails. Although a few nematodes or mites may not harm the snails, under optimum conditions, the numbers of these animals may reach tremendous amounts. When that happens, snails are overwhelmed and usually become dormant and although the large ones may survive, the small snails, particularly the juveniles, soon die. In heavily infested cultures, one may find otherwise empty snail shells with numerous nematodes or mites inside.

If a snail container is heavily infested with nematodes or mites, there is no practical way to get rid of them. The infestations usually return, even after washing the snails from infested containers in tap water, picking them clean under a stereomicroscope, and then placing them in new containers with fresh material. In all likelihood, some of the nematodes and the mites remain inside the shells or on the snails' bodies. Therefore, the best cure is prevention from the outset. As discussed above, either use composted potting soil or heat-treated forest soil. If one doubts the cleanliness of commercial soil, it too should be heated before use.

Before placing newly collected snails in their containers, wash and clean their shells with an old toothbrush under tap water. Small shells may be cleaned by briefly dipping them in a small container of water. Also, if the snails are fed lettuce or other leafy vegetables, make sure the vegetables are free from mud that may harbor nematodes. To prevent any unwanted organisms from feeding on rotting leftover food, promptly remove it. Do not mix the snails from infested containers with those from uninfested containers.

Some of my snail containers have also been infested with very small flies and their larvae, and these were difficult to eliminate (see also Gray *et al.* 1985). Although neither the flies nor their larvae seemed to be harming the snails, they were nevertheless a nuisance, requiring frequent changes of containers. The recommendations given above to prevent nematode and mite infestations also should help prevent the fly infestations.

23.6 LITERATURE CITED

Archer, A. F. 1937. Vivarium methods for the land Mollusca of North America. *In:* J. G. Needham, ed., *Culture Methods for Invertebrate Animals*. Comstock Publishing Co., Ithaca, New York. Pp. 527-529. [Reprint by Dover Publications, Inc. 1959].

Arias, R. O. and H. H. Crowell. 1963. A contribution to the biology of the gray garden slug. *Bulletin of Southern California Academy of Sciences* **62**: 83-97.

Atkinson, J. W. and M. Balaban. 1997. Size-related change in feeding preference in the carnivorous land snail *Haplotrema concavum* (Pulmonata: Stylommatophora). *Invertebrate Biology* **116**: 82-85.

Batts, J. H. 1957. Anatomy and life cycle of the snail *Rumina decollata* (Pulmonata: Achatinidae). *Southwestern Naturalist* **2**: 74-82.

Carmichael, E. B. 1937. Culture methods for *Limax flavus*. *In:* J. G. Needham, ed., *Culture Methods for Invertebrate Animals*. Comstock Publishing Co., Ithaca, New York. Pp. 529-531. [Reprint by Dover Publications, Inc. 1959].

Cowie, R. H. and A. J. Cain. 1983. Laboratory maintenance and breeding of land snails, with an example from *Helix aspersa*. *Journal of Molluscan Studies* **49**: 176-177.

Desbuquois, C., L. Chevalier, and L. Madec. 2000. Variability of egg cannibalism in the land snail *Helix aspersa* in relation to the number of eggs available and the presence of soil. *Journal of Molluscan Studies* **66**: 273-281.

Dundee, D. S. 1970. Introduced Gulf Coast molluscs. *Tulane Studies in Zoology and Botany* **16**: 101-115.

Elwell, A. S. and M. J. Ulmer. 1971. Notes on the biology of *Anguispira alternata* (Stylommatophora: Endodontidae). *Malacologia* **11**: 199-215.

Emberton, K. C. 1994. Partitioning a morphology among its controlling factors. *Biological Journal of the Linnean Society* **53**: 353-369.

Fain, A. 2004. Mites (Acari) parasitic and predaceous on terrestrial gastropods. *In:* G. Barker, ed., *Natural Enemies of Terrestrial Molluscs*. CABI Publishing, Cambridge, Massachusetts. Pp. 505-524.

Fisher, T. W. 1974. Miscellaneous observations on *Rumina decollata* (Linnaeus, 1758) (Achatinidae) and a request. *Veliger* **16**: 334-335.

Frest, T. J. and R. S. Rhodes. 1982. *Oxychilus draparnaldi* in Iowa. *Nautilus* **96**: 36-39.

Gray, J. B., R. A. Kralka, and W. M. Samuel. 1985. Rearing of eight species of terrestrial gastropods (order Stylommatophora) under laboratory conditions. *Canadian Journal of Zoology* **63**: 2474-2476.

Grimm, F. W. 1974. Techniques for rearing land snails. *In:* M. K. Jacobson, ed., *How to Study and Collect Shells*, 4th Ed. American Malacological Union. Wrightsville Beach, North Carolina. Pp. 102-104.

Kingston, N. 1966. Observations on the laboratory rearing of terrestrial molluscs. *American Midland Naturalist* **76**: 528-532.

Krull, W. 1937. Rearing terrestrial snails. *In:* J. G. Needham, ed., *Culture Methods for Invertebrate Animals*. Comstock Publishing Co., Ithaca, New

York. Pp. 526-527. [Reprint by Dover Publications, Inc. 1959].

Mapes, C. R. and W. H. Krull. 1951. Collection of the snail *Cionella lubrica*, and its maintenance in the laboratory. *Cornell Veterinerian* **41**: 433-444.

McCracken, G. F. and P. F. Brussard. 1980. Self-fertilization in the white-lipped land snail *Triodopsis albolabris*. *Biological Journal of the Linnean Society* **14**: 429-434.

Morand, S., M. J. Wilson, and D. M. Glen. 2004. Nematodes (Nematoda) parasitic in terrestrial gastropods. *In:* G. Barker, ed., *Natural Enemies of Terrestrial Molluscs*. CABI Publishing, Cambridge, Massachusetts. Pp. 525-558.

Pearce, T. A. and A. Gaertner. 1996. Optimal foraging and mucus-trail following in the carnivorous land snail *Haplotrema concavum* (Gastropoda: Pulmonata). *Malacological Review* **29**: 85-99.

Pearce, T. A. 1997. Interference and resource competition in two land snails: adults inhibit conspecific juvenile growth in field and laboratory. *Journal of Molluscan Studies* **63**: 389-399.

Riddle, W. A. and V. J. Miller. 1988. Cold-hardiness in several species of land snails. *Journal of Thermal Biology* **13**: 163-167.

Thompson, R. and S. Cheney. 1996 [revised November 2004]. *Raising Snails*. Special Reference Briefs Series no. SRB 96-05, Alternative Farming Systems Information Center, National Agricultural Library, USDA. Available at: <www.nal.usda.gov/afsic/AFSIC_pubs/srb96-05.htm>.

USDA. 2001. *How to Import Foreign Soil and How to Move Soil Within the United States.* U.S. Department of Agriculture, Circular Q-330.300-2, Soil (01/2001). Available at: <www.aphis.usda.gov/ppq/permits/soil/circul.pdf>.

Vail, V. A. 1978. Laboratory observations on the eggs and young of *Triodopsis albolabris major* (Pulmonata: Polygyridae). *Malacological Review* **11**: 39-46.

C. F. Sturm, T. A. Pearce, and A. Valdés. (Eds.) 2006. The Mollusks: A Guide to Their Study, Collection, and Preservation. American Malacological Society.

CHAPTER 24

MARINE GASTROPODA

DANIEL L. GEIGER

24.1 MARINE GASTROPODS, A HETEROGENOUS ASSEMBLAGE OF TAXA

The marine gastropods are the largest subgroup within the approximately 130,000 mollusk species. They comprise approximately 60,000 shelled gastropods and approximately 13,000 seaslugs out of the over 100,000 described gastropod species. The group comprises most of what has traditionally been called the prosobranchs, and all opisthobranch snails. With a few exceptions, the pulmonates are restricted to the land and the fresh water. Additional information on the biology, anatomy, and ecology of snails can be found in Götting (1974), Solem (1974), Thompson (1976), Purchon (1977), Thompson and Brown (1984), Hughes (1986), Fretter and Graham (1994), Gosliner (1994), Voltzow (1994) and Luchtel et al. (1997). This chapter mentions species and families that are not illustrated here; illustrations may be found in any general shell book such as Dance (1974), Eisenberg (1981), or Abbott and Dance (1983). Many regional volumes will show many of these forms as well (e.g., Springsteen and Leobrera 1986, Poppe and Goto 1991, Wilson 1993, 1994). For a more comprehensive listing of literature sources see Chapter 9.

24.2 TERMINOLOGY OF GASTROPOD SHELLS AND ANIMALS

One of the most prominent structures of gastropods is the shell (Figure 24.1). It can be present as the most conspicuous structure in typical snails so that the snail can either withdraw into it like in whelks, or hide under it as in the case of limpets. The shell can also be reduced so that the animal can no longer use the shell for protection; this condition is found in semislugs. True slugs do not have a shell at all as in the case of the nudibranch sea slugs (Figure 24.2M-P) and the marine pulmonate *Onchidium*. Functional aspects of shells have been summarized by Vermeij (1993). The most important functions include protection against predators, providing a somewhat buffered environment to balance extreme environmental conditions particularly in intertidal species, and sometimes also a tool in predation such as the labral teeth of muricids that assist in opening bivalves. Shells are often further modified to accommodate respiration (slit in Vetigastropoda; siphon in Buccinoidea) or reproduction (brooding chambers in Liotiidae, *Larochea*). The often-striking coloration of shells is challenging to interpret; some are cryptic (Ovulidae on gorgonian hosts), while cowry patterns that contrast with the mantle pattern are thought to confuse would-be predators. The coloration of many shells is revealed only after overgrowth and/or the periostracum has been removed.

The shell of a gastropod is usually a coiled, calcified structure. A gastropod grows by adding new shell material at the margin of the existing shell, so the earlier shell is maintained by the growing individual. This growth pattern is in marked contrast to arthropods, in which the old exoskeleton is discarded when the animal is molting, and a new one is produced by the growing individual. Shell growth is continuous in mollusks; usually fast early on in life, slowing down to almost zero growth at old age. In some groups growth occurs in marked steps that are reflected in the growth pattern of the shell (e.g., axial varices in Figure 24.1F), whereas

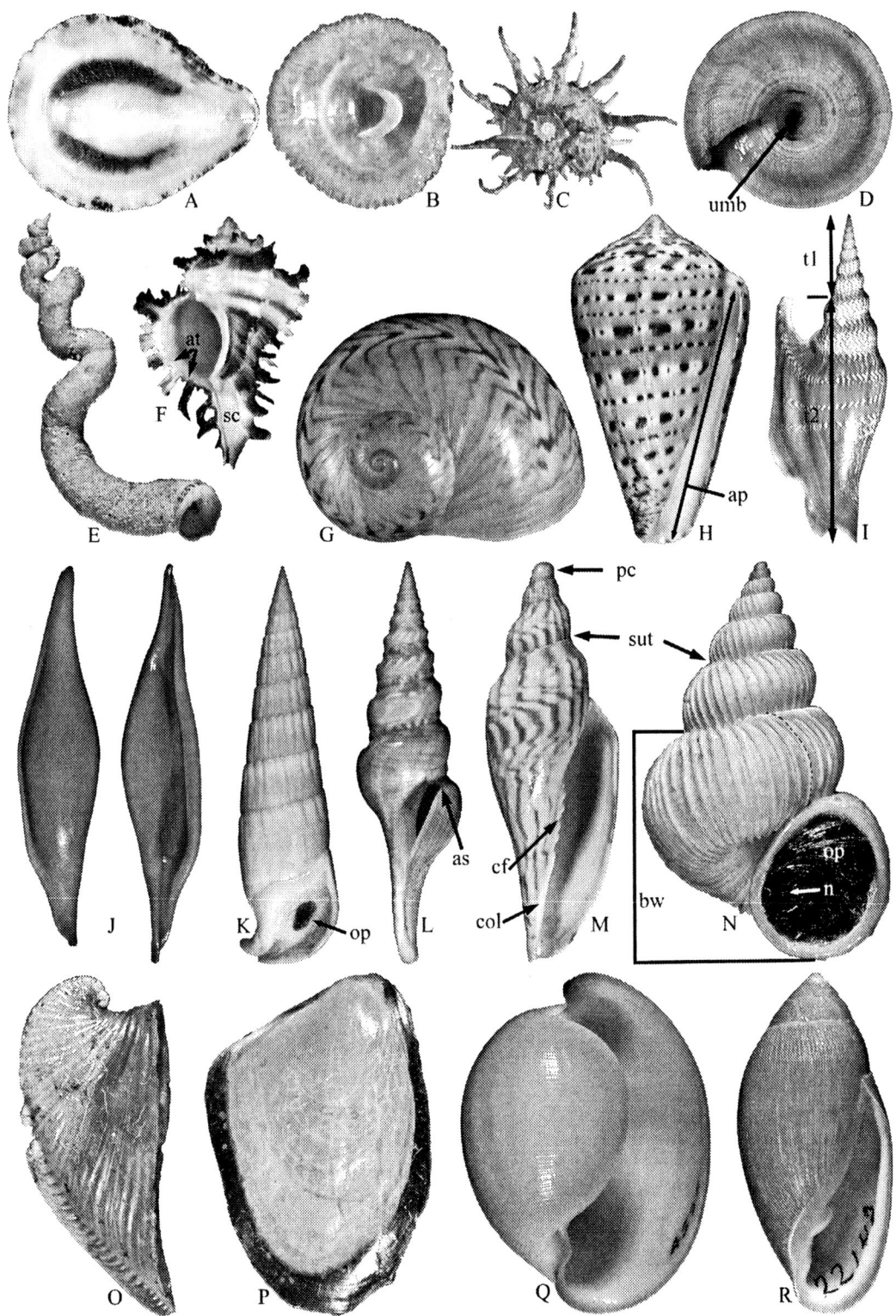

others grow in small but steady increments (Figures 24.1A, 24.1G).

In the pelagic Pseudothecosomata, the calcified structure is not homologous to that of most snails. Their "shell" is therefore called a pseudoconcha. The adult shell is not a continuation of the larval shell, but the larval shell rudiment is discarded and a new calcified structure is formed (Lalli and Gilmer 1989, Figure 24.2J).

Due to the continuous growth of the shell, discrete stages can be recognized, separated by distinct boundaries. The embryonic shell of the larva is called the protoconch. When the individual becomes a juvenile, it switches the growth pattern (teleoconch). In some cases (e.g., *Tibia* spp.), there are two post-larval growth phases namely the juvenile, sexually immature shell (teleoconch I) and the shell of the sexually mature animal (teleoconch II: Figure 24.1I).

The protoconch marks the tip of the shell from which most shells coil downwards (orthostrophic). In some cases, the shell coils upwards, leaving the protoconch in a cavity produced by the whorls (hypostrophic). In a third variation, the coiling axis of the protoconch is at a right angle with the coiling axis of the teleoconch (heterostrophic). In some groups, the coiling is in one plane (planispiral). For

Figure 24.1 Selected shells of marine snails
Maximum size of shell illustrated given. A, *Patella cochlear* Born, 1778 (Patellogastropoda: Patellidae). Cape Town, South Africa. 70 mm. Limpet shaped shell in ventral view with distinct dark, horseshoe shaped muscle scar in center of shell. B, *Cheila corrugata* (Broderip, 1834) (Caenogastropoda: Calyptraeidae). Las Hadas, Manzanillo, Colima, Mexico. 39 mm, SBMNH uncatalogued. Limpet shaped shell in ventral view with internal modification forming a shelf. C, *Angaria vicdani* Kosuge, 1980 (Vetigastropoda: Turbinidae). Bolut, Mindanao, Philippine Islands. 65 mm, SBMNH uncatalogued. Apical view of trochiform shell with axial spines. D, *Architectonica acutissima* Sowerby, 1914 (Heterogastropoda: Architectonicidae). Tosa, Shikoku, Japan. 39 mm, SBMNH uncatalogued. Ventral view of trochiform shell showing stepped umbilicus. E, *Vermicularia pellucida eburnea* (Reeve, 1842) (Caenogastropoda: Turitellidae). La Paz, Baja California del Sur, Mexico. 105 mm, SBMNH 22572. Early shell high-spired and regularly coiled, subsequently the whorls become disjunct, in which the whorls do not touch. F, *Hexaplex chicoreum* (Gmelin, 1791) (Caenogastropoda: Muricidae). Bantayan Island, Philippines. 55 mm, SBMNH uncatalogued. A rare sinistral muricid with regular axial ornamentation arranged as varices. The siphonal canal is almost closed and elongated. G, *Natica canrena*, Linnaeus, 1758 (Caenogastropoda: Naticidae). Cabo La Vela, La Guajira, Colombia. 56 mm, SBMNH uncatalogued. A smooth and polished-looking shell. H, *Conus genuatus* (Linnaeus, 1758) (Caenogastropoda: Neogastropoda: Conidae). Senegal. 70 mm, SBMNH uncatalogued, ex DLG. A typical coniform shell with narrow aperture. I, *Strombus listeri* Gray, 1862 (Caenogastropoda: Strombidae). Andaman Sea, Thailand. 139 mm, SBMNH uncatalogued. The aperture in this species is modified into a flared lip. J, *Phenacovolva rosea* (A. Adams, 1854) (Caenogastropoda: Ovulidae). Bohol, Philippines. 43 mm (left), 45 mm (right), SBMNH uncatalogued. An involute shell without a spire, in dorsal (left) and ventral (right) view. The aperture extends over the entire length of the shell. K, *Rhinoclavis fasciata* (Bruguière, 1792) (Caenogastropoda: Cerithiidae). Gibson Island, Hamilton Pass, Choiseul, Solomon Islands. 56 mm, SBMNH 4401. A high-spired shell with a recurved siphonal canal. The reduced operculum is mounted on cotton in the aperture. L, *Turricula tornata* (Dillwyn, 1817) (Caenogastropoda: Neogastropoda: Turridae). Off Phuket Island, Thailand. SBMNH 141386. The anal sulcus at the top of the aperture is distinct. M, *Fulgolaria hamillei* (Crosse, 1869) (Caenogastropoda: Volutidae). Off Honshu, Japan. 83 mm, SBMNH uncatalogued. The columella shows columellar folds or plicae. There is no umbilicus (anomphalous). N, *Epitonium imperiale* Sowerby, 1844 (Caenogastropoda: Epitoniidae). Queensland, Australia. 32 mm, SBMNH 4948. A high-spired shell with strong axial lamellae, a narrowly open umbilicus, and an operculum sealing the entire aperture. Dotted line shows start of body whorl. O, *Carinaria* cf. *australis* Quoy and Gaimard, 1833 (Caenogastropoda: Heteropoda: Carinariidae). From the stomach of a wolf fish [*Alepisaurus borealis* (Gill, 1862)] 34°N, 132°W, NE Pacific. 21 mm, SBMNH 23369. The reduced shell from a pelagic snail. P, *Dolabrifera dolabrifera* (Rang, 1828) (Opisthobranchia: Anaspidoidea: Aplysiidae). Trailer Park, San Carlos, Sonora, Mexico. 33 mm, SBMNH uncatalogued. The internal and reduced shell of a sea hare. Q, *Atys naucum* (Linnaeus, 1758) (Opisthobranchia: Cephalaspidoidea: Haminoeidae). Philippines. 46 mm, SBMNH WW4525. An only slightly reduced shell of a bubble snail. R, *Ellobium aurisjudae* (Linnaeus, 1758) (Pulmonata: Ellobiidae). Cape York, Queensland, Australia. 40 mm, SBMNH 22143. The shell of a mangrove dwelling marine pulmonate snail. Abbreviations. ap: aperture. as: anal sulcus. at: apertural teeth. bw: body whorl. cf: columellar fold. col: columella. n: nucleus. op: operculum. pc: protoconch. sc: siphonal canal. sut: suture. t1: teleoconch I. t2: teleoconch II. umb: umbilicus. SBMNH: Santa Barbara Museum of Natural History, DLG: D. L. Geiger.

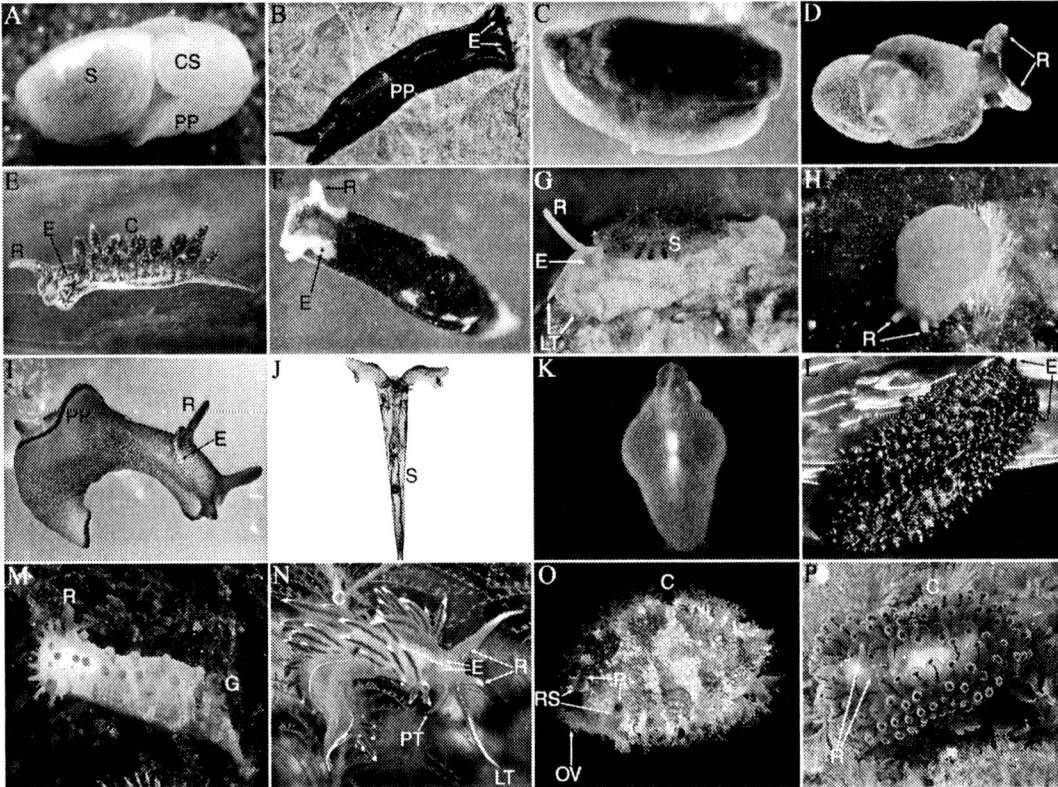

Figure 24.2 Selected sea slugs.
Cephalaspidea A - D. A, *Philine scabra* (Müller, 1776). Animal approximately 5 mm. Millport, Isle of Cumbrae, Scotland, U.K. DLG 2518. B, *Navanax inermis* Cooper, 1863. Animal approximately 4 cm. Flat Rock Point, Palos Verdes, Los Angeles, California, USA. DLG 3565. C, *Runcina coronata* (Quatrefages, 1844). Animal approximately 2.5 mm. Village Bay, Hirta, St. Kilda Archipelago, Outer Hebrides, Scotland, U.K. DLG 3121. D, *Colpodaspis pusilla* Sars, 1870. Animal approximately 5 mm. Village Bay, Hirta, St. Kilda Archipelago, Outer Hebrides, Scotland, U.K. DLG 3172. Ascoglossa E - F. E, *Hermaea (Placida) verticillata* (Alder and Hancock, 1855). Animal approximately 6 mm. On *Bryopsis* sp. (Chlorophyta). Banyuls-sur-Mer, S. France. DLG 1030. F, *Limapontia senestra* (Quatrefages, 1844). Animal approximately 3 mm. Millport, Isle of Cumbrae, Scotland, U.K. DLG 2563. Notaspidea G - H. G. *Tylodina perversa* (Gmelin, 1791), on *Verongia aerophoba* (Schmidt, 1862) (Porifera: Demospongiae). Animal approximately 5 cm, note external shell. Banyuls-sur-Mer, S. France. DLG 1095. H, *Berthella aurantiaca* (Risso, 1826). Animal approximately 5 cm, with internal shell. Ile de Bendor, Bandol, S. France. DLG 1254. Anaspidea. I, *Aplysia rosea* (Rathke, 1799). Animal approximately 3 cm, on *Ulva* sp. (Chlorophyta). Village Bay, Hirta, St. Kilda Archipelago, Outer Hebrides, Scotland, U.K. DLG 3151. Pteropoda: Thecosomata. J, *Creseis* sp. Animal approximately 2 mm. Banyuls sur Mer, S. France. DLG 2338. Gymnosomata. K, *Cliopsis krohni* Troschel, 1854. Animal approximately 4 cm. Monterey Bay Aquarium, Monterey, California, U.S.A. DLG 3975. Pulmonata. L, *Onchidiella celtica* Cuvier, 1817. Animal approximately 2 cm, on *Ulva*. sp. Erquy, Brittany, France. DLG 104. Nudibranchia M-P. Doridoidea M, *Triopha catalinae* (Cooper, 1863). Animal approximately 18 mm. W. Anacapa Island, California, U.S.A. DLG 4746. Aeolidioidea N, *Coryphella lineata* (Lovén, 1846) on *Nemertesia ramosa* (Lamarck, 1816) (Hydrozoa: Thecaphora: Plumulariidae). Animal approximately 3 cm. Clashfarland Point, Millport, Isle of Cumbrae, Scotland, U.K. DLG 2982. Dendronotoidea O, *Tritonia hombergeri* Cuvier, 1803. Animal approximately 10 cm. Millport, Isle of Cumbrae, Scotland, U.K. DLG 3017. Arminoidea P, *Janolus = Antiopella cristatus* (Delle Chiaje, 1841). Slug approximately 3 cm. Village Bay, Dun, St. Kilda Archipelago, Outer Hebrides, Scotland, U.K. DLG 3188. Abbreviations. C: Cerata. CS: cephalic shield. E: eye. G: gill. LT: labial tentacles. OV: oral veil. PP: parapodium. PT: propodial tentacles. R: rhinophore. RS: rhinophore sheath. S: shell. DLG: Author's image archive number.

those that do not coil in a plane, the coiling direction is either right handed (dextral: Figure 24.1N) or left handed (sinistral: Figure 24.1F). The handedness can be determined by holding the shell with the protoconch to the top, and observing whether the opening of the shell (aperture) is on the right or the left side of the shell. Note that the aperture is usually below the protoconch (orthostrophic shells), but may also be located above (hypostrophic shells); turning a hypostrophic dextral shell upside down will make it look superficially like an orthostrophic sinistral shell.

When the shell is composed of several whorls, the last one is usually called the body whorl (Figure 24.1N). In most species, the whorls of the shell are in contact with one another. In some groups such as worm snails, the whorls coil erratically and do not touch one another, which is referred to as disjunct (Figure 24.1E). Where the shoulder of the larger whorl makes contact with the previous whorl, the suture is formed (Figures 24.1M, 24.1N). The latter may connect smoothly, or may form a more or less distinct groove or canal.

The construction of a shell can be defined by just three variables: the translation rate, or how quickly the coiling is moving along the shell axis; the expansion rate, or how quickly the aperture grows in diameter, and the distance from the generating curve, or how far from the axis one coil is separated from the next. Models that are more complex can account for allometric growth and periodic additions such as varices (Raup 1962, 1966). The MacCoil computer program (Palmer 1996) is very instructional in learning about how shells can attain various shapes by changing just a few parameters. Some of the common overall shell shapes are described with reference to some common genera: trochiform, naticiform, patelliform, mitriform, coniform, turritelliform.

The longitudinal axis of the shell can be composed of the touching inner walls of the whorls that produce a twisted spindle or columella (*Fusinus, Vexillum*, Figure 24.1M). When those inner walls do not touch, they will leave a cavity, the umbilicus (Figure 24.1D). The inner margin of the umbilicus can be ornamented in various forms and yield several taxonomically informative characters. Occasionally, the umbilicus is covered with a thickening, the umbilical callus (*Umbonium, Neverita*).

The aperture can be modified in a number of ways. The overall shape varies from circular to slit-like. The margin can be thin and sharp (*Conus*, Figure 24.1H), or can be thickened (Cassidae, *Ellobium*, Figure 24.1R). The outline can be continuous (*Epitonium*, Figure 24.N) or interrupted by various indentations and projections. Starting from the suture in a clockwise direction on a dextral shell (with the aperture facing toward you), the first notch close to the suture is the anal sulcus (Turridae, Figure 24.1L). A similar structure, the slit, is found in many vetigastropods (Pleurotomariidae, Scissurellidae, and Emarginulinae). The two structures can be distinguished by the way the growth increments are formed in the region of the indentation. In vetigastropods, the closed slit is delineated by parallel, spiral ridges that form the selenizone or slitband, whereas in turrids the growth increments form a single, curved line over the entire axis of the whorl. The inner margin of the aperture can be lined by apertural teeth (Figure 24.1F), whereas in the lower part of the aperture a single spine (labial tooth) may project in the coiling direction of the shell (*Achanthina, Opeatostoma*). At the bottom of the aperture a second notch can be found, the siphonal canal (Figures 24.1F, 24.1K-M). It may vary from a shallow notch (*Buccinum*) to a long, drawn-out tube (*Typhis*). The inner side of the aperture may include columellar teeth, which are limited to the apertural region, or columellar folds (Figure 24.1M), which continue along the whorls of the shell (*Fasciolaria, Vexillum*).

In many coiled gastropods, the aperture can be fitted with an operculum (Figures 24.1K, 24.1N). The operculum is a flattened and hard structure attached to the foot of the snail; when the body of the snail retracts into the cavity of the shell, the operculum will cover the aperture and protect the animal from predators and desiccation. The material composition of the operculum, proteinaceous (Trochidae, *Polinices*) or calcified (Turbinidae, *Natica*), defines some major gastropod groups

Figure 24.3 Selected live marine snails.
A, *Atlanta peroni* Lesueur, 1817 (Heteropoda: Atlantidae), animal approximately 1 cm. Banyuls sur Mer, S. France. X 1991. DLG 2260. B, *Peltospira delicata* McLean, 1989 (Vetigastropoda: Peltospiridae). Animal approximately 1 cm.

that are otherwise superficially similar. Additional parameters used to characterize an operculum are its shape (round to lanceolate), the position of the nucleus (central to peripheral; Figure 24.1N), the number of whorls (few: paucispiral; many: multispiral), the surface texture (crenellated, rough, dull, shiny), and the modifications of the margin (none, frills, serrations). Most gastropods have an operculum during their larval stage. Opercula, in a shell collection, can easily become disassociated with the proper shell. Numbering the shell and its corresponding operculum with India ink or gluing (water-soluble glue) the operculum to a plug of polyester fibers placed in the aperture of the corresponding shell, may avoid this problem (Figure 24.1K). Natural cotton wool may be acidic and should be avoided.

The shells of most gastropods are composed of several layers of calcareous material. Calcium carbonate can be secreted by living organisms in two different modifications: calcite and aragonite. Both of these building materials can be shaped differentially, which allows great plasticity in the construction of gastropod shells. The shell is not secreted as pure calcium carbonate, but an internal lattice of proteins guides the crystallization of the calcium carbonate minerals in particular shapes. These processes will produce such diverse crystal types as crossed lamellar, columns, and thin platelets. A typical shell is composed of three layers: the inner hypostracum, the bulk of the shell as the ostracum, and a thin proteinaceous cover, the periostracum. These three layers are produced by three discrete folds on the margin of the mantle. Additionally, muscle attachment points are often modified as myostracum. Their rather rough surface most likely provides additional strength for muscle attachment by enlarging the surface area.

24.3 ANATOMY

Below I provide a brief outline of gastropod anatomy. For additional information see Götting (1974), Thompson (1976), Purchon (1977), Thompson and Brown (1984), Fretter and Graham (1994), Gosliner (1994), Voltzow (1994), Luchtel et al. (1997), and Beesley et al. (1998).

24.3.1 The typical marine gastropod. A typical marine snail has the following major anatomical components: head, mantle cavity, visceral hump, and foot. Figure 24.3 shows some examples of living marine snails.

9°N Vent field. 2001. Leg. Alison Green and Douglas Pace. SBMNH uncataloged, DLG 4482. C, *Cerithium vulgatum* (Brugière, 1792) (Caenogastropoda: Cerithiidae). Animal approximately 4 cm. Banyuls sur Mer, S. France, 0.5 m. II-III 1991. DLG 1036. D, *Littorina obtusata* (Linnaeus, 1758) (Caenogastropoda: Littorinidae). Animal approximately 1 cm. Erquy, Brittany, France. 1989. DLG 235. E, *Mitrella scripta* (Linnaeus, 1758) (Caenogastropoda: Columbellidae). Animal approximately 8 mm. Banyuls sur Mer, S. France. 1988. DLG 72. F, *Fasciolaria lignaria* (Linnaeus, 1758) (Caenogastropoda: Fasciolariidae). Animal approximately 6 cm. Isla Cabrera, Menorca, Balearic Islands, Spain, 5 m. VII-VIII 1989. DLG 381. G, *Thais* (*Stramonita*) *haemastoma* (Linnaeus, 1767) (Caenogastropoda: Muricidae). Animals approximately 6 cm. Ile de Bendor, Bandol, S. France, 6 m. 1989. DLG 163. Arrows indicate egg capsules deposited in oyster. H, *Lottia gigantea* (Sowerby, 1834) (Patellogastropoda: Lottiidae). Animal approximately 2 cm. Ventura Point fishing area, Palos Verdes, Los Angeles, California, U.S.A. 0.5 m. VII 4 2000. DLG 4506. I, *Haliotis tuberculata* Linnaeus, 1758 (Vetigastropoda: Haliotidae). Animal approximately 6 cm. Banyuls sur Mer, S. France. III 1992. DLG 2482. J, *Calyptraea chinensis* (Linnaeus, 1758) (Caenogastropoda: Calyptraeidae). Animal approximately 1 cm. Banyuls sur Mer, S. France. II-III 1991. DLG 1024. Arrow shows eggmass guarded by the snail. K, *Simnia spelta* (Linnaeus, 1758) (Caenogastropoda: Ovulidae) on *Eunicella singularis* (Esper, 1794) (Octocorallia: Plexauridae). Animal approximately 1 cm. Banyuls sur Mer, S. France. III 1992. DLG 2428. L, *Trivia arctica* (Solander and Humphreys, 1797) (Caenogastropoda: Triviidae). Animal approximately 1 cm. Hirta, St. Kilda Archipelago, Scotland, U.K. VII 1993. DLG 3107. M, *Lamellaria* cf. *persicula* (Linnaeus, 1758) (Caenogastropoda: Lamellariidae). Animal approximately 3 cm. Erquy, Brittany, France. 1988. DLG 105. N, Vermetid reef composed of *Petaloconchus erectus* Dall, 1888. Image shows approximately 10 m wide habitat stretch. Spanish Point near Spittal Pond, Bermuda. VII 1991. DLG 1819. O, cf. *Serpulorbis arenaria* (Linnaeus, 1758), with mucus net (arrows) used for feeding. Image width approximately 25 cm. Porros and Illa del Aire, Menorca, Balearic Islands, 6 m. VII-VIII 1989. DLG 375. P, *Serpulorbis squamigerus* (Carpenter, 1857). Image width approximately 5 cm. Cabrillo Marine Aquarium, Los Angeles, California, U.S.A. VI 1994. DLG 3864. Abbreviations: CT: cephalic tentacle. E: eye. EP: epipodium. M: mantle. MP: mantle papillae. OP: operculum. S: shell. SI: Siphon. DLG: Author's image archive number.

Head: The head contains many sense organs, most of the nervous system, and the opening of the digestive tract. The digestive tract opens through the mouth, but may be further extended by means of the proboscis. The proboscis can be used as a suction device to ingest liquid or liquefied food (preoral digestion). The radula, which usually consists of rows of chitinous teeth, is found in most snails. It is an important organ that has served as the basis for the classification of gastropods and often shows differences even between congeneric species. In many gastropods, a pair of jaws is situated in front of the radula to cut larger pieces of food to be ingested. The radula lies in the foregut, which is connected to the stomach. The sensory organs include a pair of eyes and a pair of cephalic tentacles. In addition, opisthobranchs have a pair of chemosensory organs, the rhinophores.

Mantle Cavity: Behind the head opens the dorsal mantle cavity. It serves as a respiratory chamber and as the opening for the hindgut and the gonads. The respiratory organs, gills, are found as a pair or as a single unit. Water flow through the cavity is strongly regulated. In those gastropods with a slit (Vetigastropoda), the incurrent enters the mantle cavity over the head and exits through the posterior part of the slit, taking along waste products and gametes (eggs and sperm). In the higher prosobranchs, ciliary bands in the mantle cavity help to produce a current that enters and exits over the head. In many gastropods, the incurrent opening is drawn out into an elongated tube, the siphon. Be careful to distinguish the siphon from the proboscis. In air breathing snails (Pulmonata, Littorinidae: Figure 24.3D), the gills are reduced or absent and the wall of the mantle cavity is strongly vascularized. In the amphibious patellid limpets, true gills (ctenidia) are absent and are functionally replaced by modifications of the mantle that increase the surface area (Figure 24.3H).

Visceral mass: The visceral hump contains most of the internal organs. From the foregut, the stomach takes up the ingested food and begins to digest it. Through openings in the wall of the stomach, nutrients are transported into the large digestive gland (hepatopancreas). From the stomach, the intestine continues in more or less complicated loops towards the hindgut and discharges the feces into the mantle cavity. The other large organ found in the visceral hump is the gonad. Snails can be either of a single gender (male or female, gonochoristic), or may function as both sexes (hermaphrodites), either simultaneous, as in seaslugs, or sequentially, as in slipper shells (Calyptraeidae: Figure 24.3J). The reproductive opening can simply lead to the outside of the body, or it may be modified terminally into a penis in the male, and/or a complex receptacle in the case of the female, so that internal fertilization can be achieved.

Snails have an open circulatory system meaning that the spaces through which the blood travels are not lined continuously, but are composed of the surfaces of tissues from various organs. The heart is situated just behind the gills and pumps blood into the gills, from where the oxygenated blood is distributed throughout the body and is eventually collected and channeled towards the heart. The entire visceral hump is covered by the mantle, which secretes the shell. At the margin of the shell, the mantle is differentiated into several mantle folds that produce the various layers of the shell. In some gastropods (Cypraeidae, Fissurellidae), the mantle can be drawn over the shell for camouflage or defense.

Foot: The foot serves the purposes of locomotion and attachment to the substrate. It is a highly glandular organ; it has been estimated that approximately 30% of a snail's energy expenditure goes towards the production of mucus. The animals glide over the surface by waves of muscle contraction that travel over the foot. There are several specific patterns of wave movements. At the head end, the propodium is the most mobile part of the foot. In many instances, the propodium is extended into a pair of propodial tentacles, which should not be confused with cephalic tentacles or the rhinophores of Opisthobranchia. The posterior part (metapodium) carries the operculum, can be drawn out as in Aeolidioidea, and in some groups can be willfully detached (autotomized: Harpidae). Lateral extensions of the foot, the parapodia, are drawn over the body in some groups (Ascoglossa: Elysiidae; Anaspidea: *Aplysia*). Many Vetigastropoda have a more or less elaborate set of epipodial tentacles nested between the foot and the margin of the shell (Figures. 24.3B, I).

Opisthobranchia - additional features: Many opisthobranchs do not have a shell as adults. The mantle is hence modified to be tougher and is called the notum. In dorids, it may contain calcareous spicules and often has sulfuric acid glands for protection. The posterior end in this group usually shows the conspicuous circumanal gills as a rosette of branched gills. In the remainder of the Nudibranchia (Aeolidioidea, Dentronotoidea, Arminoidea) as well as in some other opisthobranchs (Ascoglossa), the hepatopancreas is drawn out into dorsal, tentacular processes, the cerata. Note that not all cerata-bearing seaslugs are aeolids. In order to distinguish the four groups, the structure of the rhinophores is particularly helpful.

The propodium of cephalaspids is modified into a cephalic shield that covers the head, from which the name of the group has been derived. The pteropods have many unique modifications including sucker-bearing pedal lobes that evolved separately from those of the cephalopods.

24.4 THE MAJOR MARINE GASTROPOD GROUPS

The higher level of gastropod classification has been undergoing a remarkable revision over the past 15 years. The old classification of three gastropod classes - Prosobranchia, Opisthobranchia, Pulmonata - going back to Thiele (1931), has been shown to be inaccurate. In addition, the familiar division of the prosobranchs into Archaeogastropoda, Mesogastropoda, and Neogastropoda is no longer considered valid. The reasons for the changes outlined below are due to new insights into the evolutionary history of gastropods, and the premise that only monophyletic groups or clades (see Chapter 11) should be named. Many of the gastropod groups proposed by Thiele have now been shown to be non-monophyletic.

For the purpose of arranging your collection, these frequent changes in classification can be unnerving. Because some of these new arrangements are tentative, a standard system, even if not up to the latest flavor, may be quite acceptable from a pragmatic perspective. Two main sources can be considered: Vaught (1989), which is already quite dated, but encompasses all Mollusca, and Beesley *et al.* (1998), which is more current, but, does not include families missing from the Australian malacofauna. Figure 24.4 shows a gross comparison between the Thiele (1931) and the modern system of classification. The following highlights some of the major changes in gastropod classification.

24.4.1 Gastropoda. Snails are still considered by the great majority to be monophyletic; hence, Gastropoda remains the scientific name for snails. However, gastropods did not split right away into three discrete groups (Thiele's Prosobranchia, Opisthobranchia, Pulmonata), but Opisthobranchia and

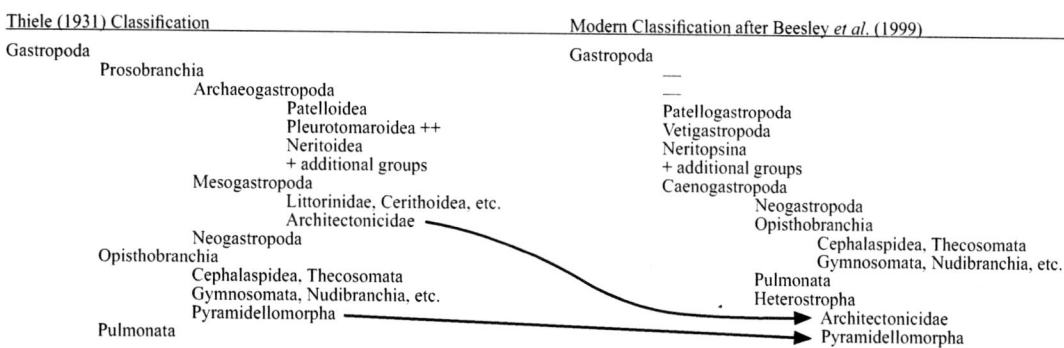

Figure 24.4 Comparisons of traditional and modern classification.
The comparison is very general and strongly condensed. The finer points of gastropod classification and its various clades are still in flux. The interested reader is referred to recent conference proceedings and relevant publications in serials. ++ = minor groups.

Pulmonata are derived from within the Caenogastropoda. Hence, there is no longer a need for Thiele's Prosobranchia as it is synonymous with Gastropoda. Opisthobranchia plus Pulmonata (="Euthyneura") do not form a clade because of fundamental differences in the nervous system; these two groups are distinct lineages. Many additional names have been introduced for the higher classification of gastropods (e.g., Eogastropoda, Apogastropoda, Orthogastropoda); for an overview of these and other names see Bieler (1992) and Ponder and Lindberg (1997).

24.4.2 Basal Gastropoda. These were termed by Thiele Archaeogastropoda. Today they are recognized to be a conglomerate of several independent lineages. The most prominent ones are Patellogastropoda (Patellidae, Lottiidae, Acmaeidae: Figure 24.1A), Vetigastropoda (Pleurotomariidae, Haliotidae, Fissurellidae, Trochoidea, ++: Figure 24.1C), and Neritopsina (Neritidae, Neritopsidae, ++); there are additional minor groups (++ means addition minor groups e.g., Seguenzina, Cocculiniformia).

24.4.3 Higher Gastropoda. The remainder of all Gastropoda is united under Caenogastropoda, i.e., Thiele's Mesogastropoda, Neogastropoda, Opisthobranchia, and Pulmonata. Caenogastropoda is composed of two Thiele suborders plus two Thiele orders. In some modern classification, some of the most basal Caenogastropoda (Cyclophoridae and Ampullariidae) are put in a distinct group, the Architaenioglossa, although there is some disagreement among phylogeneticists.

Neogastropoda (Buccinoidea, Muricoidea, Conidae, Turridae: Figures 24.1F, H, L, M) has been retained as a clade, although only as part of Caenogastropoda, and not as an equivalent unit of equal taxonomic rank.

Pulmonata has also been retained more or less as in Thiele. There are only a few marine and brackish water pulmonates, including the intertidal and subtidal Siphonariidae, the mangrove dwellers Ellobiidae (Figure 24.1R), and the Onchidiidae (Figure 24.2L, previously placed in its own group Gymnomorpha). For more details on the biology of pulmonates, see Chapter 21.

Opisthobranchia, the sea slugs, has also been retained as in Thiele with a single important exception. The members of the Opisthobranchia include: Cephalaspidea (= Bullomorpha. Figures 24.2A-C); Ascoglossa (= Sacoglossa. Figures 24.2D-F); Anaspidea (= Aplysiomorpha. Figure 24.2I); Notaspidea (= Pleurobranchomorpha. Figures 24.2G-H); Accochlidioidea; Pteropoda (Gymnosomata. Figure 24.2K; Thecosomata. Figure 24.2J); Nudibranchia (Doridoidea, Aeolidioidea, Dendronotoidea, Arminoidea. Figures 24.2M-P). Pyramidellomorpha has been removed; it has been placed in Heterogastropoda.

24.4.4 Heterogastropoda. This recently proposed group is characterized by the heterostrophic apex. Two main families are united: Architectonicidae (Figure 24.1D. formerly in the Mesogastropoda) and Pyramidellidae (formerly in the Opisthobranchia).

24.4.5 Incertae sedis: There are some other snails of uncertain affinity, particularly those that live between sand grains (interstitial fauna, meiofauna), because their bodies are highly reduced due to small size and lack features that could aid in their classification (e.g., Rhodomorpha).

24.5 HABITAT AND ECOLOGY

Marine gastropods inhabit virtually all possible habitats, including estuaries, rocky and sandy shores up to the supralittoral fringe, all continental shelf and slope areas, and the abyssal plain. They are even found in the deep-sea trenches of the hadal region. Some are found crawling between sand grains (meiofauna), and some floating (pelagic) groups are known (Pteropoda, Heteropoda). Further information on marine habitats can be found in Levinton (1995), Hogarth (1999), and Nybakken (2001).

24.5.1 Estuaries. Where fresh and sea water mix, we find estuaries. The conditions with neither pure fresh water nor typical salt water being found, combined with strong fluctuations in salinity, makes a unique habitat. Estuarine faunas are mainly composed of more typically marine organisms, with very few freshwater representatives being able to tolerate the salt in the mixed water. There are certain groups of gastropods characteristically inhabiting these areas, including hydrobiids, nassariids, and cerithioideans in the most common muddy conditions, and littorinids and some patellogastropods can be encountered on hard substrates.

24.5.2 Intertidal. The intertidal region can be subdivided with reference to the water level reached at the various tide levels. It spans from the spring tide high mark to the spring tide low mark. The splash zone, which is not fully submerged even during spring tide high water, is usually categorized as a separate unit. Within the intertidal proper, particular zones or belts are defined locally that are delimited by common organisms, such as seaweeds and barnacles along rocky shores, and worms and clams along sandy beaches.

Rocky shore: The rocky shores can be classified according to substrate type (chalk, granite, volcanic rocks), exposure (protected, battered by waves), and topography (inclination, smooth, with tide pools). Furthermore, the vertical zones defined by the presence of certain typical organisms (seaweeds, barnacles) can be determined.

Splash zone: There is very little overall gastropod diversity in the splash zone. The only regularly found snails include selected species of Littorinidae and Patellogastropoda.

Upper half of intertidal zone: This zone is characterized by an abundance of several conspicuous gastropods. Among them are the top snails (Trochoidea), nerites (Neritidae), limpets (Patellogastropoda), Littorinidae, and Lacunidae. Within the tide pools, you may encounter many of the organisms from the lower intertidal zone, but not necessarily those from the subtidal zone. The tide pools in the higher intertidal zone are not replenished with fresh seawater every day. Hence, changes in temperature and salinity due to evaporation and influx of freshwater make them a much more stressful environment for organisms inhabiting them.

Lower half of intertidal zone: As a rule, any hard substrate will be covered with organisms unless the shore is heavily exposed to open ocean conditions. A great variety of snails can be encountered, including all of those found in the higher half of the intertidal zone. Additionally, many muricids and some whelks (Buccinidae) are frequently found. Overturning rocks and checking shaded and overhanging portions of the shoreline may reveal unexpected finds such as nudibranchs, Onchidiidae, abalone, and miter shells, among the more conspicuous elements. However, those interested in micromollusks will reap great rewards collecting algal samples and brushing rocks (see below). Be sure to check in and among other organisms for snails. For instance, wentletraps (Epitoniidae) can be found associated with sea anemones, and eulimids are living on echinoderms. The tide pools are flushed with water during every tidal cycle. They act as a habitat extension for organisms otherwise only found in the subtidal area. There are some differences in temperate and tropical regions. In the former, the shore is often covered with algae, whereas in the tropics, corals are more prominent. In general, tropical faunas are also richer in the variety of species encountered.

24.5.3 Soft bottom shores. Soft substrates can be classified according to the composition of the sediment such as grain size (mud, sand, pebbles) and material (rock, coral, Foraminifera), inclination of the shore, and exposure. The composition of a soft bottom shore can vary greatly among seasons. For instance, the sand beach of Village Bay on the island of Hirta of the St. Kilda Archipelago (Outer Hebrides: Scotland, UK), is fully eroded during winter storms, only to be redeposited the following spring.

There is much less differentiation found in the gastropod fauna of soft bottoms. In the splash zone, mostly dead shells are found, but hardly any usually live there. There is a less obvious differentiation of the intertidal soft substrate, although subtle differences in the infaunal worm and bivalve fauna can be observed. The most typical snails of soft substrates are Nassariidae, Cerithioidea (Figure 24.3C), Naticidae, and Cephalaspidea. In the tropics Strombidae, Terebridae, and Conidae may be found regularly.

25.5.4 Mangroves. Mangroves constitute a specialized habitat in which the framework is composed of trees. This habitat is particularly noted for the tree snails, and few marine species are specialized for mangroves. The main families encountered are Littorinidae, Rissoidea, and Hydrobiidae. Mangroves are often encountered in estuaries (see above), but may also be found in full marine conditions.

24.5.5 Subtidal. Strictly speaking, the subtidal refers to any habitat below the intertidal, i.e., any marine habitat that is constantly covered by water. However, it is customarily understood to comprise a more restricted depth range, namely from the spring tide low mark to approximately 30 m depth. It is the well-lit area of the sea, where marine plants (algae, seagrass) can easily grow. As in the intertidal, the subtidal area is subdivided according to the substrate. In addition to the already introduced rocky and soft substrate, a third category of biogenic substrates is encountered, including coral and other reef types, seagrass meadows, and kelp forests.

Reefs: Coral reefs are the most diverse marine habitat and harbor the most diverse gastropod faunas (Dubinsky 1990, Bouchet *et al.* 2002). Virtually all gastropod families can be found in coral reefs. A few, such as the Coralliophilidae, are restricted to coral reefs because they are specialized to feed on them. A little known reef type is of special interest to the malacologist, the vermetid reefs made by worm snails (Vermetidae). There are only a few places in the world harboring this type of reef, including the US Gulf coast and Bermuda (Figure 24.3N).

Seagrass meadows: These can be found in many tropical and subtropical areas around the world, often associated with reefs (Larkum *et al.* 1989). The meadows are composed of relatives to the terrestrial grasses (Angiosperma: Monocotyledoneae: Potamogetonaceae), which are flowering plants, not algae. They are found in well-lit water usually to depths less than 20 m. They are important nursery grounds for fishes. Some snails that have adapted to this unique habitat, particularly to live on the blades of the seagrasses, include some Patellogastropoda [*Notoacmea depicta* (Hinds, 1842)], and the seahare genus *Phyllaplysia*. A diverse assemblage of trochids is also typical of seagrass meadows.

Kelp forests: Kelp forests are areas with a high standing biomass of large brown algae. These kelp algae may reach up to 50 m in height and form a dense thicket. On the algae, trochids, seaslugs, and many small gastropods (rissoids, cerithiids, triphorids) are encountered, while the rocky substrate is populated with larger snails such as fissurellids, haliotids, and patellogastropods. Kelp forests require cold water; more equatorial occurrences of kelp beds are found in deeper water than those found more towards the poles. This phenomenon is called isothermal submergence because the kelp beds are found at similar temperatures (Lüning 1990).

24.5.6 Shelf or neritic zone. The shelf is the area from the subtidal (~30 m) to the edge of the continental plate and extends to depths of 200 m. Both hard substrates and soft substrates can be encountered, although soft substrates are somewhat more common. These areas are often populated by whelks (Buccinidae) and the Turridae, which comprise a highly diverse group that become more common and more diverse with increasing depth. Most gastropod families are commonly found on the shelf, with a few exceptions such as the patellid limpets.

24.5.7 Beyond the shelf. Deep-water habitats are distinguished by major geological features. The continental slope extends from the edge of the shelf (80-200 m deep) to the edge of the abyssal plain (~4,000 m). The abyssal plain is the largest uniform expanse on our planet; it covers approximately half of the surface of the earth at depths ranging from 4,000 to 6,000 m. Deep sea trenches form the hadal region and include the greatest depths in the ocean (11,000 m). Gage and Tyler (1991) provided a very readable introduction to the deep sea. Deep-water habitats are characterized by whelks (Buccinoidea) and Turridae, but also include some families that are found only in these remote areas.

One spectacular deep-sea habitat is the deep-sea hydrothermal vents (van Dover 2000). These are found in 2,000-3,000 m depth and are noted for their unique environmental conditions. The water released from vents can reach a temperature up to 400°C compared to ambient 4°C and is rich in minerals and hydrogen sulfide. Several groups of gastropods are known only from these areas including many vetigastropods (Peltospiridae: Figure 24.3B, Lepetodrilidae, and Neomphalidae) and some caenogastropods (Provanidae).

24.5.8 Pelagic region. A few snails have abandoned their dependence on substrate and have invaded the water column of the open ocean. Lalli and Gilmer (1989) provided a fascinating account on the natural history of the various groups. They include caenogastropods (Heteropoda: Carinariidae, Atlantidae: Figure 24.3A) and opisthobranchs (Pteropoda, Aeolidioidea). A common adaptation to the free-floating (pelagic) lifestyle is the reduction of the shell. Some groups (Gymnosomata) have no shell at all, whereas others have strongly reduced it (Pseudothecosomata).

24.6 ECOLOGY

24.6.1 Food and feeding. Within the "prosobranchs," the more basal groups tend to be herbivorous. Depending on size, the full range of marine plant life is used as food, including single-celled diatoms (Scissurellidae, many juveniles), turf algae and macroalgae (haliotids and trochids), and seagrass (haliotids). A specialized form of herbivory is exhibited by the opisthobranch group Ascoglossa, in which the animals puncture individual cells of green algae and suck out the content of the cell. In some species, the part of the algal cell that engages in photosynthesis (chloroplast) is retained intact by the snail and cultured in the cerata, where the chloroplasts continue to produce sugar for the new host, and color the snail green.

For most groups of the animal kingdom, there is a snail that preys upon it: sponges (Pleurotomariidae, Doridoidea), Cnidaria (Aeolidioidea, Coralliophilidae), segmented worms (Conidae), bivalves (Naticidae), other snails (Conidae), arthropods (Muricidae), Bryozoa (Doridoidea), echinoderms (Cymatiidae), tunicates (Ovulidae), and vertebrates (Conidae). A significant number of snails parasitize other animals; the hosts include cnidarians (Epitoniidae, Architectonicidae), bivalves (Pyramidellidae), sea urchins (Eulimidae), sea cucumbers (Entoconchidae), and fishes (Marginellidae). Nassariidae are scavengers. A few groups have independently evolved filter feeding, including the slipper shells (Calyptraeidae), the worm snails (Vermetidae, Figure 24.3O), and the pelagic Thecosomata (Declerck 1995). Some highly specialized feeding modes are exhibited, for instance, by the Addisoniidae and Choristellidae, which feed exclusively on the egg cases of sharks, rays, and skates (Chondrichthyes: McLean 1985, 1992).

24.6.2 Reproduction. Most shelled gastropods have separate sexes, whereas opisthobranchs and pulmonates are simultaneous hermaphrodites. A few caenogastropods change sex during life. Some Calyptraeidae are first male and change later to be female. In the opisthobranchs the reproductive organs are arranged so that self-fertilization is difficult, which helps prevent inbreeding.

In the more basal groups, fertilization is external. Males and females release eggs and sperm into the surrounding water (broadcast spawning); fertilization occurs outside the body. The fertilized eggs can be dispersed over shorter or longer distances. The larvae hatch as trochophores or veligers and metamorphose upon contact with the appropriate substrate.

In other gastropods, fertilization may be internal and an egg mass is usually deposited. The shape of the egg mass is often characteristic; Winner (1985, 1991, 1993, 1999) provides an excellent overview of molluscan egg masses. A few species, such as the cowries, engage in parental care. The snails hatch at various stages ranging from trochophores to crawling juveniles. Often the egg masses contain nutrients in the form of unfertilized eggs, but some also use their younger, less developed siblings as a food source. In those forms that have a free veliger stage, the animals disperse at that time. Larval stages can last up to several months, during which the larvae can be passively drifted across ocean basins. These veligers are characterized by a large and highly developed protoconch that supports the rather large lobes of the velum.

Once the animals have settled, the shell will grow to adulthood. Estimates of maximum age of some larger gastropods are approximately 30-50 years.

24.6.3 Predators. Snails serve as food sources for a variety of organisms, including other snails (Baur 1992). Over evolutionary time, predation by crabs has been so significant it seems to have led to an arms race between predator and prey (Vermeij 1987). Some gastropods specialize in eating other snails, including Naticidae, Muricidae, Conidae, and some Cephalaspidea. In the intertidal zone, birds are important, and marine mammals, particularly sea otters, prey on gastropods.

Defense strategies of snails against predation include a thickened shell, varices, spines, a clampdown mechanism (limpets), camouflage, and noxious chemicals. The opisthobranchs have perfected the last strategy. Bubble snails and dorids store strongly acidic secretions in their mantle or notum, a strategy also employed by some false cowries (Triviidae).

24.6.4 Mobility and locomotion. Most species move freely either on hard substrates, soft bottoms, or within the water column. Many species are regularly found on top of other organisms (epizoans), including plants (*Phylaplysia*) and soft corals (Ovulidae). Most limpet-shaped groups move little as adults (Patellidae, Haliotidae, and Calyptraeidae). Most limpets live in high-energy environments and need to be able to attach firmly to the substrate. Some limpets go as far as to produce depressions in the substrate (homing scars) to which they will return after a feeding excursion. A few groups are permanently attached to the bottom, as in the case of worm snails (Vermetidae). Others live within the hard part of scleractinian corals (Coralliophilidae) or within the tissue of sea cucumbers (Entoconchidae).

24.7 COLLECTING

Gastropods can be collected, with appropriate permits, in most habitats and using techniques described elsewhere in this volume (see Chapters 2, 3, and 4). Snails are active organisms; some gastropods are carnivorous and some are even cannibalistic. It is a good idea to keep gastropods separate from one another. Geiger (1997) described a container system called SAmpling Container On a String (SACOS: Figure 24.5) that serves that purpose. Another good sampling container for small specimens that would otherwise fall through the meshes of most underwater collecting bags is a plastic container with a snap-on lid, from which the center of the lid has been cut out, and a piece of neoprene rubber has been affixed with a simple slit made with the cut of a knife (S. Singer, pers. comm.). You simply push the specimen through the cover, and the elastic neoprene will close the opening immediately.

Some snails can cause injury to collectors. The best-known case is the cone snails, whose venom can be fatal for humans. As a novice collecting in the tropics, familiarize yourself with cone snails, particularly with the known harmful *Conus textile* Linnaeus, 1758, *C. marmoreus* Linnaeus, 1758, *C. geographus* Linnaeus, 1758, and *C. striatus* Linnaeus, 1758. When in doubt, handle every suspicious snail as if it were a potential hazard, because some cone snails do not look like typical cones. Members of the family Strombidae (*Strombus*, *Tibia*, and *Lambis*) possess a strongly serrated operculum than can cause some minor injury. Never stick any shells in your bathing suit or in your wet suit!

A very rewarding technique to find small (< 5 mm) snails is scrubbing rocks. Find a suitable rock in the intertidal region and scrub its surface with a brush in a large bucket half full with seawater. The small shells will fall into the bucket and accumulate at the bottom. Larger rocks are usually the better ones, but beware of sea urchins, moray eels, and scorpion fishes. Then sieve the mucky water through a fine mesh screen. The small mesh insets used to rear juvenile fish in aquaria work well. They come with a collapsible frame, and are of very fine mesh that will even retain omalogyids, but will wash the silt out of the sample. Carefully remove larger algal pieces and larger pebbles and keep the remainder for live sorting or bulk preservation.

A related technique can be applied to algal samples. Stuff a 1 liter zip lock bag with any kind of algae (can be done while SCUBA diving). Then either add magnesium chloride ($MgCl_2$) to the bag, or let the bag sit in full sun light; both methods cause any epizoans such as snails to fall off the algae. Place a handful of algae into a jam jar half filled with seawater and shake it vigorously so that the water will foam. Take the algae out, inspect it for any large specimens logged in between larger branches, and sieve the contents of the jam jar through a sieve as explained above.

When collecting, take thorough field notes. There are many suitable procedures, including written descriptions or cassette recorder notes, marks on high-resolution maps, and GPS readings. Be sure that the specimens are clearly associated with the collection data. Use heavy bond paper or so-called underwater paper made of synthetic fibers to store reference numbers with wet lots and use pencil; most pens will run and become unreadable in a short time. Regular writing paper easily disintegrates when wet. Labeling, data basing, and identification is covered elsewhere in this volume (see Chapters 5 and 8).

Preservation can be carried out in many different ways. Empty shells are best washed in fresh water and then air-dried. Biogenic sediment samples, also called grunge, should also be washed in freshwater and then air-dried on newsprint, otherwise the salt from seawater will crystallize on the shell, render it unsightly, and may even destroy it in the long run. This is a particular problem with microshells. If living animals are collected, consider preserving the animal with the shell specimen. If you do not want to keep preserved animals, consider offering specimens to museums, where the material will be available to those interested in tissue samples or whole bodies (see Chapter 14). If you want to keep the animal, you should first relax them with 7% magnesium chloride solution in fresh water ($MgCl_2 \cdot 6H_2O$). This solution is isosmotic with seawater. One should add small amounts of the magnesium chloride solution at a time so one can judge how the animals are reacting to the chemical and to adjust the dosage and the timing. To test whether the animals are fully relaxed, touch a cephalic tentacle and watch for any reaction, or poke the foot with a needle. For larger specimens (> 5 cm diameter) relaxation may take 1-2 hours, but small (< 1 cm) specimens may be relaxed in just a few minutes. Some opisthobranchs evert their genital apparati and shrivel up into an unsightly ball in the narcotic (see also Chapter 2.5).

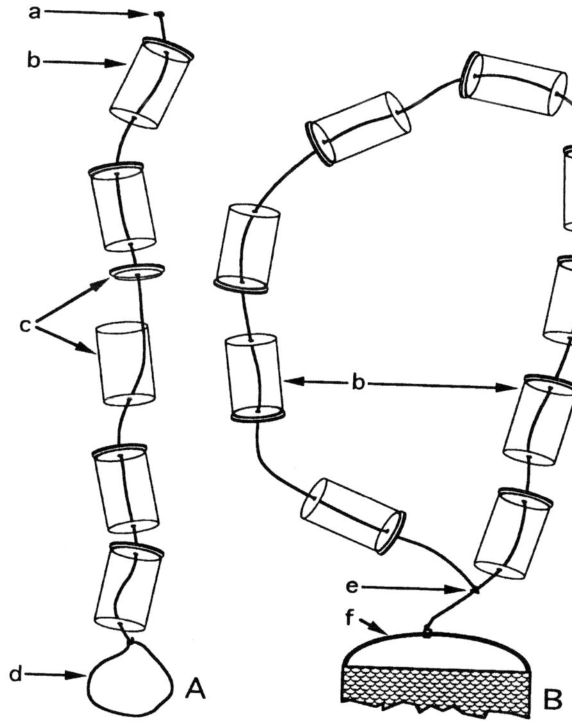

There are two main techniques for preserving animals, depending on what the purpose is. For molecular work (DNA extraction and sequencing), preserve the animal (or part thereof) in 100% ethanol. Isopropyl alcohol works as well; when in a remote location, 140 proof or higher liquor (>70% alcohol) can even do the job. The advantage of this method is that it only takes a single step and involves no strong toxins. It is mandatory for molecular work that tissue is preserved in alcohol only. However, long-term stability of the specimen is lower.

Figure 24.4 SACOS (Geiger 1997).
A. Linear model, B. Loop model. Abbreviations: a: terminal knot, b: closed film container, c: open film container, d: nylon cord, e: loop knot, f: handle of sampling net. Reproduced with kind permission of *The Festivus*.

For long-term storage and histology, fix the specimen in 5-10% formalin solution in seawater for a few hours to several days depending on the size of the specimen. The shells of large specimens or species with an operculum that seals well should be broken so that the fixative will reach the visceral hump. Alternatively, a relaxed specimen may be injected with formalin, or the gut cavity can be flushed with formalin through the mouth opening. Formalin is a suspected carcinogen; handle with care, wear gloves, and use only in a well-ventilated area. Dispose of the leftover formalin properly; institutions may have arrangements to safely and legally dispose of formalin; alternatively, bring the remainder to your local municipal toxic waste collecting agency. Transfer the specimen from the aqueous formalin solution to 80% ethanol. Contrary to earlier recommendations, there is no advantage of using lower percentage ethanol concentration (30%, 50%) in a graded series, because these weaker ethanol solutions will cause tissue swelling, and will not alter the degree of tissue shrinkage even in subsequent elevated ethanol concentrations (Glauert and Lewis 1998). Place the specimens in generous volumes of fluid, at least twice the volume of the specimens. If you have to pack specimens tightly in a small volume of liquid, change the 80% ethanol after a day to compensate for the water contained in the tissue of the animals.

Minute shells (<5 mm) need special care. The shells are very thin and are prone to dissolution in ethanol; they may be actively dissolved in the acidic formalin. Formalin can be neutralized with sodium bicarbonate (baking soda). Even under the best storage conditions, microshells can dissolve. As most features that allow identification of the species are found on the shell, such isolated bodies can no longer be identified. Therefore, it is good practice to store a subsample of wet-preserved microshells in dry conditions and to cross-reference the two subsamples. Seaslugs should be photographed prior to preservation; the coloration of the specimens fades quickly and provides important species-level characteristics (see Chapters 6 and 7).

The radula is an important anatomical part used in systematics of snails. It can easily be dissected out of larger specimens. Insert a pair of scissors or a scalpel into the snout, and cut dorsally along the roof of the foregut for 20-30% of the length of the animal. When you spread the foregut walls, you will see the radular ribbon on the floor of the foregut. It is attached to the odontophore, so you will have to use a fine spatula to lift the ribbon, and work your way posteriorly underneath the radula. Use a pair of forceps and hold on to the radula as far posterior as possible and pull gently with increasing force to pull the radula out of the radular sac. Occasionally the ribbon may break off in the middle, but because the radula is fully developed in the anterior half, this is of little consequence for the identification. The radula needs to be cleaned of the tissue adhering to it by immersion in a 10% solution of either potassium hydroxide (KOH) or sodium hydroxide (NaOH); for faster preparation time, the solution can be heated. For smaller specimens, the head or the entire animal can be dissolved. Radulae can easily be extracted from dried specimens. Remember, NaOH and KOH are extremely caustic and can dissolve protein, including human skin and the cornea; wear chemically resistant gloves and safety goggles when working with them. After a rinse in water, the radula can be mounted in Canada balsam for light microscopy, or prepared for scanning electron microscopy. For larger radulae, it may be advantageous to isolate teeth from discrete sections (central field, laterals, marginals), because in the intact radula, the sections may overlap and obscure part of the other teeth.

24.8 ACKNOWLEDGEMENTS

Charlie Sturm provided guidance and editorial support while writing this chapter. Peter Schuchert helped with tracing authorship of a hydrozoan. Christine Thacker, Kirstie Kaiser, two reviewers, and the editors read the chapter and made many suggestions for improvement.

24.9 LITERATURE CITED

Abbott, R. T. and S. P. Dance. 1983. *Compendium of Seashells*, 2nd Ed. Dutton, New York. 411 pp.

Baur, B. 1992. Cannibalism in gastropods. *In:* M. A. Elgar and B. J. Crespi, eds., *Cannibalism: Ecology and Evolution Among Diverse Taxa.* Oxford Scientific Publications, Oxford. Pp. 102-127.

Beesley, P. L., G. J. B. Ross, and A. Wells, eds. 1998. *Mollusca, the Southern Synthesis. Fauna of Australia,* Vol. 5, Parts A-B. CSIRO Publishing, Melbourne. 1234 pp.

Bieler, R. 1992. Gastropod phylogeny and systematics. *Annual Review of Ecology and Systematics* **23**: 311-338.

Bouchet, P., P. Lozouet, P. Maestrati, and V. Héros. 2002. Assessing the magnitude of species richness in tropical marine environments: exceptionally high numbers of molluscs at a New Caledonia site. *Biological Journal of the Linnean Society* **75**: 421-436.

Dance, S. P. 1974. *The Collector's Encyclopedia of Shells*. McGraw-Hill, New York. 288 pp.

Declerck, C. H. 1995. The evolution of suspension feeding in gastropods. *Biological Reviews* **70**: 549-569.

Dubinsky, Z., ed. 1990. *Coral Reefs*. Elsevier, Amsterdam. xi + 550 pp.

Eisenberg, J. M.1981. *A Collector's Guide to Seashells of the World*. McGraw-Hill, New York. 239 pp.

Fretter, V. and A. Graham. 1994. *British Prosobranch Molluscs*, Revised and Updated Ed. Ray Society, London. xix + 820 pp.

Gage, J. D. and P. A. Tyler. 1991. *Deep Sea Biology: A Natural History of Organisms at the Deep-Sea Floor*. Cambridge University Press, Cambridge. 504 pp.

Geiger, D. L. 1997. SACOS: An inexpensive and robust underwater sampling device for SCUBA divers. *Festivus* **34**: 32-34.

Glauert, A. M. and P. R. Lewis. 1998. *Biological Specimen Preparation for Transmission Electron Microscopy*. Princeton University Press, Princeton, New Jersey. xxi + 326 pp.

Gosliner, T. M. 1994. Gastropoda: Opisthobranchia. *In:* F. W. Harrison and A. J. Kohn, eds., *Microscopic Anatomy of Invertebrates*, Vol. 5 Mollusca, I. Wiley-Liss, New York. Pp. 253-355.

Götting, K.- J. 1974. *Malakozoologie*. Gustav Fischer Verlag, Stuttgart. 320 pp.

Hogarth, P. J. 1995. *The Biology of Mangroves*. Oxford University Press, Oxford. ix + 228 pp.

Hughes, R. N. 1986. *A Functional Biology of Marine Gastropods*. John Hopkins University Press, Baltimore. 245 pp.

Lalli, C. M. and R. W. Gilmer. 1989. *Pelagic Snails. The Biology of Holoplanktonic Gastropod Mollusks*. Stanford University Press, Stanford, California. 259 pp.

Larkum, A. W. D., A. J. McComb, and S. A. Shepherd, eds. 1989. *Biology of Seagrasses: a Treatise on the Biology of Seagrasses with Special Reference to the Australian Region*. Elsevier, Amsterdam. xxiv + 841 pp.

Levinton, J. S. 1995. *Marine Biology: Function, Biodiversity, Ecology*. Oxford University Press, New York. x + 420 pp.

Luchtel, D. L., A. W. Martin, I. Deyrup-Olsen, and H. H. Boer. 1997. Gastropoda: Pulmonata. *In:* F. W. Harrison and A. J. Kohn, eds., *Microscopic Anatomy of Invertebrates*, Vol. 6B Mollusca II. Wiley-Liss, New York. Pp. 459-718.

Lüning, K. 1990. *Seaweeds: Their Environment, Biogeography, and Ecophysiology*. Wiley, New York. xiii + 527 pp.

McLean, J. H. 1985. The archaeogastropod family Addisoniidae Dall, 1882; life habit and review of species. *Veliger* **28**: 99-108.

McLean, J. H. 1992. Systematic review of the family Choristellidae (Archaeogastropoda: Lepetellacea) with description of new species. *Veliger* **35**: 273-294.

Nybakken, J. W. 2001. *Marine Biology, an Ecological Approach*, 5th Ed. Benjamin Cummings, San Francisco. xi + 516 pp.

Palmer, R. A. 1996. MacCoil 1.07. Available at: <www.biology.ualberta.ca/palmer.hp/progs/MacCoil/MacCoil.htm>.

Ponder, W. T. and D. R. Lindberg. 1997. Towards a phylogeny of gastropod molluscs: an analysis using morphological characters. *Zoological Journal of the Linnean Society London* **119**: 83-265.

Poppe, G. T. and Y. Goto. 1991. *European Seashells*, Vol. 1. Christa Hemmen, Wiesbaden. 352 pp.

Purchon, R. D. 1977. *The Biology of the Mollusca*, 2nd Ed. Pergamon Press, Oxford. 560 pp.

Raup, D. M. 1962. Computer as aid in describing form in gastropod shells. *Science* **138**:150-152.

Raup, D. M. 1966. Geometric analysis of shell coiling: General problems. *Journal of Paleontology* **40**: 1178-1190.

Solem, A. 1974. *The Shell Makers, Introducing Mollusks*. Wiley Interscience, New York. 289 pp.

Springsteen, F. J. and F. M. Leobrera. 1986. *Shells of the Philippines*. Carfel Seashell Museum, Manila. 377 pp.

Thiele, J. 1931. *Handbuch der Systematischen Weichtierkunde*, Vol. 1-2. Gustav Fischer, Jena. 1154 pp.

Thompson, T. E. 1976. *Biology of Opisthobranch Molluscs*, Vol. 1. Ray Society, London. 206 pp.

Thompson, T. E. and G. H. Brown. 1984. *Biology of Opisthobranch Molluscs*, Vol. 2. Ray Society, London. 229 pp.

van Dover, C. L. 2000. *The Ecology of Deep-Sea Hydrothermal Vents*. Princeton University Press, Princeton, New Jersey. 424 pp.

Vaught, K. C. 1989. *A Classification of the Living Mollusca*. American Malacologists, Melbourne, Florida. 195 pp.

Vermeij, G. J. 1987. *Evolution and Escalation: An Ecological History of Life*. Princeton University Press, Princeton, New Jersey. xv + 527 pp.

Vermeij, G. J. 1993. *A Natural History of Shells*. Princeton University Press, Princeton, New Jersey. 207 pp.

Voltzow, J. 1994. Gastropoda: Prosobranchia. *In:* F. W. Harrison and A. J. Kohn, eds., *Microscopic Anatomy of Invertebrates*, Vol. 5 Mollusca I, pp. 111-252. Wiley-Liss, New York.

Wilson, B. 1993. *Australian Marine Shells*, Vol. 1. Odyssey, Kallaroo, Western Australia. 408 pp.

Wilson, B. 1994. *Australian Marine shells,* Vol. 2. Odyssey, Kallaroo, Western Australia. 369 pp.

Winner, B. E. 1985. *A Field Guide to Molluscan Spawn*, Vol. 1. E. B. M. Publications, North Palm Beach, Florida. 139 pp.

Winner, B. E. 1991. *A Field Guide to Molluscan Spawn*, Vol. 2. E. B. M. Publications, North Palm Beach, Florida. 94 pp.

Winner, B. E. 1993. *Life Styles of the Seashells*. E. B. M. Publications, North Palm Beach, Florida. 61 pp.

Winner, B. E. 1999. *The Sexual Behavior of Mollusks: Land-Freshwater-Sea*. E. B. M. Publications, North Palm Beach, Florida. 123 pp.

CHAPTER 25

UNIONOIDA: FRESHWATER MUSSELS

KEVIN S. CUMMINGS
ARTHUR E. BOGAN

25.1 INTRODUCTION

This chapter is a general guide to collecting, identifying, and curating freshwater mussels based upon our years of experience in the field. It is not meant to be an exhaustive guide but it does contain information learned through trial and error and the experience of others in the field.

Freshwater mollusks are perhaps the most endangered group of animals in North America (Williams *et al.* 1993). Accurate data on distribution and status are essential to freshwater mollusk conservation. Museums and private collections play an important role as repositories for specimens and associated data to document species occurrences historic and present. Without these collections, it would be impossible to ascertain or assign conservation status to species-at-risk.

Amateur collectors have provided important specimens and information throughout the years to aid in this effort. Many states have recently initiated citizen stream monitoring networks to help state and federal agencies properly protect and manage areas of high biodiversity or endemism. Carefully collected and documented collections are essential to these efforts.

This chapter mainly addresses the identification and preparation of specimens for curation, and only generally covers collecting methods. Most of the techniques and recommendations in this paper are not novel and have been used by others for many years. The following references contain excellent ideas and guidelines for specimen curation (Lewis 1868, Walker 1904, Baker 1921, 1942, Goodrich 1942, van der Schalie 1942, Stephens 1947, Stansbery and Stein 1983, Watters 1994).

25.2 ECOLOGY AND BIOLOGY

The Unionoida are found on all continents except Antarctica, but reach their greatest diversity in eastern North America where they number about 300 species.

25.2.1 Habitat. Because of their large size, freshwater mussels are some of the most conspicuous animals in freshwater ecosystems. Mussels spend most of their life partially or wholly buried in mud, sand, or gravel in permanent bodies of water. The vast majority of mussel species are found in streams, although a few are present in ponds or lakes. Although they can be found in almost any type of substrate, mussels are usually absent or rare from areas of shifting sand or deep silt.

25.2.2 Food. Freshwater mussels are filter feeders and filter about a liter of water in 45 minutes. Water enters the mussel via the incurrent aperture, comes in contact with the gills which are lined with cilia. Food items become entangled in the mucous of the animal and are passed on to the anterior of the animal where the mouth is located. The food of freshwater mussels is poorly known. Studies thus far have indicated that available foods consist of organic matter (detritus) or microscopic plants and animals (algae, protozoa, rotifers, diatoms, and desmids) suspended in the water called plankton.

25.2.3 Longevity. Mussels are long-lived, with many species living over 10 years and some reported to live over 100 years. Thin-shelled species (*Leptodea*, *Potamilus*, *Pyganodon*, *Utterbackia*, etc.) grow faster, but usually live shorter, than the thick-shelled species.

25.2.4 Predators. Freshwater mussels are an important food source for many animals including muskrat, mink, otter, fish, and some birds. Large piles of shells of freshly eaten mussels are called middens, and often contain a wide variety of species. Middens can be one of the best places to find and collect shells.

25.3 ANATOMY

25.3.1 The shell. In order to identify your specimen, you first need to know a little about the basic anatomy of a freshwater mussel shell. The anterior or front end of a mussel can be determined by the position of the umbo and pseudocardinal teeth, which are always located anteriorly. One of the features often cited in the description of a mussel is the beak sculpture. Beak sculpture is composed of the ridges or raised lines found on the umbo. Beak sculpture can vary from simple v-shaped lines to a series of wavy double-looped ridges (see Figure 25.1).

The shape, size, thickness, and color of the shell vary enormously among species of freshwater mussels, but basic structures and method of shell formation are the same for all. The shell consists of two valves held together by a dorsal ligament. Each valve is made up of three layers. The external, or outside, horny covering is called the periostracum. The periostracum protects the calcareous layers underneath from abrasion or the leaching action of acidic water. Should the periostracum become thin or wear completely away in spots, which often happens in the dorsal beak area (umbo) of old individuals, the shell in these exposed areas can become pitted or eroded. Variations in the color of the periostracum range from light yellow to green, brown, or black; shells of many species are patterned with various markings or rays. The color and patterns are important characteristics used in identifying many species.

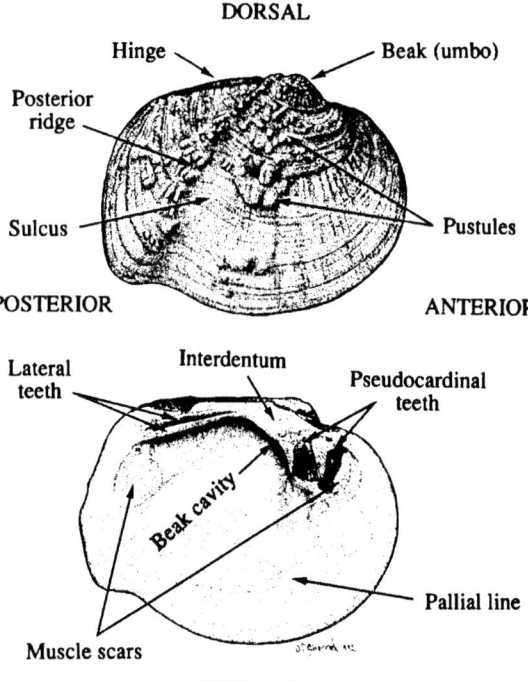

Figure 25.1 Anatomical features of a unionid shell. The major features of a freshwater mussel, as exemplified by *Quadrula quadrula* (Rafinesque, 1820). The drawing shows the exterior of the right valve (top) and the interior of the left valve (bottom). From Cummings and Mayer (1992).

Beneath the epidermis or periostracum is a thin, calcareous layer composed of vertical prisms of calcium carbonate. The third layer is referred to as the nacreous or mother-of-pearl layer. The nacre consists of a large series of thin, calcium carbonate plates that lie one upon another parallel to the surface of the shell. Some workers recognize a fourth inner layer called the hypostracum; in many species, this inner layer is often iridescent. The nacre or mother-of-pearl varies from a pure silvery white through shades of pink, salmon to dark purple.

The two outer shell layers are secreted by glands in the edge of the mantle, while the nacre, the inner layer, is produced by the surface of the mantle. Growth of the shell is achieved by the addition of material around the edge of the shell and in thickness by successive deposits of nacre over the shell's entire inner surface.

The outer surface of the shell of many kinds of mussels is completely smooth; other species possess knobs, pustules, spines, wrinkles, or folds. Often the size, combination, or arrangement of these surface structures is diagnostic of a particular species, as are the ridges or loops on the beaks.

The prominent, often raised, and/or darker concentric lines or rings on the surface of a valve indicate rest periods or stages when little or no growth took place. These lines or rings often result during the resting stage in winter, in periods of low water, or from some physical disturbance or other adverse condition. (Parmalee and Bogan 1998: 6-9).

25.3.2 Anatomy and life history. Freshwater mussels belong to the Class Bivalvia, members of which by definition have two opposing valves held together dorsally with a ligament. The animal or the soft tissue of the animal is found inside the protective valves. The body of the freshwater mussel consists of the rather thick central mass, which is attached dorsally by the dorsal muscles. Two adductor muscles act to close the valves, and the pedal retractor muscles are used to withdraw the foot.

The anterior ventral portion of this central mass is composed of the muscular foot, which is used to move or anchor the mussel in the substrate. The digestive tract, circulatory system, and reproductive organs are all housed in the central mass dorsal to the foot. On either side of this central mass and attached to the dorsal surface of the central mass are the demibranchs or gills; there are two gills on either side of the foot.

The posterior end of the animal contains the incurrent and excurrent apertures, which are composed of the edges of the mantle and pressed together to form functional siphons. Water is brought in through the incurrent aperture (located ventrally), and flows across the gills where oxygen is absorbed and exits through the dorsal excurrent aperture.

Food particles are removed from the water passing over the gills and move ventrally and anteriorly where they become attached to mucus threads. Food particles then move across the gills via ciliary action on the surface of the gills. The food particles move from the anterior end of the gills to the labial palps, then into the anterior food grove and finally into the esophagus and then to the stomach. During this whole journey, the particles are sorted; non-food particles are excluded, move off the ventral margin of the gills, and are expelled (pseudofeces).

Adult freshwater mussels live their entire lives partly embedded in the bottom of a stream or lake and they are active mostly during the warm months. Movement is accomplished by extending and contracting the foot between the valves; extension of the foot enables the mussel to wedge itself into the substratum. During periods of rest or inactivity, the mussel remains partially embedded with the valves slightly spread and the posterior end of the shell (siphons) exposed. The depth often depends on the particular species, water temperature, current, or other conditions.

Scars on the inner surface of each valve indicate points of muscle attachment. The largest muscles are the anterior and posterior adductor muscles, which hold the valves together. Anterior and posterior retractor muscles draw the foot into the shell, while the anterior protractor muscle helps to extend the foot. Additional dorsal muscles help secure the animal in the dorsal portion of the shell.

In freshwater mussels, the sexes are usually separate. Four species, the Paper Pondshell [*Utterbackia imbecillis* (Say, 1829)], Creek Heelsplitter [*Lasmigona compressa* (Lea, 1829)], Green Floater [*L. subviridis* (Conrad, 1835)], and the Lilliput [*Toxolasma parvus* (Barnes, 1823)] are hermaphroditic, a condition in which an individual possesses both male and female sex organs. Some individuals of a few other species, such as the Spectaclecase [*Cumberlandia monodonta* (Say, 1829)], in which the sexes are normally separate, occasionally exhibit this hermaphroditic condition.

When the reproductive systems are distinct, the shells of the male and female of some species within several genera, e.g., *Epioblasma, Lampsilis,* and *Villosa* (Tribe Lampsilini) are often distinct. The posterior portion of the valves of females are often more inflated and rounded than those of males. This is compensation, at least in part, for the gills, which become enlarged and distended when filled with developing eggs and glochidia (larval mussels). Sperm are released into the water, taken into the female through the incurrent aperture, and then carried to tubes in the gills where the eggs, having been previously discharged from the ovaries, are apparently fertilized. The gills or portions of the gills then serve as brood pouches, called marsupia, as well as respiratory organs.

Development of a freshwater mussel from the fertilized egg is unique, since one stage of growth of the young, called glochidia, must usually take place on the skin, gills, or fins of a fish [or, in one species, *Simpsonaias ambigua* (Say, 1825), on the common mudpuppy, *Necturus maculosus* (Rafinesque, 1818)]. Four distinct stages can be recognized in the growth of a mussel: (1) the fertilized egg; (2) the larva or glochidium in the brood pouch (gill) of the female mussel; (3) the parasitic stage on a fish (or salamander); and (4) the free-living stage with a completely formed shell.

In some species, this process is complex and involves sophisticated methods of insuring that the glochidia reach a fish host. For example, some mussels produce glochidia in packets (conglutinates) that resemble tiny fish or larval insects. These conglutinates are attractive to fish as potential prey items.

In many of the species of mussels in the Tribe Lampsilini, the edge of the mantle has been modified to resemble a fish or other prey item (Kraemer, 1970). Probably the most bizarre adaptation by freshwater mussels occurs in a few species of the genus *Hamiota*. These species develop a three- to six-foot-long (1 to 2 m) clear, gelatinous strand with all of the glochidia placed in a structure at the end (called a super conglutinate) of this tube. The super conglutinate containing the glochidia looks like a wounded minnow, floating and moving in the current (Haag *et al.* 1995).

During the parasitic stage, each glochidium remains embedded in the tissues of the host fish, changing little in size but developing many of the adult organs and structures. The length of time the glochidia remain embedded depends on the species of host fish, place of attachment, and temperature of the water; this stage may last slightly over one week or as long as six months or more. Most unionoids appear to produce a single brood each year, but *Cumberlandia* has been documented as producing two broods per year. Although each female mussel bears a very large number of glochidia, thus insuring continuation of the species, a large proportion fails to pass through the parasitic stage and, consequently, they die.

Studies have shown that glochidia from some species of mussels react to, or will parasitize, only one species of fish, while certain kinds of fish may serve as host for several species of mussels. It should be emphasized that the importance of knowing the host fish species for glochidia of freshwater mussels is essential to their effective propagation, conservation, and recovery. The ramification of this host specificity is that if the fish becomes locally extinct, the mussels can survive but not reproduce. The mussel population will eventually become locally extinct as well.

Normally, most infestations of glochidia are light and apparently do not harm or cause injury to the host fish. Upon completion of its parasitic stage, the young mussel breaks through the tissue of the fish and falls to the bottom, where it begins an independent life. With adult structures being formed, including a juvenile shell, the animal begins its growth to adulthood.

25.4 TAXONOMY OF THE UNIONOIDA

The Order Unionoida (freshwater mussels) as currently recognized contains about 900 species in approximately 170 genera. The higher level classification of freshwater mussels has been hotly debated in the past 200 years and it has been decades since the last comprehensive synthesis of mussel taxonomy (Haas 1969). It has been generally recognized that there are six families of freshwater mussels, but their placement into higher level groups (superfamilies, etc.) has been a subject of much debate. (Parodiz and Bonetto 1963, Haas

Table 25.1 Overview of the classification, diversity, and distribution of extant freshwater mussel families (from Graf and Cummings in prep.).

Taxon	Genera	Species	Distribution
Order Unionoida			Global, except Antarctica
Family Unionidae	65	374	Nearctic, south through Mesoamerica
	56	295	Palearctic, Oriental, New Guinea?
	10	38	Ethiopian
Family Margaritiferidae	3	12	Holarctic, with one species in southeastern Asia
Family Hyriidae	8	66	Neotropical
	8	27	Australasian
Family Mycetopodidae	10	50	Neotropical, and north into Mexico
Family Iridinidae	6	32	Ethiopian
Family Etheriidae	4	4	Neotropical, Ethiopian, and India
	170	898	

1969a, 1969b, Heard and Guckert 1970, Bogan and Hoeh 2000, Graf 2000, Graf and Ó Foighil 2000a, 2000b, Hoeh et al. 2001, Hoeh et al. 2002, Roe and Hoeh 2003). As currently recognized the order can be divided into two superfamilies (Unionoidea and Etherioidea). The six families contained in this order are all restricted to freshwater and this order is unique in the Class Bivalvia in having an obligate parasitic larval stage on the gills, fins, or body of a host fish or rarely some other freshwater vertebrate. A summary table of the classification and distribution of each is found in Table 25.1.

25.5 CONSERVATION ISSUES

Freshwater mussels have disappeared from many streams due to habitat changes and human activity. Some of the issues regarding the conservation of freshwater mussels are included here. More information can be found in Chapter 30.

25.5.1 Endangered species, permits, and reports.
Freshwater mollusks may be the most endangered group of animals in North America (Williams et al. 1993). Arguably, the most important task once you have determined that you want to collect freshwater mussels is to acquaint yourself with the requirements covering local, state, and federal permits. This will require the collector to be aware of which species are threatened or endangered in a particular region. It is worthwhile to take the time to learn about the local species before accidentally collecting an endangered species. Because many freshwater mussels are threatened or endangered, it is a necessity to obtain permits to collect in many states.

The permitting process can vary widely from country to country, state to state, and place to place, and laws and regulations regarding permits change often. Often there is a small fee charged in order to process the permit. It is important to check with the U.S. Fish and Wildlife Service before importing or exporting any freshwater mollusks into the U.S.A. An excellent and relatively current reference to collecting plants, animal, rocks, and fossils can be found in Wolberg and Reinard (1997).

25.5.2 Voucher specimens.
Specimens with good locality data are indispensable in documenting the historic distribution of a species. Any type of research regardless if it is a site or regional survey, taxonomic, genetic, ecological, or anatomical study, should be documented with voucher specimens. Vouchers are important for verifying the identification of specimens cited in a study. When specimens used in research are clearly designated as vouchers in a report or publication, the identifications can be verified. Eventually, all voucher specimens should be placed in an established museum for long-term preservation.

25.5.3 Number of specimens.
The purpose of the collection and the proposed long-term use of

the data dictate the number of specimens to be collected. Many collectors wish to have only a representative of a species in their collection. In this case, a single specimen would suffice. People interested in the range of variation in shell shape, size, coloration, etc. at a locality and from various locations within the range of the species would need a greater number of specimens. The critical point to remember when collecting live animals is to collect only those specimens actually needed to answer the question being researched. Do not over harvest even currently abundant species; remember the passenger pigeon was once abundant and is now extinct due to over-hunting (a kind of harvesting).

25.5.4 Live vs. dead shells. In most cases involving freshwater mussels, it is often sufficient to rely on dead shells, either from a midden, or those that have recently died and are found on the shore. Collection of live animals just for obtaining the shell should be avoided.

25.6 FIELD COLLECTING

"Collecting unionids is not at all glamorous, but is a wet and muddy activity."

Alan Solem in a tribute to Fritz Haas 1967.

25.6.1 Habitat selection - When and where to look. Freshwater mussels occur on all continents except Antarctica and in just about any habitat with permanent water (ponds, lakes, creeks, and rivers). They are rarely if ever found in seeps, springs, or wetlands, and although a few species can be abundant in ponds or lakes, they achieve their greatest diversity in streams. The best time to search for freshwater mussels is in the late summer or early fall when streams are at their lowest levels. When collecting freshwater mussels, all available habitat types (pools, runs, riffles, backwater areas, etc.) should be searched so that species restricted to those specific habitats are not overlooked.

25.6.2 Field gear. Most of the field gear needed to collect freshwater mussels can be carried in a small ice chest or cooler. None of the gear is expensive or elaborate and most of the supplies needed can be found around the house or obtained by making a trip to the local department store. The selection of field gear varies with the individual but a few of the essential items needed are discussed below.

Maps and/or Global Positioning Systems (GPS): Maps are essential to doing fieldwork and the availability of state atlases and gazetteers makes it easy to get where you are going (and know exactly where it is you were!). The best maps available for the U.S.A. are the 7.5-minute U.S. Geological Survey topographic sheets, but these are large and cumbersome to use in the field. Global Positioning Systems are becoming more widespread and are frequently used in field expeditions. GPS units can be very handy in remote areas, particularly the tropics or western North America where roads are often temporary, poorly marked, and scarce.

Field guides and notebooks: Even collectors with excellent memories forget certain details over time. That makes the taking of field notes essential to fully document collection efforts. It is customary to record ecological observations in a field notebook, many of which may have surprising relevance at a later date. Some collectors go a little overboard in recording every little detail in their notebooks, but it is difficult to predict what observations will be important in the future. Good field notes accompanied by photographs of the site are excellent ways to document a collecting trip. A field notebook is also a good place to record the number of individuals of each species found and returned to a site so that relative abundances can be documented.

Mesh bags, dip nets, pans, and viewing buckets: The most essential piece of field gear for a freshwater mussel collector is the collection bag. These can be purchased from a commercial supplier or made at home. It is strongly suggested that a breathable mesh bag be used. Avoid using plastic bags at all costs; because there are few things less forgiving than a dead mollusk stored for an undefined period of time in an airtight bag (trust us on this one). Airtight bags often result in the degradation of labels as well. An old onion or potato sack works well in a pinch and an old pillowcase could also be used effectively. The advantage of the commercially manufactured bags is that they often have drawstrings attached to secure the bag while in the field and after the collection has been made.

Plastic dishwashing pans are excellent for sorting through the vegetation or substrate samples to search for small freshwater mussels. An excellent tool to use to find and observe freshwater mussels in areas with clear water is a viewing bucket. These are usually constructed by cutting out the bottom of a 5-gallon plastic pail and replacing it with a piece of clear Plexiglas adhered with silicon sealer.

Plastic jars, film canisters, and forceps: Small plastic screw-top jars, recycled film canisters, or old medicine vials work well in the field for storing small, fragile mussels that might be too small or fragile to put in a mesh bag with larger specimens. Be sure to add a preservative to the jar, vial, or other airtight container so you do not wind up with rotten specimens.

Coolers: Specimens are often collected and returned to the lab alive so that the fixative or long-term storage fluid can be selected at a later date. To do this, a cooler is needed to transport the specimens back home safely and without mortality. Many freshwater mussels can survive in mesh bags on ice for up to a week.

Field clothes: This would seem self-explanatory, but if you are new to fieldwork, it is advisable to wear comfortable, dispensable clothes in the field. If you do much river collecting, do not wear anything you hope to wear in a social situation again. Tennis shoes, waterproof boots, or waders are recommended for collecting in streams or lakes where sharp rocks and omnipresent trash are found. Because most collecting is done in late summer or early fall when water levels are at their lowest, shorts or swim trunks are usually sufficient for collecting. However, walking through brush, stinging nettle, and poison ivy often necessitates wearing long pants or waders depending on where you collect. Waders are a necessity for collecting during the cold months.

Labels and pencils: While in the field, always add a locality label to your specimens. Never wait until you get home or back to the lab and then try to remember which label goes in which bag or vial, because invariably you will make a mistake that will result in confusion and headaches in the future. Do not use field numbers alone in lieu of complete locality data for your labels. Field numbers are fine in addition to locality data, but never alone. We cannot count the number of times we have received shells from ecologist and field biologists with a field tag that read something like A-EFWR.98-24(c). That may mean something when accompanied by a field notebook and a translator, but it is impossible to decipher by those unfamiliar with the code.

Be complete and clear when writing locality labels. The question to ask when writing a label is whether someone would be able to place a spot on a map using the data on the label that accompanied your specimens. If not, you need to sharpen your label-writing skills. The type of paper used for field labels is very important. Resistal™, Tyvek™, Rite-in-Rain™, or some kind of waterproof paper is essential. Plain notebook paper will disintegrate in water and should not be used. A pencil or India ink should be used when writing the label. Ballpoint pens should never be used.

25.6.3 Labeling. The quality of the data written on labels varies greatly. The importance of a specimen or lot is directly tied to the quality and completeness of the locality information on the accompanying label. At a minimum, a good field label should include the following:

Body of water (stream or lake and drainage): The first item written on a label should be the body of water where the specimen was found. It is also helpful to a curator or someone receiving an exchange of shells that a sufficiently large drainage name accompanies the stream or lake name. For example, there are many Sugar Creeks in the U.S.A. and unless one is familiar with the region in question, it may take looking through countless pages of an atlas to track down which Sugar Creek it actually is. If the stream is not sufficiently large enough to place on sight, it is advisable to place the drainage on the label as well. For example: Sugar Creek (Kaskaskia River Drainage). River mile designations are often handy for large rivers (i.e. the Mississippi).

Common location (distance, direction, and location): This is perhaps the most important part of the label, because even if most of the other data are missing it can be inferred from this piece of information. Ideally, the common location should include three parts; the distance from the center of town (or other permanent geographic object), the direction from the town, and the exact location. An example would be: 3.7 miles SE of Manteno just upstream Co. Rd. 1000E bridge.

Township, range, and section, UTM, latitude/longitude: Many areas have been laid out in various grids or other means of mapping out real estate to record on deeds, etc. One of the most common in the Midwestern U.S.A. is the Township, Range, and Section system. A regularly formed township consists of 36 square mile blocks arranged six across and six down. UTM stands for Universal Transverse Mercator system and is a highly precise system of coordinates, which can be found on topographic maps. Latitude and longitude are also helpful. GPS units can often assign values in either UTM or latitude and longitude.

County, state, and country: Self-explanatory, although some states (e.g., Louisiana) use parishes instead of counties and some countries use districts, townships, or provinces instead of states.

Date of collection: This would seem self-explanatory but given the Y2K problem, attention to detail when writing out a date is highly desired. Although many in the U.S.A. have been taught to hyphenate dates in the form of 4-5-56 or 4/5/56 with month first, followed by the day and then year, this convention can lead to confusion to our overseas colleagues who typically place the day in the first place followed by the month and then the year. To avoid all confusion, write out all dates unambiguously such as 5 April 1956. Write out the year in four places. Although the spelling of the month is different in other countries, it is easier to translate from one language to another than to infer from a numerical system.

Collector (field numbers): Collector's names should at least include the first two initials followed by the surname. Names associated with collections are often used to date collections because we often know the date of death of a collector and can infer that he or she was not present to collect after that date!

Map (optional, but often helpful): A photocopy of a map indicating where a particular collection was made occasionally accompanies specimens sent to a museum or on exchange. This map can be very helpful especially if the locality data that accompanied the specimens was poorly written. However, if complete locality information is included on a label, a map is not necessary.

25.6.4 Survey methods. A wide variety of techniques can be used to collect freshwater mussels. The method selected largely depends on the objective of the study. Qualitative surveys (often in the form of timed-searches) are those that document the presence/absence and the relative abundance of species at a particular site. Quantitative methods (usually in the form of linear transects or quadrat samples) are used to gather data on community and population structure. The method chosen will influence the way the data collected can be used or interpreted in future studies.

It is not our intent to suggest a particular method that should be used, but documenting how the specimens were obtained is important. Many previously published papers address various collecting methods and the strengths and weaknesses of those techniques. A good starting point is Strayer and Smith (2003) and references cited within.

Shoreline samples, hand picking, and middens: Collecting shells along the shore is something that has been done by almost anyone that has been to a beach or stream. As mentioned above, the field equipment needed to make a collection is not elaborate and neither is the most commonly used method for collecting: your bare hands. However, it does put a fair amount of strain on your back after a while. The quality of shoreline specimens is often poor due to the weathering of the shells, but simply picking up what you find along the bank can make a nice collection. Nevertheless, it is always easier to let someone else do the collecting and place the shells in a neat little pile for you. A naturally occurring pile of shells along a riverbank is termed a midden. Middens are usually made by predators such as raccoons or muskrats. Small mammals do an excellent job of finding specimens (particularly small ones) and often clean the catch as well. Some of the finest specimens we have come across in the field were collected by muskrats and deposited in middens.

Brail: A brail is a device that has been used for many years by commercial shellers and mussel survey crews. It is composed of a long bar about 8 to 16 feet (2.5 to 5 m) in length with approximately 100 thin ropes or chains hanging from the bar. Each of these ropes has from four to six wire hooks attached to the chain. These hooks are made of heavy gauge wire ending in four ends or hooks. The brail is pulled downstream from the back of a boat drifting at the speed of the current. The brail bar is held just off the bottom as the hooks drag across the river bottom. One end of the hook lodges between the valves of the mussel, the mussel closes its valves due to the irritation, and as the brail moves on, the mussel is yanked from the substrate. This method of collection is used in deeper water in large rivers but the brail is very inefficient for sampling the mussels that actually occur in the river.

Rakes and dredges: Rakes can be useful in shallow water in sandy and fine substrates to bring mollusks to the surface. Dredges such as those used in marine habitats are not easily used in freshwater. Mechanical grabs of various designs are available and useful in collecting standard samples in deeper waters. (see Chapter 3)

SCUBA and surface supplied diving: Today the most efficient method for sampling freshwater mussels in waters deeper than about one meter is with a device to supply air to a diver on the bottom of the lake or river. SCUBA and surface-supplied air diving provide the collector with the ability to remain on the bottom in deeper water for an extended period of time. These tools allow for quantitative work in otherwise inaccessible locations and the greatest ability to assess the density and diversity of the mollusks occurring in that habitat. Both of these techniques require diver certification and should not be used without adequate logistical support. (see Chapter 4)

25.7 IDENTIFICATION

"What we should have in this country is a fully illustrated manual of the Naiades such as has been published on the land and fresh-water univalves."

Charles Torrey Simpson, 1914

The above quote is as relevant today as it was over 90 years ago. Although many regional monographs have been published in recent years, much of the literature is scattered and hard to find. Many of the papers needed to identify a specimen properly are scarce, out-of-print, and expensive. The popularity of shell books by bibliophiles makes some of them out-of-reach with respect to price for all but the most affluent collector. A well-equipped library and helpful librarian are your best friends. The process of identifying freshwater mollusks consists of two basic parts: (1) compiling the literature needed to help in making the correct determination, and (2) cleaning the shells to reveal the characters used in keys or descriptions.

25.7.1 Literature. The first step in trying to identify a specimen is to gather the appropriate literature that contains information on distribution and a key or detailed description of the animal (shell or soft parts) that can be referred to while looking at a specimen. The jargon used to describe shells or their anatomy is often intimidating to the beginner. Two good references that help in that regard are Marshall (1930) and Burch (1990).

For North American freshwater mussels, there are many state or regional monographs available but many are out of date or print and in need of revision. A state-by-state listing of nearly 200 references on freshwater mussels can be found in Williams *et al.* (1993). Regional monographs on mussels have been recently published (post-1991) for Florida (in part) (Williams and Butler 1994), Vermont (Fichtel and Smith 1995), Texas (Howells *et al.* 1996), Maine (Martin 1997, Nedeau *et al.* 2000), Minnesota (Graf 1997), New York (Strayer and Jirka 1997), Tennessee (Parmalee and Bogan 1998), Kansas (Bleam *et al.* 1999) and the Rio Grande and Apalachicola river drainages (Johnson 1999, Brim Box and Williams 2000), the Midwest (Cummings and Mayer 1992), and others are in preparation.

In addition to the regional guides, there are a few published monographs on particular taxonomic groups that can be extremely valuable (e.g., Johnson 1978, Clarke 1981, 1985). Some textbooks contain helpful information for identification but they become quickly dated. McMahon and Bogan (2001) provide a general overview to the problems of identification, ecology, and physiology of freshwater bivalves (also see Pennak 1989). For information on non-North American works see Chapter 9.2.6: Unionoida.

Specimens are not always acquired through field collecting. Orphaned collections or exchanges often contain specimens that have been identified but are accompanied by outdated nomenclature that needs to be brought up to date. The essential references for tracking unionid nomenclature through the ages are Lea (1834-1874), Simpson (1900, 1914), Frierson (1927), Modell (1942, 1949), Haas (1969a, 1969b), and Burch (1975).

25.7.2 Web sites. Although web sites and addresses frequently change over time, we provide a list of four resources that may be of interest to those interested in freshwater mollusks.

Freshwater Mollusk Conservation Society: A society devoted to the advocacy for, public education about, and conservation science of freshwater mollusks, North America's most imperiled fauna <ellipse.inhs.uiuc.edu/FMCS/>

Mollusk collections of the world: This is a list of the systematic mollusk collections of the world. It includes links to the home pages of each collection and the phone number and/or e-mail address of the curator(s) and collection manager. <www.inhs.uiuc.edu/cbd/collections/mollusk_links/museumlist.html>

Freshwater mollusk researchers: This is a list of people interested in or conducting research on freshwater mollusks. It is predominantly a list of freshwater mussel researchers, but anyone with an interest in freshwater mollusks is urged to submit their name for inclusion. <www.inhs.uiuc.edu/cbd/collections/mollusk_links/uniopeoplelist.html>

Freshwater mollusk bibliography: This site is a database of over 10,000 references to papers on freshwater mollusks worldwide with an emphasis on freshwater mussels of North America. It was put together by the authors with G. Thomas Watters of the Ohio State University, Museum of Biological Diversity and Christine A. Mayer of the Illinois Natural History Survey. <ellipse.inhs.uiuc.edu:591/mollusk/>

25.7.3 Cleaning shells. After compiling the necessary literature, it is always helpful to clean the specimens to reveal characters that might be obscured by mud, marl, or algae. Harsh cleaning agents should be avoided as they may damage the shell, nacre, or the periostracum. Shells can easily be cleaned with an old toothbrush, vegetable brush, or a nylon scrub pad. Some of the shells may be covered with a thick organic crust that must be removed before any identification can be attempted.

25.8 CURATION

In order to preserve specimens for study centuries to come, it is essential that the proper materials are used to house and label the specimens. This topic is addressed more fully in Chapter 5. However, we thought it appropriate at least to cover some curation issues of importance to freshwater mussels.

25.8.1 Dry shells. Dry shells should be kept in cabinets out of direct light and preferably in a room with a constant moderate temperature and low humidity. The paper used in a dry shell collection should be acid-free bond paper or high rag content paper for long life. If the label is handwritten, the ink should be permanent and preferably black. Unfortunately, it has been a common practice in some collections (mostly private) to coat the shells with paraffin, oil (baby, linseed, etc.), or petroleum jelly to enhance the colors and presumably keep thin-shelled species from cracking. We strongly suggest that no coating be applied to the shell. Although some cracking will occur, it is preferable to the problems caused by the oils applied to the shell and dust that is attracted to them. Coatings also make it extremely difficult to write a catalogue number on the shell.

25.8.2 Wet specimens. There are a number of issues to be resolved if one decides to develop a fluid-preserved collection of freshwater mollusks. The first question is about the presumed long-term use of the collection. If the specimens are to be used for comparative anatomy, they should ideally be relaxed in as life-like position as feasible before being fixed (van der Schalie 1953, Meier-Brook 1976, Coney 1993, Araujo *et al.* 1995, Smith 1996). With the increased interest in molecular genetics, consideration should be placed on putting specimens directly in ethanol (80-95%).

Traditionally, the fixative of choice has been 10% buffered formalin (formalin is considered a carcinogen and if used, should be used with adequate ventilation). The specimens should not be left over long periods of time in formalin. About 12 hours or overnight should suffice. The specimens should be soaked in freshwater until the formalin smell is gone (usually about a day or two) and then transferred to 80% ethanol.

Some researchers that are concerned with anatomical studies contend that formalin makes the specimens hard and difficult to work with after a prolonged period of time and prefer specimens preserved in alcohol. Today both denatured ethanol and isopropyl alcohol are available in drug stores. Denatured ethanol is expensive. Isopropyl alcohol is relatively cheap and can be used if ethanol is unavailable. Distilled spirits work in a pinch. Eighty-percent ethanol is preferred by museums for long-term storage of specimens.

If the specimens are preserved in alcohol, the label should be written in permanent ink and the paper should be Resistal™ or other archival paper. Use of laser-printed labels in long-term alcohol storage is still being debated, but probably should be avoided until the issue is resolved.

Storage of wet specimens is an expensive proposition and requires a long-term commitment. See Chapter 5 for additional details regarding the selection of jars and lids. Small specimens may be more efficiently stored in shell vials closed with a polyester batting. To save shelf space, a large number of small vials can be stored in a large jar.

25.8.3 Databases. The manner in which collectors choose to keep track of the shells in their collection varies considerably. There are three common ways this information is kept: the card file, ledger, and in an electronic database. Today in the age of computers, handwritten catalogs are becoming extinct. A completed handwritten catalog is simply a numeric list of your collection. However, if the same data are entered into an electronic file, the data can be used in a variety of ways. For a more extensive treatment of using and creating a computer database, see Chapter 8.

25.9 LITERATURE CITED

Araujo, R., J. M. Remon, D. Moreno, and M. A. Ramos. 1995. Relaxing techniques for freshwater molluscs: Trials for evaluation of different methods. *Malacologia* **36**: 29-41.

Baker, F. C. 1921. Preparing collections of the Mollusca for exhibition and study. *Transactions of the American Microscopical Society* **40**: 31-46.

Baker, F. C. 1942. Collecting and preserving fresh-water snails. *American Malacological Union, Inc., News Bulletin and Annual Report* for 1941: 5-9.

Bleam, D. E., K. J. Couch, and D. A. Distler. 1999. Key to the unionid mussels of Kansas. *Transactions of the Kansas Academy of Science* **102**: 83-91.

Bogan, A. E. and W. R. Hoeh. 2000. On becoming cemented: Evolutionary relationships among the genera in the freshwater bivalve family Etheriidae (Bivalvia: Unionoida). *In:* E. M. Harper, J. D. Taylor, and J. A. Crame, eds., *The Evolutionary Biology of the Bivalvia*. Special Publication 177. Geological Society, London. Pp. 159-168.

Brim Box, J. and J. D. Williams. 2000. Unionid mollusks of the Apalachicola Basin in Alabama, Georgia and

Florida. *Bulletin of the Alabama Museum of Natural History* **21**: 1-143.

Burch, J. B. 1975. *Freshwater Unionacean clams (Mollusca: Pelecypoda) of North America* [Revised Edition]. Malacological Publications, Hamburg, Michigan. 204 pp.

Burch, J. B. 1990. Thomas Say's glossary for conchology. *Walkerana* **4**: 279-306.

Clarke, A. H. 1981. The tribe Alasmidontini (Unionidae: Anodontinae), Part I. *Pegias, Alasmidonta,* and *Arcidens. Smithsonian Contributions to Zoology* **326**: i-iii + 1-101.

Clarke, A. H. 1985. The tribe Alasmidontini (Unionidae: Anodontinae), Part II: *Lasmigona* and *Simpsonaias. Smithsonian Contributions to Zoology* **399**: i-iii + 1-75.

Coney, C. C. 1993. An empirical evaluation of various techniques for anesthetization and tissue fixation of freshwater Unionoida (Mollusca: Bivalvia), with a brief history of experimentation in molluscan anesthetization. *Veliger* **36**: 413-424.

Cummings, K. S. and C. A. Mayer. 1992. *Field Guide to Freshwater Mussels of the Midwest*. Illinois Natural History Survey, Manual 5. Illinois Natural History Survey, Champaign, Illinois. 194 pp.

Fichtel, C. and D. G. Smith. 1995. The freshwater mussels of Vermont. *Nongame and Natural Heritage Program, Vermont Fish and Wildlife Department Technical Report* **18**: 1-54.

Frierson, L. S. 1927. *A Classification and Annotated Check List of the North American Naiades*. Baylor University Press, Waco, Texas. 111 pp.

Goodrich, C. 1942. Certain remarks about labels. *Nautilus* **55**: 119-120.

Graf, D. L. 1997. Distribution of unionoid (Bivalvia) faunas in Minnesota, USA. *Nautilus* **110**: 45-54.

Graf, D. L. 2000. The Etherioidea revisited: A phylogenetic analysis of hyriid relationships (Mollusca: Bivalvia: Paleoheterodonta: Unionoida). *Occasional Papers of the Museum of Zoology, University of Michigan* **729**: 1-21.

Graf, D. L. and D. Ó Foighil. 2000a. The evolution of brooding characters among the freshwater pearly mussels (Bivalvia: Unionoidea) of North America. *Journal of Molluscan Studies* **66**: 157-170.

Graf, D. L. and D. Ó Foighil. 2000b. Molecular phylogenetic analysis of 28S rDNA supports a Gondwanan origin for Australasian Hyriidae (Mollusca: Bivalvia: Unionoida). *Vie et Milieu - Life and Environment* **50**: 245-254.

Graf, D. L. and K. S. Cummings. (In prep.). Paleo-heterodont diversity (Trigonoida + Unionoida): What we know and what we wish we knew about freshwater mussel evolution. *Zoological Journal of the Linnean Society*.

Haag, W. R., R. S. Butler, and P. D. Hartfield. 1995. An extraordinary reproductive strategy in freshwater bivalves: prey mimicry to facilitate larval dispersal. *Freshwater Biology* **34**: 471-476.

Haas, F. 1969a. Superfamilia Unionacea. *Das Tierreich (Berlin)* **88**: i-x + 1-663 pp.

Haas, F. 1969b. Superfamily Unionacea. *In:* R. C. Moore, ed., *Treatise on Invertebrate Paleontology*, Part N, Vol. 1, Mollusca. Geological Society of America and the University of Kansas, Lawrence, Kansas. Pp. 411-470.

Heard, W. H., and R. H. Guckert. 1970. A re-evaluation of the Recent Unionacea (Pelecypoda) of North America. *Malacologia* **10**: 333-355.

Hoeh, W. R., A. E. Bogan, K. S. Cummings, and S. E. Guttman. 2002. Evolutionary relationships among the higher taxa of freshwater mussels (Bivalvia: Unionoida): Inferences on phylogeny and character evolution from analyses of DNA sequence data. *Malacological Review* **31**: 123-141.

Hoeh, W. R., A. E. Bogan, and W. H. Heard. 2001. A phylogenetic perspective on the evolution of morphological and reproductive characteristics in the Unionoida. *In:* G. Bauer and K. Wächtler, eds., *Ecology and Evolution of the Freshwater Mussels Unionoida*. Ecological Studies Vol. 145. Springer-Verlag, Berlin. Pp. 257-280.

Howells, R. G., R. W. Neck, and H. D. Murray. 1996. *Freshwater Mussels of Texas*. Texas Parks and Wildlife Press, Austin, Texas. iv + 218 pp.

Johnson, R. I. 1978. Systematics and zoogeography of *Plagiola* (=*Dysnomia* =*Epioblasma*), an almost extinct genus of freshwater mussels (Bivalvia: Unionidae) from Middle North America. *Bulletin of the Museum of Comparative Zoology* **148**: 239-321.

Johnson, R. I. 1999. Unionidae of the Rio Grande (Rio Bravo del Norte) system of Texas and Mexico. *Occasional Papers on Mollusks, Museum of Comparative Zoology, Harvard University* **6**: 1-65.

Kraemer, L. R. 1970. The mantle flap in three species of *Lampsilis* (Pelecypoda: Unionidae). *Malacologia* **10**: 225-282.

Lea, I. 1834-1874. *Observations on the Genus* Unio, Vols. 1-13. Philadelphia, Printed for the Author. 1482 pp., 301 pls.

Lewis, J. 1868. Directions for collecting land and freshwater shells. *American Naturalist* **2**: 410-420.

Marshall, W. B. 1930. Former and present terms used in describing fresh-water mussels. *Nautilus* **44**: 41-42.

Martin, S. M. 1997. Freshwater mussels (Bivalvia: Unionoida) of Maine. *Northeastern Naturalist* **4**: 1-34.

McMahon, R. F. and A. E. Bogan. 2001. Mollusca: Bivalvia. *In:* J. H. Thorpe and A. P. Covich, eds., *Ecology and Classification of North American Freshwater Invertebrates*, 2nd Ed. Academic Press, New York. Pp. 331-429.

Meier-Brook, C. 1976. An improved relaxing technique for mollusks using pentabarbital. *Malacological Review* **9**: 115-117.

Modell, H. 1942. Das natürliche system der najaden. *Archiv für Molluskenkunde* **74**: 161-191.

Modell, H. 1949. Das natürliche system der najaden. 2. *Archiv für Molluskenkunde* **78**: 29-48.

Nedeau, E. J., M. A. McCollough, and B. I. Swartz. 2000. *The Freshwater Mussels of Maine*. Maine Department of Inland Fisheries and Wildlife, Augusta, Maine. 118 pp.

Parmalee, P. W. and A. E. Bogan. 1998. *The Freshwater Mussels of Tennessee*. University of Tennessee Press, Knoxville, Tennessee. 328 pp.

Parodiz, J. J. and A. A. Bonetto. 1963. Taxonomy and zoogeographic relationships of the South American naiades (Pelecypoda: Unionacea and Mutelacea). *Malacologia* **1**: 179-213.

Pennak, R. W. 1989. *Fresh-water invertebrates of the United States: Protozoa to Mollusca*, 3rd Ed. Wiley Interscience, New York. 628 pp.

Roe, K. J. and W. R. Hoeh. 2003. Systematics of freshwater mussels (Bivalvia: Unionoida). *In:* C. Lydeard and D. R. Lindberg, eds., *Molecular Systematics and Phylogeography of Mollusks*. Smithsonian Books, Washington D.C. Pp. 91-122.

Simpson, C. T. 1900. Synopsis of the naiades, or pearly fresh-water mussels. *Proceedings of the United States National Museum* **22**: 501-1044.

Simpson, C. T. 1914. *A Descriptive Catalogue of the Naiades, or Pearly Fresh-water Mussels*, Parts I-III. Bryant Walker, Detroit. xii + 1540 pp.

Smith, D. G. 1996. A method for preparing freshwater mussels (Mollusca: Unionoida) for anatomical study. *American Malacological Bulletin* **13**: 125-128.

Solem, A. 1967. The two careers of Fritz Haas. *Bulletin of the Field Museum of Natural History* **38**: 2-5.

Stansbery, D. H. and C. B. Stein. 1983. Mollusk collections at the Ohio State University Museum of Zoology. *In:* A. C. Miller, compiler, *Report of Freshwater Mollusks Workshop*. U.S. Army Engineer Waterways Experimental Station, Vicksburg, Mississippi. Pp. 94-113.

Stephens, T. C. 1947. The collection and preparation of shells. *Turtox News* for 1946 and 1947: 1-12.

Strayer, D. L. and K. J. Jirka. 1997. The pearly mussels of New York State. *New York State Museum Memoir* **26**: i-xiii + 1-113, pls. 1-27.

Strayer, D. L. and D. R. Smith. 2003. A guide to sampling freshwater mussel populations. *American Fisheries Society Monograph* **8**: 1-103.

van der Schalie, H. 1942. On collecting fresh-water mussels. *American Malacological Union, Inc., News Bulletin and Annual Report* for 1941: 9-14.

van der Schalie, H. 1953. Nembutal as a relaxing agent for mollusks. *American Midland Naturalist* **50**: 511-512.

Walker, B. 1904. Hints on collecting land and fresh-water Mollusca. *Journal of Applied Microscopy and Laboratory Methods* **6**: 2365-2368.

Watters, G. T. 1994. North American freshwater mussels. Part II. Identification, collection, and the art of zen malacology. *American Conchologist* **22**: 11-13, 18.

Williams, J. D. and R. S. Butler. 1994. Class Bivalvia, Order Unionoida, Freshwater Bivalves. *In:* M. Deyrup and R. Franz, eds., *Rare and Endangered Biota of Florida*. IV, Invertebrates. University Press of Florida, Gainesville. Pp. 53-128, 740-742.

Williams, J. D., M. L. Warren, Jr., K. S. Cummings, J. L. Harris, and R. J. Neves. 1993. Conservation status of freshwater mussels of the United States and Canada. *Fisheries* **18**: 6-22.

Wolberg, D. and P. Reinard. 1997. *Collecting the Natural World. Legal Requirements and Personal Liability for Collecting Plants, Animals, Rocks, Minerals, and Fossils*. Geoscience Press, Inc., Tucson, Arizona. 330 pp.

CHAPTER 26

NON-UNIONOID FRESHWATER BIVALVES

ALEXEI V. KORNIUSHIN (1962-2004)

26.1 INTRODUCTION

The Recent North American fauna of non-unionoid freshwater bivalves includes three families: Sphaeriidae, Corbiculidae, and Dreissenidae. Since only the first family is native, this chapter is focused on this group. The other two families are represented by a few species introduced from Europe and Asia and are described only briefly. While this chapter focuses mainly on the North American fauna, information on other regions will be presented where appropriate.

26.2 FAMILY SPHAERIIDAE: FINGERNAIL, PEA, OR PILL CLAMS

26.2.1 Species diversity and distribution. The Sphaeriidae are distributed all over the world. This family includes at least 150-200 valid species. However, the fauna of some regions is still poorly known and the systematics are not well established. Therefore, it is hard to evaluate the number of species more precisely. Herrington's (1962) review reported 34 species in the North American fauna. In the latest review (Burch 1975), the North American fauna comprises 38 Sphaeriidae species, but after the recent revision of *Pisidium punctatum* Sterki, 1865 by Korniushin *et al.* (2001), the number is 39.

The European fauna is almost as rich as North America and includes 25 to 30 species, though the status of some forms is in dispute. Some Russian reviewers recognized many more species (Korniushin 1996), but this splitting approach is not accepted by West European specialists. Fourteen species are common for both continents, five of them, *Sphaerium corneum* (Linnaeus, 1758), *Pisidium amnicum* (Müller, 1774), *P. henslowanum* (Sheppard, 1825), *P. supinum* Schmidt, 1850, and *P. moitessierianum* Paladilhe, 1866 are considered to be introduced from Europe to North America (Burch 1975, Grigorovich *et al.* 2000), and one species, *Musculium transversum* (Say, 1816), was apparently introduced from North America to Europe (Ellis 1978). Some European species in North America might be overlooked, being described as local species or forms.

The holarctic fauna of sphaeriids should be considered as rich. The fauna of the Ethiopian region, which is relatively well-studied (Mandahl-Barth 1988), comprises 29 species. Other regions have poorer fauna.

Most of the North American sphaeriid species have a transcontinental distribution, but some are more restricted (Burch 1975). The highest species diversity of sphaeriids in North America is observed in the Great Lakes region (Burch 1975). Three species, *Sphaerium nitidum* Westerlund, 1876, *Pisidium idahoense* Roper, 1890, and *P. conventus* Clessin, 1877 are distributed only in the northern part of the continent, possibly with isolated populations in the mountains in more southern locations. *Sphaerium fabale* (Prime, 1852) and *S. patella* (Gould, 1850) form a pair of morphologically similar but geographically separated species, the first being restricted to the eastern part of the continent, and the second to the western part. The Central American species *Eupera cubensis* (Prime, 1865) and *Pisidium punctiferum* (Guppy, 1867) are present in the southern states of the U.S.A. One species, *Pisidium ultramontanum* Prime, 1865 is restricted to several localities in Oregon and northern California.

In the Eastern Hemisphere, similar separation characterizes the European *S. solidum* Normand, 1844 and the Siberian *S. asiaticum* (Martens, 1864). Other species with limited distribution are known from the ancient lakes: Lake Ochrid in the Balkans, Lake Baikal in Siberia, Lake Biwa in Japan, Lake Tanganyika in Africa, and Lake Titicaca in South America.

26.2.2 Classification. According to Burch (1975), the family is divided in three subfamilies: Euperinae with the genera *Eupera* and *Byssanodonta*, Sphaeriinae with the genera *Sphaerium* and *Musculium*, and Pisidiinae with a single genus *Pisidium*. The *Eupera* are restricted to the tropical regions of Africa and America, with the greatest diversity in South America. *Byssanodonta* is a monotypical South American genus. The remaining genera are cosmopolitan and their ranges coincide with the range of the family.

The generic classification is hardly final, since heterogeneity of anatomical characters was found in all the named genera, and a phylogenetic analysis, which is necessary to construct the natural system of the group, is now in progress. It is also possible that at least Euperinae, with their very peculiar morphology and mode of reproduction, deserve the status of a separate family.

Subgeneric classification is even more uncertain, since none of the suggested classifications takes into account the entire global fauna. Burch (1975) divided the North American *Sphaerium* species into two subgenera - *Sphaerium* s. str. (in the strict sense), and *Herringtonium*. He divided *Pisidium* into three subgenera - *Pisidium* s. str, *Cyclocalyx*, and *Neopisidium*.

26.2.3 Morphological and biological peculiarities. The species of this family are the smallest

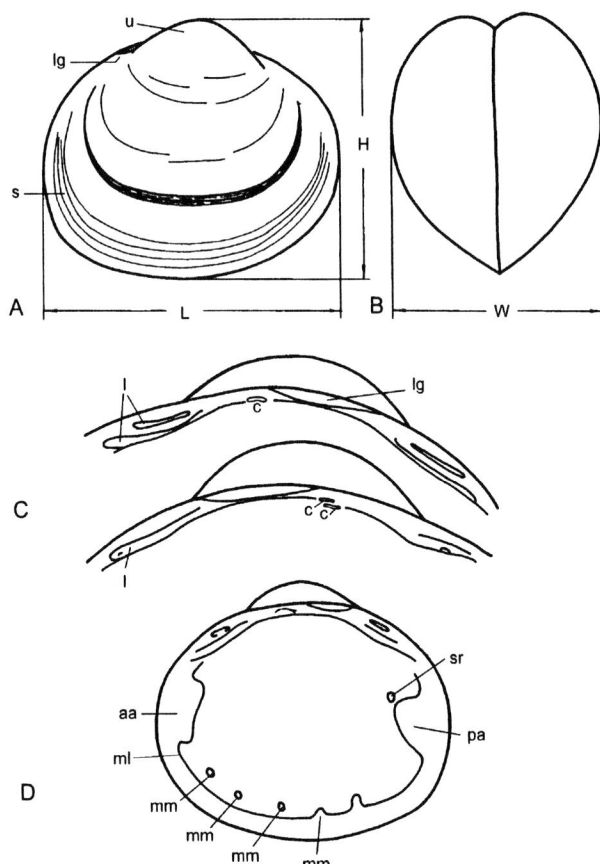

Figure 26.1 Shell of *Sphaerium corneum*.
A. View from outside, B. Front view, C. Hinge of left (lower) and right (upper) valves, D. View from inside. aa- scar of anterior adductor, c- cardinal teeth, l- lateral teeth, lg- ligament, ml- mantle line, mm- scars of mantle muscles, pa- scar of posterior adductor, s- striae, sr- scar of siphonal retractor, u- umbo; standard measurements: L- shell length, H- shell height, W- shell width (two valves).

freshwater bivalves and, probably, some of the smallest bivalves in general. Some of the adult animals (*Pisidium moitessierianum* for example) do not exceed 2.5 mm, though most of the species are somewhat larger, from 3 to about 15 mm long. The largest North American species, *Sphaerium simile* (Say, 1816), may reach 20 mm. The European species *Sphaerium rivicola* (Lamarck, 1818) is even bigger; it can range up to 25 mm long.

The sphaeriid shell (Figure 26.1A, B) is usually round or oval, but in some species may be trigonal or tetragonal (trapezoid). The position of the umbo

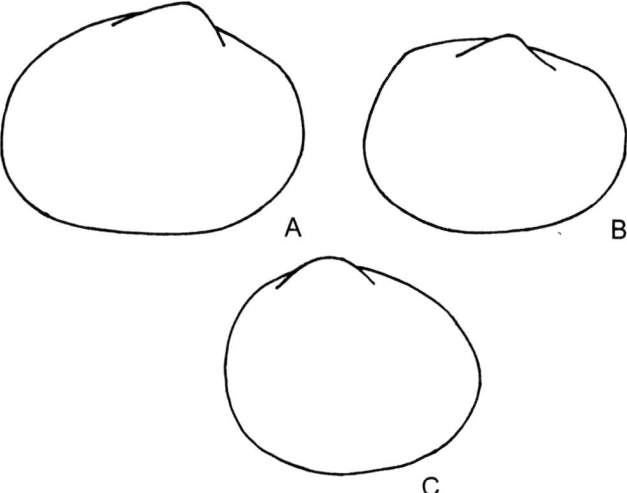

Figure 26.2 Shells of Sphaeriidae.
A. *Eupera*, B. *Musculium*, and C. *Pisidium* (external view of right valve, diagrammatic).

may be anterior (*Eupera*), central (*Sphaerium* and *Musculium*) or posterior (*Pisidium*) (Figure 26.1A, 26.2). The outer surfaces often bear concentric sculpture: ribs (larger elements with intervals of about 0.5 mm) or striae (fine sculpture with intervals 0.025-0.1 mm). Tiny pores (3-5 µm in diameter) are usually noticeable on the internal surface. The scars of the two adductors are well marked, while the smaller scars of the siphonal retractors are sometimes noticeable, lying near those of the posterior adductors or being attached to them (Figure 26.1D). The mantle line is usually weakly marked, and never forms a sinus; the scars of mantle muscle bundles may be visible above this line.

The sphaeriid hinge is typically heterodont and consists of cardinal, anterior lateral, and posterior lateral teeth (Cox *et al.* 1969). The left valve bears two cardinal teeth and two lateral teeth, one anterior and one posterior; the right valve has one cardinal and two lateral teeth on each (anterior and posterior) side (Figure 26.1C). The lateral teeth are markedly reduced in the Australian taxon *Musculium lacusedes* (Iredale, 1943), and all hinge teeth are reduced in the *Byssanodonta*.

The ligament is visible and external in some species (e.g., *Sphaerium simile, Musculium transversum* (Say, 1829), see Figure 26.3A). In a majority of the taxa, it is covered dorsally by a thin layer of calcified material and thus not visible (Figure 26.3B). In species of the subgenus *Pisidium (Odhneripisidium)*, the ligament is introverted: deeply submerged between valves (Figure 26.3C).

The soft body structure (Figures 26.4, 26.5) is characterized by some peculiar features that may be associated with the small size of these mollusks. The mantle usually forms two ventral fusions dividing the exhalant (anal), inhalant (branchial), and pedal openings. Species of *Eupera, Byssanodonta, Sphaerium,* and *Musculium* have tubular exhalant and inhalant siphons, separate (in the first two genera), or partially joined to each other (Figure 26.4). The broad (funnel-like) exhalant siphon and the absence of the inhalant siphon (Figure 26.5) characterize the genus *Pisidium*. Branchial opening may be retained [e.g., *P. amnicum, P. casertanum* (Poli, 1791)] or absent (e.g., *P. moitessierianum, P. conventus*).

The pedal opening is usually a long slit, but in some species (*Eupera* spp., *P. subtruncatum* Malm, 1855, *P. milium* Held, 1836), it is shortened because of progressive development of the presiphonal (perisiphonal) suture - mantle fusion dividing inhalant opening from the pedal slit. In contrast to

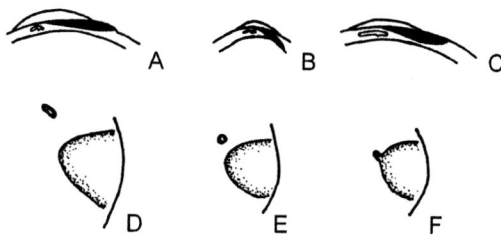

Figure 26.3 Ligament types (ligament black) and posterior muscle scars (stippled) in Sphaeriidae.
A. External ligament, B. Introverted ligament, C. Enclosed ligament, D. Scars of siphonal muscles and posterior adductors widely separated, E. Scars adjacent, F. Scars merged.

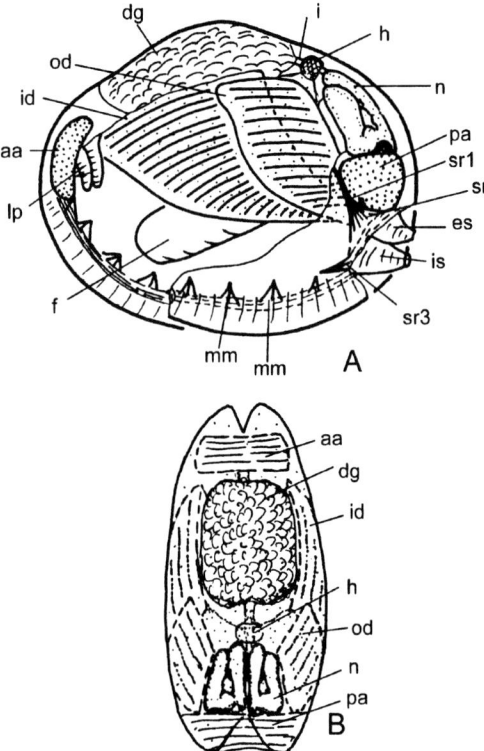

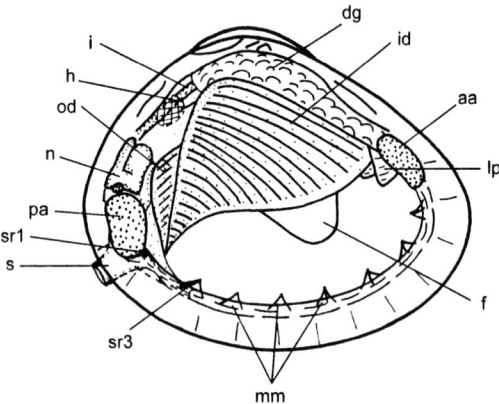

Figure 26.4 Gross anatomy of *Sphaerium corneum*.
A. Side view, B. Dorsal view. aa- anterior adductor, dg- digestive gland, es- exhalant siphon, f- foot, h- heart (ventricle), i- intestine, id- inner demibranch, is- inhalant siphon, lp- labial palps, mm- mantle muscles, n- nephridium, od- outer demibranch, pa- posterior adductor, s- siphon, sr1, sr2, and sr3- siphonal retractors.

Figure 26.5 Gross anatomy of *Pisidium* sp., side view. Abbreviations the same as in Figure 26.4.

The gills are characterized by a certain reduction of the outer demibranch (the outer demibranch is always lower than the inner one). The most profound reduction is a characteristic of *Pisidium*: in all species, except *P. idahoense*, the outer demibranch has only one lamella and does not overlap the inner one, and in many species (mainly in those lacking inhalant mantle opening), it disappears completely. As in the majority of filter-feeding bivalves, gills of sphaeriids have two basic functions: respiration and collecting food. The reduction of the size of gills definitely leads to the diminution of their respiratory and food-collecting surface and is correlated, therefore, with the size of the animal.

corbiculid or venerid bivalves in which the siphons are contracted by two broad muscle bands, the siphonal muscles of sphaeriids form several pairs of retractors (Figures 26.4, 26.5), hence the absence of the pallial sinus.

The foot is usually well developed and extendible (capable of great extension). It is a burrowing organ in the majority of species. In contrast, the especially long, motile, worm-like foot of *Sphaerium corneum* gives this animal the ability to creep on aquatic plants. A functional byssal gland is found in the larvae of some species (Mackie *et al.* 1974), however, adult animals retain a byssus only in *Eupera* and *Byssanodonta*. The latter peculiarity is apparently associated with their mode of life (see below).

The configuration of the nephridium (kidneys) is a taxonomic character. The excretory organs of sphaeriids (nephridia) are progressively developed, in contrast to the typical voluminous bivalve nephridium. The excretory organ in this group is a long specifically packed tube. Progressive development of the nephridium can be seen as an adaptation to the freshwater environment, which is hypoosmotic and needs intensification of the osmoregulatory function usually carried out by excretory organs.

All sphaeriids are hermaphrodites. The gonads are relatively big, extended to the dorsal side of the animal in *Eupera* and *Byssanodonta*, and small, placed behind the foot in the other genera. Some sphaeriid species are distinguished by remarkably

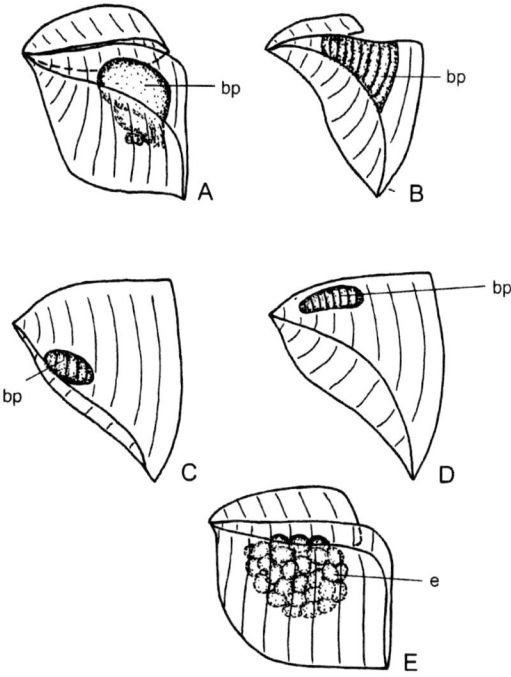

Figure 26.6 Brooding in sphaeriids, diagrammatic.
A. *Eupera*, B. *Sphaerium*, C-E. *Pisidium*, different location of a pouch. bp- brood pouch, e- eggs.

high chromosome numbers (Burch and Huber 1966, Burch *et al.* 1998), which in all probability resulted from polyploidy. The only diploid species known is *Sphaerium corneum*. However, the family is still poorly studied in this respect.

The most remarkable feature of sphaeriids is the incubation of their young in the gills of the parent animal. *Eupera* and *Byssanodonta* produce large eggs rich in yolk, which develop between the lamellae of the inner demibranch (Figure 26.6A). In the other genera, specialized organs for breeding, brood pouches or marsupia, are produced by inner demibranch filaments (Figure 26.6B-E). In *Sphaerium* and *Musculium*, several pouches develop in each gill simultaneously and parents release their young successively, while in *Pisidium*, formation of a new pouch is possible only after release of the previous brood.

Sphaeriids usually have one or two breeding seasons per year. Those from habitats with stable temperature, such as deep-water lake populations or tropical species, may reproduce year round. Details for particular species are provided by Heard (1965, 1977), Meier-Brook (1970), and Holopainen and Hanski (1986). Incubation of larvae in brood pouches may last from one month (in the smallest pisidia) up to several months.

After release from pouches, the young continue their growth lying freely in the gill (extra-marsupial larvae), and then leave the parent through its siphons. Compared to bivalves with veliger larvae, sphaeriids are characterized by low fecundity. The number of offspring released by one parent in one breeding season does not exceed several dozen. *Eupera* and *Musculium* species, as well as *Pisidium amnicum*, are characterized by the highest fecundity (sometimes more than 50 young per gravid adult), most *Sphaerium* and *Pisidium* species release up to 20 young, and *P. conventus* may incubate simultaneously only 2 to 3 larvae.

It is notable, that newly released *Sphaerium* already have mature gonads (the young begin to produce their own eggs and sperm even at the end of the incubation period) and soon form the first brood pouches (Heard 1977). This peculiarity was not found in *Musculium* and *Pisidium*. On the other hand, *Musculium* species are characterized by relatively rapid growth and reach their final size in 60-70 days, while *Sphaerium* need 4 to 8 months to mature and *Pisidium* about a year. The life span of *Musculium* species is about a year; in *Sphaerium* and *Pisidium* it is up to 5 years. Populations and species from northern countries are characterized (on the average) by slower growth and a longer life span (Holopainen and Hanski 1986).

Most of the sphaeriids are borrowing. The larger species (*Sphaerium*, *Musculium*, and *Pisidium amnicum*) only partly submerge in sediments, with their siphons stretching into the water. They are typical filter feeders and consume mainly phytoplankton (Lopez and Holopainen 1987). Most of these species are apparently not very mobile, except *Sphaerium corneum*, which often occurs openly, on submerged vegetation (personal observations) and can move almost as fast as snails crawl.

Small *Pisidium* species are typically completely submerged in bottom sediments, dwelling in long blind burrows, and their feeding mode is defined as interstitial suspension-feeding (Lopez and Holopainen 1987). Interstitial bacteria probably form the main part of their diet. Some species of sphaeriids may occur among submerged leaves (*Musculium lacustre* (Müller, 1774), *Pisidium casertanum*) or mosses (*P. obtusale* (Lamarck, 1818), *P. milium*). North American *Eupera* combine burrowed (infaunal) and attached (epifaunal) modes of life, while some South American *Eupera* and the species of *Byssanodonta* are strictly epifaunal and attach by a byssus to rocks, stones, or other hard substrata (Ch. Ituarte, personal communication).

26.2.4 Habitats. Sphaeriid clams occur in a great variety of habitats: from large rivers and lakes to springs, peat bogs and temporary pools. The highest species diversity is characteristic of lakes, ponds, and small rivers. Springs are usually colonized by one species - *Pisidium casertanum* (in Europe *P. personatum* Malm, 1855 may be also found) which is usually quite abundant. Some species show clear preference to particular habitats (e.g., *P. idahoense*, *P. lilljeborgi* (Clessin, 1877), and *P. conventus* - to lakes, *P. supinum* - to rivers), while others may live in different types of habitats. Those that live in a variety of habitats seem to be less sensitive to variations in oxygen and calcium content in the water. However, they apparently cannot live in anaerobic conditions and therefore avoid pools with a deep layer of liquid ooze or decaying debris.

Some species, e.g., *Sphaerium occidentale* (Lewis, 1856), *Musculium lacustre*, and *P. casertanum* can survive temporary drying up within a thin layer of water or even in wet soil or leaves. As a rule, sphaeriids do not tolerate salinity higher than 0.03% Cl, and only a few species can withstand higher salinity; *P. casertanum*, *P. henslowanum*, *P. nitidum* Jenys, 1832, and *P. subtruncatum* Malm, 1855 are the most tolerant in Europe and may live in the lowest courses of rivers (Kuiper and Wolf 1970).

Being burrowers, most sphaeriids avoid coarse sediments (rocks and gravel), as well as hard sand, preferring fine sand, muddy sand, and mud. Some amount of detritus seems to be necessary for them, probably because of their feeding on saprotrophic bacteria (Lopez and Holopainen 1987). Therefore, they are usually rare in the main stream of rivers or along the beaches exposed to waves and numerous in more quiet places, where some mud or debris can be accumulated.

Species such as *Sphaerium occidentale*, *Musculium lacustre*, *Pisidium casertanum*, and *P. ventricosum* Prime, 1851 (in Europe *P. obtusale*) are often abundant in small forest lakes or streams with bottoms covered with dead tree leaves. Only several species (*Sphaerium striatinum* (Lamarck, 1818) and *P. dubium* (Say, 1816); in Europe *S. rivicola*, *S. solidum*, and *P. amnicum*) show clear preference to lotic environments and coarser sediments (sand or even gravel). Despite their ability for dispersal by birds and water insects (Mackie 1979), sphaeriids are rarely found in artificial pools not connected with natural streams (personal observation).

26.2.5 Means of collection and preparation. Only the largest species of fingernail clams can be collected by hand. Smaller specimens should be washed out of the sediment with a sieve. Sediment can be taken by a shovel (in shallow water), a dipnet, or a drag dredge (in deeper water). A kitchen sieve can be used for washing, but better results are achieved when the sieve has a flat bottom (for example see the Walker Dipper, Chapter 21.4, and other devices in Chapter 2.3). As follows from the above description of habitats, the places where some mud and debris are accumulated should be searched most carefully. Dead leaves, submerged moss, or weeds (including roots) should also be collected and washed in a sieve.

Small forceps with tapered soft tips are necessary to pick up specimens (which may be very small and fragile) from the sieve. Animals can be taken to the lab, for further preparation, in tubes or jars containing water, wet cottonwool, grass, or moss (the latter is recommended for the very fragile shells of *Musculium* and some *Sphaerium* which may be damaged by shaking in water). When convenient, the whole sample of mud, leaves or vegetation can be put in a plastic bag and brought to the lab for washing or sorting.

Dry shells for conchological collections can be kept in boxes; small specimens should be additionally packed in small glass tubes or gelatin capsules. Do not put them directly on cottonwool, as it may be hard to remove them from the cotton fibers. Specimens for anatomical study should be preserved in 80% ethanol. Neutralized formalin can be also used, but it is not recommended, since it can make the tissues too rigid. The fixative does not penetrate between tightly joined valves; therefore, the animals should be relaxed before fixation.

Common narcotizing agents (menthol or barbiturates) can be used, but are not very effective for sphaeriids (exact concentration and time of relaxation should be determined experimentally for each species). Good results may be had by submerging the animals in hot (but not boiling) water for several minutes (such material may be good for anatomical preparation, but not available for DNA analysis). See Chapter 2.5 for more on relaxing mollusks.

If the valves remain closed, one of them should be broken in order to let the fixative in. Special techniques of fixation should be used for histology, electron microscopy, etc. When material for DNA sequencing is needed, a stronger alcohol (at least 95% ethanol) should be used, and the shells should be broken in order to ensure fast fixation.

Since internal shell characters (hinge, sometimes muscle scars) are necessary for species identification, it is necessary to open the valves. If dried remnants of the soft body hinder opening, the specimen should be carefully heated in a 5% solution of alkali (sodium hydroxide or potassium hydroxide), which dissolves the tissues and facilitates opening. If alkali solution is unavailable, heating in water may also be tried (see Chapter 5.9 for some additional methods for softening dried tissue).

In order to expose the soft body for anatomical study, the adductor muscles and mantle edge should be carefully detached from the shell with a needle or thin scalpel or razor blade. Most of the organs are well seen laterally (as in Figure 26.4A), but the form of the nephridium may be clearer when looking from the dorsal side (as in Figure 26.4B).

Details of the gill structure are seen much better if the mantle sheets are removed. On the other hand, the length of the presiphonal (perisiphonal) suture and the arrangement of mantle muscles are better seen from the inner side of the mantle, therefore it is necessary to separate the mantle from the body. Separation of gills is recommended to expose brood pouches formed on their inner surface. Since the smallest pouches may be covered by the inner (ascending) gill lamella, the latter should be removed if the exact number of brood pouches, their stage of development, and number of embryos are to be determined (for example, for life history studies).

Some standard histological procedures of staining and mounting specimens were suggested for microscopic investigation of mantle and gills (Korniushin 1995). While the literature on sphaeriid chromosomes is scarce and contains no detailed description of methods, some approaches developed for chromosomes of *Corbicula* (Okamoto and Arimoto 1986) can be recommended. Recently, the group has been involved in molecular studies (Park and Ó Foighil 2000, Cooley and Ó Foighil 2000), and the methods of such study can be obtained from the published literature.

26.2.6 Keeping specimens alive in the laboratory.
Living specimens of sphaeriids can be safely kept in a refrigerator for several weeks or even months. It is best to place them in wet cottonwool or moss rather than in water. Most of the species can live in an aquarium without any special demands (personal observations). However, this may be problematic for species preferring lotic environments (e.g., *Sphaerium rivicola* or *S. striatinum*).

26.2.7 Practical applications.
So far, the sphaeriids do not seem to be of any practical importance. While many of them contain trematode larvae (personal observations), none have been reported to be intermediate hosts of human parasites. Being small and having a short life span, they are not as suitable for environmental monitoring as unionids

are. Apparently, some species may be indicators of good (or bad) water quality (for example in Europe, *P. amnicum* and *P. pulchellum* Jenys, 1832, while in America they are yet to be determined). It is necessary to include this group in current ecological surveys, since some species are apparently vulnerable because of their limited distribution and specific ecological demands. They may suffer from water pollution, change of habitats, and other negative consequences of human activity.

26.3 FAMILY CORBICULIDAE: THE ASIATIC CLAM

Only one species of this family, *Corbicula fluminea* (Müller, 1774), is reliably identified in North America as being introduced before 1938 (Britton and Morton 1979). The presence of a second, not yet identified species, is seriously suspected by many malacologists (Morton 1986; Siripattrawan *et al.* 2000). The type locality of *C. fluminea* is Canton, China (Araujo *et al.* 1993), and it is considered to be widely distributed in South East Asia (Morton 1986). Many other *Corbicula* species have been described from different Asian countries, Africa, and Australia; their taxonomic status is still disputed, and some modern reviewers (Morton 1986) recognize only two species, namely *C. fluminea* and *C. fluminalis* (Müller, 1774) (type locality - Euphrat River in Mesopotamia). Introduced populations of *C. fluminea* have been found also in South America (Buenos Aires), and both above mentioned species were introduced to western Europe (see Morton 1986 and Araujo *et al.* 1993 for reviews). *Corbicula* is reported to be a serious pest in its introduced range (Morton 1979a - cited from Morton 1986).

The morphology and biology of *Corbicula* was extensively described by Britton and Morton (1979), and only the most remarkable characters are presented here. They differ from sphaeriids in their much larger size (normally up to 35 mm long, but in some populations up to 60 mm). The sculpture is much coarser (Figure 26.7A), and the arrangement of ribs may be a taxonomic character (about 10 pronounced ribs per 10 mm in *C. fluminea*, and more than 15 in *C. fluminalis*). The hinge of each

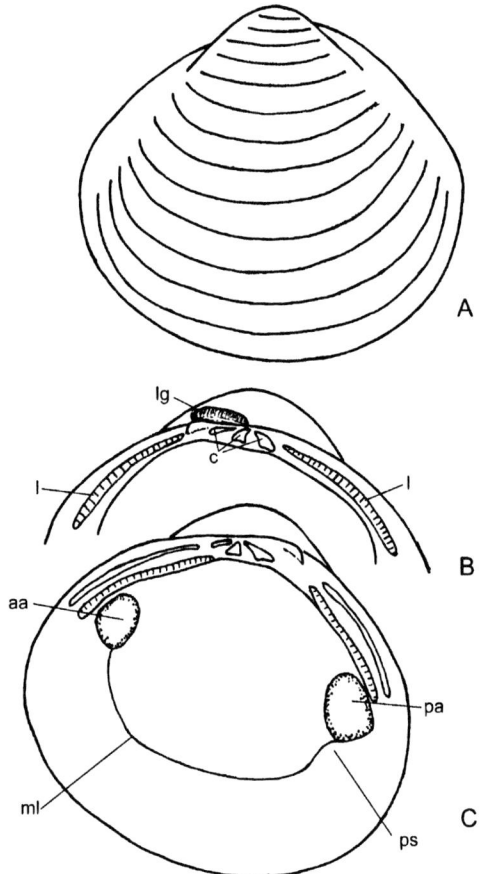

Figure 26.7 Shell of *Corbicula fluminea*.
A. View from outside, B. Hinge of the left valve, C. Right valve from inside. aa- scar of anterior adductor, c- cardinal teeth, l- lateral teeth, lg- ligament, ml- mantle line, pa- scar of posterior adductor, ps- pallial (mantle) sinus.

valve includes 3 cardinal teeth, and long lateral teeth (the same set as in *Sphaerium*) that bear small denticles (Figures 26.7B, C). Pores similar to that of sphaeriids are noticeable on the inner surface. The mantle line is clear and smooth (without any extending muscle scars), and a small pallial sinus is usually noticeable (Figure 26.7C); in some other corbiculid genera, e.g., in the South American *Neocorbicula*, the mantle sinus is pronounced.

The siphons of *Corbicula* are relatively short and their openings are fringed with papillae (Figures 26.8A, B). Form and pigmentation of siphons, as

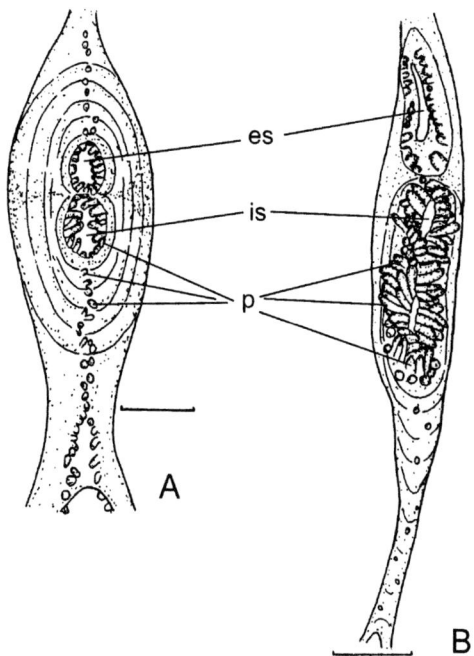

Figure 29.9 *Corbicula fluminea* gill
Gill of *C. fluminea* from inside, with developing larvae.

Figure 26.8 Some anatomical structures in *Corbicula*.
A. Siphons in *C. fluminea* (outside view), B. Another type of siphons in *Corbicula moltkiana* Prime, 1878 (a species from Sumatra). es - exhalant siphon, is - inhalant siphon, p - papillae.

well as number and arrangement of papillae were suggested as species diagnostic characters (Britton and Morton 1979, Harada and Nishino 1995), but many Asian species and forms are not yet studied in this respect. The siphonal muscles form two broad bands, as in marine venerid bivalves.

Each gill consists of two almost equally developed demibranchs (the outer one is somewhat lower). The gonad is big and branched, and produces a great number of oocytes. *Corbicula fluminea* is hermaphroditic and probably androgenic (Siripattrawan *et al.* 2000). The eggs are incubated within the inner demibranch without any brood pouches (Figure 26.9).

The larvae are released with a straight-hinged (D-shaped) shell, but an already well developed creeping foot. Even longer incubation of young takes place in *Neocorbicula limosa* (Maton, 1809); the larvae of this species reach, in the gills, an apparently advanced stage of development, and several broods may develop in one parent animal simultaneously (Ituarte 1994, Dreher Mansur and Meier-Brook 2000).

The other mode of reproduction, dioecious, releasing free-swimming larvae without incubation, was reported for some Asian taxa. Modern Japanese authors (Harada and Nishino 1995) recognize one of them as a distinct species *Corbicula japonica* (Prime, 1864), but Morton (1986) assigned all non-incubating forms to *C. fluminalis*. An intermediate mode, release of creeping larvae without incubation, is known in *C. sandai* Reinbardt, 1878, restricted to the Japanese lake Biwa (Harada and Nishino 1995). Both *C. japonica* and *C. sandai* are diploid (Okamoto and Arimoto 1986). *Corbicula fluminea* lives 3 years, and *C. japonica* up to 9 years (Morton 1986).

In America, *Corbicula* can be found in rivers, including the low courses and reservoirs, but generally not in estuaries. It occurs both in lotic and lentic conditions, mainly embedded in coarse sediments (sand and gravel). In contrast, the non-incubating Asian forms are estuarine (brackish water). The broad ecological range, correlated with a variety of reproduction modes, makes the Corbiculidae (and the genus *Corbicula* in particular) a good model for studying the evolutionary adaptation of bivalves to freshwater environments.

The methods and equipment for collecting *Corbicula* are basically the same as suggested for sphaeriids. Handling them is definitely simpler because of their larger size. Special methods for chromosome studies are described in Okamoto and Arimoto (1986), and the DNA sequencing data are summarized by Siripattrawan *et al.* (2000). Given the on-going dispute on the taxonomy of various *Corbicula* forms, further studies on their morphology (including anatomy and the chromosomes), DNA sequences, and biology are of great importance.

26.4 FAMILY DREISSENIDAE: ZEBRA MUSSELS

Dreissena polymorpha (Pallas, 1771) appeared in North America in 1988 (Hebert *et al.* 1989 and Ludyanskiy *et al.* 1993). A second European species, *D. bugensis* Andrusov, 1897 (sometimes called quagga mussel) has also been reported (Mills *et al.* 1996). *Dreissena* can be immediately recognized by their triangular shell (Figure 26.10), somewhat similar to that of *Mytilus*, and attached mode of life. The byssus is very well developed, and the foot in adult specimens is rudimentary. The anterior adductor muscles are reduced; the posterior adductor and the pedal muscle are strong and form large scars (Figure 26.10). Two siphons are present, and both demibranchs of each gill are almost equally developed. A big massive gonad is placed at the base of the foot. In contrast to other freshwater bivalves, *Dreissena* releases free-swimming veligers. Further information on distribution, habitats, and life cycles of American populations can be obtained from the above mentioned papers.

26.5 EDITORIAL NOTE

Alexei Korniushin, born on December 7, 1962, died on January 8, 2004. Alexei was 41 at the time of his death from an apparent heart attack. He studied at institutions in Kiev, Ukraine and St. Petersburg, Russia, and received a doctorate. At the time of his death, he was working at the Schmalhausen Institute of Zoology, National Academy of Sciences in Kiev. He was also an affiliated researcher at the Museum of Natural History in Berlin. He is survived by his wife Elena and two children.

The editors mourn the untimely passing of this student of the Sphaeriidae. We hope that this final

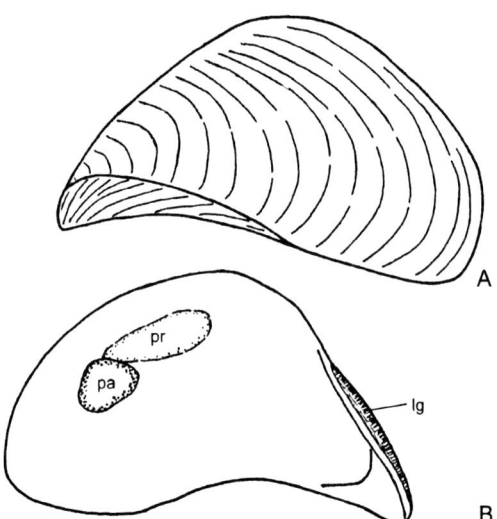

Figure 26.10 Shell of *Dreissena polymorpha*.
A - from outside, B - from inside. lg- ligament, pa- scar of posterior adductor, pr - scar of pedal retractor.

contribution from Alexei will open up the world of the Sphaeriidae to others who wish to study these taxa. While he did not have an opportunity to review the final form of this paper, only a few grammatical changes were made from the last version that he had reviewed. This paper, as well as his other publications, will serve as a lasting tribute to this eminent malacologist.

26.6 LITERATURE CITED

Araujo, R., D. Moreno, and M. A. Ramos. 1993. The Asiatic clam *Corbicula fluminea* (Müller, 1774) in Europe. *American Malacological Bulletin* **10**: 39-49.

Britton, J. C. and B. Morton. 1979. *Corbicula* in North America: The evidence reviewed and evaluated. *In:* J. D. Britton, ed., *Proceedings of the First International* Corbicula *Symposium, Texas Christian University Fort Worth, Texas, Oct. 13-15.* Texas Christian University Research Foundation, Fort Worth, Texas. Pp. 249-287.

Burch, J. B. 1975. *Freshwater Sphaeriacean Clams (Mollusca: Pelecypoda) of North America.* Malacological Publications, Hamburg, Michigan. 96 pp.

Burch, J. B. and J. M. Huber. 1966. Polyploidy in mollusks. *Malacologia* **5**: 41-43.

Burch, J. B., G-M. Park, and E-Y. Chung. 1998. Michigan's polyploid bivalves. *Michigan Academician* **30**: 351-352.

Cooley, L. R. and Ó Foighil, D. 2000. Phylogenetic analysis of the Sphaeriidae (Mollusca: Bivalvia) based on partial mitochondrial 16S DNA gene sequences. *Invertebrate Biology* **119**: 299-308.

Cox, L. R., N. D. Newell, D. W. Boyd, and 22 other authors. 1969. Bivalvia. *In:* R. C. Moore, ed., *Treatise on Invertebrate Palaeontology*, Part N, Vol. 1, Mollusca 6. University of Kansas Press, Lawrence, Kansas. Pp. 1-489.

Dreher Mansur, M. C. and C. Meier-Brook. 2000. Morphology of *Eupera* Bourguignat 1854 and *Byssanodonta* Orbigny 1846 with contributions to the phylogenetic systematics of Sphaeriidae and Corbiculidae (Bivalvia: Veneroida). *Archiv für Molluskenkunde* **128**: 1-59.

Ellis, A. E. 1978. British freshwater bivalve Mollusca. Keys and notes for the identification of the species. *Synopses of the British Fauna (New Series)* **11**: 1-95.

Grigorovich, I. A., A. V. Korniushin, and H. J. MacIsaac. 2000. Moitessier's pea clam *Pisidium moitessierianum* (Bivalvia, Sphaeriidae): A cryptogenic mollusc in the Great Lakes. *Hydrobiologia* **435**: 153-165.

Harada, E. and M. Nishino. 1995. Differences in inhalant siphonal papillae among the Japanese species of *Corbicula* (Mollusca: Bivalvia). *Publications of Seto Marine Biology Laboratory* **36**: 389-408.

Heard, W. H. 1965. Comparative life histories of North American pea clams (Sphaeriidae: *Pisidium*). *Malacologia* **2**: 381-411.

Heard, W. H. 1977. Reproduction of fingernail clams (Sphaeriidae: *Sphaerium* & *Musculium*). *Malacologia* **16**: 421-455.

Hebert, P. D. N., B. W. Muncaster, and G. L. Mackie. 1989. Ecological and genetic studies on *Dreissena polymorpha* (Pallas): A new mollusc in the Great Lakes. *Canadian Journal of Fishery and Aquatic Science* **46**: 1587-1591.

Herrington, H. B. 1962. A revision of the Sphaeriidae of North America (Mollusca: Pelecypoda). *Miscellaneous Publications, Museum of Zoology, University of Michigan*, **118**: 1-74.

Holopainen, I. J. and I. Hanski. 1986. Life history variation in *Pisidium* (Bivalvia: Pisidiidae). *Holarctic Ecology* **9**:85-98.

Ituarte, C. F. 1994. *Corbicula* and *Neocorbicula* (Bivalvia: Corbiculidae) in the Paraná, Uruguay and Río de la Plata basins. *Nautilus* **107**: 129-135.

Korniushin, A. V. 1995. Anatomy of some pill clams from Africa, with the description of new taxa. *Journal of Molluscan Studies* **61**: 163-172.

Korniushin, A. V. 1996. *Bivalve Molluscs of the Superfamily Pisidoidea in the Palaearctic Region. Fauna, Systematics, Phylogeny.* Schmalhausen Institute of Zoology, Kiev. 175 pp. [In Russian]

Korniushin, A. V., I. A. Grigorovich, and G. L. Mackie. 2001. Taxonomic revision of *Pisidium punctatum* Sterki, 1895 (Bivalvia: Sphaeriidae). *Malacologia* **43**: 337-347.

Kuiper, J. G. J. and W. J. Wolf. 1970. The Mollusca of the estuarine region of the rivers Rhine, Meuse and Scheldt in relation to the hydrography of the area. III. The genus *Pisidium*. *Basteria* **34**: 1-42.

Lopez, G. R. and I. J. Holopainen. 1987. Interstitial suspension-feeding by *Pisidium* spp. (Pisidiidae: Bivalvia): a new guild in the lentic benthos. *American Malacological Bulletin* **5**: 21-30.

Ludyanskiy, M. L., D. McDonald, and D. MacNeill. 1993. Impact of the zebra mussel, a bivalve invader. *BioScience* **43**: 533-544.

Mackie, G. L. 1979. Dispersal mechanisms in Sphaeriidae (Mollusca: Bivalvia). *Bulletin of the American Malacological Union* **45**: 17-21.

Mackie, G. L., S. U. Qadri, and A. H. Clarke. 1974. Byssus structure of larval forms of the fingernail clam, *Musculium securis* (Prime). *Canadian Journal of Zoology* **52**: 945-946.

Mandahl-Barth, G. 1988. *Studies on African Freshwater Bivalves*. Danish Bilharziasis Laboratory, Charlottenlund, Denmark. 161 pp.

Meier-Brook, C. 1970. Untersuchungen zur Biologie einiger *Pisidium*-Arten. *Archiv für Hydrobiologie, Supplement* **38**: 73-150.

Mills, E. L., G. Rosenberg, A. P. Spidle, M. L. Ludyanskiy, and Y. V. Pligin. 1996. A review of the biology and ecology of the quagga mussel (*Dreissena bugensis*), a second species of freshwater dreissenid introduced to North America. *American Zoologist* **36**: 271-286.

Morton, B. 1979. Freshwater fouling bivalves. *In:* J. D. Britton, ed., *Proceedings of the First International Corbicula Symposium, Texas Christian University Fort Worth, Texas, Oct. 13-15*. Texas Christian University Research Foundation, Fort Worth, Texas, pp. 15-38.

Morton, B. 1986. *Corbicula* in Asia - An updated synthesis. *American Malacological Bulletin Special Ed.* **2**: 113-124.

Okamoto, A. and B. Arimoto. 1986. Chromosomes of *Corbicula japonica*, *C. sandai* and *C. (Corbiculina) leana* (Bivalvia, Corbiculidae). *Venus* **45**: 194-202.

Park, J.-K. and D. Ó Foighil. 2000. Sphaeriid and corbiculid clams represent separate heterodont bivalve radiations into freshwater environments. *Molecular Phylogeny and Evolution* **14**: 75-88.

Siripattrawan, S., J.-K. Park, and D. Ó Foighil. 2000. Two lineages of the introduced Asian freshwater clam *Corbicula* occur in North America. *Journal of Molluscan Studies* **66**: 423-429.

CHAPTER 27

MARINE BIVALVES

EUGENE V. COAN
PAUL VALENTICH-SCOTT

27.1 INTRODUCTION

The marine Bivalvia are the second largest class of marine mollusks. This group includes clams, cockles, scallops, oysters, mussels, piddocks, and shipworms.

27.2 BIOLOGY

The Bivalvia are fundamentally bilaterally symmetrical mollusks in which the mantle encloses the gills (ctenidia), foot, and visceral mass, and secretes a shell in the form of two lateral valves, hinged dorsally (Figure 27.1). With the retreat of the body from direct contact with the environment, a novel mode of feeding using the ctenidia has developed.

Although the exact mode of feeding of ancestral bivalves is not known, it is likely that suspension feeding using palps and ctenidia (Figures 27. 2-4) evolved in parallel with the enclosure of the body by the mantle and shell. Modern representatives of the order Nuculoida retain several primitive morphological characters and also show a primitive form of suspension feeding. However, such forms today also possess specialized, elongate, palp appendages that can be extended from within the shell to sweep up detritus with cilia and to convey it to the mouth. The radula and other structures of the head were lost as this indirect mode of feeding developed. The protobranch ctenidia of the primitive bivalve consist of a central axis bearing, on either side, a series of flattened, ciliated filaments.

The evolution of lamellibranch ctenidia, which characterize the vast majority of modern bivalves, was accomplished by the elongation of the filaments, their folding back on themselves so that each ctenidium commonly resembles a tall, narrow W in cross-section, and the binding of adjacent filaments into extensive lamellae. Complex feeding and rejection tracts of cilia on the ctenidia transport food and pseudofeces respectively, and the palps further sort the food before passing it into the mouth. Thus equipped, bivalves have become nearly complete introverts, using their ciliated ctenidia both for respiration and for filtering food from the water. To keep in touch with their environments, they have developed sensory tentacles on the mantle edges and at the siphonal apertures, and may even possess distinctive eyes along the edges of the mantle, as in the scallops.

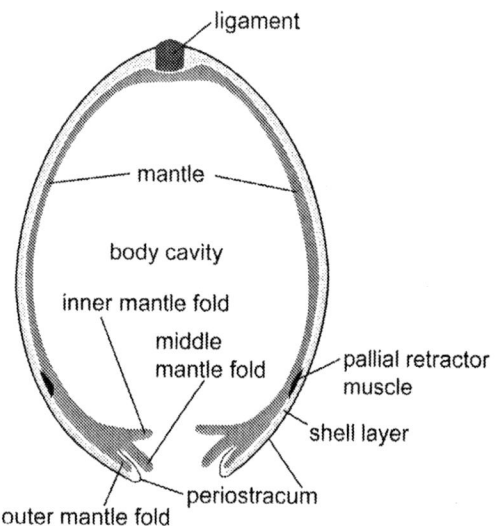

Figure 27.1 Cross section through a typical bivalve shell. After Coan *et al*. (2000).

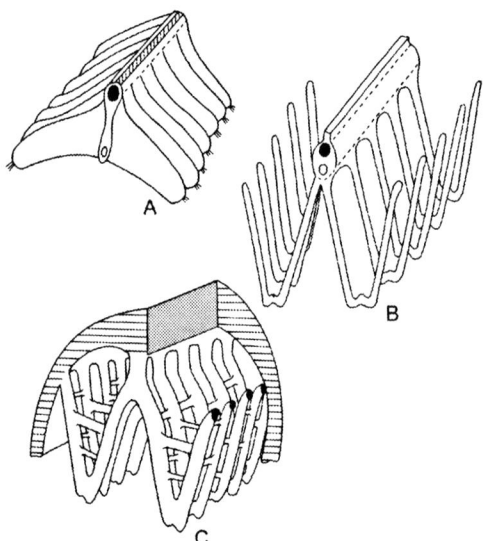

Figure 27.2 View of ctenidium filaments.
A. Protobranch ctenidium, B. filibranch ctenidium, C. eulamellibranch ctenidium. After Coan *et al.* (2000).

Ctenidial food collecting, successful as it is, has imposed certain limitations; no bivalve can lead a terrestrial existence, and enclosure of the body within the mantle precludes an active lifestyle. The ability to collect food from the water through siphons has enabled many bivalves to retreat into protected crevices or burrows in the substratum.

27.3 CLASSIFICATION OF THE BIVALVIA

Bivalves furnish splendid examples of evolutionary diversification and adaptive radiation. Easily recognizable adaptive modifications create features of taxonomic importance. Closely related groups generally show a uniformity in way of life. On the other hand, similar ways of life have also produced convergences in structure and adaptation, such as we see among distantly related genera that attach themselves to hard substrata, such as *Chama, Spondylus,* and *Crassostrea*. Certain structures, such as the hinge, often permit the recognition of evolutionary affinities despite external morphologic similarity or dissimilarity.

Modern classification of bivalves is based on a wide spectrum of characters, the most important of which are:

1. the structure of the ctenidia, including their relationship to the palps and the types of cilia on them (Figure 27.2-4);
2. the mode of life, such as burrowing, boring, attaching with a byssus, cementing to a substratum, or free-living (Figure 27.5);
3. the morphology of the shell, particularly the hinge teeth and the ligament, and the relative sizes and degree of gape of the two valves (Figure 27.6, Appendices 1C and 1D);
4. the surface sculpture of the valves (Figure 27.7, Appendices 1C and 1D);
5. the size and position of the adductor muscles pulling the shell closed, which create distinctive scars on the insides of the valves (Figure 27.8);
6. the degree of fusion of the mantle edges and the presence and morphology of siphons, reflected by frequently visible pallial scars on the insides of the valves (Figure 27.9);
7. the microstructure and mineralogy of the shell;
8. the morphology of the stomach;
9. the form of the foot and the presence of byssal attachment threads;
10. information now available through biochemical and genetic analyses.

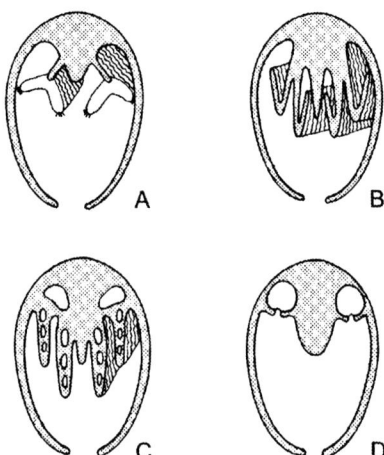

Figure 27.3 Diagrammatic cross section of the pallial cavity.
These illustrations show cross sections of the pallial cavity and the gills of four bivalves showing different arrangements of the ctenidia. A. Protobranch, B. filibranch, C. eulamellibranch, D. septibranch. After Coan *et al.* (2000).

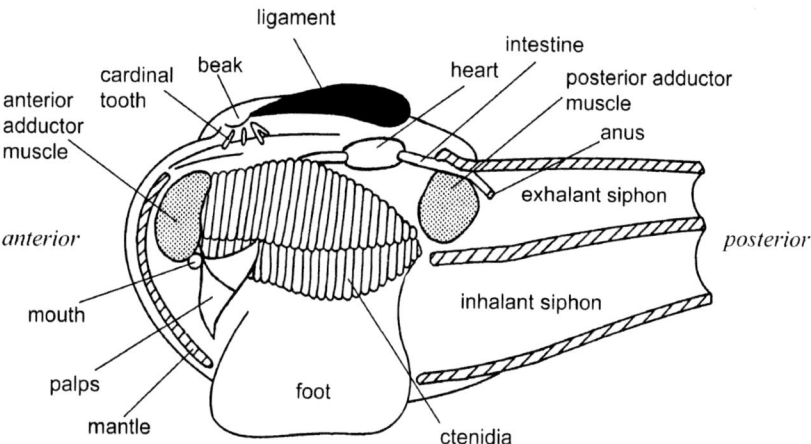

Figure 27.4 Internal anatomy of a typical heterodont bivalve.
In this diagram, the left valve and mantle lobe are removed. The siphons in this specimen are long and fused. After Coan *et al.* (2000).

A long and abundant bivalve fossil record has also enabled systematists to establish relationships and rankings that would be difficult to discern using only living forms. The number of characters now employed in bivalve classification and the degree of parallel evolution in the expression of these features has led workers to adopt names for orders not based on any one set of characters.

27.3.1 Protobranchia. A compact, probably natural group composed of two sub-groups:

Solemyoida: The genera *Solemya* and *Acharax* have primitive, protobranch ctenidia, which are the main organs of feeding and, like the Nuculoida, have a flattened foot. Solemyoida are subtidal to deep sea and infaunal or epifaunal.

Nuculoida: These usually small bivalves have protobranch ctenidia, a primitive taxodont hinge with a row of similar teeth, and they feed in large part by the palps. They are subtidal to deep sea and infaunal.

27.3.2 Pteriomorphia. Most systematists agree that this is a natural group, a conclusion based both on fossil evidence and on the overall similarity of living representatives. Most members are epifaunal, attached to surfaces by a byssus or by cementation and, as a result, the foot is reduced or entirely absent. The mantle margins are less fused than in the subclass Heterodonta. The following are some of the groups of pteriomorphs:

Arcoida: This order has the filter-feeding, filibranch ctenidium in which the elongate filaments are reflected so that each gill appears as a tall, narrow W in cross-section; adjacent filaments are united by patches of interlocking cilia to form lamellae. This is chiefly but not exclusively a tropical group and most have a taxodont hinge. Some genera in this order are *Arca, Barbatia, Anadara,* and *Noetia.*

Mytiloida: Members of this order have either filibranch or eulamellibranch ctenidia, the latter with actual bridges of tissue between adjacent filaments. The adductor muscles are unequal in size (heteromyarian). The group includes the Mytilidae (*Mytilus, Brachidontes*), the well-known mussels, which lack conspicuous hinge teeth. Most are found attached to rocks or pilings by a byssus, although some occur offshore burrowing in soft substrata.

Limoida: This group contains the family Limidae, species of which occur in the intertidal zone and offshore in water of moderate depth.

Ostreoida: Members of this group include some common families: Ostreidae (true oysters, *Ostrea, Crassostrea*), Pectinidae (scallops, *Pecten, Chla-*

mys), and Anomiidae (rock jingles or rock oysters, *Anomia*, *Pododesmus*). Only the much-enlarged posterior adductor muscle is present, a condition termed monomyarian. Both filibranch and eulamellibranch ctenidia are present in different members of the group. Many attach to the substratum by cement or byssus (sometimes calcified), whereas others are free living.

27.3.3 Paleoheterodonta. This group includes two subgroups: the Neotrigonia and the Unionoida:

Neotrigonia: These are marine species found in the Indo-Pacific region. The few extant species are representatives of a group that is more prominent in the fossil record.

Unionoida: These are the freshwater mussels. This group includes the following families: Unionidae, Margaritiferidae, Mycetopodidae, Iridinidae, Hyriidae, and Etheriidae. The Unionoida are treated in Chapter 25.

27.3.4 Heterodonta. This group includes most of the familiar clams. Ctenidia are eulamellibranchiate; the mantle margins are well fused, and elongate siphons are present in most. Distinctive patterns of hinge teeth and ligament characterize the different families:

Veneroida: Most heterodonts are members of this order, in which hinge teeth are well developed. Many veneroids are shallow to deep burrowers. Families include the Lucinidae, Thyasiridae, Ungulinidae, Carditidae, Chamidae, Galeommatidae, Lasaeidae, Neoleptonidae, Cardiidae, Veneridae, Petricolidae, Tellinidae, Donacidae, Psammobiidae, Semelidae, Solecurtidae, Solenidae, Pharidae, and Mactridae. Two additional families are Corbiculidae and Pisidiidae (Sphaeriidae), both found in freshwater (see Chapter 26).

Myoida: In this group, burrowing and boring are characteristic ways of life; most have long siphons, and the hinge has few teeth, although some have large projecting chondrophores. Families include

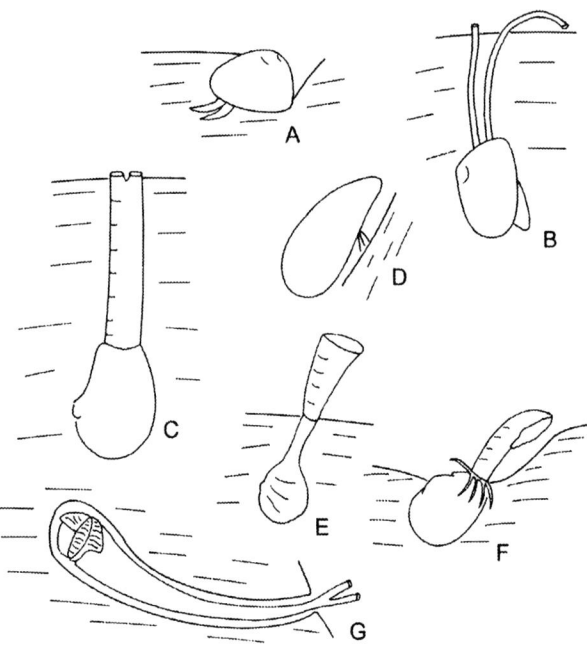

Figure 27.5 Bivalve feeding orientation.
A. Infaunal nuculid deposit feeder, B. Infaunal tellinid with separate siphons, C. Deeply infaunal heterodont with fused siphons, D. Epifaunal mytilid attached by byssus, E. Shallow infaunal cuspidarid with enlarged inhalant siphon used to suck in prey, F. Shallow infaunal poromyid with large hood used to capture prey, G. Boring teredinids in wood. After Coan *et al.* (2000).

Myidae, Corbulidae, Hiatellidae, Pholadidae, and Teredinidae (shipworms). Recent evidence suggests that this group is polyphyletic.

27.3.5 Anomalodesmata. Members of this group have siphons and many burrow into the substratum though some relatively common intertidal members are rock crevice nestlers; the shells are generally thin, and many are nacreous within. The hinge teeth are inconspicuous or absent. This group also includes the offshore, deepwater, carnivorous Septibranchia and the strange watering pot shells, Clavagellidae (*Penicillus*), as well as the Lyonsiidae, Pandoridae, Periplomatidae, and Thraciidae.

27.4 COLLECTING BIVALVES

Marine bivalves exploit a wide range of habitats, from intertidal rocks to subtidal sands and deep-sea muds. Some bivalve groups burrow deeply into the

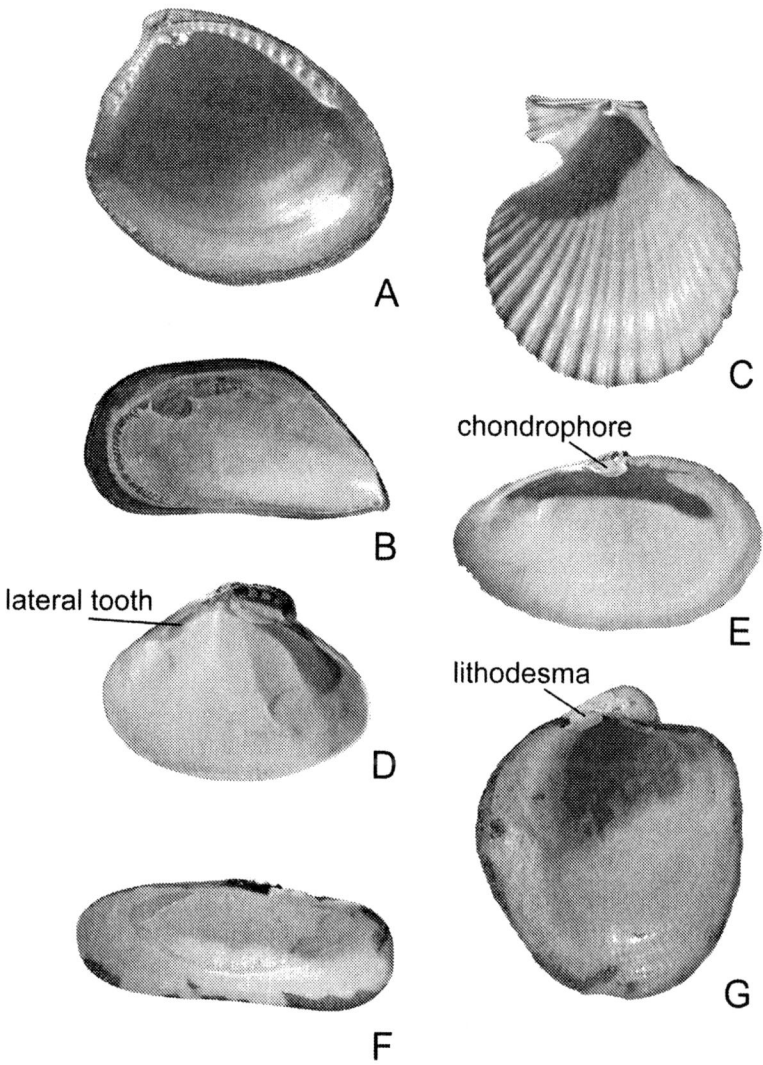

Figure 27.6 Bivalve hinge types.
A. Taxodont, rows of similar, simple, interlocking teeth [*Acila castrensis* (Hinds, 1843), SBMNH 123586, Santa Barbara County, California, U.S.A.], B. Dysodont, week teeth close to the umbones [*Mytilus galloprovincialis* Lamarck, 1819, SBMNH 44195, Ventura County, California, U.S.A.], C. Isodont, lateral tubercles and sockets [*Chlamys hastata* (Sowerby, II, 1842), SBMNH 23489, Orange County, California, U.S.A.], D. Heterodont, teeth usually large, including up to three cardinal teeth, lateral teeth may be present [*Tivela stultorum* (Mawe, 1823), SBMNH 17616, Baja California, Mexico], E. Asthenodont, cardinal teeth usually absent, replaced by large chondrophores or resilifer [*Mya arenaria* Linnaeus, 1758, SBMNH 49773, Bolinas Bay, Marin County, CA, U.S.A.], F. Anodont, true teeth absent in adults, or as amorphous tubercles [*Cyrtodaria kurriana* Dunker, 1861, SBMNH 345358, Beaufort Sea, 70°56.2'N, 153°8.4'W], G. Lithodesmodont, hinge without true teeth but may have a calcareous lithodesma ventral to, and cradling, the ligament [*Policordia jeffreysi* (Friele, 1879), SBMNH 345287, Arctic Ocean, 78°12'N, 129°58.7'E]. SBNMH = Santa Barbara Museum of Natural History. After Coan *et al.* (2000).

sediments, and still others bore into rock or wood. Listed below are a variety of collecting techniques that can be used in the many habitats where bivalves live (see Chapters 2 and 3 for additional information on collecting).

27.4.1 Beach drift. Many near-shore species are easily collected in beach drift, especially after storms and times of high wave activity. Micro species (< 5 mm) can be found in sandy beach drift lines, and in shell deposits around tidal pools and crevices along rocky shores. Collecting "shell gravel" and sorting it under a microscope or magnifying glass will often reveal many micromollusks.

27.4.2 Wood and rock. Some bivalves burrow into rock or wood. Rock-dwelling bivalves, and the nestlers that later utilize their holes, can be obtained by breaking up the rock. However, collectors should obtain such species from rock that has been broken loose by storms and washed up on the beach. Intertidal areas with suitable rocky substrata are far too uncommon for destructive collecting techniques.

Wood boring teredinids can be obtained from old pilings and other wood that has been submerged in the ocean. The usually small bivalve shells may be removed by teasing the wood apart, taking care not to crush the valves. In some, wood infestation is indicated by calcium carbonate lined borings.

27.4.3 Soft sediments. Epifaunal and shallow infaunal bivalves that live in soft substrata are easily collected by hand with shovels, rakes or guns, and with dredges and grabs. A variety of clam guns, rakes, and shovels can be found at many sporting goods stores. Clam guns (long cylindrical tubes), in particular, allow the extraction of deep burrowing bivalves in muddy sediments (Figure 2.1F).

27.4.4 Dredging. Boat dredging offers the best collection method for offshore infaunal and epifaunal bivalves. Skoglund (1990) gives an excellent overview of small boat dredging, along with plans for a small portable dredge. Grab samplers (e.g., Smith-McIntyre, Van Veen, and Ponar) or box corers provide another method for collecting infaunal bivalves but require a much larger vessel to take samples (see Chapter 3, Remote Bottom Sampling).

27.4.5 Commercial fishing. Excellent collections of small rare offshore bivalves have been obtained by sorting the stomach contents of bottom feeding fishes obtained from commercial fishermen. Dragboats obtaining bottom-feeding fishes also occasionally obtain large, rare bivalves in their nets (see Chapter 2.4.5 *Ex pisce* Collecting and Chapter 2.4.11 Commercial Fishing Boats).

27.4.6 Commensal species. Other specialized habitats for marine bivalves that are often overlooked include those living commensally (e.g., *Neaeromya*, *Mysella*, *Rochefortia*) on hermit and sand crabs, segmented and peanut worms, or attached to other invertebrates or in their burrows. Often small bivalves are found on the abdomen and telson of burrowing crustaceans. Some polychaetes, such as the chimney worm *Diopatra cuprea* (Bosc, 1802), collect small mollusks, including small bivalves, to construct their tubes.

27.5 PRESERVING AND STUDYING BIVALVES

Examination of live bivalves can yield valuable information on the ciliary and mantle currents of the animal. One can use the relaxation methods listed below and this should allow the bivalve to gape open slightly. Use a single edge razor or scalpel carefully to cut the adductor muscles. One valve can then be carefully removed, along with part of the mantle, giving an excellent view of the live animal. Carmine red stain in powder form or graphite dust (found at bicycle shops) placed on the ctenidia is used to observe ciliary currents under a dissecting microscope or magnifying glass.

In some benthic studies, bivalves are immediately preserved in formalin without proper relaxation. While this yields acceptable hard parts for identification, it is difficult to examine the anatomy of the animal, as many species "clam-up" tightly. In addition, it can be difficult to open these specimens to observe the diagnostic hinge dentition and muscle scars. There is no single technique for relaxing all species of bivalves. We have had varying results relaxing bivalves by adding menthol crystals, magnesium chloride (7% in seawater) or dilute ethyl alcohol (approximately 3%) to native seawater until the specimens are not responsive to touch. We have also had success allowing cool-water species to warm to room temperature until specimens are non-responsive, and similarly cooling warm-water species. Members of the family Veneridae can be especially difficult to relax, and we have had to resort to a combination of the above methods to relax them properly (also see Chapter 2.5).

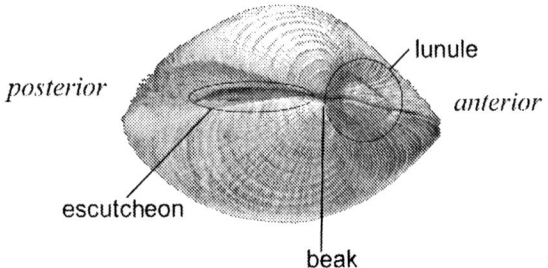

Figure 27.7 External structure of a typical heterodont bivalve. After Coan *et al.* (2000).

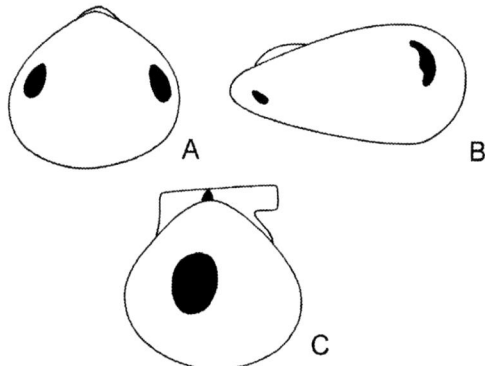

Figure 27.8 Schematic view of adductor mussle scars.
A. Isomyarian bivalve, B. Anisomyarian bivalve, C. Monomyarian bivalve. After Coan et al. (2000).

When a specimen is nonresponsive to touch, it can then be fixed in 5% buffered formalin (buffered by adding a handful of sodium borate [Borax] per gallon sample, not scientific but the reality of fieldwork). Formalin should be buffered to prevent decalcification, especially of small shells that can very rapidly dissolve. Most bivalve specimens should not be allowed to remain in formalin for more than 48 hours or severe damage can occur to the shell. Formalin is a carcinogen that should be used only outdoors or in well-ventilated conditions. After fixation, bivalves should be rinsed and soaked in distilled water to remove any formalin residue. Finally, they should be placed in 80% ethyl alcohol for long-term preservation. If DNA studies are anticipated, formalin should not be used. Instead, the specimens should be placed directly into 95% alcohol, although this makes the specimens poorer candidates for histological examination.

If a specimen is dried (e.g., museum dry collections), no attempt should be made to pry or cut the shell valves apart. Instead, the entire dried specimen should be rehydrated in warm water. The valves will eventually separate without damaging hinge structures. This can take just a few minutes with small, dried specimens or several days for larger material (see Chapter 5.9 for additional comments on reconstituting dried tissue).

Perhaps the most difficult process in examining a bivalve is safely opening tightly closed (and frequently fragile) valves to observe details of the hinge, pallial line and muscle scars. Depending on the final deposition of the specimens, one of the following two techniques should provide acceptable results. With either method, one must be sure to examine the outside characters (ligament, lunule, escutcheon, etc.) of the specimen first, as these may be unobservable once the specimen is opened.

If there is no need to examine the soft parts of the bivalve, the specimen can be placed in a dilute solution (approximately 10% to 50%) of household bleach (Clorox®, sodium hypochlorite) in distilled water. Usually after 15 minutes for small shells, longer for larger specimens, all soft tissue will dissolve allowing easy access to the inside of the shell. This method destroys all soft tissue and the periostracum.

To preserve the soft tissue and periostracum of the specimen, one can open a bivalve shell by using a single edge razor blade. Carefully place the specimen, ventral side up, in a stiff adhesive (e.g., poster mount) or clay. Then, while carefully holding the specimen with forceps, gently slice the blade into the opening between the valves. With small bivalves, one must accomplish this task while viewing the specimen through a dissecting microscope. While this technique takes practice, it is possible to open specimens as small as 1 mm in length with little or no damage. This method is usually not effective with species that have a heavily crenulated ventral margin.

External observations should include shell sculpture, ligament (if external), beak size and orientation, lunule and escutcheon (if present), and periostracum. Internal shell observations should include type and number of teeth, type of resilifer (if present), pallial line and sinus (if present), and adductor muscle scars.

In smaller specimens, it is often difficult to observe the pallial line and muscle scars. Dyeing the shell in crystal violet allows easier examination of these features. Place the specimen in a solution of crystal violet (no definite concentration, just mix a small amount maybe 3 grams in 30-50 ml of water and dispense from a dropper bottle) for approximately 15 minutes, or until the pallial line becomes more visible. Crystal violet stain can be removed from the shell by soaking it in ethyl alcohol (ethanol).

For anatomical study of preserved specimens, one shell valve must be removed to observe the body. It can be difficult, or not possible, to remove the tissues without damaging the shell. If several preserved specimens are available, and they are small, it is possible to decalcify the shell in 5% EDTA (ethylenedinitrilotetraacetic acid). This will allow one to obtain tissues with their organ systems intact. Large histologically fixed or fresh specimens may be studied either by a vertical cut with a razor along the antero-posterior axis, bisecting the tissues and leaving one half in each valve, or by carefully severing the connections between one valve and the mantle and adductor muscles. Small fresh material is best relaxed (see above), and then placed in an acidic fixative, such as Davidson's solution (a mixture of glacial acetic acid, formalin, and ethanol), which will fix the tissues while decalcifying the shell.

27.6 FURTHER STUDY

The literature dealing with the Bivalvia is vast. The following are some useful references.

27.6.1 Identification. For identifying marine bivalves the following books will prove helpful: Olsson (1961), Tebble (1966), Habe (1977), Scarlato (1981), Lamprell and Whitehead (1992), Oliver (1992), Poppe and Goto (1993), Lamprell and Healy (1998), Coan *et al.* (2000), and Qi (2004). Coan *et al.* (2000) also contains an extensive listing of bivalve references by biogeographical provinces.

27.6.2 Biology. For information on biological aspects of the Bivalvia see Giese and Pearse (1979), Carter and Lutz (1990), Jorgensen (1990), Morton (1990), Morse and Zardus (1997), Kas'ianov *et al.* (1998), and Morton *et al.* (1998).

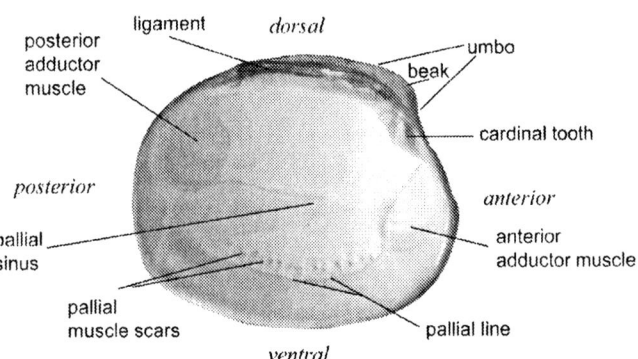

Figure 27.9 Interior view of a typical heterodont bivalve shell. After Coan *et al.* (2000).

27.6.3 General and Miscellaneous References. For information on systematics, paleontology as well as some general biological topics see Haas (1929-1955), Dechaseaux (1952), Franc (1960), Cox *et al.* (1969), Yonge and Thompson (1978), Vokes (1980, 1990), Morton (1996), Johnston and Haggart (1998), and Harper *et al.* (2000).

27.7 ACKNOWLEDGEMENT

We would like to than the Santa Barbara Museum of Natural History for permission to use the figures in this chapter. They were originally published in Coan *et al.* (2000).

27.8 LITERATURE CITED

Carter, J. G. and R. A. Lutz. 1990. Bivalvia (Mollusca). *In:* J. G. Carter, ed., *Skeletal Biomineralization: Patterns, Processes and Evolutionary Trends*, Vol. II, Atlas and Index. Van Nostrand Reinhold, New York. Pp. 5-28, pls. 1-121.

Coan, E. V., P. Valentich Scott, and F. R. Bernard. 2000. *Bivalve Seashells of Western North America; Marine Bivalve Mollusks from Arctic Alaska to Baja California.* Santa Barbara Museum of Natural History, Santa Barbara, California. vii + 764 pp.

Cox, L. R. and 24 other authors. 1969. Part N [Bivalvia], Mollusca 6, Vols. 1-3. *In:* R. C. Moore, ed., *Treatise on Invertebrate Paleontology.* Geological Society of America and University of Kansas, Lawrence, Kansas. 1224 pp.

Dechaseaux, C. 1952. Classe des lamellibranches (Lamellibranchiata Blainville 1816). *In:* J. Piveteau, ed., *Traité de Paléontologie*, Vol. 2. Masson, Paris. Pp. 220-364.

Franc, A. 1960. Classe des bivalves. *In:* P.-P. Grassé, ed., *Traité de Zoologie; Anatomie, Systematique, Biologie,* Vol. 5, Fas. 2. Masson, Paris. Pp. 1845-2133.

Giese, A. C. and J. S. Pearse. 1979. *Reproduction of Marine Invertebrates,* Vol. 5, Molluscs: Pelecypods and Lesser Classes. Academic Press, New York. xvi + 369 pp.

Haas, F. 1929-1955. Bivalvia. *In:* H. G. Bronn, ed., *Klassen und Ordnungen des Tierreichs* (3 - Mollusca) (3 - Bivalvia) [1(1): 1-176 and 1(2): 1-292, 1 April 1929; 1(3): 177-384, 1 March 1931; 1(4): 385-544 and 1(4-II): I-1 to I-41, 1933; 1(5): 545-704, 1 April 1934; 1(6): 705-864, 1 Dec. 1934; 1(7): 865-984 and 1(7-II): II-1 to II-20, 31 Dec. 1935; 2(1): 1-208, 1 Aug. 1937; 2(2): 209-466, 24 May 1938; 2(3): 467-678, June 1941; 2(4): 679-909, 1955)].

Habe, T. 1977. *Systematics of Mollusca in Japan. Bivalvia and Scaphopoda.* Zukan-no-Hokuryukan, Tokyo. xiii + 372 pp. + [4] pp. errata, 72 pls.

Harper, E. M., J. D. Taylor, and J. A. Crame. 2000. The Evolutionary Biology of the Bivalvia. *Geological Society of London Special Publication* **177**: i-vii, 1- 494 pp.

Johnston, P. A. and J. W. Haggart, eds. 1998. *Bivalves: An Eon of Evolution.* University of Alberta, Alberta, Canada. xiv + 461 pp.

Jorgensen, C. B. 1990. *Bivalve Filter Feeding: Hydrodynamics, Bioenergetics, Physiology and Ecology.* Olsen & Olsen, Fredensborg, Denmark. vi + 140 pp.

Kas'ianov, V. L., G. A. Kriuchkova, V. A. Kulikova, and L. A. Medvedeva, edited by D. L. Pawson. 1998. *Larvae of Marine Bivalves and Echinoderms.* Smithsonian Institution Libraries, Washington, D.C. viii + 288 pp.

Lamprell, K. L. and J. M. Healy. 1998. *Bivalves of Australia, Vol. 2.* Backhuys, Leiden. 288 pp.

Lamprell, K. L. and T. Whitehead. 1992. *Bivalves of Australia, Vol. 1.* Crawford House, Bathurst, New South Wales. xiii + 182 pp.

Morse, M. P. and J. D. Zardus. 1997. Bivalvia. In: F. W. Harrison and A. J. Kohn, eds., *Microscopic Anatomy of Invertebrates* 6A (II). Wiley-Liss, New York. Pp. 7-118.

Morton, B., ed. 1990. *Proceedings of a Memorial Symposium in Honour of Sir Charles Maurice Yonge (1899-1986), Edinburgh, 1986.* Hong Kong University, Hong Kong. viii + 355 pp.

Morton, B., R. S. Prezant, and B. Wilson. 1998. Class Bivalvia. In: P. L. Beesley, G. J. B. Ross, and A. Wells, eds., *Mollusca: The Southern Synthesis. Fauna of Australia,* Vol. 5. CSIRO Publishing, Melbourne, Australia. Pp. 195-429.

Morton, B. S. 1996. The evolutionary history of the Bivalvia. In: J. D. Taylor, ed., *Origin and Evolutionary Radiation of the Mollusca.* Oxford University, Oxford. Pp. 337-359.

Oliver, P. G. 1992. *Bivalved seashells of the Red Sea.* National Museum of Wales, Cardiff, Wales and Hemmen, Wiesbaden. 330 pp.

Olsson, A. A. 1961. *Mollusks of the Tropical Eastern Pacific, Particularly from the Southern Half of the Panamic-Pacific Faunal Province (Panama to Peru).* Panamic-Pacific Pelecypoda. Paleontological Research Institution, Ithaca, New York. 574 pp., 86 pls.

Poppe, G. T. and Y. Goto. 1993. *European Seashells.* Vol. II (Scaphopoda, Bivalvia, Cephalopoda). Hemmen, Wiesbaden, Germany. 221 pp., 32 pls.

Qi Zhongyan, ed. 2004. *Seashells of China.* China Ocean Press, Beijing. viii + 418 pp. 193 plates. [Covers 605 bivalve species].

Scarlato, O. A. 1981. *Dvustvorchatye Molliuski Ume-Rennykh Shirot Zapadnoi Chasti Tikhogo Okeana. [Bivalve Mollusks of the Temperate Latitudes of the Western Part of the Pacific Ocean].* Akademiia Nauk SSSR, Zoologicheskii Institut, Opredeliteli po Faune SSSR 126: 480 pp., 64 pls.

Skoglund, P. 1990. Small boat dredging. *Festivus* **22**: 106-109.

Tebble, N. 1966. *British Bivalve Seashells. A Handbook for Identification.* British Museum (Natural History), London. 212 pp., 12 pls.

Vokes, H. E. 1980. *Genera of the Bivalvia: A Systematic and Bibliographic Catalogue (revised and updated).* Paleontological Research Institution, Ithaca, New York. xxvii + 307 pp [see also Vokes, 1990].

Vokes, H. E. 1990. Genera of the Bivalvia: A systematic and bibliographic catalogue - addenda and errata. *Tulane Studies in Geology and Paleontology* **23**: 97-120.

Yonge, C. M. and T. E. Thompson, organizers. 1978. Evolutionary systematics of bivalve mollusks. *Philosophical Transactions of the Royal Society of London* (B) **284**: 199-436.

CHAPTER 28

THE MARINE AQUARIUM: A RESEARCH TOOL

BEATRICE WINNER
CHARLES F. STURM

28.1 INTRODUCTION

[This chapter originally appeared as Winner (1993). It was adapted and expanded by the junior author.]

Any aquarist can tell you how delightful it is to observe marine animals. What prompted me to write down my observations was an article published in *Hawaiian Shell News* quite a number of years ago, stating that information was needed by malacologists regarding the diet and behavior of mollusks. I thought at that time that I could help in this way. I have spoken to shell collectors about interesting happenings that have taken place in their aquaria, and I have encouraged them to write them down. This chapter might stir you to do the same.

Remember that the habits of a mollusk in the field can differ from the habits observed in an aquarium, particularly concerning diet. Aquarists that keep invertebrates must sometimes guess what they prefer to eat. It is often a trial and error situation. Littoral species such as periwinkles may stay above the water line. If so, it is necessary to splash water on them from time to time. Intertidal species such as chitons that are exposed to low and high tides and confined to rocks will not always stay below the water line in an aquarium. Species that hide under rocks at low tide but are less confined to rocks will climb from time to time above the water line. Species that consistently live below the water can take days to several weeks to adjust to an aquarium. The larger the animal, the longer it takes to adjust. Those mollusks found in deeper waters may require aquaria with more sophisticated equipment.

Mollusks can deteriorate from poor environmental conditions such as incorrect temperature, salinity, diet, and pollution, both in the field and in an aquarium. I have seen at Oyster Pond on the Bahamian island of Eleuthera at least two species that are reduced in size, presumably due to environmental conditions.

Collecting living mollusks has made it possible for me to observe egg-laying habits. These behaviors captured my interest to the point that I gathered enough material to photograph and write two field guides (Winner 1985, 1991). Years of watching snails spawn in the aquarium have convinced me that most gastropods seek a smooth surface on which to lay eggs. They generally spawn on the sides of the aquarium making it convenient to collect, photograph and observe them under a microscope.

Mollusks may behave differently in the field and in the aquarium. For example, *Melongena corona* (Gmelin, 1791), whose eggs are found on rocks, laid them on the glass in my aquarium. Marginellids normally lay eggs on clam shells or on other egg masses, yet in the aquarium the eggs are laid on the sides of the glass. *Nassarius obsoletus* (Say, 1822) when they spawned, it was always on the glass of my aquarium. Where do they lay eggs in the field?

Some species that I observed in the field were not particular about where they spawn. Eggs could be found on empty bottles, tin cans, plastic bags, wood, or whatever was convenient at the time of egg laying. Nevertheless, in the aquarium, these same species laid eggs on the glass.

My most frequent collecting areas were the beaches at Phil Foster Park, and Peanut Island, both in Riviera Beach, Florida. Other much less frequently collected areas throughout the years were Marco Island and Sanibel Island, Florida. I recorded many activities, such as what they ate, when and how the food was eaten, and any unusual behavior they performed. Much remains unknown about these animals. If you are interested in maintaining an aquarium and studying mollusks, please record your observations and consider publishing them. The amateur malacologist can make significant contributions to the understanding of the behavior, feeding habits, and development of mollusks.

28.2 AQUARIA

The marine and freshwater aquaria that I kept for years are not the typical ones that you see in pet stores or those that most people keep as a hobby. Mine were strictly for study and I was not concerned whether they were attractive or not. I never used artificial plants and other aquarium decorations.

My marine aquaria had sand bottoms, simulating the invertebrate environment as much as possible. Critters that are sand dwellers prefer the sandy bottom to store-bought gravel. I would collect sand from the beach at low tide, wash it until it was clean, and placed it not less than four inches (100 mm) deep in the aquarium. Sand collected at the beach has a coating of algae and bacteria which helps to maintain a more realistic environment. At times, I would filter salt water for critters I might be raising in petri dishes. I never was concerned about heating my aquarium or trying to keep it at an even temperature. As I mentioned previously, I tried to simulate the animals' environment, and in its natural setting, the temperature varied.

In some of my large aquaria, I had various invertebrate species, and in others, I kept only one species. The aquaria with multiple species are fun to watch. You can see who gets along with whom, who eats whom, and who eats what. The reason for having only one species in an aquarium was in the hope that I could view egg laying as well as feeding habits, and I have often been successful.

28.2.1 Water. In preparing water for your aquarium, you must consider several factors. The pH of the water should be in the range of 8-8.4. The specific gravity of the water should be from 1.02-1.03, lower for brackish water species, higher for ocean dwellers. You can measure specific gravity easily with a hydrometer. The water should be low in nitrates, have a calcium concentration of 400-450 parts per million, and have an alkalinity of 2.5-3.5 milliequivalents per liter. Finally, there should be no copper ions as these are toxic to marine invertebrates. Though this balancing act may seem daunting, most pet stores sell simple kits to test water quality. You do not need to be an analytical chemist to maintain a salt-water aquarium.

If you live by the ocean, getting water is easy. You have but to walk down to the shore with a bucket or bottle and collect some water. You can probably use the water without testing it, especially if it is from an area that is supporting a healthy community of organisms and seems to be unpolluted. If you do not live near the shore, you must obtain salt water by other means. There are formulae to follow to make salt water (Jones 1972: 471, Presnell and Schreibman 1997: 466). These are fairly involved requiring you to mix many different salts and trace elements. On the other hand, you can take the easy way out and purchase a salt mix. These can be obtained at most pet stores and all you need to do is add the required amount of water.

If you maintain an aquarium for a period of time, some of the water will evaporate. You should not replace evaporated water from a salt-water tank with salt water. If you do, you will raise the salinity, and over time, the water will become too salty to support life. Evaporated water should be replaced with distilled or de-ionized water.

In general, larger tanks are easier to maintain. The larger volume of water and chemicals tend to buffer changes better than in a smaller tank. However, you do not want to use a tank that is so large that you have difficulty finding your shells. The tank should be well lit but avoid direct sunlight. Make sure that you do not put too much food in the tank, as it will eventually degrade the quality of the water. Adding too much food is a larger concern for smaller tanks as there is a smaller margin of error than in a larger tank. Put in just enough food to feed your specimens and remove the excess. Finally, you will need to filter and aerate the water.

28.2.2 Plants. The plants in an aquarium help to recycle animal waste products. I kept living plants in my marine aquarium, and if the animal I collected was found on a plant, I brought both it and the plant home. It is sometimes difficult to keep plants alive. Those that are found in the shallows do best in the aquarium. I discovered that some plants had to be replaced frequently. You will have to experiment with different plants to see which do best for you. A little exposure to sunlight is a help and if possible, keep aquariums close to but not directly in sunlight. Algal buildup is a clue that your tank is receiving too much light.

28.2.3 Rocks. In my aquaria, the rocks I used were carefully chosen at low tide. I prefer smooth rocks because those with crevices may hide microscopic crab larvae that grow quickly. Suddenly your specimens start to disappear; everything the crabs capture, they eat. I have had this happen to me several times and finally discovered how the crabs had entered my aquaria. In one aquarium, I had some tiny newly hatched fish that I had caught with a net. I also had a small crab. This crab perched itself on a rock and when a tiny fish went by, it quickly grabbed it with its claw. An astonishing feat! Having fish in an invertebrate tank works well. It seems to improve the chemical balance of the water, both fresh and marine.

If you do not want to deal with an aquarium, and are still interested in observing small mollusks, I suggest shallow, wide-mouthed containers or jars with a layer of good clean sand. An occasional blast of air pumped in the mini containers is all that is required. A small pump purchased in an aquarium shop can take care of several containers at one time.

An important thing to remember is that the smaller the aquarium, the easier it is to foul by giving your animals too much food. Be sure to remove uneaten food as soon as your mollusks have finished eating, in both the mini- and larger aquaria.

28.2.4 Food. Mollusks may be herbivores, carnivores, or omnivores. I found that some species reported to be herbivores turned out to be omnivorous, that is, they can eat plants or animals. Raw shrimp was the most convenient food to feed carnivores. It can be purchased in pet shops, or collected in the field by dragging a hand net along the bottom in grassy areas. If you collect more than you need they can be frozen for future use.

Bivalves such as the common *Chione cancellata* (Linnaeus, 1767) are also a good source of food and can be collected at low tide. One of the easiest to collect is the coquina clam (*Donax variabilis* Say, 1822). They can be found on the beaches in some areas, and some people make a clam broth from them. Store-purchased clams can also be used. Any of these can be frozen and used when needed.

Small crabs are an excellent food source and can be collected by snorkeling in the shallows where they are found under rocks and can be scooped up with a net. Store-purchased crabs can also be used, cut into small portions, and the balance frozen. My objection to crabs is that they can foul your aquarium very easily. The shells do not dissolve and so must be removed.

One of the errors often made in feeding mollusks is "overfeeding." I cannot emphasize this enough. Overfeeding pollutes the water and can very easily kill your specimens. It must be remembered that these small mollusks eat very tiny bits of food, and it is not necessary to feed them daily. It is sufficient to feed them every few days and this lessens the chance of fouling the water. Remember that they rarely die of starvation. This trial and error situation is both challenging and fun. Commercial foods for invertebrates can be bought in pet shops, and you might want to try them. I preferred to feed them natural sea life that they would ordinarily find in the field so that I could see exactly what they like to eat. There are exceptions to daily feedings. The herbivorous Notarchidae (side-gilled slugs) ate all day and all night in my aquarium. I had to release them because almost daily I had to go to the Intracoastal Waterway to get algae. One species in particular, having no food, ate its own eggs.

When planning on setting up an aquarium, speak to other aquarists. If there are none that you know, strike up a conversation with someone at the pet supply store. While they may not have experience in maintaining mollusks, the principles are similar to those for tropical fish. For more information regarding maintaining an aquarium, see Skomal (1997), Tullock (1998), and Tunze (1999).

28.2.5 Permits and conservation. When collecting specimens, you will need to consider whether a collecting permit is required. Information on agencies that grant permits will be found in Chapter 31.

When collecting, try not to collect more specimens than you need for your project. Also, when you are finished studying the specimens you should consider if the specimens need to be kept as voucher specimens or if they can be released. If you are keeping the animals as voucher specimens, you will find the narcotization techniques mentioned in Chapter 2.5 useful. If you are returning the animals to their native habitat, you should return them to where they were collected or as close to there as possible.

28.3 SELECTED SPECIES OBSERVATIONS

28.3.1 *Corbula contracta* Say, 1822 [Contracted Corbula]. Using a hand dredge at Little Munion Island in Lake Worth, Florida, I found several of these gray, quarter-inch (6 mm) bivalves. They are shallow burrowers and both valves are about the same size except that the posterior ventral margin of one valve overlaps the edge of the other. They produce a fine rubbery, gelatinous thread, called a byssus, which is almost an inch (25 mm) long and has tiny bits of stone and broken shells attached to it. The thread extends from the center and edge of the bivalve and anchors the bivalve to the substrate. In the process of transporting them to the petri dish, they became detached from the thread. Sometime later when I looked at them again I was surprised to see that they had again attached themselves to the dish with this fine thread. Some of the other corbulids had attached themselves to the threads of other corbulids so that they were anchored with the same bits of stone and broken shells. These little shells had a white foot to propel themselves around. They lived in my miniature tank for a while.

28.3.2 *Janthina globosa* Swainson, 1822 [Elongate Janthina]. There are about four species of *Janthina* in Florida, and all are commonly called the "purple snail." These pelagic violet snails are blown onto beaches after a storm. It was after one of these storms that I was able to collect several of them alive from Riviera Beach, Florida. I also collected living *Velella velella* (Linnaeus, 1758) [By-the-Wind-Sailor, a cnidarian, see Chapter 29.2.4], on which *Janthina* feed.

Whenever I stirred the water in the aquarium and the By-the-Wind-Sailors neared the *J. globosa*, the snails would catch the *Velella* with their propodium and begin to eat it, an interesting event to watch. By-the-Wind-Sailors have a chambered pneumatophore (float) and a sail extending above the surface of the water. The *Janthina* forms a "bubble float" which is created with air cells by means of viscous mucus. When blown ashore they are doomed to die, as they cannot crawl back to the sea.

28.3.3 *Polinices lacteus* (Guikling, 1834) [Milk Moonsnail]. My writing and slide presentations about this snail received much attention from malacologists. Looking into my aquarium one day, I discovered a naticid in the center of a grooved circle. I watched this sand dweller crawl below the sand and disappear, but then I noticed the sand was moving at the center of the grooved circle. Emerging from the edge of the circle was the beginning of a sand collar (the egg case for *Polinices*). The snail worked feverishly for twenty minutes to push it completely above the sand. During this period, it appeared twice, perhaps to see if the collar was completely out. The collar, composed of sand, consisted of three whorls. It was soft to the touch when first laid, but became firm and slippery within three hours. Another twenty-four hours later, the collar had lost its slick texture.

These sand collars and those of other species of naticids were often found at Phil Foster Park, Riviera Beach, Florida. The gelatinous matter that the eggs are embedded in picks up the grains of sand. In this particular species, the snail picks up the coarse grains of sand. Thus, the texture of the sand collar is rougher than that of *Neverita duplicata* (Say, 1822) or *Naticarius canrena* (Linnaeus, 1758), although they are found in the same area. The weight of the eggs and gelatinous matter causes the difference. Sand collars never fail to amaze the people who find them on the beaches at low tide.

28.3.4 *Ficus communis* Röding, 1798 [Atlantic Figsnail]. A four-inch shell that I collected for my aquarium would not eat and I surely thought that it would die. I applied the same method that I used to persuade *Pleuroploca gigantea* (Kiener, 1840) to eat. I placed a piece of fresh-dead shrimp under its proboscis and it soon extended its radula and ate the shrimp. I had to do this a few times before it started eating on its own.

Probably all gastropods will eventually eat on their own, but I believe that coaxing helps them to eat sooner. When I placed this shell in the aquarium with a *P. gigantea*, I was disheartened to see that *P. gigantea* ate it. *Ficus communis* has been reported from shallow subtidal waters to depths beyond 3,000 meters.

28.3.5 *Cymatium labiosum* (Wood, 1828) [Lip Triton]. On Sept. 4, 1981, I captured a specimen that had a chipped lip, and by September 15th the lip was completely repaired, a total of eleven days. As with *Cymatium nicobaricum* (Röding, 1798), I was not lucky enough to see it eat. *C. labiosum* did not do well in my aquarium.

28.3.6 *Melongena corona* (Gmelin, 1791) [Crown Conch]. I collected seven small *Melongena corona* from the west coast of Florida and placed them in a one-gallon (4 liter) aquarium by themselves. I found that they would eat clams, oysters, mussels, crabs, shrimps, fish, and almost any other animal tissue. Shrimp were their main diet, and after many months in captivity, I was able to feed them pieces of shrimp by hand. I would take a *M. corona* out of the aquarium, hold it by its shell in one hand, and with my other hand feed it shrimp. However, if my hand accidentally touched the animal, it would withdraw into its shell immediately. They are interesting to watch and make good pets.

Egg capsules were laid on the glass of the aquarium. The female required almost an hour to release one capsule. I observed an unusual behavior while a female was laying her capsules. Another one climbed upon her back, extended its proboscis completely around each side of her, (I never realized how much the proboscis could stretch), then over each egg capsule that she had laid, starting with the one furthest away. When its proboscis reached the egg she was in the process of expelling, the second conch then pushed its proboscis in and out and completely around the capsule that was still in her gonopore with such force that it turned the proboscis red with each insertion, and finally forced the egg completely out of the gonopore. When the capsule was released, they both fell to the bottom of the aquarium. It appeared as if the animal was helping her to release the egg; I cannot even imagine why. This happened almost every time a capsule was being released. In researching this species, I was not able to find anything regarding this episode though it might be activity relating to fertilization.

I found *M. corona* and its egg capsules to be very hardy. On one of my trips to Sarasota, Florida, I found large egg capsules at low tide (larger animals produce larger egg capsules) and placed them in a plastic bag. Days later, I ran across the plastic bag that I had forgotten to take care of, and decided to wash them as they were covered with mud. After cleaning these egg capsules, I placed one under the microscope and was surprised to find that there was still movement in the capsules. I could see veligers swimming around; they had survived in the plastic bag with no air or water.

28.3.7 *Nassarius vibex* (Say, 1822) [Bruised Nassa].

Having kept a crab trap in the saltwater Earman River, North Palm Beach, Florida, I often found that the bait in the trap would be covered with mud dwelling nassariids. I collected them for my aquarium from time to time. I never saw them crawling around and so I thought that they were beneath the sand. One day I headed for the dock from my apartment by way of the patio, which faces the water. I was bare footed. I felt something under my foot and stooped to pick it up, it was a nassariid. I found others, all near the patio screen door. It was apparent that they had crawled out of the aquarium, across the carpet and out to the patio. They were dead, as they were unable to crawl out the door to the water. I repeatedly placed nassariids in the aquarium to observe this behavior, and each time they would climb out and head for the water. The distance traveled by these snails was 18 feet (6 m). It appears they have a homing instinct.

When food was placed in the tank, they would scurry out of the sand quickly in search of the food using their strong sense of smell. *Nassarius* is a scavenger and will help to keep an aquarium clean. It can and will eat almost any organic material placed in the aquarium. Its proboscis has a sucking action that allows it to cling tightly to food even when food is lifted out of the water.

28.3.8 *Leucozonia nassa* (Gmelin, 1791) [Chestnut Latirus].

These are shallow-water snails and can be collected from tide pools. The shell is brown and the animal is red. Having raised one from an egg to a juvenile, I found that both shell and animal go through interesting color changes.

Having watched a set of egg capsules carefully for weeks, I decided to remove an egg from its capsule at twenty-eight days. It had already developed a shell with three whorls while in the capsule. The apex and second whorl were a beautiful brilliant red, the third was a transparent tan, and the aperture was light brown. Two eyes could be seen through the thin tan shell, and the animal possessed a white foot. At thirty-two days, the apex turned from red to white, the second whorl was still red, and the third (the body whorl) turned from tan to almost brown in color. The following day the animal began to turn from white to light tan. At thirty-four days, the apex of the shell began to turn tan, and ten days after that, the entire shell turned brown. At thirty-six days, the animal changed from a very light tan to pink. To my disappointment, the animal died on day 51 and I was unable to record how long it takes the animal to turn red.

Besides this juvenile, I had several other *Leucozonia nassa* in my aquarium. I found them to be carnivores and bivalves were their favorite food. They do not drill holes in bivalves as some snails do, but pry them open with their foot. I have seen two *L. nassa* pull a bivalve apart. They also ate cut shrimp, and like almost all snails, ate at night.

28.3.9 *Fasciolaria tulipa* (Linnaeus, 1758) [True Tulip].

Tulip shells are quite active in the aquarium. They have been reported as aggressive carnivores, and my observations are as follows: I saw a *Fasciolaria tulipa* capture a limpet by holding the limpet's shell with its foot, the fleshy part of the limpet was away from the tulip's foot. It then extended its proboscis between the limpet's shell and the animal, turning the limpet slowly around a few millimeters at a time. These movements were convulsive and each time it turned the limpet, fluid was exuded. Actually, what it was doing was cutting the animal away from its shell, and when it was completely released, it began to eat it.

I placed another limpet with its fleshy side exposed on the bottom of the aquarium. The *F. tulipa* turned it over in order to grab it by its shell. It then proceeded to loosen the animal away from its shell with its proboscis as it had previously, and when it was completely released from its shell, feasted on it. *F. tulipa* must be carefully watched, as it is a carnivore and will eat other invertebrates that are placed in the aquarium.

28.3.10 *Prunum apicinum* (Menke, 1828) [Common Atlantic Marginella].

This little snail, to me, is the most interesting of all snails to observe, particularly in their eating habits. I observed them for more than five years, starting with twelve marginellids and eventually ending with more than eighty as a result of mating in the tank. They are a hardy species and extremely easy to breed. I had one aquarium just for marginellids.

These carnivorous sand dwellers rise from below the sand to forage. When given pieces of food, a snail will crawl over it, hesitate, then hold the food with its foot, continue on, and then carry it below the sand. This is a peculiar sight to see as a large lump is formed at the end of the foot, sometimes larger than the snail itself. If the food is too massive for one snail to carry, several will engage in a feast above the sand. They eat most anything that is alive, freshly dead, or frozen. I was never able to detect whether they had a preference for one type of food over another, with the exception of one particular bivalve that they will not eat, which is mentioned below.

They were fed bivalves such as *Donax variabilis*, *Chione cancellata*, store-bought *Mercenaria mercenaria* (Linnaeus, 1758), and many other species. The one bivalve that they would not eat is *Lucina pensylvanica* (Linnaeus, 1758). I presented the marginellids with these lucinid clams several times and even opened a clam for them, but it was rejected. When I withheld food from them, they ultimately ate a very small portion of lucinid, most likely out of extreme hunger.

I observed marginellids carrying *Donax variabilis*, a bivalve larger than itself, below the sand. An odd observation was when one took a *D. variabilis* to the top of the aquarium, wedged it to the side of the glass with its foot and, holding it in this manner, proceeded to eat it. When it had its fill of the clam, it crawled away leaving the clam still attached to the glass. *D. variabilis* were collected from Sanibel Island, and placed in my freezer in North Palm Beach so that food was available whenever needed.

On one occasion, I placed a large store-bought *Mercenaria mercenaria* clam, more than two inches in diameter, in the aquarium with the intentions of opening it later. The clam had been out of the water for some time, so that when I placed it in the aquarium, it opened quickly giving the marginellids an opportunity to attack it. The following day I saw the clam was almost closed and that the siphon of a marginellid was extending out of the clam. I thought that the marginellid was trapped inside the clam. I decided to cut the clam's muscle to allow it to remain open, and when I did, out came four more marginellids.

On another occasion, I cracked a living *Bulla striata* Bruguière, 1792 to see if they would eat it, and eat it they did, while it was still alive in its cracked shell. I cracked a *Cerithium litteratum* (Born, 1778), and it too was consumed while it was still alive. Limpets were defenseless and easy prey for the marginellids, as the animal is completely exposed on one side. I crushed small fresh crabs that they pounced upon, and in short order, left empty crab shells.

Another interesting occurrence took place when I placed an *Oliva sayana* Ravenel, 1834 into the aquarium. Before it could crawl below the sand, one of the marginellids climbed on it and grabbed its propodium, the anteriormost section of its foot. The olive went berserk! It tried to go below the sand but could not. It tried to move forward but made little progress. It swayed from side to side, desperately trying to rid itself of the marginellid. Then other marginellids started to climb on it. The first marginellid never did let go and finally all the other marginellids viciously ganged up on it and began to eat the olive with gusto. The marginellids won the battle. This olive was more than two inches (50 mm) in length.

Adult marginellids have a thick outer lip with four brown spots, two in the center and one on each end. The body whorl has three diffuse brown bands. By raising this species, I discovered that occasionally some had developed thin shells, some lacked the brown body whorl, and some were more slender than others were. There were some lighter in color and some darker, yet a common resemblance always existed although there was variation. Juveniles do not resemble the adults; their shells are transparent so that the attractive white and black tissue is seen through their lucid shells.

Although there were rocks and plants in the aquarium, they chose to lay eggs on the glass and other smooth surfaces. The average number of eggs in a cluster was five to eight. I discovered that the eggs were laid in communal fashion and a cluster of eggs would be right next to other clusters of eggs. One day as I was cleaning out the debris from the bottom of the aquarium, I picked up a small clear plastic tube that was lying on the bottom. It measured a quarter of an inch (6 mm) in diameter and three inches (75 mm) long. I counted seventy-two egg capsules on this plastic tube.

One particular group of egg capsules that I had carefully observed for days all hatched except for one. While activity was always visible in the transparent capsules, this last remaining one showed no activity for several days. I decided to remove it from the glass so that I could observe it under the microscope. When I punctured the capsule, to my astonishment this inactive juvenile scurried away swiftly and happily. I delivered a baby marginellid.

28.3.11 *Hydatina physis* **(Linnaeus, 1758) [Brown-Lined Paperbubble].** This taxon is sometimes known in Florida as *Hydatina vesicaria* (Lightfoot, 1786). In twenty-two years of snorkeling at Phil Foster Park, Riviera Beach, Florida, I never found or heard of anyone finding this species in the area. I was flabbergasted when I found a colony at, of all places, the swimming beach on 2 May 1983.

I saw fifteen damaged shells, two badly crushed. There were two living individuals with broken lips. One had a part of its foot cut off, possibly by a crab or fish. I counted 35 egg masses of various sizes, with the attachment strings of the ribbons firmly attached into the sand. The shells varied in size, as did the egg masses, so I assume that the size of the egg ribbon corresponds with the size of the snail.

I took the two with broken lips home to observe in my aquarium, and on the following day, they laid eggs in an upside down position. They died twenty-four hours after egg laying. Others with whom I have spoken, who have had *H. physis* in their aquarium, have told me that the snails did not survive for long. According to the literature, they eat polychaete worms.

28.3.12 *Bulla striata* **Bruguière, 1792 [Striated Bubble].** To find these sand dwellers, I snorkeled at low tide, probing the sand with my fingers all around spaghetti-like egg masses that were attached to seagrass. Using this technique, I collected several for my aquarium, in order to observe their activities and feeding habits. Seldom did I find them above the sand.

A large meal can prevent a snail from completely withdrawing into its shell and occasionally one could be found in the field in this state. Now and then, this would happen in the aquarium with body extending from the shell so that they were unable to crawl around or go below the sand. This creates a tempting meal for crabs because the snail is exposed and cannot protect itself.

In the aquarium, as in the field, they were seldom seen during the day, coming out only at night to feed. If there was no seagrass or algae to eat, they would eat shrimp, showing evidence that they are omnivores. They possess a radula (rasp-like organ equipped with teeth), as well as grinding gizzard plates to process food. After feasting, they returned below the sand.

These bubble shells laid egg strings in the aquarium at night and I would assume they do the same in the field. They are quite common in the intracoastal waters near Peanut Island and Phil Foster Park, near Riviera Beach, Florida, so I was able to observe them for several years.

The snails are extremely timid and non-aggressive. They will not attack another animal. Feeding them by hand was not possible. Their eyes are deeply set in their head, a characteristic of snails that burrow in the sand. They exude immense amounts of mucus that sand particles adhere to, thus can look like part of the sandy environment. The colors of both the animal and the shells are shades of mottled black and brown. The colors blend well with the environment and the animal can go unnoticed while it grazes algae off the rocks.

28.3.13 *Haminoea elegans* (J. E. Gray, 1825) [Elegant Glassy Bubble].
In May 1984 while snorkeling in Lake Worth, Florida, I came across a colony of *H. elegans*. I collected eighteen to take home so I could observe their activities. I placed them in a tank by themselves. A few days later, I was surprised to see my one-gallon aquarium filled with white, ribboned egg masses. They had laid eggs in a communal fashion on the glass of the aquarium. They died shortly afterwards, although I had supplied them with algae.

28.3.14 *Stylocheilus longicauda* (Quoy and Gaimard, 1825) [Longtail Seahare].
These little seahares, averaging two inches in size, were found in the field with *Bursatella leachii pleii* Rang, 1828 [Ragged Seahare], and I originally thought them to be juvenile *Bursatella*. I took home a few, less than an inch (25 mm) in size, for study. When I looked at one under the microscope, I was surprised to find its true identity. This is a very attractive animal, with thin black stripes on a pale yellow background and iridescent round blue spots surrounded by an orange border. They ate algae and were easy to maintain in the aquarium. Thin brown eggstrings were laid on the glass of the aquarium and several times, I observed them eating their own eggs.

28.4 LITERATURE CITED

Jones, G. E. 1972. Table 62-Artificial seawater. *In:* P. L. Altman and D. S. Dittmer, eds., *Biology Data Book*, Vol. 1, 2nd Ed. Federation of American Societies for Experimental Biology, Bethesda, Maryland. Pp. 471-472.

Pressnell, J. K. and M. P. Schreibman. 1997. *Humanson's Animal Tissue Techniques*, 5th Ed. John Hopkins University Press, Baltimore, Maryland. xix + 572 pp.

Skomal, G. 1997. *Setting up a Saltwater Aquarium: An Owner's Guide to a Happy, Healthy Pet*. Howell Book House, Simon & Schuster Macmillan Company, New York. 126 pp.

Tullock, J. H. 1998. *Your First Marine Aquarium*. Barron's Educational Series, Inc. Hauppauge, New York. 79 pp.

Tunze, A. 1999. *Saltwater Aquarium*. Barron's Educational Series, Inc., Hauppauge, New York. 65 pp.

Winner, B. E. 1985. *A Field Guide to Molluscan Spawn*, Vol. I. E. B. M. Publications, North Palm Beach, Florida. 139 pp.

Winner, B. E. 1991. *A Field Guide to Molluscan Spawn*, Vol. II. E. B. M. Publications, North Palm Beach, Florida. 94 pp.

Winner, B. E. 1993. *Life Styles of the Seashells*. E. B. M. Publications, North Palm Beach, Florida. 61 pp.

CHAPTER 29

AN INTRODUCTION TO SHELL-FORMING MARINE ORGANISMS

LUCIA M. GUTIERREZ

29.1 INTRODUCTION

For any invertebrate zoologist, reading the zoological verses by the great zoologist Walter Garstang (1951) may prove to be a rich and profound experience. To me it was a point of departure for what would become a fascination with marine invertebrates, and especially mollusks. The field of invertebrate zoology is vast, and invertebrates play a major and vital role in most ecological systems. Even though undergraduate and graduate courses in invertebrate zoology have remained popular in colleges and universities, it appears that the availability of existing invertebrate taxonomists specializing either in a few taxa or in the fauna of a limited geographic area has continued to decline (Winston 1988, 1999). For many groups of other more obscure organisms, there simply may not be a systematist available.

The predominance of invertebrates in virtually every imaginable environment on Earth is certainly unquestionable. Most of the 250,000 marine species described so far are known from shallow depths (intertidal and oceanic continental shelf) and well-studied coasts (Barnes and Hughes 1982). It is reassuring to know that the exploration of new regions and habitats can still result in a large number of invertebrate species being discovered and described. Spectacular body patterns, peculiar shapes, and bright colors of most marine invertebrates provide an inexhaustible opportunity for specialization, with some collectors becoming taxonomic authorities on certain groups (Winston 1999). For centuries, naturalists, collectors, museum staff, university students, and children have enthusiastically and keenly catalogued invertebrate specimens found throughout the world. In spite of this, our knowledge of the taxonomy of marine organisms, and more so of marine invertebrates, is far from complete as most of them are hidden from our sight in an environment inhospitable to humans (Winston 1999).

Marine invertebrates have adapted to a variety of different living conditions; from rocky shorelines, flat pavements, sharp-edged rocks or cliffs, down to a sandy shore, flourishing coral reefs, or estuaries where mud mixes with sand. Invertebrates exhibit a remarkable diversity of form and habit. There is something special about each marine invertebrate animal and how it interacts with its habitat. Generally, with the exception of certain taxa (e.g., flatworms, nemerteans, sipunculans, and echiurans), the diversity of invertebrates is usually manifest in external morphology, not in the minute detail of internal systems.

Contrary to popular opinion, all invertebrates possess a skeleton system, even soft-bodied types. Aquatic or soil dwelling soft-bodied invertebrates may possess a type of skeleton referred to as a hydrostatic skeleton. Organisms with hydrostatic skeletons do not have solid supportive skeletons, but rather an incompressible fluid in which muscles exert pressure to serve functions in locomotion. Earthworms, for example, possess hydrostatic skeletons. Other types of skeletons are present in hard-bodied marine invertebrates, which may possess internal (endoskeleton) or external (exoskeleton) skeletons that exhibit a variety of morphologies and functions. Both types usually have organic and inorganic components. Invertebrate skeletons may be of the non-articulating type as in the complex rigid tests of sea urchins and sand dollars, or of the articulating type such as in the exoskeleton of arthropods. In mollusks, the hard outer shells can be articulating either as in bivalves or a rigid one-piece shell as in gastropods. In the broadest sense, environmental pressures have greatly shaped all aspects of molluscan shell formation from overall appearance to the smallest details of strength, structure, and color. The calcareous molluscan shell represents one very widespread way of building an external skeleton (Vermeij 1993). Considering that calcium carbonate, or limestone, is a good material for a skeleton due to its resistance to corrosion, it is not surprising that similar structures are found in marine invertebrates belonging to different phyla.

The present chapter is intended as an introduction to those marine invertebrates that shell collectors are most likely to encounter during shell collecting trips and may, consequently, be misidentified as being mollusks or molluscan in origin. The majority of the external morphological descriptions included below are based on my own observations of mostly live specimens or of specimens from my private collection from the west and east coasts of North America, the Hawaiian Islands, and the Northwestern Hawaiian Islands. I include descriptions of calcareous algae, foraminiferans (mineralized protozoa), coralline sponges, stony and soft corals, tube-dwelling polychaetes, lampshells (brachiopods), barnacles, mud shrimps, sea urchins, and other echinoderms as they all possess structures composed partly or fully of calcium carbonate. I also include crustaceans and other arthropods, even though a completely different form of exoskeleton characterizes them.

This chapter is certainly not conceived to be a comprehensive manual that addresses nearly all taxa. I recommend Brusca and Brusca (1990) for more in-depth factual information on (1) functional body architecture, (2) gross internal anatomy, (3) developmental patterns and life history strategies as they relate to adult lifestyles, and (4) evolution and phylogenetic relationships. The reader should consult a general invertebrate zoology textbook to become familiar with the taxonomic hierarchy of the animals presented in this chapter. Some may consider invertebrate zoology textbooks and laboratory manuals too detailed in their taxonomic descriptions, but on a positive note, such books emphasize the diversity of form and function among invertebrates and often acknowledge the evolution of these groups. One such book is by Buchsbaum et al. (1987), *Animals Without Backbones*.

29.2. TAXONOMIC GROUPS

29.2.1 Protozoa. Under certain classification schemes, protozoans are no longer considered animals and reside in a different kingdom, the Protista. In this group, mineralized microorganisms, known as Foraminifera (forams), are a large and ecologically important group that typically has shells (tests) of organic, agglutinate, or rarely siliceous material (Figure 29.1). The test is usually microscopic and has a variety of shapes. Some forams may have a series of chambers of increasing size, with a main aperture in the largest chamber. Shells of certain planktonic forams occur in such high numbers that the tests of dead individuals sinking to the bottom provide a major and important contribution to marine sediments. Large stretches of the ocean floor are covered by foraminiferan ooze. Such sediments are restricted to depths shallower than about 3,000 to 4,000 m, as calcium carbonate dissolves under high pressure.

Shell enthusiasts are turning to collecting sand samples to search for the dead shells of new and often spectacular species. Collectors of micromollusks sorting through sand with the aid of a dissecting microscope ought to be aware of forams, as they can at times constitute a large component in marine sediments. Most forams with calcareous multi-chambered shells range from 5 to 300 µm and are easily identified. Some forams are visible to the naked eye. Many forams, including benthic forms, have coiling tests that could at times be mistaken for an odd-looking snail.

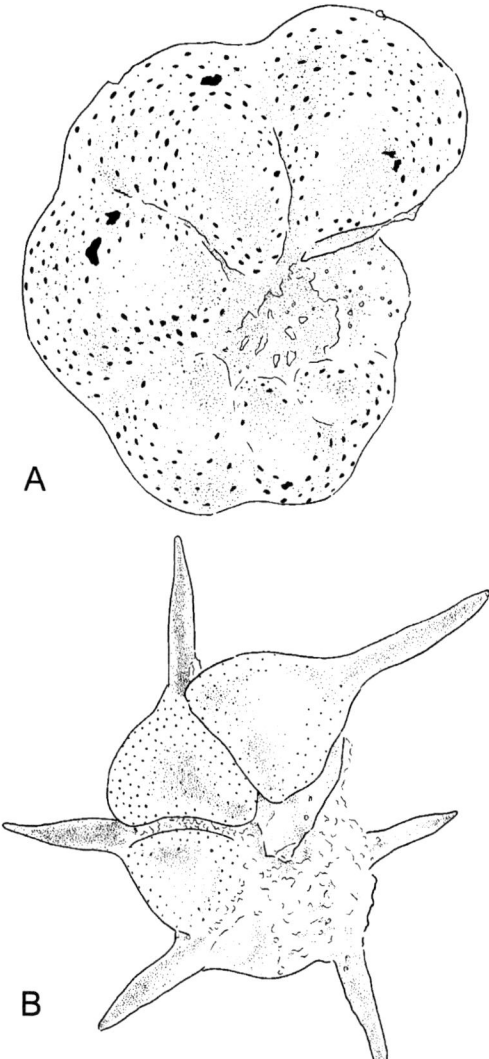

Figure 29.1 Protozoa.
A. *Trochammina* sp. (Protozoa: Foraminifera), 0.3 mm.
B. *Hantkenina* sp. (Protozoa: Foraminifera), 0.5 mm.

In some parts of the world, foraminiferan oozes of some species are many meters thick and have been mined as gravel used in walkways and roads, as in the Island of Bali. Most of the stones used to build the great pyramids of Egypt are foraminiferan in origin (Murray 1973). Some of the world's chalk, limestone, and marble are composed largely of calcareous foraminiferan tests. In the tropics, the foraminiferan *Homotrema rubrum* (Lamarck, 1816) forms bright, red, and round calcareous growths several millimeters in diameter at the base of corals. *Homotrema rubrum* is so common in Bermuda that their skeletons are responsible for the island's famous pink beaches. An excellent picture of this foram can be found on the Internet at <www.reefkeeping.com/issues/2003-07/rs/index.php>.

29.2.2 Calcareous algae. Some marine species of calcareous (coralline) algae deposit calcium carbonate in their tissues as they grow. The coralline green alga *Halimeda* (Figure 29.2) deposits calcium carbonate within a thin layer of live green tissue on the outside. The body (thallus) consists of numerous segments and when the tissue dies the segments separate. Calcareous accumulation in the body discourages grazers such as fish, sea urchins, and others from eating it. The remnants of *Halimeda* are important to the formation of coral reefs, as they accumulate as white small disks (about 4 mm diameter), spherical, or tear-shaped beads, and are often encrusted by other organisms such as sponges and sea mosses (also called bryozoans). Encrusting coralline red algae

Figure 29.2 Alga.
Thallus and root system of the calcareous green alga *Halimeda* sp. (Chlorophyta: Halimedaceae), 100 mm.

[e.g., *Porolithon gardinieri* (Foslie) Foslie, 1909, *Lithothamnion* spp., *Hydrolithon* spp.] grow on hard substrates such as dead coral fragments, and are known to deposit considerable amounts of calcium carbonate. They usually grow over sediments as it builds up on dead corals, cementing the sediment in place, or form loose nodules that roll with the surf. Live coralline algae never have polyps like stony corals do. Dead encrusting algae often resembles coral rubble with no visible holes of any kind, and has a smooth appearance. Due to this smooth appearance and its calcium carbonate look, it would be easy to mistake any piece of rubble with encrusting coralline alga as being molluscan in origin as the algae would be devoid of pigmentation and appear shelly.

29.2.3 Porifera. Porifera (sponges), which are mostly marine, live as filter feeders attached to hard substrates, and can adopt a variety of shapes and sizes due to the structural support of transparent siliceous or opaque calcareous spicules. Shapes can vary from round masses, to branching, and to tubular forms that may reach large sizes. Sponges are often referred to as fouling organisms. Sponges can encrust rocks or dead corals or create thin channels through calcium carbonate of corals or snail shells. For example, *Clione limacine* (Phillips, 1774), a species of boring sponge, bore into coral or snail shells by making holes about 1 mm diameter in shells. Other sponges like the Venus'-flower-basket [*Euplectella speciosa* (Quoy and Gaimard, 1833)] live anchored in deep-water sediments and are characterized by fused glass spicules forming a lace-like skeleton. For coralline sponges, or sclerosponges, a calcium carbonate skeleton forms beneath the body of the sponge, which contains siliceous spicules and spongin [e.g., *Ceratoporella nicholsoni* (Hickson, 1911)]. They are inhabitants of overhangs and occur in caves in tropical reef areas and can be distinguished from corals by the smoother appearance of the external surfaces and the presence of many holes (Figure 29.3). These holes allow the passage of seawater through the live sponge.

29.2.4 Cnidaria. The phylum Cnidaria comprises jellyfishes, sea anemones, stony corals, precious corals, and sea fans. The basic cnidarian life cycle alternates between two adult body forms, the polyp (Figure 29.4A) and the medusa. Polyps are sessile while medusae are free-swimming. Many cnidarians such as anemones, stony corals, and soft corals, have lost the medusa stage and exist only as polyps. Solitary and colonial stony corals secrete skeletons of calcareous material, the latter forming coral reefs in shallow tropical seas with the aid of symbiotic algae (zooxanthellae). On the contrary, the hard skeleton of precious corals and sea fans are made mostly of a flexible horny protein and contribute little to reef formation. Corals sold as souvenirs in aquarium and shell shops are in fact the skeletons of live corals whose living polyps have been removed. Jellyfishes and allies still retain both life cycle forms, and do not ordinarily form calcareous structures.

Sea anemones are solitary polyps that usually do not secrete skeletons; however, some spe-

Figure 29.3 Porifera.
This specimen is a figure of an unidentified member of Porifera: Calcarea, the calcareous sponges. Average size of calcareous sponges is 4-10 cm.

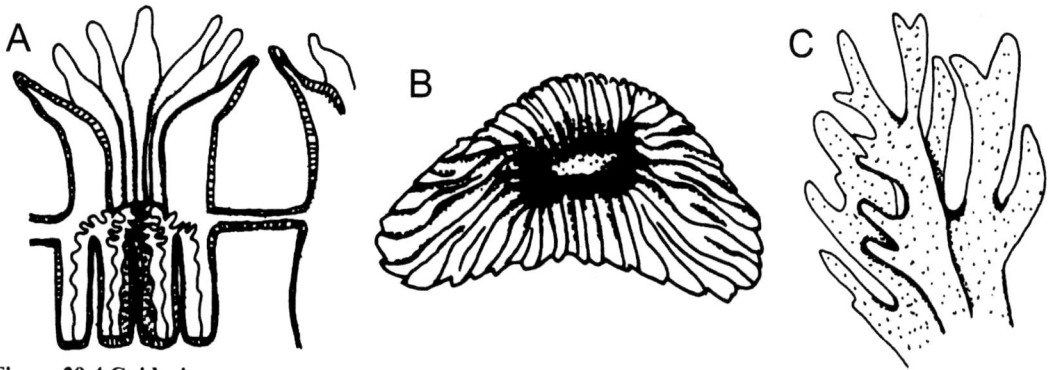

Figure 29.4 Cnidaria.
A. Cross section of an individual coral polyp belonging to a reef-forming species, showing feeding tentacles (Cnidaria: Anthrozoa), approximately 0.5 mm. B. Dead calcareous skeleton of *Fungia* sp., a non-reef forming solitary coral (Cnidaria: Anthrozoa), 70 mm in length. C. The branching calcareous skeleton of a hypothetical hydrocoral (Cnidaria: Hydrozoa), 75 mm.

cies are known to produce calcareous structures (Dunn *et al.* 1980). Cloak anemones of the genus *Adamsia* (e.g., *Adamsia sociabilis* Verrill, 1882) exhibit an extreme case of mutualism that results in the formation of a flexible anemone shell that can be mistakenly described and classified as flexible gastropod shells. Initially, cloak anemones wrap themselves around small gastropod shells occupied by hermit crabs and the pedal disc secretes a chitinous cuticle over the shell, often dissolving the original gastropod shell over time. The anemone simply grows to provide the hermit crab with a living protective cnidarian shell (carcinoecium), that is flexible and coiled. The hydrozoan colony *Janaria mirabilis* Stechow, 1921 secretes a shell-like casing inhabited by hermit crabs that resembles the structure formed by *Hippoporidra calcarea* (Smitt, 1873) (a bryozoan). These 'odd' shells can be mistakenly described and classified as gastropod shells.

Stony corals are anemone-like polyps that are distributed in all tropical and sub-tropical oceans. In stony corals, the colony of polyps forms delicate to massive calcareous (aragonite) exoskeletons perforated by pores of one size. They can be recognized not only by their minute monomorphic (one type) individual cups with raised rims but also by their internal spaces that are divided radially by thin ridges. Dried pieces of coral skeletons are often found on the beach shore.

Fire corals, which are not true corals, may also form large calcareous exoskeletons that are coral-like, distinguishable from stony corals by having skeletons covered with pinholes and lacking individual cups and raised rims. The common name fire coral is a tribute to their very strong stinging cells when alive. Two kinds of fire corals are commonly known, the milleporine and stylasterine hydrocorals (Figure 29.4C).

Like true corals, millepore hydrocorals (e.g., *Millepora alcicornis* Linnaeus, 1758) may assume a great variety of shapes from massive to stony encrustations, but usually form a flattened, erect, or laterally extended colony on top of reefs. The milleporinid exoskeletons are mustard-yellow to brown with white tips or edges. They have a characteristic pattern of small perforations of two sizes, large holes that house the feeding polyps, surrounded by a circle of smaller holes that house defensive polyps. Occasionally, the surface of the millepore skeleton has rounded bumps covered by lacy openings. These lacy openings overlie pits called ampullae, in which the sexual individuals of the colony reside.

Stylasterine hydrocorals also form erect or encrusting calcareous skeletons that are often brightly colored (purple, red, or yellow). A definite pattern of small holes of two sizes is visible. Usually there is a single star-shaped pit that has one or more prominent projections towards the edges. Each of the points of the star is also a pit, but shallower than the center, and may or may not contain projections called dactylostyles. Lacy openings are scattered on the skeleton and are usually covered by a prominent hemispherical roof.

Two additional examples of soft corals that produce calcium carbonate skeletons are the blue corals and pipe organ corals. Blue corals include the single genus *Heliopora* and usually form calcareous skeletons from encrusting to branching forms in coral reefs in the Pacific and Indian Oceans. Two distinguishing characteristics of *Heliopora* are that the surface of the calcareous skeleton is perforated by hollow cylinders of two sizes, and that iron salts are responsible for the amazing blue color of this coral. This coral also lacks individual cups and raised rims. Pipe organ corals are also found in the Indo-Pacific [e.g., *Tubipora musica* (Linnaeus, 1758)], as well as in the Atlantic and Pacific coasts [e.g., *Clavularia modesta* (Verrill, 1874)]. *Tubipora musica* has large green polyps and a red skeleton. The skeleton forms an encrusting sheet or network of vertical hollow tubes periodically fused together by horizontal plates. The polyps are located in the hollow vertical tubes, which may resemble the entwining network of vermetid snails.

This is a good section to mention scleractinian corals like the unusual genus *Fungia* (Figure 29.4B) and other solitary corals (*Cycloseris* spp.). These corals are non-reef forming (i.e., ahermatypic) and are often found leading solitary existences, and like another hermatypic corals, utilize the energy produced by symbiotic algae. The majority of fungiids are solitary corals, but colonies of individuals are also found. In this family, which is distributed from the west coast of Central America through the whole Indo-Pacific to the Red Sea and as far south as South Africa, the calcareous skeleton consists of elongated septa running from the center of the polyp towards the edge. On the upper side, the skeleton has septal teeth and on the lower side, it has costal spines. The calcareous skeleton of these corals could be mistaken for being molluscan in origin.

29.2.5 Bryozoa. Bryozoa, or sea mosses, are colonial invertebrates that secrete delicate skeletons of a variety of shapes on seaweed, rocks, shells, and other surfaces (Figure 29.5). The colonies may be encrusting or take the form of tufts of crusty lace. Other bryozoans may take more unconventional shapes. An unusual bryozoan is the Texas Longhorn, the skeletal structure built by a colony of bryozoans of the genus *Hippoporidra*. This skeleton looks like two straight long horns pointing in opposite directions that arise from a coiling spiral in the middle. In the middle, this curiosity of nature houses a small hermit crab [*Pylopagurus corallinus* (Benedict, 1892)]. *Hippoporidra* sp. has a granulated surface with erected little cups, with round openings, that are found periodically on the skeleton. Specimens have been found at 64 m deep in the North Atlantic Ocean, off the east coast of Florida. Similar structures with additional ornamentation are secreted by *Hippoporidra edax* (Busk, 1859). For pictures of these amazing skele-

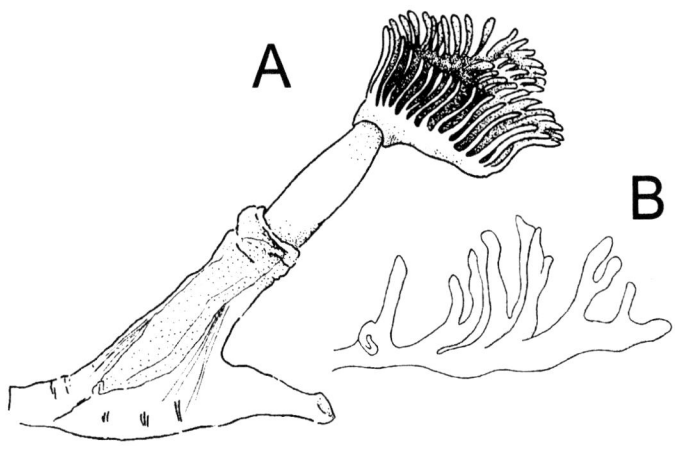

Figure 29.5 Bryozoa.
A. Bryozoan polyp, approximately 0.5-1 mm. B. Bryozoan colony, approximately 10 cm.

29.2.6 Annelida. Almost all segmented marine worms (annelids, phylum Annelida) are polychaetes and occupy a great variety of habitats. In comparison to other annelids, polychaetes have high species diversity and many different body morphologies. Some polychaetes build and live in a variety of tubes. Two families of polychaetes, the serpulids [e.g., *Serpula vermicularis* Linnaeus, 1767 and *Spirobranchus giganteus* (Pallas, 1766)] and spirorbids (e.g., *Spirorbis bifurcates* Knight-Jones 1978)] produce calcium carbonate tubes (Figure 29.6), which are cemented on rocks and other hard surfaces. Polychaete tubes are secreted by a pair of large glands near the base of the worm's crown. These tubes are sometimes misidentified as tubes belonging to small vermetid snails.

In general, tubes from vermetid snails are calcareous, three-layered tubes, with the inner layer being glossy. There may be a spiral protoconch present. Often there is a spiral (appears longitudi-

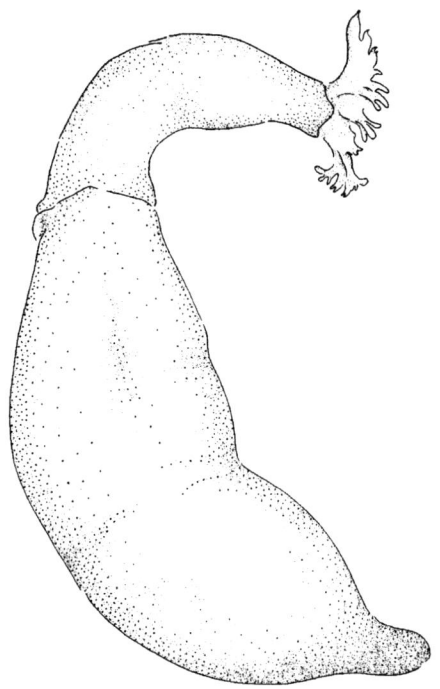

Figure 29.7 Sipuncula.
A worm from the Phylum Sipuncula, commonly known as peanut worms. They range in size from 10-60 mm.

nal) or a cancellate pattern to the shell's surface. Polychaetes build two-layered tubes of unadorned mud, sand, or parchment (often decorated with sand, shell fragments, algae, or hydroids) or of hardened calcium carbonate. A live vermetid snail will often have an operculum at the opening of the tube, whereas a tube-dwelling polychaete may have a funnel shaped operculum that arises from a modified radiole (Keen 1961, ten Hove 1994).

As tube dwelling polychaetes, both serpulids and spirorbids bear a tentacular crown, as in feather duster worms, in addition to a funnel-shaped operculum that can be pulled into the end of the small calcareous tube when the worm withdraws. Serpulids form moderately straight tubes whereas spirorbids form tubes that coil to the right (dextral or right-handed) or left (sinistral or left-handed). In Hawaii, sea frost that occurs in tide pools, deeper reefs, and on harbor pilings is actually masses of white intertwining tubes of the serpulid *Salmacina dysteri* (Huxley, 1855) (Hoover 1998).

Figure 29.6 Annelida.
A live spirorbid polychaete (Annelida: Polychaeta) showing a dextrally coiled calcareous tube, feeding crown, and, operculum that closes the shell. The spiral shell is approximately 4 mm in diameter.

29.2.7 Sipuncula. Sipuncula, which are exclusively marine, are often called peanut worms. Peanut worms commonly bury their soft, nonsegmented bodies in muddy sediments, hard calcareous substrata, or drifting wood. Some sipunculids can also inhabit abandoned gastropod shells and polychaete tubes. Any amateur shell collector ought to be aware of the existence of peanut worms as they could be mistaken for a shelled mollusk if residing inside an empty gastropod micro-shell or for an aplacophoran (non-shelled mollusk, see Chapter 16). However, peanut worms have a sausage-shaped body that is sometimes covered by a well-developed cuticle often bearing papillae, warts, or spines (Figure 29.7). Peanut worms also have an elongated tentacle (introvert) that can be retracted into the remaining portion of the body. It is when the introvert is retracted and the body is turgid that some resemble a peanut, hence the vernacular name.

29.2.8 Arthropoda. The spectacular radiation of arthropods (phylum Arthropoda) is recognized in the huge variety of animals that has invaded all types of environments. Crustaceans, insects, and the extinct trilobites are arthropods. A characteristic of arthropods is that they have an external, jointed, and rigid body covering composed of a cuticle made of chitin and protein produced by the epidermis. The exoskeleton provides support and maintenance of body shape besides acting as a protective covering against physical injury and physiological stress. The hard cuticle, which is inflexible except at the joints, imposes limitations on growth and size. As they grow, arthropods periodically shed or molt the old exoskeleton and deposit a new larger one. Molting is regulated by hormones and occurs in such a way that the old cuticle splits and the animal can wriggle free and pull itself out.

Horseshoe crabs (Figure 29.8A) are regarded as living fossils and are widely represented in the fossil record. Only four species of modern horseshoe crabs are known, one in the Gulf of Mexico and along the Atlantic coasts [*Limulus polyphemus* (Linnaeus, 1758)], while the other three occur in Southeast Asia [*Tachypleus gigas* (Müller, 1785), *T. tridentatus* (Leach, 1819), and *Carcinoscorpius rotundicauda* (Latreille, 1802)]. All four species are similar in terms of ecology, morphology, and serology. Horseshoe crabs are not true crabs. They inhabit shallow marine waters, generally on sandy bottoms where they crawl or burrow just beneath the surface, preying on clams and other small soft-bottom invertebrates. Horseshoe crabs do not have a discrete head. The carapace (head and thorax) is horseshoe-shaped and is fused to the abdomen at the end of which there is a large spike-like tail (telson).

Members of the class Crustacea (e.g., barnacles, mussel shrimps, beach hoppers, fish lice, fairy shrimps, krill, shrimps, crabs, and lobsters) are morphologically diverse and have adapted for life in water, with most species living in the ocean. Crustaceans possess appendages specialized for feeding, crawling, and swimming. The chitinous skeleton of crustaceans is further mineralized by the deposition of calcium carbonate in the outer layers of the cuticle. The joints are very flexible and the cuticle at the joints is much reduced and unhardened.

For many years in the early 1800s, barnacles were classified with mollusks, because heavy calcareous shell plates usually enclose their bodies. Louis Agassiz (born 1807-died 1873) described a barnacle as "a little shrimp-like animal standing on its head in a limestone house and kicking food into its mouth" (in Hoover 1998). Barnacles are abundant suspension feeders that usually attach to hard substrates in shallow marine waters around the world. Some have specifically adapted to live attached to floating objects such as seaweeds, driftwood, and nektonic marine animals (e.g., whales and sea turtles), while others are often found on shells and exoskeletons of various errant invertebrates (e.g., crabs and gastropods). Barnacles can be grouped as gooseneck and acorn barnacles, parasitic, boring, benthic crawlers, burrowers, or planktonic. This chapter discusses goose and acorn barnacles only as they are most common and are most often encountered by mollusk collectors out in the field. Other types of barnacles, which are less ubiquitous and less likely to be encountered by the amateur shell collector, are not covered in this chapter.

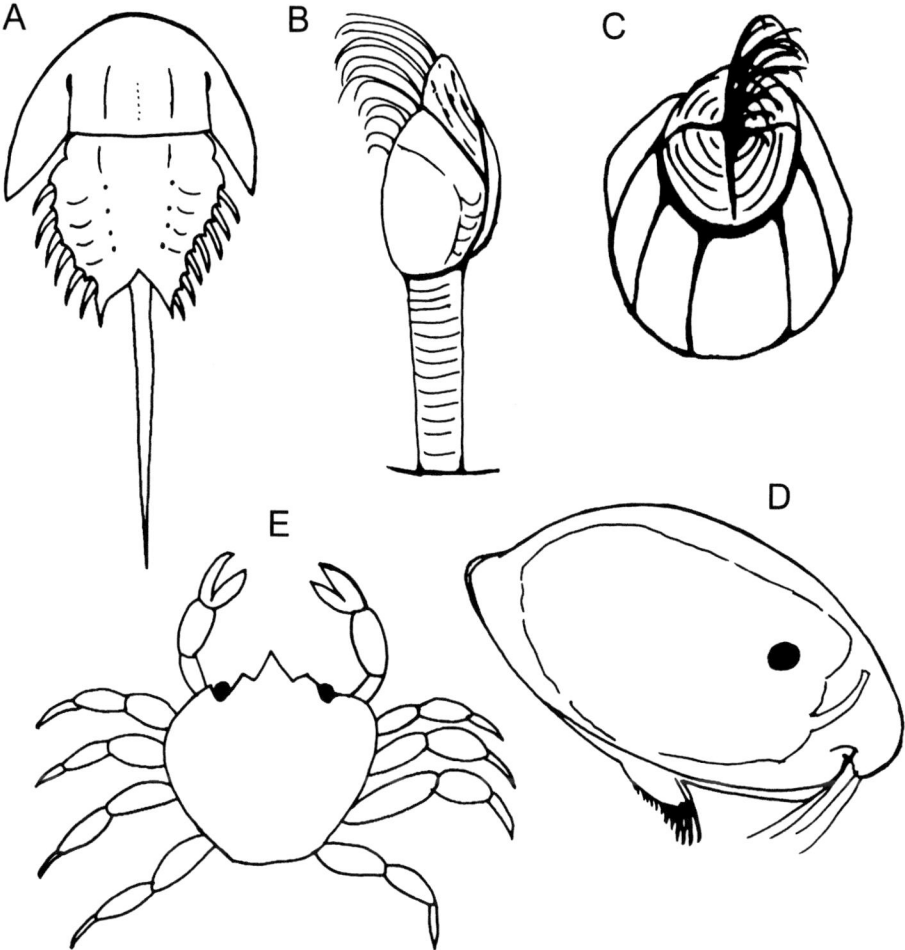

Figure 29.8 Arthropoda.
A. Dorsal view of *Limulus*, a horseshoe crab (Arthropoda: Merostomata) common in the Gulf of Mexico, 50 cm in length. B. Side view of a live gooseneck barnacle of the genus *Lepas* (Arthropoda: Maxillopoda) showing feeding tentacles, bivalved plates and a fleshy stalk called the peduncle, 100 mm. C. Dorsal view of a live acorn barnacle (Arthropoda: Maxillopoda) showing complete skeletal plates and feeding tentacles, 25 mm. D. Side view of an ostracod (Arthropoda: Ostracoda), a bivalved crustacean, 0.1-3 mm. E. Dorsal view of the molt of a decapod crab (Arthropoda: Malacostraca), 75 mm.

In gooseneck barnacles (Figure 29.8B), the body is divided into two regions, one covered by calcium carbonate valves and the other a fleshy stalk called the peduncle. Both regions can vary among species. For example, in *Pollicipes pollicipes* (Gmelin, 1790) and in *Pollicipes polymerus* Sowerby, 1833 the peduncle is often covered with a number of small calcareous scales, and in the goose barnacle *Lepas anserifera* Linnaeus, 1767, the stalk may reach up to 7-10 cm long. Stalked barnacles often drift the seas worldwide on the underside of ships, and on flotsam and jetsam.

Conversely, acorn barnacles (Figure 29.8C) are completely enclosed by a rigid calcareous shell on top of which are four shell plates that open and close to allow the feathery filtering appendages to feed. The rigid shell is composed of a series of calcium carbonate plates strongly appressed to each other, appearing fused. The plates can be smooth, slightly ribbed vertically, and colorfully marked. Some species of barnacles that attach to the skin and teeth of various groups of whales are host specific, attaching only to whales; other barnacles are more versatile and are also found attached to sea turtles. For example, *Cryptolepas rhachianecti*, Dall, 1872 occurs exclusively on the gray whale whereas *Coronula diadema* (Linnaeus, 1787) occurs on several species of whales. Turtle barnacles of the genus *Chelonibia* [e.g., *Chelonibia testudinaria* (Linnaeus, 1787), *Chelonibia caretta* (Spengler, 1790)] usually attach to many animals, but in Hawaii, it specializes on the Green Turtle *Chelonia mydas* (Linnaeus, 1758).

Seed shrimp or mussel shrimp are small crustaceans also known as ostracods (Ostracoda). They range in size from 0.1 to 3 mm in length and their exoskeletons resemble the shells of small bivalve mollusks (Figure 29.8D). Mussel shrimps are common benthic inhabitants of freshwater and marine habitats. Two valves that are hinged dorsally enclose the animal, and are variously hardened with deposited calcium carbonate. The body lacks segmentation and has a maximum of two appendages behind the head. Depending on the habitat, the outer surfaces of the valves may be smooth or variously sculpted. The extreme delicate shells of mussel shrimps are easily recognized by the lack of concentric growth rings found in bivalve mollusks.

In vernacular terms, nearly every decapod crustacean may be recognized as some sort of crab, hermit crab, shrimp, or lobster. Decapods (Decapoda) feature five pairs of walking legs, and the first pair is modified into pincers for obtaining food (Figure 29.8E). Most decapods are known to be scavengers. Old decapod exoskeletons that covered the entire external surface of the crab, even its eyes and mouthparts, are discarded after molting and are frequently found on shorelines. In some male crabs, the abdomen is visible as a flat V-shaped plate but in females, it is expanded and U-shaped.

Hermit crabs are abundant members of the intertidal fauna around the world. Most hermit crabs usually hide their long, soft, and asymmetrical abdomen in empty gastropod shells of all sizes or other empty houses not of their own making. These little crabs coil their tails inside the empty shells and hold on from within with their specialized last pair of legs. Hermit crabs rapidly retreat into the shell when threatened or disturbed. Some cover their borrowed shells with sea anemones or sponges. The coconut crab of the tropical Pacific and Indian oceans (*Birgus latro* Linnaeus, 1767) is the largest land-dwelling hermit crab. Adults do not use shells because their abdomen is leathery and tough, only small young coconut crabs inhabit shells.

29.2.9 Brachiopoda. At first sight, lampshells (phylum Brachiopoda) may be easily confused with bivalve mollusks, and at one point, brachio-

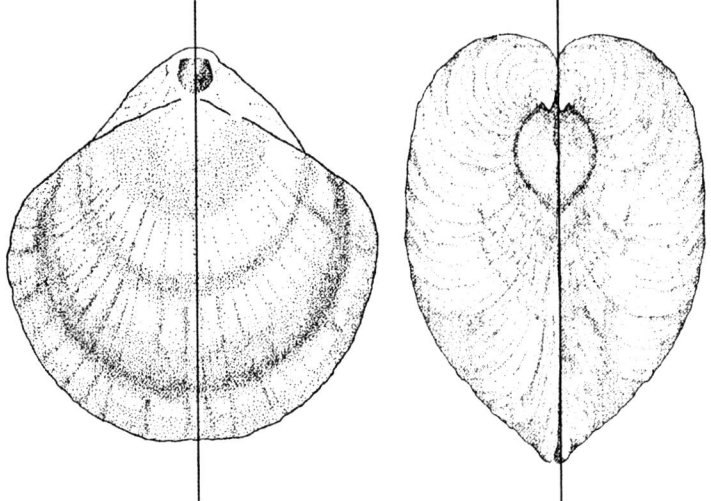

Figure 29.9 Brachiopoda - Bivalvia Symmetry.
Brachiopods (image on left) show bilateral symmetry along a dorsal-ventral plane that divides the shell into right and left mirror images as seen in *Kraussina rubra* (Pallas, 1776) [Brachiopoda: Articulata], 27mm dorsal-ventral length. Bivalvia (image on right) generally show bilateral symmetry between the two valves as seen in *Venus verrucosa* Linnaeus, 1758 [Mollusca, Bivalvia], 40 mm width.

pods were allied with mollusks. This view was held until late in the nineteenth century. There are approximately 335 living marine species, with most adult animals measuring from 4 to 6 cm. They thrive on practically every continental shelf and have been recorded from nearly all oceanic depths. Lampshells are solitary animals that inhabit burrows in loose sand or are directly cemented to the substratum by a fleshy pedicle. The pedicle is an outgrowth of the body wall arising from the posterior area of the ventral or larger valve (usually uppermost in the living animal). Brachiopods resemble bivalve mollusks because the body is enclosed within a pair of external bivalve shells. The shell contains the visceral organs and an extensive crown of hollow, ciliated tentacles surrounding the mouth (the lophophore). Brachiopod shells are oriented dorso-ventrally, and frequently bear perforations and spines, while bivalves have lateral valves.

An important difference between bivalves and brachiopods is that the shells in bivalves are typically hinged together dorsally by an elastic ligament and shell-teeth. The bivalve body is enclosed by two valves of similar size, between one valve oriented to the right and the other valve oriented to the left thus showing bilateral symmetry. The brachiopod body is enclosed between a pair of dorso-ventrally oriented valves, which are usually unequal in size and can be attached to one another posteriorly by a tooth-and-socket hinge (articulate lampshells, Figure 29.9) or simply by muscles (inarticulate lampshells, Figure 29.10).

The majority of brachiopods in modern seas belong to a group known as articulate lampshells [e.g., *Kraussina rubra* (Pallas, 1776), Figure 29.9]. Their shells are composed of scleroprotein and calcium carbonate. The two valves are unequal in size and posteriorly attached to one another by a tooth and socket hinge. Inarticulate lampshells (e.g., *Glottidia* and *Lingula*, Figure 29.10) have valves of organic composition that may consist of interleaved layers of chitin and calcium phosphate or low magnesium calcium carbonate. The anterior body region consists of two shells of equal size strongly held together by muscles (unhinged), and the posterior body consists of the long, fleshy pedicle. Inarticulate brachiopods are common in some near-shore habitats of the Gulf of Mexico.

29.2.10 Echinodermata. Sea urchins, sand dollars, sea biscuits, heart urchins, sea stars, and sea cucumbers are all grouped as the phylum Echinodermata (Figure 29.11). Echinoderms are widely distributed in all oceans at all depths and display many modifications of the echinoderm body plan. It is in urchins and sand dollars (class Echinoidea) that the internal skeleton forms a rigid test composed of crystals of calcium carbonate with the individual plates tightly bound to each other in most species. The empty tests of these animals are seen more often than the living animals.

Sea urchins (Figure 29.11A) graze and live on the sea bottom or solid surfaces, and move by means of tube feet and movable spines. The solid tests of sea urchins are composed primarily of calcium carbonate in the form of calcite, with trace amounts of magnesium carbonate. Sea urchin tests may be globose, or round to ovoid. A live sea urchin has hundreds of tube feet and movable spines that can usually be long to short, slender sharp spines (e.g., the Long-Spined urchin *Diadema paucispina* Agassiz, 1863), club-like [e.g., the Red Pencil urchin *Heterocentrotus mammillatus*

Figure 29.10 Brachiopoda Side view of the inarticulate brachiopod *Lingula* sp. showing calcareous valves and a fleshy pedicle. (Brachiopoda: Inarticulata). The valves are 25 mm long.

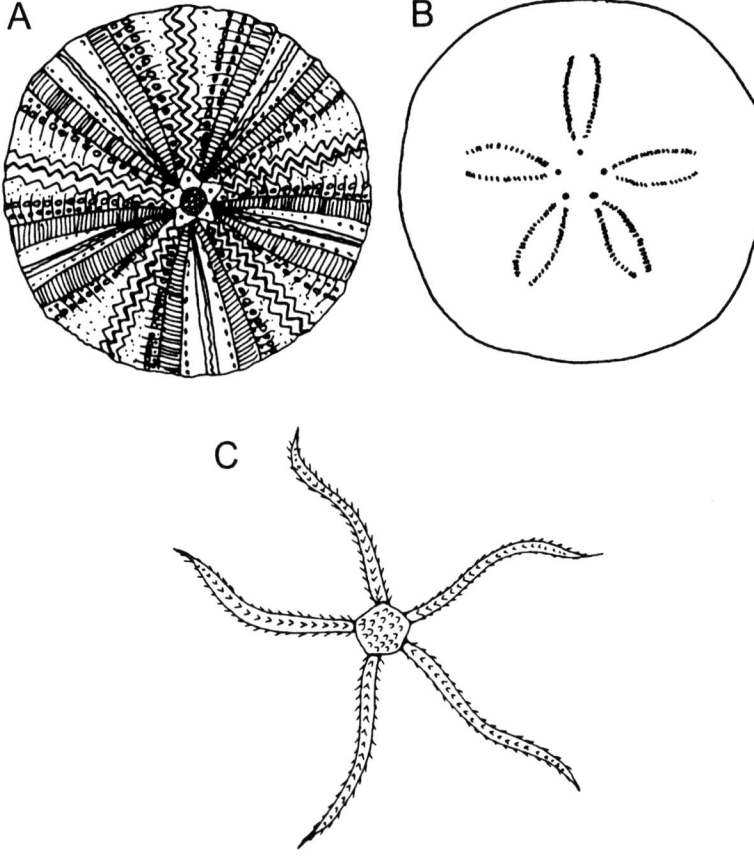

Figure 29.11 Echinodermata
A. Aboral view of a sea urchin's test with spines and tissue removed showing 20 individual rows of plates fused into a solid piece. (Echinodermata: Echinoidea), 7.5 cm. B. Aboral view of the dead skeleton of a sand dollar with the 5 distinguishing petaloid marks. (Echinodermata: Echinoidea), 10 cm. C. Dorsal view of a brittle star (Echinodermata: Ophiuroidea), 15 cm.

below the surface, or partially burrowing by keeping the posterior part of the body extended at an angle above the surface. Heart urchins, sea biscuits, and sand dollars do not have round tests as sea urchins do, and the movable spines are usually short and hair-like. These echinoderms have adapted to live on the bottom by having flattened bodies. The ovoid or somewhat elongated tests of heart urchins typically have a sunken portion on the frontal part and somewhat resemble the shape of a heart. Sand dollars (Figure 29.11B) burrow and crawl in clean sands. Their flat tests are thin and have a little hole near the center of the body on the bottom (oral surface). In heart urchins and sand dollars, the upper surface of the tests displays five beautiful petaloid marks with double holes forming a distinctive pattern. Specialized tube feet protruded through these pores when the animal was alive. Sand dollars are conspicuous animals on sandy bottoms along the southern United States and the Caribbean [e.g., *Mellita quinquiesperforata* (Leske, 1778)].

(Linnaeus, 1758), or massive blunt ones (e.g., the Ten-Lined urchin *Eucidaris metularia* (Lamarck, 1816)]. The test is composed of fused plates that can be easily seen if the urchin is cleaned of spines and tissue. Twenty vertical rows that are symmetrical are arranged around a top-bottom (i.e., oral-aboral) axis. In each test are two types of alternating plates, those with two rows of tiny double holes on each side and those without. All plates have numerous bumps of different sizes where the spines articulate on ball and socket joints.

Other echinoderms inhabit soft sediments, either living on the surface, completely burrowing well

29.3 FLOTSAM AND JETSAM

Sometimes you will be collecting and come across items that may resemble mollusks but are not mollusks. Above were delineated several groups of invertebrates that might cause confusion. In addition, there are several other items that bear mentioning.

The first items are pieces of plastic, especially Styrofoam packing particles. Some of these superficially resemble the valves of a bivalve mollusk. Others may resemble gastropods. If they have been floating around for a while, they may be covered with other marine growth and this camouflage may make them more difficult to identify. Encrusted pieces of plastic straw and tubing might be mistaken for vermetid gastropods.

Another item is the shell of nuts such as the pistachio. These may be mistaken for bivalves. Sea beans, floating nuts of several different types of plants, may also be mistaken for bivalves.

Sometimes fossils will wash up on shore and some of these may be mistaken for mollusks. Some people believe that the ear bone from cetaceans superficially resembles a cowry shell of the genus *Cypraea*.

If you find something that is strange and cannot be identified, it is often wise to carry it out and determine what it is later.

29.4 CONCLUSIONS

Marine invertebrate animals exist in enormous variety and in habitats such as sand and rubble bottoms, tide pools, brackish pools, mangroves, and shallow reef flats. To an invertebrate zoologist, more often than not, discovering the dead skeleton of a marine invertebrate or a shell of a gastropod in any of such habitats may be a noteworthy experience. For the amateur shell collector, it may mean an unfamiliar shell suitable to add to his/her collection or, perhaps, the misidentification of a calcareous structure. I hope that this introduction of marine invertebrates that produce calcareous skeletal components will be both instructive and enjoyable. The above descriptions should give you a sense of which other organisms produce calcareous structures.

29.5 ACKNOWLEDGMENTS

I extend special thanks to Jennifer Dreyer and Amanda E. Zimmerman for their outstanding artwork. Ms. Dreyer completed Figures 29.2, 4, 6, 8, 10, and 11 while Ms. Zimmerman drew Figures 29.1, 3, 5, 7, and 9. I thank two anonymous reviewers for their invaluable comments on this manuscript and Dr. Thomas Dix for lending me bryozoan specimens. I am grateful to Dr. Charles Sturm for providing comments, advice, and patience.

29.6 LITERATURE CITED

Barnes, R. S. K. and R. N. Hughes. 1982. *An Introduction to Marine Ecology*. Blackwell, London. 351 pp.

Brusca, R. C. and G. J. Brusca. 1990. *Invertebrates*. Sinauer Associates, Inc., Sunderland. 922 pp.

Buchsbaum, R., M. Buchsbaum, J. Pearse, and V. Pearse. 1987. *Animals Without Backbones*, 3rd Ed. University of Chicago Press, Chicago. 527 pp.

Dunn, D. F., D. M. Devaney, and B. Roth. 1980. *Stylobates*: A shell forming sea anemone (Coelenterata, Anthozoa, Actiniaria). *Pacific Science* **34**: 379-388.

Garstang, W. 1951. *Larval Forms and Other Zoological Verses*. Basil Blackwell, Oxford. 98 pp. [Reprinted in 1985 by the University of Chicago Press].

Hoover, J. P. 1998. *Hawai'i's Sea Creatures: A Guide to Hawai'i's Marine Invertebrates*. Mutual Publishing, Honolulu. 366 pp.

Keen, A. M. 1961. A proposed reclassification of the gastropod family Vermetidae. *Bulletin of the British Museum (Natural History) Zoology* **7**: 183-213.

Murray, J. W. 1973. *Distribution and Ecology of Living Benthic Foraminiferids*. Crane, Rusak, and Company. New York. 274 pp.

ten Hove, H. A. 1994. The dualistic relation between molluscs and serpulid tube-worms. In: M, Coomans-Eustatia, R. Moolenbeek, W. Los, and P. Prins, eds., *De Horen en Zijn Echo. Stichting Libri Antilliani*, Zoologisch Museum Amsterdam. Zoologisch Museum Amsterdam, Amsterdam, pp. 65-70. [In English].

Vermeij, G. 1993. *A Natural History of Shells*. Princeton University Press, Princeton, New Jersey. 216 pp.

Winston, J. E. 1988. The systematists' perspective. *Memoirs of the California Academy of Sciences* **13**: 1-6.

Winston, J. E. 1999. *Describing Species, Practical Taxonomic Procedure for Biologists*. Columbia University Press, New York. 512 pp.

CHAPTER 30

CONSERVATION AND EXTINCTION OF THE FRESHWATER MOLLUSCAN FAUNA OF NORTH AMERICA

ARTHUR E. BOGAN

30.1 INTRODUCTION

The rivers, streams, and lakes of North America historically were home to a very diverse freshwater molluscan fauna containing approximately 1019 species (Tables 30.1-3). This fauna reached its greatest diversity in the southeastern U.S.A. (Neves et al. 1997). To be able to understand fully the current state of the fauna we have first to establish the early historic levels of freshwater molluscan diversity. The description and illustration of this fauna began with the illustration of unionid shells from Virginia by Martin Lister in 1685. However, Thomas Say was the first American biologist to examine and publish descriptions of new species of freshwater mollusks of North America. Say was followed by Rafinesque, Barnes, Conrad, Lea, Tryon, Simpson, Baker, Frierson, and Haas to present only a few of the actors. Parmalee and Bogan (1998) provide a short history of the development of the taxonomy of the Unionoida in North America. Burch (1989) provides the most recent overview and introduction to the freshwater gastropod classification. Today, we recognize about 344 taxa of freshwater unionoid bivalves (Table 30.2) and 675 taxa of freshwater gastropods based initially on Turgeon et al. (1998) but supplemented with taxa from the recent papers of R. Hershler and F.G. Thompson (Table 30.3).

30.2 UNIONOIDA

Freshwater bivalves in the two families Unionidae and Margaritiferidae are the largest aquatic invertebrates in North America and represent a major portion of the invertebrate biomass in lakes, rivers, and streams. An understanding of their biology, longevity, ecology, and behavior is critical to establishing their role as valuable water quality indicator organisms. We have observed some of the more amazing features of their biology only within the last few years, a time of rapidly declining freshwater molluscan species populations.

Unionoid bivalves are unique among the representatives of the Class Bivalvia in having an obligate parasitic life stage on the gills or fins of a host fish and in one case the external gills of the mudpuppy (*Necturus*). We know the host fishes for only a small percentage of the unionoid species of North America (Watters 1994). A good overview of the larval forms of Unionoida has been provided by Wächtler et al. (2001). Mechanisms for the dispersal of the glochidia by the female onto host fish ranges from releasing the glochidia as a cloud as a potential host fish swims past to some remarkable strategies involving mimicry. Gravid females in the genus *Lampsilis* typically develop mantle extensions ventral and posterior to the incurrent aperture. These extended mantle structures are in the shape of a small fish, usually minnows (Cyprinidae) or darters (Percidae) and are pigmented to resemble fish, including distinct eyespots, dorsal and caudal fins with character-

Table 30.1 Freshwater molluscan diversity in North America north of Mexico.

	Gastropods	Bivalves	Total
Families	14	5	19
Genera	94	57	151
Species	675	344	1,019

Table 30.2 Diversity of North American freshwater bivalves.

Family	Genera	Species
Margaritiferidae	2	5
Unionidae	48	295
Dreissenidae	2	3
Corbiculidae	1	2
Sphaeriidae	4	39
Total	57	344

Table 30.3 Diversity of North American freshwater gastropods.

Family	Genera	Species
Acroloxidae	1	1
Ampulariidae	2	4
Ancylidae	4	13
Bithyniidae	1	1
Hydrobiidae	43	298
Lymnaeidae	10	58
Neritidae	1	1
Physidae	4	43
Planorbidae	12	47
Pleuroceridae	7	160
Pomatiopsidae	1	6
Thiaridae	2	3
Valvatidae	1	11
Viviparidae	5	29
Total	94	675

istic movements (Kraemer 1970). These mantle modifications mimic fish and are presumed to act as lures for piscivorous fish, which also act as host species for the glochidia of these species. These lures apparently entice the potential host fish close enough that the gravid female is able to quickly release a cloud of glochidia in the face of the fish and with any luck, the fish takes in water full of glochidia and some of them attach to the gills of the fish. Recently conglutinate (gelatinous) packages full of glochidia have been reported in various shapes; mimicking larval fish in *Ptychobranchus occidentalis* (Conrad, 1836), an aquatic dipteran (Chironomidae) with red head capsule and sticky tail in *Ptychobranhus greenii* (Conrad, 1834), and some producing conglutinate packages resembling fish eggs. Probably the most interesting and bizarre structure is reported as being produced by four species of *Hamiota* from rivers tributary to the Gulf of Mexico. Haag *et al.* (1995) reported the production of a long gelatinous tube with all of the glochidia placed in the distal end in a package, pigmented externally to resemble a minnow. This minnow mimic is held in the current and moves in the current like a sick or wounded fish and acts as bait for a predator and presumably the host fish.

An overlooked aspect of the biology of these bivalves is their long life span, 30 to 70 years for many species, with *Margaritifera margaritifera* (Linnaeus, 1758) having a life span of 132 to about 200 years (Bauer 1992, Zuiganov *et al.* 2000). This long lifespan is recorded in the shell of the animal, thus providing a detailed history of the local water chemistry and water temperature (Mutvei and Westermark 2001).

Naturalists remarked on the clarity of the streams and rivers from the earliest settlement and exploration of eastern North America. However, the problems with the fouling and pollution of these same rivers also were recognized early.

Sir Charles Lyell (1849: 256) recorded the transformation of Piedmont streams from clear to turbid as exemplified by his comments on the changes seen in the two major tributaries of the Altamaha River, Georgia, during December 1845:

"As our canoe was scudding through the clear waters of the Altamaha, Mr. Couper mentioned a fact which shows the effect of herbage, shrubs, and trees in protecting the soil from the wasting action of rain and torrents. Formerly, even during floods, the Altamaha was transparent, or only stained of a darker color by decayed vegetable matter, like some streams in Europe which flow out of peat mosses. So as late as 1841, a resident here could distinguish on which of the two branches of the Altamaha, the Oconee or Ocmulgee, a freshet had occurred, for the lands in the upper country (Piedmont) drained by one of these (the Oconee) had already been partially cleared and cultivated, so that that tributary sent down a copious supply of red mud, while the other (the Ocmulgee) remained clear, though swollen. But no sooner had the Indians been driven out, and the woods of their old hunting ground begun to give way before the ax of the new settler, than the Ocmulgee also became turbid." Lyell did not witness this transformation himself but there appears to be little reason to doubt the story in light of the other evidence presented. These conditions contrast with those of today.

Higgins (1858: 551) writing slightly later on the Mollusca found in the vicinity of Columbus, Ohio observed:

> "eleven of the fluviatile shells are also extinct, or nearly so. This remarkable decrease and extinction among the mollusca [sic], may, to a great degree be accounted for, when we consider the immense change which the surface of the country has undergone. The change of a wilderness into a highly cultivated country, the immense area of forest which as yielded to the plow; the decrease in the volume of the water in our rivers and creeks, the total change of vegetation and change of climate from moist to dry, have each had their influences upon the character and increase of the Mollusca of this vicinity."

Rhoads (1899) continued the documentation of the environmental destruction and modification in his remarks on the decimation of the freshwater bivalve fauna of the Monongahela River above Pittsburgh, Pennsylvania:

> "Above the city of Pittsburgh the Monongahela is bordered for the greater part of its navigable length with factories, furnaces, refineries, mines, and oil and gas well, whose refuse products are continually draining in to the river. The sewage of the towns on this river is also a factor in its pollution. Great as this pollution may appear, it is not likely that it would cause the death of many mussels and fish, which now no longer exist in the lower half of the Monongahela, if the waters had their free course; but the damming of the river has so concentrated this sewage during low water that the imprisoned animals have no relief from the free flow of the current nor means of escape from the limits of the dammed area."

Ortmann (1909) reported the decimation of the unionoid, fish, and crayfish fauna of western Pennsylvania. He recognized the worst damage to the fauna was done by pollution, both industrial and commercial, including sewage, acid mine drainage, oil well brines, and "the worst are the *oil refineries*, which discharge into the water chemicals which are utterly destructive of life." (Ortmann 1909: 98).

Ortmann (1918) reported the decimation of the aquatic fauna in the Pigeon River in East Tennessee due to the wood pulp and paper mill effluents. These were all rather localized events or impacts. Van der Schalie (1938) decried construction of the dams on the Tennessee River and major tributaries by the Tennessee Valley Authority and the impending loss of the Tennessee River's aquatic faunal diversity.

30.3 AQUATIC GASTROPODA

The freshwater gastropod fauna of North America is composed of about 675 species in 14 families (9 operculate and 5 pulmonate; Tables 30.3 and 30.4) but our knowledge of the status of the freshwater gastropods is much more limited than for the bivalves. The greatest diversity of freshwater gastropods occurred in the Mobile Bay Basin followed by the Tennessee River Basin (Bogan *et al.* 1995, Neves *et al.* 1997). The freshwater gastropod fauna has lost 6.2 % of its total taxa to extinction and only 4.2% of the total taxa are accorded legal status (Table 30.4). Out of a total of 675 freshwater gastropod taxa reported for North America, 42 are presumed to be extinct, 38 in the Mobile Bay Basin alone (Bogan *et al.* 1995, Neves *et al.* 1997). The 42 extinct gastropods include four extinct genera of aquatic gastropods: *Clappia* (2 species, Hydrobiidae); *Gyrotoma* (6 species, Pleuroceridae); *Neoplanorbis* (4 species, Planorbidae) and *Amphigyra* (1 species, Planorbidae) (Bogan *et al.* 1995, Neves *et al.* 1997, Turgeon *et al.* 1998), all of which were endemic to the Coosa River in the Mobile Bay Basin. These species were lost when the big-river shoal habitat was impounded and covered with deep standing water and silt. A number of other pleurocerid gastropod species are persisting on the clean swept shoal areas below the dams on the Tennessee River. Robison and Allen (1995) reported the extinction of an additional three species of Hydrobiidae due to the covering of their home range with a reservoir.

30.4 DETERMINATION OF STATUS

The first step to understand the status of a species and a local fauna is based on all available distributional data. It is only after a good, thorough survey for a species throughout its known range that the

Table 30.4 Status of North American freshwater mollusks (see Appendix 30.1).

	Diversity	Endangered	Threatened	Extinct
Unionoid bivalves	300	62	8	35
Freshwater gastropods	675	16	4	42
Total	975	78	12	77

status of a species can be determined. This modern survey data coupled with the historic reports and the historic museum specimens will either document continued health or the decline and loss of a species. In some areas of the U.S.A., the archaeological record can supplement the museum record giving time depth to understanding the dynamics of the molluscan fauna, in some cases 6,000 years (Bogan 1990).

30.5 FACTORS CAUSING THE DECLINE IN SPECIES

The freshwater mollusks, both the gastropods and bivalves, are declining due to habitat modification and destruction. Dams and pollution in its varied forms, including acid mine drainage, sewage treatment plant effluent, sedimentation due to numerous factors, sand and gravel dredging, all contribute to the destruction of habitat. Damming of major rivers has had a major impact on the unionoid fauna and on the obligate unionoid host fishes, due to changes in water quality and loss of habitat. Instream gravel mining, dredging, stream channelization, and the often associated headcutting, has eliminated stable aquatic mollusk habitat. Acid mine drainage, and various point and non-point pollution sources also continue to decimate local unionoid populations.

A new threat to the continued survival of unionoid and pleurocerid taxa is the introduction, in the mid-1980's of the zebra mussel, *Dreissena polymorpha* (Pallas, 1791) and to a lesser extent the Quagga Mussel, *Dreissena bugensis* Andrusov, 1897. These small, byssally-attached bivalves cover and smother native mussels and larger gastropods. Another introduced mollusk, the New Zealand Mudsnail, *Pyrgulopsis antipodarum* (Gray, 1853) has been introduced into the Snake River basin and is growing to populations of 200,000-500,000 animals per square meter. They are having a major impact on the local ecology. Another source of introductions that is new but may present a growing problem is the introduction of native species of unionids to locations outside of their historic range (e.g., Bogan *et al.* 2002).

Another threat to the continued health of freshwater mollusks is the introduction of native molluscivorous fish outside of their native range. These include the Blue Catfish, *Ictalurus furcatus* (Lesueur, 1819), Flathead Catfish *Pylodictus olivaris* (Rafinesque, 1818), and introduction of the Asian molluscivore, the Black Carp, *Mylopharyngodon piceus* (Richardson, 1846).

30.6 EXTINCTION

We have considered the freshwater molluscan faunal diversity of North America and the concentration of the diversity in the southeastern U.S.A. We have seen that environmental perturbations have resulted in the decline and demise of many of the species. Before dealing with the levels of decline, some terms must be clearly defined. The term extinction is often widely and incorrectly used. A species becomes extinct, when the last living specimen of that organism perishes. There are no longer any living specimens of that species. Extinction has been confused with extirpation. Extirpation is the loss of a local population or group of animals from within the total range of a species. When a local population dies out, it is extirpated, but the species continues to exist in other parts of its range. When the last passenger pigeon died in the St. Louis Zoo, the species became extinct.

Another question arises in defining extinction; when is a species extinct? What are the data that support the confirmation that a species is extinct? Some people have expressed the belief that a species should not have been collected or documented for a minimum period of 50 years. Since many of our species live on and in the bottom of major rivers it is even more difficult to categorically state when a particular species is in fact extinct. The species is listed for this reason as presumed extinct and by using a minimum time of 50 years since it was last collected or documented before it can be reasonably considered extinct. This also assumes that there have been recent, thorough, credible surveys in the known habitat of the species at various times in the recent past.

Parmalee and Bogan (1998) have defined endangered as: "this status at the state level includes peripheral forms which may be common in another part of the range, but whose continued existence within the political boundaries of the state is in danger of extirpation. At the national level, this status means the organism is in danger of extinction and is included on or being considered for the U.S. List of Endangered Fauna and Endangered and Threatened plant species of the U.S.A., under the Endangered Species Act of 1973." They have also defined threatened as "this status at the state level includes forms which are likely to become endangered in the foreseeable future if certain conditions are not met. This includes forms which exhibit a considerable decrease in numbers beyond normal population fluctuations or a documented range contraction, but are not yet considered endangered. At the national level, this applies to the Endangered Species Act of 1973."

The first attempts to assess the status of the freshwater molluscan fauna in North America began in 1970. Stansbery (1970, 1971) reported that 11 species of unionid bivalves had not been found since 1900 and were to be presumed extinct and that 120 taxa were to be considered rare or endangered. Stansbery (1970) considered the information on seven families of freshwater gastropods to be too insufficient to determine their status. He noted the Pleuroceridae were declining but listed a single species as extinct. Heard (1970) presumed three Gulf Coast species of gastropods to be extinct while Athearn (1970) listed 69 aquatic gastropod species as rare or endangered in the southeast. Taylor (1970) added ten more species of aquatic gastropods to this growing list. Stansbery (1971) listed 82 species of rare and endangered aquatic gastropods and presumed a single species within the Pleuroceridae to be extinct.

With the passage of the Clean Water Act (1972) and the Endangered Species Act (1973), the U.S. Fish and Wildlife Service began evaluating species survival status and listed some aquatic mollusks as being threatened or endangered. Increased interest and the listing of species stimulated various states to begin treating freshwater bivalves as locally extirpated, threatened, or endangered. The American Fisheries Society listed 13 unionoid taxa as presumed extinct and 30 as federally endangered (Turgeon et al. 1988). Bogan (1993) reviewed the evidence for extinctions for freshwater bivalves worldwide and listed 19 species of North American Unionoida as presumed extinct, but found evidence for extinction of freshwater bivalves outside of the U.S.A. only in Israel, which lost three taxa when a river dried up. Neves (1993) listed 21 species of unionoids as presumed extinct. Turgeon et al. (1998), citing the U.S. Fish and Wildlife Service, listed 35 unionids as presumed extinct, 57 as endangered, and 6 taxa as threatened. Williams et al. (1993) provided a revised evaluation of the status of the North American unionoid fauna and listed 12% of the unionoids as presumed extinct, 43% as listed or to be listed as endangered or threatened and an additional 25% as declining. At this time, none of the North American unionoid genera has become extinct.

Bogan (2001) examined the current precipitous decline and extinction specifically in North American freshwater bivalves and addressed the problem; an extinction wave in the making. Are we facing a major wave of extinction and have we seen only the beginning of the wave of extinctions soon to be visited on the freshwater bivalve fauna of North America?

30.7 ACTIVITIES LEADING TO A REVERSAL OF TRENDS

The aquatic environment began to respond to positive actions due to the Clean Water Act and the attention brought by the Endangered Species Act.

Over the past 30 years, the health of the aquatic ecosystem has become more visible and the results of reducing the effects of pollution have become very obvious. Many conservation groups have grown up and expanded their efforts both at the local level as well as nationally and internationally. The Nature Conservancy (TNC) has been instrumental in purchasing land for conserving habitat as well as developing riparian buffer zones along streams and river. TNC and the U.S. Fish and Wildlife Service have been working with local landowners to fence cattle, a major source of pollution, out of streams. Corporations and towns have been developing ordinances addressing the setting of riparian stream buffers, and stabilization of river and creek banks. The improvement of the quality due to improved water treatment and stopping industrial effluent (e.g. Allegheny and upper Ohio River in Pennsylvania.) are all having a dramatic positive effect on the local fauna (e.g. Locy *et al.* 2002).

The perception of the need for conservation of freshwater mollusks has recently been addressed by a symposium of the IUCN (World Conservation Union formerly the International Union for the Conservation of Nature) resulting in a volume on the conservation biology of mollusks (Kay 1995). More recently, the Conchological Society of Great Britain and Ireland hosted an international symposium *Molluscan conservation: a strategy for the 21st Century* (Bogan 1998 and Killeen *et al.* 1998). Seddon *et al.* (1998) provided a list of strategies gleaned from the symposium on the conservation of mollusks. These include:

1. To improve the public awareness of the need to maintain mollusks as a biological resource.
2. To reduce the confusion about the numbers of, and names of mollusks.
3. To improve the information on threatened mollusks so as to enable prioritization for their management.
4. To maintain the skill and staffing to identify mollusks.
5. To insure access to molluscan collections in museums.
6. To reduce the number of accidental/deliberate introductions of alien species.
7. To improve communication about mollusks that should be protected
8. To improve knowledge on the life-cycles/breeding biology of mollusks to provide a better basis for successful implementation of management plans.

30.8 CONCLUSIONS

Recent efforts in the U.S.A. to expand riparian buffer zones, control mine runoff, close point sources of pollution, and otherwise improve water quality are enabling the return of fish and freshwater mollusks to formerly polluted waters. However, the return of the long-lived mussels is a slow process and may take as long as 100 years, making it impossible to predict whether more species will be lost than preserved and thus, how devastating a tragedy this extinction wave will be.

30.9 ACKNOWLDGMENTS

Dr. Terry Ferguson, Wofford College, Spartanburg, SC assisted with the Lyell citation. Many people over the years have assisted me by sending literature and listening to my rantings. This is a product of many years and my friends are thanked for their time, assistance, and patience.

30.10 LITERATURE CITED

Athearn, H. D. 1970. Discussion of Dr. Heard's paper. Symposium on Endangered Mollusks. *Malacologia* **10**: 28-31.

Bauer, G. 1992. Variation in the life span and size of the freshwater pearl mussel. *Journal of Animal Ecology* **61**: 425-436.

Bogan, A. E. 1990. Stability of Recent unionid (Mollusca: Bivalvia) communities over the past 6000 years. *In:* W. Miller III, ed., *Paleocommunity Temporal Dynamics: The Long-Term Development of Multispecies Assemblages.* The Paleontological Society Special Publication No. 5. The Paleontological Society, Lawrence, Kansas. Pp. 112-136.

Bogan, A. E. 1993. Freshwater bivalve extinctions: Search for a cause. *American Zoologist* **33**: 599-609.

Bogan, A. E. 1997. The silent extinction. *American Paleontologist* **15**: 2-4

Bogan, A. E. 1998. Freshwater molluscan conservation in North America: Problems and practices. *In:* I.

J. Killeen, M. B. Seddon, and A. M. Holmes, eds., *Molluscan Conservation: A Strategy for the 21st Century. Journal of Conchology*, Special Publication No. 2. Conchological Society of Great Britain and Ireland, London. Pp. 223-230.

Bogan, A. E. 2001. Extinction wave in the making. *In:* A. Bräutigam and M. D. Jenkins, eds., *The Red Book: The Extinction Crisis Face to Face*. IUCN-International Union for Conservation of Nature and Natural Resources and CEMEX, S.A., Mexico City. Pp. 138-139.

Bogan, A. E., S. A. Ahlstedt, and P. W. Parmalee. 2002. Exotic freshwater bivalves found in the Nolichucky River, East Tennessee. *Ellipsaria* **4**: 9.

Bogan, A. E., J. M. Pierson, and P. Hartfield. 1995. Decline in the freshwater gastropod fauna in the Mobile Bay Basin. *In*: E. T. LaRoe, G. S. Farris, C. E. Puckett, P. D. Doran, and M. J. Mac, eds., *Our Living Resources: A Report to the Nation on the Distribution, Abundance, and Health of U.S. Plants, Animals, and Ecosystems*. U.S. Department of Interior, National Biological Service, Washington, D.C. Pp. 249-252.

Burch, J. B. 1989. *North American Freshwater Snails*. Malacological Publications. Hamburg, Michigan. 365 pp.

Haag, W. R., Butler, R. S., and P. D. Hartfield. 1995. An extraordinary reproductive strategy in freshwater bivalves: Prey mimicry to facilitate larval dispersal. *Freshwater Biology* **34**: 471-476.

Heard, W. H. 1970. Eastern freshwater mollusks (III). The south Atlantic and gulf drainages. *Malacologia* **10**: 23-31.

Higgins, F. 1858. A catalogue of the shell-bearing species of Mollusca, inhabiting the vicinity of Columbus, Ohio, with some remarks thereon. *Twelfth Annual report of the Ohio State Board of Agriculture with an abstract of the Proceedings of the County Agricultural Societies, to the General Assembly of Ohio* for 1857: 548-555.

Kay, E. A. ed. 1995. The conservation biology of Molluscs. *Occasional Paper of the IUCN Species Survival Commission* **9**: 1-81.

Killeen, I. J., M. B. Seddon, and A. M. Holmes, eds. 1998. *Molluscan conservation: A strategy for the 21st Century. Journal of Conchology*, Special Publication No. 2. Conchological Society of Great Britain and Ireland, London. 320 pp.

Kraemer, L. R. 1970. The mantle flap in three species of *Lampsilis* (Pelecypoda: Unionidae). *Malacologia* **10**: 225-282.

Locy, D., T. Proch, and A. E. Bogan. 2002. *Anodonta suborbiculata* (Say, 1831) added to the freshwater bivalve fauna of Pennsylvania. *Ellipsaria* **4**: 10.

Lyell, C. 1849. *A Second Visit to the United States of North America*, Vol. 1. Harper and Brothers, New York. 273 pp.

Mutvei, H. and T. Westermark. 2001. How environmental information can be obtained from naiad shells. *In:* Bauer and K. Wächtler, eds., *Ecology and Evolutionary Biology of Freshwater Mussels, Unionoida* Ecological Studies, Vol. 145, G. Springer Verlag, New York. Pp. 367-379.

Neves, R.J. 1993. A state-of-the-unionids address. *In:* K. S. Cummings, A. C. Buchanan, and L. M. Koch, eds., *Conservation and Management of Freshwater Mussels. Proceedings of a Symposium, 12-14 October, 1992, St. Louis, Missouri*. Upper Mississippi River Conservation Committee, Rock Island, Illinois. Pp. 1-10.

Neves, R. J., A. E. Bogan, J. D. Williams, S. A. Ahlstedt, and P. D. Hartfield. 1997 [March 1998]. Status of aquatic mollusks in the southeastern United States: a downward spiral of diversity. *In:* G. W. Benz and D. E. Collins, eds., *Aquatic Fauna in Peril: The Southeastern Perspective*. Special Publication No. 1. Southeast Aquatic Research Institute, Lenz Design and Communications, Decatur, Georgia. Pp. 43-86. [Published May 1998].

Ortmann, A. E. 1909. The destruction of the fresh-water fauna in western Pennsylvania. *Proceedings of the American Philosophical Society* **48**: 90-110.

Ortmann, A. E. 1918. The nayades (freshwater mussels) of the Upper Tennessee drainage. With notes on synonymy and distribution. *Proceedings of the American Philosophical Society* **57**: 521-626.

Parmalee, P. W. and A. E. Bogan. 1998. *The freshwater mussels of Tennessee*. The University of Tennessee Press, Knoxville, Tennessee. 328 pp.

Rhoads, S. N. 1899. On a recent collection of Pennsylvanian mollusks from the Ohio River system below Pittsburgh. *The Nautilus* **12**: 133-138.

Robison, H. W. and R. T. Allen. 1995. *Only in Arkansas. A study of the Endemic Plants and Animals of the State*. University of Arkansas Press, Fayetteville, Arkansas. 121 pp.

Seddon, M. B., I. J. Killeen, P. Bouchet, and A. E. Bogan. 1998. Developing a strategy for molluscan conservation in the next century. *In:* I. J. Killeen, M. B. Seddon, and A. M. Holmes, eds., *Molluscan Conservation: A Strategy for the 21st Century. Journal of Conchology, Special Publication* No. 2. Conchological Society of Great Britain and Ireland, London. Pp. 295-298.

Stansbery, D. H. 1970. Eastern freshwater mollusks. (I). The Mississippi and St. Lawrence River systems. American Malacological Union Symposium on Rare and Endangered Mollusks. *Malacologia* **10**: 9-22.

Stansbery, D. H. 1971. Rare and endangered freshwater mollusks in eastern United States. In: S. E. Jorgensen and R. E. Sharp, eds., *Proceedings of a Symposium on Rare and Endangered Mollusks (Naiads) of the U.S. Region 3*. Bureau of Sport

Fisheries and Wildlife, U.S. Fish and Wildlife Service. Twin Cities, Minnesota. Pp. 5-18.

Taylor, D. W. 1970. Western freshwater mollusks. *Malacologia* **10**: 33.

Turgeon, D. D., A. E. Bogan, E. V. Coan, W. K. Emerson, W. G. Lyons, W. L. Pratt, C. F. E. Roper, A. Scheltema, F. G. Thompson, and J. D. Williams. 1988. *Common and Scientific Names of Aquatic Invertebrates from the United States and Canada: Mollusks.* American Fisheries Society, Special Publication 16. American Fisheries Society, Bethesda, Maryland. 277 pp.

Turgeon, D. D., J. F. Quinn, Jr., A. E. Bogan, E. V. Coan, F. G. Hochberg, W. G. Lyons, P. M. Mikkelsen, R. J. Neves, C. F. E. Roper, G. Rosenberg, B. Roth, A. Scheltema, F. G. Thompson, M. Vecchione. and J. D. Williams. 1998. *Common and Scientific Names of Aquatic Invertebrates from the United States and Canada: Mollusks.* American Fisheries Society Special Publication 26, 2nd Ed. American Fisheries Society, Bethesda, Maryland. 536 pp.

van der Schalie, H. 1938. The naiades (freshwater mussels) of the Cahaba River in northern Alabama. *Occasional Papers of the Museum of Zoology, University of Michigan* **392**: 1-29.

Wächtler, K., M. C. Dreher-Mansur, and T. Richter. 2001. Larval types and early postlarval biology in Naiads (Unionoida). *In:* G. Bauer and K. Wächtler, eds., *Ecology and Evolutionary Biology of Freshwater Mussels, Unionoida. Ecological Studies*, Vol. 145. Springer Verlag, New York. Pp. 93-125.

Watters, G. T. 1994. An annotated bibliography of the reproduction and propagation of the Unionoidea (Primarily of North America). *Ohio Biological Survey Miscellaneous Contributions* **1**: 1-158.

Williams, J. D. and A. E. Bogan. 1998. Endangered and threatened species in North America. Appendix V. *In:* D. D. Turgeon, J. F. Quinn, Jr., A. E. Bogan, E. V. Coan, F. G. Hochberg, W. G. Lyons, P. M. Mikkelsen, R. J. Neves, C. F. E. Roper, G. Rosenberg, B. Roth, A. Scheltema, M. J. Sweeney, F. G. Thompson, M. Vecchione, and J. D. Williams, eds., *Common and Scientific Names of Aquatic Invertebrates from the United States and Canada: Mollusks.* American Fisheries Society Special Publication 26, 2nd Ed. American Fisheries Society, Bethesda, Maryland. Pp. 329-338.

Williams, J. D., A. E. Bogan, and R. J. Neves. 1998. Possibly extinct Mollusks of North America. Appendix III. *In:* D. D. Turgeon, J. F. Quinn, Jr., A. E. Bogan, E. V. Coan, F. G. Hochberg, W. G. Lyons, P. M. Mikkelsen, R. J. Neves, C. F. E. Roper, G. Rosenberg, B. Roth, A. Scheltema, M. J. Sweeney, F. G. Thompson, M. Vecchione, and J. D. Williams, eds., *Common and Scientific Names of Aquatic Invertebrates from the United States and Canada: Mollusks.* American Fisheries Society Special Publication 26, 2nd Ed. American Fisheries Society, Bethesda, Maryland. Pp. 315-318.

Williams, J. D., M. L. Warren, Jr., K. S. Cummings, J. L. Harris, and R. J. Neves. 1993. Conservation status of the freshwater mussels of the United States and Canada. *Fisheries* **18**: 6-22.

Zuiganov, V., E. San Miguel, R. J. Neves, A. Longa, C. Fernández, R. Amaro, V. Beletsky, E. Popkovitch, S. Kalluzhin, and T. Johnson. 2000. Life span variation of the freshwater pearl mussel: A model species for testing longevity mechanisms in animals. *Ambio* **29**: 102-105.

APPENDIX 30.1

List of Extinct, endangered, and threatened freshwater mollusks based on Williams and Bogan (1998) and Williams *et al.* (1998) and modified to include additional data from Robison and Allen (1995) and the U.S. Fish and Wildlife Service.

CLASS BIVALVIA

Presumed extinct freshwater bivalves

Alasmidonta mccordi Athearn, 1964 Coosa Elktoe AL

Alasmidonta robusta Clarke, 1981. Carolina Elktoe NC, SC

Alasmidonta wrightiana (Walker, 1901) Ochlockonee Arcmussel FL

Elliptio nigella (Lea, 1852) Winged Spike AL, GA

Epioblasma arcaeformis (Lea, 1831) Sugarspoon AL, KY, TN

Epioblasma biemarginata (Lea, 1857) Angled Riffleshell AL, KY, TN

Epioblasma flexuosa (Rafinesque, 1820) Leafshell AL, IL, IN, KY, OH, TN

Epioblasma florentina florentina (Lea, 1857) Yellow Blossom AL, KY, TN

Epioblasma haysiana (Lea, 1834) Acornshell AL, KY, TN, VA

Epioblasma lenior (Lea, 1842) Narrow Catspaw AL, TN

Epioblasma lewisii (Walker, 1910) Forkshell AL, KY, TN

Epioblasma personata (Say, 1829) Round Combshell IL, IN, KY, OH

Epioblasma propinqua (Lea, 1857) Tennessee Riffleshell AL, IL, IN, KY, OH, TN

Epioblasma sampsonii (Lea, 1861) Wabash Riffleshell IL, IN, KY

Epioblasma stewardsonii (Lea, 1852) Cumberland Leafshell AL, KY, TN

Epioblasma torulosa gubernaculums (Reeve, 1865) Green Blossom TN, VA

Epioblasma torulosa torulosa (Rafinesque, 1820) Tubercled Blossom AL, IL, IN, KY, OH, TN, WV

Epioblasma turgidula (Lea, 1858) Turgid Blossom AL, AR, TN
Lampsilis binominata Simpson, 1900 Lined Pocketbook AL, GA
Medionidus mcglameriae van der Schalie, 1939 Tombigbee Moccasinshell AL
Pleurobema altum (Conrad, 1854) Highnut AL, GA
Pleurobema avellanum Simpson, 1900 Hazel Pigtoe AL
Pleurobema bournianum (Lea, 1840) Scioto Pigtoe OH
Pleurobema chattanoogaense (Lea, 1858) Painted Clubshell AL, GA, TN
Pleurobema flavidulum (Lea, 1861) Yellow Pigtoe AL
Pleurobema hagleri (Frierson, 1900) Brown Pigtoe AL
Pleurobema hanleyianum (Lea, 1852) Georgia Pigtoe AL, GA, TN
Pleurobema johannis (Lea, 1859) Alabama Pigtoe AL
Pleurobema murrayense (Lea, 1868) Coosa Pigtoe AL, GA, TN
Pleurobema nucleopsis (Conrad, 1849) Longnut AL, GA
Pleurobema rubellum (Conrad, 1834) Warrior Pigtoe AL, GA, TN
Pleurobema troschelianum (Lea, 1852) Alabama Clubshell AL, GA, TN
Pleurobema verum (Lea, 1861) True Pigtoe AL
Quadrula tuberosa (Lea, 1840) Rough Rockshell TN, VA

Endangered species
Alasmidonta heterodon (Lea, 1829) Dwarf Wedgemussel CT, DC, DE, MA, MD, NC, NH, NJ, NY, PA, VA, VT; Canada: NB
Alasmidonta atropurpurea (Rafinesque, 1831) Cumberland Elktoe KY, TN
Alasmidonta raveneliana (Lea, 1834) Appalachian Elktoe NC, TN
Amblema neislerii (Lea, 1858) Fat Threeridge FL, GA
Arkansia wheeleri Ortmann and Walker, 1912 Ouachita Rock Pocketbook AR, OK
Cyprogenia stegaria (Rafinesque, 1820) Fanshell AL, IL, IN, KY, OH, PA, TN, VA, WV
Dromus dromas (Lea, 1834) Dromedary Pearlymussel AL, KY, TN, VA
Elliptio steinstansana R. I. Johnson and Clarke, 1983 Tar River Spinymussel NC
Epioblasma brevidens (Lea, 1831) Cumberlandian Combshell AL, KY, TN, VA
Epioblasma capsaeformis (Lea, 1834) Oyster Mussel AL, KY, TN, VA
Epioblasma florentina curtisi (Utterback, 1916) Curtis Pearlymussel AR, MO
Epioblasma florentina florentina (Lea, 1857) Yellow Blossom AL, KY, TN
Epioblasma florentina walkeri (Wilson and Clark, 1914) Tan Riffleshell KY, TN, VA

Epioblasma metastriata (Conrad, 1838) Upland Combshell AL, GA, TN
Epioblasma obliquata obliquata (Rafinesque, 1820) Catspaw AL, IL, IN, KY, OH, TN
Epioblasma obliquata perobliqua (Conrad, 1836) White Catspaw IL, IN, KY, MI, OH
Epioblasma othcaloogensis (Lea, 1857) Southern Acornshell AL, GA, TN
Epioblasma penita (Conrad, 1834) Southern Combshell AL, GA, MS
Epioblasma torulosa gubernaculums (Reeve, 1865) Green Blossom TN, VA
Epioblasma torulosa rangiana (Lea, 1838) Northern Riffleshell IL, IN, KY, MI, OH, PA, WV; Canada: ON
Epioblasma torulosa torulosa (Rafinesque, 1820) Tubercled Blossom AL, IL, IN, KY, OH, TN, WV
Epioblasma turgidula (L ea, 1858) Turgid Blossom AL, AR, TN
Fusconaia cor (Conrad, 1834) Shiny Pigtoe AL, TN, VA
Fusconaia cuneolus (Lea, 1840) Finerayed Pigtoe AL, TN, VA
Hemistena lata (Rafinesque, 1820) Cracking Pearlymussel AL, IL, IN, KY, OH, PA, TN, VA
Lampsilis abrupta (Say, 1831) Pink Mucket AL, AR, IL, IN, KY, LA, MO, OH, PA, TN, VA, WV
Lampsilis higginsii (Lea, 1857) Higgins Eye IA, IL, MN, MO, WI
Lampsilis streckeri Frierson, 1927 Speckled Pocketbook AR
Lampsilis subangulata (Lea, 1840) Shinyrayed Pocketbook AL, FL, GA
Lampsilis virescens (Lea, 1858) Alabama Lampmussel AL, TN
Lasmigona decorata (Lea, 1852) Carolina Heelsplitter NC, SC
Lemiox rimosus (Rafinesque, 1831) Birdwing Pearlymussel AL, TN, VA
Leptodea leptodon (Rafinesque, 1820) Scaleshell Mussel AL, AR, IA, IN, KY, MI, MO, MS, OH, OK, SD, TN, WI
Medionidus parvulus (Lea, 1860) Coosa Moccasinshell AL, GA, TN
Medionidus penicillatus (Lea, 1857) Gulf Moccasinshell AL, FL, GA
Medionidus simpsonianus Walker, 1905 Ochlockonee Moccasinshell FL, GA
Obovaria retusa (Lamarck, 1819) Ring Pink AL, IL, IN, KY, OH, PA, TN, WV
Pegias fabula (Lea, 1838) Littlewing Pearlymussel AL, KY, NC, TN, VA
Plethobasus cicatricosus (Say, 1829) White Wartyback AL, IL, IN, KY, OH, TN
Plethobasus cooperianus (Lea, 1834) Orangefoot Pimpleback AL, IL, IN, KY, OH, PA, TN
Pleurobema clava (Lamarck, 1819) Clubshell AL, IL, IN, KY, MI, OH, PA, TN, WV

Pleurobema collina (Conrad, 1837) James Spinymussel VA, WV
Pleurobema curtum (Lea, 1859) Black Clubshell AL, MS
Pleurobema decisum (Lea, 1831) Southern Clubshell AL, GA, MS, TN
Pleurobema furvum (Conrad, 1834) Dark Pigtoe AL
Pleurobema georgianum (Lea, 1841) Southern Pigtoe AL, GA, TN
Pleurobema gibberum (Lea, 1838) Cumberland Pigtoe TN
Pleurobema marshalli Frierson, 1927 Flat Pigtoe AL, MS
Pleurobema perovatum (Conrad, 1834) Ovate clubshell AL, GA, MS, TN
Pleurobema plenum (Lea, 1840) Rough Pigtoe AL, IL, IN, KY, OH, PA, TN, VA
Pleurobema pyriforme (Lea, 1857) Oval Pigtoe AL, FL, GA
Pleurobema taitianum (Lea, 1834) Heavy Pigtoe AL, MS
Potamilus capax (Green, 1832) Fat Pocketbook AR, IA, IL, IN, KY, LA, MN, MO, MS, OH, OK, WI
Ptychobranchus greenii (Conrad, 1834) Triangular Kidneyshell AL, GA, TN
Quadrula cylindrica strigillata (Wright, 1898) Rough Rabbitsfoot TN, VA
Quadrula fragosa (Conrad, 1835) Winged Mapleleaf AL, IA, IL, IN, KY, MN, MO, OH, TN, WI
Quadrula intermedia (Conrad, 1836) Cumberland Monkeyface AL, TN, VA
Quadrula sparsa (Lea, 1841) Appalachian Monkeyface KY, TN, VA
Quadrula stapes (Lea, 1831) Stirrupshell AL, MS
Toxolasma cylindrellus (Lea, 1868) Pale Lilliput AL, TN
Villosa perpurpurea (Lea, 1861) Purple Bean GA, TN, VA
Villosa trabalis (Conrad, 1834) Cumberland Bean AL, KY, TN, VA

Threatened species
Margaritiferidae
Margaritifera hembeli (Conrad, 1838) Louisiana Pearlshell LA

Unionidae
Elliptio chipolaensis (Walker, 1905) Chipola Slabshell AL, FL
Elliptoideus sloatianus (Lea, 1840) Purple Bankclimber AL, FL, GA
Lampsilis altilis (Conrad, 1834) Finelined Pocketbook AL, GA, MS, TN
Lampsilis perovalis (Conrad, 1834) Orangenacre Mucket AL, MS
Lampsilis powellii (Lea, 1852) Arkansas Fatmucket AR

Medionidus acutissimus (Lea, 1831) Alabama Moccasinshell AL, GA, MS, TN
Potamilus inflatus (Lea, 1831) Alabama Heelsplitter AL, LA, MS

CLASS GASTROPODA
Presumed extinct species
Pleuroceridae
Elimia brevis (Reeve, 1860) Short-spire Elimia AL
Elimia clausa (Lea, 1861) Closed Elimia AL
Elimia fusiformis (Lea, 1861) Fusiform Elimia AL
Elimia gibbera (Goodrich, 1922) AL
Elimia hartmaniana (Lea, 1861) High-spired Elimia AL
Elimia impressa (Lea, 1841) Constricted Elimia AL
Elimia jonesi (Goodrich, 1936) Hearty Elimia AL
Elimia lachryma (Reeve, 1861) AL
Elimia laeta (Jay, 1839) Ribbed Elimia AL
Elimia macglameriana (Goodrich, 1936) AL
Elimia pilsbryi (Goodrich, 1927) Rough-lined Elimia AL
Elimia pupaeformis (Lea, 1864) Pupa Elimia AL
Elimia pygmaea (H. H. Smith, 1936) Pygmy Elimia AL
Elimia vanuxemiana (Lea, 1843) Cobble Elimia AL
Elimia varians (Lea, 1861) Puzzle Elimia AL
Gyrotoma excisa (Lea, 1843) Excised Slitshell AL
Gyrotoma lewisii (Lea, 1869) Striate Slitshell AL
Gyrotoma pagoda (Lea, 1845) Pagoda Slitshell AL
Gyrotoma pumila (Lea, 1860) Ribbed Slitshell AL
Gyrotoma pyramidata (Shuttleworth, 1845) Pyramid Slitshell AL
Gyrotoma walkeri (H. H. Smith, 1924) Round Slitshell AL
Leptoxis clipeata (H. H. Smith, 1922) Agate Rocksnail AL
Leptoxis compacta (Anthony, 1854) Oblong Rocksnail AL
Leptoxis crassa crassa (Haldeman, 1842) Boulder Snail AL
Leptoxis foremani (Lea, 1843) Interrupted Rocksnail AL
Leptoxis formosa (Lea, 1860) Maiden Rocksnail AL
Leptoxis ligata (Anthony, 1860) Rotund Rocksnail AL
Leptoxis lirata (H. H. Smith, 1922) Lirate Rocksnail AL
Leptoxis melanoides (Conrad, 1834) Black Mudalia AL
Leptoxis occultata (H. H. Smith, 1922) Bigmouth Rocksnail AL
Leptoxis showalterii (Lea, 1860) Coosa Rocksnail AL
Leptoxis torrefacta (Goodrich, 1922) AL
Leptoxis vittata (Lea, 1860) Striped Rocksnail AL

Hydrobiidae
Clappia cahabensis Clench, 1965 Cahaba Pebblesnail AL
Clappia umbilicata (Walker, 1904) Umbilicate Pebblesnail AL
Pyrgulopsis nevadensis (Stearns, 1883) Corded Pyrg NV
Pyrgulopsis olivacea (Pilsbry, 1895) Olive Marstonia AL
Somatogyrus amnicoloides Walker, 1915 Ouachita Pebblesnail AR

Somatogyrus crassilabris Walker, 1915 Thicklipped Pebblesnail AR
Somatogyrus wheeleri Walker, 1915 Channelled Pebblesnail AR

Planorbidae
Amphigyra alabamensis Pilsbry, 1906 Shoal Sprite AL
Neoplanorbis carinatus Walker, 1908 *classification uncertain* AL
Neoplanorbis smithi Walker, 1908 *classification uncertain* AL
Neoplanorbis tantillus Pilsbry, 1906 *classification uncertain* AL
Neoplanorbis umbilicatus Walker, 1908 *classification uncertain* AL

Endangered species
Viviparidae
Campeloma decampi (Binney, 1865) Slender Campeloma AL
Lioplax cyclostomaformis (Lea, 1841) Cylindrical Lioplax AL, GA, LA
Tulotoma magnifica (Conrad, 1834) Tulotoma AL

Pleuroceridae
Leptoxis anthonyi (Redfield, 1854) Anthony Riversnail AL, GA, TN
Leptoxis plicata (Conrad, 1834) Plicate Rocksnail AL

Hydrobiidae
Antrobia culveria Hubricht, 1971 Tumbling Creek Cavesnail MO

Lepyrium showalteri (Lea, 1861) Flat Pebblesnail AL
Pyrgulopsis bruneauensis Hershler, 1990 Bruneau Hot Springsnail ID
Pyrgulopsis idahoensis (Pilsbry, 1933) Idaho Springsnail ID
Pyrgulopsis neomexicana (Pilsbry, 1916) Socorro Springsnail NM
Pyrgulopsis ogmorhaphe (F. G. Thompson, 1977) Royal Springsnail TN
Pyrgulopsis pachyta (Thompson, 1977) Armored Marstonia AL
Tryonia alamosae Taylor, 1987 Caliente Tryonia NM

Valvatidae
Valvata utahensis Call, 1884 Desert Valvata ID, UT

Lymnaeidae
Lanx new species Banbury Springs Limpet ID

Physidae
Physella natricina Taylor, 1988 Snake River Physa ID

Threatened species
Hydrobiidae
Taylorconcha serpenticola Hershler, Frest, Johannes, Bowler, and Thompson, 1994 Bliss Rapids Snail ID

Pleuroceridae
Elimia crenatella (Lea, 1860) Lacy Elimia AL
Leptoxis ampla (Anthony, 1855) Round Rocksnail AL
Leptoxis taeniata (Conrad, 1834) Painted Rocksnail AL

CHAPTER 31

ISSUES IN MARINE FISHERIES AND CONSERVATION

PATRICK BAKER

31.1 INTRODUCTION

The tale of marine mollusk conservation, particularly since the Endangered Species Act (1973), is very different from that for freshwater or terrestrial mollusks. Freshwater mollusks in particular are among the most imperiled groups of organisms in North America. Williams *et al.* (1993), for example, conservatively estimated that 21 of the nearly 300 described species of unionoid clams (freshwater pearly mussels) in the United States were extinct, and more have since been added to that sad list. That is more than the total known historic marine mollusk extinctions worldwide. Whether this difference is real or only perceived is unknown, but the unstated perception is that the sea is too vast, and its lowly inhabitants too numerous, for humans to impact as much as they have done in freshwater (Williams *et al.* 1993, Neves *et al.* 1997, Carlton *et al.* 1999). As a consequence, the majority of freshwater mollusks in the United States are currently managed with the goal of conserving species, but the story of marine mollusk conservation is still mainly that of fishery management.

31.2 HISTORICAL ATTITUDES TOWARDS MARINE MOLLUSKS

Native Americans on all coasts valued marine mollusks as reliable and abundant food sources. Favored sites for cooking and processing mollusks from a nearby source were used for generations, and the remains of these sites are known as shell middens or shell mounds. Oyster shell middens reached spectacular proportions, and survive today as uplands in otherwise low-lying areas of the Gulf of Mexico coastline. Shell middens remain along all coastlines and many are protected by federal law, but they are vulnerable to natural erosion.

Marine mollusks entered other aspects of the lives of Native Americans, as tools, ornamentation, currency, and cultural or religious objects. *Wampum* was beads made by Native Americans of whelks (Melongenidae) and quahog clams (*Mercenaria mercenaria* Linnaeus, 1758). Wampum was known among Europeans mainly as a mode of economic exchange, but had additional significance to Native Americans, as reviewed by Scozzari (1995). In the Pacific Northwest, Native Americans harvested tusk shells [*Antalis pretiosum* (Sowerby, 1860)] (Figure 31.1) solely for currency - they had no value as food - and the presence of these distinctive shells in archeological sites throughout western North America has been used to infer ancient trade routes

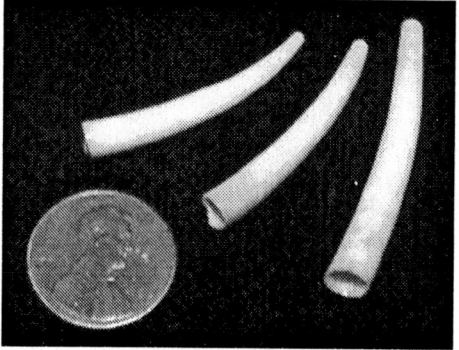

Figure 31.1 *Antalis pretiosum.*
Antalis pretiosum (Sowerby II, 1860), the tusk shell harvested by Northwestern Native Americans for use as currency [Florida Museum of Natural History 63627 (Photo by author)].

Figure 31.2 *Strombus gigas*.
Adult and juvenile queen conch, *Strombus gigas* Linnaeus, 1758, from the Florida Keys [Florida Museum of Natural History 8601/123714 (Photo by author)].

(Barton 1994). In southern Florida and the Caribbean, the queen conch (*Strombus gigas* Linnaeus, 1758) (Figure 31.2) was used for tools, currency, and adornment, as were abalones (*Haliotis* spp.) (Figure 31.3) in western North America (Cox 1962, Carstarphen 1982).

The first European settlers, in contrast to Native Americans, regarded marine mollusks as starvation rations. "Mussels" were eaten only by minorities and the poor, and Claasen (1994) believes this attitude explains why freshwater bivalves, also consumed only by minorities and the impoverished, came to their common name of "mussels" even though they do not closely resemble marine mussels. Pismo clams were once harvested by teams of horses and plows, to be used as animal feed. In Florida and the Caribbean, queen conch, which are now costly delicacies, were initially harvested by Europeans to bait fish traps. California abalones, another modern gourmet item, were ignored by European and Hispanic settlers, and the list continues. Oysters were the single exception to this low regard towards shellfish, and were exported to London prior to the 1770s, although even oysters had to overcome an earlier aversion by the first British settlers.

Attitudes towards seafood changed as immigrants themselves moved up the sociological food chain and brought their seafood preferences with them. Eastern oysters (*Crassostrea virginica* Gmelin, 1791) (Figure 31.4) again led the way, and during the 19th Century, "Blue Points," "Chingaroras" and other terms referred to oysters from favored locales. Other 19th Century molluscan fisheries included the Olympia oyster [*Ostrea conchaphila* (Carpenter, 1857)] (Figure 31.4) on the West Coast, New England softshell clams (*Mya arenaria* Linnaeus, 1758), hard clams (*Mercenaria mercenaria*) in the Mid-Atlantic, and bay scallops (*Argopecten irradians* Lamarck, 1819) from both New England and the Carolinas (Brooks 1891, Field *et al.* 1910, Baker 1995, Bleyer 2002). During this period, observers and harvesters were aware that local shellfish beds could be over-exploited and numerous local ordinances were passed by the early 18th Century to

Figure 31.3 *Haliotis sorenseni*.
An endangered white abalone, *Haliotis sorenseni* Bartsch, 1940, Newport Bay, California [Florida Museum of Natural History 121447 (Photo by author)].

Figure 31.4 Oysters.
Examples of the eastern oyster, *Crassostrea virginica* (Gmelin, 1791) (left) and the Olympia oyster, *Ostrea conchaphila* (Carpenter, 1857), from Virginia and British Columbia, respectively. Both specimens are market size (Photo by author).

limit harvest. Many of these early "conservation" efforts were appeals by one group of harvesters to exclude harvesters from other localities or ethnicities, while others were broader limits on harvest season, the number of harvesters, and wasteful harvest practices (Kochiss 1974, Wennersten 1981, McCay 1998, Bleyer 2002).

For the most part, however, the sea and its fisheries were treated as unlimited resources, endlessly renewable. The "boundless sea" (William Shakespeare, *Sonnet LXV*) is a part of our cultural conscience. We cannot see across the ocean, or under it, and it can be difficult to understand how we could deplete anything in it. Marine molluscan resources seemed "inexhaustible" to early resource managers (Brooks 1891). Clam and oyster shells were so abundant that they were (and still are) used as industrial, construction, and roadway material. Despite limited official recognition that shellfish stocks could be depleted, the wider perception was that it was just a local problem, and there were plenty of shellfish in the next bay or river (Kochiss 1974, McCay 1998).

Molluscan resources were not inexhaustible and, by the 20th Century, this had become more apparent. Oyster stocks declined or disappeared on both Atlantic and Pacific coasts. The commercial scallop industry in Massachusetts lasted a mere thirty years before going into decline at the end of the 19th Century. Hand harvesting techniques depleted near-shore abalone stocks in California and Mexico in about the same amount of time (Brooks 1891, Field *et al.* 1910, Cox 1962, Baker 1995). Some rudimentary conservation measures were in place - for example, in the latter part of the 19th Century, Maryland required sub-market sized oysters to be returned to the oyster grounds. Compliance was lacking (Brooks 1891), and the concerns of a handful of fishery managers were largely ignored until about the end of the century. It was not until the first part of the 20th Century that modern notions of seasons, take limits, size limits, and gear restrictions began to be combined into what we now call fishery management.

31.3 MARINE MOLLUSK CONSERVATION IN THE 20th CENTURY

William J. Hargis, Jr., fishery biologist and former director of the Virginia Institute of Marine Science, noted a distinctive character of a declining fishery, using Chesapeake Bay oysters as an example. Since little concern is given to a robust fishery, he declared, the health of a fishery can be measured inversely by the number of biologists studying it (pers. comm.). The study of fisheries became a serious science in the 20th Century and the reason, as noted by Dr. Hargis, was the failure of one fishery after another.

The Industrial Revolution had two important implications for marine resources. First, human impacts upon natural resources increased sharply, both by virtue of the rapid expansion of industry and the human population, and by the addition of new resource stressors. For example, Brooks (1891) noted that steam-powered vessels with dredges, a technological advance over hand-powered dredges or tongs, rapidly depleted oyster stocks in some regions. The expansion of dikes and resultant changes in water flow in the Mississippi Delta were blamed for oyster declines in Louisiana. The advent of hard-hat diving exposed new abalone stocks to exploitation. Industry also produced more pollution than ever before, with pulp mills and oil wells, for example, indicted for shellfish mortalities on the Pacific and Gulf of Mexico coasts, respectively (Cox 1962, Mackin and Hopkins 1962, Baker 1995).

seashore became a top destination, and millions of recreational shellfish harvesters added new pressure to what had been commercial or subsistence fisheries - particularly on the Pacific coast and in New England. In California, for example thousands of people flocked daily to Pismo Beach while, in New England, the clambake became a cultural institution

---

"Well here we are, Pismo Beach, and all the clams we can eat!" Thus spoke Bugs Bunny (Warner Bros.©), on those rare occasions when he remembered to take a left at Albuquerque. The Pismo clam, *Tivela stultorum* (Figure 31.5), is named after the best-known locality in California for collecting it, which in turn takes its name from Rancho Pismo, an early land grant when California was under Mexican rule.

The colorful Pismo clam was one of California's first important commercial and recreational mollusks, and ranked third in the state's molluscan fisheries, following abalone and oysters, in the period between the World Wars. The Pismo clam was also the first California mollusk to be regulated, and the history of its exploitation and management mirrors that of many other mollusks on the Pacific coast.

Like many other modern delicacies, Pismo clams were first harvested by European settlers to feed hogs and chickens. Around the beginning of the 20th century, though, settlers began eating the clams themselves. In 1911, a daily limit (200 clams) and a minimum size limit were established. In 1915, the daily limit was lowered to 50 clams, and in 1917, some beaches were closed seasonally. The minimum size limit varied for a few years, but eventually settled on 5 inches (12.6 cm) across the shell in most areas. In contrast, the bag limit continued to decline, to 10 clams in 1949, where it remains as of 2005. Other protection strategies were also developed. In 1929, a Pismo clam no-take sanctuary was established. In 1931, a California sports fishing license was required to take Pismo clams, and in 1947, all commercial harvesting of Pismo clams was halted.

The Pismo clam fishery is currently managed by a complex set of regulations. Only hand tools may be used to harvest Pismo clams, and sub-legal clams must be returned unharmed. In addition to sports fishing license requirements for personal use and a ban on commercial harvests, there is a closed season (May through August) in part of its range. Clams north of San Luis Obispo have a larger minimum size requirement than those taken to the south, and three areas have now been designated as no-take sanctuaries.

Figure 31.5 *Tivela stultorum*. Pismo clam, *T. stultorum*, from Pismo Beach, California [Florida MNH 266701 (Photo by author)].

The California Fish and Game Commission is charged with the delicate balance of conserving both the species and a sports fishery. For Pismo clams, at least, their approach seems to be successful, and clamming is still listed as a tourist activity in Pismo Beach visitor information.

Sources: Herrington (1929), Fitch (1950), California Fish and Game Commission (2002).

The Industrial Revolution also permitted the rise of a large middle class. Ordinary people could afford to invest in leisure activities and partake in vacations. Destinations that were once considered rugged were now considered scenic; activities such as hunting and fishing that were once done for survival were now done as recreation. The

(Neustadt 1992). Thus, not only were more people exploiting molluscan resources, but more people noticed when they declined.

Shellfish regulations existed before the 20th Century, but were primarily local, with no specific agencies or enforcement behind them. This began to change in the 20th Century, although not equally in all parts of the United States. Most marine mollusks of interest are coastal species, and as such, are regulated by states, not the federal government. How states choose to do this is strongly affected by historical factors that vary among regions. For example, in New England and New York, marine mollusks have been regarded as a public resource since Colonial times, but the intertidal and coastal waters where the shellfish reside are controlled by townships, not the state (Kochiss 1974). The same was originally true in the Mid-Atlantic region, but state governments have since assumed regulatory authority (Wennersten 1981, McCay 1998).

In the western United States, the state government assumed regulatory authority for fisheries early in the 20th Century. As on the Atlantic coast, fish and shellfish were regarded as public commodities, with two notable exceptions. In Washington State, most inland (e.g., Puget Sound) coastlines and the shellfish they contain are regarded as the private property of the adjacent landowner. The state regulates shellfish harvests, but does not imply the public right to harvest those shellfish. In Hawaii, most shellfish harvests are conducted by the native Hawaiians, who also control much of the shorelines where these harvests are conducted.

In most of the southeastern United States and the Gulf of Mexico, geography both protected molluscan resources and delayed management. The low, malarial swamps that stretched with few breaks from the southern Chesapeake Bay to Texas were long inhabited only by a hardy few souls, many of whom intentionally limited contact with the outside world. Louisiana was a notable exception to this, but even there, no statewide shellfish surveys had been conducted by the latter half of the 20th Century (Mackin and Hopkins 1962). Florida has comparatively well developed shellfish regulations, but in surrounding states, fewer molluscan species are regulated, and those to a lesser degree, than in New England or Pacific states.

Another contrast between Pacific and Atlantic states is the use of recreational regulations. From Alaska to California, recreational shellfish harvesters are limited to precise numbers of shellfish. In Oregon, for example, a clam digger is allowed twelve gaper clams, thirty-six softshell clams, fifteen razor clams, seventy-two mussels, and so forth. The same sorts of numerical limits apply even in Alaska, with its huge resource-to-user base. As fisheries come under pressure, the first response by regulatory agencies is to reduce the bag limit, and continue reducing it as needed. Fisheries in this region are closed only as a last resort, and the closures then tend to be permanent, as has happened for most abalone fisheries (see Table 31.4, for information on state regulations).

On the Atlantic seaboard, shellfish managers are less willing to set low recreational bag limits, and most Maine townships do not set recreational limits at all. Maryland allows 250 hard clams and one bushel of oysters per day, Florida allows one five-gallon bucket of bay scallops per user during the open season, and neighboring states have comparable regulations. Rather than reduce bag limits when the fishery is perceived to be threatened, resource managers are likely to simply close the fishery until it recovers, as Florida does regionally for its bay scallop fishery (Greenawalt 2002; see also Tables 31.2 and 31.4).

Regardless of the regulatory history, most marine mollusks are still regulated as user resources, not species. Since the rise of microbiology, shellfish harvest closures are more likely to be for human health reasons than for species conservation. Florida is the only state to acknowledge shell collecting or the aquarium trade in its regulations, although some other states, such as California and Oregon, place broad limits on the harvest of non-food intertidal invertebrates, including most gastropods. A growing number of states have also established marine reserves, where the collection of all invertebrates is curtailed or prohibited, and these and other trends probably represent a change in perceptions about marine mollusks that will be carried into the 21st Century. Of course, it could be argued that fishery conservation is as viable as species conservation when it comes to preserving a species; certainly this has been the approach taken by Duck Unlimited and salmon restoration programs like the Salmon and Trout Enhancement Program (STEP).

The federal government was late getting into fishery conservation, but the Magnuson-Stevens Fisheries Act, as amended in 1996, mandated regulation for a number of species by the National Marine Fisheries Service (NMFS), a division of the National Oceanic and Atmospheric Administration (NOAA). Through the development and implementation of Fishery Management Plans, NMFS administers fisheries in the "Continental Shelf," defined as waters outside the 12 mile (19.3 km) territorial limits of the United States, but within the 200 mile (322 km) Exclusive Economic Zone. Under certain circumstances, however, such as in the National Marine Sanctuaries and in U.S. territories, near shore waters are also regulated by NOAA.

Just six mollusks were regulated by NOAA in 1996: three species of abalone (*Haliotis rufescens* Swainson, 1822, *H. corrugata* Wood, 1828, and *H. kamtschatkana* Jonas, 1845), queen conch (*Strombus gigas*), ocean quahogs [*Arctica islandica* (Linnaeus, 1767)], and surf clams [*Spisula solidissima* (Dillwyn, 1817)]. Since then, Fishery Management Plans have been added for the sea scallop [*Placopecten magellanicus* (Gmelin, 1791)] in the Atlantic and the weathervane scallop [*Patinopecten caurinus* (Gould, 1850)] off Alaska, two species of squid [*Illex illecebrosus* (Lesueur, 1821) and *Loligo pealeii* (Lesueur, 1821)] in the Atlantic, and a third squid (*Loligo opalescens* Berry, 1911) off California, although squid are generally regulated within an integrated fishery management plan for several pelagic species. As this chapter was being written, a fishery management plan was being developed for the calico scallop [*Argopecten gibbus* (Linnaeus, 1758)] of the southeast United States where the fishery collapsed shortly prior to the Sustainable Fisheries Act (Moyer *et al.* 1993).

Only in the past few years has it been officially recognized that not all marine mollusks are suitable for fishery status. The white abalone of southern California (*Haliotis sorenseni* Bartsch, 1940), which will be discussed later, was federally listed as endangered, and was the first marine mollusk to gain that dubious status in the United States. It will probably not be the last, however. Four other abalones are listed as species of concern by the NMFS, harvest of any abalone species is banned in Washington State, and the queen conch (*Strombus gigas*) is completely protected in Florida (Tables 31.1-31.4).

Industrial-scale fishing has been present since the 1950s and provides much needed protein to hundreds of millions of people, but it is so effective that we are in the process of driving many species towards extinction - unthinkable to earlier generations who saw the sea as too vast to impact (Parfit 1995). Only recently have the federal government and international treaties begun to address the costs that come with cheap seafood. Pelagic squid fisheries, for example, are monitored by the National Marine Fisheries Service for *bycatch*, or the accidental take of non-target species.

Most large-scale molluscan fisheries are benthic, and some form of a benthic dredge is usually used, with predictable impacts on other biota. The Maryland softshell clam fishery represents an extreme case, using hydraulic dredges that entirely disrupt the sediments, and any associated community, down to a depth of about 20 cm (The hydraulic dredge industry in Maryland is currently restricted to one portion of Chesapeake Bay, well offshore). Benthic impacts are not limited only to molluscan fisheries, but mollusks are heavier than fish or crustaceans, with their massive carbonate shells. Many are infaunal, so they require heavier, more disruptive gear to harvest. As a result, these and other fishing grounds, as well as coastal benthic ecosystems around the world, are now nearly featureless plains, flattened by decades of pounding by a variety of fishing gear. The impact of fishing gear cannot easily be addressed, because there is no other effective way to harvest many species - particularly mollusks. The creation of marine sanctuaries, both through state efforts and the National Marine Sanctuary Program, approaches the issue of industrial fishing in the most direct way, by simply banning it in designated refuges. Mollusks are seldom the target of such sanctuaries, but are protected along with the rest of the species (Duff and Brownlow 1997).

Maritime user groups, from commercial fisheries to recreational divers, fishers, and collectors, are converging on the realization that the sea is not boundless after all. Our ability to cross the Pacific Ocean by jet aircraft in only hours contributes in some way to this new understanding, but so have the gloomy statistics accompanying virtually every marine fishery, and the visibly increasing numbers of beachgoers, recreational fishers, and SCUBA divers. In combination with a growing body of scientific evidence for decline or even extinction of marine mollusks, political will to develop and apply non-fisheries conservation for marine mollusks has also arisen. The bias is still towards fishery species, however. No attempt has been made to determine whether non-fishery species are in need of protection, despite the fact that several have become extinct (Carlton 1993, Carlton et al. 1999). If a snail lacks value as a fishery, or the charisma of a seal, it is difficult to arouse official or public sympathy for its plight.

31.4 DECLINING, COLLAPSED, OR EXTINCT: CONSERVATION ISSUES FOR MARINE MOLLUSKS

Management options for marine fisheries in decline are fundamentally different from more familiar conservation efforts applied to endangered species (or populations) such as the spotted owl or freshwater pearly mussels. In the United States, the legally defined terms *threatened* and *endangered* are starting to be applied to marine fishery species, but fisheries managers still use the terms *declining* and *collapsed*, respectively, to mean somewhat the same things. A declining fishery is one in which catch-per-unit effort must increase to produce the same yield. Most fishery populations, however, fluctuate naturally. A fluctuation may be interpreted as a fishery decline, or a fishery decline may be mistaken for natural fluctuation.

Using the weight of historical evidence, one might generalize that all heavily exploited fisheries, molluscan or otherwise, are in decline. A well documented decline is the red abalone (*H. rufescens*) in California, which remains a fishery but on a greatly decreased scale. Other cases are more difficult to interpret. For example, inter-annual variation of the Pacific weathervane scallop (*P. caurinus*) south of Alaska is so extreme that there is an effective fishery only at irregular intervals, and the remainder of the time the scallop persists at apparently natural low densities (Nunez 1988). If a decline were to occur in this fishery, how would we know?

A collapsed fishery is one that has attained a new stable population level below what is profitable to harvest, throughout much or all of its former range. The California sardine, for example, was the foundation of wealth for Monterey, California, and was the colorful backdrop for John Steinbeck's *Cannery Row* (Viking Press 1945). Only a few years after the publication of *Cannery Row*, the sardine fishery collapsed, and never recovered its former level. A fishery collapse typically follows a period of intense exploitation, but why the fishery remains collapsed is seldom understood. In the case of the sardine, fishery pressure may have accelerated a natural decline due to cyclic (decade-scale) fluctuations in climate. Fishery collapses are becoming more common in the ocean, and the term *commercially extinct* has recently been used in that context (Torres and Sullivan-Sealey 2000).

Mollusks have not been immune to fishery collapse, despite having some of the highest fecundities of any marine organisms. Well known collapses include bay scallops and eastern oysters along parts of the Atlantic coast of North America, Olympia oysters and abalone on the Pacific coast, and queen conch in the Caribbean (Marshall 1947, Rothschild et al. 1994, Baker 1995, Torres and Sullivan-Sealey 2000). While all of these collapses occurred during periods of exploitation, some were associated with extraneous environmental processes. A blight that nearly wiped out eelgrass [*Zostera marina* (Linnaeus, 1753)] in New England and the Mid-Atlantic was associated with the simultaneous decline of the bay scallop, although some scallops persisted in areas without eelgrass (Marshall 1947). Oyster diseases in Delaware and Chesapeake bays have been blamed for the near-eradication of the eastern oyster in those regions, but these collapses occurred during and after a century of intense fishing pressure (Brooks 1891, Rothschild et al. 1994, Ford and

Tripp 1996). In addition, the question of whether humans introduced any of those diseases - eelgrass or oyster - has never been definitively addressed, but seems likely in at least some cases (Carlton and Mann 1996).

Extinctions are the same whether one is speaking about fisheries or species, but biologists recognize two categories of extinction. The first is species extinction, when no members of a species remain alive anywhere on Earth. Well documented for freshwater mollusks, modern species extinctions seem to be rare among marine mollusks; some exceptions will be discussed below. The other category is the extirpation of a population, or as resource managers now prefer, distinct population segment (DPS) or evolutionarily significant unit (ESU) (Moritz 1994, Waples 1995). A DPS or ESU is important because it is believed to contain unique genetic information that may be important for the continued survival of the species as a whole. Additionally, an ESU may be uniquely adapted to a particular habitat. Bowen (1998) describes and critiques ESUs and the underlying evolutionary concepts, and King et al. (1998) overviews gene-based conservation in mollusks in particular. The term extirpation is sometimes preferred over extinction for the loss of an ESU but not an entire species.

Management Units (MUs) are populations with significant divergence of genotype frequencies, regardless of the phylogenetic distinctiveness of the genotypes (Moritz 1994). Differences in genotype frequencies denote populations with independent demographic trajectories (Wright 1931, Slatkin 1987). Hence, MUs are grounded in principles of population biology and theory. MUs often correspond to "stocks" in fisheries management, and are the fundamental units of wildlife management: reproductively isolated populations.

The Evolutionarily Significant Unit (ESU) is a category above MU, describing a population or groups of populations that are evolutionarily distinct, as indicated by substantial divergence in ecological, morphological, or genetic traits (see Waples 1995). The ESU concept is valuable because it allows wildlife managers to circumvent taxonomic issues (such as controversial subspecies definitions) and focus on the preservation of major evolutionary subdivisions within species. One common criterion for ESUs is diagnostic genetic differences among populations, indicating evolutionary depth (see Bernatchez 1995). However, several reports have expressed dissatisfaction with the ESU because some populations that probably merit conservation resources do not qualify under current definitions (see Barlow 1995, Stauffer et al. 1995, Pennock and Dimmick 1997, Karl and Bowen 1999).

The Distinct Population Segment (DPS) is the lowest category to hold legal protection under the U.S. Endangered Species Act. The DPS incorporates many of the same criteria as the ESU: a DPS must be exceptional in a way that indicates evolutionary divergence or novelty (Waples 1995, Pennock and Dimmick 1997).

In the United States, extinctions of marine mollusks have received little notice compared to their freshwater counterparts. The first documented historic extinction of a marine invertebrate was the Atlantic eelgrass limpet, [*Lottia alveus alveus* (Conrad, 1831)] in New England and southeast Canada (a separate subspecies, *L. a. parallela* (Dall, 1914), persists in the North Pacific). The decline occurred following a catastrophic die-off of eelgrass in the region; an essential habitat to this specialized species. It is possible that the extinction was a natural event, but it is also possible that the disease was introduced by human activity (Carlton et al. 1991, Carlton and Mann 1996)

Several other marine mollusk extinctions have since been documented in the United States. A horn snail, *Cerithidea fuscata* Gould, 1857, was last reliably reported from its only known habitat, San Diego Bay, California, in 1935. A geographically restricted habitat is more typical of freshwater than marine mollusks, which may be why freshwater mollusks and *C. fuscata* have proven vulnerable to extinction. Habitat modification of San Diego Bay is implicated in the case of *C. fuscata*. A limpet of uncertain generic status, "*Collisella*" *edmitchelli* (Lipps, 1963), which was restricted to the southern Channel Islands of California, has not been recorded since 1861. The extinction of a Florida sea hare, *Phyllaplysia smaragda* Clark, 1977, is regarded as tentative by Carlton et al. (1999), pending further research (Carlton 1993, Carlton et al. 1999).

Mollusks lack the charisma of sea otters or pelicans; they are much less likely to receive public sympathy - and corresponding political support - if imperiled. Worldwide, only a handful of threatened marine mollusks have received significant public attention; examples include giant clams (Tridacnidae) of the Indo-Pacific and the Triton's trumpet snail [*Charonia tritonis* (Linnaeus, 1758)] from the same region. The International Union for the Conservation of Nature (IUCN) listed ten marine mollusks as critically endangered, endangered, or vulnerable in 2004; all but two of these are either giant clams (Tridacnidae) or cone snails (*Conus*), popular with aquarists and shell collectors. CITES (Convention on International Trade in Endangered Species) currently lists all giant clam species and the queen conch, *S. gigas*, in Appendix II. The date mussel *Lithophaga lithophaga* (Linné, 1758) is also listed in Appendix II but probably because the fishery for this rock-boring species is so destructive (CITES 2004).

Marine mollusk population (or ESU) extirpations have apparently also occurred in North America. For example, while the sea hare *P. smaragda* (above) may still occur somewhere in the northern Caribbean, it is almost certainly extinct from all of Florida. The fact that it has not reappeared in many decades suggests either that the species is extinct everywhere else, or that the Florida animals comprised a discrete unit, not part of a metapopulation maintained by larval transport in oceanic currents. The bay scallop (*Argopecten irradians*) disappeared from Maryland and Virginia 70 years ago, and despite the fact that it persists to the north and south, has never reappeared in those states. Habitats of the Olympia oyster (*O. conchaphila*) are highly discrete, and in some of these, the species has been extirpated in historical times (Baker 1995, Tarnowski and Homer 1999).

An important difficulty in understanding population extinctions is that we usually do not know how or when populations arose, and how discrete they were before extirpation. Was the lost "population" an Evolutionarily Significant Unit, or merely a geographical aggregation? In the scenario of most concern, a population was reproductively isolated long enough to evolve a unique genetic response to local conditions, and its extinction means a permanent loss of genetic information. It is also possible, however, that what appears to be a discrete population is actually a subpopulation - one unit of a metapopulation - but that gene flow among subpopulations is so infrequent it may not be sufficient to restore a population over the timeframe (years and decades) relevant to wildlife management strategies.

In a hypothetical example, imagine an intertidal snail that occurs on both the mainland and a remote island, and reproduces independently in both habitats. In a subpopulation scenario, oceanic conditions that occur only once or twice a century carry numerous larvae from one habitat to the other, swamping any incipient genetic divergence (Figure 31.6). In this scenario, what appears to be two separate populations is actually a single population from an evolutionary standpoint. Extirpation of the island snails would result in little genetic loss to the species, and re-stocking might reasonably be done using mainland snails. This is not to say that the extirpation of the island snails is acceptable, nor that re-stocking is easily achieved. Nonetheless, in this scenario, the loss of unique genetic information is one less thing about which a fishery manager has to worry.

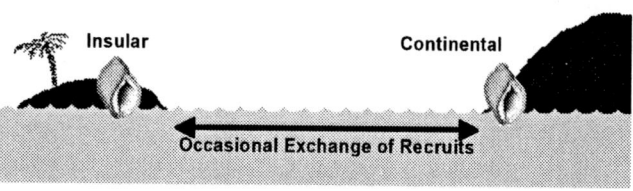

Figure 31.6 Subpopulations (Management Units).
Island (insular) snails and mainland snails are usually isolated, but on rare occasions larvae from one location recruit to the other, providing just enough exchange to prevent the development of unique local genotypes.

In a population scenario, the snails in the two habitats are completely reproductively isolated. Random mutations followed by genetic drift will result in unique genotypes, and it is also possible that natural selection in response to local conditions will produce populations that are uniquely adapted to insular and continental habitats. If the island population was extirpated, true genetic loss to the species would occur. A fishery manager must first try to prevent the loss of this unique genotype. According to current conservation practices, efforts must be made to make sure no relict populations of the island snail remain, prior to re-stocking, and if they do, the focus should be on population restoration, not replacement.

The importance of ESUs and the risk of hybridization has become one of the conservation focuses in western North America for landlocked and anadromous fishes. The topic is also beginning to be discussed for endangered freshwater mollusks (King et al. 1998, Roe and Lydeard 1998). For marine mollusks, conservation has barely reached the stage where concern for native species, let alone populations, takes precedence over the immediate needs of fisheries or aquaculture (Carlton 1979, Carlton and Mann 1996, Robinson and Johnson 1997, Colson and Sturmer 2000). In some of the most recent efforts under consideration, however, population genetic integrity is a crucial part of the solution.

Why are marine mollusks, with their enormous reproductive capacity, vulnerable to human pressure? For some, just as for freshwater or terrestrial organisms, the answer is habitat limitation. For example, seagrass is optimal habitat for bay scallops, (although not necessarily essential; see Marshall 1947), and the decline in eelgrass in the 1930s resulted in corresponding declines in bay scallops. To oysters, optimal habitat is more than merely a hard place to settle, but the proper topography as well. Centuries of fishing pressure in Chesapeake Bay have reduced former oyster reefs from high-relief intertidal structures to low-relief subtidal areas. When a high-relief intertidal reef was artificially restored in one branch of Chesapeake Bay, the result was strong oyster recruitment and growth (Bartol et al. 1999). For many species, habitat may not seem limiting, but there are also possible changes in habitat, or *ontogenetic niche shifts*, between juveniles and adults. It is now known that some mollusks recruit in one habitat as newly metamorphosed postlarvae, and then relocate to the adult habitat as juveniles. The postlarval habitat is known for only a few bivalve species, and until we know the habitat for all life stages, we cannot rule out habitat limitation.

Other marine mollusks are vulnerable to fisheries because of their long life cycles. Bay scallops in Florida reach the fishery in only a year and usually do not live much longer, but some species take much longer to reach fishery or reproductive size. Geoduck clams [*Panopea abrupta* (Conrad, 1849)] and knobbed whelk [*Busycon carica* (Gmelin, 1791)] take a decade to reach full size or sexual maturity. The ocean quahog (*Arctica islandica*) takes 15-20 years to recruit to the fishery and can live to at least 70 years. Long-lived species cannot replace themselves as fast as a fishery can remove them (Zettler et al. 2001, Greenawalt 2002).

The most fundamental vulnerability of many marine mollusks may be their reproductive strategy of broadcasting spawning. Reproductive success in broadcast spawners is subject to what is known as the Allee Effect, or density dependence (Allee 1931). Gametes that are broadcast into the water column have a short functional lifespan, sometimes measured in seconds. If the currents are wrong or a mate is too far away, the gametes loose viability before fertilization can occur. Some gastropods and most bivalves cannot migrate to a common mating ground, but must spawn from where they are, regardless of whether other potential mates are nearby or spawning at the same time. Unless two oysters are within centimeters of each other, spawning simultaneously, fertilization success will be negligible. The Allee Effect is not limited to broadcast spawners but has been implicated for species such as the queen conch (*Strombus gigas*). The implications are that any time a fishery reduces the density of mollusks dependent upon broadcast spawning, the remaining population is at risk of reproductive failure (Stoner and Ray-Culp 2000).

In the following sections, fishery concerns, management, and conservation of several important groups of mollusks are discussed. This is not intended to be a comprehensive review, and in the interest of space, many groups are omitted; I apologize if I have omitted your favorite mollusk. The discussion is focused on the United States, with some examples from Canada, but the same issues for marine mollusk conservation occur worldwide.

31.5 NATIVE OYSTERS ON TWO COASTS

The United States and Canada share two native commercial oysters - one species on each seaboard. Populations of both oysters have declined, following fishery exploitation and industrial development, and both have subsequently been the targets of conservation and restoration attempts. Management methods for these two species have differed greatly, however, and serve as an illustration of how historical and cultural differences between east and west coasts, of the United States in particular, can influence marine molluscan conservation.

The northwestern Atlantic basin has a rich oyster (Ostreidae) fauna, with at least five species making it into the United States. Of these, the most ecologically and economically important oyster is the eastern oyster, *Crassostrea virginica*, which occurs in estuaries from the Gulf of St. Lawrence to the Gulf of Mexico. Its distribution in coastal waters is nearly continuous, except at the northern and southern extremes of its range. The eastern oyster is a relatively large species, with a heavy shell often more than 20 cm in length. Oysters cement permanently to hard substrata, and in most areas, the favored substrate is shells of other oysters, living or dead. Generations of oysters attaching to each other create three-dimensional structures called reefs or bars. Reefs created by *C. virginica* can be hundreds of meters or even kilometers in length, and are believed to enhance biodiversity of both sessile and mobile estuarine organisms (Cake 1983, Coen *et al.* 1999, Luckenbach *et al.* 1999). Newell (1988) estimated that a century ago, oysters in Chesapeake Bay could, by their suspension-feeding on plankton, filter a volume of water equal to the entire bay in only three days.

In contrast, the northeastern Pacific Basin, north of Mexico, has a single described oyster species, the Olympia oyster, *Ostrea* (*Ostreola*) *conchaphila*, which occurs in estuaries from Alaska to Mexico. Its distribution is highly discontinuous, particularly along the outer coast from Washington to Point Conception, California, where it is known from only ten estuaries. The Olympia oyster is thin-shelled and much smaller than the eastern oyster, seldom exceeding 7 cm in shell length. This species forms reef-like aggregations, but its ecology is much less studied than for eastern oysters (Baker 1995).

31.5.1 History of eastern oyster fishery and decline.
The eastern oyster was a food staple of coastal Native Americans, and the shell middens they created, over generations of fishing the same waters, reached the proportions of small islands in some regions. Unlike most shellfish, oysters in New England and the mid-Atlantic were consumed by early British settlers, and there was also a long history of oyster exploitation by French immigrants in Louisiana. Oysters were also harvested purely for their shells, used for road construction or for the production of lime. The harvest of oysters for their shells only was recognized as wasteful, and banned in some states in the 18th Century. These various early forms of exploitation were blamed by residents and government officials for depletion of local oyster stocks. In fact, many southern New England oyster stocks were depleted prior to the 19th Century, and stocks in other areas were impacted soon after (Mackin and Hopkins 1962, Kochiss 1974, McCay 1998).

More waves of immigrants arrived, and seafood preferences and harvesting techniques broadened. By the 19th Century, well-developed commercial fisheries occurred in areas such as Malpeque Bay (Prince Edward Island), Long Island (New York), Raritan, Barnegat, and Delaware Bays (New Jersey), Chesapeake Bay (Maryland and Virginia), Apalachicola Bay (Florida), and Barataria and Terrebonne Bays (Louisiana). In Virginia, the oyster industry gained thousands of new recruits when former soldiers of the Confederacy sought to make a living in Chesapeake Bay. The southeastern United States, however, from North Carolina to eastern Florida, was a comparatively minor production region during this period (Mackin and Hopkins 1962, Kochiss, 1974, McCay 1998).

On the Atlantic and Gulf of Mexico coasts, fishery resources were traditionally recognized as a common property of the local (e.g. township or county) users, but the definition of *local user* was a source of conflict. In Maryland and Virginia in particular, fighting between rival groups of "watermen" led to bloodshed and state intervention. Conflicts over oysters are too broad of a topic to be covered adequately here, but have been covered in several excellent books (Kochiss 1974, Wennersten 1981, McCay 1998). Virginia, Maryland, and other states subsequently codified the concept of public fishing grounds. For example, the Baylor Survey Grounds in Virginia are defined areas of traditional oyster production, preserved for common use by all oyster harvesters. It should be noted that while it is used as a conservation tool, the Baylor Survey was conceived purely as a legal and political effort. In Chesapeake Bay, control of oyster production was eventually taken over by the states, but in New York and New England, it mostly remains in the control of townships to this day (Baylor 1894, Kochiss 1974, Wennersten 1981). Not all eastern oyster production has remained in the public domain, and in many states, certain oyster production areas may be privately owned or leased (McKenzie 1996, McCay 1998).

Declines in eastern oyster abundance were noted prior to the 19th Century. Brooks (1891) was among the first researchers to sound the alarm, for the Maryland portion of Chesapeake Bay, but stated that the decline had been ongoing for some time. This point is also made by McCay (1998), who cites numerous early warnings by locals and fishery observers that oyster resources were being depleted. In Louisiana, Moore (1899) suggested that not only harvesting but also habitat alteration (the construction of levees) were to blame for widespread oyster declines. In the resource-rich waters of Chesapeake Bay and the Mississippi Delta, the industry itself was slower to be impacted because harvesting technology improved, oyster prices increased, or the harvesters simply moved on to new beds.

Not only were oysters declining, but so was their habitat - the three-dimensional bar or reef created from the shells of many generations of oysters attached to each other. For most of the fishery's history in all areas, shells of the harvested oyster, plus the shells they were attached to, were left in large piles near processing plants - modern shell middens - either to be discarded or to be used for construction. This meant that years and generations of harvesting the same reefs planed these three-dimensional structures down to low-relief bars, which were then vulnerable to siltation (Kennedy and Sanford 1999).

After World War II, with the growth of population and industry in coastal areas, the rate of oyster fishing picked up but worse was yet to come. In the late 1940s, the oyster parasite *Perkinsus marinus* (Mackin, Owen, and Collier, 1950) (Dermo) was reported from the Gulf of Mexico, but its real impacts occurred after the mid-1950s, when it was reported from Delaware Bay, and subsequently, Chesapeake Bay. At about the same time, a second parasite, *Haplosporidium nelsoni* (Haskin, Stauber, and Mackin, 1966) (MSX) appeared in the same area, and has since spread from Maine to Florida. Either disease typically kills the oyster before it reaches market size (Ford and Tripp 1996). These diseases may have been introduced by human activity (e.g. shellfish transfers to the Atlantic coast) (Sindermann 1993, Carlton and Mann 1996). Regardless of their origin, these two diseases, combined with continued harvesting and other factors, reduced oyster reefs and oyster harvests to a tiny fraction of their former level, at least from Chesapeake Bay north. Newell (1988), in his study on the filtering capacity of Chesapeake oysters, indicated that modern oyster populations would take close to a year to filter the entire Bay. Since the residence time of water in Chesapeake Bay is far less than a year, this means that the ecological impact of modern oyster populations on the water column is negligible. Other diseases, notably the so-called Juvenile Oyster Disease, have hit eastern oyster populations in other parts of the species range.

Oysters south of Chesapeake Bay, including the Gulf of Mexico, are less affected by MSX or

Dermo. MSX is mostly absent from the Gulf of Mexico, and Dermo, while present, is seldom implicated in major mortality events. Oyster harvests in southern areas have remained relatively stable since the 1950s, and as a consequence of declines in other areas, southern sources now supply most of the market for eastern oysters (Newell 1988, Rothschild et al. 1994, Ford and Tripp 1996).

31.5.2 Management and conservation of eastern oysters.

Academic recognition of the need for oyster conservation followed the opus simply titled *The Oyster*, by William Brooks (1891), a professor at Johns Hopkins University in Baltimore, Maryland. The first two thirds of the book were devoted to biology and culture of the eastern oyster, but the last third described the already dramatic decline of oyster populations in Maryland, and promoted the need for conservation and management. Prior to this, it was generally recognized that oysters were a limited commodity, but apart from setting harvest seasons, official responses were limited to laws against "poaching" by one group of users against another (Wennersten 1981, McCay 1998).

More scientific or official steps towards conservation or management gradually followed. Studies like the Baylor Grounds Survey in Virginia (Baylor 1894) and the Louisiana oyster beds report (Moore 1899) established the extent of surviving areas of oyster production. In the 20th Century, following the rise of new harvest technologies, gear restrictions were added to harvests from public grounds. Gear restrictions have two purposes: to level the playing field for all entrants to the fishery, and to control harvest efficiency. The most famous of these gear restrictions are in Maryland, where oysters from public grounds may be harvested by dredges, but the vessels involved must be powered by sail alone. These vessels, known as *skipjacks*, are now mainly historical relics, along with the oyster reefs. In Virginia, by contrast, powered vessels may be used, but the gear is limited to hand-held tongs. In both cases, the intent was to strike a compromise between a harvest method that could provide a living to watermen and preserving the fishery from overexploitation. It might have worked, but fishing on top of other stressors, such as disease, all but eliminated the Chesapeake oyster fishery by the late 1980s (Rothschild et al. 1994).

Oyster fishery managers have more than fishery restrictions as a management tool, and as the fishery has dwindled, management has become more aggressive. Proactive methods can be divided into three main types: silt removal, oyster bar augmentation, and "seed" transfers. Silt, from a variety of natural or anthropogenic sources, can cover oyster bars, particularly those flattened by years of oyster removal, and prevent the settlement of juvenile oysters (e.g., Butler 1949). Silt removal is usually done with a towed harrow or similar implement that either blows the silt off to one side or lifts some of the shells above the silt, thus exposing them to natural settlement of juvenile oysters. Oyster bar augmentation is similar in principle, but consists of taking oyster shells from one location, preferably a non-productive site, and placing them on a productive area, again, to provide substratum ("cultch") for natural settlement. In some cases, fossil shell, clamshell, or even artificial cultch is used (Mann et al. 1990). "Seed," or juvenile oyster transfers consist of relocating juvenile oysters from habitats were they settle well to areas where they grow well, based either on established local knowledge of such areas or fishery studies (McKenzie 1996, Luckenbach et al. 1999)

Numerous attempts have been made to augment or replace the eastern oyster fishery with aquaculture. In Connecticut, Virginia, and other states, private grounds or leases for oyster "culture" were established, but they generally relied upon wild settlement or relocation of wild oysters. Culture of other (nonindigenous) oyster species was also suggested, and pilot oyster hatcheries were developed, but the culture of oysters in the Mid-Atlantic has never become a commercial reality (Kochiss 1974). Sociological and political factors played a role in this; during the 1980s, the well-organized Virginia oyster harvesting lobby ("watermen") opposed efforts to introduce a disease-resistant nonindigenous oyster, fearing their market share would be lost to a cultured species. In contrast, the much less-organized northern quahog (*Mercenaria mercenaria*) fishery in Virginia put up little resistance to clam culture, and Virginia is now a leading cultured clam producer. A similar dichotomy was observed in Florida; oyster culture and clam culture were promoted simultaneously as a source of income for fishers displaced by a gill net ban. Fisheries for both shellfish species already existed but while clam culture has since thrived in Florida, oyster culture met resistance from the oyster fishery and failed to take hold (Colson and Sturmer 2000). In both Virginia and Florida, the historical importance of the oyster fishery, plus a justifiable fear that the development of an oyster culture industry would spell the end of the wild oyster fishery, have impeded acceptance of aquaculture as a means of producing oysters. Only in the northern part of its range, particularly in New England and the Maritime Provinces of Canada, has oyster aquaculture become established. Not coincidentally, these areas have also been receptive to the culture of nonindigenous oysters (Kochiss 1974).

From a conservation perspective, social resistance to oyster culture in the Mid-Atlantic and Gulf of Mexico has its advantages. Nonindigenous oysters (namely, the commercially successful Pacific oyster, *Crassostrea gigas*) have not been introduced to possibly compete with natives, nor have local culturists supplied the market for fresh oyster products. This has forced resource managers to concentrate instead on ways to preserve and restore natural, native eastern oyster resources. This focus on natural populations has, in turn, fueled more ecological studies on the eastern oyster than on any bivalve in the world. Oysters are recognized not only as a resource, but also as *ecosystem engineers* that alter the environment around them in ways beneficial to the ecosystem as a whole (Cake 1983, Newell 1988, Rheinhardt and Mann 1990, Jones *et al.* 1997, Coen *et al.* 1999). In combination with their economic value, this makes natural oyster beds a prime candidate for preservation in the face of human pressure. Additionally, oyster bar augmentation has now been taken to the next logical level with the successful re-creation of realistic oyster reefs in areas where they had once existed (Bartol *et al.* 1999). Since the early 1990s, however, social resistance to oyster culture in Chesapeake Bay has eroded along with its political base in the oyster fishing industry, and Maryland and Virginia are again considering the introduction of a disease-resistant nonindigenous oyster.

The human population in coastal areas of the Gulf of Mexico and southeastern United States is growing rapidly and will put continued pressure on oyster resources, directly or indirectly (Baker *et al.* 2003). It is very possible that the trend seen elsewhere, a decline or collapse of oyster fisheries, will also be observed in these areas. Once that happens, the only eastern oysters to reach the palates of discriminating diners will come from Maine, Canada, or other areas where aquaculture has replaced the fishery.

In the Mid-Atlantic, the historic center of the American oyster industry, we are seeing a transition of attitudes and approaches to conservation of the eastern oyster. From a common resource that supported a rugged population of watermen, the industry has dwindled, and its supporters fallen away, until almost all that remain are the managers and researchers who tried to sustain the fishery since the time of Brooks (1891). It is now they, not the watermen, who tell the oyster's tale to the public, and it is they who have emphasized the oyster as ecosystem engineer, rather than a mere restaurant delicacy. Should this new group of stewards succeed in restoring the oyster to its former dominance in the ecosystem, will the oyster fishery also return? Will a new generation of watermen be willing to deal with the rugged life on the oyster bars, combined now with the demands of a state bureaucracy more protective of the oysters than the harvesters? It is difficult to guess, but given the examples of other resource industries elsewhere in the United States, from logging to cattle ranching, the answer is likely to be no.

31.5.3 History of Olympia oyster fishery and decline. By the time of the California Gold Rush (1849), oysters were a delicacy among the immigrants. Hang Town Fry, so named because it was reputed to be a favored last meal among condemned convicts, included eggs and oysters - in this case, the small Olympia oyster native to California. There was no pre-existing fishery, and this was the West, where natural resources were bestowed freely by a benevolent federal government. The fishery was open to whoever wished to enter it and in whatever manner they chose. The opinions of the surviving Native Americans were not consulted.

Olympia oysters attach to each other, but the aggregations they form are not as robust or three-dimensional as those of eastern oysters. It was a simple matter (albeit, still hard work) for the harvesters to simply rake up Olympia oysters off the mud. Unfortunately, this form of harvest removed almost all habitat for re-settlement, because of the lack of exposed layers of shell beneath. Once harvested, the oyster fishery did not always recover. San Francisco Bay was the first to be fished out, although silt from gold mining and other disturbances probably played a role as well by destroying habitat (Carlton 1979). From San Francisco Bay, the fishery worked its way south to Mexico and north through California, Oregon, and Washington. Local extirpations occurred in a number of estuaries in all three states over the course of the next century, although this was concurrent with massive disturbance in the form of siltation, dredging, and filling, so it is difficult to blame only fishing. An oyster fishery also existed in British Columbia, but the distribution of the species was not as precisely documented as further south, and it is unclear whether any local extirpations occurred (Baker 1995, Baker et al. 1999).

By the early 19th Century, most easily exploited Olympia oyster stocks were gone. Attempts to introduce the eastern oyster, *C. virginica*, were for the most part unsuccessful, although numerous other species were accidentally introduced in the process (Carlton 1979). To meet West Coast demand, private growers in Washington State, where tidelands may be privately owned to this day, modified a French oyster culture system using concrete dikes to pond water at low tide. This system relied upon natural settlement of oysters, but growing conditions and predators could be controlled. The dike system was also tried in British Columbia and northern California, but did not become widely used except for the southwest Puget Sound of Washington. Here, oyster culture provided a living for a number of growers for several decades and modern shellfish aquaculture companies in Washington date their beginnings to the Olympia oyster (Baker 1995).

The Olympia oyster culture industry of Puget Sound collapsed due to high mortalities and reproductive failure during the 1950s. Pollution from pulp mills was generally accepted as the source of the problems but, by the time the problem had been mitigated, the oyster industry had moved on. The Pacific oyster, *C. gigas*, was imported from Japan, and even though it took decades to develop local hatcheries for this species, the industry was sustained by annual imports of juvenile oysters, or "seed," direct from Japanese waters. Again, this period was associated with the accidental introduction of many other species, including some serious oyster pests (Carlton 1979, Baker 1995). The Pacific oyster is roundly despised by many American epicureans, compared to the eastern oyster (*C. virginica*) or the Olympia oyster. Most consumers, however, are content with the fast-growing Pacific oyster, which attains sizes larger than the eastern oyster, and it is certainly a more commercially attractive product that the tiny Olympia oyster. The Olympia oyster remains a commercial product, but ranks near the bottom of all commercial oyster species in total value and is strictly a specialty product. A simple comparison of tissue mass reveals why; a standard 1-gallon (U.S.) shipping container holds 80-140 shucked Pacific oysters, but 1600-2000 shucked Olympia oysters. Olympia oyster populations have recovered somewhat in southwest Puget Sound, which remains the primary source, but California, Oregon, and British Columbia also produce small quantities (Baker 1995).

The rapid demise of the Olympia oyster fishery and the subsequent rise of oyster aquaculture have led to a stark contrast in regional perceptions of oysters, compared to the Atlantic seaboard. In California and Oregon, all oysters (native and non-indigenous) are assumed to be private property, and harvest, even on public land, is forbidden. Public harvest in Washington is permitted, but only on clearly designated public lands (Tables 31.2 and 31.4). Other shellfish, from abalone to clams, are considered to be in the public domain if they are on public land, even though some of them are also cultured. In Chesapeake Bay or the Gulf of Mexico, such exclusion from the natural resource would be unacceptable.

31.5.4 Management and conservation of Olympia oysters. Coastal communities on the Pacific coast, with the exception of the few surviving Native American communities, are young compared to their counterparts in the Atlantic seaboard. No shellfish industries or traditions had a chance to develop before most Olympia oyster populations were fully exploited. By the time state and federal agencies began taking an interest in shellfish resources, oyster culture attempts were already underway. Federal and state efforts were focused on assisting the oyster industry, either by importing eastern oysters (which failed to become established in most areas), or by assisting in the development of Olympia oyster culture (Carlton 1979, Baker 1995).

With the near-demise of the Olympia oyster culture industry in Puget Sound and the rise of the Pacific oyster industry, official interest in the Olympia oyster waned. Regulations against public harvest of any oyster, in any locality, were subsequently erected, presumably because of the difficulty (to the law enforcement officer) of determining the origin of oysters once they were collected. Bans against harvests of wild Olympia oysters were probably erected simply because of the difficulty in distinguishing between the native Olympia oyster and small cultured Pacific oysters. Almost no specific efforts towards Olympia oyster conservation, apart from monitoring the waning culture industry, were made for most of the second half of the 20th Century.

Several trends may have come together at the same time to renew interest in the Olympia oyster. Olympia oyster commercial production began to rise slowly in the 1980s. The spectacular success of the non-indigenous zebra mussel in the Great Lakes focused attention on non-indigenous aquatic species, bivalves in particular, including the Pacific oyster. The reverse side of non-indigenous species management is native species conservation, and on the Pacific coast, this included the Olympia oyster.

Finally, recent research on the role of oysters as *ecosystem engineers* (Newell 1988, Jones *et al.* 1997, Coen *et al.* 1999) provoked interest in the Olympia oyster for reasons other than a fishery.

Robinson and Johnson (1997) undertook a unilateral effort to "restore" the Olympia oyster in Oregon, by relocating stock from the only surviving population (Yaquina Bay) to several other isolated estuaries. In at least one case, this relocation effort appeared to have re-established a self-sustaining population where one had previously died out (Netarts Bay). In another location (Coos Bay), a population had already been established, probably via the translocation of cultured Pacific oysters from other estuaries (Baker *et al.* 1999). Olympia oysters were also introduced to estuaries with no historical or fossil record of that species, although there is no evidence that any of these transplants established a new population.

The above anecdote brings up a significant hurdle facing managers who wish to conserve or restore the Olympia oyster. Unlike the eastern oyster, the Olympia oyster distribution is highly discrete through much of its range, and isolated localities may be genetically distinct populations. Isolated populations, it follows, may be uniquely adapted to that locality, and thus be Ecologically Significant Units (ESUs - see Section 31.4). It may be inappropriate to use oysters from one ESU to restore a population in another habitat, if a remnant ESU in the target habitat still persists, because of the risk of genetic dilution. For this reason, conservation efforts for the Olympia oyster remain in the conceptual stage (Baker *et al.* 1999).

31.6 FISHERIES COLLAPSE: ABALONES AND QUEEN CONCH

Bivalves dominate American molluscan fisheries, but the most costly fishery products come from gastropods - abalone on the Pacific coast and queen conch in the Caribbean. Abalones (*Haliotis* spp.) are free-spawning, releasing millions of microscopic eggs into the plankton. Queen conch (*Strombus gigas*) deposit egg cases, like most gastropods, but large females still produce large numbers of eggs, each of which hatches into a free-swimming veliger larva. Given this fecundity, one would expect high fishery resilience of these species. The reality is that starting from huge fisheries less than a century ago, only two species of abalone are still fished to a limited extent; the remaining species have been harvested to the point of fishery collapse. The queen conch is completely protected in Florida, the only U.S. state in which it occurs.

Up to fifty species of abalone occur worldwide and seven of these are found in California, the center of the U.S. fishery. The common gastropod spiral is collapsed into a large, ear-shaped shell, and the Latin name for the genus, *Haliotis*, means, roughly, *sea ear*. The abalone has no operculum and most of the ventral surface of the animal is a huge foot that clings to the rocky substrate, where it grazes on algae. It is this foot that is the primary target of the fishery, with retail prices for wild specimens approaching $100 per pound. Abalone shells are strong, pearly, and attractive, and were used by indigenous peoples around the world for tools, ornaments, and even currency. The shells of a few large species, notably the New Zealand *paua* [*Haliotis iris* (Gmelin, 1791)], are still used for jewelry today, and several species produce rare but prized pearls (Landman *et al.* 2001).

Despite this wide regard for abalones, the first Spanish and U.S. settlers to California ignored the abundant species found along that coast. Chinese immigrants arriving for the Gold Rush were the first to recognize the large gastropods for the delicacy they were, and began harvesting them in both California and Mexico. Early harvest methods were crude, but they were sufficient to effectively eliminate red, green, and black abalone (*H. rufescens* Swainson, 1822, *H. fulgens* Philippi, 1845, and *H. cracherodii* Leach, 1814, respectively) from near-shore localities in southern California in just a few years. The Mexican government enacted some harvest limits in 1880, and local and state Californian governments began regulating the fishery in 1901 and 1911, respectively (Cox 1962).

Diving technology expanded the abalone fishery further from shore and to new species. Hardhat diving, with a heavy brass helmet and a surface air compressor, was introduced by Japanese immigrants in the 1920s, but the fishery remained focused on red abalone. World War II halted the fishery for a few years but, after the war, "frogman" technology was applied to the fishery. A surface air compressor was still used, but the diver was more mobile, and the fishery expanded to include pink abalone, *H. corrugata* Wood, 1828. The California fishery began at Monterey, but after World War II moved south, and almost all catch came from San Luis Obispo County (Cox 1962). After the spread of SCUBA technology in the 1960s and 1970s, even more areas and species became accessible. Offshore populations of green and black abalone again became important to the fishery. Pink abalone numbers declined rapidly after 1970 and by 1985 were no longer a significant part of the fishery.

The red abalone has always been a large component of the industry and is the only species still harvested in California. In the early 1970s, however, green abalone dominated the industry, peaking at nearly 10,000 short tons of meat in 1971. This fishery level could not be sustained; by 1977 it had fallen under 1000 tons and it slowly declined after that. Black abalone harvests resumed in 1970 after nearly a century of being fished out of near-shore waters. The fishery peaked in 1973 and then declined but remained important until the late 1980s. In 1985, Withering Syndrome, tentatively identified as the bacterium *Xenohaliotis californiensis* Friedman, Andree, Beauchamp, Moore, Robbins, Shield, and Hedrick, 2000, caused mass mortality of wild black abalone. This led to a complete collapse of that fishery by 1993, when harvests were prohibited. Black abalone populations have since recovered slightly, but remain at densities orders of magnitude below historic levels (Moore *et al.* 2001).

The most dramatic decline has been that of the white abalone (*H. sorenseni*). A comparatively deep-water species largely restricted to California's Channel Islands, it was caught sporadically after 1955, when a 1933 ban was lifted. The fishery surged in 1969, peaked at around 1300 short tons, and then declined steadily, until by 1980, it was essentially absent from the fishery. Few realized the extent of the decline, however, and a sport fishery was allowed to remain until 1995. The fishery was so efficient that researchers surveying the same area that supported up to 30,000 white abalone in the 1970s could find only three animals in 1993. Subsequent surveys failed to find any, and the fishery was permanently closed in 1996. So severe was the white abalone decline that imminent extinction of the species was feared. In 2001, the National Oceanic and Atmospheric Administration made history by listing the white abalone as the first U.S. federally endangered marine invertebrate. The species is not yet extinct, fortunately, but remains in critically low numbers.

California banned harvests of green and black abalone in 1996. The entire fishery south of San Francisco, which included the once-productive San Luis Obispo and Channel Islands, was closed in 1997, and in 2000, the fisheries for pinto abalone (*H. kamtschatkana*) and flat abalone (*H. walallensis* Stearns, 1899), which had never been large, were closed. In California and in Oregon, the only other state to permit harvests of red abalone, the fishery is currently open only to recreational fishing, with restrictive bag and size limits. The fishery for pinto abalone in British Columbia followed the same trend as in California, resulting in a permanent closure in 1990. In Alaska, the pinto abalone fishery was sporadic until 1977, when systematic harvest methods were brought to bear. The Alaska pinto abalone fishery peaked immediately, and has slowly declined since, but remains open. California recently adopted a comprehensive recovery plane for abalone, which depends almost exclusively on fishery closures and monitoring; re-stocking with hatchery-reared abalone is not planned.

Ironically, it may be that the entire California abalone fishery was an artifact to start with, created by prior human predation on the main natural predator of adult abalones. The sea otter [*Enhydra lutris* (Linnaeus, 1758)] was nearly eradicated by hunters prior to the California Gold Rush, but a small pocket survived undetected in central California. In the 1960s, sea otters began to be locally abundant again, and their predatory impact on abalone was dramatic. By the end of the 1960s, a rapid decline in the fishery began in areas with sea otters, and the current belief of fishery managers is that healthy sea otter populations and any sort of abalone fishery cannot co-exist (Cox 1962, Ault 1985).

The queen conch, *Strombus gigas*, is a more typical marine snail than the abalone in some regards, possessing a spiraled shell and an operculum. The characteristic flared aperture lip does not form until the animal nears maturity. Like abalone, conch shells were valued by indigenous peoples for tools, ornamentation, and currency. Conch pearls are prized by collectors today (Carstarphen 1982, Landman *et al.* 2001). True conchs (Strombidae) should not be confused with other large snails sometimes termed *conch*, such as whelks of the family Melongenidae.

Queen conch fisheries by Europeans date back to the 17th Century, first as fish bait, and later as dried rations. The U.S. fishery is not as well documented as that of the abalone, but was originally important throughout the Caribbean, including the Florida Keys. Florida Key residents are sometimes referred to as "conchs," a term that reflects the past importance of the colorful shellfish to the region. The fishery remains valuable in other parts of the Caribbean, but most areas have experienced severe fishery declines, leading to closures in Mexico and elsewhere (Carstarphen 1982, Torres and Sullivan-Sealey 2000).

The Florida queen conch fishery, which extended from the Dry Tortugas to Biscayne Bay, is as old as those elsewhere in the Caribbean, but the well-documented portion of the fishery picked up rapidly in the 1960s, following the Florida real estate and tourism boom. Both meat and shells of the queen conch were targeted. In 1975 the commercial fishery was closed in response to declining stocks and, in 1986, the recreational fishery was closed. The initial goal of the Florida closure was to permit fishery stocks to recover on their own but, by 1993, it was apparent that this was not going to happen as quickly as hoped. Stoner and Ray-Culp (2000) suggest that density dependent reproduction is at least partly to blame; at low densities, queen conch simply fail to breed. In 1993, queen conch management became proactive, and using recently developed culture technologies, Florida began stocking areas with juvenile queen conch. To date, stocking efforts have yielded few encouraging results, in Florida or elsewhere. One problem may be pollution. There is a negative correlation with spawning success in near-shore conchs, which are exposed to eutrophication and other anthropogenic inputs, compared to offshore specimens. Another more fundamental issue may be the transport of planktonic larvae on oceanic currents. Some authors argue that Florida queen conch actually recruited from now-depleted stocks elsewhere in the Caribbean (Iversen and Jory 1997, Glazer and Quintero 1998, Hawtof et al. 1998).

For abalone and queen conch, the future seems to be replacement of the fishery with aquaculture. Aquaculture techniques are well-developed for abalone, not only in California but in many parts of the world, and the market is now dominated by aquaculture products. We may one day see abalone affordable to the ordinary consumer, just as the retail price of farm-reared salmon has plunged in recent years. Queen conch aquaculture, despite decades of efforts, is in its infancy. No major technological hurdles appear to exist, and current prices for conch certainly support such ventures. No one is counting on restoration of a commercial fishery for either abalone or queen conch in the foreseeable future. Even if abalone or queen conch numbers recover significantly, the sport fishing industry is now larger, better equipped, and more politically active than ever before, and could easily take whatever portion of the stock managers designate as a fishery (Glazer 2001, Department of Fisheries and Oceans 2002).

The historic listing of the white abalone as federally threatened represents a shift in mollusk conservation, from fishery conservation to species conservation. If enough of the voting public comes to recognize abalone and queen conch as animals, just as are California sea otters and Florida manatees, they may eventually be valued for more than table fare or decorative shells.

31.7 SCALLOPS

Scallops (Pectinidae) are a diverse group of bivalves with numerous representatives in the United States, and include valuable fisheries. So-called deep-sea scallops include the large and colorful sea scallop (*Placopecten magellanicus*) off the Atlantic coast south to North Carolina and the similar-size weathervane scallop (*Patinopecten caurinus*) off the Pacific coast from Alaska to California. Inshore fisheries include the bay scallop (*Argopecten irradians*) along parts of the Atlantic and Gulf of Mexico coasts, and the oyster-like purple-hinged rock scallop [*Crassadoma gigantea* (Gray, 1825)] along the Pacific Coast. Two small scallops, *Chlamys hastata* (Sowerby, 1842) and *C. rubida* (Hinds, 1845), make up a minor fishery off the Pacific coast. In this section, only the sea scallop, weathervane scallop, and bay scallop will be discussed.

The sea scallop is the single most valuable mollusk fishery in the U.S. and, while landings over past 70 years have been more variable than for surf clams (*S. solidissima*) or ocean quahog (*A. islandica*), the next most valuable species, they have not dropped below 2000 metric tons annually since World War II. This is in contrast to the weathervane scallop, for which the fishery south of Alaska is so variable that most years it simply does not exist, yet is productive in other years. NMFS Fishery Management Plans exist for both species, but while overfishing is a common concern for the sea scallop, the weathervane scallop fishery is too poorly understood to know whether conservation concerns exist. Given the rapid expansion of the weathervane scallop fishery since the 1980s, the possibility for over-exploitation certainly exists (Mullen and Moring 1986, Shirley and Kruse 1995, Nunez 1988).

Bay scallops, which have long been considered to comprise multiple separate populations, present a more complex scenario. They are semelparous (spawning only once) in most populations and are thus especially vulnerable to population collapses. Bay scallops were originally harvested for fertilizer in New England but became a fishery in the 19th Century and, in only a few decades, were depleted to fishery collapse in many areas. Scallops were not a valuable enough resource to continue harvesting at low densities and stocks recovered when the fishery ceased. In the 1930s, however, an eelgrass disease destroyed most of the habitat for scallops from Virginia north and populations of bay scallops again collapsed. Incidentally, this was also when the eelgrass limpet, *Lottia alveus alveus*, went extinct. Fishing pressure was not responsible for this bay scallop collapse, although it has been suggested that the eelgrass disease was accidentally introduced by humans. Bay scallops recovered again in New England, and comprise a recreational fishery in Massachusetts, but were extirpated in Maryland and Virginia, and did not reappear for over 60 years. From North Carolina to Florida, bay scallop populations are under intense recreational fishing pressure, leading to regional closures in some states (Field *et al.* 1910, Fay *et al.* 1983, Marshall 1947, Carlton *et al.* 1991, Wilber and Gaffney 1997, Greenawalt, 2002).

Although restoration of marine shellfish by transplanting has a mixed record of accomplishment, this has not stopped attempts to restore bay scallop populations in this manner. Attempts have been made, in Virginia and Florida, to plant cultured bay scallops. Transplanting has not been clearly demonstrated to be effective to date, but harvest closures have been followed by fishery restoration in some areas (Blake 1996, Tarnowski and Homer 1999, Greenawalt 2002). Aquaculture of the bay scallop has been attempted since the 1960s but, in the 1980s, China began large-scale aquaculture production of this species. Today it supplies most of the market for the adductor muscles (Castagna 1975, Yan *et al.* 2000).

For most scallops, only the adductor muscle is consumed, but this is a wasteful and purely cultural bias. The entire scallop is as edible as any other mollusk. It is eaten, much as are whole mussels or clams, in most parts of the world. In the United States, much of the small fishery for spiny and pink scallops (*Chlamys* spp.) on the Pacific coast goes to a restaurant-based market for whole product; a comparable product, using the bay scallop is still struggling for market acceptance in the eastern U.S.

If such a market existed, aquaculture for the bay scallop in the U.S. might be a commercially viable venture (Tarnowski 2005).

31.8 CLAMS

Clams, as opposed to oysters, scallops, and mussels, are free-living bivalves living in soft sediments that draw water and nutrition through tubular mantle extensions known as siphons to biologists and "necks" to clam diggers. Clams are a functional group, rather than taxonomic. Around twenty clam species are regulated in the United States, and an additional half-dozen species are taken by local fisheries. Only a few of these species will be discussed here.

The most valuable clams, in terms of total landings in the U.S., are the surf clam (*Spisula solidissima*) and the ocean quahog (*Arctica islandica*), both of which tend to end up in clam chowder and clam strips. Although both species make up minor recreational fisheries in New England, the commercial fisheries are well offshore in the North Atlantic, and regulated by NOAA Fishery Management Plans. Both fisheries have been relatively stable, but the ocean quahog in particular, is probably not sustainable. The ocean quahog, known locally as the mahogany clam for its dark brown periostracum, takes decades to reach full size, and may live seventy years or more (Zettler *et al.* 2001). It does not follow that the species is endangered, though; unlike for abalone, the value of a single ocean quahog is low, and the fishery industry will not pursue low densities or small pockets of clams.

The oldest U.S. clam fisheries, not including those developed by Native Americans, are those for the hard clam or northern quahog (*Mercenaria mercenaria*) and the softshell or steamer (*Mya arenaria*) in New England and the Mid-Atlantic. Although habitat for both species has been lost to coastal filling or areas that are closed for pollution or sanitation reasons, fisheries for both have remained relatively robust. The dramatic declines observed among eastern oysters, bay scallops, and abalone have, for the most part, not occurred for these two clams. Consequently, neither species is subject to strict conservation measures throughout most of their ranges, although Fegley (2001) questions whether this incaution is warranted for hard clams. Diseases, such as QPX in hard clams, have inhibited some local efforts to manage populations. Currently, aquaculture in Virginia and Florida supplies most of the market for the smaller-sized (more expensive) hard clams, but the softshell clam industry remains strictly a fishery (Newell and Hidu 1986, Eversole 1987, Roegner and Mann 1991, Kraeuter et al. 1998, Colson and Sturmer 2000).

On the Pacific coast, a somewhat larger number of clams are sought, but only one, the geoduck clam [*Panopea abrupta* (Conrad, 1849)], is a major fishery in the U.S. The geoduck is the world's largest burrowing clam and mature specimens cannot retract into their shells. It is primarily subtidal and lives buried a meter (3 ft) in the sediment. It was not until the application of SCUBA technologies in 1970 that a commercial fishery began for this species. This U.S. fishery rapidly expanded to several thousand metric tons annually but, for the past decade, has been less than 1000 tons annually. Given that the clam takes a decade to reach maturity and can live for decades more, there are legitimate conservation concerns. Commercial-scale poaching for this valuable species (worth up to $100 per pound) is a serious problem but it is believed that unexploited stocks survive in various refugia, especially in British Columbia. Aquaculture techniques for the geoduck have been developed in both the U.S. and British Columbia, where the species also forms a fishery, but the market remains dominated by fishery products (Department of Fisheries and Oceans 2001).

The geoduck, however, has achieved what few other mollusks have, and that is a place in the hearts and lore of the local populace. The clam's unusual size, bizarre and frankly phallic appearance, and peculiar name probably all contribute to this phenomenon. Just as mispronunciation of "quahog" will mark you as an outsider on Long Island, so will a phonetic pronunciation of "geoduck" in the Puget Sound region. Plush geoduck dolls are sometimes available for sale in Seattle's Pike Place Market. Such notoriety is of great value when the time comes to seek the political will to impose fishery restrictions or other conservation measures.

The wide, sandy beaches of the Pacific coast support important local fisheries for two additional clam species. The Pismo clam [*Tivela stultorum* (Mawe, 1823)] is restricted to California and Mexico, and is taken only recreationally. The Pacific razor clam [*Siliqua patula* (Dixon, 1789)], prized for its sweet flesh, is sought from northern California to Alaska, but the center of the sport fishery is two beaches flanking the mouth of the Columbia River, in Oregon and Washington. Their habitat, high-energy sandy beaches, is exposed for only a few hours of a few days a month, often at dawn or dusk. To live in this environment, razor clams must be able to re-bury themselves in seconds, and to dig rapidly to escape pounding waves or predatory birds. This ability also allows them to escape rapidly from the clam digger, by digging down out of reach, so harvesting razor clams is not a leisurely activity.

Razor clams became a major recreation fishery around 1900, with catches peaking in the 1970s. Over 300,000 clam diggers were involved annually in Washington alone, and catches could not be sustained. Oregon and Washington began reducing bag limits and enacting closed seasons, but declines continued. In the 1980s, a presumably bacterial disease (NIX) caused 90% mortalities in Washington. Mortalities since then have been less severe, but the disease still causes mortalities, although mainly in northern Oregon and southern Washington. The commercial fishery relocated to British Columbia, mainly in Haida Gwaii (Queen Charlotte Islands). In the 1990s, domoic acid, a serious toxin associated with harmful algal blooms, began appearing in razor clams, prompting regular testing and emergency closures. In Washington, Native Americans began exercising a federally recognized right to harvest razor clams outside of their reservations, putting further pressure on the species. Bag limits have declined further. Despite these problems, a popular sport fishery for razor clams still exists (Elston 1986, Simons and Ayres 1991).

31.9 MUSSELS: EXCEPTIONS TO THE RULE?

Many of the above tales of molluscan fisheries have been of decline or collapse. Not so the marine mussels. Until recently, all of the edible mussels in the United States have been in the genus *Mytilus*, but the only readily identifiable species is *M. californianus* Conrad, 1837, the California sea mussel, which is found from Alaska to Mexico. *M. edulis* Linnaeus, 1758, the blue mussel, was originally thought to be native to all northern oceans, but has since been recognized as at least three morphologically identical species. *M. edulis* is native to both coasts of the North Atlantic, *M. galloprovincialis* Lamarck, 1819 is native to Europe but is introduced to the Pacific coast of North America, and *M. trossulus* Gould, 1850 appears to be native to the North Pacific and portions of the North Atlantic. Other native mussels are mostly small or inedible (McDonald and Koehn 1988, Newell 1989).

Mussels, regardless of species, are the easiest of mollusks to harvest. They cling to the substrate with strong byssal threads, which are easily severed with hand tools. Most *Mytilus* species occur intertidally in huge beds, and in the Netherlands, blue mussels are simply scraped off the tide flats with industrial equipment (Saier 2002). Mussels are popular in most cultures where they occur, and are widely eaten on both coasts of North America. Mussel "culture" is one of the mainstays of marine aquaculture, but it too, is at least partly a fishery. Mussel culturists do not culture larvae, as do most other shellfish culturists, but instead rely upon settlement onto their culture rafts from natural mussel populations (Newell *et al.* 1991).

Despite their vulnerability to harvest, mussels have proved remarkably resilient fisheries. The evidence of this can be seen in the literature: there are thousands of biological or ecological references to *Mytilus*, but compared to other important marine fishery species, there are relatively few fishery papers on mussels. As William J. Hargis pointed out for oysters, fishery biologists do not study a species much until it is in decline. Almost all mussel fishery declines or closures are directly related to human health concerns about the water the mussels grow in, not about the mussel populations themselves. Conservation concerns about mussel harvests are as likely to be about ancillary impacts on other species as on mussels (Saier 2002). There is no simple explanation for why mussel fisheries remain robust, while oyster, scallop, and other fisheries are in decline. Perhaps we should study mussels from a fishery perspective, if only to determine how we have failed to undermine their populations as we have for so many other marine mollusks.

31.10 CEPHALOPODS

The United States is nearly unique, among nations with major fisheries, in ignoring cephalopods as a major table fare; initial fisheries were for bait, and bait is still one of the primary markets for squid in the U.S.A. Calamari is gaining popularity in Italian and seafood restaurants, but the vast majority of the U.S. fishery goes to specialty markets or is consumed and sold abroad. Squid (*Loligo* spp.) are regulated as a sport fishery in several states (Table 31.1), but bag limits are generous and the majority of the fishery is in federal waters. Pelagic squid are typically dispersed in small aggregations, but gather in huge numbers to spawn, and are targeted at that time. This makes them vulnerable to overexploitation and, in the late 1990s, the National Marine Fishery Service concluded that *L. pealei* Lesueur, 1821 from North Carolina northward was being overexploited. The fishery was subsequently closed in 2001 and again in 2002. On the Pacific coast, impacts of the fishery on stocks are still being assessed (Cadrin 2000, Washington Fish and Wildlife 2003).

The Caribbean reef squid [*Sepioteuthis sepioidea* (de Blainville, 1823)], which occurs in Florida, is not regulated, nor is the neon flying squid [*Ommastrephes bartrami* (Lesueur, 1821)], which occurs off the Pacific coast from Mexico to Alaska. Both are fishery species in other regions, and the neon flying squid in particular is the target of an experimental fishery. The brief squid [*Lolliguncula brevis* (de Blainville, 1823)] is an abundant coastal species from New York to South America, and is often harvested for bait, but is not widely eaten in its U.S. range.

Octopods (*Octopus* spp.) are more tightly regulated by states and provinces than are squids, with generally low take limits or closures and strict gear limits. Industrial-scale fishing methods are inappropriate for octopuses, which live among rocks or other structures, but because octopuses are large, territorial predators, a well-developed hand fishery can have significant impacts. The advent of SCUBA technology has placed most large octopus within reach of a specialty fishery, and there is a modest commercial fishery on the Pacific coast and in Florida, the only Atlantic seaboard state with specific octopus regulations (Table 31.1). Unfortunately, while octopus fisheries are widespread, there appears to be little research on fishery stocks or sustainability.

Cephalopods differ from the majority of gastropods and bivalves in ways important to their conservation or management. They are predators, whereas many commercially important gastropods and all bivalves feed on algae or plankton. This tends to make cephalopod fisheries less productive per unit area than other molluscan fisheries, and thus more vulnerable to exploitation. Second, cephalopods are semelparous (spawning only once), while all but a few marine gastropods and bivalves are iteroparous, with populations consisting of multiple age classes. Loss of an entire cohort of abalone is bad; loss of an entire cohort of octopus is catastrophic. A third issue, particularly for octopus conservation or regulation, is the difficulty in distinguishing species, and in most parts of the world, probably more species exist than are currently recognized (Voss *et al.* 1998, Herb 2001).

31.11 UNREGULATED AND OVERLOOKED MOLLUSKS

So far in this chapter we have covered a tiny fraction of the thousands of species of marine mollusks that occur on our shores. Some of these species, overlooked by regulators and conservationists, are nonetheless fishery species, or have been in the past (Table 31.1). Just because they are unregulated, however, does not mean that serious conservation issues do not exist for these mollusks.

Some molluscan species are unregulated because the fishery collapsed before regulators took an interest. Bean clams (*Donax gouldi* Dall, 1921) were once an important local fishery in southern California before World War II, but the fishery collapsed, for reasons unknown, and while populations of the bean clam have at least partially recovered, the fishery has not. Calico scallops (*Argopecten gibbus*) were an important fishery off the coast of Georgia and Florida in the 1980s, but collapsed, possibly permanently, before the reauthorization of the Magnuson-Stevens Fishery Act in 1993. Disease was blamed, but the collapse occurred during a period of commercial exploitation. Sunray venus clams [*Macrocallista nimbosa* (Lightfoot, 1786)] were briefly the target of a dredge fishery in west Florida in the 1960s and 1970s, but this fishery was not sustained. Currently, the sunray venus forms only a minor recreational fishery and is not specifically regulated (Gibbons 1964, Godcharles and Jaap 1973, Moyer *et al.* 1993).

Other edible mollusks are probably not currently regulated for the same reason mussels and clams took generations to become common table fare in America: the ethnic groups that target them have simply not moved up the socio-economic ladder enough to make their seafood mainstream. A prime example of this is ark clams (Arcoida), which occur from Cape Cod to the Caribbean. The group is also collectively called *blood clams* due to the presence of extracellular hemoglobin, which colors tissue and blood. Most *Norte Americanos* prefer their clams pasty white but, in parts of Latin America, ark clams are a delicacy and they are also eaten widely in Southeast Asia. Sporadic efforts have been made in the U.S. to tap into this market, targeting blood arks (*Anadara* spp.) and the ponderous ark [*Noetia ponderosa* (Say, 1822)]. A minor fishery exists in Virginia but stocks cannot keep up with demand and there are fears of overharvesting. The ponderous ark and one of the blood arks, *Anadara ovalis* (Bruguière, 1789), are currently being investigated as aquaculture species (Baker *et al.* 2001, Walker and Gates 2001).

Unregulated gastropods are not exempt from fishery impacts, either. In California, the large abalones get all the press, however, limpets are also harvested. The owl limpet (*Lottia gigantea* Sowerby, 1834) and the great keyhole limpet [*Megathura crenata* (Sowerby, 1834)] are targeted for their large size. Both are actually regulated in California under a combination of general take limits for all intertidal invertebrates and no-take zones, but enforcement is lacking in most areas. The result, at least for the owl limpet, has been not only a decline in abundance, but also an interesting adaptive, or possibly evolutionary response; mature animals are smaller in impacted areas than in either museum collections or in strictly enforced no-take zones (Roy *et al.* 2003).

California, Florida, Oregon, and Washington all have rich near-shore faunas, and ideal conditions for casual collectors, whether it is warm water or alluring tide pools. All four of these states have recognized the possibility that casual collecting can deplete some species, especially mollusks. Thus, along the outer coasts of California, Oregon, and Washington, and in parts of southern Florida, there are general limits on invertebrates, as noted above for California. Research on harvest sustainability is lacking for most species. Law enforcement agencies do not have the ability to keep track of hundreds of species so general limits solve just a few conservations problems. These limits may still be too broad for some species, like the large limpets noted above, so some refinement may be necessary. In addition, all four states have a network of state parks, state reserves, and federal parks or reserves that are designated no-take areas (Tables 31.1 and 31.4).

31.12 CONCLUSION

Marine mollusks have not been noted for the spectacular rate of decline and extinction seen in freshwater species, but declines are widespread nonetheless, and catastrophic in a few cases. Until recently, marine mollusks were regulated as fisheries (if at all) rather than species, and this is still the case for most. There are promising trends, however, two of which I will reiterate here. One is the growing number of marine parks and sanctuaries, within which mollusks are regulated as part of the total benthic community and are harvested sparingly or not at all. Given the taxonomic diversity and often cryptic habitat of many mollusks, broad protection is not only the most practical solution but also the most ecosystem-friendly. Marine aquaculture also offers relief to wild mollusk populations by replacing fisheries with farmed products. Aquaculture is often criticized as non-sustainable but most mollusks are different, as argued by Shumway *et al.* (2003). The most valuable gastropods - abalone and conch - are herbivorous and do not require feed processed from wild-caught fish. Bivalve nutrition is even simpler - they simply feed on natural concentrations of phytoplankton in the water flowing past their culture area. No human activities are without some impacts on the environment, but shellfish aquaculture may provide a way to have our conch and eat it, too.

31.13 LITERATURE CITED

Allee, W. C. 1931. *Animal Aggregations. A Study in General Sociology*. University of Chicago Press, Chicago, Illinois. 431 pp.

Ault, J. S. 1985. Species profiles: Life histories and environmental requirements of coastal fishes and invertebrates (Pacific Southwest) - black, green, and red abalones. *U.S. Fish and Wildlife Service Biological Reports* **82**(11.32): 1-19.

Baker, P. 1995. Review of ecology and fishery of the Olympia oyster, *Ostrea lurida*, with annotated bibliography. *Journal of Shellfish Research* **14**: 501-518.

Baker, P., D. Bergquist, and S. Baker. 2003. *Oyster Reef Assessment in the Suwannee River Estuary*. Final Report to Suwannee River Water Management District, Live Oak, Florida. 34 pp.

Baker, P. and R. Mann. 1997. The postlarval phase of bivalve mollusks: A review of functional ecology and new records of postlarval drifting of Chesapeake Bay bivalves. *Bulletin of Marine Science* **61**: 409-430.

Baker, P., N. Richmond, and N. B. Terwilliger. 1999. Re-establishment of a native oyster, *Ostrea conchaphila*, following a natural local extinction. *In:* J. Pederson, ed., *Marine Bioinvasions*. MIT Sea Grant, Cambridge, Massachusetts. Pp. 221-231.

Baker, S. M., L. N. Sturmer, and J. Baldwin. 2001. *Preliminary Investigation of Blood Ark,* Anadara ovalis, *and Ponderous Ark,* Noetia ponderosa, *Culture to Initiate Diversification for the Hard Clam,*

Mercenaria mercenaria, *Aquaculture Industry*. U.S. Department of Agriculture Special Research Grants, Aquaculture.

Bartol, I. K., R. Mann, and M. Luckenbach. 1999. Growth and mortality of oysters (*Crassostrea virginica*) on constructed intertidal reefs: Effects of tidal height and substrate level. *Journal of Experimental Marine Biology and Ecology* **237**: 157-184.

Barlow, G. W. 1995. The relevance of behavior and natural history to evolutionary significant units. *In:* J. L. Nielsen, ed., *Evolution and the Aquatic Ecosystem: Defining Unique Units in Population Conservation. American Fisheries Society Symposium* 17. American Fisheries Society, Bethesda, Maryland. Pp. 169-175.

Barton, A. J. 1994. *Fishing for Ivory Worms: A Review of Ethnographic and Historical Recorded* Dentalium *Source*. M.A. Dissertation, Simon Fraser University, Vancouver, Canada. 176 pp.

Baylor, J. B. 1894. *Method of Defining and Locating Natural Oyster Beds, Rocks, and Shoals*. Oyster Records, Board of Fisheries of Virginia. [Set of pamphlets for each Tidewater county in Virginia].

Bernatchez, L. 1995. A role for molecular systematics in defining evolutionary significant units in fishes. *In:* J. L. Nielsen, ed., *Evolution and the Aquatic Ecosystem: Defining Unique Units in Population Conservation. American Fisheries Society Symposium* 17. American Fisheries Society, Bethesda, Maryland. Pp. 114-132.

Blake, N. J. 1996. *Demonstration of Large-Scale Reintroduction of the Southern Bay Scallop to Tampa Bay, Florida*. Tampa Bay National Estuary Program Technical Publication #13-95. 28 pp.

Bleyer, W. 2002. The oyster was their world. *In: Long Island: Our Story*, Chapter 6, City and Suburb. Newsday, Melville, New York <www.lihistory.com>.

Bowen, B. W. 1998. What is wrong with ESUs?: The gap between evolutionary theory and conservation principles. *Journal of Shellfish Research* **17**: 1355-1358.

Brooks, W. K. 1891. *The Oyster*. Johns Hopkins University Press, Baltimore, Maryland. 230 pp.

Butler, P. A. 1949. An investigation of oyster producing areas in Louisiana and Mississippi damaged by flood waters in 1945. *U.S. Department of Interior, Fish and Wildlife Service, Special Scientific Report Fisheries* **8**: i-vi + 1-27.

Cadrin, S. X. 2000. *Status of the Fishery Resources of the Northeastern United States. Species Synopses - Longfin Inshore Squid*. National Oceanic Atmospheric Administration, National Marine Fisheries Service. <www.nefsc.noaa.gov>.

Cake, E. W., Jr. 1983. *Habitat Suitability Index Models: Gulf of Mexico American oyster*. U.S. Fish and Wildlife Service. FWS/OBS-82/10.57.

California Fish and Game Commission. 2002. *2002 Ocean Fishing Regulations Booklet*. Department of Fish and Game, Sacramento, California. 34 pp.

Carlton, J. T. 1979. *History, Biogeography, and Ecology of the Introduced Invertebrates of the Pacific Coast of North America*. Ph.D. Dissertation, University of California, Davis, California. 904 pp.

Carlton, J. T. 1993. Neoextinctions of marine invertebrates. *American Zoologist* **33**: 499-509.

Carlton, J. T., J. B. Geller, M. L. Reaka-Kudla, and E. A. Norse. 1999. Historic extinctions in the sea. *Annual Review of Ecology and Systematics* **30**: 515-538.

Carlton, J. T. and R. Mann. 1996. Transfers and worldwide introductions. *In:* V. S. Kennedy, R. E. I. Newell, and A. F. Eble, eds., *The Eastern Oyster*. Maryland Sea Grant, College Park, Maryland. pp. 691-706

Carlton, J. T., G. J. Vermeij, D. R. Lindberg, D. A. Carlton, and E. C. Dudley. 1991. The first historical extinction of a marine invertebrate in an ocean basin: the demise of the eelgrass limpet *Lottia alveus*. *Biological Bulletin* **180**: 72-80.

Carstarphen, D. 1982. *The Conch Book*. Banyan Books, Miami, Florida. 75 pp.

Castagna, M. 1975. Culture of the bay scallop, *Argopecten irradians*, in Virginia. *Marine Fisheries Reviews* **37**: 19-24.

Claassen, C. 1994. *Washboards, Pigtoes, and Muckets: Historic Musseling in the Mississippi Watershed*. Society of Historical Archaeology, Tucson, Arizona. 145 pp.

CITES (Convention on International Trade of Endangered Species). 2004. Electronic update to *CITES Handbook*. Secretariat of the Convention on International Trade of Endangered Species of Wild Fauna and Flora, 2001. Sadag Imprimerie, Bellegarde-sur-Valserine, France. 350 pp. <www.cites.org/eng/app/index.shtml> [accessed Jan. 2005].

Coen, L. D., M. W. Luckenbach, and D. L. Breitberg. 1999. The role of oyster reefs as essential fish habitat: A review of current knowledge and some new perspectives. *American Fisheries Society Symposium* **22**: 438-454.

Colson, S. and L. Sturmer. 2000. One shining moment known as Clamelot: the Cedar Key story. *Journal of Shellfish Research* **19**: 477-480.

Cox, K. W. 1962. California abalones, Family Haliotidae. *California Fish and Game Bulletin* **118**: 1-133.

Department of Fisheries and Oceans. 2001. *Pacific Science Advice Review Committee Stock Status Reports. Invertebrates. C6 (1999-2001)*. Canada Department of Fisheries and Oceans, Pacific Region. <www.pac.dfo-mpo.gc.ca/sci/psarc/SSRs/invert_ssrs_e.htm>

Department of Fisheries and Oceans. 2002. *Fisheries Management - Pacific Region. Shellfish and Invertebrates*. Canada Department of Fisheries Oceans,

Pacific Region. <www.pac.dfo-mpo.gc.ca/ops/fm/shellfish/default_e.htm>

Duff, J. A. and R. Brownlow. 1997. National Marine Sanctuaries Act. *Water Log* **17**: 7-9.

Elston, R. A. 1986. An intranuclear pathogen (nuclear inclusion X (NIX)) associated with massive mortalities of the Pacific razor clam, *Siliqua patula*. *Journal of Invertebrate Pathology* **47**: 93-104.

Eversole, A. G. 1987. Species profiles: life histories and environmental requirements of coastal fishes and invertebrates (South Atlantic) - hard clam. *U.S. Fish and Wildlife Service Biological Report* **82**(11.75): 1-33.

Fay, C. W., R. J. Neves, and G. B. Pardue. 1983. Species profiles: Life histories and environmental requirements of coastal fishes and invertebrates (Mid-Atlantic) - bay scallop. *U.S. Fish and Wildlife Service Biological Report* **82**(11.12): 1-17.

Fegley, S. R. 2001. Demography and dynamics of hard clam populations. *In:* J. N. Kraeuter and M. Castagna, eds., *Biology of the Hard Clam*. Elsevier, New York. Pp. 383-422.

Field, G. W., J. W. Delano, and G. H. Garfield. 1910. *A Report upon the Scallop Fishery of Massachusetts*. Wright and Potter State Printers, Boston, Massachusetts. 150 pp.

Fitch, J. E. 1950. The Pismo Clam. *California Fish Game* **36**: 285-312.

Ford, S. E. and M. R. Tripp. 1996. Diseases and defense mechanisms. *In:* V. S. Kennedy, R. E. I. Newell, and A. F. Eble, eds., *The Eastern Oyster*. Maryland Sea Grant, College Park, Maryland. Pp. 581-660.

Gibbons, E. 1964. *Stalking the Blue-Eyed Scallop. Foraging Our Native Seacoasts for Food and Pleasure*. McKay Co., New York. 332 pp.

Glazer, R. A. 2001. *Queen Conch Stock Restoration*. Florida Marine Research Institute, St. Petersburg, Florida. 7 pp.

Glazer, R. A. and I. Quintero. 1998. Observations on the sensitivity of queen conch to water quality: implications for coastal development. *Proceedings of the Gulf and Caribbean Fisheries Institute* **50**: 78-93.

Godcharles, M. F. and W. C. Jaap. 1973. Exploratory clam survey of Florida nearshore and estuarine waters with commercial hydraulic dredging gear. *Florida Depatment of Natural Resources, Professional Paper Series* **21**: 1-77.

Greenawalt, J. M. 2002. *Mortality Estimates and Distributional Patterns of the Southern Bay Scallop along the Gulf Coast of Florida*. M.S. Dissertation, University of Florida, Gainesville, Florida. 52 pp.

Hawtof, D. B., K. J. McCarthy, and R. A. Glazer. 1998. Distribution and abundance of queen conch, *Strombus gigas*, larvae in the Florida Current: Implications for recruitment to the Florida Keys. *Proceedings of the Gulf and Caribbean Fisheries Institute* **50**: 94-103.

Herb, H. A. 2001. *A Systematic Review of the Shallow-Water Octopuses (Cephalopoda: Octopodidae) of the Fiji Islands*. M.A. Dissertation, University of Florida, Gainesville, Florida. 163 pp.

Herrington, W. C. 1929. The Pismo Clam. Further studies of its life history and depletion. *California Division of Fish and Game Bulletin* **18**: 1-69.

Iversen, E. S. and D. E. Jory. 1997. Mariculture and enhancement of wild populations of queen conch (*Strombus gigas*) in the western Atlantic. *Bulletin of Marine Science* **60**: 929-941.

Jones, C. G., J. H. Lawton, and M. Shachak. 1997. Positive and negative effects of organisms as physical ecosystem engineers. *Ecology* **78**: 1946-1957.

Karl, S. A. and B. W. Bowen. 1999. Evolutionary significant units versus geopolitical taxonomy: molecular systematics of an endangered sea turtle (Genus *Chelonia*). *Conservation Biology* **13**: 990-999.

Kennedy, V. S. and L. P. Sanford. 1999. Characteristics of relatively unexploited beds of the eastern oyster, *Crassostrea virginica*, and early restoration programs. *In:* M. W. Luckenbach, R. Mann, and J. A. Wesson, eds., *Oyster Reef Habitat Restoration: A Synopsis and Syntheses of Approaches*. Virginia Institute of Marine Sciences Press, Gloucester Point, Virginia. Pp. 25-46.

King, T. L., E. C. Pendleton, and R. F. Villella. 1998. Gene conservation: Management and evolutionary units in freshwater bivalve management - introduction to the proceedings. *Journal of Shellfish Research* **17**: 1351-1353.

Kochiss, J. M. 1974. *Oystering from New York to Boston*. Published for Mystic Seaport, Inc., by Wesleyan University Press, Middletown, Connecticut. 251 pp.

Kraeuter, J. N., S. E. Ford, R. Smolowitz, D. Leavitt, and L. M. Ragone. 1998. QPX, a protistan parasite of hard clams (*Mercenaria mercenaria*) and its importance to rehabilitation efforts (Abstract). In: *Proceedings of the International Conference on Shellfish Restoration*. Hilton Head, South Carolina. p. 62.

Landman, N. H., P. M. Mikkelsen, R. Bieler, and Bronson. 2001. *Pearls: A Natural History*. Harry N. Abrams, Inc. and American Museum of Natural History, New York. 230 pp.

Luckenbach, M. W., R. Mann, and J. A. Wesson, eds. 1999. *Oyster Reef Habitat Restoration: A Synopsis and Syntheses of Approaches*. Virginia Institute of Marine Science Press, Gloucester Point, Virginia. 366 pp.

McCay, B. J. 1998. *Oyster Wars and the Public Trust: Property, Law, and Ecology in New Jersey History*. University of Arizona Press, Tucson, Arizona. 246 pp.

McDonald, J. H. and R. K. Koehn. 1988. The mussels *Mytilus galloprovincialis* and *M. trossulus* on the

Pacific coast of North America. *Marine Biology* **99**: 111-118.

McKenzie, C. L., Jr. 1996. Management of natural populations. *In:* R. I. E. Newell, V. S. Kennedy, and A. F. Eble, eds., *The Eastern Oyster, Crassostrea virginica.* Maryland Sea Grant, College Park, Maryland. Pp. 707-721.

Mackin, J. G. and S. H. Hopkins. 1962. Studies on oyster mortality in relation to natural environments and to oil fields in Louisiana. *Publications of the Institute of Marine Science, Texas* **7**: 1-131.

Mann, R., B. J. Barber, J. P. Whitcomb, and K. S. Walker. 1990. Settlement of oysters, *Crassostrea virginica* (Gmelin, 1791) on oyster shell, expanded shale, and tire chips in the James River, Virginia. *Journal of Shellfish Research* **9**: 173-175.

Marshall, N. 1947. An abundance of bay scallops in the absence of eelgrass. *Ecology* **28**: 321-322.

Moore, H. F. 1899. Report on the oyster beds of Louisiana. *Report of the Commissioner, U.S. Commission of Fish and Fisheries* for 1898: 45-100.

Moore, J. D., T. T. Robbins, R. P. Hedrick, and C. S. Friedman. 2001. Transmission of the Rickettsiales-like prokaryote "*Candidatus Xenohaliotis californiensis*" and its role in withering syndrome of California abalone, *Haliotis* spp. *Journal of Shellfish Research* **20**: 867-874.

Moritz, C. 1994. Defining 'evolutionary significant units' for conservation. *Trends in Ecology and Evolution* **9**: 373-375.

Moyer, M. A., N. J. Blake, and W. S. Arnold. 1993. An acetosporan disease causing mass mortality in the Atlantic calico scallop, *Argopecten gibbus* (Linnaeus, 1758). *Journal of Shellfish Research* **12**: 305-310.

Mullen, D. M. and J. R. Moring. 1986. Species profiles: Life histories and environmental requirements of coastal fishes and invertebrates (North Atlantic) - sea scallop. *U.S. Fish and Wildlife Service Biological Report* **82**(11.67): 1-13 pp.

Neustadt, K. 1992. *Clambake: A History and Celebration of an American Tradition.* American Folklore Society, University of Massachusettes Press, Amherst, Massachusetts. 227 pp.

Neves, R. J., A. E. Bogan, J. D. Williams, S. A. Ahlstedt, and P. W. Hartfield. 1997. Status of aquatic mollusks in the southeastern United States: a downward spiral of diversity. *In:* G. W. Benz and D. E. Collins, eds., *Aquatic Fauna in Peril: The Southeastern Perspective.* Special Publication 1, Southeast Aquatic Research Institute, Cohutta, Georgia. Pp. 43-85.

Newell, C. R. and H. Hidu. 1986. Species profiles: life histories and environmental requirements of coastal fishes and invertebrates (North Atlantic) - softshell clam. *U.S. Fish and Wildlife Service Biological Report* **82**(11.53): 1-17 pp.

Newell, C. R., H. Hidu, B. J. McAlice, G. Podniesinski, F. Short, and L. Kindblom. 1991. Recruitment and commercial seed procurement of the blue mussel *Mytilus edulis* in Maine. *Journal of the World Aquaculture Society* **22**: 134-152.

Newell, R. I. E. 1988. Ecological changes in Chesapeake Bay: Are they the result of overharvesting the American oyster, *Crassostrea virginica*? *In:* M. P. Lynch and E. C. Krome, eds., *Understanding the Estuary: Advances in Chesapeake Bay Research. Proceedings of a Conference 29-31 March 1988, Baltimore.* Maryland Chesapeake Research Consortium Publication 129. Chesapeake Research Consortium, Solomons, Maryland. Pp. 536-546.

Newell, R. I. E. 1989. Species profiles: Life histories and environmental requirements of coastal fishes and invertebrates (North and Mid-Atlantic) - blue mussel. *U.S. Fish and Wildlife Service Biological Report* **82**(11.102): 1-25.

Nunez, J. D. 1988. *The Fishery Biology of the Weather-Vane Scallop [*Pecten (Patinopecten) caurinus *Gould, 1850] in Oregon coastal waters.* M.S. Dissertation, Oregon State University, Corvallis, Oregon. 160 pp.

Parfit, M. 1995. Diminishing returns: Exploiting the ocean's bounty. *National Geographic* **188**: 2-37.

Pennock, D. S. and W. W. Dimmick. 1997. Critique of the evolutionary significant unit as a definition for "distinct population segments" under the U.S. Endangered Species Act. *Conservation Biology* **11**: 611-619.

Rheinhardt, R. D. and R. Mann. 1990. Temporal changes in epibenthic fouling community structure on a natural oyster bed in Virginia. *Biofouling* **2**: 13-25.

Robinson, A. and Johnson, J. 1997. Native oyster restorations in Oregon (Abstract). *Journal of Shellfish Research* **16**: 337.

Roe, K. J. and C. Lydeard. 1998. Species delineation and the identification of evolutionarily significant units: lessons from the freshwater mussel genus *Potamilus* (Bivalvia: Unionidae). *Journal of Shellfish Research* **17**: 1359-1363.

Roegner, C. G., and R. Mann. 1991. Hard clam. *In:* S. L. Funderburk, J. A. Mihursky, S. L. Jordan, and D. Riley, eds., *Habitat Requirements for Chesapeake Bay Living Resources,* 2nd Ed. Living Resources Subcommittee, Chesapeake Bay Program, Annapolis, Maryland. Pp. 5.1-5.17.

Rothschild, B. J., J. S. Ault, P. Goulletquer, and M. Héral. 1994. Decline of the Chesapeake Bay oyster population: A century of habitat destruction and overfishing. *Marine Ecology Progress Series* **111**: 29-39.

Roy, K., A. G. Collins, B. J. Becker, E. Begovic, and J. M. Engle. 2003. Anthropogenic impacts and historical decline in body size of rocky intertidal gastropods in southern California. *Ecology Letters* **6**: 205-211.

Saier, B. 2002. Subtidal mussel beds in the Wadden Sea: threatened oases of biodiversity. *Wadden Sea Newsletter* 2002 - **1**: 12-14.

Scozzari, S. 1995. The significance of wampum to seventeenth century Indians in New England. *Connecticut Review* **17**: 59-69.

Shirley, S. M. and G. H. Kruse. 1995. Development of the fishery for weathervane scallops, *Patinopecten caurinus* (Gould, 1850), in Alaska. *Journal of Shellfish Research* **14**: 71-78.

Shumway, S. E., C. Davis, R. Downey, R. Karney, J. Kraeuter, J. Parsons, R. Rheault, and G. Wikfors. 2003. Shellfish aquaculture - in praise of sustainable economics and environment. *World Aquaculture* **34**: 15-17.

Simons, D. and D. L. Ayres. 1991. Struggle for survival - the Pacific razor clam and disease (Abstract). *Journal of Shellfish Research* **10**: 278.

Sindermann, C. J. 1993. Disease risks associated with importation of nonindigenous marine animals. *Marine Fisheries Review* **54**: 1-10.

Slatkin, M. 1987. Gene flow and the geographic structure of natural populations. *Science* **236**: 787-792.

Stauffer, J. R., N. J. Bowes, K. R. McKaye, and T. D. Kocher. 1995. Evolutionary significant units among cichlid fishes: The role of behavioral studies. *In:* J. L. Nielsen ed., *Evolution and the Aquatic Ecosystem: Defining Unique Units in Population Conservation. American Fisheries Society Symposium* 17. American Fisheries Society, Bethesda, Maryland. Pp. 227-244.

Stoner, A. W., and M. Ray-Culp. 2000. Evidence for Allee effects in an over-harvested marine gastropod: Density-dependent mating and egg production. *Marine Ecology Progress Series* **202**: 297-302.

Tarnowski, M. 2005. Chapter 8.4. Status of shellfish populations in the Maryland Coastal Bays. *In:* C. Wazniak, D. Goshorn, M. Hal, D. Blazer, R. Jesien, D. Wilson, C. Cain, W. Dennison, J. Thomas, T. Carruthers, and B. Sturgis, eds., *Maryland's Coastal Bays: Ecosystem Health Assessment 2004*. Document DNR-12-1202-0009. Maryland Department of Natural Resources, Annapolis, Maryland. Pp. 8.52-8.73.

Tarnowski, M. L. and M. L. Homer. 1999. Re-introducing the bay scallop *Argopecten irradians* into Chincoteague Bay, MD. (Abstract). *Journal of Shellfish Research* **18**: 315.

Torres, R. E. and K. M. Sullivan-Sealey. 2000. Shell midden surveys as source of information about fished queen conch (*Strombus gigas*) populations: A case study in Parque Nacional Del Este, Dominican Republic. *Proceedings of the Gulf and Caribbean Fisheries Institute* **53**: 143-153.

Voss, N. A., M. Vecchione, R. B. Toll, and M. J. Sweeney, eds. 1998. Systematics and biogeography of Cephalopods. *Smithsonian Contributions to Zoology* **586**: 1- 599. [2 volumes].

Walker, R. L. and K. W. Gates. 2001. *Survey of Ark, Anadara ovalis, Anadara brasiliana, and Noetia ponderosa, Populations in Coastal Georgia*. Georgia Sea Grant GAUS-S-01-002. 34 pp.

Waples, R. S. 1995. Evolutionarily significant units and the conservation of biological diversity under the Endangered Species Act. *In:* J. L. Nielsen, ed., *Evolution and the Aquatic Ecosystem: Defining Unique Units in Population Conservation. American Fisheries Society Symposium* 17. American Fisheries Society, Bethesda, Maryland. Pp. 8-27.

Washington (Department of) Fish and Wildlife. 2003. *Fishing and Shellfishing*. Washington Department of Fish and Wildlife, Olympia, Washington <www.wdfw.wa.gov/fishcorn.htm>.

Wennersten, J. R. 1981. *The Oyster Wars of Chesapeake Bay*. Tidewater Publications, Centreville, Maryland. 147 pp.

Wilbur, A. E. and P. M. Gaffney. 1997. Mitochondrial DNA variation and population structure of the bay scallop, *Argopecten irradians* (Abstract). *Journal of Shellfish Research* **16**: 329-330.

Williams, J. D., M. L. Warren, Jr., K. S. Cummings, J. L. Harris, and R. J. Neves. 1993. Conservation status of freshwater mussels of the United States and Canada. *Fisheries* **18**: 6-22.

Wright, S. 1931. Evolution in Mendelian populations. *Genetics* **16**: 97-159.

Yan, J., H. Sun, and J. Fang. 2000. The present status, some problems and developing countermeasures in the cultivation of bay scallop in China. *Marine Fisheries Research/Haiyang Shuichan Yanjiu* **21**: 77-80.

Zettler, M. L., R. Bonsch, and F. Gosselck. 2001. Distribution, abundance, and some population characteristics of the ocean quahog, *Arctica islandica* (Linnaeus, 1767), in the Mecklenberg Bight (Baltic Sea). *Journal of Shellfish Research* **20**: 161-169.

TABLES
Summary of marine mollusk regulations.

A growing number of states regulate the harvest of marine invertebrates in general, including mollusks not covered by other regulations, although effective enforcement may be lacking. In addition, there are a growing number of federal and state preserves, in which the take of all marine life or specific groups (e.g., intertidal invertebrates) is limited or banned.

Table 31.1 Marine gastropods and cephalopods for which U.S. state or federal regulations exist.
Omissions or errors in this table are possible due to unclear or missing information in state regulations. States in **boldface** prohibit harvest of the taxon listed, except by special permit or by aquaculturists.

Family	Latin Name	Common Name	States Regulated
Gastropoda			
Halitiotidae	*Haliotis* spp. [1]	abalone	**CA HI* OR WA**
	Haliotis kamtschatkana	pinto abalone	AK **CA OR WA**
	Haliotis rufescens	red abalone	CA OR **WA**
Patellidae	*Cellana* spp. [2]	opihi	HI
Trochidae	*Trochus* spp.	top shells	**HI***
Strombidae	*Strombus gigas*	queen conch	**FL**
Natacidae	*Polinices lewisii*	moon snail	CA, WA
Muricidae	*Ocinebrinellus inornatus*	Japanese oyster drill	WA[3]
Melongenidae	*Busycon carica* [4]	knobbed whelks, "conch"	CT DE MA MD
	Busycoptus canaliculatus [4]		
	B. contrarium	whelks, "conch"	CT DE MA MD
Buccinidae	*Buccinum undatum*	waved whelk	ME NH
	Colus stimpsoni	Stimpson's whelk	NH
"Nudibranchs"	many taxa [5]	sea slugs, sea hares	FL WA
Cephalopoda			
Loliginidae	*Loligo opalescens*		CA* WA
	Loligo pealei [6]		ME to NC*
Ommastrephidae	*Illex illecebrosus* [6]		ME to NC*
Octopodidae	*Octopus* spp. [7]	octopus	AK CA FL OR WA
	Octopus cyanea and *O. ornatus*	he'e	HI

*Nonindigenous
[1] Except as noted for *H. rufescens* and *H. kamtschatkana*, harvest of all abalone is prohibited.
[2] All Hawaiian intertidal limpets are regulated collectively, but primary species harvested are *C. exarata* and *C. sandwicensis*.
[3] Transfer of this nonindigenous pest between marine waters is banned.
[4] In all states but DE, *Busycon* and *Busycoptus* are regulated collectively where sympatric.
[5] Opisthobranchia "sea slugs" and similar forms are regulated collectively
[6] Federally regulated only.
[7] Numerous species, seldom specifically defined, but primary species harvested include *O. dolfleini* and *O. vulgaris* on Pacific and Atlantic coasts, respectively.

Table 31.2 Marine bivalves for which U.S. state or federal harvest regulations exist.
Omissions or errors in this table are possible due to unclear or missing information in state regulations. States in **boldface** prohibit harvest of the taxon listed, except by special permit or by aquaculturists.

Family	Latin Name	Common Name	States Regulated
Mytilidae	*Mytilus* spp. [1]	blue mussels, sea mussels	CA CT DE ME NH NY OR RI WA
Pteriidae	*Pinctada margaritifera*	black-lipped pearl oyster	**HI**
Pectinidae	*Argopecten irradians*	bay scallop	CT MA NJ NY RI NC SC FL
	Argopecten ventricosus	speckled scallop	**CA**
	Chlamys hastata, C. rubida [2]	spiny scallop, pink scallop	WA
	Crassadoma gigantea	rock scallop	AK CA OR WA
	Euvola diegensis	San Diego scallop	**CA**
	Patinopecten caurinus	weathervane scallop	AK WA
	Placopecten magellanicus	sea scallop	MA NH
Limidae	*Lima scabra*	file shell	FL
Ostreidae	*Crassostrea* spp. [2, 3]	cupped oysters	**CA* HI* OR* WA***
	Crassostrea virginica	Eastern oyster	AL CT DE FL GA LA MA ME MS NC NH NJ NY RI SC TX VA WA*
	Ostrea conchaphila	Olympia oyster	**CA OR** WA
	Ostrea edulis	European flat oyster	**CA* OR* WA***
Veneridae	*Arctica islandica*	mahogany quahog	ME NH NY
	Mercenaria mercenaria, M. campechiensis [2]	hard clams, quahogs	AL CA* CT DE FL GA **HI*** MA ME MS NC **NH** NJ NY RI SC TX VA
	Protothaca staminea	Pacific littleneck clam	AK CA OR WA
	Saxidomus giganteus	butter clam	AK CA OR WA
	Tivela stultorum	Pismo clam	CA
	Venerupis philippinarum	Manila clam	WA*
Tellinidae	*Macoma brota* and *M. nasuta*	sand clam, bentnose clam	WA
Cardiidae	*Clinocardium nuttallii*	basket cockle	CA OR WA
Mactridae	*Spisula solidissima*	surf clam	MA ME NH NJ NY RI
	Tresus capax, T. nuttallii [2, 4]	gaper clam, horse clam	AK CA OR WA
Pharidae	*Siliqua patula, S. alta*	razor clams	AK CA OR WA
	Ensis directus	jackknife clam	CT NH
Myidae	*Mya arenaria*	softshell clam	CA* CT MA ME MS NH NJ NY OR* RI WA*
Hiatellidae	*Panopea abrupta*	geoduck	AK CA WA
Pholadidae	*Penitella penita* [5]	piddocks	OR
	Zirfaea pilsbryi [5]	piddocks	OR
	Barnea subtruncata [5]	piddocks	OR

*Nonindigenous
[1] *M. californiensis, M. edulis, M. galloprovincialis,* and *M. trossulus,* regulated collectively where sympatric. *M. galloprovincialis* is nonindigenous in CA, OR, and WA.
[2] Regulated collectively where sympatric.
[3] *C. araikensis, C. gigas,* and *C. sikamea,* all cultured and nonindigenous.
[4] A third, recently described, sympatric species, *T. allomyax,* is unrecognized by state agencies.
[5] All species of Pholadidae (rock-boring clams) are regulated collectively.

Table 31.3 Examples of unregulated edible or commercial near-shore marine mollusks.
These mollusks are not specifically regulated in U.S. waters, although non-specific regulations (e.g., collecting limits on all intertidal invertebrates or coastal no-take preserves) may affect these species. Errors in this table are possible due to unclear or missing information in state or local regulations.

Family	Latin Name	Common Name	U.S. Range
Gastropoda			
Lottiidae	*Lottia gigantea*	owl limpet	CA
Fissurellidae	*Megathura crenata*	great keyhole limpet	CA
Thaididae	*Stramonita haemastoma*	rock shell, oyster drill	NC to TX
Melongenidae	*Melongena corona* [1]	crown conch	FL AL
	Busycotypus spiratus	pear whelk	NC to TX
Fasciolariidae	*Pleuroploca gigantea* [1]	Florida horse conch	NC to TX
Bivalvia			
Arcidae	*Anadara* spp. [2]	blood ark	MA to TX
Noetiidae	*Noetia ponderosa*	ponderous ark	VA to TX
Mytilidae	*Perna perna*	brown mussel	TX*
	Perna viridis	green mussel	FL* GA*
Pinnidae	*Atrina rigida, A. serrata*	pen shells	NC to TX
Veneridae	*Chione elevata* [3,4]	cross-barred venus	NC to FL
	Macrocallista maculata [4]	calico clam	NC to FL
	Macrocallista nimbosa [4]	sunray venus	NC to TX
Cardiidae	*Trachycardium* spp.	cockles	NC to TX
Donacidae	*Donax gouldii*	beanclam	CA
	Donax fossor, D. variabilis	coquina clams	NY to TX
Mactridae	*Rangia cuneata*	rangia clam	NY* VA-TX
Psammobiidae	*Nuttallia nuttallii*	California mahogany clam	CA
	Nuttallia obscurata	purple mahogany clam	WA* OR*
Pholadidae	*Cyrtopleura costata*	angelwing clam	NJ to TX
Cephalopoda			
Loliginidae	*Lolliguncula brevis*	brief squid	NJ to TX
	Sepioteuthis sepioidea	Caribbean reef squid	FL

*nonindigenous
[1] regulated in Mexico federal waters
[2] *A. brasiliana, A. floridana, A. ovalis*, and *A. transversa*, depending on locality
[3] Listed as *C. cancellata* in most references.
[4] North Carolina regulations are vague on the definition of "clam"

Table 31.4 State regulatory agencies and contacts for marine mollusk regulations, as of 2005.
Regulations, Internet addresses (URLs), and state agencies are subject to change.

State	Regulatory Agencies	Contact Information (URL)
Alabama	Department of Conservation and Natural Resources Marine Resources Division	<www.dcnr.state.al.us>
Alaska	Department of Fish and Game [1]	<www.adfg.state.ak.us>
California	Department of Fish and Game, Marine Region	<www.dfg.ca.gov/mrd>
Connecticut	Department of Agriculture, Bureau of Aquaculture and Laboratory Services [2]	<www.state.ct/doag>
Delaware	Division of Fish and Wildlife	<www.dnrec.state.de.us/fw>
Florida	Fish and Wildlife Conservation Commission Marine Fisheries	<marinefisheries.org>
Georgia	Coastal Resource Division Department of Natural Resources	<www.gadnr.org>
Hawaii	Department of Land and Natural Resources Division of Aquatic Resources	<www.state.hi.us/dlnr/dar>
Louisiana	Department of Wildlife and Fisheries Marine Fisheries Division	<www.wlf.state.la.us>
Maine	Department of Marine Resources [2]	<www.maine.gov/dmr>
Maryland	Department of Natural Resources, Fisheries Service	<www.dnr.state.md.us/fisheries>
Massachusetts	Division of Marine Fisheries [2]	<www.state.ma.us/dfwele/dmf>
Mississippi	Department of Wildlife, Fisheries and Parks	<www.mdwfp.com>
New Hampshire	Fish and Game Department [2]	<www.wildlife.state.nh.us>
New Jersey	Division of Fish and Wildlife	<www.state.nj.us/dep/fgw>
New York	Department of Environmental Conservation Bureau of Marine Resources [2]	<www.dec.state.ny.us/website/dfwmr/marine>
North Carolina	Department of Environmental and Natural Resources Division of Marine Fisheries	<www.ncfisheries.net>
Oregon	Department of Fish and Wildlife, Marine Resources Program	<www.dfw.state.or.us/mrp>
Rhode Island	Department of Environmental Management Division of Fish and Wildlife [2]	<www.dem.ri.gov/programs/bnatres/fisheries>
South Carolina	Department of Natural Resources Marine Resources Division	<www.dnr.state.sc.us>
Texas	Parks and Wildlife	<www.tpwd.state.tx.us>
Virginia	Marine Resource Commission	<www.mrc.state.va.us>
Washington	Department of Fish and Wildlife [1]	<wdfw.wa.gov>

[1] Shellfish regulations vary regionally.
[2] Specific shellfish regulations may be set by individual townships.

APPENDICES

APPENDIX 1: Morphological Features of Gastropod and Bivalve (Pelecypod) Shells
    Appendix 1A. Structural features of gastropods
    Appendix 1B. Form of gastropod shells
    Appendix 1C. Morphological features of pelecypod shells
    Appendix 1D. Descriptive terms applied to pelecypod shells

Appendix 1 is reproduced from Raymond C. Moore, Cecil G. Lalicker, and Alfred G. Fischer. 1952. *Invertebrate Fossils*. McGraw-Hill Book Company, Inc., New York, NY. xiii + 766 pp. Figures 8-5, 8-6, 10-5, and 10-6 are here reproduced as Appendices 1A through 1D. They are reproduced with the kind permission of the McGraw-Hill Companies.

APPENDIX 2
    Expanded Table of Contents

GLOSSARY

APPENDIX 1

Appendix 1A. Structural features of gastropods

Anterior (22). Direction towards head of gastropod, in spiral shells towards aperture.
Aperture (3). Opening of shell, out of which part of the gastropod body may extend.
Apex (14). Point of beginning of shell growth, at tip of spire or cone.
Axis (43). Imaginary center line around which a shell coils.
Basal fasciole (8). Tract of more or less strongly inflected growth lines on the base of a shell, marking the location of a siphonal notch.
Base (4). Part of shell surface at extremity opposite apex.
Body whorl (2). Last-formed single complete loop of a spiral shell.
Callus (38). Generally thickened deposit of shell substance on anterior part of inner lip of aperture, partly or wholly covering umbilicus; it is part of inductura.
Canal (12). Semitubular anterior extension of aperture, enclosing siphon; at least slightly open along side, not closed like a pipe.
Carina (30). Spiral keel on exterior of whorl, generally at edge of shelf.
Columella (31). Solid or perforate pillar formed by inner walls of a conispiral shell.
Columellar fold (33). Spiral elevation on columella produced by localized thickening of shell.
Columellar lip (25). Part of inner lip of aperture adjoining columella.
Costa (5). Coarse threadlike thickening of shell running spirally or axially.
Digitation (48). Finger-like outward projection of outer lip of aperture.
Dip (18). Deviation of suture from a plane normal to the shell axis.
Growth line (40). Marking on shell parallel to apertural margin, denoting a former position of aperture.
Gutter (19). Groove or canal at posterior extremity of aperture, in some gastropods marking location of anal outlet.
Heterostrophy (35). Abrupt change in type of coiling between nucleus and later-formed part of shell.
Inductura (36). Layer of lamellar shell material along inner lip of aperture or extending over shell surface beyond outer lip, characterized by smooth surface; includes callus.

Inner lip (20). Margin of aperture adjacent to next-to-last whorl; may include parietal lip and columellar lip.
Labral outline (39). Shape of outer lip in view normal to aperture.
Labral profile (27). Shape of outer lip in view parallel to edge of aperture.
Lunula (46). Crescentic growth line on selenizone.
Neck (7). Constricted anterior part of body whorl of some gastropods, exclusive of canal.
Nucleus (34). Embryonic gastropod shell, commonly consisting of one to four whorls.
Operculum (13). Horny or horny-calcareous plate carried on posterior part of foot, used to close aperture when gastropod withdraws into its shell.
Ornamentation (28). Raised or depressed markings of shell surface other than growth lines.
Outer lip (21). Edge of aperture on side away from next-to-last whorl.
Parietal lip (24). Part of inner lip which adjoins next-to-last whorl.
Periphery (6). Part of whorl farthest from shell axis.
Peristome (37). Margin of whole aperture.
Posterior (15). Direction backward from head of gastropod, in spiral shells toward apex.
Ramp (44). Sloping surface of a whorl next below a suture.
Rib (23). Well-marked linear elevation of shell surface, larger and broader than a costa.
Selenizone (45). Sharply defined band parallel to coiling of whorls, which bears crescentic growth lines denoting a notch or slit in outer lip.
Shelf (29). Subhorizontal part of whorl surface next to a suture, bordered on side toward periphery of whorl by a sharp angulation or by a carina.
Shoulder (47). Salient angulation of a whorl parallel to coiling.
Sinus (49). Reentrant in outer lip with nonparallel sides.
Siphonal notch (9). Reentrant at junction of outer and columellar lips occupied by siphon.
Slit (26). More or less deep, parallel-sided reentrant in outer lip, which gives rise to a selenizone.
Spire (1). Coiled gastropod shell exclusive of body whorl.
Spiral angle (10). Angle formed by lines tangent to two or more whorls on opposite sides of shell; inasmuch as lines tangent to all whorls of the spire may define a curve, spiral angle is commonly determined by drawing straight-line tangents to lowermost whorls of spire.

Appendices

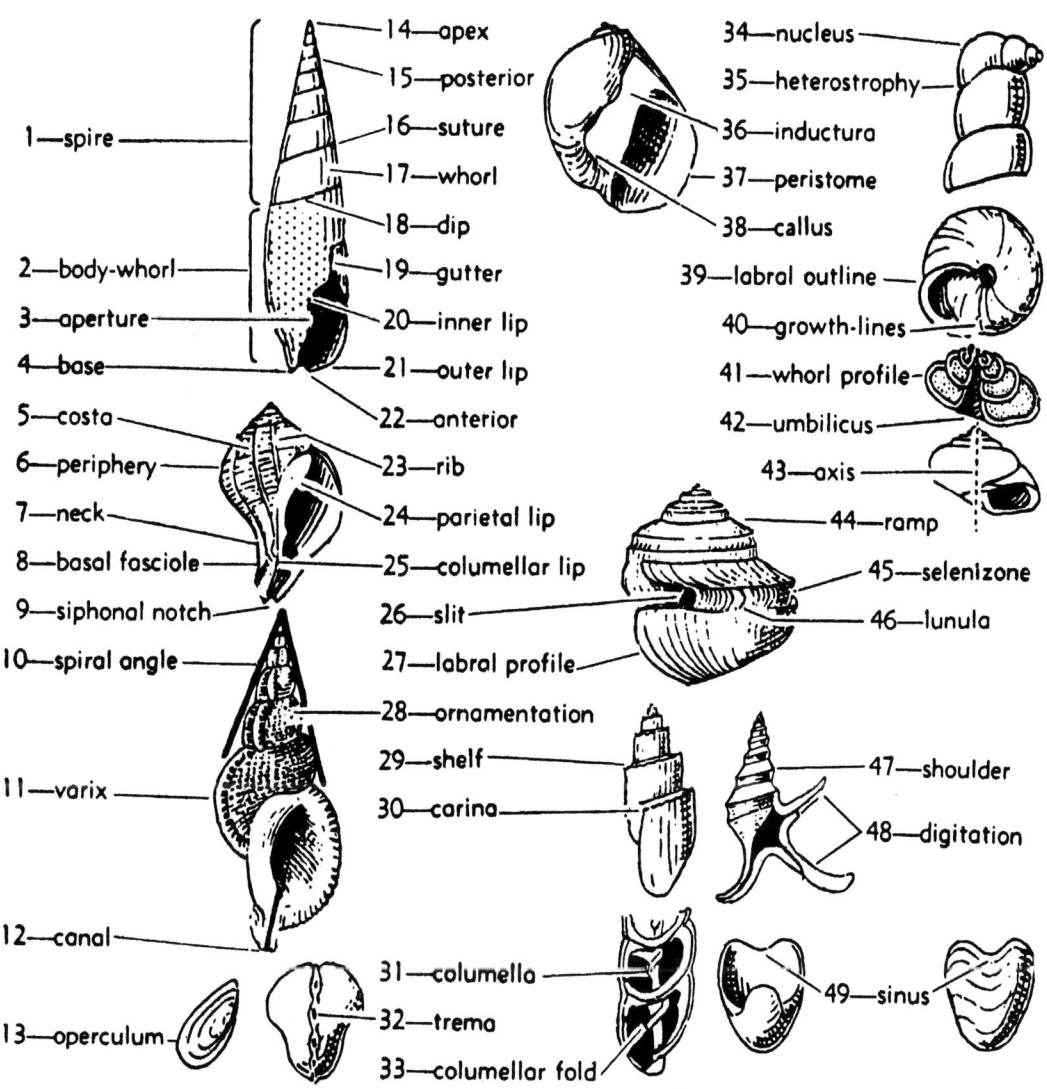

Suture (16). Spiral line of junction between surfaces of any two whorls; includes external sutures on outer side of shell and umbilical sutures within umbilicus.

Trema (pl. tremata, 32). Perforation of shell, generally formed by periodic closure of a slit, but occurring also at apex of some cap-shaped shells.

Umbilicus (42). Central cavity of a shell formed by walls on inner sides of whorls; most common is basal umbilicus of orthostrophic conispiral shells but also included are apical umbilicus of convolute and hyperstrophic shells and lateral umbilici of planispiral (isotrophic) shells.

Varix (11). Ridge, flange, or row of spines parallel to growth lines and marking modification of shell at former position of aperture.

Whorl (17). Single complete loop of a spiral shell.

Whorl profile (41). Transverse contour of surface of a whorl in a plane intersecting the axis of coiling; differs from labral profile and labral outline.

Appendix 1B. Form of gastropod shells

Advolute (5). Whorls in contact but not distinctly embracing.

Biconical (23). Spire and base both having moderately elevated conical form.

Conical (22). Spire and body whorl having evenly confluent sides which form a straight-walled cone, base flattened.

Conoidal (24). Spire and body whorl forming a cone which is distinctly steeper-sided near the base than at the apex.

Convolute (14). Outer whorls so deeply embracing inner ones that latter are nearly or quite invisible externally.

Dextral (1). Right-handed; in apertural view with apex directed upward, aperture is on right side of shell; in apical view, shell coils in a clockwise direction.

Discoidal (3). Having shape of a wheel or disc; spire extremely low conical, flat, or shallowly concave, and base also having one of these shapes.

Evolute (4). Loose-coiled, whorls not in contact.

Extraconical (17). Spire and body whorl forming a cone distinctly steeper-sided near apex than toward base.

Fusiform (25). Spindle-shaped; largest in middle and sharp-pointed at both extremities.

Holostomatous (9). Peristome continuous, not interrupted anteriorly by siphonal notch or canal.

Hyperstrophic (19). Spire depressed instead of elevated; a hyperstrophic dextral shell may be identical in form to an orthostrophic sinistral shell, and a hyperstrophic sinistral shell likewise may correspond to an orthostrophic dextral shell; identification is based on organization of soft parts.

Involute (12). Outer whorls slightly to strongly embracing inner whorls but not completely.

Multispiral (7). Spire composed of numerous whorls.

Obconical (26). Reversed-cone shape, base strongly conical and spire nearly flat.

Orthostrophic (2). Spire slightly to strongly elevated; as measured by form of whorl center lines, this embraces shells in which upper sides of whorls are all tangent to a single plane.

Ovoid (15). Egg-shaped, apical and basal parts of shell somewhat evenly rounded.

Patelliform (20). Low cap-shaped, like shell of *Patella*, noncoiled.

Paucispiral (16). Spire composed of very few whorls.

Planispiral (13). Coiling in a plane, with part of shell on one side of a plane the mirror image of other.

Pupaeform (6). Elevated ovoid, like shell of *Pupa*, in which late-formed whorls have decreasing radii of curvature.

Sinistral (10). Left-handed; in apertural view with apex directed upward, aperture is on left side of shell; in apical view, shell coils counter-clockwise.

Siphonostomatous (27). Peristome discontinuous, interrupted anteriorly by a siphonal notch or canal.

Trochiform (21). Sides of shell evenly conical, base flat, like shell of *Trochus*.

Turbinate (8). Top-shaped, generally with rounded base.

Turreted (18). Very high-spired shell with flat or gently rounded base, like shell of *Turritella*.

Umbilical shoulder (11). Angulation of whorls at margin of umbilicus and within it.

Appendices

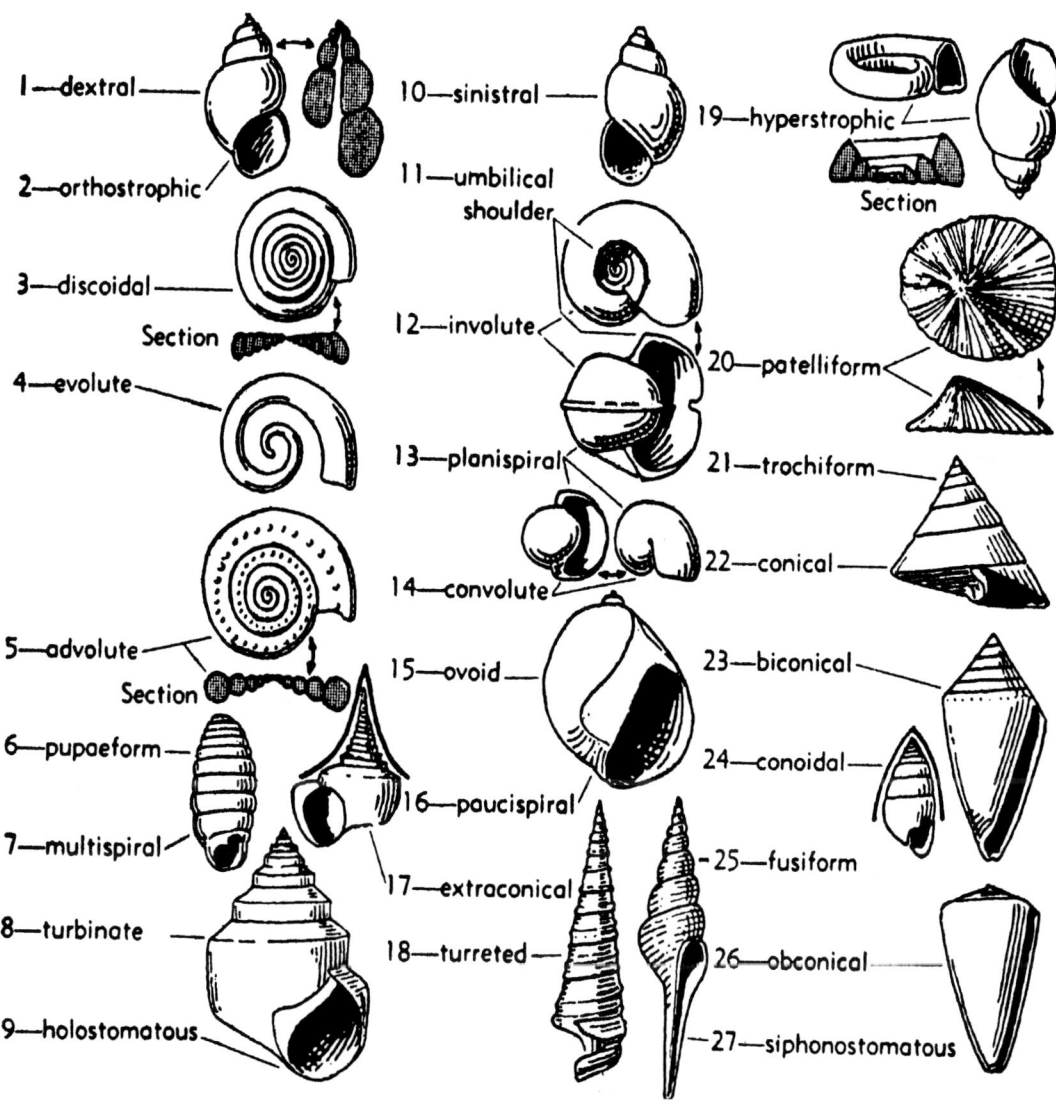

Appendix 1C. Morphological features of pelecypod shells

Adductor scar (30). Impression on inside of valve made by attachment of muscle which functions for closure of valve.
Anterior (5). Part of shell containing mouth; beaks of most pelecypods point forward.
Auricle (17). Forward or backward projection of shell along hinge line in some pelecypods; also called ear.
Auricular sulcus (44). Groove on shell exterior separating auricle from remainder of valve.
Beak (2). More or less sharp-pointed projection at the initial point of shell growth, located along or above hinge line.
Buttress (25). Ridge on inner surface of a valve which serves as support for part of hinge.
Byssal notch (18). Indentation on anterior edge of some shells for protrusion of threadlike attachment called byssus; most common in pectinoid shell on right valve, which is lowermost, allowing protrusion of the small foot without opening valve widely.
Byssal sinus (42). Indentation beneath auricle of left valve of pectinoid shells on anterior margin.
Cardinal area (40). Plane or curved surface between beak and hinge line, generally distinguished from remainder of valve exterior by sharply angulated border.
Cardinal teeth (28). Projections vertical or oblique to hinge line directly beneath or closely adjacent to beak; they fit into sockets of opposite valve.
Chevron groove (39). Narrow depressions on cardinal area having an inverted V shape, marking ligament attachments.
Chondrophore (24). Relatively prominent internal spoon-shaped structure, which holds an internal ligament (resilium).
Costa (22). Radial ridge on shell surface formed by thickening of shell.
Dorsal (1). Direction toward part of shell containing hinge line.
Escutcheon (11). Depressed plane or curved area along hinge line behind beak, corresponding to posterior part of cardinal area.
Gape (47). Anterior or posterior space between edges of valves when ventral margins are in contact.
Growth line (6). More or less obscure concentric lines parallel to shell margin, marking successive advances of edge of shell.
Height (3). Distance from dorsal to ventral margin measured normal to length.
Hinge line (9). Edge of valve along dorsal margin which is in permanent contact with opposite valve.
Hinge plate (27). Internal surface adjacent to hinge line along which hinge teeth project.
Hinge teeth (14). Projections from hinge plate for articulation of valves.
Lamellar layer (35). Generally innermost part of pelecypod shell, consisting of microscopically thin sheets of calcite or aragonite separated by layers of conchiolin.
Lateral teeth (15). Projections from hinge plate nearly parallel to hinge line, situated in front or behind cardinal teeth.
Left valve (43). Shell on left side of the antero-posterior axis; among pelecypods which characteristically lie on one side, the left valve is typically uppermost in some (pectinoids) but lowermost in others (oysters and many pachyodonts).
Length (4) Distance from anterior to posterior margin at farthest points or measured parallel to hinge line.
Ligament area (26). Portion of surface along hinge line to which ligament is attached.
Ligament groove (38). Linear depression in cardinal area or ligament area marking attachment of ligament fibers.
Lunule (10). Depressed plane or curved area along hinge line in front of beak, equivalent to anterior part of cardinal area.
Muscle scar (19). Generally depressed (less commonly raised) area on inner surface of shell, marking attachment place of muscle.
Myophore (46). Plate or rodlike structure on inside of shell for attachment of muscle.
Ostracum (36). Calcareous structure composing all of pelecypod shell except thin outer conchiolin layer (periostracum).
Pallial line (32). Linear depression on inside of pelecypod shell along ventral side, marking inner margin of thickened mantle edges.
Pallial sinus (31). Inward deflection of posterior part of pallial line, defining space for retraction of siphons.
Periostracum (33). Outer thin layer of conchiolin, which in many pelecypods covers calcareous ostracum.
Plane of commissure (13). Surface approximately coinciding with valve margins.
Plica (21). Radially disposed ribs formed by fold that involves entire thickness of shell.
Posterior (29). Direction or part of shell toward position of anus and siphonal opening; in most pelecypods opposite to inclination of beak.
Prismatic layer (34). Outer part of ostracum in many pelecypods, consisting of closely spaced polygonal prisms of calcite.
Prodissoconch (37). Earliest-formed part of shell; generally preserved at tip of beak.
Resilifer (41). Portion of hinge plate which bears internal ligament (resilium); generally a simple shallow pit.
Resilium (23). Part of ligament below level of valve margins and under compression instead of tension.

Appendices

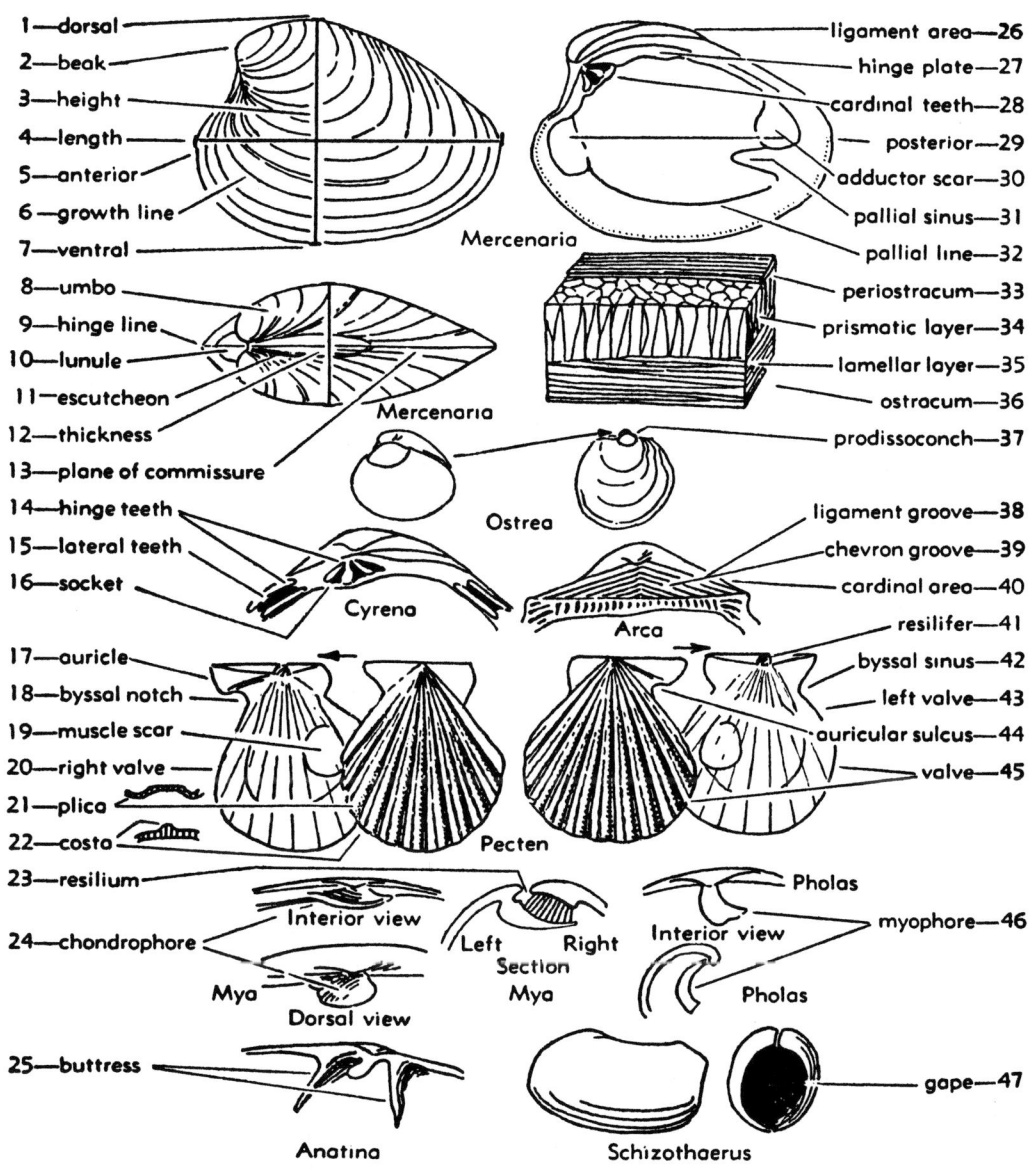

Right valve (20). Shell on right side of antero-posterior axis; among pelecypods which lie on their sides, it is generally lowermost in pectinoids and uppermost in oysters and pachyodonts.

Socket (16). Depression in hinge plate for reception of a hinge tooth of opposite valve.

Thickness (12). Maximum dimension of pelecypod shell measured normal to plane of commissure.

Umbo (8). Very strongly convex part of valve adjacent to beak.

Valve (45). Part of shell lying on either side of hinge line.

Ventral (7). Direction or part of shell lying opposite the hinge line; generally located lowermost in pelecypods which move about freely.

Appendix 1D. Descriptive terms applied to pelecypod shells

Acline (13). Shell having neither forward nor backward obliquity, mid-line of umbo being normal to hinge line.
Alate (32). Shell characterized by possession of wings or auricles.
Alivincular (18). Type of external ligament having greatest length transverse to plane of commissure.
Amphidetic (43). Ligament located along hinge line both in front and behind beak.
Anisomyarian (4). Adductor muscle scars conspicuously unequal.
Auriculate (11). Shell possessing auricles; equivalent to alate.
Cancellate (48). Shell surface marked by subequal concentric and radial markings.
Carinate (27). Shell surface marked by sharp-angled edge extending outward from beak.
Compressed (1). Transversely flattened shell having small thickness.
Concentric (28). Shell surface marked by ridges parallel to shell margin.
Costate (12). Shell bearing radial ribs formed by localized thickening.
Desmodont (40). Type of shell characterized mainly by prominence of internal ligament.
Dimyarian (35). Valves having two adductor scars, whether equal or unequal.
Divaricate (47). Shell surface marked by two sets of parallel lines which meet at a distinct angle.
Duplivincular (19). Ligament composed partly of fibrous (compressional) tissue and partly of lamellar (tensional) tissue.
Dysodont (39). Shells mainly characterized by absence or near absence of hinge teeth and narrow external ligament.
Edentate (17). Lacking hinge teeth.
Equilateral (3). Anterior and posterior halves of valve subequal and nearly symmetrical.
Equivalve (2) Right and left valves subequal and comprising mirror images of one another except for hinge structure.
Gaping (31). Part of valve margins not in contact when other parts are pulled tightly together.
Heterodont (41). Characterized by hinge teeth of distinct type – cardinals beneath beak, and laterals in front or behind or both.
Inequilateral (6). Anterior and posterior parts of valve unequal and lacking symmetry.
Inequivalve (9). Opposite valves dissimilar in size or shape or both.
Isodont (15). Characterized by two subequal prominent hinge teeth on one valve and corresponding sockets in the other.
Isomyarian (36). Having two adductor muscles of approximately equal size.
Monomyarian (46). Having only one adductor muscle, originally posterior but tending to be central in position.
Multicostate (21). Surface marked by costae which increase by intercalation or bifurcation.
Multivincular (38). Ligament chiefly formed by successive bands of fibrous (compressional) tissue.
Mytiliform (24). Slipper-shaped, like the genus *Mytilus*.
Opisthocline (14). Shell having backward obliquity, approach along mid-line to beak pointing backward.
Opisthodetic (37). External ligament located behind beaks.
Opisthogyral (33). Beaks turned backward instead of projecting forward.
Orbicular (20). Shell subcircular in outline.
Pachydont (42). Thickened specialized teeth, typically developed in coral-like rudistids.
Parivincular (44). Ligament having long axis parallel to hinge line, consisting mainly of lamellar (tensional) tissue.
Perforate (45). Valve (right) characterized by rounded opening for passage of byssus.
Plicate (22). Shell radially folded to form ribs.
Produced (7). Shell much elongated in one direction.
Prosocline (10). Having forward obliquity, approach to beak along mid-line of shell inclined forward.
Prosogyral (8). Beaks directed forward.
Quadrate (23). Shell rectangular in outline.
Radial (26). Surface marked by costae or plicae diverging from beak.
Rhomboidal (5). Shell outline rhomb-shaped.
Rostrate (29). Having prominent beaks.
Schizodont (16). Having prominent diverging or bifurcate hinge teeth.
Taxodont (34). Characterized by more or less numerous subequal hinge teeth, generally arranged in a row.
Trigonal (25). Shell outline subtriangular.
Truncate (30). Edge of shell, generally posterior, having a chopped-off appearance.

Appendices

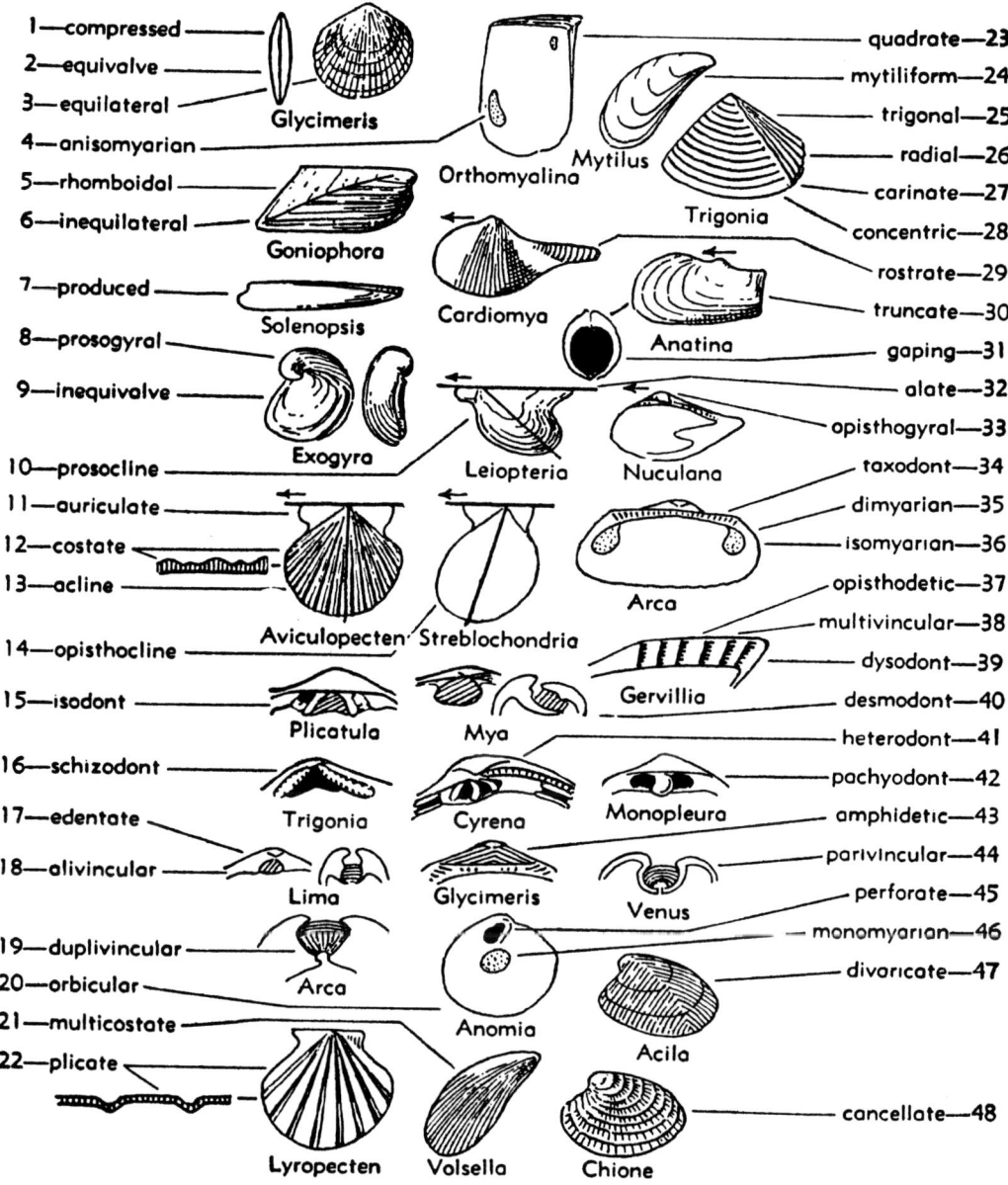

APPENDIX 2

FRONTISPIECE ii
PREFACE vii
TABLE OF CONTENTS ix
CONTRIBUTORS xi
CHAPTER 1 THE MOLLUSKS: INTRODUCTORY COMMENTS 1
    1.1 INTRODUCTION 1
    1.2 CHAPTER REVIEWS 3
    1.3 ACKNOWLEDGMENTS 6
    1.4 LITERATURE CITED 7
CHAPTER 2 FIELD AND LABORATORY METHODS IN MALACOLOGY 9
    2.1 INTRODUCTION 9
    2.2 COLLECTING BASICS 9
        2.2.1 Purchasing 9
        2.2.2 Trading 10
        2.2.3 Self-collecting 10
    2.3 COLLECTING EQUIPMENT 10
        2.3.1 Allison Scoop 11
        2.3.2 Ferriss Hoe 11
        2.3.3 Davis Rake Drag 11
        2.3.4 Screens, Dippers, and Nets 11
        2.3.5 van Eeden Scoop 12
        2.3.6 Shovels 12
        2.3.7 Clam Tube 12
        2.3.8 Hammer 12
        2.3.9 Bales Hook 12
        2.3.10 Water Pumps 13
        2.3.11 Bags and Collecting Containers 13
        2.3.12 Glass-bottomed Bucket 13
        2.3.13 Lights for Night Collecting 14
        2.3.14 Forceps/Tweezers 14
        2.3.15 Loupes 14
        2.3.16 Thread 14
        2.3.17 Tide Tables 14
        2.3.18 Miscellaneous Items 14
    2.4 FIELD COLLECTING TECHNIQUES 15
        2.4.1 Land Snails 15
        2.4.3 Freshwater Mollusks 15
        2.4.3 Marine Mollusks 15
        2.4.4 SCUBA Diving 16
        2.4.5 *Ex pisce* Collecting 16
        2.4.6 Sea Stars (Starfish) 17
        2.4.7 Tidal Pools 17

2.4.8 Traps 17
2.4.9 Navigational Buoys 17
2.4.10 Marine Grasses and Algae 18
2.4.11 Commercial Fishing Boats 18
2.4.12 Specific Methods for Select Groups 19
2.4.13 Ecological Considerations 21
2.5 TECHNIQUES FOR NARCOTIZING MOLLUSKS 21
2.6 TECHNIQUES FOR MARKING AND TAGGING MOLLUSKS 22
2.7 TECHNIQUES FOR CLEANING AND PRESERVING MOLLUSKS 23
    2.7.1 Boiling 23
    2.7.2 Hooks and Pins 24
    2.7.3 Flushing with Water 24
    2.7.4 Preservation of Tissue 24
    2.7.5 Formaldehyde 25
    2.7.6 Cleaning Shells with Insects 25
    2.7.7 Vacuum Pumps 25
    2.7.8 Microwave Ovens 25
    2.7.9 Cleaning the Exterior Shell Surface 26
    2.7.10 Ultrasonic Cleaner 27
    2.7.11 Walnut Shell Blasting 27
    2.7.12 Techniques for Specific Groups 27
    2.7.13 Coating Shells with Preservatives 28
    2.7.14 Acid Treatment 29
2.8 LITERATURE CITED 29

CHAPTER 3 REMOTE BOTTOM COLLECTING 33
3.1 INTRODUCTION 33
3.2 DREDGING 33
    3.2.1 Boats 34
    3.2.2 Dredge Line 35
    3.2.3 Types of Dredges 35
    3.2.4 Dredging Operations 37
    3.2.5 Locating Where you Are 37
    3.2.6 Taking Care of the Hauls 38
3.3 OTHER METHODS OF REMOTE BOTTOM SAMPLING 38
    3.3.1 Grabs 38
    3.3.2 Box Cores 39
    3.3.3 Tangle Nets 40
    3.3.4 Bail Hooks 40
    3.3.5 Dip Nets, Walker Dipper, *etc.* 40
3.4 LITERATURE CITED 40

CHAPTER 4 SNORKELING AND SCUBA DIVING 41
4.1 INTRODUCTION 41
4.2 LEARNING TO SNORKEL 41
4.3 EQUIPMENT NEEDED TO SNORKEL 42
4.4 SCUBA DIVING 43
4.5 EQUIPMENT NEEDED FOR SCUBA DIVING 43
4.6 DRAWBACKS TO SCUBA DIVING 43
4.7 UNDERWATER HAZARDS 43

Appendices

    4.8 CONCLUSIONS 44
    4.9 LITERATURE CITED 44
CHAPTER 5 ARCHIVAL AND CURATORIAL METHODS 45
    5.1 BASIC PRINCIPLES 45
    5.2 DANGERS TO A COLLECTION 45
        5.2.1 Acid, Temperature, and Humidity 46
        5.2.2 Light 48
        5.2.3 Pests 48
        5.2.4 Shock and Abrasion 48
    5.3 PAPER 49
    5.4 INKS AND COMPUTER PRINTERS 50
    5.5 VIALS AND JARS 50
        5.5.1 Glass Decay 51
        5.5.2 Vials, Jars, and Closures 51
    5.6 PLASTICS 52
    5.7 CONSOLIDANTS AND ADHESIVES 53
    5.8 RECORDS 53
    5.9 WET COLLECTIONS 54
    5.10 CONCLUSIONS 55
    5.11 LITERATURE CITED 56
    APPENDIX 5.1 GENERAL AND ARCHIVAL LABORATORY SUPPLIERS 57
CHAPTER 6 DIGITAL IMAGING: FLATBED SCANNERS AND DIGITAL CAMERAS 59
    6.1 INTRODUCTION 59
    6.2 PRIMER ON DIGITAL IMAGING 59
        6.2.1 Pixels, Resolution, and Image Size 59
        6.2.2 Color Depth and Graphic Types 60
    6.3 DIGITAL *VS.* FILM PHOTOGRAPHY 61
    6.4 USES OF DIGITAL PHOTOGRAPHY IN MALACOLOGY 61
    6.5 INPUT DEVICES 62
        6.5.1 Scanners 63
        6.5.2 Digital Cameras 63
        6.5.3 Other Sources 63
    6.6 THE FLATBED SCANNER: AN AFFORDABLE SOLUTION TO SHELL PHOTOGRAPHY.
    63
        6.6.1 How to Scan Shells with a Flatbed Scanner 64
    6.7 DIGITAL CAMERAS 66
        6.7.1 Desirable Functions on a Digital Camera (or Buying Tips) 66
        6.7.2 Recommended Accessories 67
        6.7.3 Tips on Photographing Shells 67
    6.8 OUTPUT DEVICES 68
    6.9 BASICS OF DIGITAL IMAGE EDITING 68
    6.10 CONCLUSIONS 70
    6.11 RESOURCES ON DIGITAL IMAGING AND MALACOLOGY 70
    6.12 ACKNOWLEDGMENTS 71
    6.13 LITERATURE CITED 71
CHAPTER 7 APPLIED FILM PHOTOGRAPHY IN SYSTEMATIC MALACOLOGY 73
    7.1 INTRODUCTION 73
    7.2 LIGHT 73

    7.2.1 Light Intensity 73
    7.2.2 Estimating Exposure Correction 74
    7.2.3 The Color of Light 75
7.3 EQUIPMENT 76
    7.3.1 Body 76
    7.3.2 Motor drive 78
    7.3.3 Lenses 78
    7.3.4 Focus and Depth of Field 79
    7.3.5 Filters 80
    7.3.6 Flashes 82
    7.3.7 Monopods, Tripods, Copy Stands 82
7.4 FILM 84
    7.4.1 Types of Film 84
    7.4.2 Specialty Films 84
    7.4.3 Color print or Slide Film? 85
    7.4.4 Comparative Metrics 85
    7.4.5 Professional and Consumer Film 87
    7.4.6 Storage of Films 87
    7.4.7 Error of Reciprocity 87
7.5 REPRODUCTION PHOTOGRAPHY 87
    7.5.1 Even Illumination 88
    7.5.2 Surface Texture 88
    7.5.3 Shadow-Free Illumination 89
    7.5.4 3D-Objects, Top View 89
    7.5.5 3D Objects, Lateral View 91
    7.5.6 Glossy Objects 91
    7.5.7 Microphotography 91
7.6 ULTRAVIOLET (UV) PHOTOGRAPHY 91
    7.6.1 UV reflectance photography 92
    7.6.2 UV fluorescence photography 92
7.7 INFRARED (IR) PHOTOGRAPHY 92
7.8 PHOTOGRAPHY THROUGH GLASS AND WATER 93
    7.8.1 Optics and Geometry 93
    7.8.2 Aquarium Set-up 94
    7.8.3 Outdoor Applications 94
7.9 UNDERWATER (UW) PHOTOGRAPHY 95
    7.9.1 Range of Equipment 95
    7.9.2 Photography in Water 96
7.10 STORAGE AND ARCHIVAL CONSIDERATIONS 97
    7.10.1 Processing 97
    7.10.2 Mounting 98
    7.10.3 Labeling 98
    7.10.4 Storage 98
7.11 ACKNOWLEDGMENTS 99
7.12 LITERATURE CITED 99
CHAPTER 8 COMPUTERIZING SHELL COLLECTIONS 101
8.1 INTRODUCTION 101
8.2 CHOOSING A PROGRAM 101

Appendices

8.3 DATABASE CONVENTIONS ... 102
8.4 SUGGESTED FIELDS ... 103
    8.4.1 Specimen ... 104
    8.4.2 Identification ... 105
    8.4.3 Classification ... 106
    8.4.4 Locality ... 106
    8.4.5 Collecting Event ... 107
8.5 DATABASE STRUCTURES ... 108
8.6 TESTING AND USING THE SYSTEM ... 109
8.7 ACKNOWLEDGMENTS ... 110

CHAPTER 9 THE MOLLUSCAN LITERATURE: GEOGRAPHIC AND TAXONOMIC WORKS ... 111

9.1 INTRODUCTION TO THE MALACOLOGICAL LITERATURE ... 111
    9.1.1 Iconographies ... 111
    9.1.2 Monographs ... 111
    9.1.3 Nomenclators ... 111
    9.1.4 Handbooks ... 112
    9.1.5 Classic Books ... 112
    9.1.6 General Books ... 112
    9.1.7 Journals ... 112
    9.1.8 Separates ... 112
    9.1.9 Miscellaneous Printed Works ... 112
    9.1.10 Internet Resources ... 112
9.2 REGIONAL AND TAXONOMIC GUIDES ... 113
    9.2.1 Mollusks - General References and Research Tools ... 114
    9.2.2 Marine Biogeographic Zones ... 114
    9.2.3 Terrestrial and Freshwater Biogeographic Zones ... 115
    9.2.4 Marine Molluscan Groups ... 118
    9.2.5 Terestrial Gastropods ... 120
    9.2.6 Freshwater Mollusks ... 121
9.3 MALACOLOGICAL BIBLIOGRAPHY ... 122

CHAPTER 10 TAXONOMY AND TAXONOMIC WRITING: A PRIMER ... 147

10.1 INTRODUCTION ... 147
10.2 REFERENCING A SPECIES ... 147
    10.2.1 Basics ... 147
    10.2.2 Further Considerations ... 147
    10.2.3 New Names ... 148
    10.2.4 Uncertainty Statements ... 148
10.3 DESCRIPTION OF A NEW TAXON ... 148
    10.3.1 Quality ... 148
    10.3.2 Publication Platform ... 149
    10.3.3 Illustrations ... 149
    10.3.4 How to Recognize a New Species ... 150
    10.3.5 What is a Species? ... 150
10.4 ANATOMY OF A SPECIES OR GENUS DESCRIPTION ... 151
    10.4.1 Title and Abstract ... 151
    10.4.2 Introduction ... 151
    10.4.3 Material and Methods ... 151

10.4.4 Systematics 152
10.4.5 New Genus 152
10.4.6 New Species 152
10.5 REVISIONS 153
10.5.1 Genus 153
10.5.2 Species 154
10.6 THE ICZN CODE 155
10.6.1 History 155
10.6.2 Aim (Articles 1-3) 155
10.6.3 Principles (Articles 3-4) 156
10.6.4 Publication (Articles 7-9) 156
10.6.5 Availability (Articles 10-20) 156
10.6.6 Publication Date (Articles 21-22) 156
10.6.7 Validity (Articles 23-24) 156
10.6.8 Formation of Names (Articles 25-34) 156
10.6.9 Family, Genus, and Species Level Names (Articles 35-49) 157
10.6.10 Authorship (Articles 50-51) 157
10.6.11 Homonymy (Articles 52-60) 157
10.6.12 Types (Articles 61-76) 157
10.6.13 The Commission (Articles 77-90) 157
10.6.14 Code of Ethics (Appendix A) 157
10.7 NOMENCLATURE, TAXONOMY, AND CLASSIFICATION 157
10.8 ACKNOWLEDGMENTS 158
10.9 LITERATURE CITED 158
CHAPTER 11 CLADISTICS AND MOLECULAR TECHNIQUES: A PRIMER 161
11.1 INTRODUCTION 161
11.2 CLADISTICS - WHAT IS IT? 161
11.2.1 The Root, the Ingroup, and the Outgroups 162
11.2.2 Parsimony 163
11.2.3 Distances 164
11.2.4 Maximum Likelihood 164
11.2.5 Branch-and-Bound and Heuristic Methods 165
11.2.6 Consensus 165
11.2.7 Bootstrapping and Jackknifing 165
11.2.8 Long Branch Attraction 166
11.3 HOW TO DO DNA STUDIES 166
11.3.1 Collecting 167
11.3.2 Processing the Sample 167
11.3.3 What are the Results? 170
11.4 LITERATURE CITED 171
CHAPTER 12 ORGANIZATIONS, MEETINGS, AND MALACOLOGY 173
12.1 INTRODUCTION 173
12.2 ORGANIZATIONS AND MEETINGS 173
12.3 THE INTERNET 177
12.3.1 MOLLUSCA LIST 178
12.3.2 CONCH-L 178
12.3.3 UNIO 178
12.3.4 PaleoNet 178

Appendices

    12.3.5 NHCOLL-L 179
    12.3.6 PERMIT-L 179
  12.4 SUMMARY 179

CHAPTER 13 MUSEUMS AND MALACOLOGY 181
  13.1 INTRODUCTION 181
  13.2 MUSEUMS AND SOCIETY 181
    13.2.1 Education 181
    13.2.2 Collections 182
    13.2.3 Research 183
  13.3 UNITED STATES AND CANADIAN MUSEUMS 183
  13.4 LITERATURE CITED 187

CHAPTER 14 DONATING AMATEUR COLLECTIONS TO MUSEUMS 189
  14.1 INTRODUCTION 189
  14.2 WHAT MAKES A SPECIMEN VALUABLE TO A MUSEUM? 190
  14.3 HIERARCHY OF USES 190
    14.3.1 Research or Systematic Collection 190
    14.3.2 Specimens for Exchange 190
    14.3.3 Exhibits 190
    14.3.4 Education 190
    14.3.5 Specimens for Sale 191
    14.3.6 Crafts 191
    14.3.7 Specimens to be Discarded 191
  14.4 WHAT AND HOW MUCH DATA? 191
  14.5 IS SPECIMEN QUALITY IMPORTANT? 192
  14.6 SOFT PARTS 193
  14.7 REMOVING THE SOFT PARTS FROM A SHELL 193
  14.8 PRESERVATIVES FOR SOFT PARTS 193
  14.9 INCREASING SPEED AND LIKELIHOOD OF DONATIONS BEING INCORPORATED INTO MUSEUMS 194
    14.9.1 Establish the Fate of the Collection ahead of Time 194
    14.9.2 Associate Data with Specimens 194
    14.9.3 Contact a Museum 195
    14.9.4 Donate Money or Time 195
    14.9.5 No Special Conditions 195
  14.10 SUMMARY 196
  14.11 LITERATURE CITED 196

CHAPTER 15 FOSSIL MOLLUSKS 197
  15.1 INTRODUCTION 197
  15.2 FOSSIL MOLLUSKS 197
  15.3 COLLECTING 200
    15.3.1 In the Field 200
    15.3.2 Transporting Material Home 201
    15.3.3 Cleaning and Preparing Fossils 202
  15.4 LOCALITY AND STRATIGRAPHIC DATA 203
  15.5 IDENTIFYING FOSSILS 203
  15.6 AMATEUR ACTIVITIES AND OPPORTUNITIES 204
  15.7 LITERATURE CITED 205

CHAPTER 16 APLACOPHORA 207
   16.1 INTRODUCTION 207
   16.2 ORGANIZATION 207
   16.3 ECOLOGY 208
   16.4 COLLECTION TECHNIQUES 209
   16.5 EXAMINATION 209
   16.6 LITERATURE CITED 210

CHAPTER 17 MONOPLACOPHORA 211
   17.1 INTRODUCTION 211
   17.2 BIOLOGY 212
      17.2.1 Shell 212
      17.2.2 Locomotion 212
      17.2.3 Digestion and Diet 213
      17.2.4 Gas Exchange and Water Balance 213
      17.2.5 Reproduction and Development 214
      17.2.6 Substratum 214
   17.3 ZOOGEOGRAPHIC DISTRIBUTION 214
      17.3.1 Fossil Species 214
      17.3.2 Living Species 214
   17.4 COLLECTING AND STORAGE TECHNIQUES 214
      17.4.1 Recent Monoplacophora 214
      17.4.2 Fossil Monoplacophora 216
   17.5 LITERATURE CITED 216

CHAPTER 18 POLYPLACOPHORA 217
   18.1 INTRODUCTION 217
   18.2 GROSS MORPHOLOGY 217
      18.2.1 The Plates (Valves) 217
      18.2.2 The Perinotum (Girdle) 218
      18.2.3 The Ctenidia (Gills) 219
   18.3 ANATOMY 219
      18.3.1 The Radula 220
      18.3.2 The Digestive System 220
      18.3.3 Nervous System and Sensory Organs 220
      18.3.4 The Body Musculature 221
   18.4 REPRODUCTION, LIFE HISTORY, AND ORGANOGENESIS 222
   18.5 HABITAT 223
   18.6 COLLECTION AND PREPARATION 223
   18.7 POLYPLACOPHORAN PHYLOGENY 224
   18.8 SELECTED WEB RESOURCES 225
   18.9 LITERATURE CITED 225

CHAPTER 19 SCAPHOPODA: THE TUSK SHELLS 229
   19.1 INTRODUCTION 229
   19.2 BIOLOGY 230
      19.2.1 Shell 230
      19.2.2 Mantle Cavity 230
      19.2.3 Foot 231
      19.2.4 Feeding 231
      19.2.5 Circulation and Excretion 231

19.2.6 Nervous System and Sensation ... 231
19.2.7 Reproduction and Development ... 232
19.3 ECOLOGY ... 232
19.3.1 Distribution ... 232
19.3.2 Interspecific Associations ... 233
19.4 SAMPLING AND CURATION ... 234
19.4.1 Collection ... 234
19.4.2 Maintenance ... 234
19.4.3 Preservation and Storage ... 235
19.5 MISCELLANEOUS ... 235
19.6 ACKNOWLEDGMENTS ... 235
19.7 LITERATURE CITED ... 236

CHAPTER 20 CEPHALOPODA ... 239
20.1 INTRODUCTION ... 239
20.2 TAXONOMY, BIOLOGY, ECOLOGY, AND BEHAVIOR ... 239
20.2.1 Basic Biology ... 239
20.2.2 Behavior ... 240
20.2.3 Taxonomic Overview ... 241
20.3 COLLECTING TECHNIQUES ... 243
20.3.1 Teuthids ... 243
20.3.2 Sepiolids ... 243
20.3.3 Octopuses ... 243
20.4 AQUARIUM CULTURE ... 244
20.5 PREPARATION ... 244
20.5.1 Documentation ... 244
20.5.2 Preservation for Genetic Research ... 245
20.5.3 Fixation and Preservation for Morphological Study ... 245
20.6 FIXATION AND PRESERVATION PROTOCOLS ... 245
20.6.1 Narcotization ... 246
20.6.2 Octopuses and Small Squids ... 246
20.6.3 Large Squids ... 246
20.7 IDENTIFICATION ... 248
20.8 WEB RESOURCES ... 249
20.9 SUMMARY ... 249
20.10 ACKNOWLEDGMENTS ... 249
20.11 LITERATURE CITED ... 249

CHAPTER 21 FRESHWATER GASTROPODA ... 251
21.1 INTRODUCTION ... 251
21.2 BIOLOGY AND ECOLOGY ... 251
21.3 CONSERVATION ... 253
21.4 COLLECTING TECHNIQUES ... 253
21.5 PREPARATION AND STORAGE ... 255
21.6 IDENTIFICATION ... 256
21.7 AQUARIUM CULTURE ... 257
21.7.1 *Pomacea* ... 257
21.7.2 *Bellamya* (or *Cipangopaludina*) ... 257
21.7.3 *Melanoides tuberculata* ... 257
21.7.4 *Helisoma trivolvis* ... 257

    21.7.5 *Physa*...258
  21.8 SUMMARY 258
  21.9 WEB RESOURCES..258
  21.10 LITERATURE CITED...258
CHAPTER 22 TERRESTRIAL GASTROPODA...261
  22.1 INTRODUCTION 261
  22.2 BIOLOGY OF LAND SNAILS..261
  22.3 WHERE TO FIND LAND SNAILS 263
    22.3.1 General Considerations..263
    22.3.2 Macrohabitat Requirements 264
    22.3.3 Microhabitats 265
  22.4 FIELD METHODS AND EQUIPMENT..267
    22.4.1 Methods, General Considerations..267
    22.4.2 Visual Search 268
    22.4.3 Leaf Litter and Soil Sampling..268
    22.4.4 Transporting Soil..269
    22.4.5 Stream Drift 269
    22.4.6 Trapping 270
    22.4.7 Vacuuming, Sweeping, and Beating 270
    22.4.8 Mark-Release-Capture Method..270
    22.4.9 Containers, Shipping, and other Equipment 271
    22.4.10 Other Considerations 272
  22.5 RECORD KEEPING IN THE FIELD 275
  22.6 PROCESSING AND STORING SAMPLES..276
    22.6.1 Laboratory Recovery of Snails from Leaf Litter and Soil 276
    22.6.2 Cleaning and Preserving Empty Shells...278
    22.6.3 Preserving Soft Parts 278
    22.6.4 Vials and Closures...280
    22.6.5 Labeling, and Keeping Good Records 280
    22.6.6 Storing and Display..280
  22.7 HOW TO IDENTIFY LAND SNAILS...281
  22.8 LITERATURE CITED..282
CHAPTER 23 REARING TERRESTRIAL GASTROPODA..287
  23.1 INTRODUCTION 287
  23.2 A SURVEY OF THE LITERATURE...287
  23.3 SHORT TERM MAINTENANCE OF LAND SNAILS 288
  23.4 REARING NORTH AMERICAN WOODLAND SNAILS 288
    23.4.1 Containers 288
    23.4.2 Substrate...288
    23.4.3 Food 289
    23.4.4 Moisture, Temperature, and Light...290
    23.4.5 Maintenance of Eggs and Juveniles 291
  23.5 SNAIL HEALTH 291
  23.6 LITERATURE CITED..292
CHAPTER 24 MARINE GASTROPODA 295
  24.1 MARINE GASTROPODS, A HETEROGENEOUS ASSEMBLAGE OF TAXA...........295
  24.2 TERMINOLOGY OF GASTROPOD SHELLS AND ANIMALS 295
  24.3 ANATOMY 301

Appendices

    24.3.1 The Typical Marine Gastropod 301
  24.4 THE MAJOR MARINE GASTROPOD GROUPS 303
    24.4.1 Gastropoda 303
    24.4.2 Basal Gastropoda 304
    24.4.3 Higher Gastropoda 304
    24.4.4 Heterogastropoda 304
    24.4.5 Incertae Sedis 304
  24.5 HABITAT AND ECOLOGY 304
    24.5.1 Estuaries 304
    24.5.2 Intertidal 305
    24.5.3 Soft Bottom Shores 305
    25.5.4 Mangroves 305
    24.5.5 Subtidal 306
    24.5.6 Shelf or Neritic Zone 306
    24.5.7 Beyond the Shelf 306
    24.5.8 Pelagic Region 307
  24.6 ECOLOGY 307
    24.6.1 Food and Feeding 307
    24.6.2 Reproduction 307
    24.6.3 Predators 308
    24.6.4 Mobility and Locomotion 308
  24.7 COLLECTING 308
  24.8 ACKNOWLEDGMENTS 310
  24.9 LITERATURE CITED 310
CHAPTER 25 UNIONOIDA: FRESHWATER MUSSELS 313
  25.1 INTRODUCTION 313
  25.2 ECOLOGY AND BIOLOGY 313
    25.2.1 Habitat 313
    25.2.2 Food 313
    25.2.3 Longevity 314
    25.2.4 Predators 314
  25.3 ANATOMY 314
    25.3.1 The Shell 314
    25.3.2 Anatomy and Life History 315
  25.4 TAXONOMY OF THE UNIONOIDA 316
  25.5 CONSERVATION ISSUES 317
    25.5.1 Endangered Species, Permits, and Reports 317
    25.5.2 Voucher Specimens 317
    25.5.3 Number of Specimens 317
    25.5.4 Live *vs.* Dead Shells 318
  25.6 FIELD COLLECTING 318
    25.6.1 Habitat Selection - When and Where to Look 318
    25.6.2 Field Gear 318
    25.6.3 Labeling 319
    25.6.4 Survey Methods 320
  25.7 IDENTIFICATION 321
    25.7.1 Literature 321
    25.7.2 Web Sites 322

| 25.7.3 Cleaning Shells 322
| 25.8 CURATION 322
| 25.8.1 Dry Shells 322
| 25.8.2 Wet Specimens 323
| 25.8.3 Databases 323
| 25.9 LITERATURE CITED 323
CHAPTER 26 NON-UNIONOID FRESHWATER BIVALVES 327
| 26.1 INTRODUCTION 327
| 26.2 FAMILY SPHAERIIDAE: FINGERNAIL, PEA, OR PILL CLAMS 327
| 26.2.1 Species Diversity and Distribution 327
| 26.2.2 Classification 328
| 26.2.3 Morphological and Biological Peculiarities 328
| 26.2.4 Habitats 332
| 26.2.5 Means of Collection and Preparation 332
| 26.2.6 Keeping Specimens Alive in the Laboratory 333
| 26.2.7 Practical Applications 333
| 26.3 FAMILY CORBICULIDAE: THE ASIATIC CLAM 334
| 26.4 FAMILY DREISSENIDAE: ZEBRA MUSSELS 336
| 26.5 EDITORIAL NOTE 336
| 26.6 LITERATURE CITED 336
CHAPTER 27 MARINE BIVALVES 339
| 27.1 INTRODUCTION 339
| 27.2 BIOLOGY 339
| 27.3 CLASSIFICATION OF THE BIVALVIA 340
| 27.3.1 Protobranchia 341
| 27.3.2 Pteriomorphia 341
| 27.3.3 Paleoheterodonta 342
| 27.3.4 Heterodonta 342
| 27.3.5 Anomalodesmata 342
| 27.4 COLLECTING BIVALVES 342
| 27.4.1 Beach Drift 343
| 27.4.2 Wood and Rock 343
| 27.4.3 Soft Sediments 344
| 27.4.4 Dredging 344
| 27.4.5 Commercial Fishing 344
| 27.4.6 Commensal Species 344
| 27.5 PRESERVING AND STUDYING BIVALVES 344
| 27.6 FURTHER STUDY 346
| 27.6.1 Identification 346
| 27.6.2 Biology 346
| 27.6.3 General and Miscellaneous References 346
| 27.7 ACKNOWLEDGMENTS 346
| 27.8 LITERATURE CITED 346
CHAPTER 28 THE MARINE AQUARIUM: A RESEARCH TOOL 349
| 28.1 INTRODUCTION 349
| 28.2 AQUARIA 350
| 28.2.1 Water 350
| 28.2.2 Plants 351

  28.2.3 Rocks...351
  28.2.4 Food..351
  28.2.5 Permits and Conservation 352
 28.3 SELECTED SPECIES OBSERVATIONS 352
  28.3.1 *Corbula contracta* Say, 1822 [Contracted Corbula]..352
  28.3.2 *Janthina globosa* Swainson, 1822 [Elongate Janthina] 352
  28.3.3 *Polinices lacteus* (Guikling, 1834) [Milk Moonsnail]......................................352
  28.3.4 *Ficus communis* Röding, 1798 [Atlantic Figsnail] 353
  28.3.5 *Cymatium labiosum* (Wood, 1828) [Lip Triton] 353
  28.3.6 *Melongena corona* (Gmelin, 1791) [Crown Conch].......................................353
  28.3.7 *Nassarius vibex* (Say, 1822) [Bruised Nassa] 354
  28.3.8 *Leucozonia nassa* (Gmelin, 1791) [Chestnut Latirus] 354
  28.3.9 *Fasciolaria tulipa* (Linnaeus, 1758) [True Tulip]..354
  28.3.10 *Prunum apicinum* (Menke, 1828) [Common Atlantic Marginella] 354
  28.3.11 *Hydatina physis* (Linnaeus, 1758) [Brown-Lined Paperbubble] 356
  28.3.12 *Bulla striata* Bruguière, 1792 [Striated Bubble]..356
  28.3.13 *Haminoea elegans* (J. E. Gray, 1825) [Elegant Glassy Bubble]....................357
  28.3.14 *Stylocheilus longicauda* (Quoy and Gaimard, 1825) [Longtail Seahare] 357
 28.4 LITERATURE CITED...357
CHAPTER 29 AN INTRODUCTION TO SHELL-FORMING MARINE ORGANISMS 359
 29.1 INTRODUCTION 359
 29.2 TAXONOMIC GROUPS 360
  29.2.1 Protozoa 360
  29.2.2 Calcareous Algae 361
  29.2.3 Porifera..362
  29.2.4 Cnidaria...362
  29.2.5 Bryozoa...364
  29.2.6 Annelida..365
  29.2.7 Sipuncula...366
  29.2.8 Arthropoda 366
  29.2.9 Brachiopoda 368
  29.2.10 Echinodermata 369
 29.3 FLOTSAM AND JETSAM 370
 29.4 CONCLUSIONS..371
 29.5 ACKNOWLEDGMENTS..371
 29.6 LITERATURE CITED...371
CHAPTER 30 CONSERVATION AND EXTINCTION OF THE FRESHWATER MOLLUSCAN FAUNA OF NORTH AMERICA...373
 30.1 INTRODUCTION 373
 30.2 UNIONOIDA...373
 30.3 AQUATIC GASTROPODA...375
 30.4 DETERMINATION OF STATUS...375
 30.5 FACTORS CAUSING THE DECLINE IN SPECIES...376
 30.6 EXTINCTION 376
 30.7 ACTIVITIES LEADING TO A REVERSAL OF TRENDS 377
 30.8 CONCLUSIONS..378
 30.9 ACKNOWLDGMENTS 378
 30.10 LITERATURE CITED...378

APPENDIX 30.1 380
CHAPTER 31 ISSUES IN MARINE CONSERVATION 385
   31.1 INTRODUCTION 385
   31.2 HISTORICAL ATTITUDES TOWARDS MARINE MOLLUSKS 385
   31.3 MARINE MOLLUSK CONSERVATION IN THE 20th CENTURY 387
   31.4 DECLINING, COLLAPSED, OR EXTINCT: CONSERVATION ISSUES FOR MARINE.....
   MOLLUSKS 391
   31.5 NATIVE OYSTERS ON TWO COASTS 395
      31.5.1 History of Eastern Oyster Fishery and Decline 395
      31.5.2 Management and Conservation of Eastern Oysters 397
      31.5.3 History of Olympia Oyster Fishery and Decline 398
      31.5.4 Management and Conservation of Olympia Oysters 400
   31.6 FISHERIES COLLAPSE: ABALONES AND QUEEN CONCH 400
   31.7 SCALLOPS 403
   31.8 CLAMS 404
   31.9 MUSSELS: EXCEPTIONS TO THE RULE? 406
   31.10 CEPHALOPODS 406
   31.11 UNREGULATED AND OVERLOOKED MOLLUSKS 407
   31.12 CONCLUSION 408
   31.13 LITERATURE CITED 408
APPENDICES 417
APPENDIX 1: Morphological Features of Gastropod and Bivalve (Pelecypod) Shells 418
   Appendix 1A Structural Features of Gastropods 418
   Appendix 1B Form of Gastropod Shells 420
   Appendix 1C Morphological Features of Pelecypod Shells 422
   Appendix 1D Descriptive Terms Applied to Pelecypod Shells 424
APPENDIX 2: Expanded Table of Contents 427
GLOSSARY 441

GLOSSARY

Abyssal: Oceanic depths from 2000 to 6000 m.

Adductor muscle: One of the muscles used to close a Bivalvia shell. Most species have anterior and posterior adductor muscles; others have just a single adductor muscle.

Aestivation: A period of inactivity seen in some land snails usually during prolonged dry or hot periods; the aperture is often covered by dried mucous, also known as the epiphram, to retard the loss of moisture.

Aperture: An opening, such as the opening of a gastropod shell.

Aragonite: The form of calcium carbonate most commonly used by mollusks, with an unstable crystal structure, and a specific gravity of 2.9.

Bathyal: Oceanic depths from 200 to 2000m.

Bauplan: An idealized and generalized body plan.

Benthos (adj.: benthic): The bottom of a body of water; e.g. ocean floor, lakebed, river bottom.

Blastomeres: Cells formed by cleavage from a fertilized egg.

Blastopore: An opening, in the gastrula stage of the embryonic development of mollusks (and many other animals) that is destined to become the mouth.

Blastula: An early stage of embryonic development, usually consisting of a hollow ball of cells, preceding the gastrula larval stage.

Calcite: A form of calcium carbonate with a specific gravity of 2.7.

Carnivore: Flesh eating animal.

Circumpharyngeal nerve ring: A ring of nerves and ganglia (the brain) that form a circular structure surrounding the pharynx.

Coeloblastula: The hollow cavity found in the blastula.

Commensalism: A form of symbiosis where individuals of two different taxa live together, one deriving a benefit from the association (usually food), while the other is not affected by the association.

Commissure: A seam, where the two valves of Bivalvia come together.

Conchiolin: The organic substance that forms the thin outer layer of mollusks known as the periostracum.

Crystalline style: In the stomach, a structure that rotates and rubs against the gastric shield aiding in mixing and grinding food particles, and releasing digestive enzymes. Found in conchiferan mollusks (i.e., not chitons or Aplacophora), but not all members of Conchifera have a crystalline style.

Ctenidium (pl.: ctenidia): A respiratory organ (specialized gills) unique to mollusks; in Bivalvia it also serves in food gathering.

Demibranch: When filaments of the gills have two branches, each branch is a demibranch. Present in gills of Bivalvia and Cephalopoda.

Detritus: Dead and decaying litter, which is derived from plants, and covers the environmental surface (e.g. ocean bottom, forest floor).

Diploid: Having paired chromosomes.

Epifaunal: Animals that live upon (as opposed to within) the substrate of a body of water. They may be attached by a byssus or be free-living.

Escutcheon: An area of the posterior-dorsal bivalved shell associated with the ligament.

Euryhaline: A term applied to organisms that tolerate a wide range of salinity.

Foraminifera: A group of Protozoa that form shells, usually of calcium carbonate. Though most are microscopic, some are visible to the naked eye.

Four d (4d): Mollusks and most other protostomes have determined or fixed cell fates during early embryo development. All mesoderm is derived from the 4d (four little d) cell, formed at the sixth division after fertilization. After the first two cleavages, the more or less equally sized cells are termed A, B, C, D. The third cleavage produces unequal sized cells; the smaller cells are 1a, 1b, 1c, 1d. The fourth cleavage of the larger cells produces the 2a, 2b, 2c, and 2d cells. The fifth cleavage produces the 3a, 3b, 3c, and 3d cells. The sixth cleavage of the larger cells produces 4a, 4b, 4c, and 4d.

Gamete: Specialized reproductive cells; eggs and sperm.

Ganglion (pl.: ganglia): A structure composed of nerve cell bodies.

Gastric shield: a hardened cuticular plate near the anterior end of the stomach, made of protein and chitin with a roughened, rasp-like surface. The crystalline style rotates against the gastric shield aiding in mixing and grinding food particles, and releasing digestive enzymes.

Gastrula: The mass of embryonic cells following the blastula. During the formation of the gastrula, cells that are fairly superficial migrate to where they will develop into future organs.

Glochidium (pl.: glochidia): A modified larva of a freshwater mussel (Unionidae)

Gonopore: Where the reproductive system opens to the outside of the body.

Hadal: Oceanic depths greater than 6000 m.

Hemocoel circulation: The hemocoel is a coelom (cavity) surrounding the heart. Mollusks do not have a system of arteries, veins, and capillaries. When the blood is pumped out of the heart, it passes into sinuses that carry it to the various organs. The blood collects in other sinuses and is then returned to the heart.

Hermaphrodite: An individual that produces both male and female gametes.

Homeobox Genes: genes that provide the identity of particular body regions in a developing embryo by determining where body structures will develop.

Infaunal: Animals that live within the sediment at the bottom of a body of water.

Labial Palps: In Bivalvia, folds of tissue on either side of the mouth that help to bring food to the mouth.

Lecithotroph: An egg rich in yolk.

Lentic: A freshwater habitat that is characteristically calm, e.g. ponds, lakes, swamps.

Ligament: An elastic structure linking the two valves in Bivalvia.

Littoral: The shoreline zone where the water meets the land.

Lotic: A freshwater habitat that is characteristically running, e.g. brooks, streams, and rivers.

Lunule: An area of the anterior-dorsal bivalved shell that is circular to heart shaped and separated from the rest of the shell by a ridge.

Macrophyte: A large macroscopic plant; often used in reference to kelp and other seaweeds.

Mantle: An organ that covers the body of a mollusk. It functions in producing the shell and periostracum. In air-breathing gastropods (Pulmonata), it serves in a respiratory function as well.

Mantle Cavity: The space within the mantle where the gills, osphradia, excretory, and genital openings are found.

Mesoderm: Embryonic cells that develop into muscle, blood, and connective tissue.

Metamerism: Repetition of body segments or repeated organ systems. Seen in Annelida, but mollusks do not have true metamerism. Those mollusks having repeated organ systems, such as the Polyplacophora, Monoplacophora, and some Aplacophora, have pseudometamerism.

Microphagous: Feeding on small particles or small prey (microorganisms).

Molluscivore: Feeding on mollusks.

Mutualism: A form of symbiosis in which individuals of two different taxa live together, both deriving a benefit from the association.

Nacre: The iridescent or shiny layer that lines the inner surface of some mollusks such as Unionidae (freshwater mussels) and Ostreidae (oysters).

Nephridium (pl.: nephridia): A renal organ in mollusks, a kidney.

Nidamental gland: Provides the outer coating for eggs and sometimes the gelatinous part of the egg mass.

Omnivore: An animal that feeds on both plants and animals.

Operculum (pl.: opercula): A plate that wholly or partly covers the aperture of some gastropods. It may be horny or calcareous.

Osphradium (pl.: osphradia): Sense organs found in some mollusks, usually in the mantle cavity. These structures can detect certain chemicals in the water surrounding the mollusk.

Ovoviviparous: A condition in which the female holds the eggs in her uterus until they hatch, so her offspring emerge as juveniles rather than as eggs.

Pallial line: In Bivalvia, a fine linear impression, on the inner surface of the shell, made by the mantle edge. The line runs from one adductor muscle to the other.

Pallial sinus: An invagination of the pallial line where the siphons pass through. This is found in the posterior end of the shell.

Palps: see labial palps

Parasitism: A form of symbiosis in which individuals of two different taxa live together, one deriving a benefit and the other being harmed by the association.

Parthenogenesis: Development of an organism from an egg that did not require fertilization. The offspring are genetically identical to the parent.

Pectinoid: A member of the superfamily Pectinoidea (Bivalvia): examples include *Pecten*, *Chlamys*, *Lima*, and *Spondylus*.

Pelagic: Open water. It is the water column in the marine environment extending from the air-sea interface down to the benthos.

Pelecypoda: A synonym of the Class Bivalvia.

Pericalymmal larva: a barrel-shaped larva with prominent apical tuft. It propels itself with transverse bands of cilia and a patch of accessory locomotory cilia. In the later stages, the larvae become laterally compressed as the shell develops

Pericardium: A sac surrounding the heart. The molluscan pericardium is unique to mollusks.

Periostracum: The proteinaceous outer shell layer.

Polyploidy: Having three or more of each type of chromosome.

Prismatic Layer: A cross section of a molluscan shell will usually reveal several layers. The prismatic layer is often the middle layer, between the outer (periostracum) and inner (often nacreous) layers.

Propodium: The leading edge of the gastropod foot.

Protostome: A major group of animals including Mollusca in which, during development, the mouth forms from the blastopore, and the anus forms from a second opening.

Pseudofeces: After bivalves remove the desirable food particles through suspension feeding, the remaining material, bound by mucus, is ejected by reverse flow. This material is called pseudofeces.

Pseudometamerism: The appearance of being segmented, when in fact true segmentation does not occur, as in some mollusks.

Piscivorous: An animal that eats fish. Some Conidae are piscivorous.

Radula (pl.: radulae): A rasp-like structure, found in the oral cavity of most mollusks (except Bivalvia) that is used in feeding. It usually looks like a ribbon with small tooth-like structures attached to it in rows and columns.

Refugium (pl.: refugia): A location where major environmental changes have not occurred and therefore the biota of an earlier period persists.

Resilifer: A recess on the hinge plate that contains the resilium (internal portion of the ligament).

Riffle: part of a stream with shallow, fast-flowing water followed by deep, slow-flowing water.

Schizocoel coelomic cavities: A coelom is a cavity formed by a splitting of the embryonic mesoderm (connective tissue). In Mollusks, as development progresses, this coelom becomes reduced to just cavities surrounding heart, kidneys, and gonads (reproductive organs).

Semislug: A snail partially covered by an external shell, into which the animal cannot retreat. A condition of some land snails and some sea slugs (e.g., Notaspidea).

Shell: A hard calcareous structure found encasing mollusks. In some mollusks, such as Bivalvia, this structure encloses most or all of the animal. In other mollusks, such as some Gastropoda (slugs and semislugs) the shell is greatly reduced or vestigial.

Sipuncula: A phylum of marine worms, often called peanut worms, considered to be one of the possible sister groups to mollusks.

Spicule: A small calcareous projection found in some mollusks [Aplacophora and on the girdles of some Polyplacophora (chitons)].

Stenocalymma: a specialized larval form of some Aplacophora intermediate between pericalymma and trochophore larvae

Symbiosis: Individuals of two different taxa living in close association. Forms of symbiosis can be described as commensalistic, mutualistic, and parasitic.

Teeth: There are three uses for this term in malacology. 1) Projections from the hinge of a bivalve shell that fit into sockets on the opposing shell. 2) Projections seen on the inner surface of the aperture in some gastropods that might function as anti-predator barriers. 3) Structures on the radula that give it its rasp-like character.

Tetraneury: A nervous system, found in mollusks, that is composed of two sets of paired nerves, one dorsal and the other ventral.

Torsion: A feature of all gastropods: an early developmental 180° twisting of the shell, mantle, and visceral mass with respect to the head and foot. Torsion results in the nervous system being twisted or crossed and in the products of excretion and reproduction being discharged from a position above the animal's head.

Trochophore: early larval stage of most mollusks. The animal is more or less shaped like a toy top and has an equatorial band of cilia and an apical tuft that provides locomotion as well as help with feeding. The trochophore often develops into a veliger larva.

Uropod: A structure, found in Malacostraca (shrimp, crabs, and lobsters) that forms part of the swimming apparatus of the tail. Some Bivalvia attach to this structure.

Veliger: late larval stage of many mollusks, derived from the trochophore. The more or less spherical animal has a pair of large lobes, each being known as a velum, which help with swimming and feeding. The larval shell, the protoconch, is formed at this stage. Eventually the animal transforms into a juvenile by metamorphosis.

Visceral Mass: The soft tissues of mollusks that comprise the internal organs (digestive and reproductive) of a mollusk.

Printed in the United States
56735LVS00003B/1